TRADITIONAL AMERICAN FARMING TECHNIQUES

A ready reference on all
phases of agriculture for farmers of
the United States and Canada

FRANK D. GARDNER

FOREWORD BY JAMES R. BABB

INTRODUCTION BY MARYJANE BUTTERS

THE LYONS PRESS

Guilford, Connecticut
An imprint of The Globe Pequot Press

First Lyons Press edition, 2001

Originally published in 1916 by L.T. Myers as *Successful Farming*

The Lyons Press is an imprint of The Globe Pequot Press.

10 9 8 7 6 5 4 3 2 1

Printed in the United States of America

ISBN: 978-1-59921-079-7

Library of Congress Cataloging-in-Publication Data is available on file.

TRADITIONAL AMERICAN FARMING TECHNIQUES

A Ready Reference on all Phases of Agriculture for Farmers of the United States and Canada

Including

SOILS, MANURES, FERTILIZERS, LIME, DRAINAGE, IRRIGATION, TILLAGE, FIELD CROPS, CROP ROTATIONS, PLANT IMPROVEMENT, MEADOWS AND PASTURES, WEEDS—VEGETABLES, VEGETABLE FORCING, MUSHROOM CULTURE, MEDICINAL PLANTS—ORCHARDING, SMALL FRUITS, NUT CULTURE, FARM WOODLOT, FLORICULTURE—LIVESTOCK, DAIRYING, FEEDS AND FEEDING, ANIMAL DISEASES—FARM MANAGEMENT, RECORDS AND ACCOUNTS, MARKETS AND MARKETING, LAND RENTAL, LABOR—FARM BUILDINGS, FENCES, ENGINES, TRACTORS, MACHINERY, SANITATION—PLANT DISEASES, INSECTS, FUNGICIDES, INSECTICIDES—DOMESTIC ECONOMY, HOUSING, CLOTHING, EDUCATION, INFORMATION—USEFUL TABLES, COMPOSITION OF PRODUCTS, FEEDING STANDARDS, WEIGHTS, MEASURES.

By FRANK D. GARDNER

Professor of Agronomy, Pennsylvania State College and Experiment Station

With special chapters written expressly for this book by the following authorities

AGEE, ALVA, Director Agricultural Extension, N. J.
ANTHONY, E. L., Instructor in Dairying, Pa.
BAKER, MRS. CECIL, Textile Specialist, Ill.
BLASINGAME, R. U., Professor of Agricultural Engineering, Ala.
BUCKLEY, SAMUEL, Professor Veterinary Science, Md.
CAUTHEN, E. F., Associate in Agronomy, Ala.
COCHEL, W. A., Professor of Animal Husbandry, Kansas.
COOK, M. T., Plant Pathologist, N. J.
CORBETT, L. C., Horticulturist, U. S. Department Agriculture.
COWELL, A. W., Professor of Landscape Gardening, Pa.
DARST, W. H., Assistant Professor of Agronomy, Pa.
EVVARD, J. H., Swine Specialist, Ia.
GARMAN, H., Entomologist and Zoologist, Ky.
GRINDLEY, H. S., Professor Animal Nutrition, Ill.
GOLDTHWAITE, MISS NELLIE E., Dean of Women, N. H.
HUGHES, E. H., Assistant Professor of Animal Husbandry, Mo.
HUMPHREY, GEORGE C., Professor of Animal Husbandry, Wis.
KAINS, M. G., Professor of Horticulture, Pa.
KILPATRICK, M. C., Instructor in Poultry Husbandry, Ohio.
LARSON, C. W., Professor of Dairying, Pa.
McNESS, G. T., Tobacco Expert, Texas.
MOON, F. F., Professor of Forestry, N. Y.
NOLL, C. F., Associate in Agronomy, Pa.
PUTNEY, F. S., Assistant Professor of Dairying, Pa.
REED, C. A., Nut Culturist, U. S. Department Agriculture.
SEARS, F. C., Pomologist, Mass.
STEWART, J. P., Pomologist, Pa.
STOCKBERGER, W. W., Plant Psychologist, U. S. Department Agriculture.
STONE, TOM, Instructor in Animal Husbandry, O.
TOMHAVE, W. H., Professor of Animal Husbandry, Pa.
WADE, C. W., Extension Specialist, Mich.
WARE, H. M., Mushroom Specialist, Pa.
WATTS, R. L., Dean and Director, Pa.
WEBBER, H. J., Dean and Director, Tropical Agriculture, Cal.
WOOD, W. B., Assistant Entomologist, U. S. Department Agriculture.
WORK, PAUL, Superintendent of Vegetable Gardening, N. Y.

PROFUSELY ILLUSTRATED WITH EXCLUSIVE DRAWINGS AND PHOTOGRAPHS

FOREWORD

I little knew, back when I was an editor at The Lyons Press looking for something agricultural to seed the spring list, and slipped from the shelf my well-thumbed and precisely titled copy of Rolf Cobleigh's *Handy Farm Devices and How to Make Them*, just what fruits this act of archival plunder might bear.

Lyons brought out a facsimile edition, for which I wrote a brief introduction to lend context and poke gentle fun at my fellow innocents from the homesteading flurry of the early 1970s. And, as publishers are wont to say, copies began flying off the shelves.

Six years have passed, and they're still flying off the shelves, along with a dozen subsequent titles in the same vein, none of which I had anything at all to do with but all of which share the common goals of bringing to modern readers with rural inclinations the forgotten wisdom of their forebears.

Yet books of the golden age of American agriculture are more than just nostalgic novelties. To latter-day homesteaders who aspire to do what seems impossible now but that three generations back was taken for granted—which is to say earn a living from the soil, or if you will, farm successfully—they are of immense value.

Consider this: Between the American Revolution and the Civil War, most Americans were farmers and most were unsuccessful, if you measure success by how many families farmed a piece of land long enough to make it prosper. Between the Civil War and World War I, far fewer Americans were farmers but, measured by the same yardstick, most were successful. Many factors contributed to this agricultural renaissance, but chief among them was the growing store of knowledge then being widely disseminated by agricultural researchers and college experiment stations nationwide. In other words, by combining rapidly developing science with centuries of accumulated wisdom, farmers were finally learning how to farm.

And then along came the dust bowl and the Great Depression, and for reasons beyond the power of knowledge to contravene, this legion of prosperous family farmers mostly went broke. In the aftermath of World War II, industrial agriculture arrived, and by investing in ever more and ever larger machinery and supercharging crops and sterilizing soils with ever more chemicals, the remaining farms grew ever larger and grew ever more and ever cheaper food while employing ever fewer people. Nowadays, as any number of socio-economists will gravely tell you, practically no one in America is a farmer and those who remain are either starvation-wage serfs of agribusiness multinationals or are failed farmers awaiting rescue by a Willie Nelson concert.

But even as the conventional family farmer becomes an endangered species and rural areas lose population nationwide, a different breed of small farmer is finding a prosperous niche in out-of-the-way places from Maine to Oregon, Wales to Bordeaux—not as global agribusiness ciphers churning out industrial food but as artisans nurturing real food meant for people who care about what they eat and will pay for it, people who understand that 49-cents-a-pound factory chicken has hidden costs ranging far beyond its cardboard flavor.

These new-age artisanal farmers have their gurus—John Jeavons, John Seymour, and Eliot Coleman come to mind—all of whom write well and usefully on subjects from raising raddichio to double-digging a raised bed. But as thoughtful and diligent as it is, the new farming literature is missing something important: an unbroken thread that runs from the time of writing straight back to the earliest days of agriculture.

This is what the best of the books from the golden age of farming offered, and TRADITIONAL AMERICAN FARMING TECHNIQUES is the best I know. I bought my copy at a yard sale thirty years ago, and I still find something useful every time I open it. Which, in growing season, is almost daily.

It begins, as any book presuming to call itself TRADITIONAL AMERICAN FARMING TECHNIQUES must, with the soil—its composition, its structure, what fertility means and how to maintain it by feeding the soil, not the plants.

There are chapters on the composition and fertilizer constituents of various manures, how to store and handle them, make them go further and ensure they deliver the maximum advantage to the soil. How to supplement fertility with commercial fertilizers ranging from comparatively benign old-fashioned chemicals—superphosphate, nitrate of soda, muriate of potash—to the outright organic: cottonseed meal, tankage, dried blood, phosphate rock, bone meal, and wood ash. How to deal with pernicious weeds, insect pests, and diseases using homemade methods that still work fine today, such as soaps, tobacco extracts, and lime-sulphur washes.

There is much on how to improve pastures and meadows, make hay with hand-, horse-drawn, and tractor-towed machinery, and how to hand-cure alfalfa to preserve the maximum possible protein. How to reduce feed costs by adding alfalfa pasturage and feeding root crops such as turnips and mangles. How to make the best butter, including the ripened-cream ambrosia now tasted only in frighteningly expensive restaurants. How to keep, breed, feed, and manage all manner of livestock, from horses and mules, hogs and poultry, to the small-scale dairy herd and homestead honey production.

For seed-savers, there is extensive information on selectively breeding varieties perfectly adapted to your local conditions. And for fans of heritage varieties? Well, *all* the varieties detailed in this book, from potatoes to corn to tomatoes and oranges, are heritage varieties.

And of course there's a complete exploration of vegetable gardening, along with such entertaining exotica as anise, ginseng, wormwood, belladonna, and digitalis, not to mention all kinds of fruits, from temperate staples such as strawberries and pears through tropical favorites such as pineapples, oranges, mangoes, and bananas. For those so inclined, in these additive-laden and condemnatory times, there is even a chapter on growing and curing the finest tobacco.

And what complete farm manual from pre–World War I days would be truly complete without cautioning the farm housewife to avoid extremes of fashion and to wear what is appropriate, for the "ball gown is immodest if worn on the street, and the bathing suit shocking in the parlor." Words to live by, now as in 1916.

JAMES R. BABB
SEARSPORT, MAINE

PREFACE

This book is written for farmers. It makes a popular appeal to all men engaged in farming. It will also be of interest to the student of agriculture and the prospective farmer. It is designed to be a handy reference on the whole range of agriculture in the United States and Canada. Technical terms and lengthy discussions have been avoided.

Ages of farm experience and a few generations of agricultural research have given us a vast store of practical knowledge on tilling the soil and raising crops and animals. This knowledge is scattered through many volumes on different phases of the subject, in experiment station bulletins, agricultural journals and encyclopedias. The important facts on which the most successful farming is based are here brought together in orderly and readable form. Not only are directions given for the management of the soil and the raising of crops and livestock, but the business of farming is fully discussed, showing why some achieve success and why others fail.

The subject-matter is arranged in ten parts of a number of chapters each, and by referring to the Table of Contents any subject may be quickly found. References are freely given at the close of each chapter. Each chapter has been prepared by a specialist in the subject presented. The name of the author appears at the beginning of each chapter. Those unacknowledged have been prepared by myself.

The illustrations have been secured from many sources. Due credit has been given these.

Special acknowledgment is due the publishers of this volume for its conception, and for many helpful suggestions in the presentation of its subject-matter.

Acknowledgment is also due Professor E. L. Worthen and Professor R. S. Smith, both of The Pennsylvania State College, for helpful suggestions and criticisms on soils and crop rotations. I wish also to especially acknowledge the valuable editorial assistance of my wife in the preparation of the manuscript.

FRANK D. GARDNER.

INTRODUCTION

In the past two decades, since I first settled on my five acres at the end of a dirt road in Idaho, I've somehow managed to survive and call myself an "organic farmer." My ability to stay put and pay my bills in those early years, when I was a single mother with two young children, went beyond dogged determination and stick-to-it-ness; farming, for me, became religion.

In the last few years, whenever I encountered an obstacle and muttered to myself, "There must be a way" . . . to get this pasture looking better, to give better ventilation to my chickens, to take my two beehives that seem weak and combine them, to choose the perfect dairy cow . . . I've reached for "the good book," this book in your hands right now that I call my "farm bible." This epic manual was written in an era when almost all farmers in North America were still organic, small-scale farmers; when four-fifths of the farms in the United States had a dairy cow out back and two-thirds still made their own butter; when cows grew up on abundant pastures, not corporate feedlots; when chickens were free-range, not caged; when eating a strict organic diet didn't require a trip to a health-food store; when the best-tasting carrot was simply a carrot and not a "certified organic" carrot.

Daily farm life (including Sundays) on my piece of Eden went something like this: I'd find a new weed or bug or one of my fruit trees would get a cankerous eruption or my daughter's pig would find a way out of its pen again and I'd say, "Run to the barn and get the scripture on fences." Scripture? Farming for me has, no doubt, been an act of faith. So much so that rapture, farm romance, and humor were the only way I could keep going at times.

This book, Traditional American Farming Techniques, has it all and more, and it is still a ready reference for the modern-day organic farmer. (It's easy enough to ignore the brief chapter on insecticides and the occasional mention of the then-newfound wonder, arsenic—the very beginning of an industry that would some day give farmers a challenge far greater than the vagaries of weather and the availability of water.)

And for sure, some of it is nostalgic novelty. I'm not likely to build a farm limekiln, even though for two winters I did use a window box instead of a refrigerator. The instructions for creating a window box in this book not only are doable, the drawing itself is worthy of framing. If you love old line drawings and photos, this book is for you.

And when it comes to gender, it's definitely outdated. In 1916, women were "farm housewives." Ninety years later, we're the fastest-growing group of people buying and operating small farms. Although the number of American farms has dropped 14 percent in the last twenty-five years, the number of farms run by women has increased 86 percent, according to the U.S. Department of Agriculture. But even the outdated gender stuff serves a useful purpose—a reminder that we womenfolk have come a long way! For example, a textile and clothing expert in the chapter on home economics says farm women must do "the planning, the buying, perhaps the making, as well as care for the clothes"—aka shopping, sewing, and laundry. Our 1916 expert warns us: "The ball gown is immodest if worn on the street, and the bathing suit shocking in the parlor."

So from hogs to handiwork, hearth to home, everything "4-H" you'll need to run a farmstead is here—things like how to build up your soil or how to deal with pernicious weeds and insect pests using down-home methods that still work today.

It was here that I read about "sweeping" a field every ten days to control thistle. We visited an old John Deere boneyard, brought home and then repaired a retired sweep, hooked it to our tractor, and that was the end of thistle in an entire two-acre field. Afterwards, our sun-baked, worn-out soil needed rebuilding. Again, this book gave me new ideas for how to store and handle manures, make them go further, and kick-start the rebirth of tired soil with old-fashioned soil amendments ranging from benign supplements like lime and phosphate rock to organic regulars like wood ash and manure.

And my favorite section? The one on how to make better butter! The ultimate in farm delicacies is perfectly ripened butter made using the cream you've coaxed from the udder of your very own cow (a farm pet as adorable as a clucking hen that you've trained to help with breakfast).

Speaking of pets, we still joke about the captioned photo "Hogging Down Corn" every August when our corn patch ripens. In 1916 it meant turning loose your pet swine onto your cornfields after you've "downed" (harvested) the stalks, in a form of natural rototilling. For us, it means eating nothing for dinner but fresh, hot ears of corn cooked over a campfire and then brushed with melted butter (using bunches of fresh, bundled sprigs of rosemary as brushes).

There are tried-and-true techniques for improving pastures and meadows, making hay by hand, even details on how to hand-cure alfalfa for maximum protein, the perfect tip for small farmers who don't have access to a lot of working capital.

For fruit lovers, everything from apples to pomelos is discussed in detail. Did you know that strawberries can be a good companion groundcover crop to peach trees? Or that pears will grow too fast if planted in soil that is fertile? (In other words, if you have a troublesome clay area, plant a pear tree.) Beyond fruit trees, the discussion on ornamental trees is divided into those best suited for lining lanes and those for lawns, screens, hedges, and windbreaks.

When it came time for my son to take over milking my cow "Chocolate," I assigned him some reading material from our library in the barn that included the chapter on dairy herd management. The description on how to manipulate a cow's "glands" using the Hegelund method, thought in 1916 to produce more milk, is still the best description I've come across. In perfect detail, the three stages of manipulation are detailed: "When no more milk is obtained by the first manipulation, the hind quarters are milked by placing a hand on the outside of each quarter, likewise with fingers spread and turned upward, but with the thumb just in front of the hind quarter. The hands are lifted and grasped into the gland from behind and from the side, after which they are lowered to draw the milk." Typical of the detail found in this book, a total novice could be sitting by the side of a cow for the first time, book in hand, and learn how to milk her—or a goat or sheep for that matter.

For seed savers, there is extensive information on selectively breeding varieties perfectly adapted to your locale. And for zealots rapturous about heirloom varieties, there are details on seed saving. Who would have thought back then that by the turn of the century we would save seed not only for economic reasons but also in a desperate attempt to rescue certain varieties from extinction? Seed saving and planting heirloom seeds has become a deeply spiritual act—one that has converted thousands of forward-thinking gardeners and farmers into hellfire-and-brimstone evangelists.

If you're a land-poor farmer longing for your place in the country, take heart—a new kind of farm is on the horizon, and the demand for local, hands-on food, or "food with a face," is rising. New kinds of markets are opening up. Consumers are listening and shopping more with their hearts instead of their pocketbooks. The value of food has taken on a new meaning. I had a conversation recently with a bookkeeper that might prove helpful to anyone lusting for land. (I've yet to meet a person who hasn't, at some point in his or her life, had a farm fantasy.) A wannabe farmer in her early fifties, this bookkeeper is constantly on the prowl for the perfect piece of land. She's also a watercolor artist. Her search for land seemed to be taking its toll on her artwork, her weekends off, her inner calm. I offered this bit of advice: If you can't yet cultivate your own back forty, cultivate instead your farm fantasies. Give them free rein, letting them grow with abandon, like weeds gone wild. With books like this one, your fantasies can be a sensory journey almost as good as actually being there. In the meantime, do your homework—plant a garden. Learn how to eat zucchini a hundred different ways! And while you're waiting for your farming stars to align, sit back and enjoy the journey, read, take notes, buy tools. Give yourself fully to the farmstead life you hold in your heart until the day it starts in your hands.

MaryJane Butters

CONTENTS

BOOK I. SOILS AND SOIL MANAGEMENT

CONTENTS

CONTENTS

CONTENTS

CONTENTS

BOOK VI. FARM BUILDINGS AND EQUIPMENT

BOOK VII. FARM MANAGEMENT

CONTENTS

CONTENTS

BOOK VIII. PLANT AND ANIMAL DISEASES, INSECT ENEMIES AND THEIR CONTROL

CONTENTS

TRADITIONAL AMERICAN FARMING TECHNIQUES

BOOK I

SOILS AND SOIL MANAGEMENT

CHAPTER 1

SOIL CLASSIFICATION AND CROP ADAPTATION

The thin layer of the earth's surface known as the "soil and subsoil" supports all vegetation and makes it possible for the earth to sustain a highly developed life. The prosperity and degree of civilization of a people depend in a large measure on the productivity and utilization of this thin surface layer of the earth's crust. From it come the food supply and the materials for clothing and to a considerable extent the materials for housing of mankind.

Soils are Permanent.—The soil is indestructible, and according to the great laws of nature, it should be capable of supporting generation after generation of men each living on a slightly higher plane than the preceding. This necessitates a system of agriculture that is permanent, and one that will foster and maintain the productivity of the soil. Each man who owns and cultivates land owes it to his fellow-men to so cultivate and fertilize the soil that it will be left to his successor in as good or even better condition than it was during his occupation. In return, his fellowmen should make it possible for him to secure a living without resorting to soil robbery. A faulty system of soil management that permits a decline in soil productivity will ultimately be just as injurious to the men indirectly dependent upon the soil as it is to those actually living on the land.

The soils of the United States and Canada are a great asset, and one over which man has relatively large control. Intimately associated with this great asset are two other resources, namely, the atmosphere that envelops the earth and the sunshine that reaches it. Little can be done, however, to control these assets, but with the surface of the earth man can do much as he pleases.

What Farmers Should Know.—Every farmer should have a thorough knowledge of the soil on his own farm. In this and following chapters, the soil and its properties as related to the business of farming will be discussed chiefly from the standpoint of the farmer. The practical farmer expects cash compensation for the intelligent care he gives to his land. He should be able to distinguish between the essentials and non-essentials in the science of the soil. He should know that all soils may be made productive, but this cannot always be done at a profit. Soils on which men, by the exercise of intelligence and reasonable industry, cannot make more than a meager living, should not be cultivated. They should revert to nature or be devoted to forestry. There is some land that has been cleared of its virgin growth and come under the plow that should never have been farmed. There are farms, once productive, that have been robbed of fertility and neglected until they are no longer fit for occupation. There are also some types of farming in some localities, once profitable, that are not paying under the changed economic conditions. These are some of the more acute problems that call for a fuller knowledge of the soil than we have previously possessed. The following chapters in Book I will deal with the essentials in a non-technical manner.

ROCK WEATHERING AND THE PROCESS OF SOIL FORMATION.[1]

It is hoped it may all be profitable reading for any one engaged in the business of farming.

The Science of the Soil.—In recent years science has been directed towards the soil in search of new truths. The reasons for methods of tillage, crop rotations, use of manures, need for lime and many other things have been explained. Soils are being classified and mapped. Crop adaptation is being studied. Field experiments with fertilizers and cultural methods are being conducted extensively in every state in the Union. As a result of all this activity, much progress has been made and we now have a voluminous literature relating to the soil. The subject is recognized as vital to successful farming everywhere, because the soil is the foundation of all agriculture.

[1]Courtesy of E. P. Dutton & Co., N. Y. From "The Soil," by Hall.

How Soils are Formed.—Many agents are active in the formation of soils. Among these may be mentioned changes in temperature, the mechanical action of wind and water, the solvent action of water, and the action of bacteria, fungi and the higher forms of plants.

The manner of formation gives rise to two general classes of soil known as (1) residual soils and (2) transported soils. Residual soils are those formed from rocks like those on which they rest, while transported soils are those carried some distance either by the movement of glaciers, or by moving water in the form of streams and tides, or by the action of the wind.

Weathering and Disintegration.—Rocks absorb more or less water. Low temperatures cause a freezing of the water, which exerts a pressure approximating one ton per square inch. This ruptures the rocks, and the process repeated many times every year gradually reduces the portion subjected to these changes in temperature to fragments. Little by little rocks are thus reduced to soil. On the immediate surface the change in temperature between night and day causes expansion and contraction which also tends to sliver off particles of rock. The movement of soil particles as the result of wind and rain also tends to wear down the surface and break off minute particles that contribute to the process of weathering and disintegration.

In addition to this the vegetation which gradually secures a foothold develops into larger plants, the roots of which penetrate the crevices, exerting a pressure which still further moves and often ruptures the already weakened rocks or fragments thereof. In this way, through generations, the soils are gradually formed and become incorporated with the decomposed vegetation that gradually accumulates on and near the surface. As a further aid to the process of weathering and disintegration we find numerous worms and insects that burrow into the soil, living on the organic matter and living plants. These not only move particles of soil from place to place but carry the organic matter down into the soil.

The rain which falls upon the soil is also a factor in soil formation. When thoroughly wet the soils expand and when quite dry they contract and little fissures open in the surface. A succeeding rain washes the fine surface particles and organic matter into the fissures and causes a gradual mixture of these two essential parts of the soil solids.

Decomposition.—The processes of weathering and disintegration result in a change in the physical properties of the soil without necessarily changing the character of the compounds. Decomposition, on the other hand, generally results in the formation of new compounds. The processes of decomposition are technical and we will not undertake to discuss them.

What is the Soil?—The soil consists of three principal parts, namely, solids, a liquid and gases. The solids consist of the minerals and the organic matter mingled with them. The liquid is the soil water in which is dissolved small quantities of various soil solids. The gases consist chiefly of the air intermingled with various quantities of other compounds, such as carbon dioxide, marsh gas, etc.

The soil and subsoil include all material to the depth to which plant roots distribute themselves. It, therefore, constitutes a wide range of material, both in depth and character. It may be deep or shallow, loose or compact, wet or dry, coarse or fine in texture, having all degrees of variation in its physical, chemical and biological properties.

The Soil Solids.—The solid part of the soil consists of the minerals and organic matter. In practically all soils the minerals form ninety-five per cent or more of the solids. The exception to this would be the peat and muck soils, which may contain as much as eighty per cent or more of organic matter. The mineral matter of the soil consists chiefly of the minute particles or fragments of the mother rock from which the soil has been derived. In case of residual soils this will correspond in a large degree to the rock formation generally found beneath the soil and subsoil at varying depths. In transported soils the mineral particles, having been transported either by water, glaciers, or wind, may have come from different sources, and will generally show a greater diversity in character. It is significant, however, that the minerals of all soils contain all the essential mineral elements for plant growth, although these may vary widely in their relative proportions.

The minerals of the soil are sparingly soluble in the soil water and the solubility is influenced by a number of factors that will be discussed in a subsequent chapter. It is fortunate that this solubility takes place very slowly, otherwise soils would be dissolved and disappear in the drainage waters too rapidly, and the waters of the earth would become too saline to be used by plant and animal life. Loss of the mineral constituents takes place by leaching. The drainage waters from land always contain a very small quantity of many of the elements of which the soil is composed. Nitrogen, the most valuable decomposition product of the organic matter of the soil, is most rapidly leached away in the form of nitrates. Likewise, lime slowly disappears from the body of the soil. Limestone soils, formed from the disintegration and decomposition of limestone rocks, sometimes ninety per cent or more carbonate of lime, generally contain not more than one-half of one per cent of carbonate of lime. The rate of leaching corresponds in a large measure to the rainfall of the region. In regions of sparse rainfall very little leaching takes place, and the soil solution frequently becomes so concentrated that the soils are known as alkali soils. Such soils are either bare of vegetation or produce only crops that are tolerant of alkali. The soils of arid regions are as a rule very productive when placed under irrigation.

The Soil Fluid.—This consists of water in which is dissolved minute quantities of the different minerals of the soil together with organic products and gases. The soil solution moves through the soil by virtue of gravity and capillarity. The water from rain passes downward by gravity. The rate of downward movement depends on the size of the little passageways through the soil. In fine-textured, compact soils it is often very slow. The depth to which it penetrates depends upon the character of the subsoil or underlying strata. It is frequently intercepted by impervious layers, and consequently in times of excessive rainfall the soil becomes saturated and water accumulates on the surface. It then seeks an escape by passage over the surface and often carries with it portions of the soil. thus becoming a destructive agent in soil formation. In dry periods the surface of the soil loses its water through direct evaporation and through the consumption of water by the plants growing in the soil. This should be replaced by the water in the subsoil which returns to the surface by capillarity. The distance through which capillary water will rise is measured by a few feet. The height of rise is greatest in case of fine-textured soils, but in this type of soil the rate of movement is slowest. The rate of movement in sandy soils is much more rapid, but the height of rise is much less.

Gases of the Soil.—The soil atmosphere consists of air and the gases resulting from decomposition of the organic solids in the soil. The dominant gas is carbon dioxide, which, dissolved in water, increases the solvent action of the water and helps to increase the available plant food. The movement of the gases in the soil is affected by changes in temperature which cause an expansion and contraction of their volume. It is also affected by the movements of soil water. As the water table in the soil is lowered air enters and fills up all spaces not occupied by water. The movement is also facilitated by changes in barometric pressure and by the movement of the air over the surface of the soil. Just as a strong wind blowing over the top of a chimney causes a strong draft in the chimney, so does such a wind cause a ventilation of the soil and increases the circulation of the air within the soil.

The roots of most economic plants require oxygen and this is secured in properly drained and well aerated soils from the soil atmosphere. When soils are filled with water the plant roots have difficulty in getting the required supply of oxygen and the growth of the plant is retarded. A proper aeration of the soil is necessary to the development of microscopic organisms that live in great numbers in the soil and play an important part in making available the mineral constituents necessary for the higher forms of plants. It is essential that farmers understand the movement of water and air in the soil in order that they may do their part in bringing about that degree of movement that is essential to the highest productivity of the soil. Drainage, cultivation and the judicious selection of the crops grown are some of the means of influencing the movement of water and air in the soil.

Soil Classification.—Science is classified knowledge. In order that there may be a science of the soil it becomes necessary to classify soils.

Such a classification should meet the needs of an enlightened agriculture. The first classification of the soils of the United States and Canada to be put into extensive use was that devised by the Bureau of Soils of the United States Department of Agriculture, and used extensively in the soil survey of the United States during the past sixteen years. This classification is based upon factors that can be recognized in the field, and has for its ultimate aim the crop adaptation and management of the soil.

Soil Surveys.—"A soil survey exists for the purpose of defining, mapping, classifying, correlating and describing soils. The results obtained are valuable in many ways and to men of many kinds of occupation and interests. To the farmer it gives an interpretation of the appearance and behavior of his soils, and enables him to compare his farm with other farms of the same and of different soils. The soil survey report shows him the meaning of the comparison and furnishes a basis for working out a system of management that will be profitable and at the same time conserve the fertility of his soil. To the investor, banker, real estate dealer or railway official it furnishes a basis for the determination of land values. To the scientific investigator it furnishes a foundation knowledge of the soil on which can be based plans for its improvement and further investigation by experiment. To the colonist it furnishes a reliable description of the soil."

Soils of the United States.—"For the purposes of soil classification the United States has been divided into thirteen subdivisions, seven of which, lying east of the Great Plains, are called soil provinces, and six, including the Great Plains and the country west of them, are known as regions.

"A soil province is an area having the same general physiographic expression, in which the soils have been produced by the same forces or groups of forces and throughout which each rock or soil material yields to equal forces equal results.

"A soil region differs from a soil province in being more inclusive. It embraces an area, the several parts of which may on further study resolve themselves into soil provinces.

"Soil provinces and soil regions are essentially geographic features."* The soils in a province are separated into groups. Each group constitutes a series. A soil series is divided finally into types. The type is determined by texture. The texture may range from loose sands down to the heaviest of clays. All types in a soil region or province that are closely related in reference to color, drainage, character of subsoil and topography and are of a common origin, constitute a group or series of soils. A soil type is, therefore, the unit in soil classification. "It is limited to a single class, a single series and a single province."*

Classification by Texture.—The soil type of a particular series is

* That which is enclosed in quotation marks is quoted from U. S. Bureau of Soils Bulletin No. 96, "Soils of the United States."

based on soil texture and is determined in the laboratory by separating a sample into seven portions, or grades. Each portion contains soil particles ranging in diameter between fixed limits. This process constitutes a mechanical analysis. In such an analysis the groups and their diameters are as follows:

Groups.	Diameter in mm.	Number of Particles in 1 Gram.
1. Fine gravel	2.000–1.000	252
2. Coarse sand	1.000–0.500	1,723
3. Medium sand	0.500–0.250	13,500
4. Fine sand	0.250–0.100	132,600
5. Very fine sand	0.100–0.050	1,687,000
6. Silt	0.050–0.005	65,100,000
7. Clay	0.005–0.000	45,500,000,000

Fifteen types of soil are possible within any soil series. The relative proportions of the several soil separates, given in the table above, determine the type. The twelve most important of these are known as

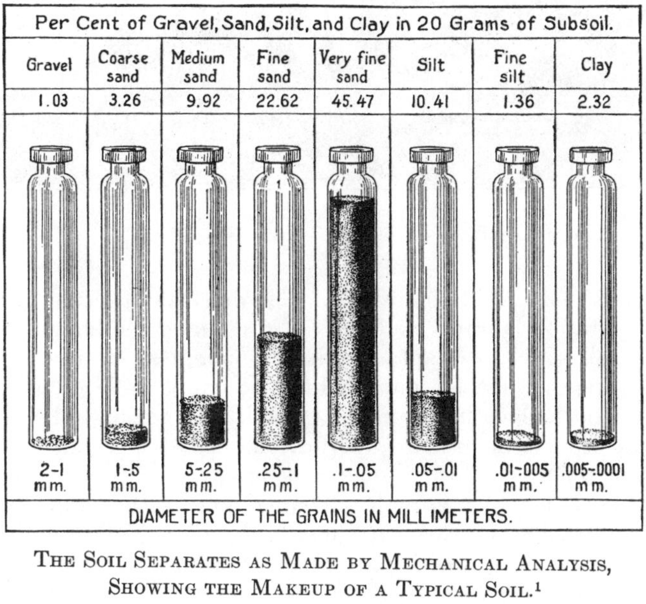
Per Cent of Gravel, Sand, Silt, and Clay in 20 Grams of Subsoil.
Gravel
Coarse sand
Medium sand
Fine sand
Very fine sand
Silt
Fine silt
Clay
1.03
3.26
9.92
22.62
45.47
10.41
1.36
2.32
2-1 mm.
1-.5 mm.
.5-.25 mm.
.25-.1 mm.
.1-.05 mm.
.05-.01 mm.
.01-.005 mm.
.005-.0001 mm.
DIAMETER OF THE GRAINS IN MILLIMETERS.

The Soil Separates as Made by Mechanical Analysis, Showing the Makeup of a Typical Soil.[1]

coarse sand, medium sand, fine sand, coarse sandy loam, medium sandy loam, fine sandy loam, loam, silt loam, clay loam, sandy clay, silt clay and clay. They range from light to heavy in the order named, and, except

[1] Courtesy of Orange Judd Company. From "Soils and Crops," by Hunt and Burkett.

as influenced by presence of organic matter, their water-holding capacity varies directly with the increase in fineness of texture, the sand having the smallest water-holding capacity and the silty clays and clays the largest.

In classifying soils in the field the soil expert determines the type by the appearance and feel of the soil. He takes numerous samples which are sent to the laboratory where they are subjected to a mechanical analysis in order to verify his judgment and field classification.

The accompanying map shows the extent and location of the several soil provinces and regions in the United States.

Crop Adaptation.—That certain soils under definite climatic conditions are best adapted to certain plants is obvious to anyone who has studied

INSPECTING AND SAMPLING THE SOIL.

different soils under field conditions. The marked variation in the character of vegetation is often made use of in defining the boundaries of soil types and soil series. Adaptation is also manifest in the behavior of cultivated crops. Among our well-known crops tobacco is the most susceptible to changes in character of soil, and we find that a specific type of tobacco can be grown to perfection only on a certain type of soil, while a very different type of tobacco demands an entirely different type of soil for its satisfactory growth. The red soils of the Orangeburg series in Texas will produce an excellent quality of tobacco, whereas the Norfolk series with gray surface soil and yellow subsoil, occurring in the same general locality, gives very unsatisfactory results with the same variety of tobacco. This difference in the tobacco is not due to the texture of the soil, since soil of the same texture can readily be selected in both of these series. The most casual observer cannot fail to distinguish the difference between the Norfolk and Orangeburg soils, as manifested chiefly in their color.

The question of crop adaptation, therefore, becomes exceedingly important, and success with a crop in which quality plays an important part will be determined to a large extent by whether or not it is produced on the soil to which it is by nature best adapted.

Variety tests of wheat afford further illustration of crop adaptation. In Illinois the wheat giving the highest yield on the black prairie soil of the central and northern part of the state is Turkey Red, but this variety when grown on the light-colored soil in the southern part of the state yielded five bushels per acre less than the variety Harvest King. It is evident, therefore, that if Turkey Red, which was demonstrated to be the best variety at the experiment station, had been planted over the wheat-growing region of the southern part of the state, farmers of that region would have suffered a considerable loss. In Pennsylvania and North Carolina Turkey Red has been grown in variety tests, and found to be one of the lowest yielding varieties. For example, the yield in North Carolina, as an average of four years, was only 8.4 bushels per acre as compared with 13.5 bushels for Dawson's Golden Chaff. At the Pennsylvania Station the yield for two years was 26.5 bushels per acre for Turkey Red and 37.5 bushels for Dawson's Golden Chaff.

Similar observations have been made relative to varieties of cotton and varieties of apples. There is no doubt but that the question of varietal adaptation, with reference to all of the principal crops, is important, and it should be the business of farmers in their community to ascertain the varieties of the crops grown which are best adapted to local conditions.

Dr. J. A. Bonsteel, born and reared on a New York farm, and for fifteen years a soil expert in the U. S. Bureau of Soils, prepared for the *Tribune Farmer* in the early part of 1913 a series of articles on "Fitting Crops to Soils." The following is a portion of his summary and is a concise statement of the soil adaptation of the fifteen leading crops in the northeastern part of the United States.

"**Summary of Soil Adaptedness.**—Summarizing, briefly, the facts stated in the articles and derived from a large number of field observations made in all parts of the northeastern portion of the United States, we see:

"First.—Clay soils are best suited to the production of grass. They are suited to the growing of wheat when well drained and of cabbages under favorable local conditions of drainage and market. Oats may be grown, but thrive better upon more friable soils.

"Second.—Clay loam soils are especially well suited to the growing of grass, wheat, beans and cabbages, the latter two only when well drained.

"Third.—Silt loam soils produce wheat, oats, buckwheat, late potatoes, corn, onions and celery. The last two crops require special attention to drainage and moisture supply to be well suited to silt loam soils.

"Fourth.—Loam soils, which are the most extensively developed of any group in the Northeastern states, are also suited to the widest range of crops. These are wheat, oats, corn, buckwheat, late potatoes, barley, rye, grass, alfalfa and beans.

"Fifth.—The sandy loam soils are best suited for the growing of barley, rye, beans, early potatoes, and, under special conditions of location near to water level, of onions and celery.

"Sixth.—Sandy soils are best adapted to the early potatoes grown as market garden or truck crops, and to rye.

"This summary takes into consideration only the texture of the soil and its adaptations under fair conditions of drainage, organic matter content and average skill in treatment.

"Yet the articles have called special attention to certain other features than those of soil texture. Otherwise, the specific naming of the different loam soils would not have been given.

"The noteworthy lime content of the soils of the Dunkirk, Ontario, Cazenovia, Dover and Hagerstown loams has been made evident as a basis for the profitable growing of alfalfa, since the plant is known to be particularly sensitive to the amount of lime contained in the soil.

"Similarly the production of the late or staple potato crop has been noted upon soils which are particularly well supplied with organic matter as in the case of the Caribou loam and the Volusia loam. Other loams and silt loams produce good crops of potatoes upon individual farms where there is an unusually good supply of organic matter in the soil, but not on portions of the other types not so well supplied. Good organic matter content is rather a general characteristic of a good potato soil and is found on the types named.

"Beans may be grown upon a large number of different soils if the farmer is satisfied with average crops. But the best bean crops are secured from soils which are well supplied both with organic matter and with lime. Hence, the Clyde loam and clay loam and the soils of the Dunkirk series are among the best bean soils.

"It is still impossible to state precisely what varieties of the different crops are best suited to a particular soil, yet I hope to see the time when there will be special breeding of staple crops to meet the different conditions which prevail upon different soils. Some time there will be strains of wheat, of corn, of oats, of alfalfa and of other field crops which have been developed for generations upon a specific type of soil and which excel all other strains of the crop for that soil. This is inevitable in time, since the characteristics of plants may be fixed by growing them under the same conditions of soil and climate for many plant generations.

"There are certain broad generalizations in crop adaptation which are very generally known, but may profitably be stated again.

"The friable loam is the great soil texture of the temperate, humid regions, possessing the broadest crop adaptations, and usually the most permanent natural fertility of all soils.

"As any departure is made from the loam texture there is a restriction in the number of the different crops which may be grown upon this type, and frequently in the yields of the common crops, which may be expected, The crop range in number of kinds best grown usually decreases in both directions, becoming decidedly limited at a rapid rate in the case of more sandy soils, and at a less rapid rate in the case of the clay loams and clays. This expresses moisture control. It has been more difficult to control moisture in the sandy soils than in the clay loams and clays. Irrigation is the answer to the difficulty with the sands, and drainage with the clays.

"Leguminous crops of all descriptions are particularly favored by a high lime content in both soil and subsoil.

"Soils well supplied with organic matter atone for some other soil deficiencies in texture and structure.

"Compacted layers of any kinds beneath the surface soil are unfavorable to crop production. This applies to compacted subsoil, due to shallow plowing, as well as to actual 'hard-pan.'

"Good soil management always increases the range of crops which may be grown as well as the amounts harvested. Man's ingenuity may be used profitably to overcome nature's deficiencies.

"**Eastern Soils Not Worn Out.**—Finally, I wish to state as a result of years of observation under widely varying circumstances of soil study and of farming:

"I. That the soils of the Northeastern states are in nowise 'worn out' or seriously depleted of anything essential to good crop production with the local exception of organic matter in the surface soil.

"II. That the majority of soils of the Northeastern states are capable of producing average crops or greater if given fair treatment, especially when the proper crops for the climate and the soil are selected for planting and others are discarded.

"III. That soils which have been called 'worn out' have frequently revived within a period of five years or less of good farming methods, until their yields equaled or exceeded any production before known upon that soil.

"IV. That the best methods of crop growing and of soil management now practiced by the best farmers of the Northeastern states would, if made general in their application, more than double the total cropping ability of the improved lands now in use.

"V. That the market facilities of the Northeastern states are now and will continue to become more and more favorable to the intensive use of land and to the man who uses each acre for the crop or group of crops best suited to his soil and climate.

* * * * * * * * *

"To the young farmers who are to carry on the great work of redeeming land and of feeding people I have just one more thing to say. Study the fundamental principles, which are true in Asia or the United States;

true today and for the centuries to come; true for all crops and for all seasons. The details of modifying these principles of agriculture, experience alone can teach you."

SOIL ADAPTATION OF FIFTEEN CROPS COMMON TO NORTHEASTERN STATES.

CROPS.	SOILS BEST SUITED TO.	WAYS OF MODIFYING SOILS TO FIT CROPS.	FERTILIZERS TO APPLY.
Wheat.	Clay and silt loams containing considerable lime. Surface soil friable. Subsoils of same nature, but heavier and more compacted.	Use manure liberally. Practice rotation with leguminous crops. Apply moderate amounts of lime.	Principally phosphatic fertilizers containing small amounts of nitrogen and potash.
Oats.	Wide adaptation. Loams or heavy loams rather fine in texture best. Avoid dry sands. Plenty of humus desirable.	Apply manure to crop preceding. Turn under green manure. Plow only moderately deep. Seed early in spring. Prepare land thoroughly.	Always use some form of phosphate, preferably acid phosphate or basic slag. Use small amounts of potash, usually muriate.
Rye.	Well-drained, sandy loams give the longest, brightest straw and largest crops of grain. Will do fairly well on lighter and poorer upland soils.	Smaller amounts of humus necessary. Will grow on more acid soils than wheat or oats. Fine general utility crop.	About same as wheat. Little lime needed.
Barley.	Well-drained fertile loam. Intermediate between rye and oat soils. Heavy loams give best yields. Sandy loams give brighter grain. Avoid clay on account of lodging and too light sand because of drought.	Requires moderate amount of humus. Avoid too rich soils on account of lodging. Good drainage essential.	About same as oats.
Buck-wheat.	Moderately friable loam, underlain by compacted but well-drained loamy subsoils.	Will do well on rather poor, thin hill lands, because of power to loosen pulverized soil. Prepare land thoroughly, providing organic matter. Good drainage necessary.	Complete fertilizer.
Potatoes.	Sandy or sandy loam preferably for early crop. Silt loam or loam best for late. Avoid clay and clay loams.	Thorough drainage essential. Abundant organic matter needed. Grow in rotation and turn under green manures.	Apply large amounts of fertilizer high in potash. Small amounts of nitrogen for late crops. More on sandy soils. Avoid liming immediately ahead of potatoes.
Corn.	Loam or silt loam, with heavier subsoil at least ten inches below surface. Where seasons are short, sandy or gravelly loams give larger yields, because of earlier maturity.	Well-drained, moisture-holding lands. Turn under good grass sod or preferably clover sod. Apply barnyard manure to previous crop if possible.	Use 200 to 500 lbs. of fertilizer containing 3 to 4 per cent of nitrogen, 8 to 12 per cent phosphoric acid, 3 to 4 per cent potash.
Clover and Timothy Hay.	Loam or clay loam best. Heavy soils retain moisture best. Avoid too compacted clays or hardpans. Timothy: Loam or well-drained clay loam or clay.	Use stable manure on preceding crop. Apply lime in most cases. See that both surface and subsoil are well drained. Prepare land very thoroughly for seeding.	Stable manure best fertilizer; 100 to 300 lbs. an acre of complete fertilizer. High in nitrogen (8 to 10 per cent). Gives good results.
Alfalfa.	Very fertile, well-drained, alkaline soils. Strong loams containing limestone best. Avoid shallow soils and hardpans near surface.	Drain soil thoroughly. Standing water fatal to alfalfa. Apply lime liberally. Inoculate soil.	Top dress with stable manure or with 300 to 400 lbs. of acid phosphate or 400 to 600 lbs. basic slag, or 200 lbs. or more of steamed bone meal an acre.
Beans.	Wide range of soils. Best results on types not more coarse grained than sandy loam or more compacted than clay loam. Lime-bearing soils best.	Must be well drained and well supplied with organic matter. If soils do not contain limestone give moderate application of lime.	Fertilize with 200 to 300 lbs. an acre of mixture containing 2 per cent nitrogen, 8 to 12 per cent phosphoric acid, 4 to 6 per cent potash. Use stable manure.
Apples.	Fairly deep, well-drained loams and clay and silt loams with fair proportion of sand in surface soil. A heavy subsoil retentive of moisture, but not impervious to water.	See that soils are thoroughly drained. Apply moderate amounts of manure. Plow under leguminous cover crop. In general give thorough cultivation in early part of the season.	Depends on soils and variety. On heavier soils none may be needed except stable manure, which is always best. Experiment with commercial fertilizers.

SOIL ADAPTATION OF FIFTEEN CROPS COMMON TO NORTHEASTERN STATES (*Continued*).

CROPS.	SOILS BEST SUITED TO.	WAYS OF MODIFYING SOILS TO FIT CROPS.	FERTILIZERS TO APPLY
Cabbage.	Heavy loam or silt loam, with retentive subsoils. Muck soils generally well suited if not too loose.	See that soil is well supplied with organic matter. Apply lime liberally to surface of soil. Grow crop in rotation.	Apply complete fertilizer, high in potash and moderately high in nitrogen, in liberal amounts.
Celery.	Muck soils best adapted. Silty river flood plains and silty or fine silty uplands, high in organic matter, will do.	Soil must be moist, but well drained and well supplied with organic matter. Lime and salt both affect celery favorably.	Fertilize heavily with stable manure where possible. Large amounts of commercial fertilizer, rich in nitrogen, can be applied profitably.
Onions.	Sandy loam just above water level, protected from overflow and well supplied with moisture. Strong, well-drained muck land tilled two or three years.	Must be well drained. Large amounts of organic matter necessary. Lime gives good results. Crop rotation or alternation desirable.	Stable manure and high grade commercial fertilizers must be abundantly supplied for continued large yields.
Tobacco.	Many grades of soil from light silt to heavy loams suitable, depending on grade of leaf desired.	Must be well drained. High in organic matter. Very thoroughly prepared soil and constant cultivation necessary.	Depends on kind of soil and type of leaf being grown. Usually requires large amounts of potash derived from sulphate. Liming usually thickens leaf and makes it harsh.

Following the plan of Dr. Bonsteel, the author has gone carefully through the soil literature of the United States and summarized the crop adaptations, the means of modifying soils and the fertilizers to apply to them. This is given for the leading crops by regions as follows: (1) The North Central region, covered mostly by the Glacial and Glacial lake soils lying between Pennsylvania and the Dakotas, and north of the Ohio and Missouri Rivers; (2) the South Central and South Atlantic Coast region, comprising Delaware, Maryland, Virginia, West Virginia, Kentucky and the Cotton Belt; (3) the Plains and Mountain region west of the 97th meridian of longitude; and (4) the Pacific Coast region, including the three coast states and most of Nevada.

The following is a summary of the leading crops adapted to soils of the North Central region:

Sand.—Good for very early truck and small fruits; fair for sugar beets and poor for small grains. May be kept in grass to prevent drifting.

Sandy Loam.—Good for tobacco, truck, apples, beans, root crops, fruit, and fair for hay, small grains and corn.

Loam.—Good for general crops, truck and fruit.

Silt Loam.—Finest corn soil; good for small grains, hay, fruit, tobacco and heavy truck, such as cabbage.

Clay Loam.—Best wheat soil; good for corn, oats, rye, barley, grass, clover, alfalfa and fruit.

Clay.—Good for hay, small grains, export tobacco, some fruit and small fruit. (For continuation see next page.)

The following is a summary of the leading crops adapted to soils of the South Central and South Atlantic Coast region:

Sand.—Adapted to earliest vegetables, some fruits and some varieties of grapes. Small grains may be grown, but do better on heavier soils.

SOIL ADAPTATION OF THE LEADING CROPS OF THE NORTH CENTRAL REGION.

CROPS.	SOILS BEST SUITED TO.	WAYS OF MODIFYING SOILS TO FIT CROPS.	FERTILIZERS TO APPLY.
Corn.	Loam or silt loam. Deep soil with heavy subsoil. For short season, sandy loam.	Well-drained moisture-holding lands. Turn under good grass or clover sod. Apply barnyard manure.	Phosphoric acid and legumes. Use lime on sour soils.
Wheat.	Clay or silt loam. Deep soil well supplied with humus. Subsoil, heavier clay.	Rotate with legumes and hoed crops. Add organic matter as manure or green manure when available.	Small to moderate amounts of fertilizers high in phosphoric acid, and with small amounts of nitrogen and potash. For western portion, phosphoric acid only.
Oats.	Any soil but light sand. Loam or silt loam best. Good supply of humus desirable.	Should follow hoed crops, usually corn. Prepare seed bed by disking, seed early, drilling preferable.	Manure or fertilizer should be applied to preceding crop. On poor soils, small amounts of phosphorus and nitrogen may be used.
Rye.	Sandy loam or loam; must be well drained.	Good crop for poor land; will stand considerable acid.	About same as wheat. Does not need much lime.
Barley.	Loam to clay loam. Clay causes lodging. Heavy soils give larger yields; light soils brighter straw.	Moderate amounts of humus. Must be well drained. Too rich soils will cause lodging.	About same as oats.
Buckwheat.	Loam with well-drained loamy subsoil.	Good pulverizer, hence will do well on rather poor soil. Good drainage essential. Add organic matter.	A minor crop, seldom fertilized. Small amounts of complete fertilizer advised for poor soils.
Potatoes.	Sandy loam or loam; avoid heavy soils.	Fall plow; use winter cover crop and turn under. Grow in rotation. Thorough drainage needed.	Do not lime immediately before potatoes. Apply fertilizer high in potassium.
Hay, Clover, Timothy.	Wide variety of soils. Loam to clay loam best.	Drain land, top dress with manure; small applications spread uniformly.	Top dress beginning of second year with small amounts of complete fertilizer high in nitrogen.
Alfalfa.	Rather heavy soil but must be deep and well drained.	Plow deep and inoculate soil.	Use good supply mineral fertilizer and lime.
Beans.	Sandy loam and clay loam best.	Apply manure and drain.	Moderate amounts complete fertilizer high in phosphoric acid and potash. Apply lime.
Apples.	Loamy soil best; must be quite deep and well drained. Avoid poor air drainage.	Sow to cover crop, preferably to legume in fall; plow under in spring and cultivate clean during early summer.	Depends on soil. On good soils, none needed for several years. Experiment.
Heavy Truck—Cabbage, Celery, etc.	Heavy loams or muck soils, high in organic matter.	Use plenty of stable manure.	Complete fertilizer high in nitrogen. Also lime.
Other Truck—Lettuce, Radishes, etc.	Light soils, sandy for very early markets; sandy loam and loam for later crops.	Must be prepared to irrigate sand. Apply lots of manure. Rotation desirable.	High grade complete fertilizer. High nitrogen content for leaf crops, as lettuce.
Tobacco.	For "bright" cigarette tobacco, sand; for wrapper, sandy loam; for filler and export grade, heavier soils.	Prepare soil thoroughly and cultivate frequently. Must have high organic content and be well drained for best results.	Avoid lime, as it thickens leaf. Kind of fertilizer depends on the soil. Usually large amounts of potassium sulphate.
Plums, Cherries, Small Fruits.	Sand and sandy loam. Provide for good air drainage in order to avoid danger from frost.	Use leguminous cover crops for winter. Clean cultivation in summer.	Varies with soil and location. Experiment.

Sandy Loam.—"Bright" tobacco, mid-season truck, peanuts, forage crops and cotton and small grains to some extent.

Loam.—Cotton, tobacco, main crop truck, corn, small grains, sugar cane, fruit and small fruit, legumes for hay or cover crops, rice and nursery stock.

Silt Loam.—Cotton, tobacco, truck for canning, corn, small grains, hay and pasturage, tree and small fruits.

Clay Loam.—Cotton, export tobacco, corn, small grains, very good for grazing, fruit, rice, flax, hemp, etc.

Clay.—Rice, sugar cane, export tobacco, forage crops, hay and fruit.

Soil Adaptation of the Leading Crops of the South Central and South Atlantic Coast Region.

Crops.	Soils Best Suited To.	Ways of Modifying Soils to Fit Crops.	Fertilizers to Apply.
Cotton.	Loam or silt loam.	Fall plow, cultivate frequently, rotate with legumes.	Add manure and other forms of organic matter. Complete fertilizer.
Corn.	Any soil but very light sand and heavy clay. Best on loam.	Plow deep and rotate.	Complete fertilizer high in phosphoric acid. Also plenty of organic matter. Add lime.
Tobacco.	Varies with kind of tobacco grown. (See North Central Region.)	Frequent, careful cultivation and cover crop in winter to prevent erosion. Rotate with legume.	Do not lime light tobacco. Avoid muriate of potash in fertilizer.
Sugar Cane.	Loam to clay; best on clay loam. Soil must be rich.	Drain when needed; add organic matter.	Heavy complete fertilizer.
Truck.	Sand for extra early, loam for main crop.	Must be well drained and have abundant supply of humus.	High grade complete fertilizer.
Rice.	Clay or clay loam; heavy subsoil essential.	Must be able to flood at proper time and drain at proper time.	Plow deep and add lime.
Peaches, Plums, Cherries, Small Fruits.	Sand or sandy loam.	Use cover crops to prevent washing, legumes best.	Varies with location, climate and crop. Experiment.
Forage Crops—Millet, Sorghum, etc.	Clay loam or clay.	Plow deep, use winter cover crop.	Complete fertilizer and manure, or green manure.
Grapes.	Varies with variety from sand to clay.	Add organic matter.	Varies with soil. Experiment.
Peanuts.	Sandy loam.	Organic matter and fall plowing.	Mineral fertilizers.
Annual Legumes, Cowpeas, Soy Beans, etc.	Sandy loam to clays.	Plow deep, give good cultivation. Good for interplanting with cotton or corn.	Mineral fertilizers and lime.

Plains and Mountain Region.—Most of this region is semi-arid to arid and used largely as pasture, but where transportation and water are available, very good crops may be grown by the aid of irrigation. The following is a summary of the leading crops adapted to soils of the Plains and Mountain region:

Sand.—Is the predominating soil and care must be taken to prevent its drifting. It gives fair crops of truck, fruit, cotton, Kaffir, sorghum, wheat, oats and hay.

Sandy Loam.—Does not drift quite so badly. On it may be grown truck, fruit, cotton, Kaffir, sorghum, milo, sugar beets, wheat and alfalfa. It also gives good pasturage.

Loam.—Is about the most productive soil. It is good for broomcorn, sorghum, milo, truck, sugar beets and, in the South, cotton. In the Central States small grains and forage crops; and in the North, wheat, oats, flax and millet.

Silt Loam.—Is not quite so good as loam, but is used for about the same crops.

Clay Loam.—Is very hard to handle and not very productive. It is used for general crops and special local crops.

Clay.—Very hard to manage to prevent puddling. It is used to some extent for general crops, but chiefly for grazing.

SOIL ADAPTATION OF THE LEADING CROPS OF THE PLAINS AND MOUNTAIN REGION.

CROPS.	SOILS BEST SUITED TO.	WAYS OF MODIFYING SOILS TO FIT CROPS.	FERTILIZERS TO APPLY.
Cotton.	Loam.	Irrigate.	Manure and complete fertilizer.
Corn.	Loam to clay loam.	Plant with lister. Manure, cultivate frequently.	Fertilizer seldom used.
Small Grains.	Silt loam.	Add organic matter.	Fertilizer seldom used.
Hay and Pasturage	Most any soil with enough water.	Do not pasture too closely or when wet.	
Sugar Beets.	Sandy loam and loam.	Irrigate, plow deeply and give clean cultivation.	Complete fertilizer.
Forage Crops—Kaffir, Sorghum, Millet.	Loam best, but will grow in wide range of soils.	Plow deeply, give thorough cultivation. Do not plant too early.	
Alfalfa.	Sandy loam to clay.	Plow deeply; irrigate. Seed and light crops of hay produced without irrigation.	

Pacific Coast Region.—This region is in most places almost arid. With the aid of irrigation it becomes one of the garden spots of the country. The following is a summary of the leading crops adapted to soils of the Pacific Coast region:

Sand.—Used for early truck, figs, stone fruits, citrus fruits and some of the small fruits. It requires large amounts of water and frequent cultivation to conserve moisture.

Sandy Loam.—Used for most of the fruits grown in this region, also grapes, small fruits, alfalfa and, to some extent, general crops. This soil is quite light and requires much the same care as sand.

Loam.—Used for fruit, late truck, small fruit, grapes, hops, hay and general crops.

Silt Loam.—Used for fruit (including citrus fruit), small fruit, heavy truck, English walnuts.

Clay Loam.—Used for fruit, small fruit, truck for canning, and general crops. This soil is much used in southern California for citrus groves and lima beans.

Clay.—Grains and hay, some heavy truck and tree fruit.

Soil Adaptation of the Leading Crops of the Pacific Coast Region.

Crops.	Soils Best Suited To.	Ways of Modifying Soils to Fit Crops.	Fertilizers to Apply.
Truck.	Sandy loam for early; silt or clay loam for late.	Add lots of organic matter.	Depends on crop and soil.
Fruit.	Any soil; loam or silt loam best for most fruits.	Practice clean cultivation to prevent evaporation. Add organic matter.	Varies with kind of fruit.
Grapes.	Sandy loam or loam.	Same as for fruit.	Complete fertilizer.
Small Fruit.	Sandy loam to silt loam.	Same as for fruit.	Experiment.
English Walnut.	Silt loam.	Cultivate clean in dry season, but grow cover crop in rainy season, and plow under.	
General Crops Grains, Hay.	Any of the heavier soils.	Give soil thorough preparation before planting and cultivate wherever possible.	Complete fertilizer.

Aids to Solution of Soil Problems.—The soil survey conducted by the Bureau of Soils of the United States Department of Agriculture, in co-operation with the various state departments of agriculture or agricultural experiment stations, is now extended into many counties in every state. Two kinds of surveys have been made: (1) that known as the reconnoissance soil survey, in which detailed mapping is not undertaken (it consists chiefly in mapping the soil series); and (2) a detailed county survey showing the location and extent of each soil type. The results of this work are issued as government reports, accompanied by colored maps outlining the soils. In these reports the soils are fully described and their crop adaptations stated. Much other valuable data pertaining to agricultural conditions, climate and soil requirements are also given. These reports are available to all farmers living in the districts in which the surveys are made. They may be secured either through the local senator or representative, or directly from the National Department of Agriculture. In some cases the state experiment station or state department of agriculture will be able to supply them.

The detailed county surveys will enable any one in such an area to ascertain the types of soil on his farm. If there is any doubt in this particular on the part of the farmer, he can submit samples of his soil to his state experiment station, and by giving the exact location of his farm, the authorities at the station will be able to advise him not only as to the type of his soil, but in a general way can give him facts concerning crop adaptation and the treatment most likely to bring good results.

Samples of soil should accurately represent the field from which taken. Samples should be taken to the depth of plowing in not less than ten places in the field. These may be put together and thoroughly mixed. A pound of this mixture sent to the experiment station by parcel post will meet the requirements. It is frequently desirable also to send a sample of the subsoil. If there is no great hurry it will be better to write to the experiment station first and ask for instruction on collecting and sending samples.

A SOIL AUGER.[1]

The soil auger is most convenient for taking soil samples. It consists of an ordinary $1\frac{1}{2}$-inch wood auger having the shank lengthened and the threaded screw and sharp lips removed. Any blacksmith can do the work in a few minutes. The accompanying figure shows a three-foot auger with gas pipe handle. For a farmer's use the wooden handle will serve just as well. If an auger is not available, a square-pointed spade will serve very well for taking samples. Dig a hole to the depth of plowing, having one perpendicular side, then cut from the perpendicular side a slice of uniform thickness from top to bottom. This repeated in ten or more places in the field will give a sample representing the soil accurately.

Because of the difficulty on the part of the experiment station authorities in giving definite advice at long range, some of these institutions now employ experts who travel about the state, inspect farms and consult with farmers relative to their soil problems as well as other problems of the farm. By such inspection these men are able to advise more definitely than can be done by letter.

In the last few years another innovation for the benefit of the farmers has been introduced, namely, the providing of the county farm adviser, who is located within a county permanently and who soon becomes familiar with the agricultural problems of his restricted territory. Through these sources the farmer can always secure able assistance in the solution not only of his soil problems, but of all problems that concern his business.

REFERENCES

"Soils: How to Handle and Improve Them." Fletcher.
"Soils." Lyon and Fippin.
"Soils." Burkett.
Pennsylvania Agricultural Expt. Station Bulletin 132. "Soils of Pennsylvania."
Canadian Dept. of Agriculture Bulletin 228. "Farm Crops."
Farmers' Bulletin No. 494, U. S. Dept. of Agriculture. "Lawn Soils and Lawns."

[1] Courtesy of The Macmillan Co., N. Y. From "How to Choose a Farm," by Hunt.

CHAPTER 2

PHYSICAL, CHEMICAL AND BIOLOGICAL PROPERTIES

Texture of Soil.—Texture pertains to the size of the mineral particles that make up the body of the soil. In the laboratory, texture is determined by a mechanical analysis. This is described in Chapter 1. The clay portion of a soil will range anywhere from a fraction of one per cent to as high as fifty per cent of the body of the soil. The particles of clay are so small that they can be seen only by the use of a high-power microscope. When clay is thoroughly mixed with water the particles will remain in suspension for several days. It is this clay that is chiefly responsible for the turbid condition of the streams of water flowing from the land after heavy rains. Clay, when thoroughly wet and rubbed between the thumb and finger, has a smooth, greasy feel.

The silt may also range from a very small percentage to sixty per cent or more of the body of the soil. It forms the group of particles next larger than clay. It produces practically no perceptibly gritty feel when wet and rubbed between the thumb and finger. Silt particles will remain in suspension in water for only a short time, seldom more than one-half hour.

The various grades of sand consist of particles very much larger than those of either clay or silt, and can be seen with the naked eye. The percentage of sand in soils like that of clay and silt varies between wide ranges. Sandy soils may contain seventy-five per cent to ninety per cent of the different grades of sand. All of the sandy soils give a distinctly gritty feel when the wet soil is rubbed between the thumb and finger.

Water-Holding Capacity of Soils.—The texture of the soil is very important and determines in a large degree the water-holding capacity of the soil, the rapidity of movement of water and air in the soil, the penetration of plant roots, ease of cultivation and, above all, the crop adaptation of the soil. Texture is determined by the relative amounts of the particles that fall into the several groups mentioned. The textural effect is modified by the structure of the soil (discussed later) and its content of organic matter.

The larger the proportion of fine particles, such as clay and silt, the greater is the surface area of these particles in a unit volume of soil. In a well-drained soil all gravitational water passes away and only capillary water is retained. This capillary water consists of very thin films of water adhering to the surface of the soil particles and surrounding them in such a way as to make a continuous film of water in the soil. Through this continuity of the film, water moves by capillarity from a point where the films are thickest to a point where they are thinner, tending always to equality in the thickness of the film, but gradually becoming thinner as the distance from the source of water increases.

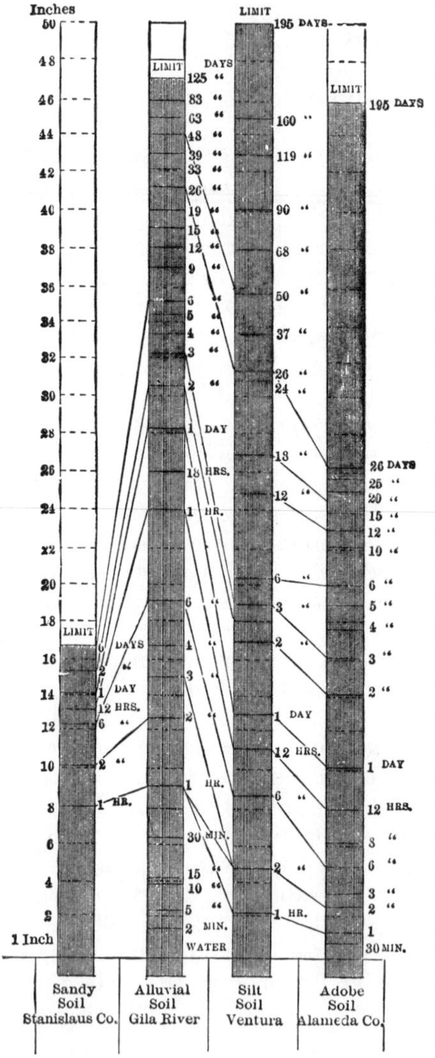

RATE AND HEIGHT OF CAPILLARY RISE OF WATER IN SOILS OF DIFFERENT TEXTURE.[1]

It is evident, therefore, that the fine-textured soil will hold much more water than the one consisting largely of sand. Such a soil can supply crops with more water than a sandy soil, and such a soil is adapted to grass, wheat and other plants having fibrous roots that do not penetrate to great depths.

If a glass tumbler is filled with water and emptied, a thin film of the liquid adheres to the surface. This will equal only a fraction of one per cent of the weight of the tumbler. If the tumbler can be pulverized into a very fine powder and the particles saturated with water and allowed to drain, they may hold water to the extent of ten to fifteen per cent of the weight of the glass. This change in the water-holding power is the result of pulverization and especially of the increase of the exposed surface which is brought in contact with the liquid. The finer the degree of pulverization the larger the percentage of water the glass particles will retain. So we find that soils of very fine texture will sometimes hold as much as forty per cent of their weight of water, while some of the coarse, sandy soils will not hold more than four or five per cent of their weight of water. This water-holding capacity of the soil is also modified by its content of organic matter. Organic matter

[1] Courtesy of The Macmillan Company, N. Y. From "Soils," by Hilyard.

will absorb from two to four times its own weight of water. The sponge best illustrates the capacity of organic matter to absorb and hold water.

Water Movement in Soil.—The movement of water in the soil is influenced chiefly by soil texture. In soils of coarse texture the water moves very freely. Drainage is rapid and the soils dry soon after rains so that tillage operations may soon be resumed. On such soils there is generally little loss of time during the period when they need tillage. On very heavy soils, that is, on those consisting chiefly of clay and silt particles, the movement of water within the body of the soil is exceedingly slow. Drainage is difficult, and where the land is level and the substratum is dense, underdrainage is often required in order to make the soils productive. In sandy soils the rainfall penetrates and descends rapidly through the soil body. In this kind of soil leaching is rather rapid. Rain penetrates heavy soils very slowly, and if the rainfall is rapid, its passing from the surface of the soil causes severe erosion. Furthermore, a large proportion of the rainfall is thus lost and in no way benefits the growing plants. On the part of the farmer it therefore becomes essential so to plow and cultivate the fine-textured, heavy soil as to increase its penetrability and facilitate the movement of air and water and the penetration of roots as much as possible. In case of the very sandy soil it is often advisable to do just the reverse. Applications of lime, which tend to cement the particles together, and of organic matter to fill up the interspaces, and compacting the soil by rolling to reduce the spaces, are often resorted to. Where land has a high value it may even pay to add clay to a sandy soil in order to improve its physical properties. On the other hand, it may sometimes pay to add sand to a very heavy, clay soil. Such practice, however, is justifiable only in case of land of high value when used for intensive cropping.

Absorption of Fertilizers.—The absorptive power of the soil is also proportional to the surface area of the particles within a unit volume. Soils of fine texture are, therefore, capable of absorbing and holding much larger amounts of fertilizers than those that are sandy. This is very important in connection with the application of fertilizers. It is also true that the soil absorption is much stronger for some substances than it is for others, and this will often determine the time of application of fertilizers. The absorptive power of the soil is less marked for nitrogen, either as ammonia or nitrates, than it is for either potash or phosphorus. Consequently, nitrogenous fertilizers should be used in quantities just sufficient to meet the needs of the crop, and applied just preceding the time at which the crop most needs it. In view of this fact, surface applications of nitrogen are often effective, since the downward movement of the material in the soil soon brings it into the region of root activity.

Potash and phosphorus are, however, absorbed and held much more tenaciously by the soil particles, and are not subject to severe loss by leaching. Liberal applications of potash applied to the surface of the soil to which large amounts of water were applied by irrigation were found to have penetrated to a depth of only about three inches in the course of as many months. This suggests that such fertilizers should be distributed in that zone of the soil where root activity is most marked, in order that the plants may utilize the fertilizer as fully as possible. All of this has a bearing upon the fertilizer practices which will be discussed in a subsequent chapter.

Plasticity and Ease of Cultivation.—Soils of fine texture are very plastic when wet, and clay soils in this condition tend to adhere to cultural implements, wheels of vehicles and the feet of animals. Such soils should not be tilled when they are wet. The movement of the soil particles upon one another when in this condition causes them to be cloddy and hard when they dry out. It furthermore gives rise to what is known as puddling, and prevents the free movement of water and air through the soil. This is well illustrated by a clay road in the spring when wagons pass over it and form ruts while it is in a wet condition. These ruts will often become filled with water, which escapes only by evaporation, none of it finding its way through the soil below. The fine-textured soils, when not well supplied with organic matter, tend to run together and become very compact and difficult to cultivate. This condition can be alleviated to a certain extent by avoiding tillage operations when too wet, and also by the application of organic matter in the form of manure or green manuring crops. Likewise, this condition is improved by the application of lime, which causes a flocculation of the soil particles; that is, causes them to gather into little groups with larger spaces between these groups.

The sandy soils and those containing a liberal amount of sand are less affected by rains, are more easy of cultivation and do not call for as great precautions in their tillage. Such soils when wet do not adhere to cultural implements and the feet of animals as do the heavy soils, and the roads made of such soil are often as good or better immediately after rains than they are when in a dry condition.

Texture Affects Crop Adaptation.—Heavy clay soils and those containing large amounts of silt are generally best adapted to the grasses such as timothy, blue grass, orchard grass and redtop, and to wheat, rye and what is commonly known as the heavy truck crops, such as cabbage, tomatoes and asparagus. The soils known as loam, which are of medium texture, are better adapted to such crops as corn, oats, barley, buckwheat, peas, beans, clover and potatoes. The soils of light texture, known as fine sand and sandy loams, are also well adapted to potatoes, beets and all tuber and root crops, and are also extensively used for the early truck crops, such as spinach, lettuce, early potatoes, early peas, etc. Some of the very lightest sands, such as are found in certain parts of Florida, are especially adapted to the growing of pineapples. In general, the pomaceous fruits, such as apples and pears, will do well on fairly heavy soils, while the stone fruits, such as peaches, cherries and plums, succeed better on soils that are lighter in texture and better drained. In fact, peaches will often succeed admirably on shaly ridges and mountains in the Piedmont Plateau.

Texture Affects Tillage.—Soil texture so influences the cost of tillage that it often determines the crop to be grown. Crops that require a great deal of tillage and hand work, such as sugar beets, are more economically

THE EASE OF SEED-BED PREPARATION DEPENDS ON CONDITION OF SOIL.[1]

grown on soils of light texture, because of the greater ease of weeding and tillage. Even though these light soils under intensive cultivation may require considerable expenditure for fertilizers, the additional cost thus entailed is generally more than offset by the saving in labor.

Structure of the Soil.—The structure of the soil pertains to the arrangement of the soil particles within the body of the soil in much the same way that the arrangement of the bricks in a building determines the style of architecture. In all soils of fine texture it is good soil management to strive to obtain a granular structure. This consists of a grouping of the soil particles into small groups or granules. A good illustration of

[1] Courtesy of Doubleday, Page & Co., Garden City, N. Y. From "Soils," by Fletcher.

a granular structure is found in what is known as buckshot land. Such a soil when plowed breaks up into small cubical fragments an eighth of an inch to a quarter of an inch in size. The granular structure facilitates the circulation of the air and soil moisture, permits easier penetration by plant roots and lessens the difficulty of cultivation.

Granular Structure.—The granular structure may be improved by tillage. Every time the soil is plowed, cultivated, disked or harrowed, it is pulverized and broken up into particles, each formed of a larger or smaller number of grains. Granular structure is also improved by good drainage. When the body of the soil is saturated or completely filled with water the soil particles move with little resistance and tend to arrange themselves into a compact mass. This fact is taken advantage of in filling excavations, and when the soil is returned to the excavation water is turned into it in order that it may settle compactly, so that when once filled no depression will occur at the surface. Soils that are thoroughly underdrained seldom, if ever, become saturated, so that there is no opportunity for the soil particles to arrange themselves in this compact mass. Consequently, a soil of this character when once drained gradually assumes the granular structure through plowing and cultivation, together with the penetration of the roots of the plants and the work of insects and worms. This is further facilitated by the thorough drying of the soil in periods of prolonged drought.

The process of alternate freezing and thawing also has an influence on structure. As the water in the soil solidifies it expands and causes an elevation of the soil, making it more porous. As it thaws and the water again becomes liquid the soil does not fully return to its original position, and consequently its tilth is improved.

Granulation Improved by Organic Matter.—Granular structure is also improved by the addition of organic matter to the soil either as barnyard manure or the residues of crops turned under. The organic matter incorporated with the soil occupies spaces that would otherwise be occupied by soil particles, and upon its gradual decay it leaves small cavities which separate small groups of soil particles. Plant roots are also influential in improving the structure of the soil, first, by an actual moving of the soil particles due to the enlargement of the roots as they grow; and, second, by the gradual decay of these roots, which leaves minute channels in the soil through which air and water find free passage. Earthworms open channels of considerable depth, and also incorporate in the soil the organic matter upon which they live.

Good Tilth Important.—It is common to speak of the soil as having a good or poor tilth. A soil in good tilth means that it is in good physical condition, or that it has a granular structure that makes it the best possible home for the plants to which it is adapted. The degree of granulation desired will be determined to considerable extent by the character of crop that is planted. Corn and potatoes, demanding a rather open soil, call for a loose seed-bed in which granular structure is accentuated. Wheat, rye, clover and the grasses, on the other hand, demand a rather compact, fine-grained seed-bed, and, therefore, do not demand an equal degree of granulation.

Solubility of Soil Minerals.—Plants take their mineral food only when it is in solution. This necessitates a degree of solubility of the essential plant food minerals that will meet the maximum needs of the plants. The solubility of the soil particles depends upon a number of factors, and is a rather complex process. In pure water the solubility is very slight, but as the water of the soil becomes impregnated with carbonic acid gas, organic compounds and mineral compounds, these all exert an influence on the degree of solubility of other mineral constituents. Solubility is also markedly influenced by temperature. This fact is well recognized by the housewife, who by heating dissolves sugar in water until it becomes a syrup; so the solubility of the soil minerals is increased by a rise in soil temperature.

Rate of Solubility Depends on Texture and Kind of Minerals.—The rate of solubility is approximately in proportion to the surface of the particles on which the solvent acts. Consequently, we find as a rule larger amounts of plant food in solution in soils of fine texture than we do in soils that are coarse in texture. This doubtless accounts for the practice of the more extensive use of fertilizers on sandy soils. It is also true that the different minerals have varying degrees of solubility, some being far more soluble than others. The limestone particles in a soil mass are much more readily soluble than the quartz, and, consequently, lime disappears from the soil. Plant roots also have an influence upon solubility by means of certain excreta given off by the roots. Since, therefore, carbon dioxide, organic compounds and plant roots increase the solubility of the soil particles, it is plain to be seen that the incorporation of organic manures with the soil and the production of good crops tend always towards a more productive soil, except in so far as the minerals of the soil are exhausted through plant removal.

Soil Bacteria Increase Solubility.—The bacteria of the soil are also instrumental in increasing the solubility of the soil minerals. Since, for their greatest activity, bacteria require proper sanitary conditions, such as aeration, a neutral soil medium and organic matter as their food, it will be seen that fertile soils encourage increased numbers of bacteria, which in turn make for increased fertility. It is, therefore, essential for the tiller of the soil to understand the various factors which enter into soil productivity, and to perform his part in encouraging the development of those which are beneficial and discouraging those which may be destructive.

Rapid Solubility Results in Loss of Fertility.—The rate of the solution of soil minerals should not far exceed the needs of the crops grown, lest there be an unnecessary loss of plant food through leaching and the consequent hastening of the impoverishment of the soil. Except in very sandy soils, in the practice of bare fallowing of soils, and in the Southern states where land is left without cover-crops, there is very little danger, however, in this regard.

Chemical Composition of Soils.—The soil has long been an intricate problem for the chemist. Many years of research have been spent in an endeavor to determine through chemical analysis not only the composition of the soil but its power to produce crops and its need for fertilizers. The chemist has little difficulty in determining the absolute amounts of the essential plant food constituents in the soil, although the process is rather long, tedious and costly. Unfortunately, such analyses seldom indicate the relative fertility of different soils, and tell us comparatively little as to the present fertilizer needs of them. The chemist has also endeavored to devise methods of analysis that will determine the amounts of available plant food present in the soil. For this he has used different solvents of varying concentrations in an endeavor to imitate the plant in its extraction of the elements from the soil. So far, however, such methods have met with comparatively little success, and we are, therefore, obliged to conclude that, as a rule, a chemical analysis of the soil is of very little help to the farmer. This statement admits of certain exceptions. If the analyst finds that the total potash or phosphorus content of a soil is very small, it at once indicates that this soil is either immediately in need of the deficient element or soon will become so. It is also true that, when the physical conditions of the soil are good, the drainage satisfactory and unusually large amounts of the essential elements are present, the soils are, or may easily be made, productive without the addition of plant food.

The above statements should not be construed to mean that the chemist should cease to put forth his best efforts in the solution of unsolved soil problems; but in its present status, it is not worth while for the farmer to ask for a complete chemical analysis of his soil, or to go to the expense of having a commercial chemist make such an analysis for him. Chemical analyses are useful and helpful to the scientist and soil expert, and are to be encouraged as a help in the advancement of our knowledge of soils.

Availability Important.—In the majority of cases it is important that the farmer know how to increase the availability of plant food in the soil. This question has been partly analyzed in the preceding topic on solubility of soil minerals. In general, however, the farmer may increase availability by deep plowing, thorough tillage, the incorporation of organic matter and soil drainage. The best measure of soil fertility or available plant food is the growth that plants make upon any particular soil. Not only is the degree of growth an indication of fertility, but likewise the color of the plants, the manner of growth and the proportion of vegetative parts to seeds or fruits are often indicative of the presence or absence of particular elements. The first essential to profitable crops is the production of a healthy and vigorous plant. Added to this is a high degree of fruitfulness. A deficiency in phosphorus may not prevent a satisfactory development of the plant, but may seriously curtail the production of seed. This is often illustrated in the case of wheat which makes a rank growth of straw and a comparatively small yield of wheat. The absence of available nitrogen is often indicated by the yellow color of the foliage.

The form in which the elements are combined may influence the quality of the product. This is illustrated in tobacco when the application of muriate of potash causes a poor burning quality of the leaf that is to be used for cigars. Better results with a cigar tobacco are secured when the potash is applied in the form of sulphate or carbonate. Furthermore, the essential plant food constituents dominate in the development of certain parts of the plant or in the performance of certain vegetative functions. For example, potash is believed to be largely instrumental in the development of starch, and fertilizers for starch-producing plants, such as potatoes, generally contain a high percentage of potash. It is believed also that the color of fruits is controlled to a certain extent by the presence or absence of certain essential elements, such as potash or iron.

Elements Essential to Plants.—The essential elements of plant food may be grouped as follows: First, those obtained from air and water, consisting of oxygen, hydrogen and carbon; second, those constituents that are frequently deficient in soils and are supplied through the use of commercial fertilizers, namely, nitrogen, phosphorus and potassium; the third group is not likely to be deficient as elements of plant food. These consist of calcium, magnesium, sulphur and iron. In this group calcium and magnesium in the carbonate form may become so deficient that soils become sour, in which case the practice of applying lime is advisable. The five other elements commonly present and fitting into a fourth group are silicon, aluminum, sodium, chlorine and manganese.

Soil Bacteria.—Bacteria are microscopic plants. They are composed chiefly of protoplasm, and differ from higher plants in that they contain no chlorophyll. Bacteria are generally single-celled, and they are so small that it would require about one and one-half millions brought together in a mass in order to be visible to the naked eye. These small plants are omnipresent. Soils are teeming with millions upon millions of them. They are present in the air and in the water of the lakes and rivers, and occur on all vegetation and are present in the foods we eat. These minute organisms were unknown until the high power microscope was invented a comparatively short time ago. They play a very important part in all life processes. More than a thousand species of bacteria have already been identified and described, and new species are being discovered every day.

Bacteria Make Plant Food Available.—The bacteria of the soil are of great importance in preparing plant food for our ordinary farm, garden and orchard crops. They are instrumental in making nitrogen available for higher plants. They also bring about availability of the mineral constituents of the soil. It is essential for the farmer to understand that the bacterial flora of the soil is important, and that the multiplication of these bacteria is generally to be encouraged. It is also well to know that there are two great classes of bacteria: first, those that thrive best in the presence of plenty of air, from which they obtain oxygen; and second, those that thrive best with little air and even in the total absence of oxygen. These classes are spoken of as aerobic and anaerobic bacteria, respectively. The first class, or those thriving best with plenty of air, are made up generally of the beneficial forms, and these dominate in the more productive soils. They require for their life and rapid multiplication food in the form of organic matter, although many forms live directly on the mineral elements of the soil. They need moisture and are dormant or may die when the soil remains long in a very dry condition. They must have air and this is facilitated by the tillage of the soil.

Nitrogen Increased by Bacteria.—Soil bacteria have no greater function in soils than the conversion of organic nitrogen into ammonia, nitrites, and finally nitrates, thus making the nitrogen available for higher plants. Nitrogen is the most expensive element that farmers have to purchase in a commercial form. It costs about twenty cents per pound, or three times as much as granulated sugar. Nitrogen is present in the air in great quantities, and it is chiefly through various forms of bacteria that the higher plants are able to secure the necessary supply. Among the bacteria instrumental in this process are the numerous species that are found in the nodules on the roots of the various leguminous crops. For ages legumes, such as clovers, have been recognized as beneficial to the soil, as shown by the increased growth of the non-leguminous crops that follow. Not until the discovery of these bacteria in the nodules on the roots of legumes (about one-fourth century ago) was it understood why legumes were beneficial.

The species of bacteria that occur in the nodules on the roots of one leguminous crop is generally different from that occurring on a different leguminous crop, although there are a few exceptions to this rule. The same species of bacteria occur on the roots of both alfalfa and sweet clover, but a different species is characteristic of red clover, and one species cannot be successfully substituted for another. It is, therefore, essential to use the right species when attempting to inoculate soil artificially for a particular leguminous crop. The different species of bacteria for the leguminous crops will be discussed under each of those crops in chapters which follow.

There are also species of bacteria living in the soil, not dependent directly upon legumes, which have the power of abstracting free nitrogen from the air and converting it into forms available for general farm crops.

Bacteria Abundant Near Surface.—The soil bacteria are most abundant in the plowed portion of the soil. Their numbers greatly diminish as the depth increases, and disappear entirely at a depth of a few feet. It is generally believed that direct sunshine is destructive to practically all forms of bacteria. Consequently, we find few living bacteria immediately at the surface of a dry soil. In the practice of inoculating soils, therefore, it is recommended that the bacteria be distributed on a cloudy day or in the morning or evening when there is little sunshine, and that the inoculation be at once thoroughly mixed with the soil, by disking or harrowing.

Barnyard manures are always teeming with myriads of bacteria, and the practice of applying such manure adds many bacteria to the soil. Bacteria are most active during the warmer portions of the year, and most of them are dormant when the temperature of the soil falls below the freezing point. Those instrumental in nitrification are very inactive when the soil is cold and wet and become exceedingly active in mid-summer when the temperature of the soil is comparatively high, when plant growth in general is most active and when nitrogen is most needed by growing crops. This is a fortunate coincidence, since it enables the higher plants to utilize the nitrates made available at that particular season by bacteria. If nitrification through the bacteria were equally rapid during periods when farm crops made little growth, a great loss of nitrogen would occur through leaching of the soil. The freezing of the soil does not destroy bacteria, as a rule, but simply causes them to be temporarily dormant.

REFERENCE

"The Soil." Hall.

CHAPTER 3

FERTILITY AND HOW TO MAINTAIN

Fertility Defined.—The fertility of a soil is measured by its capacity to produce an abundant growth of the crops to which the soil and climate of the region are adapted. Fertility is not dependent upon a single factor, but requires the presence and co-ordination of a number of factors acting in unison. The fertility of the soil is, therefore, dependent, first, upon the presence of a sufficient supply of the necessary plant-food elements in an available form; second, upon an adequate water supply to convey these elements in solution to the roots of the plants; third, upon sufficient warmth to promote plant growth; fourth, upon the presence of sufficient air to meet the needs of the roots for oxygen. A fertile soil will, therefore, generally consist of the ordinary soil minerals reduced to a fine state of subdivision, incorporated with more or less organic matter, and containing a sufficient supply of air, water and soil bacteria.

Vegetation an Index to Fertility.—The best index to soil fertility is the growth and condition of plants produced by the soil. On a virgin soil, either in timbered regions or on the prairies, the species of plants and their conditions of growth have long been recognized as indications of the character and value of the soil. In general, such trees as apple, ash, basswood, black walnut, burr oak, crab-apple, hard maple, hickory and wild plum, are indicative of good soil. On the other hand, where beech, chestnut, hemlock, pine or spruce dominates the forest growth, the soils are likely to be comparatively poor. White oak and beech are frequently found growing together in considerable abundance. If the white oak predominates the soil may be considered fairly good, but if beech predominates it may be looked upon with suspicion, and will probably prove to be a poor soil.

Herbaceous plants in the same manner are a good indication of the fertility of the soil. For example, in regions where alfalfa, Canada thistle, bindweed, clover, corn, cockle-burr, Kentucky blue grass, quack grass, ragweed and wheat grow well, the soils are generally found to be fertile. On the other hand, the predominance of buckwheat, Canada blue grass, the daisy, five-finger, oats, paint-brush, potatoes, redtop, rye, sorrel and wild carrot, indicate soils relatively poor.

In general, legumes indicate a good soil, although in case of the wild legumes there are some exceptions to this. Soils on which the grasses predominate are generally better than those given over largely to the growth of sedges. The sedges in general indicate wet soils. Golden-rod is a common weed having a wide habitat. It grows on both poor and good soils. The character of growth of this plant will suggest whether or not the soil is good or poor. On good soil it will have a rank and vigorous growth. The same may be true with other plants, but where nature is allowed to run her course and the law of "the survival of the fittest" has free sway, those plants naturally best adapted to the region are the ones which will ultimately predominate.

It should not be understood that any one species of plant should be relied upon to indicate whether or not a soil is good or poor, but when one takes into consideration all the vegetation present, one can then judge quite accurately as to the relative strength or fertility of the soil.

Drainage Reflected in Character of Vegetation.—The condition of the soil with reference to drainage is, of course, a modifying factor. Swamp soils, for example, are adapted only to those plants that can grow in the presence of an excess of moisture. So long as soils are in a swampy condition they are unsuited to agricultural crops, and in that condition may be considered unproductive. A good system of artificial drainage may change the whole aspect and cause them to be transformed into highly productive farm soils. Indeed, the establishment of a drainage system under such conditions would ultimately cause the disappearance of the native vegetation and encourage the encroachment of an entirely different set of plants. Then, again, climate is a modifying factor, and certain plants are found in regions of continuous warm climate that are not found where cold winters prevail.

Lime Content and Acidity Related to Plants.—The predominance of chestnut trees as above indicated suggests a poor soil and one low in lime content. Chestnut trees are not found on limestone soils, and the limestone soils in general are considered among the most fertile. Such plants as the huckleberry, blueberry, cranberry and wintergreen are seldom found on soils well supplied with lime. Redtop, while often indicative of a poor soil, will grow luxuriantly on a fertile soil. It is also very tolerant of soil acidity and an excess of moisture. It has a wide adaptation and is often grown as a hay crop on poor soils.

The presence of an abundance of sorrel, plantain and moss in cultivated fields is indicative of the condition of the soil, although it may have no relation to the soluble plant food present. Such plants generally indicate an acid soil, and call for the application of lime to encourage the growth of clover. Sorrel, like clover, is generally benefited by lime, but it is more tolerant of soil acidity than clover, and on an acid soil the clover disappears and the sorrel takes its place. Red clover is less tolerant of soil acidity than alsike clover. Many farmers make it a practice to mix these two species of clover. On neutral soils the red clover will always dominate and the alsike will scarcely be noticeable. But if the acidity of the soil approaches the limit for red clover, then the alsike will predominate, and this predomination is very noticeable when the crop comes into blossom.

Vegetation and Alkali.—In the irrigation districts of the semi-arid regions of the United States the character of vegetation often enables one to determine at a glance whether or not the soils are too alkaline for the production of staple crops. This fact is taken advantage of and serves as a great aid to the soil expert in the mapping of alkali soils. The predominance of sage bushes and rabbit's foot indicates freedom from alkali, while such plants as greasewood, mutton sass and salt grasses show at once that the soils are highly impregnated with alkali salts.

Color of Soils Related to Fertility.—Another index to soil fertility is the color of the soil. It cannot always be explained just why a certain color is indicative of fertility or otherwise, but there seems to be a comparatively consistent relationship between color and degree of fertility. Nearly all black soils are fertile, while those that are of an ashy hue or have a yellowish cast are generally poor. The chocolate-colored soils, the red soils and those of a brown color are, as a rule, fairly fertile. The farmer, as well as the soil expert, soon learns that color is a good index relative to soil fertility.

It is wise, however, to look further than merely on the surface of the soil or the character of the vegetation. Subsoil is also very important in connection with fertility. There are regions where the surface soil is black and where the subsoil immediately beneath is of a light-colored, tenacious clay, so nearly hardpan that the soils are not productive for any considerable range of general farm crops, although they may be well adapted to grass.

Maintenance of Fertility.—Soils are permanent. They constitute the most important asset of the nation. Their maintenance through rational systems of farming is essential. Nature has made for increased soil fertility, but unfortunately the occupation of the soil by man has often resulted in soil robbery and a decline in productivity. This serious fault should be remedied.

Fertility Lost by Plant Removal.—Loss of soil fertility by plant removal is legitimate. Such loss must ultimately be replaced, either by the return of the residues of crops thus removed in the form of unused portions or by-products and the excreta of the animals that consume the crops, or by the purchase of the different elements in commercial fertilizers. In rational systems of farming the removal of plant food through the removal of crops is not to be considered undesirable, and such removal should result in sufficient profits to enable the soil loss to be replaced at a cost less than the profits received through the crops grown. In the preceding chapter we found that of the mineral elements potassium and phosphorus are the only ones likely to become exhausted to such a degree as to necessitate replacement. As a matter of fact, potash occurs in large quantities in most soils, and the problem of the future seems to be largely the adoption of methods that will bring about its availability. Many soils, however, contain phosphorus in such small amounts that in a short time the supply will be so nearly exhausted as to necessitate the return of this element to the soil in some commercial form. In some soils it is already necessary for most profitable crop production.

Loss by Erosion.—The loss of soil fertility by erosion is more serious than the loss by plant removal. In this way there is not only a loss of plant food but a loss of a portion of the soil body itself. The millions of tons of finest soil particles and organic matter carried annually to the ocean by the rivers of the United States are a monument to careless soil management. This waste may be witnessed everywhere. The removal of the most fertile part of the soil is not only a loss to the soil, but is often a menace to navigable streams which are filled up with this material. An enormous expenditure on the part of our national government is necessary in dredging them out and making them again navigable. This erosion also becomes a menace to our great city water supplies, necessitating expensive filter plants to remove the suspended matter and purify the water. It also frequently does damage to other land subject to overflow, and on which the deposits may be left.

The great problem, therefore, seems to be the control of the rain that falls upon the land. A portion of this may pass over the surface, carrying with it small amounts of the surface, which in the course of time has been largely exhausted of plant-food elements. This loss should be accompanied by a renewal of the soil from below. The addition of new soil below should keep pace with the removal from the surface if permanent soil fertility is to be maintained. The remainder of the rainfall should find its way into the soil. A portion of this may pass off into the drainage waters, removing certain soluble material that without such drainage might accumulate in the course of centuries to the detriment of plant growth. Another portion should return to the surface, bringing with it the soluble constituents of the soil and leaving them near the surface for the use of growing plants.

Preventing Soil Erosion.—Water escaping from the soil by means of underdrainage never carries with it any of the soil material other than the slight portions that are soluble. It is, therefore, essential to establish systems of farming that will enable a large proportion of the rainfall to penetrate the soil; and to remove the excess of water by underdrainage when nature fails to provide such a system. Erosion may be largely prevented on most farms by deep plowing and by keeping the soil covered as much as possible with growing crops or their remains. Deep plowing encourages an increased penetration of the rainfall and, therefore, reduces the amount passing over the surface of the soil. The presence of growing plants retards the movement of surface water and holds back the soil particles. An abundance of roots in the soil helps to hold it together and prevent erosion. The application of barnyard and green manures also retards erosion. In some places terracing the soil to prevent erosion becomes necessary, but it is a costly and cumbersome method and not to be recommended where other and cheaper methods can be used.

Lands that are steep and subject to erosion should be kept covered with vegetation as fully as possible. Such lands should not be plowed in the fall and allowed to lie bare through the winter.

Farming Systems that Maintain Fertility.—Systems of farming which provide for a return of the largest possible proportion of the plant-food constituents removed in crops are those that most easily maintain the fertility of the soil. It is, therefore, evident that livestock farming in general is least exhaustive of soil fertility, provided the excreta of the animals are carefully saved and returned to the soil. In the rearing of animals for meat, about ninety per cent of the plant food consumed by the animals is voided in the liquid and solid excreta. If this is carefully saved and returned to the soil, depletion of soil fertility will be exceedingly slow.

In dairy farming, where the milk is sold, a somewhat larger proportion of the plant food elements is sold from the farm. Even here the total amount is relatively small, and may be offset by the plant food in concentrates purchased for the dairy. If the milk is fed to pigs and calves and only the butter is sold, the exhaustion in the long run will be no greater than in meat production. It is, therefore, evident that the type of farming is closely related to the maintenance of soil fertility, and those types which permit a maximum sale of cash crops cause the largest direct removal of plant food from the farm. All types of livestock farming, therefore, come closest to maintaining permanent fertility.

In new countries it is not an uncommon practice for farmers to dump the manure from stables into a nearby stream in order to get rid of it. It is also a common practice to burn stacks of straw and the stubble of the field in order that the soil may be freed of rubbish and easily plowed and cultivated. Such practices are to be condemned, for in the long run they encourage soil depletion. Where land is cheap and fertile and labor expensive, the immediate returns from applying manure may not justify the cost of its application, but in a long term of years it will prove profitable. A farmer should be far-sighted enough to calculate what the result will be in the course of a lifetime. There should be more profit in the removal of fifty crops in as many years where fertility has been maintained or increased, and where the crop yields have increased, than there is in the removal of fifty crops with a constantly decreasing yield. In the first case the land is left in good condition for the succeeding generation; in the second case, in bad condition.

Deep Plowing Advisable.—Fertility of the soil is generally improved by increasing the depth of plowing. It is a common observation that in regions of good farming where farmers are prosperous, the soil is generally plowed to a depth of seven to ten inches. In many portions of the South we find the one-mule plow that barely skims the surface of the soil, and accompanying this we have the unsuccessful farmer. Plowing is an expensive operation. It is estimated that the power required annually to plow the farm land of the United States exceeds that used in the operation of all the mills and factories in the country.

There is a limit to the profitable depth of plowing, and numerous experiments indicate that it is seldom profitable to plow deeper than eight to ten inches. There doubtless are some exceptions to this found in case of the production of intensive crops or the occasional deep plowing for the preparation of a deep-rooted crop like trees or alfalfa. Deep plowing increases fertility by increasing the area of pulverized soil in which the roots of the plants find pasturage. Such plowing increases the aeration of the soil, encourages the multiplication of bacteria to a greater depth in the soil, and results in increased availability of plant food. Deep plowing also incorporates the organic matter applied as manure or as the stubble of the preceding crop in a deeper stratum of soil, thus increasing its water-holding capacity. Deep plowing also increases the penetration of rainfall and provides for greater storage of it. This provides a larger water supply for the growing crops in periods of drought.

Tillage is Manure.—Cultivation of the soil, and especially the intertillage of crops, such as corn, potatoes and truck crops, aids in maintaining fertility: first, by conserving soil moisture; second, by more thorough aeration of the soil; third, by a fuller incorporation and distribution of the organic matter with the mineral matter; and fourth, by the destruction of weeds which consume plant food and water to the detriment of the crop grown.

Rotations are Helpful.—Crop rotations also help to maintain fertility. By means of rotating crops the soil may be occupied for longer periods of time than when one crop is planted year after year on the same soil. The roots of different crops, having very different habits, occupy somewhat different zones in the soil. A shallow-rooted crop may be advantageously followed by a deep-rooted one. One takes the major portion of its plant food from near the surface and the other from a somewhat lower stratum. All crops do not use mineral constituents in the same proportion. One which demands large amounts of nitrogen may appropriately follow one which has the power of gathering nitrogen from the air. For example, corn appropriately follows clover, the corn benefiting by the nitrogen left in the soil by the roots and stubble of the clover crop.

Rotations Reduce Diseases.—Rotations also make for fertility by checking the epidemics of plant diseases and the depredations of insects. As a rule, a plant disease is common only to one crop and where that one crop is grown year after year on the same soil the disease increases until finally the crop must be abandoned. Many of the insect pests of crops either live permanently in the soil or have but little power of migration. These likewise prey upon certain crops and do not bother others, and the rotation of crops prevents serious injury by them. While these do not add plant food to the soil, their absence increases the growth of crops, which means the same thing.

Cover-Crops Prevent Loss of Fertility.—Cover- or catch-crops may be grown greatly to the benefit of the soil. Cover-crops consist of any suitable plants occupying the soil when the money crop is not in possession. They make growth during the cool season of the year, take up plant food as it is made available, and hold it in plant form, where it may be returned to the soil when such a crop is plowed under. In this way it prevents the loss of soil fertility by direct soil leaching and converts mineral plant food into an organic form which upon decay is more readily available than it previously was. Such a crop also adds organic matter to the soil, increasing its power for holding water and being generally beneficial. Good examples of cover-crops are crimson clover or a mixture of rye and winter vetch seeded in corn late in the summer and occupying the soil during the winter. Such crops do not at all interfere with the growth and maturity of the corn. They make most of their growth in the late fall and early spring and may be plowed under in ample time for planting a crop the following year. Such crops are adapted especially to the South, where the winters are mild and freezing of the soil is slight, while erosion and leaching are marked. This practice is quite common with truck farmers, as cover-crops may be seeded after the removal of a truck crop.

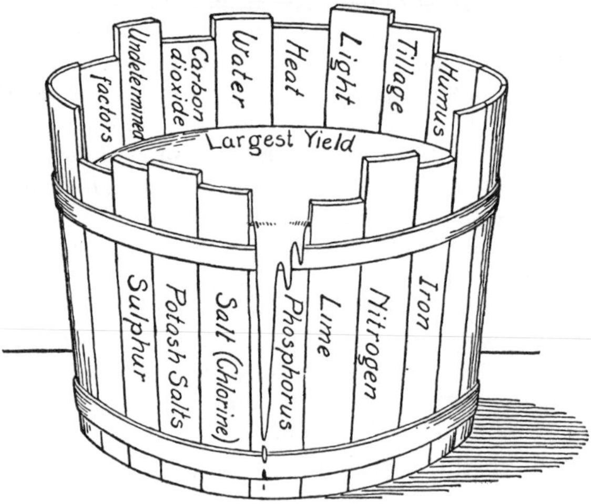

SOIL FERTILITY BARREL.[1]
Illustrating the limiting factor in crop production.

Legumes Increase Soil Nitrogen.—Of all the crops instrumental in increasing soil fertility, none equal the legumes, for these alone have the power, through the instrumentality of bacteria residing in the nodules on their roots, to extract free nitrogen from the air. While such crops are richer in protein than the non-legumes, yet at the same time they leave in the roots and stubble a large amount of nitrogen which is available for non-legumes. A crop rotation which does not have at least one leguminous crop every four or five years is decidedly faulty.

[1]Courtesy of the Wisconsin Agricultural Experiment Station.

Drainage Increases Fertility.—Fertility is increased by drainage, especially underdrainage, which lowers the water table, increases aeration, and causes plant roots to go deeper in the soil. The amount of plant food that plants can secure is approximately proportionate to the volume of the soil to which they have access. Drainage virtually deepens the soil.

Manure is the Best Fertilizer.—Manures increase fertility by the direct addition of plant food and by increasing the organic matter of the soil. Manures increase the water-holding capacity of the soil, improve its physical condition, introduce various forms of bacteria and encourage the multiplication of desirable bacteria.

Commercial Fertilizers Add Plant Food Only.—Commercial fertilizers increase fertility by the direct addition of the plant food elements they contain, but, as a rule, have very little if any other effect. Commercial fertilizers are expensive and call for an intimate knowledge of the requirements of the soil and the form and availability of the constituents in the fertilizer. The factors above mentioned in relation to soil fertility will be more fully discussed under the several chapters pertaining to them, which follow.

The Limiting Factor.—There is always a limiting factor in crop production, and it is the business of the farmer to ascertain his limiting factor or factors. In many cases the limiting factor in the growth of a crop will be the supply of water. This may be a deficient supply or it may be an excess. If water is the limiting factor it may be due to a low rainfall during the crop season and the low storage capacity of the soil. The farmer has no control over the rainfall, but he should endeavor to increase the water storage capacity of his soil by such means as are economical. Deeper plowing, the addition of organic matter, thorough tillage to conserve soil moisture or the application of water in the form of irrigation are all of them means to such an end. If the limiting factor is due to an excess of water, thus preventing plant growth, the problem becomes one of land drainage and the removal of the water.

The limiting factor may be a deficiency in phosphorus. This being the case, it is important that the farmer know the truth in order that he may supply the deficiency by the application of a phosphatic fertilizer. When the limiting factor or deficiency has been supplied, something else may then become a limiting factor. For example, the limestone soils of Pennsylvania are generally deficient in phosphorus. Such soils, when cropped with a four-year rotation of corn, oats, wheat and mixed clover and timothy, will show a steady decline in crop yields if no manures or fertilizers are applied. Experiments with fertilizers on limestone soil and for the crops mentioned show that when nitrogen alone is applied it has no effect. Potash applied alone is likewise ineffective. When phosphorus is applied there is a marked increase in the yield of crops. Phosphorus, however, will not fully maintain the fertility of the soil. Its yield will decline, but not so rapidly as when nothing is applied. When the need for phosphorus is met, then potash becomes the limiting factor, and large applications of potash may be used in connection with phosphorus with profitable returns. In this way there will always be a limiting factor in crop production. The farmer should ascertain the limiting factors in his crop production, and then supply them most economically. He may find that there are several limiting factors, and that these will vary from

Soil Fertility Plats, Pennsylvania Agricultural Experiment Station.

On left, 200 pounds per acre muriate of potash every other year.
In center, dried blood containing 24 pounds nitrogen and dissolved bone-black containing 48 pounds phosphoric acid.
On right, dried blood containing 24 pounds nitrogen, muriate of potash 200 pounds.

time to time; so the problem of soil fertility is a never-ending problem with which the farmer will always have to contend.

Fertility an Economic Problem.—Soil fertility is a problem of far-reaching economic importance. The principal items of expense in general crop production are labor of men and horses, equipment, seeds and land rental. These cost no more for a productive acre than for one of low productivity. In fact, the productive soils are generally plowed and cultivated at less cost of time and energy than those of low productivity. Every hundredweight of product over that required to meet the cost of production is profit.

REFERENCES

"Conservation of Natural Resources." Van Hise.
"Soil Fertility and Permanent Agriculture." Hopkins.
"Soil Management." King.
"First Principles of Soil Fertility." Vivian.

"Soils and Soil Fertility." Whitson and Walston.
"The Fertility of the Land." Roberts.
Kentucky Agricultural Expt. Station Bulletin 191. "Teachings of Kentucky Agricultural Expt. Station Relative to Soil Fertility."
Farmers' Bulletins, U. S. Dept. of Agriculture:
342. "Conservation of Soil Resources."
406. "Soil Conservation."
421. "The Control of Blowing Soils."
446. "The Choice of Crops for Alkali Land."

CHAPTER 4

COMMERCIAL FERTILIZERS

A careful study of the condition of farming in the United States shows that the supply of barnyard and stable manure is not adequate to maintain the fertility of the soil. The need for commercial fertilizers is, therefore, apparent and real, although the amount required in conjunction with natural manures may be comparatively small.

It is desirable to use commercial fertilizers on many farms and the practice is becoming more general each decade. This is but natural, since there is a constant flow of soil fertility towards the cities. The rapid increase in the city population and the consequent increase in food consumption at those points cause a constantly increasing drain upon the soil fertility of the farms.

Object and Use of Commercial Fertilizers.—The object of manuring the soil, whether with stable manure, green manure or commercial fertilizers, is to increase its crop-yielding capacity. In order to justify the practice the resulting increase in products must be more than sufficient to offset the cost of manures or fertilizers applied. This increase need not necessarily be secured the first year after the application, but should be secured in the current and succeeding crops, and should give a net profit on the capital and labor so expended.

The first noteworthy use of commercial fertilizers in the United States was in 1848. In that year there was imported 1000 tons of guano. This was followed the succeeding year by twenty times that quantity. From that date the importation steadily increased until 1880, when it reached its maximum and began to decline because of a failing supply of guano. Other materials, such as sodium nitrate from Chile and the potash salts from Germany, have taken the place of the guano. These, together with the development of our phosphate mines, the use of cottonseed meal and the utilization of slaughter-house by-products, have met the continually increasing demand for commercial fertilizers by our farmers. According to census reports, the expenditures for fertilizers in the United States during the past four census-taking years have been as follows:

Year.	Value.
1879	$28,500,000.00
1889	38,500,000.00
1899	54,750,000.00
1909	112,000,000.00

There seems to be little doubt but that this rate of increase in the use of fertilizers will continue for some time to come. The subject is one of much economic importance to farmers, and one which has received much time and attention on the part of investigators in the agricultural experiment stations of all the older agricultural states. Agricultural literature now contains a vast amount of data setting forth the results of experiments with fertilizers on different types of soil and for different crops, but there is still much to be learned relative to the subject. We will always have an acute fertilizer problem. This is due to the constantly changing conditions of the soil, resulting primarily from changed agricultural practices and especially from the treatment of the soil, which will gradually change its relationship to crops.

What are Commercial Fertilizers?—In discussing the subject of fertilizers the terms manures, complete and incomplete manures, fertilizers, chemical fertilizers, commercial fertilizers, natural fertilizers, artificial fertilizers, indirect fertilizers, superphosphates, etc., are used, and there is often misunderstanding of the meaning of some of these terms. Fertilizers are first divided into natural and artificial. The former include all the solid and liquid excrement of animals and green manuring crops when plowed under for the benefit of the soil. Artificial fertilizers include all commercial forms of fertilizers. These are sometimes called prepared fertilizers and chemical fertilizers, but are becoming more generally known as commercial fertilizers. A complete fertilizer contains the three essential plant-food constituents, nitrogen, phosphorus and potassium. An incomplete fertilizer contains only one or two of these. All animal manures are complete fertilizers. Green manures are likewise complete.

A fertilizer is said to be indirect when it contains none of the essential plant-food elements, but in some way acts on the soil so as to increase the availability of plant food in the soil or increase crop growth. Lime, gypsum, salt and numerous other substances have been found to have this action and would be classed as indirect fertilizers.

The terms high-grade and low-grade are also applied to fertilizers. These terms, however, are not well defined. High-grade fertilizers generally contain large amounts of plant food per ton, while low-grade fertilizers contain relatively small amounts. Another distinction that is sometimes made is that fertilizers manufactured out of high-grade constituents, such as nitrate of soda, acid phosphate and muriate or sulphate of potash, are considered high-grade fertilizers regardless of the percentage of the elements. A high-grade fertilizer always costs more per ton than a low-grade one, but it is generally true that the elements in such a fertilizer come cheaper to the farmer than they do in a low-grade material. Whether it is more economical to purchase high-grade or low-grade material is an important question, but the answer is not difficult. All fertilizers should be bought on the basis of their content of available plant food, and it is merely a problem in arithmetic to calculate the relative cost of the elements in different grades of fertilizer.

Where are Fertilizers Secured?—Fertilizer materials are to a large extent gathered from different parts of the world, and are either treated to increase availability or combined into mixed fertilizers before being offered to the farmer. Fortunately the fertilizing element most needed in the soils of the United States and Canada, namely, phosphorus, is secured chiefly from extensive deposits of phosphate rock in Florida, South Carolina and Tennessee and a few other states. This supply is supplemented to some extent by bone phosphate, which comes chiefly from the slaughter-houses of the country; also by basic slag, a by-product of steel manufacture.

The potash salts are secured almost exclusively from the extensive potash mines in Germany. Potash salts come to us in different forms. Most of them have been manipulated and more or less purified. The one most extensively used is known as muriate of potash and is a chloride of potassium (KCl). Sulphate of potash and carbonate of potash are used to a somewhat less extent. In addition to these we have some of the crude potash salts, such as kainite and manure salt. A comparatively new source of potash suitable for commercial fertilizers has been found in the extensive kelp groves in the Pacific Ocean off the coast of the United States and Canada. As yet these have not been extensively used as a commercial source of potash.

Nitrogen is available chiefly in the form of nitrate of soda, which comes from Chile. We also have sulphate of ammonia, an extensive by-product from coke ovens and from the manufacture of artificial gas. As yet the nitrogen escaping from coke ovens is not all transformed into sulphate of ammonia. There are also organic forms of nitrogen, chief of which are cottonseed meal, dried blood, tankage, fish scrap, guano, castor pomace, together with small amounts of horn, hair, feathers and wool waste.

Carriers of Nitrogen.—Nitrate of soda ($NaNO_3$) contains 15 per cent of nitrogen. It is readily soluble in water, and nitrogen in this form is immediately available for plants. It should be applied in small quantities and not long prior to the time plants most need their nitrogen supply.

Sulphate of ammonia $(NH_4)_2SO_4$ contains 20 per cent of nitrogen. Like nitrate of soda, it is quick acting, but for most crops the ammonia must first be converted into the nitrate form before it can be utilized. Some crops, however, can utilize ammonia as such. Sulphate of ammonia is not leached from the soil quite as rapidly as nitrate of soda, but nevertheless it should not be applied in larger amounts than are necessary, nor far in advance of the needs of the crop.

Cottonseed meal is another source of nitrogen which is extensively resorted to in the cotton belt. It contains from 3 to 8 per cent of nitrogen, with an average of about 6.8 per cent. It is not wholly a nitrogenous fertilizer, since it also contains an average of 2.9 per cent phosphoric acid and 1.8 per cent potash. The nitrogen in cottonseed meal being in an organic form, is rather slowly available. Availability is gradually brought about through decomposition. The nitrogen thus resulting is, therefore, distributed through a considerable period of time. It is often used as a part of the nitrogen supply for crops with a long growing season.

Dried blood is also an organic source of nitrogen, containing on an average 10 per cent of this element. It is easily decomposed and somewhat more available than nitrogen in cottonseed meal.

Tankage contains nitrogen in variable quantities, ranging from 5 to 12 per cent. It may also contain from 7 to 20 per cent of phosphoric acid. The nitrogen in tankage is slowly available.

Forms of nitrogen that have more recently found their way into the market are cyanamide and lime nitrate. These are manufactured products in which the nitrogen is secured directly from the air through certain chemical and electrical processes. The nitrogen in these forms is not so available as that in nitrate of soda or sulphate of ammonia, although it is considered more readily available than most of the organic forms.

Phosphorus.—This constituent is available in the form of acid phosphate, which contains 14 to 16 per cent of phosphoric acid or 6 to 7 per cent of phosphorus. Most of the phosphorus is in an available form. Acid phosphate is made by treating a given bulk of finely pulverized phosphate rock with an equal weight of crude commercial sulphuric acid. The reaction that takes place makes the phosphorus available. It is this material that is chiefly used in the manufacture of complete commercial fertilizers. Phosphoric acid costs from four to five cents per pound in acid phosphate, depending on location and size of purchases. (As this goes to press, prices have advanced 25 to 30 per cent. This advance is probably temporary.)

There is now an increased tendency to make direct use of the raw rock phosphate in a finely pulverized form. Such rock contains the equivalent of 28 to 35 per cent of phosphoric acid, but it is in an insoluble form and can be economically used only on soils that are well supplied with organic matter or in conjunction with barnyard or stable manure and green manure crops. The general use of raw rock phosphate has not been advisable on the soils of the eastern and southern part of the United States. On the other hand, the raw rock phosphate has given good results on the prairie soils of Indiana, Illinois, Iowa and some other states. The cost of phosphoric acid in this form is equivalent to two cents per pound or a little less.

Basic slag, sometimes known as Thomas Phosphate, is a by-product of steel mills which is finely ground and used as a source of phosphorus. It is similar to raw rock phosphate, slightly more available and contains the equivalent of 15 to 18 per cent of phosphoric acid.

There are two types of bone meal on the market, raw bone and steamed bone. The raw bone is fresh bone which has been finely ground. Raw bone contains about 20 per cent of phosphoric acid and 4 per cent of nitrogen. Bone which has the fat and gelatin removed by extracting with steam contains only about 1 per cent of nitrogen and 22 to 28 per cent of phosphoric acid. The steamed bone is more finely ground than the raw bone, and since the fat and gelatin are removed it decomposes more rapidly and is, therefore, more readily available as plant food. While the phosphorus in both forms of bone is largely insoluble, it is nevertheless more readily available than that in rock phosphate.

Potassium.—Muriate of potash (KCl), the chief source of potash, contains the equivalent of about 50 per cent of potash (K_2O). It is the most common purified potash salt, consisting chiefly of potassium chloride. It is a very satisfactory source of potash for all crops excepting tobacco and potatoes. This form, on account of its contents of chlorine, causes a poor burn in tobacco used for smoking purposes. The chlorine is supposed to be slightly detrimental to starch formation, and for this reason the sulphate and carbonate of potash are considered superior for potatoes.

Potassium sulphate also contains the equivalent of 50 per cent of potash (K_2O). Kainite a low-grade material contains about 12 per cent of potash.

Wood ashes are also a source of potash. They contain about 6 per cent of this constituent, together with about 2 per cent of phosphoric acid and a large amount of lime. The availability of the potash in ashes is rated as medium.

Forms of Fertilizer Materials.—It is the common experience of farmers and investigators that the different carriers of nitrogen, phosphorus and potassium behave differently on different soils, in different seasons and with different crops. Most fruit and tobacco growers recognize the difference in the different forms of potash although it is not clearly understood why these differences occur.

Under present fertilizer regulations dealers are required to state only the percentage of the plant-food constituents in the fertilizers they offer for sale. It would be a wise provision if in addition to this they were required to state the source of the constituents as well as the percentage. This is especially important as relates to nitrogen, which varies widely in its availability, depending on its source. Many materials containing essential elements are nearly worthless as sources of plant food because the form is not right. Plants are unable to make use of these materials because they are unavailable. Materials that do not show wide variation in composition and in which the constituents are practically uniform in their action, may be regarded as standard in the sense that they can be depended upon to furnish practically the same amount and form of a constituent wherever secured. Among such standard materials may be considered nitrate of soda, sulphate of ammonia, acid phosphate, muriate of potash, sulphate of potash and carbonate of potash.

Relative Value of Fertilizer Ingredients.—A practical point, and one of importance to the farmer, is a reliable estimate of the relative value and usefulness of the various products that enter into commercial fertilizers. The relative rate of availability of a constituent in one carrier as compared with its availability in another is the point at issue. This determines the advantage or disadvantage of purchasing one or the other at ruling market prices. As yet definite relative values for all fertilizing materials have not been worked out. Furthermore, it is recognized that they never can be worked out for conditions in general, because of the wide latitude in the conditions which affect availability. This problem is attacked by what is known as vegetative tests; that is, tests which show the actual amounts of the constituents taken up from various substances by plants when grown under identical conditions. With nitrogenous fertilizers, for example, the results so far obtained indicate that when nitrogen in nitrate of soda is rated at 100 per cent, that in blood and cottonseed meal are equal to about 70 per cent, that in dried and ground fish and hoof meal at 65 per cent, that in bone and tankage at 60 per cent, and for leather and wool waste may range from as low as 2 per cent to as high as 30 per cent.

The Composition of Fertilizers.—In the purchase of mixed fertilizers consumers should demand that they be accompanied by a guarantee. This is essential because the purchaser is unable to determine the kind and proportion of the different materials entering into the mixture, either by its appearance, weight or smell.

At present most of the states have on their statutes, laws regulating the manufacture and sale of commercial fertilizers. These require that the composition be plainly stated on the original packages of fertilizer. The law also provides for the analysis of samples collected at any point and the publication of these analyses either by the state departments or by the state experiment stations. Such publications set forth the name of the brand of fertilizer and the name of the dealer or manufacturer, together with a statement of the analysis as given by the manufacturer as compared with that found by the official analysis. Infringements of the law relative to its provisions call for punishment generally by fines. Under such a system of regulation there is now little danger of the farmer being cheated in the purchase of fertilizers so far as their composition is concerned.

What Analyses of Fertilizers Show.—The difference between a good and inferior fertilizer is shown by a chemical analysis, providing it is carried far enough to show both the amount and form of the constituents present. An analysis of a fertilizer which shows that the nitrogen is present chiefly as nitrates, the phosphorus as acid phosphate and the potash as muriate of potash at once stamps such a fertilizer as being made up of high-grade materials. On the other hand, if the nitrogen is found largely in an organic form and the phosphorus in an insoluble form, it is evident that the materials used are low-grade forms, and result in a slow-acting and sometimes unsatisfactory fertilizer.

Commercial vs. Agricultural Value of Manures.—Agricultural value and commercial value as applied to fertilizers are not synonymous and should not be confused. The agricultural value is measured by the value of the increase in crops secured through the use of the fertilizer. The commercial value is determined by the trade conditions. It is based upon the composition of the fertilizer and the price per pound of the different forms of the several constituents that enter into it. Commercial value is merely a matter of arithmetic. Agricultural value varies greatly and depends upon a number of factors, among which the knowledge of the farmer plays no small part.

Mechanical Condition.—The mechanical condition of a commercial fertilizer deserves consideration by the farmer. The degree of pulverization controls the rate of solubility to no small extent. The finer the pulverization the more thorough can be the distribution made in the soil. The greater the number of points at which there are particles of fertilizer in the soil, the more rapid will be the solution and the diffusion of the plant-food material. Mechanical condition is also important from the standpoint of distribution through fertilizer drills. The material should be in what is known as a drillable condition. It should not only be thoroughly pulverized, but also should be sufficiently dry to feed through the mechanism of the drill at a uniform rate. Wet, sticky material clogs up the drill and causes faulty distribution.

High-Grade vs. Low-Grade Fertilizers.—Thousands of tons of low-grade fertilizer are bought by farmers because the price is low, when, as a matter of fact, the same money invested in a lesser amount of high-grade fertilizer would have given them better results. Low-grade fertilizers, as a rule, contain varying amounts of filler or inert matter. This sometimes constitutes as much as one-half the weight of the fertilizer. It costs just as much to provide bags and handle this material as it does the more active portion. Furthermore, the farmer pays for the bags and freight on this worthless material. At the same time, he hauls it from the railway station to his farm, unloads it and afterwards applies it to his fields with much more expenditure of time and effort than would be required for a smaller amount of high-grade material containing equally as much plant food.

Use of Fertilizers.—The most economical use of commercial fertilizers is secured only when a systematic crop rotation is practiced and the soil is maintained in good physical condition and well supplied with organic matter and moisture. The soil should contain sufficient lime to prevent the accumulation of acids, so that legumes such as clover will thrive. Every crop rotation should have a suitable legume occurring once every third to fifth year. The presence of legumes will lessen the necessity for nitrogen in the fertilizer. It is estimated that nitrogen can be secured through the growing of legumes at a cost of approximately four cents per pound, whereas it costs fifteen to twenty cents when purchased in a commercial form.

Value of Crop Determines Rate of Fertilization.—Crops are divided into two classes with reference to the use of commercial fertilizers. The first class includes those crops having a comparatively low money value, such as hay and the general grain crops. Because of the low money value it is possible to apply only small amounts of fertilizer profitably. It is also necessary that the crops use as large a proportion of the applied material as possible. The cropping system should be arranged so as to utilize the residues of previous applications. As a rule it is wise to purchase very little nitrogen for such crops, since their needs can generally be met by growing suitable legumes in the rotation. In the temperate climate of the United States and Canada, east of the 100th meridian, red clover is the crop best adapted for this purpose, although there are other clovers and annual legumes that may meet local conditions better. In the southern part of the United States cowpeas, soy beans, Lespedeza clover, crimson clover and some other legumes are best suited for this purpose. West of the Mississippi River alfalfa will pretty fully meet the needs of the soil for nitrogen. Ordinarily it will be grown several years in succession.

Valuable Products Justify Heavy Fertilization.—The second class of crops includes those having a high money value per acre and for which large applications of high-grade fertilizers may be economically used. Among such crops may be mentioned tobacco, cabbage, early peas, spinach, asparagus and even early potatoes. Because of the high money value of these crops a larger investment in fertilizers may be more than paid for, even though the percentage increase in yield is no greater than when fertilizers are applied to crops of low money value. In growing early truck crops, especially when grown along the lower portion of the Atlantic seaboard or in the southern states, the truck farmer who can get his product into the northern markets earliest is the one who receives the fancy prices. Such markets call for products of high quality, and quality in many cases is determined by the rate of growth. In such crops as lettuce, radishes, spinach, etc., succulence and tenderness of the product are essential. These qualities, together with earliness, are often determined not only by the time of planting and the character of soil on which the crops are grown, but also by the character of the fertilizer used. We, therefore, find such farmers using fertilizers that are readily soluble and well supplied with available nitrogen. Nitrogen tends to accelerate vegetative growth and to give quality to early vegetables. It is not unusual to find truck farmers applying as much as a ton per acre of a high-grade fertilizer. The crop grown may use a comparatively small portion of the constituents applied. This calls for a rotation of crops on the part of such a farmer so that other and less valuable crops may follow and be benefited by the residual effect of the fertilizer.

A strict classification of crops into the two classes mentioned is impossible. Conditions which would place a crop in one group in one locality may place it in the other group in a distant locality. The high price of a crop is in some cases determined by location. For example, the early strawberries and early potatoes of the South that reach northern markets very early are often worth five to ten times as much per unit as are the late strawberries and late potatoes grown in the North and at some distance from markets.

Character of Fertilizer Related to Soil.—In general, fertilizers that stimulate the production of seeds and fruit should be used on rich lands. On poor land the elements that force vegetative growth combined with those that mature fruit may be used. High-grade phosphates in a readily available form hasten maturity and increase the proportion of fruit. This is well illustrated in the fertilizer plats at the Ohio and Pennsylvania Experiment Stations. As the oats and wheat approach maturity on these plats the visitor is at once impressed with the earlier period of ripening of those grown on plats treated with acid phosphate. Nitrogen tends to a prolonged growth of the crop and retards maturity. The grain on the plats treated with liberal applications of nitrogen matures a week or ten days later than on the phosphate-treated plats.

In the use of fertilizers one should distinguish between a large increase of crop and a profitable increase, and this will be determined chiefly by the value of the crop grown. In general there will be an increase in yield accompanying an increase in the amount of fertilizer used, but it is a fact that the first unit of application, that is, the first two hundred or four hundred pounds per acre, will give a relatively larger return than the second or third unit, and there will always be a place where an added unit will give a return, the value of which will be no greater than the cost of the unit of fertilizer. It is most profitable to stop before one reaches this point in the application of fertilizers.

Finally, the purchaser of fertilizers should bear in mind that the composition of the fertilizer and availability of its constituents, its mechanical condition, the economy of its purchase and application are all factors that bear directly upon the economy of its use. This calls for a knowledge of the requirements of the soil and the crops grown.

What the Farmer Should Know.—Commercial fertilizers are valuable mainly because they furnish nitrogen, phosphoric acid and potash. In some cases they may act as stimulants, but their chief function is to supply available plant food. The returns will be approximately in proportion to their content of such constituents, when the selection is so made that it meets the needs of the soil and crops to which applied. The agricultural value of these constituents depends largely upon their chemical form, and these forms must be contained in products of well-defined character and composition. They may be purchased as such from both dealers and manufacturers. The farmer may put them together in proportions to meet his own needs, if he is competent to do so.

The farmer should know the deficiency of the soil on his farm. He should also know the requirements of the plants with which he deals. He may secure these facts in a general way from the state experiment station, but the details can best be ascertained by actual field tests by the farmer himself on his own farm. Such tests do not necessitate carefully laid out plats of a definite size. Farmers, as a rule, do not have the time and patience to do much experimenting, neither do they have the training, experience and facilities for such work; but any farmer may make a fair comparison of two or more kinds of fertilizers, or he may test the efficiency of any fertilizer ingredient, such as nitrogen, potash or phosphorus, on his soil. This can be done by applying a different character of fertilizer through his fertilizer drill, whether it be attached to the corn planter, the potato planter or to the grain drill, to a definite number of rows running clear through the field. This, if marked at one end of the field by stakes, is easily and readily compared at harvest time with the rows on either side treated with the usual fertilizers or in the usual way. Much can often be determined by observation, but more definite results are obtained by measuring the product of a certain number of rows specially treated, as compared with an equal number adjacent treated in the usual way.

A rapid growth and a dark-green color of foliage indicate the presence of an ample supply of nitrogen in the soil. If the rank growth is accompanied by a watery appearance it suggests a deficiency of phosphoric acid. If plants make a stunted growth under normal conditions of sunshine, temperature and water supply, and mature unduly early, it indicates sufficient phosphoric acid in the soil, and suggests that nitrogen or perhaps potash may materially improve the crop. Potash fertilizers are of special benefit in case of tobacco, beets and the legumes.

The user of commercial fertilizers should place his main dependence upon those that have given him best results. New brands or modified mixtures should be tried on a small scale and in an experimental way until it has been demonstrated that they are better and more economical to use than his old standby. Emphasis should also be placed upon the importance of a systematic use of fertilizers. This can be accomplished through a definite cropping system and a definite scheme of manuring and fertilizing worked out in such a way as best to meet the needs of the soil and crops. It should take into account the fullest possible utilization of the home and local supplies of manure. For example, it is found that the general farm crops in Pennsylvania are most frequently grown in a rotation consisting of corn, oats, wheat and two years of mixed clover and timothy hay. On limestone soils such crops call for a scheme of treatment about as follows: For the corn, 6 to 10 loads of manure per acre should be applied and supplemented with 200 pounds acid phosphate; to the oats following the corn, no fertilizer except when the soil is poor, in which case 150 to 200 pounds per acre of acid phosphate may be used; to the wheat, 350 pounds per acre of acid phosphate, 100 pounds muriate of potash and 50 pounds of nitrate of soda should be applied; the clover following the wheat calls for no fertilizer, but the timothy during the second year the land is in grass may be profitably treated with a complete fertilizer consisting of 150 pounds of acid phosphate, 150 pounds nitrate of soda and 50 pounds muriate of potash, applied broadcast early in the spring just as the grass starts to grow. Such a scheme of treatment makes a place for all the manure on the average farm and provides for the application of the fertilizer where it will be most fully used and give the largest returns.

A similar scheme of treatment will be found to fit various localities

EFFECT OF TOP DRESSING MEADOWS WITH COMMERCIAL FERTILIZER.

On left, average yield, 2060 pounds cured hay per acre.
On right, average yield, 3637 pounds cured hay per acre.
Grass on right, top dressed early each spring with 350 pounds per acre of 7-7-7 fertilizer. Average of four consecutive years.

in all states. The details will be determined by local conditions, and frequently they have already been worked out for various localities either by the experiment station of the state or by farmers. It is, therefore, important that every farmer become informed on the best practice for his locality.

How to Determine Needs of Soil.—The fertilizer needs of a soil are best determined by applying to the soil and for the crops grown different kinds and combinations of fertilizers. This puts the question directly to the soil, and the crops give the answer by their growth and condition. Such soil tests with fertilizers have proven more practicable and satisfactory than any others thus far devised. A chemical analysis of the soil is thought by many to enable the farmer or the soil expert to judge as to the character of the fertilizer needed. This, however, is not the case, and such chemical analyses are as a rule of very little help in this respect. The chief difficulty with this method lies in the fact that such analyses do not determine the availability of the plant food present. Another method which is fairly satisfactory is to make pot tests with the soil in question and for the crops to be grown. Such tests may frequently be completed in a shorter period of time than can field tests. They are not, however, so satisfactory as field tests because the crops are not grown under field conditions.

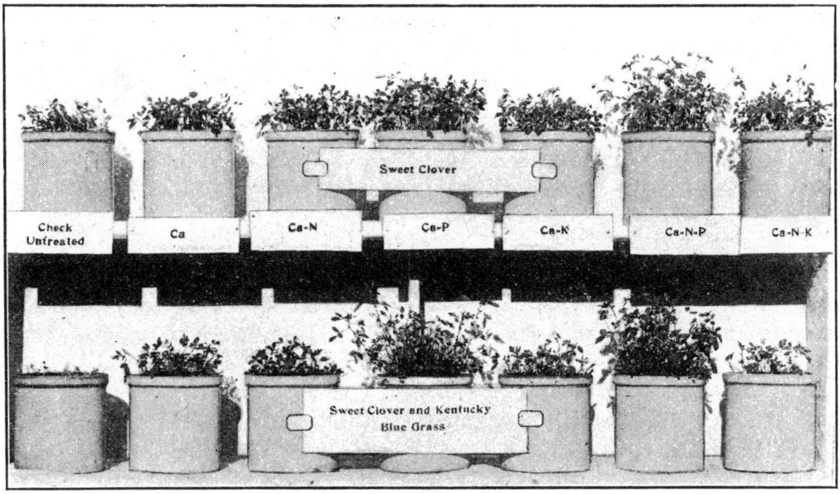

Effect of Fertilizers on the Growth of Sweet Clover.

Soil from virgin cut-over land in Pennsylvania.
Ca—Lime. N—Nitrogen. P—Phosphorus. K—Potash.

Effect Modified by Soil and Crop.—The fertilizer to be used is determined both by the needs of the soil and the crop grown. A commercial fertilizer is beneficial chiefly because of the plant-food elements it supplies. Its best action is accomplished when the soil is in good physical condition and when there is a good supply of moisture and organic matter. The effect of a fertilizer under one set of soil conditions may be reversed when the conditions are materially changed. Under favorable conditions, for example, nitrification in the soil might proceed with sufficient activity to supply a certain crop with all the nitrogen needed for normal growth. The following season being cold and accompanied by an excess of moisture might result in slow nitrification, and this might materially diminish the growth of the crop. In one case nitrogen in a readily available form would be much more beneficial than in the other. In the same way the results obtained on one farm might not be duplicated on the adjacent farm, although the soil is of the same formation and type, difference in the previous cropping or management of the soil being responsible for the difference in results.

Which is the Best Fertilizer to Use?—This question is a pertinent one, and is often asked by practical farmers. A definite answer can seldom be given. The consumer of fertilizers can best answer it by tests such as above suggested. In a general way, however, the consumer should select those fertilizers which contain the largest amount of plant food in suitable and available forms for the least money. Until a rational scheme of fertilizer treatment has been established it is safest to depend upon high-grade fertilizers used in rather limited amounts. Low-grade materials and elements in slowly available form may prove cheaper for certain soils and crops, but their use involves a larger risk, especially for the farmer who is not well informed on the subject. For soils poor in humus, nitrogenous fertilizers will generally be advisable. For those well supplied with humus, phosphates and potash generally give best results.

Needs of Different Soils.—Since the fertilizer is determined by both soil and crop, the needs of the soil can be determined only in a rather general way. There is no definite statement that will hold under all conditions. A particular soil type in one locality may be greatly benefited by a certain fertilizer, while the same type in another neighborhood may have quite a different requirement.

Heavy soils generally respond to phosphates. Sandy soils are more likely to need potash and nitrogen, while clay soils are generally well supplied with potash. There are some exceptions to this rule.

Experiments at various experiment stations show that soils vary widely in their fertilizer requirements. The results in one locality may be inapplicable in another. Acid soils respond to application of lime and generally to available phosphates. Marshy soils, especially those consisting chiefly of muck or peat, are generally in need of potash and sometimes phosphoric acid and lime. The prairie soils are as a rule deficient in phosphorus, and on such soils the insoluble phosphates are economically used. The need for lime is frequently determined by the failure of clover and the encroachment of sorrel and plantain. Potassium is likely to be needed in soils that have long been exhaustively cropped, especially if hay and straw have been sold from the land as well as the grain.

Crop Requirements.—Crops differ in their fertilizer requirements. This difference is due to the purpose for which the crop is grown, to the length of the growing season required by the crop, and to the period of the season when it makes its chief growth; also to the composition of the crop. It is also influenced by the character of the root systems. Plants which grow quickly generally need their food supply in a readily available form. Those which grow slowly and take a long time to mature can utilize the more difficultly available forms of plant food. These facts explain why plants differ in their requirements.

Fertilizers for Cereals and Grasses.—The cereals and grasses (Indian corn excepted) are similar in habits of growth and are distinguished by having extensive, fibrous root systems. They require comparatively long periods of growth, and this enables them to extract mineral food from comparatively insoluble sources. As a rule, however, these crops make the major portion of their vegetative growth during the cool part of the growing season. During this period nitrification is comparatively slow; consequently, such crops need readily available nitrogen and respond to fertilizers containing some nitrogen. This demands the application

Effect of Commercial Fertilizer on Wheat on a Poor Soil.

A complete fertilizer on the left, no fertilizer in center.

of nitrogen in a readily available form, preferably just at the beginning of vegetative growth in the spring.

Legumes Require No Nitrogen.—The clovers, peas, beans, vetches and in fact nearly all the crops that belong to the family of legumes have the power under proper soil conditions to utilize free nitrogen from the air; consequently, such crops require no nitrogen in the fertilizer. They use relatively more potash than most other forage crops; consequently, the mineral fertilizers with a rather high proportion of potash are generally most beneficial. Corn is a rather gross feeder, and since it makes the major portion of its vegetative growth in the warmer portion of the growing season when nitrification is especially active, it seldom pays to apply much nitrogen to it. Furthermore, corn is able to make use of relatively insoluble phosphorus and potash.

Available Forms Best for Roots.—Root and tuber crops are generally regarded as a class that, because of their habits of growth, are unable to make extensive use of the insoluble minerals; hence, their profitable growth requires plenty of the readily available forms of fertilizing constituents. Nitrogen and potash are especially valuable for mangels and beets, while phosphates and potash together with small amounts of nitrogen are generally used for both white and sweet potatoes.

Slow-Acting Fertilizers Suited to Orchards and Small Fruits.—Orchard trees are as a rule slow growing and do not demand quick-acting fertilizers. In old orchards that are large and are top dressed it may, however, be good practice to use the readily soluble forms of plant food in order that it may be carried into the soil by rainfall and brought in contaot with the zones of root activity. Where orchards are manured from the beginning, and especially where they are inter-tilled, barnyard manure and the more difficultly soluble forms of fertilizers may be economically used.

The fertilizer requirements of small fruits are similar to those of orchard fruits. As a rule smaller fruits make a more rapid growth; consequently, heavier applications of soluble fertilizing constituents may be used.

Nitrogen Needed for Vegetables.—The market garden crops, and especially those grown for their vegetative parts, demand rather liberal applications of available nitrogen. The higher the value of the crop per unit of weight, the larger are the applications of nitrogen that may be used economically. In such crops as early cabbage, beets, peas, etc., earliness and quality are of prime importance. To be highly remunerative such crops must be harvested early; in other words, they must be forced. At this period of the year decomposition processes in the soil are not especially active. For this reason an abundance of available nitrogen is demanded.

Fertilizers for Cotton.—Perhaps no crop has been subjected to more experiments with fertilizers than cotton. Cotton is a plant that responds promptly and profitably to judicious fertilization. Such fertilization should hasten the maturity of the crop. This tends to increase the climatic area in which cotton may be grown. In recent years it has become of great importance in connection with the cotton boll weevil. This insect multiplies rapidly throughout the season, its numbers becoming very great in the latter part of the season. It feeds on the cotton bolls. When the bolls are matured early, the insects being less numerous at that season, a larger proportion of the bolls escapes infestation than when they mature late. The most judicious proportions of nitrogen, soluble phosphoric acid and potash in a complete fertilizer for cotton has not been determined with entire accuracy. Those for Georgia are nitrogen 1, potash 1, phosphoric acid $3\frac{1}{2}$; for South Carolina, nitrogen 1, potash $\frac{3}{4}$, phosphoric acid $2\frac{1}{4}$; and for general use nitrogen 1, potash 1, phosphoric acid $2\frac{3}{4}$ or 3 will perhaps approximate reasonable accuracy.

The amount of fertilizer which may be profitably used varies widely with the season, nature of soil and other circumstances. On an average the maximum amounts indicated for Georgia are nitrogen 20 pounds, potash 20 pounds, phosphoric acid 70 pounds; those for South Carolina, nitrogen 20 pounds, potash 15 pounds, phosphoric acid 50 pounds.

Miscellaneous Fertilizer Facts.—Wheat, to which a moderate amount of manure has been applied, will not need additional nitrogen. In most cases the manure can be profitably supplemented with phosphoric acid, and on some soils a small amount of potash may be included. When the wheat field is seeded to clover and grass which is to be left down for hay, the phosphoric acid and potash in the fertilizer should be increased somewhat.

Oats as a rule receive no commercial fertilizer. On soils low in fertility small applications of readily soluble nitrogen and phosphoric acid applied at seeding time are advisable. Winter oats, grown mostly in the South, are generally fertilized with light applications of phosphorus and potash when seeded in the fall, and are top dressed with nitrate of soda in the spring.

For tobacco, barnyard manure occupies a leading position as a fertilizer, both because of its cheapness and effectiveness. When manure is not available in sufficient quantities commercial fertilizers are frequently resorted to. In fact, the manure is often supplemented with commercial fertilizers. This crop generally requires a complete fertilizer. Cottonseed meal is frequently used as a source of nitrogen for tobacco. However, manure is not used for bright tobacco and only very small amounts of cottonseed meal are used.

When nitrogen is required by a crop having a long growing season it is generally advisable to combine it in two forms, one readily available as nitrate of soda or sulphate of ammonia, the other in an organic form, as dried blood or cottonseed meal. Where nitrate of soda is depended upon entirely, two or more applications may be given during the growing season. This is applicable to open, leachy soils, but is not essential on heavy soils.

Effect of Fertilizers on Proportion of Straw to Grain.—The proportion of straw to grain is influenced by season, soil and character of fertilizer. At the Pennsylvania Experiment Station, in a test extending through many years, it was found that for twenty-four different fertilizers applied there were produced 52 pounds of stover for each 70 pounds of ear corn. The average proportion for seven complete fertilizers was 55.4 pounds stover to 70 pounds corn. Barnyard manure gave 47.6 pounds stover to 70 pounds corn, while a complete fertilizer containing dried blood gave 58 pounds stover to 70 pounds corn. In case of oats, the largest relative yield of straw was from barnyard manure. The average for twenty-four different fertilizers was 45 pounds straw per bushel of oats. The average for seven complete fertilizers was 42 pounds straw per bushel of oats. In general, the proportion of straw will be increased by an abundance of nitrogen, while the proportion of grain will be increased by liberal supplies of phosphoric acid.

This is a matter of considerable practical importance in the growing of both oats and wheat. There is often such a marked tendency for these crops to produce vegetative growth that the straw lodges before maturity. This makes harvesting of the crops with machinery difficult. It smothers out the clover and grasses that are sometimes seeded with them. Lodging also prevents satisfactory filling of the heads of grain and maturing of

Soil Fertility Plats, Pennsylvania Agricultural Experiment Station.

On left, 320 pounds land plaster.
Center, no fertilizer.
On right, dissolved bone-black, containing 48 pounds phosphoric acid and muriate of potash 200 pounds.

the kernels. A properly balanced fertilizer or the proper proportion of available constituents in the soil for these crops, therefore, is essential.

Principles Governing Profitable Use of Fertilizers.—Definite rules relative to amount and character of fertilizer for soils or crops cannot be laid down, but there are certain principles that should always be taken into consideration in connection with the use of fertilizers. In general, the higher the acre value of the crop grown the larger the amount of fertilizer that can be profitably used. This is a principle that will hold even though the same percentage increase from a definite investment in fertilizer is secured.

Another principle which always holds is that each additional unit of fertilizer gives a smaller increase in crop growth than the preceding one; consequently, the lower the money value of the crop the smaller the amount of fertilizer that can be profitably used. This principle is well illustrated in an experiment with fertilizers used in different amounts on cotton at the Georgia Experiment Station. In this experiment a fertilizer valued at about $20 per ton was applied in amounts valued at $4, $8 and $12 per acre respectively. As an average of three years with these applications the increase in lint and seed, respectively, resulting from the applications were valued at $10.11, $15.69 and $21.17, the percentage of profit on the investment in fertilizers being 153, 96 and 76 for the three amounts respectively. These results coincide with the principle above stated. In the above experiment the increase in yield of seed cotton for 400 pounds of fertilizer was 281 pounds. The increase for 800 pounds was not twice 281, which would be 562, but was only 436 pounds. The increase for 1200 pounds was not three times 281, which would equal 843, but was only 588 pounds. The smallest amount of fertilizer produced the largest return on the capital invested in fertilizer, although the largest amount made the largest aggregate profit. In this case each $4 invested brought a return greater than the actual investment, and it is evident that it might have been possible to add another $4 worth of fertilizer and still further increase the total profit per acre, although the percentage return on the investment would have been reduced still further. The fertilizer, however, is only part of the investment, since the rent of land and cost of labor and seed are comparatively large items.

If a planter has $1400 to invest in the growing of cotton and the rent of land, seed, labor and every expense connected with the cost of cultivation and picking aggregated $28 per acre, he can plant fifty acres. If his profit without fertilizer is $3 per acre, it will aggregate $150, or $10\frac{2}{3}$ per cent on the investment. On the basis of the above experiment and with the same capital, how much will he be justified in reducing his acreage in order to purchase fertilizers?

By inspection we find:

Acres.	Cost of Growing One Acre.	Total Cost.	Profit per Acre.	Total Profit.	Per Cent on Investment.
50	$28.00	$1,400.00	$3.00	$150.00	10.7
43.75	32.00	1,400.00	9.11	398.56	28.4
38.9	36.00	1,400.00	10.69	415.84	29.7
35	40.00	1,400.00	12.17	425.00	30.3

The increased cost per acre represents the addition of fertilizers to the amount of $4, $8 and $12 and is justified up to the $12 limit where the maximum profit of $425 is secured. By growing 35 acres well fertilized, his percentage profit on capital invested is $30\frac{1}{3}$ instead of $10\frac{2}{3}$ where no fertilizer was to be used.

When to Apply Fertilizers.—The time at which to apply commercial fertilizer will be determined by the needs of the crop, kind of fertilizer, rate of application, character of soil and subsoil, convenience of the farmer and the economy in applying. Plenty of plant food should be within reach of the plants when growth is rapid. Fertilizers that are readily lost by leaching should not be applied long before needed. Heavy applications may be divided into two or three portions and applied as needed. On heavy soils with retentive subsoils leaching is slight. On sandy soils it may be pronounced. As a rule it will be economy to apply small and moderate applications of fertilizer just prior to or at seeding time. Most planting and drilling machinery is now supplied with fertilizer attachments. These provide for the proper distribution of the fertilizer at the time of seeding or planting without much additional labor. So long as the amount which is distributed in the immediate vicinity of the seed is not sufficient to interfere with germination and early growth, the method is satisfactory. If the concentration of the soil solution in contact with seeds equals the concentration within the cells of the seeds, they will be unable to absorb water from the soil. This may prevent germination and cause the seed to rot. For this reason it is never wise to apply large applications in this way. Such applications should be applied some time in advance of seeding or planting in order that the fertilizers may have become uniformly disseminated through the soil. Another method in common use is to broadcast a portion of the fertilizer and mix it with the soil by harrowing. The remainder is then applied through the fertilizer attachment of the seeding machinery. As previously noted, soluble nitrates may be advantageously applied just at the time when the growing crop is most in need of available nitrogen. This is especially applicable on sandy, leachy soils. So far as danger of loss is concerned, the potash and phosphorus may be applied at almost any time.

Readily soluble fertilizers are preferable for the top dressing of grass land, and should be applied very early in the spring, just as the grass is starting to grow. Early application is necessary because the growth demands it early in the season, and also because the fertilizer must be carried into the soil by rains in order to be brought into contact with the roots.

Organic fertilizers, and especially manure, are best applied some time in advance of seeding. The early stages of decomposition frequently give rise to deleterious compounds. These should have time to disappear before the crop is started.

Methods of Application.—The manner of applying fertilizer depends on a number of conditions, especially the kind of fertilizer, the amount to be used, the character of the crop and the method of its tillage. It is a good practice to distribute the potash and phosphoric acid in that portion of the soil where the root activity of the crop grown is most abundant. In case of inter-tilled crops this will generally be in the lower two-thirds of the plowed portion of the soil. The surface two inches are so frequently cultivated during the early period that roots are destroyed. At other seasons it is likely to be so dry that roots cannot grow in it. Plant food does little good so long as it remains at the surface. It is not so essential to put the soluble nitrates in this lower zone because there is a great tendency for them to pass downward in the soil.

Where very small applications are used it is often thought advisable to deposit the fertilizer with the seed or plant in order that it may have an abundance of plant food at the very outset. This method stimulates the plant in its early stages of growth. It is probably more applicable to crops that are seeded or planted very early when the ground is cold and bacterial activity is slow.

In the cotton belt there are two methods of applying fertilizers. Experiments at the Georgia Experiment Station have shown that the method known as "bedding on the fertilizer" has given better results than applying the fertilizer through the fertilizer drill at time of seeding cotton. In the first method the fertilizer is distributed over the bottom of a furrow in which the cotton is planted one week or ten days later. The second method deposits the fertilizer in close proximity to the seed at planting time. As an average of four years the per cent profit on the investment in fertilizer was 48 when applied with the seed and 90 when "bedded on the fertilizer."

Purchase of Fertilizers.—The concentrated high-grade fertilizer materials necessarily command a higher price than low-grade materials and those containing small amounts of plant food. As a rule the high-grade materials are the cheapest. The inexperienced farmer is too much inclined to purchase fertilizers chiefly on the ton basis, without regard to the amount or form of plant-food constituents they contain. He should bear in mind that he is not buying mere weight, but that he is paying for one or more of the plant-food constituents, and those fertilizers that are richest in plant food will generally supply these ingredients at the lowest cost per unit. This is obvious from what has been previously said relative to the costs of manufacturing, handling and shipping fertilizers. It is well also to consider the relative economy of retail versus wholesale rates on fertilizers. The more hands a fertilizer passes through the greater will be its cost when it reaches the consumer. Each dealer must of necessity make some profit on his transaction. Small shipments and small consignments call for higher freight rates and additional labor in making out bills and collecting accounts. These all entail increased expense.

There is now an increased tendency on the part of farmers to co-operate in the purchase of fertilizers. As a rule the character of fertilizer that best meets the needs of a farmer in a particular locality will in general be a good fertilizer for his neighbors. It is possible for neighbors to combine and purchase their fertilizers in carload lots directly from the manufacturer, saving the profit of the middleman and getting carload freight rates which will very materially reduce the cost of the fertilizers laid down at their railway stations. Such co-operation in buying will generally lead to a discussion of the merits of the different brands of fertilizers, and in this way the purchase is generally based upon the combined judgment of the co-operating farmers instead of on an individual farmer. If by chance a diversity of crops and soils of the neighborhood is such that different brands are required, there will be no difficulty in having several brands shipped in the same car.

It is also wise to purchase early and avoid the rush which often causes a delay in shipments in the rush season. Then, too, early orders enable the farmer to plan more definitely relative to his fertilizer needs and give more careful consideration to the brand most likely to meet his needs. In this way he is enabled to receive and haul his fertilizer to his farm at a time when the field work does not demand the time of himself and teams.

It is also well to consider the relative advantages of buying mixed fertilizers as compared with the unmixed goods. In the nature of things the manufacturer with his well-equipped plant should be able to mix fertilizers more thoroughly and economically than the farmer. This, however, is not always done, since the farmer can frequently utilize labor for mixing fertilizers when it would otherwise be unemployed. The advantages of buying unmixed goods are that the farmer can make the mixture that in his judgment will best meet his needs. He may not be able to secure on the market just such a mixture. Furthermore, it will enable him to make different mixtures and try them on his soil and for his crops with the view of gaining information relative to the character of fertilizer that will best meet his future needs.

Home Mixing of Fertilizers.—The home mixing of fertilizers demands on the part of the farmer a fair knowledge of fertilizers and the needs of soils and crops. Without this, he had probably best depend upon ready mixed goods such as are recommended for his conditions. Furthermore, much will depend upon whether or not he can purchase a fertilizer the composition of which, in his judgment, is what he should have, and also whether or not there would be much saving in buying unmixed goods when the additional labor of mixing is taken into account. Such a practice is likely to be economical only when the fertilizers are used rather extensively. Where only a few hundred pounds are used by the farmer it will generally not be advisable for him to attempt to mix his own fertilizer.

So far as the mechanical process is concerned, fertilizers can be mixed by the farmer on the farm very satisfactorily. It does not require a mechanical mixer, although this may be economical when it is done on a large scale. When the unmixed goods are in good mechanical condition, as they should be, definite weights or measures of the different constituents may be placed on a tight barn floor and shoveled over a number of times until the mixture takes on a uniform color. It is advisable to empty not more than 400 to 600 pounds at one time. It can be more thoroughly mixed in small quantities. A hoe and square-pointed shovel are best suited for the mixing. A broom and an ordinary 2 by 6 foot sand screen with three meshes to the inch are all that are necessary. This assumes that the fertilizer comes in bags of definite weight, and that by putting in one bag of one ingredient and two or three of another, etc., a proper proportion can be secured. Greater exactness can, of course, be obtained by using platform scales and weighing roughly the amounts of the different kinds that are brought together. It is suggested that the most bulky ingredient be placed at the bottom of the pile and the least bulky on top. After it is mixed with a shovel and hoe it should be thrown through the screen. This removes all lumps and perfects the mixing. The lumps, should there be any, should be crushed before they are allowed to go into the next mixing batch. After thorough mixing the material will be ready to return to the bags. It can be hauled to the field when needed.

It is well to remember that most fertilizers absorb moisture, increase in weight and later on dry out and become hard. It is, therefore, wise to keep them in a building which is fairly dry.

The following list of fertilizer materials, together with the percentage of the several ingredients which they contain, is given as an aid to those making home mixtures of fertilizers:

List of Materials Used in Home-Mixing of Fertilizers.*

Name of Material.	Nitrogen, per cent.	Phosphoric Acid, per cent.	Potash, per cent.	Availability.
Nitrate of soda	15	0	0	Very quick
Sulphate of ammonia	20	0	0	Quick
Dried blood	10	0	0	Medium
Tankage (meat)	7.4	10	0	Slow
Tankage (bone)	5	15	0	Slow
Ground bone	2.5	23	0	Slow
Acid phosphate, 14 per cent	0	14	0	Quick
Acid phosphate, 12 per cent	0	12	0	Quick
Dissolved bone-black	0	15	0	Medium
Basic slag	0	15	0	Slow
Rock phosphate	0	18–30	0	Very slow
Muriate of potash	0	0	50	Quick
High-grade sulphate of potash	0	0	50	Quick
Kainite	0	0	12	Quick
Wood-ashes	0	2	6	Medium

REFERENCES

"Manures and Fertilizers." Wheeler.
"Fertilizers." Voorhees.
"Fertilizers and Crops." Van Slyke.
New York Expt. Station Bulletin 392. "Fertilizer Facts for the Farmer."
South Carolina Expt. Station Bulletin 182. "Potash."
Texas Expt. Station Bulletin 167. "Commercial Fertilizers and Their Use."
Farmers' Bulletins, U. S. Dept. of Agriculture:
388 "Incompatibles in Fertilizer Mixtures."
398. "Commercial Fertilizers in the South."

* From the Farmers' Cyclopedia.

CHAPTER 5

Barnyard, Stable and Green Manures

Barnyard and stable manure consists of the solid and liquid voidings of the farm animals mixed with various kinds and amounts of bedding. The term stable manure designates manure just as it comes from the stable in its fresh state. Yard manure applies to that which has accumulated or been kept for some time in piles in the barnyard. Fresh manure means that which is only a few hours or, at most, a few days old. The term rotted manure is used to designate that which has gone through considerable fermentation and is more or less disintegrated. The term mixed manure applies to that of the different species of farm animals when brought together in the same manure heap.

Manure an Important Farm Asset.—The manure of farm animals is the most valuable by-product of American farms. Numerous tests and analyses have been made to determine the amount and composition of both the liquid and solid excrements for different classes of farm animals. The average yield of fresh manure and its content of essential plant-food constituents, together with the yearly value of these, is given in the following table for different classes of animals. The calculations in this table are based on the composition of the solid and liquid excrements given in a subsequent table in this chapter. The plant-food constituents are valued as follows: nitrogen eighteen cents a pound, phosphoric acid four cents a pound, potash five cents a pound.

Average Yield and Yearly Value of Fresh Manure of Farm Animals, Exclusive of Bedding.

Kind of Livestock.	Amount of Manure Yearly, pounds.	Pounds of Ingredients Yearly.			Yearly Value.
		Nitrogen.	Phosphoric Acid.	Potash.	
Cow	28,000	124.	50.	132.	$30.92
Horse	15,000	96.	42.	81.	23.01
Pig	3,000	14.4	9.54	11.4	3.54
Sheep	1,140	11.02	4.75	9.88	2.67
Poultry	30	.414	.15	.123	.087

The following table gives the numbers of the different classes of farm animals in the United States according to the census of 1910, together with the calculated value of manure for each class, the calculations being based upon the valuation of manure given in the preceding table. In case of cattle, the valuation has been reduced, the reduction being based on the relative numbers and values of milch cows as compared with all other cattle.

ANIMALS IN THE UNITED STATES IN 1910 AND ESTIMATED VALUE OF THEIR MANURE.

Class.	Number of Animals.	Value of Manure.	
		Per Head.	Total.
Horses	27,618,242	$23.00	$635,219,566.00
Cattle (all kinds)	63,682,648	23.00*	1,464,700,904.00
Swine	59,473,636	3.54	210,536,671.00
Sheep and goats	55,868,543	2.67	149,169,010.00
Poultry	295,880,000	.087	25,741,560.00
Total value			2,485,367,711.00

Manure is valuable because: (1) it contains the three essential elements of plant food, namely, nitrogen, phosphorus and potassium; (2) it furnishes organic matter which is converted into humus in the soil and materially improves the physical condition, water-holding capacity and chemical and bacterial activities in the soil; (3) it introduces beneficial forms of bacteria in the soil and these multiply and become increasingly beneficial as their numbers increase.

As a Source of Plant Food.—The composition of manure varies with the kind of animals producing it, the age of animals and the amount and quality of the feed they consume. The manure consists of the solid excrements and the liquids or urine. These differ in their composition. The urine is the most valuable part of the excreta of animals. The average mixed stable and barnyard manure contains approximately ten pounds nitrogen, six pounds phosphoric acid and eight pounds potash in each ton of manure. The solid portions consist chiefly of the undigested portions of the food consumed, together with the straw or bedding that has been used in the stables. The solid portions contain approximately one-third of the total nitrogen, one-fifth of the total potash and nearly all of the phosphoric acid voided by animals. The urine contains about two-thirds of the total nitrogen, four-fifths of the potash and very little of the phosphoric acid. The elements found in the urine are insoluble. They are not immediately available as food for plants, but become so more quickly than the constituents in the solid portions.

Of the nitrogen in barnyard manure, that in the urine will be most readily available; that in the finely divided matter of the feces will be more slowly available; and that in the bedding will be most slowly available. For this reason the availability of the nitrogen in manure when applied to the soil is distributed throughout a comparatively long period. Availability will vary greatly with the nature and treatment of the manure.

* Estimated value based on relative numbers and values of milch cows and all other kinds of cattle.

Experiments at several experiment stations show that the nitrogen in manure is much less readily available than that in either nitrate of soda or sulphate of ammonia. Because of this fact, barnyard manure when used for certain truck crops is sometimes supplemented with available forms of nitrogen. In such cases it is not advisable to mix the chemical forms of nitrogen with the manure. Such mixture is likely to result in a loss of available nitrogen through denitrification in the manure pile. It is best, therefore, to apply the chemical form of nitrogen by itself, preferably some time after the manure has been applied.

Physical Effect of Manures.—Barnyard and stable manure improves the physical condition of heavy soils by increasing their tilth and making them easier to cultivate. It improves loose, sandy soils by holding the particles together and increasing the water-holding capacity. It, therefore, has the reverse effect on these two extremes of soil.

Manure tends to equalize the supply and distribution of water in the soil and renders the soil less subject to erosion and injury by winds. Experiments conducted by Professor King at the Wisconsin Experiment Station show that manured land contained eighteen tons more water per acre in the upper foot of soil than similar land unmanured, and thirty-four tons more in the soil to a depth of three feet.

Biological Effect of Manure.—Farm manures introduce into the soil a variety of bacteria and ferments. These help increase the supply of available plant food. Barnyard manure sometimes causes denitrification in the soil. By this process, nitrogen is set free in a gaseous form and may escape. This is likely to be most serious as a result of changing nitrates in the soil into other forms and therefore reducing the available nitrogen supply. Experiments show that this occurs only in exceptional cases and generally when unusually large applications of manure have been made. On the other hand, experiments in considerable number indicate that applications of manure may actually favor nitrification and aid in the formation of nitrates. At the Delaware Experiment Station it was found that soil liberally manured and producing hay at the rate of six tons per acre contained several times as many bacteria as were found in the same soil which had but little manure and was producing hay at the rate of about one ton per acre.

The Value of Manure.—The value of manure depends: (1) upon the class of animals by which it is produced; (2) upon the age of the animals producing it; and (3) upon the character of feed from which produced. Animals that are used for breeding purposes or for the production of milk or wool retain a larger proportion of the plant-food constituents of the food they consume. This will be found in their products, whether it be the young animals to which they give birth or the milk or wool produced by the cow and sheep respectively. Young animals that are making rapid growth use a portion of the plant-food constituents, and this is built into the tissues and bones of such animals. Old animals that have ceased to grow and animals that are being fattened void practically all of the plant-food constituents in their excrements. For this reason the manure from different classes of animals varies considerably in its plant-food constituents.

Mature animals, neither gaining nor losing in weight, excrete practically all of the fertilizer constituents in the food consumed. Growing animals may excrete as little as 50 per cent of such constituents. Milch cows excrete 65 to 85 per cent; fattening and working animals 85 to 95 per cent. As regards the value of equal weights of manure under average farm conditions, farm animals stand in the following order: poultry, sheep, pigs, horses, cows. At the Mississippi Experiment Station young fattening steers excreted on an average 84 per cent of the nitrogen, 86 per cent of the phosphoric acid and 92 per cent of the potash in the food consumed. At the Pennsylvania Experiment Station, cows in milk excreted 83 per cent of nitrogen, 75 per cent of phosphoric acid and 92 per cent of the potash of their food. The amount of manure produced per thousand pounds of live weight of animals also varies with the class of animals, as well as with the method of feeding and the character of the feed consumed. Sheep and hogs produce the smallest amount of manure, but yield manure of the greatest value per ton. Cows stand first in the amount of manure produced, but rank lowest in the quality of manure.

Horse Manure.—Horse manure is more variable in its composition than that of any other class of farm animals. This is due to the fluctuation in the amount and character of the feed given to the horse, depending on whether he is doing heavy or light work, or whether he is idle. Horse manure is drier than that from cattle, and generally contains more fibrous material. It ferments easily, and is, therefore, considered a hot, quick manure. When placed in piles by itself it ferments rapidly and soon loses a large part of its nitrogen in the form of ammonia. Because of its dry condition and rapid fermentation the temperature of the manure pile becomes very high, causing it to dry out quickly. This results in what is commonly called fire-fanging. To prevent this, horse manure should be mixed with cold, heavy cow or pig manure, or the piles of horse manure should be compacted and kept constantly wet in order to reduce the presence of air and consequent rapid fermentation. The quality of horse manure makes it especially valuable for use in hotbeds, for the growing of mushrooms and for application to cold, wet soils. Horse manure is more bulky than that of any other class of farm animals and weighs less per cubic foot.

Cattle Manure.—Cow and steer manure contains more water than that from other domestic animals. It is ranked as a cold manure, and has the lowest value, both from the standpoint of its plant-food constituents and its fertilizing value. The average cow produces 40 to 50 pounds of dung or solid manure, and 20 to 30 pounds of urine per day.

Hog Manure.—The manure from hogs is fairly uniform in its composition, and is considered a cold, wet manure. It ferments slowly. Hogs of average size produce 10 to 15 pounds of manure daily, and the manure is somewhat richer than that from the preceding classes of animals, chiefly because swine are fed more largely on rich, concentrated foods.

Sheep Manure.—Sheep manure is drier and richer than that from any of the domestic animals except poultry. It ferments easily and acts quickly in the soil. It keeps well, however, when allowed to accumulate in pens where it is thoroughly tramped by the animals. It is especially valuable for use in flower beds or for vegetables where quick action is desired. An average sheep produces about four to five pounds of manure daily.

Poultry Manure.—Poultry manure is the richest of farm manures. It is especially rich in nitrogen, which is due to the fact that the urinal secretions are semi-solid and are voided with the solid excrements. It ferments easily, giving rise to the loss of nitrogen, and is very quick acting when placed in the soil. It keeps best when maintained in a fairly dry condition, and should be mixed with some absorbent or preservative. Ground rock phosphate, gypsum or dry earth are good materials for this purpose. Mixing with slaked lime, ashes or any alkaline material should be avoided. These cause a liberation of ammonia, resulting in a loss of nitrogen.

The following table gives the average total production of solid and liquid excrements per year of the different classes of animals, together with their percentage of water, nitrogen, phosphoric acid and potash.

Average Yield and Composition of Fresh Excrements of Farm Animals.*

Dung—Solid Excrements.	Excreted per Year, pounds.	Water, per cent.	Composition.		Potash, per cent.
			Nitrogen, per cent.	Phosphoric Acid, per cent.	
Cows	20,000	84.0	0.30	0.25	0.10
Horse	12,000	76.0	0.50	0.35	0.30
Pigs	1,800	80.0	0.60	0.45	0.50
Sheep	760	58.0	0.75	0.60	0.30
Hen	30	48.6	1.38	0.50	0.41

Urine—Liquid Excrements.	Excreted per Year, pounds.	Water, per cent.	Composition.		Potash, per cent.
			Nitrogen, per cent.	Phosphoric Acid, per cent.	
Cows	8,000	92.0	0.80	Trace	1.4
Horse	3,000	89.0	1.20	Trace	1.5
Pigs	1,200	97.5	0.30	0.12	0.2
Sheep	380	86.5	1.40	0.05	2.0

* This table taken from Volume Five, Farmers' Cyclopedia.

Miscellaneous Farm Manures.—In addition to the manure from farm animals there is a variety of materials that may be available as manure on many farms. It is well to utilize these as far as possible. Among those most commonly met with are night-soil, leaf-mould and muck or peat. Night-soil is best used when mixed with some good absorbent, such as loam, muck or peat, and composted. Muck and peat are terms used to designate accumulations of vegetable matter that are frequently found in marshes, swamps and small ponds. Such material varies greatly in its composition, and is especially valuable for its content of nitrogen, and for its physical effect upon the soil. Leaf-mould pertains to decayed accumulations of leaves frequently found in considerable quantities in forested areas. It is especially valuable for some classes of garden truck and flowers, but is ordinarily too costly because of the difficulty of gathering it in any considerable quantities.

Value of Manure Influenced by Quality of Feed.—The plant-food content of manure is almost directly in proportion to the plant-food constituents contained in the feeds from which it comes. Thus, concentrated feeds high in protein, such as cottonseed meal, wheat bran and oil cake, produce manure of the highest value. Ranking next to these are such feeds as alfalfa and clover hay and other legumes. The cereals, including corn and oats together with hay made from grasses, rank third, while manure from roots is the lowest in plant-food constituents and fertilizing value. Not only will the plant-food constituents be most abundant in the manure from the concentrates, but it is likely also to be more readily available than that produced from roughage.

These facts are important in connection with the selling of cash crops and purchasing such concentrates as cottonseed meal and bran. One who buys cottonseed meal as a fertilizer gets only its fertilizing value. If it is purchased for feeding purposes, one may secure both its feeding value and practically all of its manurial value. The relative price, therefore, of cash crops and purchased concentrates as feed is only one phase of the exchange problem. Such concentrates produce manure having a much higher value than that from the cash crops. This should be considered in connection with the exchange.

The table on next page shows the pounds of fertilizer constituents in one ton of different agricultural products. It indicates the exchanges which might, therefore, be effected with advantage.

The feeding value of a ton of wheat bran does not differ materially from that of a ton of shelled corn. The difference in its feeding value affects the nutritive ratio rather than the energy value. By exchanging one ton of corn for an equal weight of wheat bran, there would be a gain to the farm of 21 pounds of nitrogen, 46 pounds phosphoric acid and 24 pounds of potash, as shown by the above table. At usual prices for the fertilizer constituents, this gain would amount to not less than $6 worth of plant food. With an exchange of milk or potatoes for similar concentrates, the gain would be still more striking.

Amount and Character of Bedding Affects Value of Manure.—Straw is a by-product on most farms, and is best utilized as bedding for animals. In this way the plant-food constituents are not only all returned to the soil from whence they originally came, but the straw becomes an absorbent and prevents the loss of the liquids in the manure. Straw utilized in this way is probably more valuable than it would be if applied directly

MANURIAL CONSTITUENTS CONTAINED IN ONE TON OF VARIOUS FARM PRODUCTS.

Farm Product.	Manurial Constituents.		
	Nitrogen, pounds.	Phosphoric Acid, pounds.	Potash, pounds.
Timothy hay	19.2	7.2	25.2
Clover hay	39.4	8.0	35.0
Alfalfa hay	53.2	10.8	49.2
Cowpea hay	49.6	13.2	47.2
Corn fodder, field cured	17.2	7.2	21.4
Corn silage	8.4	2.4	6.6
Wheat straw	8.6	2.6	14.8
Rye straw	10.0	5.8	15.8
Oat straw	13.0	4.4	24.4
Wheat	34.6	19.2	7.0
Rye	32.4	16.2	10.4
Oats	36.2	15.4	11.4
Corn	29.6	12.2	7.2
Barley	39.6	15.4	9.0
Wheat bran	51.2	58.4	31.4
Linseed meal	108.6	37.6	26.2
Cottonseed meal	142.8	61.8	36.4
Potatoes	7.0	3.2	11.4
Milk	10.2	3.4	3.0
Cheese	90.6	23.0	5.0
Live cattle	53.2	37.2	3.4

as such to the soil. In the manure it is intermingled with the solid and liquid excrement, and inoculated with the bacteria in the voidings of animals, which facilitates its decomposition in the soil. Straw contains less plant food than an equal weight of dry matter in manure. An abundance of straw, therefore, used as bedding tends to dilute the manure and slightly reduce its value per ton. This, however, is not a logical objection to its use on the farm, although it might become so on the part of the farmer who is purchasing barnyard manure from outside sources, providing, of course, that no distinction in price is made in accordance with the concentration or dilution of the manure.

Some farmers use a great abundance of straw for bedding their animals. It is not, however, deemed good practice to use more than is sufficient to keep the animals clean and absorb and retain all of the liquids.

A superabundance of bedding gives rise to a bulky, strawy manure that must be used in large quantities in order to be effective, and frequently results at the outset in denitrification and unsatisfactory results.

MODERN CONVENIENCE FOR CONVEYING MANURE FROM STALLS TO MANURE SPREADER.[1]

In a general way, it is estimated that the amount of bedding used for animals should equal approximately one-third of the dry matter con-

ABSORBENT CAPACITY OF 100 POUNDS OF DIFFERENT MATERIALS WHEN AIR DRY

Nature of Absorbent.	Liquid Absorbed, pounds.
Wheat straw	220
Oat straw	285
Rye straw	300
Sawdust	350
Partly decomposed oak leaves	160
Leaf rakings	400
Peat	500
Peat moss	1,300

[1] Courtesy of The Pennsylvania Farmer.

sumed. This, however, will vary greatly, depending on the absorbent power of the bedding used and the character of the feed the animals receive. It will also depend on whether or not the absorbent material is thoroughly dry when used. When bedded with ordinary oat and wheat straw, it is estimated generally that cows should each have about 9 pounds of bedding, horses $6\frac{1}{2}$ pounds and sheep $\frac{3}{4}$ pound. The table on preceding page shows the approximate absorbent capacity of various materials used as bedding.

The figures in the table are only approximate, and will vary considerably under different conditions. They are supposed to represent the amount of liquid that will be held by 100 pounds of the substances mentioned, after twenty-four hours of contact.

Aside from the absorbent power of bedding, its composition is also of some importance, and the following table gives the average fertilizer constituents in 2000 pounds of different kinds of straw.

FERTILIZER CONSTITUENTS IN 2000 POUNDS OF VARIOUS KINDS OF DRY STRAW.

	Nitrogen, per cent.	Phosphoric Acid, per cent.	Potash, per cent.
Wheat	11.8	2.4	10.2
Wheat chaff	15.8	14.0	8.4
Oats	12.4	4.0	24.8
Rye	9.2	5.6	15.8
Barley	26.2	6.0	41.8
Barley chaff	20.2	5.4	19.8
Buckwheat hulls	9.8	1.4	10.4

Methods of Storing and Handling.—The value of manure is also determined by the manner in which it is stored, the length of time it remains in storage and its manipulation in the storage heap. Manure is a very bulky material of a comparatively low money value per ton. Its economical use, therefore, demands that the consequent labor be reduced to the minimum, especially in those regions where labor is high-priced. Where manure is to be protected from the elements, it calls for comparatively inexpensive structures for the purpose.

When different kinds of animals are kept, it is advisable to place all the manure together so that the moist, cold cow and pig manure may become thoroughly mixed with the dry, hot horse and sheep dung. In this way each class of manure benefits the other. Where the manure is deposited in a barnyard in which the animals run, the swine are frequently allowed to have free access to the manure pile, from which they often get considerable feed which would otherwise be wasted. Such feed consists of the undigested concentrates fed to the horses and cattle. Swine thoroughly mix the different kinds of manure, and when it is thoroughly compacted by the tramping of the animals, fermentation is reduced to the minimum. If it is protected from rains and sufficient absorbent material has been used in the bedding, loss is comparatively small.

When horse manure is placed by itself, it ferments very rapidly and soon loses its nitrogen. Such fermentation can be materially reduced by compacting the manure pile thoroughly and applying sufficient water to keep it constantly wet. This same rapid decomposition and loss of nitrogen will take place in case of mixed manures if they are neither compacted nor wet, although loss will not be so rapid.

The use of covered barnyards for protecting manure has in recent years met with much favor in some portions of the country.

Losses of Manure.—A practice too common in many sections is to

PILES OF MANURE STORED UNDER EAVES OF BARN, SHOWING HOW LOSS TAKES PLACE.[1]

throw the manure out of stable doors and windows, and allow it to remain for a considerable length of time beneath the eaves of the barns. This not only exposes it to direct rainfall, but also subjects it to additional rain collected by the roof of the building. Under these conditions the leaching of the manure and the consequent loss is very great. Where manure piles remain long under these conditions, it is sometimes doubtful whether the depleted manure is worth hauling to the field. Certainly this is a practice to be condemned. Both the mineral constituents and organic matter are carried off in the leachings.

Experimental Results.—Experiments at the Cornell Experiment Station where manure remained exposed during six summer months showed a percentage loss for horse manure as follows: gross weight 57

[1] Courtesy of Doubleday, Page & Co., Garden City, N. Y. From "Soils," by Fletcher.

per cent, nitrogen 60 per cent, phosphoric acid 47 per cent, potash 76 per cent; for cow manure the loss was: gross weight 49 per cent, nitrogen 41 per cent, phosphoric acid 19 per cent, potash 8 per cent. The rainfall during this period was 28 inches. This shows an average loss for the two classes of manure of more than one-half in both weight and actual plant-food constituents.

By similar observations at the Kansas Station, it was found that the waste in six months amounted to fully one-half of the gross weight of the manure and nearly 40 per cent of its nitrogen.

The New Jersey Experiment Station found that cow dung exposed to the weather for 109 days lost 37.6 per cent of its nitrogen, 52 per cent of its phosphoric acid and 47 per cent of its potash. Mixed dung and urine lost during the same period of time 51 per cent of its nitrogen, 51 per cent of phosphoric acid and 61 per cent of potash. Numerous other experiments along the same line could be cited, giving essentially the same results. These experiments leave no doubt as to the large loss incurred in negligent methods in the management of manure, and emphasize the importance of better methods of storing manure.

The estimated annual value of the manure from all animals in the United States as given in the table in the first part of this chapter is $2,485,367,711. There is no means of ascertaining what proportion of all manure is deposited where it can be collected. For present purposes we will assume that one-half of it is available for return to the land. Assuming that one-third of this is lost because of faulty methods of storage and handling, the loss from this source would be valued at $414,-227,952. The enormous loss sustained by American farmers through negligence in the care, management and use of manure emphasizes the importance of the subject and the great need of adopting economic methods in its utilization.

How to Prevent Loss.—Some of the methods of preventing loss have already been suggested. Under most conditions this is best accomplished by hauling the manure soon after its production directly to the field. This has become a common practice in many localities. It is economical from a number of viewpoints. It saves labor, obviating the extra handling incurred when the manure is first dumped in the yard and afterwards loaded on wagons to be taken to the field. It keeps the premises about the barns and yards clean at all times; reduces offensive odors due to decomposition of manure; and reduces in the summer time breeding places for flies. The most important saving, however, is in the actual value of the manure, which in this way has sustained no loss due to decomposition and leaching.

Absorbents vs. Cisterns.—Losses frequently occur both in the yard and stable, due to a direct and immediate loss of the liquid portions of the manure. This is overcome either by the use of an ample supply of absorbent in the way of bedding or by collecting the liquid manure in a cistern. The cistern method of saving liquid manure is of doubtful economy in this country. The expense of cisterns and the trouble of hauling and distributing, together with the care which must be exercised to prevent loss of nitrogen by fermentation of the liquid when it stands long, are all valid objections to such provisions. It is possible under intensive farming and with cheap labor that liquid manure might be thus saved and utilized for crops that respond to nitrogenous fertilizers. Best results with manure demand that the liquid and solid portions be applied together. It is the consensus of opinion that the best general practice is to save the liquid by the use of absorbents.

Since nitrogen frequently escapes as ammonia, certain absorbents for gases, such as gypsum, kainite, acid phosphate and ordinary dust, have been recommended. As direct absorbents, however, these are of doubtful value, although some of them are effective, first, in reducing the fermentation, and second, in actually reinforcing the manure by the addition of plant-food constituents.

Sterilization.—Preservatives have also been suggested in the nature of substances that will prevent fermentation and thus reduce losses. Bisulphide of carbon, caustic lime, sulphuric acid and a number of other substances have been tested for this purpose. However, anything that will prohibit fermentation destroys the bacteria of the manure, and such destruction may more than offset the saving in plant-food constituents. Furthermore, most of these materials are rather costly, and the benefits derived are not equal to the expense incurred.

Reinforcing Manures.—A number of substances have been used to reinforce manure. The one most beneficial and economical is either acid phosphate or rock phosphate. This is undoubtedly due to the fact that phosphorus is the element most frequently needed in the soils, and that manure is inadequately supplied with it. The following table, showing results obtained at the Ohio Experiment Station by reinforcing manure with different substances, gives direct evidence as to the relative merits of such substances:

Value of Manure, Average 15 Years.—Rotation: Corn, Wheat, Clover (3 Years).

Treatment.	Nothing.	Gypsum.	Kainite.	Floats.	Acid Phosphate.
Return per ton:					
Yard manure	$2.55	$3.04	$2.93	$3.54	$4.10
Stall manure	3.31	3.56	3.97	4.49	4.82

It is evident from the above table that all the materials used have more or less increased the value of the manure, as determined by the value of increase in crops obtained from each ton when applied once in a three years' rotation of corn, wheat and clover. The value per ton of manure is based on the average farm price of the crops produced. It is also evident from the table that stall manure gave in every instance a larger return per ton than did yard manure, and that floats and acid phosphate proved by all odds the best reinforcing materials. While acid phosphate reinforcement gave the largest return per ton of manure, the floats proved about equally profitable from the investment standpoint.

In localities where phosphorus is the dominant soil requirement, the reinforcement of manure with acid phosphate at the rate of about forty pounds to each ton of manure is a most excellent practice. The manner of applying the phosphate may be determined by conditions. It will frequently be found convenient to apply this material to the manure in

Spreading Manure from Wagon, Old Way.[1]

the stalls or stables each day at the rate of about one pound for each fully grown cow, horse or steer, and in lesser amounts for the smaller animals. There is probably no place in which the raw rock phosphate is likely to give better results than when used in this way as a reinforcement to manure.

Economical Use of Manure.—The most economical use of manure involves a number of factors. It is the opinion of both chemists and farmers that manure and urine should be applied to the soil in its freshest possible condition. If this is true, manure should be hauled from the stable or barnyard to the field as soon as it is made. As previously indicated, this method reduces to the minimum the cost of handling and has several additional advantages. Well-rotted manure may be more quickly available to plants, less bulky and easier to distribute, and weight for

[1] Courtesy of Doubleday, Page & Co., Garden City, N. Y. From "Soils," by Fletcher.

weight may give as much or larger returns than fresh manure. There are, however, only a few conditions under which its use can be superior to that of using fresh material. The rotted manure may be used for intensive crops when availability is important, and especially on land where weeds, entailing hand work, become a serious problem. In fresh manure the weed-seeds that may have been in the feeds are likely to be largely viable, and give rise to trouble in the field. Thorough fermentation generally destroys the viability of weed-seeds in manure.

To Which Crops Should Manure be Applied?—Next to time of hauling may be considered the crops to which manure can be most advantageously applied. Direct applications of fresh manure are thought to be injurious to the quality of tobacco, to sugar beets and to potatoes. It should, therefore, not be applied to these crops directly. It may be applied to the crop preceding, or decomposed manure may be used. As a rule, manure should be applied directly to the crop in the rotation having the longest growing season, or the greatest money value. For example, in a rotation of corn, oats, wheat and mixed grasses, corn not only has the longest growing season, but also the greatest food and cash value. It is, therefore, considered good practice to apply the manure directly to the corn. Since the benefits of manure are distributed over a number of years, the crops which follow will benefit by its residual effect.

To What Soils Should Manure be Applied?—Character of soil may also determine where the manure should be applied. If mechanical condition is a prime consideration, fresh manure may be applied to heavy, clay soils and well-rotted manure to light, sandy soils. On the other hand, the sandy soils in a favorable season are more likely to utilize coarse manure to advantage than heavy soils. In such soils decomposition will proceed more rapidly, thus rendering available the plant-food constituents of the manure. On sandy soils manure should be applied only a short time before it is likely to be needed, in order to prevent the danger of loss by leaching. On heavy, clay soils the benefits from applying fresh manure are likely to be rather slight the first year, because of slow decomposition of the manure. This, however, is not serious, because in such soils the plant food as it becomes available is held by the soil with little or no loss.

Climate Affects Decomposition.—Climate may also be a factor influencing the use of fresh manure. In a warm, damp climate it matters little whether the manure is fresh or well rotted when applied. Under such conditions decomposition in the soil is sufficiently rapid to make fresh manure readily available. The character of season may also be a factor determining the relative merits of fresh and rotted manure. In a very dry season excessive applications of fresh manure show a tendency to burn out the soil, and this is more marked in light, sandy soils than in the heavy soils. Furthermore, heavy applications of strawy manure plowed under when the soil is dry will destroy the capillary connection between the upper and lower soils, thus preventing a rise of the subsoil water for the benefit of the newly planted crop. This occasionally results in a crop failure and the condemnation of the use of fresh manure.

Eroded Soil Most in Need of Manure.—In a general way, any kind of manure should be applied to those portions of the farm the soil of which is most in need of manure. Marked differences in the organic content of the soil in different parts of fields are often manifest. This most frequently is the result of slight erosion on the sloping portions. It is a good practice to apply manure to these portions in an effort to restore them to their original fertility. Such areas without special attention tend to deteriorate rapidly. The addition of manure improves the physical condition of the soil, increases its absorptive power for rain and lessens erosion. In this way, not only is the soil benefited, but deterioration through erosion is checked.

Rate of Application.—The rate of applying manure is also important and will determine the returns per ton of manure. Farmers in general do not have sufficient manure to apply in large quantities to all of their land. This gives rise to the question as to whether or not heavy applications shall be used on restricted areas and for certain crops, or whether the manure shall be spread thinly and made to reach as far as possible. Some German writers speak of 18 tons per acre as abundant, 14 tons as

VALUE OF MANURE. AVERAGE 30 YEARS.

Rotation: Corn,* Oats, Wheat,* Clover, Timothy (Four Years).

Treatment, One Rotation.	Value of Four Crops.	Return per Ton of Manure.	Return per Ton over 12 per Acre.
Nothing	$60.02	....	....
Manure 12 tons	88.91	$2.41	....
Manure 16 tons	89.62	1.85	$0.18
Manure 20 tons	92.68	1.63	.33
Manure 12 tons and lime 2 tons	92.22	2.68	....

moderate and 8 tons as light applications. They recommend 10 tons per acre for roots, 20 tons per acre for potatoes. In England, at the Rothampsted Experiment Station, 14 tons yearly for grain was considered heavy. In New Jersey 20 tons per acre for truck is not infrequently used. Such applications are, however, unnecessarily large for general farm crops and for the average farm.

At the Pennsylvania Experiment Station the average results for a period of thirty years in a four-crop rotation when manure was used at the rate of 12, 16 and 20 tons per acre during the rotation, show that the largest return per ton of manure was secured with the lightest application.

* Manure applied to these crops only.

The manure in this case was applied twice in the rotation; 6, 8 and 10 tons per acre to the corn, the same amounts to the wheat and none to either the oats or grass.

The returns per ton of manure are based on a valuation of crops as follows: Corn 50 cents a bushel, oats 32 cents a bushel, wheat 80 cents a bushel, hay $10 a ton, and oat straw, wheat straw and corn stover $2.50 per ton.

A similar experiment at the Ohio Experiment Station covering a period of eighteen years has also shown the largest return per ton of manure in case of the smaller applications. The results are given in the following table:

VALUE OF MANURE. AVERAGE 18 YEARS.

Rotation: Corn,* Oats, Wheat,* Clover, Timothy (Five-year Rotation).

Treatment, One Rotation.	Return per Ton of Manure.	Return per Ton over 8 per Acre.
Manure 8 tons	$3.17	....
Manure 16 tons	2.41	$1.75

Rotation: Potatoes, Wheat,† Clover (Three Years).

Treatment, One Rotation.	Return per Ton of Manure.	Return per Ton over 8 per Acre.
Manure 4 tons	$3.47	....
Manure 8 tons	2.58	$1.69
Manure 16 tons	2.15	1.72
Manure 8 tons	3.30	....

Methods of Applying Manure.—A uniform rate and even distribution of manure are essential. This can be most economically effected by the use of a manure spreader. It does the work better than it can be done with a fork, and at a great saving of labor. While a manure spreader is rather an expensive implement, it will be a paying investment on any farm where 60 tons or more of manure are to be applied annually. It is a common practice in most parts of the country to apply manure to a grass sod and plow it under. In many cases manure is also applied to corn land and land that has been in small grain, to be followed by other or similar crops. While it is the consensus of opinion that the manure applied in this way will give best results, there is some question as to whether or not more of it should not be applied in the form of a top dressing.

Top Dressing vs. Plowing Under.—At the Maryland Experiment

* Manure applied to these crops only.

† Manure applied to wheat, except in second 8 tons application, which went on potatoes.

The Modern Manure Spreader.[1]
This machine spreads the manure evenly and requires but one man to drive and operate it.

[1] Courtesy of The International Harvester Company, Chicago.

Station both fresh and rotted manure were applied before and after plowing. For fresh manure the average of two crops of corn showed a gain of 10.9 bushels per acre in favor of applying after plowing. For the wheat which followed the corn the gain was two bushels per acre. Where rotted manure was compared in the same way there was practically no difference in the yield of corn, and about one bushel gain for wheat in favor of applying after plowing. In this experiment the fresh manure under both conditions and for both crops gave yields considerably above that produced by the rotted manure.

Another experiment in which the manure was plowed under in the spring as compared with plowing under in the fall gave results with corn and wheat favorable to plowing under in the spring. This is in harmony with the preceding experiment, and suggests that manure applied to the surface, and allowed to remain for some time in that position, benefits the soil and results in a better growth of crops than when it is plowed under immediately. The subject is one worthy of further consideration and experimentation. It is not an uncommon opinion, however, among practical farmers that top dressing with manure is more beneficial than plowing it under, and it is quite a common practice to top dress grass lands and wheat with manure.

In the South, where manure is very scarce, it is frequently applied in the hill or furrow at planting time. This entails a good deal of hand labor, but it is probably justifiable where labor is as cheap as it is there. The manner of applying small applications concentrates the manure in the vicinity of the plants and stimulates growth during the early portions of the season.

The Parking System.—The cheapest possible way of getting manure on the land is by pasturing the animals, or allowing them to gather their own feed. This, of course, is an old and universal practice in case of pastures, and is becoming more popular as indicated by the practice of hogging off corn, and other annual crops. This is spoken of as the parking system. It has a disadvantage that in certain classes of animals the manure is not uniformly distributed. It is more applicable for sheep and swine than it is for the larger animals.

Distribution of Benefits.—The benefits of manure are distributed over a number of years. This often gives rise to difficulty in case of the tenant farmer who rents a farm for only one year and without assurance that he will remain for more than that length of time. He hesitates to haul and apply the manure, knowing that his successor will receive a considerable part of its benefits. Under average conditions it is estimated that the first crop after manure is applied will receive about 40 per cent of its benefits; the second crop 30 per cent; the third crop 20 per cent; and the fourth one the remaining 10 per cent. This distribution of the benefits of manure is used in cost accounting in farm crops. The accuracy of the distribution is doubtless crude. and would vary greatly for different crops and different soils, and would also be influenced by the character of the manure and its rate of application.

GREEN MANURES

Green manuring consists of plowing under green crops for the benefit of the soil. The practice results in increasing the organic matter in the soil. If legumes are used for this purpose the nitrogen content of the soil may also be increased. Preference should be given to legumes for this reason. The choice of a crop for green manuring purposes will depend on a number of factors. Other things equal, deep-rooted crops are preferable to those having shallow root systems. Plants with deep roots gather some mineral constituents from the subsoil and upon the decay of the plants leave them in the surface soil in an organic form. Deep-rooted plants are also beneficial because they improve the physical condition of the subsoil. In general, crops that will furnish the largest amount of humus and nitrogen-bearing material for the soil should be selected.

When is Green Manuring Advisable?—The practice of plowing under crops for the benefit of the soil is not justified in systems of livestock farming where the crops can be profitably fed and the manure returned to the soil. There are many localities, however, where the farming systems are such that but little manure is available to supply the needs of the soil. Under such conditions green manuring crops are often resorted to with profit. They are especially to be recommended in case of sandy soils low in organic matter, and for heavy soils in poor physical condition. In addition to serving the purposes above mentioned, green manuring crops, if properly selected, occupy the soil at seasons when it would otherwise be bare of vegetation and subject to erosion. They also prevent the loss of nitrogen by leaching. This is later made available for other crops as the green manures decompose in the soil.

Green manuring is most applicable on fruit and truck farms. It is quite extensively practiced in orchards during the early life of the trees. It is also economical in the trucking regions where the winters are mild.

Objections to Green Manuring.—The objections to green manuring lie chiefly in the fact that green manure crops are grown and plowed under for the benefit of the soil and no direct immediate return is secured. The green manuring crops generally take the place of money crops. When it is possible to grow legumes and feed them to livestock with profit, the stubble and roots of such crops, together with the manure which they will afford, make possible nearly as rapid improvement of the soil as is the case when the whole crop is plowed under. Whether or not a green manuring crop should be fed or plowed under must be determined by the cost of harvesting and feeding, together with the cost of returning the manure, as compared with the returns secured in animals or animal products in feeding it.

Principal Green Manuring Crops.—The principal crops grown in the United States for green manuring purposes are red clover, alfalfa, alsike clover, crimson clover, cowpeas, Canada peas, soy beans, vetch, velvet bean, Japan clover, sweet clover and bur clover. In addition to these, beggar weed, peanuts and velvet bean are also used in the South. These are all legumes, and are decidedly preferable to non-legumes under most conditions where green manures can be used. In the North, where the winters are severe, rye and occasionally wheat are used for this purpose. Buckwheat, which is a summer annual, is also sometimes used.

RYE TURNED UNDER FOR SOIL IMPROVEMENT.

When heavy green manuring crops are turned under allow two weeks or more to elapse before planting succeeding crop.

The characteristics and the requirements for these crops will be discussed in Book II of this work.

On poor soils lime and the mineral fertilizers may be used with profit in the production of a green manure crop. This will stimulate the crop to a greater growth, and when it decays in the soil the elements applied will again become available for the crop that is to follow.

The composition of the legumes used for green manuring varies considerably, depending upon local conditions, character of soil and the stage of maturity when plowed under. The table on next page shows the composition as determined by the average of a number of analyses, and gives the fertilizing constituents in pounds per ton of dry matter for both tops and roots in the crops indicated.

In connection with the analyses as shown in this table, it should be borne in mind that all of the mineral constituents come from the soil, and that it is not possible to increase these by the growing of green manuring crops. The only possible benefit in this respect is the more available form that may result as the green manuring crops decompose. The only real additions to the soil will be in the form of organic matter and nitrogen. It is, therefore, essential to select those crops that will give the largest increase in those two constituents.

FERTILIZING MATERIALS IN 2000 POUNDS OF DRY SUBSTANCE.

Plant and Part.	Nitrogen, per cent.	Phosphoric Acid, per cent.	Potash, per cent.
Alfalfa, tops	46.	10.8	30.4
Alfalfa, roots	41.	8.6	9.6
Cowpeas, tops	39.2	10.2	38.6
Cowpeas, roots	23.6	11.	23.2
Crimson clover, tops	42.6	12.4	27.
Crimson clover, roots	30.	9.4	20.4
Common vetch, tops	59.9	14.2	53.7
Common vetch, roots	43.8	15.8	23.6
Red clover, tops	47.	11.6	42.8
Red clover, roots	54.8	16.8	16.4
Soy bean, tops	43.6	12.5	33.6
Soy bean, roots	21.	6.8	13.4
Velvet bean	50.2	10.6	76.8

The cultivated crops, such as corn, potatoes, tobacco, cotton and some of the heavier truck crops, generally follow a green manuring crop to better advantage than crops that are broadcasted or drilled and do not require cultivation. It is good practice to plow under green manuring crops two weeks or more in advance of the time of seeding the crop which is to follow. Lime applied to the surface before the crop is turned under will tend to hasten decomposition and neutralize acids which are generally formed. The more succulent the crop when turned under, the greater the tendency to acid formation.

REFERENCES

"Fertilizers and Manures." Hall.
"Farm Manures." Thorne.
"Barnyard Manure, Value and Use." Edward Minus, Dept. of Agriculture, Cornell University, Ithaca, N. Y.
Michigan Expt. Station Circular 25. "Composition and Value of Farm Manure."
Michigan Expt. Station Circular 26. "Losses and Preservation of Barnyard Manure."
Ohio Expt. Station Bulletin 246. "Barnyard Manure."
Purdue Expt. Station Bulletin 49. "Farm Manures."

CHAPTER 6

Lime and Other Soil Amendments

Soils Need Lime.—Lime is an essential element of plant food. Many plants are injured by an acid condition of the soil. Soil acidity is most cheaply corrected by one of the several forms of lime. The beneficial effects of liming have been demonstrated by the agricultural experiment stations in a dozen or more of the states. Observations by farmers in all of the Eastern and Southern States, and in the Central States as far west as the Missouri River, show that on many of the farms soils are sour. This sourness of the soil is due to a deficiency of lime, and often occurs in soils originally rich in lime.

Lime Content of Soils.—Soils vary greatly in their original lime content. Some have very little lime to begin with. Others, such as the limestone soils, are formed from limestone rocks, some of which were originally more than 90 per cent carbonate of lime. The lime content of soils is determined by treating them with strong mineral acids. This removes all of the lime from the soil, and the content is then determined chemically. The following table shows the lime content of a number of typical soils in different parts of the United States:

Lime Content ($CaCO_3$) per Acre 7 Inches of Soil in Some Typical Soils of the United States.

Soil Type.	State.	Production.	Lime Content, pounds.
Leonardtown loam	Maryland	Very low	2,500
Orangeburg sandy loam	Alabama	Low	3,500
Orangeburg fine sandy loam	Texas	"	4,650
Cecil clay	North Carolina	"	5,000
Norfolk loam	Maryland	"	8,575
Oswego silt loam	Kansas	"	14,275
Hagerstown loam	Tennessee	Medium	14,275
Miami sand	Ohio	"	34,650
Miami silt loam	Wisconsin	High	32,500
Porters black clay	Virginia	"	59,250
Marshal loam	Minnesota	"	66,750
Podunk fine sandy loam	Connecticut	"	83,575
Fresno fine sandy loam	California	"	125,250
Huston clay	Alabama	"	1,000,750

How Soils Lose Lime.—The greatest loss of lime from the soil is due to leaching. Lime is slowly soluble in the soil solution, and is carried downward by the gravitational movement of the soil water. The rate of loss of lime in this way depends both upon the rate of solubility and the rate of underground drainage. The fact that drainage waters and well waters in all regions where lime is abundant in the soil are highly charged with it is an indication of the readiness with which lime is lost from the soil in this way.

In limestone soil regions the water generally finds its way into underground drainage channels, and few surface streams occur. Very little of it passes over the surface. This explains why limestone soils become deficient in lime. The presence of an abundance of humus in the soil may retain lime in the form of humates, and reduce its loss.

Lime is also removed in farm crops. The amount of removal in this way depends on the yield and character of crops removed, together with the amount that is returned in manures and other by-products. Legumes contain much more lime than non-legumes, and, therefore, cause a more rapid reduction in the lime of the soil.

Lime Requirements of Soils.—The character of vegetation is a good index to the lime requirement of soils. When red clover fails or when alsike clover does better than red clover, it indicates a sour soil. The presence of redtop, plantain and sorrel also indicates a sour soil. In traveling over the country from the Missouri River to the Atlantic seacoast, the acidity of the soil is indicated by the presence of these weeds.

Farmers who are troubled with failure of clover and by the encroachment of the above-mentioned weeds, may feel reasonably sure that their soils need lime. If these signs leave doubt in the mind of the farmer, he can further test his soil by the use of neutral litmus paper. Five cents worth of neutral litmus paper purchased at the drug store will enable him to make tests of many samples of soil. This is conveniently done by collecting small samples of soil to the usual depth of plowing at a number of points in the field in question. The soils should be made thoroughly wet, preferably with rain water or water that is not charged with lime. A strip of the litmus paper brought in contact with the soil and allowed to remain for fifteen or thirty minutes will turn red if the soil is sour. The intensity of the change of color will in a measure indicate the degree of sourness.

Upon request, most of the state experiment stations will test representative samples of soil and advise concerning their lime requirements. The laboratory method determines approximately the amount of lime required to neutralize the soil to the usual depth of plowing.

Crops Require Lime.—Some crops are more tolerant of soil acidity than others. Of our staple farm crops, common red clover is about the least tolerant of such a condition. The staple crops that draw most heavily on the soil for a supply of lime are those first affected by soil acidity. They are also the least tolerant of soil acidity, and are usually most responsive to applications of lime. The clovers contain much more lime and magnesia than the cereals and grasses. The following table gives the average lime and magnesia content as carbonates in a ton of the more general farm crops. Notice the large amounts in clover and alfalfa. Common red clover contains more than alsike clover. It is less tolerant of soil acidity than the latter.

AVERAGE LIME AND MAGNESIA (EQUIVALENT TO $CaCO_3$ AND $MgCO_3$) IN 2000 LBS. OF THE FOLLOWING CROPS.

(Calculated from von Wolff's Tables on the Basis of 15 per cent Moisture.)

Produce.	Pounds of Carbonates as		
	Calcium $CaCO_3$.	Magnesium $MgCO_3$.	Total.
Timothy hay	6.00	2.77	8.77
Wheat (grain and straw)	6.50	6.23	12.73
Corn (grain, cobs and stover)	8.68	8.66	17.34
Oats (grain and straw)	10.40	9.00	19.40
Clover hay (alsike)	49.00	21.47	70.47
Clover hay (red)	73.00	27.01	100.01
Alfalfa hay	91.00	13.16	104.16

Tolerance to Acidity.—Numerous tests at the Pennsylvania Experiment Station show that when the lime requirement of the soil is 1500 to 1700 pounds of burnt or caustic lime per acre seven inches of soil, red

THE GROWTH OF RED CLOVER ON AN ACID SOIL AS AFFECTED BY LIME.[1]

A sour soil is unfriendly to clover. Lime will overcome the difficulty.

clover fails. This is equivalent to from 2700 to 3000 pounds of carbonate of lime or crushed limestone. A lime requirement of 500 to 1000 pounds per acre does not seriously interfere with the growth of red clover. In ordinary farm practice the acidity seldom becomes sufficiently marked to affect noticeably the cereals and grasses, although these may be indirectly

[1] Courtesy of The Pennsylvania Agricultural Experiment Station.

affected by the failure of clover. On experimental plats where ammonium sulphate has been used, the acidity has become so marked that all of the crops in the rotation are directly affected. The degree of tolerance of these crops is in the following order: oats, wheat, corn and red clover; the last being the least tolerant of soil acidity.

At the Rhode Island Experiment Station, Wheeler has made extensive tests of the tolerance of plants to soil acidity, and the relative benefits of applying lime. The following table shows the plants falling into three classes: first, those benefited by lime; second, those but little benefited by lime; third, plants usually or frequently injured by lime.

Lime as Affecting Growth of Plants

Plants Benefited by Liming.

- Alfalfa
- Asparagus
- Balsam
- Barley
- Beets (all kinds)
- Beans
  - Bush
  - Golden Wax
  - Horticultural Pole
  - Red Valentine
- Cabbage
- Cantaloupe
- Cauliflower
- Celery
- Cherry
- Clover
  - Red
  - White
  - Alsike
  - Crimson
- Cucumber
- Currant
- Dandelion
- Eggplant
- Elm, American
- Emmer
- Gooseberry
- Hemp
- Kentucky Bluegrass
- Kohl-rabi
- Lentil
- Lettuce (all kinds)
- Linden, American
- Martynia
- Mignonette
- Nasturtium
- Oats
- Okra (Gumbo)
- Onion
- Orange
- Pea
  - Canada
  - Common
  - Sweet
- Pansy
- Parsnip
- Peanut
- Pepper
- Plum (Burbank-Japan)
- Pumpkin
- Quince
- Raspberry (Cuthbert)
- Rhubarb
- Salsify
- Salt-bush
- Sorghum
- Spinach
- Squash
  - Summer
  - Hubbard
- Sweet Alyssum
- Timothy
- Tobacco
- Turnip
  - Flat
  - Swedish
- Upland Cress
- Wheat

Plants but Little Benefited by Liming.

- Bent, Rhode Island
- Carrot
- Chicory
- Corn, Indian
- Redtop
- Rye
- Spurry

Plants Usually or Frequently Injured by Liming.

- Apple*
- Azalea†
- Bean
  - Velvet
  - Castor
- Birch, American White
- Blackberry
- Chestnut†
- Cotton
- Cowpea*
- Cranberry
- Flax
- Grape, Concord*
- Lupine
- Phlox (Drummondi)*
- Peach*
- Pear*
- Radish
- Raspberry (Black-cap)
- Rhododendron†
- Sorrel
  - Common
  - Sheep
- Spruce, Norway
- Tomato*
- Zinnia*

* These under certain conditions are benefited by liming.
† These have not been tested at the Rhode Island Station.

Crops benefited by lime were not only increased in size, but were ready for market earlier than where lime was omitted. Tobacco was improved in the character of its ash by the use of lime.

Lime is most beneficial in promoting the growth of legumes. This results in building up the nitrogen supply and general fertility of the soil.

Sources of Lime.—The principal source of lime is in the limestone rocks and deposits that occur in great abundance in many sections of the country. There are probably no states in which limestone formations do not occur, although there are sometimes considerable sections including a number of counties in which limestone deposits are not accessible.

Deposits of marl occur in certain localities. They vary greatly in composition and lime content. Marl is generally in good physical condition for application to the soil, and some of it contains phosphorus and potash.

BEETS GROWN WITH AND WITHOUT LIME.[1]

Oyster shells that accumulate in large quantities in sea-coast localities where oyster farming is carried on forms another valuable source of lime. Wood-ashes are about one-third actual lime. Three tons of wood-ashes are, therefore, equal to one ton of pure burnt lime. Unleached ashes contain 5 to 7 per cent of potash, and 1 to 2 per cent of phosphoric acid, which materially increases their value for use on land. When ashes are leached, most of the potash is lost, but the lime content is somewhat increased.

There are a number of forms of spent lime, which is a by-product of different manufacturing establishments that use lime. Among these may be mentioned dye-house lime, gas-house lime, lime from tanneries, waste lime from soda-ash works, and waste lime from beet-sugar factories. The value of these varies widely, and it is impossible to make a definite statement concerning their value. They can frequently be secured at no cost other than the hauling. Whether or not they are worth hauling depends upon circumstances. Frequently, they contain much water, are in poor physical condition and will be more expensive in the long run than to purchase first-class lime in good mechanical condition. Their

[1] Courtesy of International Agricultural Association, Caledonia, N. Y.

value can be determined only by examination by the chemist or by actual field test.

Gypsum or land plaster is frequently used on land, and while it will supply calcium as a plant food, it has little or no effect in correcting soil acidity.

The rock phosphates and Thomas slag, used as sources of phosphorus, contain considerable lime, and their liberal use may obviate the necessity for applying lime to the soil.

Forms of Lime.—Lime (CaO) does not occur in nature. It is prepared by heating limestone ($CaCO_3$) in kilns; 100 pounds of pure limestone thus heated loses 44 pounds of gas known as carbon dioxide (CO_2), and results in 56 pounds of lime. This 56 pounds of lime may be slaked with water and will combine with enough water to make 74 pounds of hydrated lime. Therefore, 1120 pounds of pure lime equals 1480 pounds of pure hydrated lime, which equals 2000 pounds carbonate of lime or pure pulverized limestone. When lime and hydrated lime are exposed to the air they slowly combine with the carbon dioxide of the air until finally reverted to the original form of carbonate of lime. The only difference between the original lime rock and completely air-slaked lime is that of fineness of subdivision, the one being in the form of large rock masses and the other a very fine powder. It is this fine state of subdivision that makes air-slaked lime valuable to apply to the soil. If the raw limestone could be made equally fine it would be just as good as air-slaked lime for the same purpose. If used in generous amounts it need not be so fine as air-slaked lime, but in order to be prompt and effective, pulverized limestone should be so fine that 90 per cent will pass through a 100-mesh screen. Where abundant and cheap, larger amounts of coarser material may be used because of the considerable amounts of finely divided active material it carries. The coarse portion may become available in later years. Lime is generally sold in one of five forms: ground limestone, freshly burnt or lump lime, ground burnt lime, hydrated lime and air-slaked lime. Some deposits of lime are nearly pure carbonates of lime, while others contain much magnesia and are known as dolomite. The presence of magnesia slightly increases the neutralizing power of a given weight of lime.

FUNCTIONS OF LIME

Lime as Plant Food.—The absence of lime prevents a normal development of plants. Lime is, therefore, essential as a plant food. Most soils contain sufficient lime to meet the food requirements of plants. Some soils, however, may contain so little, or it may be so unavailable, that plants that are hungry for lime may suffer from a lack of it.

Chemical Action of Lime.—The chemical effect of lime on most soils is of minor importance. It varies somewhat with the form in which it is applied to the soil. Freshly burnt or caustic lime is the most active form. It may combine with certain soil elements liberating other elements such as potash, and making them available for plants. Lime in the presence of soluble phosphates will readily combine with them, forming tricalcium phosphate. This will prevent the phosphates from uniting with iron and aluminum, which gives rise to compounds less available to plants than the lime phosphates.

Physical Effect of Lime.—Clay soils are frequently improved in physical condition by the liberal application of lime. Freshly burnt lime is the most active form for this purpose. Lime causes a flocculation of the clay particles and increases the porosity of the soil. Lime, therefore, facilitates drainage, makes cultivation easier, causes an aeration of the soil and makes possible a deeper penetration by plant roots. On sandy soils burnt lime may tend to bind the particles together. This may or may not be desirable. When applied for its physical effect it is usually best to apply air-slaked lime or finely pulverized limestone to sandy soils, and to use freshly burnt lime on heavy, refractory soils well supplied with organic matter.

Lime Affects Soil Bacteria.—Certain species of bacteria are instrumental in the change of ammonia and inorganic forms of nitrogen to nitrates. This process is known as nitrification, and is promoted by the presence of lime in the soil. The process not only makes the nitrogen available, but gives rise to the development of carbon dioxide, which in turn acts upon inert plant food and makes it more readily available to plants.

Lime is also beneficial to the several forms of micro-organisms that reside in the tubercles on the roots of all legumes. This may explain why legumes are generally more benefited by lime than non-legumes.

Lime Corrects Soil Acidity.—In the vast majority of instances the chief function of lime is to correct soil acidity. Lime corrects acidity by combining with the acids formed and giving rise to neutral salts. It will seldom pay to apply lime to the soil for purposes other than this. The amount of lime to apply is, therefore, determined chiefly by the degree of acidity of the soil. In practice it is found advisable to apply more than actual lime requirements indicated by chemical methods. This is advisable because in practice it is impossible to distribute lime thoroughly and uniformly and secure its thorough mixture with the soil. Because of this lack of uniformity in distribution some of the lime applied will be ineffective and portions of the soil will not be brought in contact with lime. It is not always necessary to make the soil neutral, since most crops, even the most sensitive crops, will grow fairly well in the presence of small amounts of acids.

Sanitary Effect of Lime.—The decomposition of organic matter in the soil often gives rise to products that are injurious to plant growth. While these generally disappear in time, the presence of lime often corrects the difficulty at once. It is also believed that plant roots excrete injurious substances. Lime neutralizes these objectionable substances.

Lime also affects plant diseases. It lessens the injury of club root, which is often serious in case of turnips, cabbages and other cruciferous plants. It is found to be effective in reducing soil rot of sweet potatoes and checking the root diseases of alfalfa. On the other hand, lime tends to favor the development of potato scab, providing the germ of this disease is already in the soil. In this case it encourages the disease and becomes a menace rather than an aid. For this reason, lime is seldom recommended for potatoes. If applied in a crop rotation which contains potatoes, it is advisable to apply it just after the potato crop rather than before.

Injudicious Use of Lime.—The injudicious use of lime may prove a detriment. Lime is not a fertilizer. To depend on it alone will result in failure. In the failure to recognize these principles lies the truth of the old saying, "Lime and lime without manure makes both farm and farmer poorer."

The excessive use of burnt lime may bring about the availability of more plant food than can be utilized by crops, and cause a rapid loss of it, in which case soil depletion is hastened. It is, therefore, good farm practice to use medium to small quantities at intervals of five or six years. Little is to be gained by applying more than is sufficient to meet the present needs of the soil from the standpoint of neutralizing its acidity.

Rate of Application.—The amount of lime to apply varies with the kind of lime, the requirements of the soil and the frequency of its application. If a soil is a tenacious clay and physical improvement is desired, an application of two or three tons of burnt lime per acre may be profitable. Ordinarily, lime is applied to correct acidity and make the soil friendly to clover and other plants. The equivalent of one to one and one-half tons of burnt lime per acre applied once in each crop rotation is usually a maximum amount. In some instances 1000 pounds per acre will accomplish the desired result. The equivalent of 1000 pounds of burnt lime is between 1300 and 1350 pounds of slaked lime, or a little less than one ton of finely pulverized raw limestone. Unusually large applications have emphasized the wastefulness of such applications so far as the needs of the soil and crops are concerned, through periods of five to six years. Large applications may last much longer, but they are more wasteful of lime, and result in capital being invested without returns.

Small applications are advised for sandy soils. On such soils the carbonate form is to be preferred. Wood-ashes, because of the form of lime and the content of potash, is advised for sandy soils.

Time of Applying.—Lime in any form may be applied at any time of the year. In general farm practice it is advisable to apply lime when men and teams are available for its hauling and distribution with the minimum interference with other farm work. There are some minor precautions, however, in this connection. It is never advisable to apply caustic lime in large amounts just prior to the planting of the crop. At least ten days or two weeks should intervene between time of application and planting of the seed. The caustic effect may injure the young plants. In the soil lime is converted to the carbonate form and the caustic properties soon disappear.

Lime should usually pave the way for clover. It is well to apply lime a year or more before the seeding of clover. If this has not been done, it may be put on the land when the seed-bed is being made for the wheat, oats or other crop with which clover is to be seeded. The advantages of applying a year or two in advance of clover lie in the very thorough mixture of lime and soil resulting from the plowing and tilling of the soil.

Frequency of Application.—The frequency with which lime should be applied depends upon the character of the soil, the rate of application, the length of the crop rotation and the character of the crops grown. It may also be affected by climatic conditions and soil drainage. With good drainage and heavy rainfall the losses of lime will be large, while under reverse conditions they will be comparatively small. In crop rotations five years or more in length, one application at an appropriate place in each rotation should be sufficient. For shorter rotations one application for each two rotations may meet the needs. On soils that are extremely acid and where lime is scarce and high-priced, it may be desirable to make small applications at frequent intervals until the lime requirement of the soil is fully met. Sandy soils call for light applications at rather short intervals. On clay soils larger amounts can be used and the intervals lengthened.

Method of Applying.—Lime should be applied after the ground is plowed and thoroughly mixed with the soil by harrowing or disking. The more thoroughly it is mixed with the soil the better and quicker the results will be. It should never be plowed under, because its tendency is to work downward rather than upward in the soil. Apply lime with a spreader after the ground has been plowed. Do not drill lime in with seeds, nor mix it with commercial fertilizer, nor use it in place of fertilizer. Apply lime to meet the lime requirements of a soil, and when this has been done use manure and commercial fertilizers in the ways that have been found profitable for the crops which are to be grown, regardless of the fact that lime has been applied.

Relative Values of Different Forms of Lime.—The neutralizing effect of the different forms of lime is given under the carriers of lime on a preceding page. The question of relative money values, however, is a matter of arithmetic, and involves not only the first cost of unit weights of the different forms of lime, but includes freight rates, cost of hauling and the work of applying it to the land. In this connection the purity of the product must always be taken into account. Impurities entail the expense of freight and hauling of worthless materials, and increase the cost of the active portion of the lime. The cost of lime in any locality will depend largely on the presence or absence of limestone or some other form of lime, together with the actual cost of quarrying, crushing or burning, as the case may be.

The following figures, as given by Mr. J. H. Barron in the *Tribune Farmer*, show the relative cost of equivalent amounts of three forms of lime applied to the land in southern New York. This will serve as a method for any region.

1 ton burnt lime at railroad station	$4.00
Hauling	1.00
Cost of applying	1.50
Total cost per acre	$6.50

The high cost of applying is on account of having to slake the burnt lime before it is applied, together with the difficulty in applying it in that form.

2640 pounds hydrated lime (equivalent to 1 ton burnt lime), at $7.00 per ton	$9.24
Hauling, at $1.00 per ton	1.32
Applying, at 75 cents per ton	.99
Total cost per acre	$11.55

The increased cost per acre in using this form is due to the relatively high first cost of hydrated lime and to the additional expense of hauling 650 pounds of water content in the hydrated lime.

In case of ground limestone we have the following:

3570 pounds ground limestone (equivalent to 1 ton burnt lime), at $4.00 per ton	$7.14
Hauling, at $1.00 per ton	1.78
Applying, at 75 cents per ton	1.33
Total cost per acre	$10.25

The above costs are probably considerably above the average for most localities where lime is not too inaccessible. The relative cost of ground limestone as compared with the burnt lime is also rather high.

It is good business to purchase that form which supplies the greatest amount of active lime for the amount of money involved, providing the mechanical condition is satisfactory. In this connection it should be borne in mind that no matter in what form lime is applied to the soil, it soon reverts to its original form of carbonate of lime. The advantages in using slaked burnt lime lie chiefly in the extreme fineness of subdivision and the possibilities of more thorough distribution in the soil.

Mixing with Manure and Fertilizers.—Caustic forms of lime should not be mixed with either manure or fertilizers. Such forms in the presence of nitrogenous materials cause a loss of nitrogen in the form of ammonia. In the presence of soluble phosphates they cause a reversion to insoluble forms. It is best, therefore, to apply lime in advance of applying fertilizers, and mix it with the soil by disking or harrowing. In case of manure which is plowed under, the application of lime may follow that of manure, being applied preferably after plowing.

The pulverized raw limestone may be applied with manure, or at the time of applying fertilizers, without injurious results.

Experimental Results.—Experiments with lime at many experiment stations and on all kinds of soils show that it makes little difference what form is used, so long as it is applied in sufficient quantities to meet the lime requirements of the soil, and is thoroughly and uniformly mixed with the soil. At the Pennsylvania Experiment Station finely crushed limestone in each of three field tests extending over a number of years has proven slightly better than equivalent amounts of burnt lime. Extensive pot experiments at the same experiment station have shown that finely pulverized limestone is equally as prompt and effective in correcting soil acidity and promoting the growth of clover as equivalent amounts of caustic lime. While these tests are favorable to pulverized limestone, they are not all sufficiently decisive to justify its use at a disproportionate price. If two tons of ground limestone cost much more than one ton of burnt lime, one would ordinarily not be justified in using the former.

THE OLD WAY OF SPREADING LIME.[1]
After slaking, the piles are uniformly spread over the surface.

Where lime must be shipped some distance, the more concentrated forms are usually the cheaper.

Spreading Lime.—The practice most common in the Eastern States is to place small piles of burnt lump lime at uniform intervals over the field, the amount in each pile and the distance between piles determining the rate of application. If the lime is to be spread promptly, about one-half pail of water should be applied to each pile, and then covered lightly with earth. This facilitates slaking, and the lime will be ready for distribution in a comparatively short time. In other instances the piles are allowed to remain without either wetting or covering with earth until weather conditions bring about complete slaking. Long periods of

[1]Courtesy of W. N. Lowry, Student.

rainy weather frequently prove disastrous by puddling the lime and causing it to get into bad physical condition.

Another method is to place the burnt lump lime in large stacks at the end of the field, and allow them to remain for several months until air slaked. From these stacks the lime is hauled either by wagon, manure spreader or lime spreader, and applied to the field. When the lime contains lumps the manure spreader gives best results in distribution. By screening, a lime spreader or fertilizer spreader with large capacity may be used with good results. Whatever method is used, an effort should be made to obtain uniform distribution at the desired rate at the minimum cost of time and labor. When slaked lime is spread with the lime spreader,

A MODERN LIME SPREADER IN OPERATION.[1]

a canvas may be attached to the spreader which will reach to the ground, and by tacking a strip at the lower edge to cause it to drag on the ground, the disagreeable effect of the dust is largely overcome. Goggles for the eyes and a wet sponge for the mouth may prevent some of the disagreeable effects to the operator.

In the central states where pulverized raw limestone is extensively used, both manure spreaders and lime spreaders are found satisfactory in its distribution. One successful farmer finds that the work is most cheaply and effectively done by using a short-tongue distributor hitched close behind a wagon loaded with limestone. The limestone is shoveled into the distributor as the load is drawn across the field. On loose, plowed earth four horses are required to pull the load. In this way there is no extra handling of the lime, and the distribution is completed as soon as the wagon is unloaded. Many others have had good results with the manure spreader. Several methods have been practiced with this machine.

[1] Courtesy of The Webb Publishing Company, St. Paul, Minn. From "Field Management and Crop Rotations," by Parker.

Some apply the lime and manure together. When the limestone is to be applied at the rate of three tons per acre, 600 pounds on each load of manure in case of ten loads of manure to the acre, gives the desired amount.

Another method is to put a layer of straw in the bottom of the manure spreader, set the spreader for its minimum rate of distribution, and load in the amount of lime that will give the desired rate of application. For distribution at the rate of three tons per acre, this will generally require not more than one ton.

Slaking Lime.—Lime in large quantities may be satisfactorily slaked by applying about two and one-half pails of water to each barrel of lime

A LIME CRUSHING OUTFIT SUITABLE FOR THE FARMER.[1]

as it is unloaded in the field. Eventually the whole stack should be covered with soil. In a few days all of the lime will be thoroughly slaked, and in a fine, dry condition suitable for spreading.

Crushing vs. Burning Lime.—The use of finely pulverized raw limestone has created a demand for machinery for crushing lime rock. There are now on the market quite a number of portable machines suitable for farm use. In some localities where limestone is easily accessible it can be quarried and finely pulverized with these machines at a cost of $1 to $1.50 per ton. This puts it within the reach of farmers at a moderate price.

Lime is burnt in several ways. The simplest way on the farm is to make a stack of lime rock with alternating layers of wood or coal. This is built in a conical form with an intake for air at the bottom and an opening at the top for ventilation. The stack is covered with earth and the fire lighted.

[1] Courtesy of New York Agricultural Experiment Station, Geneva, N. Y. Bulletin 400.

More effective burning is secured by burning limestone in a kiln constructed of stone or masonry. In either case the cost per ton of burning varies with the cost of fuel, the price of labor and the accessibility of

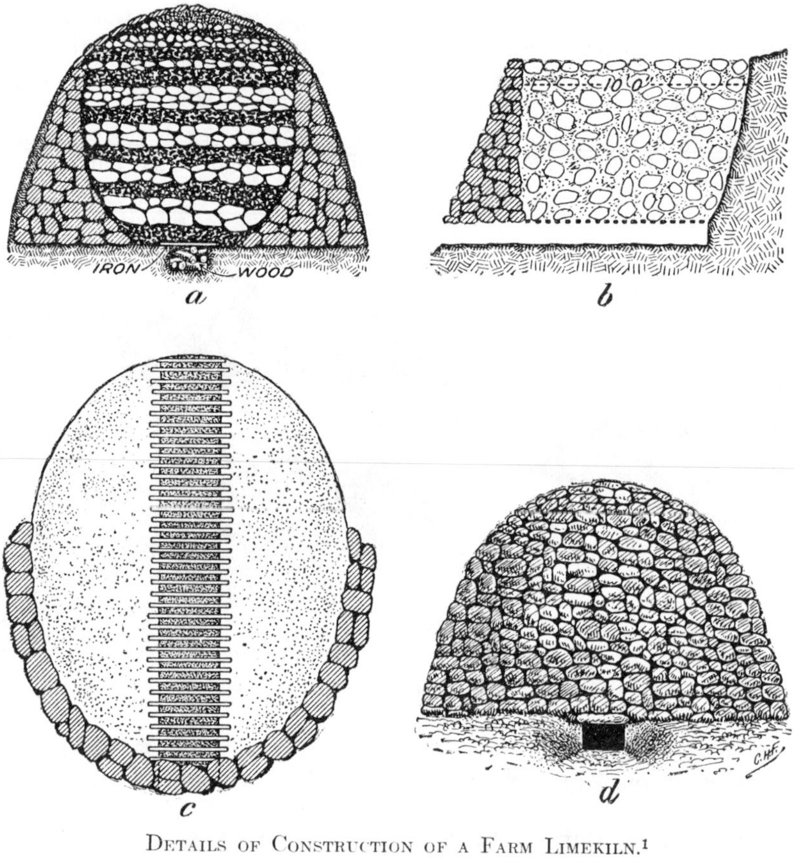

DETAILS OF CONSTRUCTION OF A FARM LIMEKILN.[1]

A—Cross section, showing layers of rock and coal. B—Longitudinal section, showing side hill used as back wall. C—Ground plan, showing trench and grate. D—Completed kiln, walled in and plastered with mud.

limestone. The minimum cost for burning, including quarrying, labor and fuel, will be about $1.75 per ton of burnt lime. In many cases it will cost much more.

REFERENCES

Alabama Expt. Station Bulletin 95. "Lime as a Fertilizer for Oats."
Iowa Expt. Station Bulletin 151. "Lime as a Fertilizer on Iowa Soils."

[1] From Farmers' Bulletin 435, U. S. Dept. of Agriculture.

Iowa Expt. Station Bulletin 2. "Bacteriological Effects of Lime."
New Jersey Expt. Station Bulletin 210. "Lime as a Fertilizer for Clover and Oats."
Ohio Expt. Station Bulletin 279. "Lime as a Fertilizer."
Pennsylvania Expt. Station Bulletin 131. "Use of Lime on Land."
Rhode Island Expt. Station Bulletin 49. "Methods of Applying Lime."
Rhode Island Expt. Station Bulletin 58. "Lime with Phosphates on Grass."
Rhode Island Expt. Station Bulletin 160. "Lime with Nitrogenous Fertilizers on Acid Soils."
Tennessee Expt. Station Bulletin 96. "Effect of Lime on Crop Production."
Tennessee Expt. Station Bulletin 109. "Lime as a Fertilizer on Tennessee Soils."
Virginia Expt. Station Bulletin 187. "Lime as a Fertilizer on Virginia Soils."
Wisconsin Expt. Station Bulletin 230. "Lime as a Fertilizer on Wisconsin Soils."
Pennsylvania State Dept. of Agriculture Bulletin 261. "Sour Soils and Liming."
U. S. Dept. of Agriculture, Bureau of Chemistry, Bulletin 101. "Lime Sulphur Wash."
Farmers' Bulletin, U. S. Dept. of Agriculture, 435. "Burning Lime on the Farm."

CHAPTER 7

SOIL WATER, ITS FUNCTIONS AND CONTROL

Water is the most abundant substance in nature. It is necessary to all forms of life. An abundant supply of moisture in the soil at all seasons of the plant's growth is essential to a bountiful harvest. Sixty to ninety per cent of all green plants consist of water. About forty per cent of the dry matter is made from water which unites with carbon to form the structure of the plant. Water is the necessary vehicle which

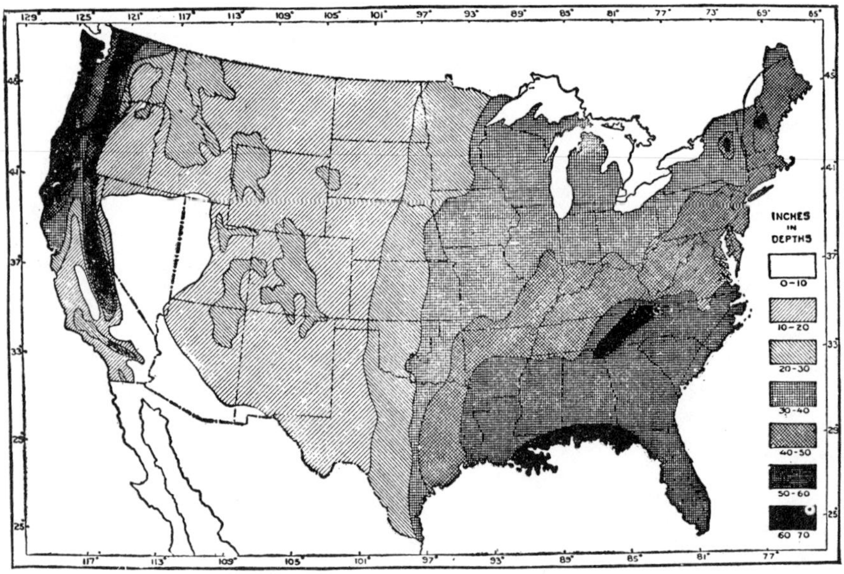

MAP SHOWING MEAN ANNUAL RAINFALL FOR ALL PARTS OF THE UNITED STATES.[1]

carries plant food to the plant, and causes it to circulate from one portion of the plant to another. When there is a deficiency of water in the soil, plant growth is checked. If the deficiency becomes sufficiently marked, plant growth ceases entirely.

Amount and Distribution of Rain.—All water comes from rains and melting snows. An acre inch of rain makes 113 tons of water. To supply the equivalent of one inch of rainfall by artificial means at 10 cents per ton of water would cost $11.30 per acre. Ten inches of rain-

[1] Courtesy of Doubleday, Page & Co., Garden City, N. Y. From "Soils," by Fletcher.

fall at the same rate would cost $113 per acre. From this it can be readily understood that artificial means of supplying plants with water must be done at a very low cost, otherwise it will not prove profitable.

The amount of rain in any region is important in connection with crop production. In all regions where the annual rainfall averages less than twenty inches, failures from insufficient moisture in the soil are frequent. The distribution of the rain is quite as important as the total annual rainfall. That which falls during the crop-growing season is more important than that which comes in the non-growing season. Consequently, there are regions of comparatively low rainfall where the distribution is so favorable that crop failures are infrequent. In other localities a large part of a good annual rainfall may come in the non-crop-growing season, and as a result, crops frequently suffer from drought. In moving from one region to another it is well to study the average rainfall and its distribution.

Amount of Water Necessary to Produce Crops.—In the processes of plant growth the amount of water transpired or given off by plants is many times greater than that used in the plant tissues. Investigations in different parts of the world and at several of the American experiment stations show that in plant growth the amount of water required to produce a pound of dry matter ranges from 200 to 700 pounds. This amount must actually pass through plants. Each ton of dry matter in alfalfa takes 700 tons of water. Each ton of dry matter in wheat required about 400 tons of water; in oats, about 500 tons; and in corn, about 300 tons. To produce three tons of alfalfa in one season requires from 16 to 17 inches of rainfall, all of which must pass through the plants. A 20-bushel crop of wheat would require about 6 inches, and 40 bushels of oats $6\frac{1}{2}$; while 50 bushels of corn would require about $8\frac{1}{2}$ inches of rainfall. For crops of the yields mentioned there should be more rainfall during the growing season than above indicated, because of the loss of water by direct evaporation from the soil, plus additional amounts that may flow from the surface if the rain falls rapidly, together with some that may pass through the soil into the underdrainage.

Transpiration by Plants.—Transpiration, or the amount of water that passes through the plant and is evaporated from the surface of the leaves, varies greatly in different localities, and is influenced by a number of factors. Transpiration takes place most rapidly during the daytime and in the presence of plenty of sunshine and warmth. During the night-time it is reduced to a very small amount. Transpiration is increased with a reduction of the humidity of the air, with rise in temperature and with intensity of sunshine. It is also increased with an increase in the movement of the air. An increase in plant food tends to decrease it, as does also a rapid growth of the plant. Transpiration is more rapid in the presence of an abundance of soil moisture than it is when the soil is dry.

Experiments at the University of Illinois by Dr. Hunt showed an increase per acre in the dry matter in corn amounting to 1300 pounds in one week in July. On the basis of requiring 300 pounds of water for each pound of dry matter, the consumption of water by the growing corn in one week would equal 1.72 inches of rain. This, of course, is for a single week in the height of the growing season, but it shows the large amount of rainfall required to meet fully the needs of a large and rapidly growing crop. It should emphasize the importance of storing in the soil the largest possible amount of available water to tide over periods of deficiency in rainfall.

Forms of Soil Water.—Water exists in the soil in three forms: (1) gravitational water, or that which is free to move through the soil under the influence of gravity; (2) capillary water, or that which is held against the force of gravity by capillary power or, as it is sometimes called, surface tension; (3) hygroscopic water, or that which adheres to the soil particles so firmly that it will not be given off, even when the soil becomes dry. Not all of the water in the soil is available for plants. Very few of our economic plants use any of the gravitational water of the soil, except as it may rise by capillarity and be used from the capillary store which it replenishes. It is also certain that plants cannot benefit from the hygroscopic water of the soil, because they are unable to get it from the soil particles by which it is so tenaciously held in this form. The capillary water is, therefore, the one form that is of importance in plant growth. The relative amounts of the three forms of water in the soil depend on a number of factors.

The amount of pore space in soils ranges from 35 to 60 per cent of the volume of the soil. When there is no underdrainage and a superabundance of rain this space may become fully occupied with water to the exclusion of air. The soil is then said to be saturated. If rains cease and underdrainage is established, the gravitational water will escape by means of the drainage channels. The amount which will escape in this way is determined chiefly by the texture of the soil and the percentage of pore space in it. The larger the pore space, the greater the amount of water that will escape in this way; the finer the texture of the soil, the larger the amount held by capillarity and the less the amount that will escape by drainage.

Capillary Water.—This is the important portion of the soil water supply. It is the form on which plants wholly depend for their water supply. Plants cannot exhaust from the soil all of the capillary water, because a portion of it will be too tenaciously held by the soil particles to be removed by the plant root hairs. The optimum, or most favorable percentage of water in the soil for plants, differs for different crops. Such crops as corn and potatoes do best with a moderate percentage of water in the soil, which gives opportunity for plenty of air. Such plants as timothy, redtop and other grasses do best when the percentage of water in the soil is somewhat higher. Field experiments have shown that when the water content of the soil is increased 25 per cent above the optimum percentage, plants begin to suffer as a result of too much moisture, and when the moisture falls 25 per cent below the optimum, they suffer from drought.

The amount of capillary water in the soil is determined chiefly by its texture. The following table shows the percentage of water held by soils ranging in texture from coarse sand to clay, when subjected to a

EFFECT OF LITTLE, MEDIUM, AND MUCH WATER ON WHEAT.[1]

centrifugal force 2940 times that of gravity. A coarse sand held only 4.6 per cent of moisture, while clay held 46.5 per cent or ten times as much. The water held under natural conditions by the several classes of soil given in the table would be much larger, but the relative amounts would be the same.

CAPILLARY MOISTURE IN SOIL.

Class.	Percentage of Clay in Soil.	Percentage of Moisture Retained against Force 2940 Times that of Gravity.
Coarse sand	4.8	4.6
Medium sandy loam	7.3	7.0
Fine sandy loam	12.6	11.8
Silt	10.6	12.9
Silt loam	17.7	26.9
Clay loam	26.6	32.4
Clay	59.8	46.5

Capillary water is also influenced to some extent by the structure of the soil, and to somewhat greater extent by its content of humus or

[1] Courtesy of The Macmillan Company, N. Y. From "Principles of Irrigation Practice," by Widtsoe.

organic matter. Soils of fine texture and those having plenty of organic matter hold the largest amount of capillary water, and are able to withstand periods of drought better than those with a lesser capacity.

Plant roots move toward the water supply in the soil, and as they withdraw water from the soil particles, water moves to those points by capillary action to replace that removed. The rate of capillary movement is slowest in soils of fine texture, and is most rapid in sandy soils. The distance through which capillary power acts on the other hand is least in sandy soils, and greatest in soils of fine texture. We find, therefore, that plant roots are most extensive in sandy soils and extend to greater depths in search of a water supply.

Gravitational Water.—Since gravitational water is but little used by plants, it becomes a menace in soils more often than a benefit. Over large areas of comparatively level land where there is an abundant rainfall, it often becomes necessary to remove the gravitational water by means of various forms of drainage. The movement of gravitational water within the soil depends chiefly on the texture and structure of the soil. The amount that needs to be removed under agricultural conditions depends chiefly on the rainfall of the region and the amount that escapes over the surface of the land. The depth to which this gravitational water should be removed will be determined chiefly by the character of crops to be grown. Seldom is it advisable to place underdrains for this purpose at a depth of less than three feet. For deep-rooted crops, such as alfalfa and orchard fruits, four feet and sometimes more is advisable.

While this form of water may be injurious to upland plants, when it exists at a depth of from four to six feet below the surface it does no harm and serves as a reservoir from which water may be drawn by capillarity to meet the losses above by evaporation and plant removal.

Hygroscopic Water.—The water which is held by the soil when a thin layer is spread out and allowed to become air dry is called hygroscopic moisture. When this soil is placed in an oven and heated to the temperature of boiling water for several hours, it loses its hygroscopic water and becomes water free. The amount of this form of water held by soils varies directly with the texture of the soil and may amount to as much as 16.5 per cent in case of clay, while in a muck soil it may be as high as 50 per cent. The percentage of hygroscopic water will also be influenced by the temperature and humidity of the air with which it comes in contact.

Water Affects Temperature of the Soil.—A requisite degree of warmth in the soil is essential to physical, chemical and biological processes that make for soil fertility. Warmth is essential to the germination of seeds and growth of plants. The chief source of warmth in the soil is from the sun. The rapidity with which a soil warms under the influence of the sun depends more largely on its water content than on any other factor. One pound of water requires four times as much heat to increase its temperature one degree as would be required by an equal weight of soil. An excess of water in the soil, therefore, greatly lessens its rate of warming. In wet soils much evaporation of water takes place at the surface. It requires more than five times as much heat to transform one pound of water from liquid to vapor as it does to raise the temperature of an equal weight of water from the freezing to the boiling point. In other words, the heat consumed in the process of evaporation is sufficient to cause a change of 900 degrees in temperature in an equal volume of water. This fact emphasizes the importance of removing surplus water by means of drainage, instead of allowing it to evaporate from the surface of the soil. An amount of evaporation sufficient to maintain a proper soil temperature in prolonged heat periods may be desirable, but excessive evaporation is undesirable in temperate latitudes, especially during the early growing season. Reduced temperature as the result of such evaporation often causes disaster during the seeding or planting season and retards the early growth of crops.

Water Storage Capacity of Soils.—Since the rains of summer are rarely fully adequate to meet the needs of growing plants, it is essential to increase the storage capacity of the soil as far as possible. For this purpose, the chief agencies are plowing, methods of tillage and the use of organic manures. Deep plowing and the incorporation of organic matter to the full depth of plowing will increase very materially the capacity of the soil for water. In conjunction with this, the soil should be so cultivated that it will receive the rainfall and thus have an opportunity for holding it. This means the maintenance of a porous surface so that rainfall will not escape over the surface until the soil has become filled with water.

Those crops endowed with the power of deep-root penetration, such as alfalfa, can draw their moisture from greater depths in the soil than shallow-rooted crops. In regions of low rainfall this amounts to the same thing as increasing the storage capacity of the surface portion of the soil.

Moisture Conservation.—The practical conservation of soil moisture is effected chiefly by preventing direct evaporation from the surface of the soil, and also by exterminating all foreign plants in the nature of weeds that tend to rob the crops of their moisture supply. Evaporation is most economically reduced to the minimum by surface tillage and the establishment of an earth mulch. The earth mulch to the depth of two or three inches is formed by periodic cultivation or a stirring of the surface of the soil so as to break the capillary action with the soil immediately beneath. The efficiency of such mulches depends largely on the perfection with which they are made. A surface mulch to be effective should consist of rather finely pulverized loose soil. This becomes dry to such an extent that the soil moisture film is discontinuous and water ceases to rise to the immediate surface. In this condition, any loss that takes place must result from the escape of water within the soil pores. A little loss will take place in this way. Such mulches must be renewed at intervals more or less frequent, depending on the rainfall and the rapidity with which the surface soil may become compacted. In the absence of rains, a well-established mulch will last for a long time. On the other hand, a comparatively light rain will spoil the mulch and establish capillary connection with the soil below.

Mulches of straw, manure and other organic materials are sometimes used. These are very effective, but are often expensive. Such mulches are most common in orchards in case of small fruits, strawberries, and sometimes for potatoes and tomatoes.

ORCHARD WELL CULTIVATED TO PREVENT EVAPORATION.[1]

Where green manuring crops which are to be followed promptly with money crops are used, it is well to take the precaution to plow these under before they have thoroughly exhausted the moisture supply of the soil. Precaution should also be taken in plowing under green manure crops and barnyard manure to avoid possibility of cutting off the capillary connection between the plowed and unplowed portion of the soil.

[1] Courtesy of The Macmillan Company, N. Y. From "Principles of Irrigation Practice," by Widtsoe.

Removing Excess of Water.—Excess of soil water pertains only to that above described as gravitational water. This may be removed by deep, open drains and by underdrains. Methods of drainage will be discussed in another topic.

On comparatively level lands where surface water often accumulates, its escape may be encouraged by so plowing the land that it will lie in slight ridges and continuous depressions. If the depressions have a continuous fall, all of the surface water will slowly escape from the land into natural drainage channels and without causing erosion.

Excess of water is sometimes removed by the use of crops, although this does not pertain to gravitational water. In most localities it is desirable to have the growth of orchard trees cease as the season draws to a close, in order that the wood may harden and withstand winter freezing. For this purpose orchards are frequently planted with crops that draw heavily on the soil moisture for the purpose of so exhausting it that the growth of the trees will be checked. This serves not only a good purpose with reference to the condition of the orchard, but produces organic matter that may be plowed under for the benefit of the soil and the trees.

LAND DRAINAGE

A wet soil is cold and late. It can seldom be plowed and tilled at the proper time. Most farm crops do not make satisfactory growth in a wet soil, and, therefore, it seldom pays to farm such land.

Wet lands, when drained, are generally above the average in fertility. Money invested in drainage seldom fails to bring good returns. In many cases the increase in crops, following drainage, has paid for its cost in one year.

Drainage Increases Warmth and Fertility of Soil.—When an excess of soil water is removed through underground drains it permits the soil to warm up rapidly under the influence of the sun; lengthens the growing season; increases the number of days during which the soil is in good condition to plow; increases aeration of the soil; encourages the deep penetration of the roots of plants, and as a result makes the plants resistant to drought. Drainage is, therefore, the first essential to soil fertility.

Improves Health Conditions.—Drainage also improves health conditions. The drainage of large areas of swampy land in the vicinity of populous districts has often been undertaken for this purpose alone and without any regard to the increased agricultural value of the land. Large portions of the prairie region when first settled were sufficiently wet to furnish abundant breeding places for mosquitoes. The great numbers of mosquitoes were not only a great annoyance, but were responsible for thousands of cases of malaria, which greatly reduced the health and efficiency of people living in that region. Tile drainage that has been so extensively established in most of that region has practically abolished breeding places for mosquitoes, and caused their disappearance to such a degree that malaria is now practically unknown in that region.

Open vs. Underground Drains.—The gravitational water in the soil may be lowered to the depth of two or three feet below the surface by open drains, but the same can be more economically effected by the installation of underground drains. Open drains waste much land, the ditches are subject to erosion and their presence interferes with cultural operations. They are also expensive to maintain, because of the necessity of annually cleaning them.

Underground or tile drains are more effective than open ones. They waste no land, require practically no outlay for annual maintenance, do not interfere with cultural operations and are permanent. The cost of excavating for underground drains is less than that for an equal length of open drains, because in the former very narrow trenches are excavated which are filled as soon as the tile is in place.

Quality of Tile.—Burned clay pipes are almost universally used for soil drains. They are made in sections, from 12 to 24 inches long, having an internal diameter ranging from 3 to 16 inches. Since the installation of underground drainage is to be permanent, care should be exercised in the selection and purchase of the tile. Only the straight, well-burned tile should be used. A well-burned tile is generally dark in color, and gives a decided ring when struck with a light metal. Formerly it was thought that such tiles should be quite pervious to water, but it is now understood that the openings at the joints are ample to admit the water from the soil as fast as it can reach the lines of tile.

Cost of Tile and Excavating.—The cost of installing underground drainage depends on the cost of the tile laid down on the land, the frequency of the underground lines of drainage as determined by the permeability of the soil to water, together with the cost of digging the trenches as determined by the ease or difficulty in excavating the soil. The cost of the tile will vary with the locality, the freight charges and the distance they must be hauled. In general, the price of the tile per 1000 feet F. O. B. cars, at the factories, will be as follows:

Size.	Price.
3 inch	$10.00–$12.00
4 "	15.00– 20.00
5 "	20.00– 27.00
6 "	27.00– 35.00
7 "	36.00– 50.00
8 "	45.00– 60.00
10 "	60.00–110.00
12 "	90.00–150.00

The cost of digging the trenches will vary greatly with the character and condition of the soil to be excavated, the skill of the digger and the prevailing cost of labor in the locality. Deep trenches cost relatively more to excavate than shallow ones, because the trenches must be wider at the top to accommodate the workman, and the earth in the bottom of the trenches is more difficult to remove. Where the soil is free from stones and hardpan, trenches are frequently excavated to the depth of three feet, and the tiles placed ready for filling the trenches, at a cost of thirty cents per linear rod. Below the depth of three feet and up to five feet, excavating under similar conditions will cost about one cent per inch per rod.

Depth and Frequency of Drains.—The depth at which to place the tile drains will be determined by the class of crops to be grown and the character of the subsoil. Three feet in depth is considered ample for most farm crops, but for orchards, alfalfa and especially deep-rooted crops, a depth of four feet is preferred. There are many localities, however, where the impervious character of the subsoil is such that tiles can be placed only twenty-four or thirty inches deep, and permit the water to enter. Even under these conditions, tile drainage is generally advisable.

The distance between lines of drain will depend chiefly on the character of the soil, with special reference to its permeability to water. A soil and subsoil that is sandy or loamy in character will frequently be satisfactorily drained with lines of tile 200 to 300 feet apart. On the other hand, a dense clay will sometimes necessitate the lines of drains being placed at intervals of not more than 30 to 40 feet. This, of course, makes underdrainage much more expensive than in the former case. The deeper the tile is placed the farther the lines may be apart.

Where land to be drained is uniformly wet, the gridiron or regular system is to be preferred. The irregular system will answer the purpose for the drainage of wet spots or sloughs. The main lines should follow approximately the natural depressions or water courses, while the laterals may run up and down the slopes. Rather long parallel lines are more economical than short ones with numerous branches.

Grades, Silt Basins and Junctions.—All lines of underdrainage should be laid with uniform grades. If the topography of the land necessitates a change in the grade, in which the grade in the lower portion of the line is less than in the upper portion, a silt basin should be placed at the point where the change of grade takes place. When the reverse is true, a silt basin is not necessary. Where laterals enter a main or sub-main which has a lesser fall than the laterals, silt basins should also be installed. Laterals should enter the main above the center of the pipe, rather than below it. All junctions should be made at an angle of about forty-five degrees up-stream. A fall of one foot in one hundred feet is considered a heavy grade. A fall of one inch in one hundred feet will give good results, although more fall than this is better. In the level prairie sections of the country hundreds of miles of tile are laid with a grade of only one-half inch in one hundred feet, and where great care is exercised in laying the tile, difficulty has seldom been encountered.

On level land a fair grade may be obtained by gradually lessening the depth of the tile from the lower to the upper end of any branch. In a drainage line 1200 feet in length a fall of one inch in each hundred feet may be obtained by having the lower end of the line $3\frac{1}{2}$ feet below the surface of the ground, and the upper end $2\frac{1}{2}$ feet below the surface, even though the land along this line is absolutely level.

The Outlet.—The first essential for a satisfactory system of underground drainage is a good outlet. The outlet must be the lowest point in the whole drainage system, and water should seldom, if ever, stand above the opening of the tile.

The outlet of the main should be protected by a screen in such a way that rabbits and other animals cannot enter. At the outlet the tiles are subject to freezing more than elsewhere in the system, as a result of which they may be broken. It is well to provide for this by using a wooden box, or an iron pipe as a substitute for the earthen tile. This should extend back from the opening six or eight feet to a position where it will not become frozen.

WATER ISSUING FROM AN UNDERGROUND DRAIN.[1]

Size of Tile.—The size of the main outlet or line is determined by the area to be drained, together with the water-shed contributary to it. Not only must we figure on removing all of the rainfall that descends directly on the land to be drained, but we must also calculate on the amount of water that reaches such land from adjacent higher land, whether as surface wash or underground seepage. The maximum amount of water necessary to remove from the land in order to effect satisfactory drainage will depend chiefly on the rainfall likely to occur in short periods of time during the growing season. It will seldom be necessary to provide for the removal of more than one-half inch of water in twenty-four hours. On this basis a system of tiles flowing at full capacity will remove rainfall at the rate of fifteen inches per month. This is much in excess of the usual rainfall in any part of the country. The removal of one-quarter inch of rainfall in twenty-four hours will generally provide adequate drainage. The size of tile required to accomplish removal of water at the above mentioned rate will be determined largely by the grades that it is possible to secure. The size of tile required is given in the chapter on "Drainage and Irrigation."

[1]Courtesy of Orange Judd Company. From "Soils and Crops," by Hunt and Burkett.

REFERENCES

"Dry Farming." MacDonald.
"Dry Farming." Widtsoe.
"Dry Farming." Shaw.
Kansas Expt. Station Bulletin 206. "Relation of Moisture to Yield of Wheat in Kansas."
Nebraska Expt. Station Bulletin 114. "Storing Moisture in the Soil."
Utah Expt. Station Bulletin 104. "Storage of Winter Precipitation in Soils."

CHAPTER 8

General Methods of Soil Management

The art of soil management consists in so manipulating the two million pounds of soil constituting the average plowed portion of each acre, that it will give the largest returns without impairing the soil. The best chance of attaining success in the art of soil management is in the hands of the man who best understands the principles underlying it. The art of soil management is the result of more than 4000 years of accumulated experience, while the science is very much a matter of yesterday. It is not to be expected that science will revolutionize the art, but it will explain why many operations are performed and will also suggest improvements in the manner of performing them. There are no definite rules relative to methods of soil tillage. The best way of performing a certain operation of soil tillage at any particular time and place is generally a matter of judgment on the part of the farmer. Accuracy in judgment on his part is greatly strengthened through knowledge of the underlying principles.

Objects of Tillage.—The chief objects of tillage are: (1) to improve the physical condition of the soil; (2) to turn under plant residues that have accumulated at the surface and incorporate them with the soil; (3) to destroy weeds; and (4) to provide a suitable seed-bed.

In recent years great changes have taken place in the methods of tillage, due chiefly to the invention and use of labor-saving implements. In this connection it is well to know the approximate duty of the cultural implements that are available. In a general way the duty of a cultural implement is obtained by multiplying the width in feet which it covers in passing over the field by 1.4. For example, a 12-inch plow will plow, on an average 1.4 acres of land per day. A harrow 6 feet in width would harrow 8.4 acres. The duty will vary somewhat with conditions, such as speed in process of operation, the length of day and percentage of time when not in actual operation. With good fast-walking teams and implements of light draft, the acreage covered per day may be somewhat increased. On the other hand, if much time is lost, if the teams are slow or if implements are of heavy draft, the acreage will be reduced. These facts are important in connection with determining the extent of equipment required to perform satisfactorily the operations on a farm of given size.

Plowing.—Plowing is the most expensive tillage operation in connection with crop production. For this reason it is important to know when it is necessary to plow the land and how deep it should be plowed, since both depth and frequency of plowing bear directly on the cost of the operation. Mold-board and disk plows are used for this purpose. Either of these implements turn the soil, pulverize it and cover rubbish. The implement to be preferred is determined largely by the character of the soil and its condition. Disk plows work best in rather dry soil. Mold-board plows are much more extensively used and will work under a wider range of soil conditions. The form of the mold-board plow varies considerably, and different forms are applicable to different purposes and different soils. The sod plow has the minimum curvature and inverts

A DEEP TILLING DOUBLE-DISK PLOW.[1]

the furrow slice with the least pulverization of the soil. The stubble or breaking plow has much more curvature of the mold board, and gives more thorough pulverization of the soil. The greater the curvature of the mold board and the more thorough the pulverization of the soil as a result of it, the heavier will be the draft. Sharpness of the share and smoothness of the plow surface tend toward lightness of draft. The presence of roots and stones may somewhat increase the draft of plows. The texture, structure and physical condition of the soil, especially with reference to its water content, greatly influence draft. The soil plows

[1] Courtesy of The Spalding Tilling Machine Company, Cleveland, Ohio.

most easily when it is in a fairly moist condition and most easily pulverized. The draft of the plow will be increased both when the soil is too wet and when it is too dry.

Coulters and jointers are both attached to plows to influence draft and improve the character of plowing. Coulters are for two purposes: (1) those which cut the roots separating the furrow slice from the unplowed land, and (2) those which cut vines and rubbish, preventing their dragging across the plow standard and clogging the plow. Rolling coulters are best for the latter purpose, while standard cutters may be equally as good for cutting the roots in the soil. The chief object of the jointer is to push the surface rubbish into the furrow so that it will be more completely covered. Sulky plows are often used instead of walking plows. The chief advantage in the sulky plow is in reducing the labor of the plowman and in more effective plowing. It is claimed that sulky plows reduce the draft of the plow by relieving the friction on the bottom and land side of the furrow. Under most favorable conditions there may be a slight reduction in draft, but under average conditions the weight of the sulky and the plowman more than offset the reduced friction.

Plowing at the same depth many years in succession often gives rise to a compacted layer just below the depth of plowing, known as plow sole or hardpan. This is a fault which may be avoided by changing slightly the depth of plowing from year to year. The plowman often looks with pride on what may be poor plowing. The furrow slice should not be completely inverted like a plank turned the other side up, but one furrow slice should lean against the previous one in such a way that the rubbish will be distributed from a portion of the bottom of the furrow nearly to the surface of the plowed ground. At the same time a portion of the furrow slice should be in direct contact with the soil below. This permits good capillary connection for a portion of each furrow slice. When there is an abundance of rubbish to be turned under, it is often wise to disk the land before plowing. This loosens the surface of the soil and causes some mixture of it with the rubbish. When plowed under in this condition it does not form so continuous a layer to cut off capillary water from below. Disking in advance of plowing in case of rather compact soil also facilitates the pulverization of the furrow slice and results in a better pulverized seed-bed.

Time of Plowing.—The best time to plow depends on many conditions. There is no particular season that will be better than other seasons under all conditions. The old maxim, "Plow when you can," is a good one to follow. Plowing done in the fall or early winter lessens the rush of work in the following spring, and under most conditions fall plowing gives better results than spring plowing. Fall plowing in temperate latitudes subjects the exposed soil to the elements and results in destruction of insects and a thorough pulverization of the soil, due to freezing and thawing. Fall plowing should neither be harrowed nor disked, but left in a rough condition in order to collect the rains and snows during the winter. This will result in storage of the winter rainfall and prevent erosion, unless by chance the land is steep and rains are very heavy. Under the latter conditions it may not be wise to practice fall plowing. In warmer latitudes plowing may be done during the winter, and when land is plowed in the autumn it should be seeded with a cover crop to prevent erosion. In the Northern states and Canada fall plowing is generally recommended, but in the South spring plowing is considered preferable. Spring plowing, unless it be very early, should be harrowed soon afterward in order to

A Badly Eroded Field.[1]

Damage of this character reflects no credit on American agriculture.

conserve soil moistures. Generally it will be found good practice to harrow towards the close of each day the land that has been plowed during the day. If the soil is rather dry and weather conditions very dry, it may be better to harrow it each half day. In case of sod and compact soil, disking in advance of plowing is advised.

Depth of Plowing.—The depth of plowing is determined by the character of the soil and the kind of crop to be grown. In general, fall plowing should be deeper than spring plowing. Deep-rooted crops call

[1] Courtesy of United States Department of Agriculture, Bureau of Soils. From "Soil Survey of Fairfield County, South Carolina."

10

for deeper plowing than shallow-rooted ones. For corn, potatoes and heavy truck crops, deep plowing is generally advised. For oats, barley, flax, millet and other spring annuals, shallow plowing generally gives as good results as deep plowing, and at a less cost. In the long run, deep plowing for most soils is to be recommended. Deep plowing increases the depth of soil from which the mass of plant roots draw moisture and plant food; it increases the water-holding capacity of the soil; it incorporates the organic matter to a greater depth in the soil; it enables the soil to receive and hold the rainfall, thus reducing erosion.

Where shallow plowing has been the practice, the depth of plowing should be increased gradually, one-half inch to one inch each year, until the desired depth has been obtained. This gives better results than increasing to the full depth at once. On virgin land with deep soil shallow plowing during the early years of cultivation may give as good results as deep plowing. Much depends on the nature of the soil, and wherever the soil at the depth of six to ten inches is compact, deep plowing and the incorporation of organic matter will improve it.

Subsoiling.—Subsoiling pertains to loosening the subsoil below the usual depth of plowing. Subsoil plows are constructed to run to a depth of sixteen to eighteen inches, with a view of loosening and slightly lifting the subsoil. It is neither turned nor brought to the surface. Such a practice is even more expensive than plowing and, consequently, more than doubles the cost of the preparation of the land for crops. While it may prove beneficial, many tests indicate that the practice does not generally pay for the expense involved. Doubtless much will depend upon the value of the land, the character of subsoil and the nature of the crops to be grown. On valuable land having impervious subsoil, and for high-priced crops, it may frequently pay. How long the benefits from subsoiling will last is determined by the rapidity with which the soil returns to its former compact condition. Heavy rains and thorough saturation with water often soon overcomes the benefits of subsoiling. As a general practice, subsoiling is not to be recommended. It might prove beneficial in semi-arid regions as a means of increasing the water storage capacity of the soil to tide over long periods of drought. In such regions the beneficial results are likely to be more lasting than where the rainfall is heavy. Both in practice and theory deep plowing is preferable to subsoiling.

Disking.—There are two forms of disk harrows: (1) having a solid disk, and (2) having a serrated disk and known as the cutaway disk. The latter is generally lighter than the former, is adapted to stony and gravelly soil and for light work. The full disk is more generally used, although in double disks both the full disk and the cutaway disk are sometimes combined in the same implement. The disk harrow stirs the soil to a greater depth than do most other forms of harrows. It is especially useful on land that has been plowed for some time and has become somewhat compacted. Fall plowing and early spring plowing, when being prepared for medium to late planted crops, should generally be gone over once or twice with the disk.

A large portion of the spring oats in the Central States are seeded on land prepared by the use of the disk and harrow, and without plowing. The disk is the most effective implement in the preparation of the seed-bed for oats. This method of preparing the land enables farmers to accomplish early seeding on a large scale. Early seeding of oats is important in connection with good yields.

Harrowing.—There are many forms of harrows varying in style of teeth, number of teeth, weight and adjustment. The steel frame harrow with levers to adjust the teeth, built in sections that are joined together, is generally preferred. The size or width of the harrow is usually determined by the number of sections it has. It is an implement of light draft, and to be effective should be used in the nick of time. Repeated harrowing is often advised (1) for the purpose of maintaining a surface mulch to conserve moisture, and (2) to destroy weeds just as they start growth. The spring-toothed harrow is effective in stony and gravelly soil, and tends to loosen the soil more than the spike-toothed harrow. The former is best for destroying weeds and loosening the soil, while the latter is preferable for soil pulverization and for covering small seeds that are broadcasted, such as clovers, grass seeds and the millets. While the harrow is generally used just prior to seeding and planting, it is found to be a good practice to harrow such crops as corn and potatoes after planting, and sometimes even after they are up. Such harrowing is often fully as effective in destroying weeds and pulverizing the soil as a good cultivation would be. It is much more rapidly and cheaply done than cultivating.

Planking or Dragging.—The plank drag is a cheap implement consisting of three or four two-inch planks fastened securely together with the edges overlapping. These may be eight to twelve feet in length. It is used for pulverizing clods and smoothing the surface of the ground. It is an effective implement to use where fine pulverization of the surface is desired, and works satisfactorily when the soil is rather dry.

Rolling.—The roller serves two chief purposes: (1) to compact the soil, and (2) to pulverize clods. The weight and size of the roller are important in this connection. Soil compacting calls for considerable weight, while pulverization demands a roller of comparatively small diameter. In recent years the corrugated roller with a discontinuous surface has come into use and is thought to be superior to the old style. It compacts the soil and yet leaves some loose soil at the surface, thus lessening direct evaporation. The roller should be used only when the soil is in dry condition and when it is desirable to encourage capillary rise of water and establish conditions favorable for the germination of seeds that lie near the surface of the soil. Rolling is most frequently resorted to in preparing the seed-bed for winter wheat. This crop calls for a compact and well-pulverized seed-bed. In the winter wheat regions the soils are frequently dry at the time winter wheat should be seeded.

A roller known as the subsurface packer has come into use in the semi-arid regions. This implement, consisting of a series of heavy disks, is so constructed as to compact the soil to a considerable depth, leaving two or three inches of loose soil at the surface. It encourages capillary rise of water without encouraging surface evaporation.

Details of a Good Seed Bed.[1]

Character of Seed-Bed.—The ideal seed-bed is determined by the character of crop to be grown. Wheat, rye, alfalfa, the clovers and most small seeds call for a finely pulverized, compact seed-bed. If these conditions are combined with a good supply of moisture these crops will make a prompt and satisfactory growth. Such crops as corn and potatoes call for a deep, loose seed-bed, and do not demand the same degree of pulverization of the soil as the crops above mentioned. Oats and barley do best with a fairly loose and open seed-bed, but demand fairly good

[1] Courtesy of The Campbell Soil Culture Publishing Co. From "Wheat," by Ten Eyck.

pulverization of the soil. As a rule, all small seeds need a seed-bed that has been thoroughly well prepared, while larger seeds, and especially those of crops that are to be inter-tilled, may be planted with less thoroughness in seed-bed preparation. The after-tillage will often overcome a lack of previous preparation.

An even distribution of seed, especially when it is sown broadcast, is essential. This, together with uniformity in germination, makes for perfection in stand of plants. The character of seed-bed is important in this connection. A well-prepared seed-bed facilitates a good stand, while a poorly prepared one often does just the reverse.

Cultivation and Hoeing.—Cultivation and hoeing pertain wholly to inter-tilled crops, such as corn, potatoes, beets, tomatoes, cabbage and a great many other garden crops. As a rule, cultivation should be sufficiently frequent during the early stages of growth to maintain a satisfactory soil mulch and destroy all weeds. This is best accomplished by cultivating or hoeing at just the right time. Weeds are easily destroyed when quite small. One cultivation at the right time is more effective than two or three cultivations when weeds have become large. As a rule, little is to be gained by inter-tillage when there are no weeds and when there is a satisfactory soil mulch. The frequency of cultivation is, therefore, largely determined by these factors. Ordinarily, nothing is to be gained by cultivating deeper than necessary to destroy weeds and maintain a good soil mulch. Two to three inches in depth is generally sufficient. Deep cultivation frequently destroys roots of the crop cultivated, much to its detriment.

Throughout most of the corn belt shallow and level cultivation is practiced. This seems to give better results than deeper cultivation or the ridging of the soil by throwing the earth toward the corn plants. Ridging the soil causes rain to flow quickly to the depressions midway between the rows, and encourages soil erosion. Level cultivation with numerous small furrows close together encourages more thorough penetration of the rain. Level cultivation makes the seeding of oats easy, as it generally follows the corn with no other preparation than the disking of the land.

Control of Weeds.—The time of plowing and the frequency and character of cultivation are related to the growth and eradication of weeds. Weed-seeds turned under to the full depth of plowing frequently lie dormant until the ground is again plowed and they are brought near to the surface. On spring-plowed land it is generally advisable to allow time for the weed-seeds to germinate, after which the small weeds may be destroyed by harrowing. Then crops may be planted with comparative safety so far as weed competition is concerned. In case of late plowing, it is advisable to plant or seed very promptly after the land is plowed in order that the crops may get ahead of the weeds.

Weeds are a great menace to crops, and especially to those that do not fully occupy the ground in their early periods of growth. Weeds compete with the farm crop plants for plant food and moisture. Where they have an equal start, they will frequently exterminate the crop unless removed promptly by cultivation. Weed destruction is most economically accomplished by hoeing and cultivating as soon as weeds have begun to grow. When such measures have been neglected and the weeds get a good start, it requires much more labor for their extermination.

Soil Mulches.—Aside from the soil mulch mentioned under the topic of cultivation and hoeing, mulches of straw, manure and other organic substances are resorted to in exceptional cases. These serve

TERRACING AS A MEANS OF PREVENTING EROSION.[1]

both to conserve soil moisture and to keep down weeds. They therefore obviate the necessity for hoeing and cultivating. Such mulches encourage capillary rise of soil moisture to the immediate surface of the ground. Furthermore, upon the decay of the mulch, organic matter and plant food are added to the soil. Such mulches are applicable only under intensive systems of farming and where the materials may be secured without too great cost.

Soil Erosion.—Soils are eroded by the rapid movement of both wind and water. Wind erosion occurs most extensively in the sandy regions

[1] From Year-Book, U. S. Dept. of Agriculture, 1913.

of the semi-arid belt, especially in western Kansas and Oklahoma. Such soil destruction calls for surface protection, either by a continuous covering of plants, or by such methods of cultivation as will prevent the movement of the surface soil. In those regions it is recommended that the plow furrows be at right angles to the prevailing direction of the wind, and that the drill rows of grain be likewise at right angles to the wind. Mulches of straw, especially in the wheat regions where straw is abundant, are also recommended. Such straw may be rolled with a subsurface packer to prevent its blowing from the soil. Under such conditions the surface soil should not be made too fine.

In the South and in southern Illinois, Iowa and Missouri, soils erode badly as result of the movement of rain water. Such erosion often results

ANOTHER WAY TO STOP EROSION.[1]

in deep and destructive gullies. These cause a direct loss of soil, and are barriers to continuous cultivation in the fields in which they occur. Such erosion should be prevented by every possible means before it proceeds far. Gullies may be stopped by the use of brush, weeds, straw and stone. These materials should be anchored in the gullies in such a way as to encourage them to fill with soil again. Deep plowing and the use of green manures, which encourage penetration of rains, help to overcome this erosion. Terracing the soil may be resorted to as a last means of preventing erosion.

Soil Injury.—Soils are frequently injured by plowing and cultivating when they are too wet. Heavy soils are more susceptible to such injury than those of a sandy nature. Such injury is often difficult to overcome. It gives rise to a puddled condition of the soil. When plowed, it turns

[1] Courtesy of The International Harvester Company.

up in hard clods which are difficult to pulverize. In this condition it requires more labor to prepare a seed-bed than if it had not been so injured.

Soils are often seriously injured by the tramping of livestock. It is unwise to allow stock to run in the fields when the soil is in a very wet condition. Hauling manure or loads of any kind across the field when the soil is too wet often results in injury to such an extent that the tracks of the wagon may be seen even after the land has been plowed and cultivated.

Time and Intensity of Tillage are Economic Factors.—The time to plow, disk harrow and cultivate is important in connection with the cost of the operations. It is essential to perform these tillage operations when the soil is in the best possible moisture condition. This enables the farmer to accomplish the desired result with the minimum amount of labor; consequently, his force of men and teams is able to properly care for the maximum acreage. It is easier and much less expensive to stir the soil at the right time and thus prevent bad physical condition than it is to change the bad physical condition to a good condition. A great deal of labor is required to reduce a hard, cloddy soil to a finely pulverized condition. As above indicated, time of cultivation in connection with weed destruction is important. The farmer who is foresighted and plans his work in such a way as to avoid undue rush at busy seasons will be the one to accomplish the various cultural operations with the minimum amount of labor.

The intensity of tillage will be determined by a number of factors, such as the price of land, the cost of labor and the value of the product grown. With cheap labor, high-priced land and a valuable product, intensive methods of tillage are applicable. On the other hand, when labor is expensive, land is cheap and products are of low value, extensive methods of tillage must be applied. It is wise to keep the soil occupied as fully as possible. This is accomplished by crop rotations and a succession of crops, one following another, throughout the growing season, so that at all times plants will be occupying the soil and gathering plant food as it becomes available.

The saving and utilization of all the manures produced on the farm is essential in this connection. It is more profitable to grow a full crop on five acres than it is to produce one-half a crop on ten acres.

In general, soil utilization and management call for a thorough understanding of the underlying principles and the adoption of methods of handling that accomplish good results without undue expense. Those practices which are injurious and those which do not make for maintenance of fertility should be avoided.

REFERENCES

"Principles of Soil Management." Lyon and Fippin.
"Crops and Methods of Soil Improvement." Agee.
"Soils." Fletcher.
"The Soil." Hall.
"Soils." Burkett.
Michigan Expt. Station Bulletin 273. "Utilization of Muck Lands."
Missouri Expt. Station Circular 78. "Control of Soil Washing."
U. S. Dept. of Agriculture Bulletin 180. "Soil Erosion in the South."

BOOK II
FARM CROPS

CHAPTER 9

Crop Improvement

By C. F. Noll

Assistant Professor of Agronomy, Pennsylvania State College

The development of varieties and strains of our farm crops which have great productiveness or superior merit in other respects is a matter of great interest to all agriculturists. Increase in yield due to natural productiveness of a variety results in a gain which is maintained year after year without additional cost of fertilizer or expense in culture. Such gains are of much economic importance, as shown by the differences secured in many variety tests. At the Pennsylvania State College Experiment Station, where varieties of various crops are tested under the same conditions, there are some which outyield others by as much as fifty per cent. Here the good yielding varieties are grown with just the same expense as the poor ones, except for the slight additional cost of handling the increase in crop. Similar results have been secured at experiment stations in nearly every state.

Plant Selection.—Crop improvement or plant breeding is often looked upon as a new thing, but ever since man has been growing plants, they have gradually been modified by seed selection. All of our cultivated plants come from wild forms, but some of them have been so changed that they could not now perpetuate their race if left to shift for themselves. Within the memory of men now living, the fruits of tomatoes have been developed from the size of a walnut to several times as large, and other changes have been effected which have made them more desirable for table use. Though plant improvement has been thus going on for ages, only within the past few decades has there been great general interest in this work, and only of late have some of the fundamental principles been understood.

Man originates to a very limited extent desirable changes in the plants with which he works. He is dependent chiefly upon changes which occur naturally, and all that he does is to take advantage of these changes and perpetuate the forms which are the most suitable for his purpose. He cannot, for example, make the pole lima beans over into the dwarf form, but when dwarf plants are found in a field of lima beans, he can save seed of these plants and perpetuate and multiply a race of dwarf lima beans.

Kinds of Variation.—No two plants are exactly alike, but most of the variations are of no significance to the plant breeder. They may be due to differences of environment, and in this case will not be inherited. If a hill of corn is heavily manured, the stalks and ears will be larger than where manure is withheld, but seed from these favored hills is not necessarily any better than seed from plants not so well fertilized. However, variations may arise which are hereditary and which may be the beginning of new varieties. When the variations are in the yield or size of plant, usually one cannot distinguish the difference between the variations which are hereditary and those which are not, except by a study of the progeny of the plants. When the variations are in color or form,

VARIATIONS IN TIMOTHY.[1]

one may have less difficulty in picking out those which could be used to develop new strains or varieties.

Hybridization.—Hybridization means the crossing of plants of different species or different varieties. It is accomplished by taking the pollen from a flower of one of the plants to be crossed and placing it on the pistil of a flower of the other. Care must always be exercised to prevent the plant-producing seed from being fertilized with its own pollen or with foreign pollen carried by the wind or by insects.

While there has been a good deal of mystery to many in regard to the crossing of plants and a disposition to regard hybrids as of superior

[1] Courtesy of The Macmillan Company, N. Y. From "Plant Breeding," by Bailey.

merit, the cross-fertilization is usually easily accomplished, and, on the other hand, the varieties produced by crossing are not necessarily of superior merit. Crossing of plants for the most part results in new combinations of parental characters. By crossing a yellow pear tomato and a large red one, one could produce a red pear tomato and a large yellow one. If a variety of wheat with bearded heads and white grains is crossed with a variety with smooth heads and red grains, there could be produced a bearded wheat with red grains and a smooth wheat with white grains. By selection and propagation the characters become fixed and give new varieties.

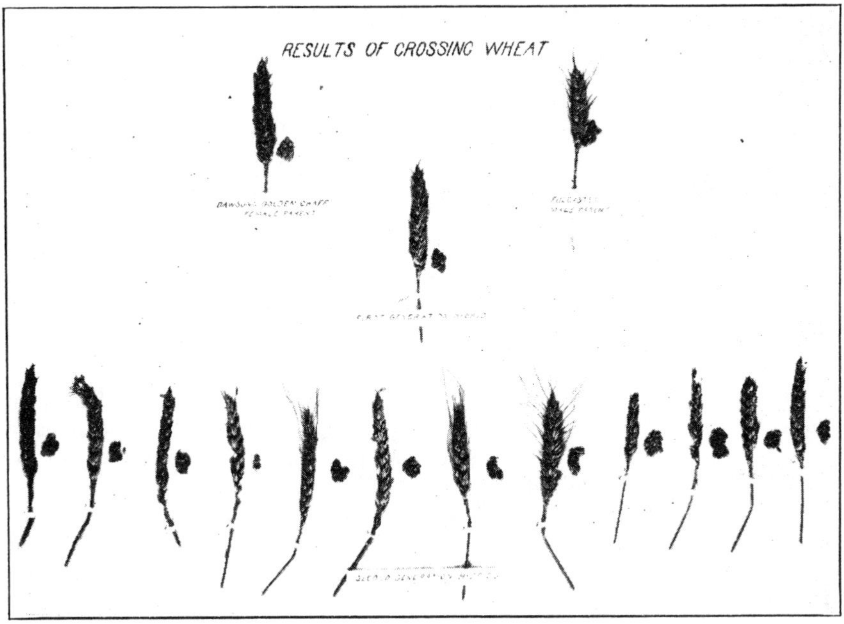

NOTE THE VARIATION IN THE SECOND GENERATION HYBRIDS.[1]

Choice of Varieties.—In attempting to improve a crop one should first endeavor to secure a first-class variety. Because of the great difference in varieties, if the poorer yielding ones were chosen and an attempt were made to improve them in productiveness, it is not likely that they could be made better than varieties already in existence.

Variety testing is a rather simple matter, but some precautions must be observed if the results are to be dependable. The main considerations are as follows:

1. The varieties should be tested on as uniform soil as possible of the kind on which the field crops are to be grown.

[1] Courtesy of Pennsylvania Agricultural Experiment Station.

2. The plats should be long and narrow rather than short and broad, and should extend across inequalities in the land rather than parallel with them.
3. A known standard variety should be planted in every third or fourth plat for comparison.
4. Trials should be conducted for a number of years and the choice of a variety based upon the average performance rather than upon the results of only one year.

Suggestions are given in this chapter for the improvement of a few crops. The methods of procedure with others would be similar, depending chiefly upon how the blossoms are fertilized and upon methods of propagation.

CORN

Special care must be exercised in the purchase of seed-corn. This crop tends to become adapted to local conditions and may not do well when removed to different localities. Especially is there likely to be a failure to mature when seed from a locality having a longer season is bought. On the other hand, a wise selection of seed should enable a farmer to adapt his corn better to his own conditions.

Most of our best known varieties have thus been developed by consistent selection of seed for a number of years on the same farm. The well-known Leaming variety was developed by J. S. Leaming in Clinton County, Ohio, by continuous selection, from a variety bought in Hamilton County, Ohio, in 1855. By selection along the same line, this variety was made very uniform. Reid's Yellow Dent, a very popular variety of a well-defined type, originated with a cross between two varieties planted in the same field by Robert Reid in 1846. The type was fixed in this case also by continuous selection. Most farmers could not do better than test a number of varieties to find a good one and then by careful selection of seed try to make it better.

The Ear-Row Method.—The most rapid improvement of corn is accomplished by some ear-row (or ear-to-row) method of breeding. There are a number of methods in use which vary in detail. By ear-row planting is meant the planting of each ear to be tested in a row by itself to determine its productiveness and other desirable qualities. The rows should be of such a length that not over half of the seed on an ear need be planted. If the rows are three and one-half feet apart and the hills three feet apart, forty-two hills will comprise approximately one one-hundredth of an acre. Five or six grains should be planted in a hill and when the corn is up, it should be thinned to three stalks per hill. Mixed seed of the variety should be planted for a check every sixth row. During the growing season the rows should be observed and desirable or undesirable characteristics noted.

Each row should be harvested separately. Since the yield of stover is of only minor importance, it does not matter whether the plants are cut or not, but they must be husked separately and the corn ears weighed. After the yields of the ear-rows have been obtained, any one of a number of methods for continuing the work may be followed. The simplest way is to take the remnants of the best ears as shown by the ear-row test, shell these together and plant in an isolated seed plat the next year. From this plat the diseased and weak stalks should be removed before the pollen is shed. Seed should be saved from the best rows in the ear-row plat for field planting the next year. The third year there should be an ear-row plat like that of the first year and the ears for this should come from the multiplying plat grown the second year. The seed of the main crop the third year should come from the multiplying plat and from the part of the field in which the seed from the ear-rows was planted. This

THE EAR-TO-ROW TEST PLAT WITH CORN HUSKED, SHOWING A METHOD USED IN ASCERTAINING WHICH SEED EARS HAVE YIELDED BEST.[1]

method provides for an ear-row plat and a multiplying plat on alternate years.

Ideals in Selection of Corn.—Besides attempting to secure greater productiveness in a variety of corn, one who would improve the crop should seek to adapt the variety in length of growing season to the locality in which it is grown. In a general way the best varieties are those which require about all of the season for development and yet can be depended upon to mature before frost.

The stalks should be of medium size and able to stand up well. The ears should be of medium height from the ground, with a rather short shank, and should droop somewhat rather than stand erect. By continuous selection for high and low ears for five years at the Ohio Experiment Station, two strains were developed from one variety with a

[1] From Year-Book, U. S. Dept. of Agriculture.

difference in height of ear of over two feet. Here the low-eared strain was the earlier and gave the greatest average yield.*

The ears should be of good size, should have medium size cobs, should be fairly uniform in color and type of kernel and should be attractive in appearance. The so-called show points of the ears are of less economic importance than was at one time thought, but corn that looks good finds the best market when sold for seed.

In the above discussion it has been assumed that the corn is grown primarily for grain. Ensilage varieties should have a rather large, leafy stalk besides a good ear, and may be somewhat later in maturing than varieties for grain.

WHEAT, OATS AND BARLEY

In the case of wheat, oats and barley, variations frequently occur within a variety which make it worth while to search for those which are better than the parent variety and to multiply them as new strains.

First Year.—Go through the field and select choice heads, taking as many as can be planted separately, remembering that really good strains are more likely to be found in a large number than in a small number.

Second Year.—Plant the seed of these heads each in a separate row. Make the rows a foot apart and plant the grains four inches apart in the row. The same number of grains should be planted from each head, which may be twenty-five or thirty.

When ripe, the number of plants in each row should be determined, the rows should be cut separately and the yield of each gotten.

Third Year.—A number of ounces of seed of each strain should now be available. If 100 head-rows has been planted the year before, about twenty of the best should be sown this year. These should be sown under field conditions and the parent variety should be sown for a check and put into every third plat. A good size of plat is sixteen feet long and two rows wide, the rows being eight inches apart and the plats ten inches apart. These grains can be planted thus side by side, for they do not readily cross-fertilize.

Fourth Year.—By the fourth year there will be enough seed for a larger plat which may be sown with a grain drill. A good plan is to shut off the middle hoe, put in a partition and sow at one time two plats of four or five drill rows each, depending upon whether the drill used has nine hoes or eleven. As before, the parent variety should be sown in every third plat. Only the most promising strains should be continued.

Fifth and Succeeding Years.—If desired, the plats may be made larger than the fourth year, but the arrangement of plats should be the same. Only those which are a decided improvement on the parent variety should be retained.

* Ohio Experiment Station Bulletin No. 282, Corn Experiments, by C. G. Williams and F. A. Welton.

Crossing of Varieties in Small Grain Breeding.—Different varieties may be crossed for new combinations of characters as discussed before. The first generation from the cross will look like one parent in respect to some characters and like the other parent in others. The seed of each plant should be kept separate and planted like the head-rows in selection work. Usually it will be found that the progeny of these parent plants are not uniform. In that case the grain from each plant must be kept separate and planted again in separate rows as before and this must be kept up until all the plants from a parent are alike in all of their visible characters. Those that are uniform should be considered pure strains, and after this the testing may proceed as with selections from the third year on.

Varieties of these crops should be improved in production first of all, but also in resistance to disease and stiffness of straw. In the case of wheat, the milling quality of the grain is important, and in oats, from the market point of view, the color of grain, white being the preferred color.

Many of the older varieties of these crops owe their origin to selections made by farmers and some to crosses. Of the varieties of wheat, Fultz was originated by Abraham Fultz in 1862 from a selection from Lancaster; Gold Coin, which was an accidental seedling variation, was selected from Deihl Mediterranean; Fulcaster, the well-known red-bearded variety, resulted from a cross between Fultz and Lancaster, made by S. M. Schindel, Hagerstown, Md.

POTATOES

Production of Seedlings.—New varieties of potatoes originate from seedlings. The seeds are produced in the true fruits, which come after the blossoms and look like little green tomatoes. These fruits or balls, as they are commonly called, are produced very sparingly and in some seasons none are seen. The seed should be sown indoors early in the spring and the young potato plants handled like tomatoes until they are set out in the field. Transplanting to pots increases their vigor. The first year few reach full development and most do not for two or more years. The seedlings, as a rule, are quite variable and few if any look just like the parents. Each should be regarded as a new variety and given a number and kept separate as long as grown.

The work is very interesting and may give varieties better than those already on the market, but most seedlings are of inferior merit.

Hill and Tuber Selection.—Potatoes vary in the hill and it is possible to improve a variety by selection of the best hills or the best tubers. It is a good practice to dig by hand a great many hills and save seed of some of the best for a seed plat the next year. This seed plat should be gone over and weak and diseased hills removed and the remainder saved for the field planting. Greater progress will be made by keeping the tubers from selected hills separate and testing these as new strains. Each should have a number by which it will always be known. The first year ten hills of each might be planted in rows side by side with the parent variety every third place. The best only should be saved and the next and succeeding years the plats may be made larger.

Tuber unit selection should start with selected tubers of the same size which are desirable in appearance and free from disease. These are each cut into four pieces, which are planted in succession, one tuber after

VARIATION IN YIELD OF POTATOES FROM SELECTED TUBERS.[1]

the other, with some space between the hills from the different tubers. When mature, the four hills from a tuber are dug together and the future selection based upon the yield of tubers and their appearance. These must be designated by numbers as in other selection work. The next year single row plats of ten or more hills each of the most promising may be planted, with the parent variety in every third plat as before.

Potatoes may be improved in productiveness, disease resistance and quality of tubers. There is a difference in susceptibility, especially to

[1]Courtesy of Pennsylvania Agricultural Experiment Station.

early and late blight, and perhaps to other serious diseases. Only strains of high market quality should be perpetuated. The tubers should be of medium size, smooth in outline, flat oval or flat oblong in shape and have shallow eyes.

Where carefully conducted, these methods of selection have resulted in the improvement of the variety.

Opportunities in Crop Improvement.—There is need in every community that at least one farmer make a specialty of producing and selling improved farm seeds. Such work is usually very remunerative, besides being of value to the whole neighborhood.

Testing of varieties and the improvement of certain crops may be made a matter of community interest, especially where there is some farmers' organization. There is also the opportunity of forming clubs or associations for crop improvement, which may be quite local or state-wide, as in the case of many state crop improvement associations now in existence.

REFERENCES

"Genetics." Walter.
"Cereals in America." Hunt.
"Plant Breeding." Bailey and Gilbert.
"Fundamentals of Plant Breeding." Coulter.
Ontario Agricultural College Bulletin 228. "Farm Crops."
Farmers' Bulletin 382, U. S. Dept. of Agriculture. "Adulteration of Forage Plant Seeds."

CHAPTER 10

The Rotation of Crops

In all of the older agricultural districts the rotation of crops is recognized as an essential to successful farming. With the prevailing price of corn, farmers on the best lands in the corn-growing belt have found it profitable to grow corn after corn for a number of years. In like manner, on the best wheat land in Minnesota, the Dakotas and Canada, wheat grown continuously has proven a profitable enterprise. In that region farmers find no good argument in favor of fencing their farms, constructing farm buildings, feeding cattle and milking cows, when they can make as much money or more by a system of farming that occupies their time for a little more than one-half the year and allows them leisure during the remainder of the year. A single crop system, while successful for a time, however, will not prove successful in the long run.

Successful farming calls not only for the best possible utilization of the soil and the maintenance of its fertility, but also demands the fullest possible utilization of the labor that is to be employed. The efficiency of the labor of men and teams on farms is measured largely by the proportion of time for which they are profitably employed. In nearly all other enterprises labor is fully and continuously employed. In order that farming may compete with other enterprises for labor, it must be likewise employed on the farm.

Rotations Defined.—A crop rotation is a succession of crops grown on the same land. A good crop rotation is a systematic succession of the three general classes of farm crops, namely, cultivated crops, grain crops and grass crops, in such a way as to give large yields and provide pasture and forage on the farm at the least expense of labor and soil fertility.

The rotation is definite when the crops recur in a fixed order, and it is a fixed rotation when they not only recur in a fixed order, but also at regular intervals. A rotation consisting of corn, oats, wheat and clover and timothy is a definite one, regardless of whether the clover and timothy remain for one, two or three years, but it becomes a fixed rotation when not only the order of the crops is named, but the length of time of each crop is also specified.

Purpose of Rotations.—A rotation of crops (1) provides for maintaining the soil in good tilth; (2) supplies organic matter and nitrogen; (3) prevents destructive outbreaks of insect pests; (4) reduces plant diseases; (5) provides for the economical destruction of weeds; (6) maintains crop yields; (7) distributes the labor of men and horses; (8) saves labor in cultivation of land; (9) keeps the soil occupied; (10) provides for a balanced removal of plant food; (11) systematizes farming; and (12) may control toxic substances.

Maintain Good Physical Condition of Soil.—Deep-rooted plants, such as alfalfa and the clovers, improve the physical condition of the subsoil as a result of root penetration. The cultivation given to inter-tilled crops, such as corn, potatoes, beets and the truck crops, improves the physical condition of the surface soil. Such frequent cultivation may tend to reduce the organic matter of the soil, but this will be largely overcome by the stubble and roots of the grasses and clovers that follow the grain crops.

Conserve Organic Matter and Nitrogen.—Extensive rotation experiments at the Minnesota Experiment Station show that standard rotations,

Dangers of Continuous Cropping.[1]

On the left is corn growing on land that has grown corn continuously for 19 years. On the right is corn in a five-year rotation. Both fields were planted on the same day to the same kind of corn. The yield on the field to the left is 27.5 bushels to the acre. The field on the right gives 61.3 bushels an acre. These are the average yields for ten years.

which include an inter-tilled crop, small grains and grasses with clover, all gave net profits. A four-years' rotation of millet, barley, corn and oats was no better than four years of continuous growing of wheat. All of these are classified as exhaustive crops. They cause a reduction in both the organic matter and nitrogen supply of the soil. Land cropped continuously to wheat, corn, potatoes or mangels for a period of ten years, showed a loss of 1100 pounds of nitrogen and 20,000 pounds of carbon per acre. In twelve standard rotations covering the same period of time, there was a gain of 300 pounds of nitrogen per acre, while the carbon and humus in the soil was maintained and in some cases increased. In the standard rotations eight tons of manure per acre were applied once during the rotation.

Provide for Extermination of Weeds.—Noxious weeds often cause a serious loss in farming. Weeds not only rob the crops of plant food and

[1] From "Farm Management" by Boss. Courtesy of Lyons and Carnahan, Chicago.

moisture, thus reducing the yield and sometimes causing absolute failure, but they entail additional labor in the process of cultivation. Many weeds grow best in certain kinds of crops. For example, mustard is a common weed in the small grain crops in the prairie states. The seeds ripen a little earlier than the grain, and in the process of harvesting are freely shattered and seed the land for the succeeding year. Where small grain is grown continuously this weed becomes a serious pest. Its extermination calls for an inter-tilled crop following the small grain. Pigweed, bindweed, foxtail and crab-grass are common in corn and potato fields, but they seldom become serious in small grain fields or in grass land; consequently, cultivated crops followed by grasses and small grains make for extermination of these weeds. Daisies, wild carrot and buckhorn are common weeds in hay fields, and generally grow worse the longer the land remains in hay. Such weeds, however, give no trouble in cultivated fields devoted to corn, potatoes, etc., and the cultivation helps to exterminate them.

Lessen Insect Depredations.—Most insect pests live upon some particular crop or a few closely related crops. A crop or related crops, grown continuously on the same land, affords an opportunity for the associated insects to multiply and become very numerous. The remedy is to plant the infested fields with a crop which will not be injured by the pest in question. Unless these insects have the power of migration they will perish for the want of suitable food or for lack of conditions suitable for multiplication.

However efficient the rotation of crops may be in the extermination of insects, some rotations may prove not only ineffective but actually disastrous. For example, land that has been long in grass sometimes becomes so infested with wire-worms as to cause a practical failure when devoted to corn. Grass affords conditions favorable to the multiplication of wire-worms, and they may live in the soil sufficiently long after the grass is plowed up to destroy a crop of corn which follows. Under such conditions fall plowing or bare fallow should precede the planting of the corn. The bill bug breeds freely in the bulbous roots of timothy, and when timothy sod is plowed late in the spring and planted to corn, this insect transfers its attention to the corn with disastrous results. Such trouble may be avoided by destroying the existing vegetation some time in advance of planting the corn. The insect under such conditions will either be starved or forced to leave the field before it is planted to corn.

Cutworms are a great menace to newly planted tobacco and many other crops, but their presence depends largely on the preceding crop. Cutworms multiply extensively only in grass land where the eggs are laid by the moths. Many similar examples could be cited, and success in preventing insect depredation by crop rotation calls for a knowledge of the life history and habits of the insect pest concerned. (See Chapter **76**: "Insect Pests and Their Control.")

Reduce Plant Diseases.—Plant diseases, like insect pests, are generally restricted to a particular crop or small group of closely related crops. The potato scab, so far as is known, is confined solely to potatoes. Its presence in the soil prevents the continuous growing of potatoes, and calls for a rotation in which the interval between successive potato crops is sufficiently long to provide for the disappearance of the disease. In a similar manner flax wilt or cotton wilt demands a rotation of crops in order to prevent the disease becoming disastrous. Bacterial diseases of tomatoes, potatoes, eggplants, cabbage and numerous other vegetables, the rusts and smuts of small grains, and many other diseases accumulate in the soil under the one-crop system. These troubles can be largely avoided and the crop-producing power of the soil maintained by intelligent systems of rotation. The most profitable system for any locality or type of farming can generally be ascertained from the state experiment station.

Improve Environment of Crop.—Aside from insect pest, plant diseases and weeds which flourish under the one-crop system to the disadvantage of the crop, there is another factor inimical to best plant growth. This consists of excreta given off by the roots of plants that accumulate in the soil to their detriment. As a rule, such excreta are not equally injurious to a different class of crops, and a rotation, therefore, lessens the injury. The excreted substances are organic in nature and are either changed in character or entirely disappear with time, so that the crop giving rise to them may be returned to the land after a year or more without injury.

Rotations Insure Returns.—The old adage, "Don't place all your eggs in one basket," applies with equal force in the production of crops. Unfavorable conditions in any locality are seldom such as to cause a failure of all kinds of crops, although a complete failure of a particular crop in a certain locality is not uncommon. A rotation of crops which includes a variety of crops, therefore, avoids complete failure.

Prevent Reduced Crop Yields.—The tillage given to a cultivated crop, such as corn or potatoes, increases the yield of the crop that follows by providing a better physical condition of the soil. In like manner legumes leave organic matter and nitrogen in the soil which is utilized to the advantage of corn or potatoes which may follow. The cultivation given crops destroys weeds to the advantage of crops which follow, and which do not receive cultivation.

Rotations Systematize Farming.—A well-planned rotation of crops enables the farmer to know definitely what is to be done each year, and makes possible an estimation of the general expenses and returns that may be expected. It also enables him to plan his work and secure his materials, such as seed, fertilizers, etc., in advance of the time they are needed.

Rotations Distribute Labor.—A good rotation of crops will enable the farmer to do a larger proportion of his own work than would be possible if the land were devoted to one crop. This enables him to utilize his own labor to the fullest possible advantage, and to reduce the expense necessary for hired labor. It is important, therefore, in selecting crops for a rotation, to select those that will compete with each other for the labor of men and teams as little as possible. The common rotation of corn, oats, wheat and hay fulfils these requirements fairly well. To illustrate, the preparation of land and seeding of oats take place in the early spring. Between the seeding time of oats and the time for planting corn there is sufficient time to prepare the land for the latter crop. The cultivation of corn will precede the harvest of hay and oats. The preparation of land for winter wheat will take place after the harvest period and prior to the harvest of corn. This fully occupies the time of the farmer during the growing season. There will sometimes be conflict between the harvest of wheat and hay, and the cultivation of corn, necessitating a little extra labor at that time.

Essentials of a Good Rotation.—A good crop rotation should contain (1) an inter-tilled crop, (2) a cash crop, (3) crops to feed, and (4) a crop to supply humus and nitrogen. All crops may be roughly classified under three heads, namely: exhaustive, intermediate and restorative. All crops, when harvested, remove from the land more or less plant food, and in this sense they are exhaustive. No crop restores to the soil any considerable amount of plant food unless it is plowed under or allowed to decay on the surface of the soil. Notwithstanding these facts, certain crops leave land in poorer condition for subsequent crops than it was before. These are designated as exhaustive crops, and include wheat, oats, rye, barley and millet. Their ill effect upon subsequent crops may be due to any one or a combination of a number of factors, among which are physical condition of the soil, injurious insects, plant diseases, reduction of soil moisture and a failure to supply either organic matter or nitrogen in any appreciable quantity.

It is wise, therefore, to select as many restorative crops as possible and so arrange the crops that these will be followed by the exhaustive crops. These two classes of crops should alternate as far as possible. In conjunction with this, one should select crops that will yield well and for which there is a demand, either for feeding on the farm or as a cash crop. The best varieties of the crops entering into the rotation should always be used. These will be determined largely by local conditions.

Sequence of Crops.—It is a good plan to follow a crop with a long growing season by one having a short growing season. This is typified when corn is followed by oats. In turn oats or barley is removed from the land in ample time for seeding winter wheat, which occupies the land for a rather long period. In this connection it is wise to provide in the rotation a place where manure may be hauled directly from stables and barnyards and applied to the fields. Where there is an abundance of manure and corn is extensively and advantageously used as feed for livestock, corn may be grown two years in succession, especially when the soil is fertile and manure is available for both the first and second crops. It is desirable that crops be arranged in such a way that the improving effects of each crop shall be regularly received and the ill effects of the exhaustive crops be systematically neutralized by the crop that follows.

Length of Rotations.—The length of crop rotations will be determined by local conditions and the character of crops grown, together with the value of land and cost of labor. Crops that are costly to establish, such as alfalfa, should occupy the land for two or more years in order to minimize the annual cost of production. The length of time that a crop remains productive is also a factor. The annual cost of seed and the preparation of the land for the crop is one-half or one-third as much if the crop is continued for two or three years respectively, as it is if allowed to remain only one year. So long as the yield is satisfactory, it generally pays to continue the crop. This tends toward a longer crop rotation.

In many localities where general farm crops prevail, a seven-year rotation is common, such for example as corn, oats, wheat and mixed clover and timothy for four years. Such long rotations with only one legume in them do not make for increased soil fertility, unless all the crops produced are fed upon the farm and the manure returned to the fields. Where cash crops dominate the type of farming, short rotations may be better. A rotation of corn, wheat and clover or of potatoes, wheat and clover affords the maximum of cash crops, while the frequency of clover in the rotation tends to maintain the nitrogen supply of the soil. Such short rotations also maintain the soil in good tilth as a result of the frequent plowing and abundant tillage.

What Crops to Grow.—The crops to be grown in a rotation will be determined by a number of factors, as soil adaptation, length of growing season, market demands, transportation facilities, and the system of farming that prevails. Aside from these facts there is another consideration that must not be overlooked. Usually it is unwise to follow a crop like tobacco, which is considered a gross feeder, with another crop such as corn having similar feeding habits. Such a practice is permissible only on very fertile soil or where the quality of the following crops is to be influenced through reduction in organic matter or available plant food. For example, coarseness in tobacco might be reduced by having it preceded by corn.

When to Apply Manure and Fertilizers.—It is generally advisable to apply barnyard manure to those crops in the rotation that have a long growing season or a high money value, or to those that are considered gross feeders, such as corn. In the absence of manure, the same rule will apply in the applications of commercial fertilizers. When manure is supplemented with fertilizers, the fertilizers are best adapted to crops of short growing season or to those influenced in quality by the character or form of a particular fertilizer ingredient. In this connection it should be borne in mind that the legumes require only mineral fertilizers and that crops that demand much nitrogen should follow the legumes.

Some Suggested Rotations.—Crops should naturally follow each other in such a way that each crop paves the way for the one that is to follow. Best results will be secured when plants are not compelled to do their part at a disadvantage. Wherever feasible, a large proportion of the product of a rotation should be food for livestock. This provides for the maintenance of soil fertility.

In the northeastern part of the United States a rotation of corn, oats, wheat and hay with various modifications dominates most of the general and livestock types of farming. By omitting oats a three-crop rotation results, which, if restricted to three years in length, makes for soil fertility, provides a cash crop and at the same time furnished an abundance of livestock food and bedding. This may be supplemented with alfalfa, thus increasing the protein supply. On soils poorly adapted to wheat this crop may be omitted and oats will take its place. In the northernmost latitudes and at higher elevations the acreage of corn will be reduced and that of oats and hay increased. Where markets are favorable and the soil is adapted to potatoes, this crop may be substituted for a portion of the corn, thus increasing the cash crops at the expense of forage.

Wheat generally proves a better crop in which to seed clover and the grasses than does oats. In most parts of this section of the country the grasses are seeded in the autumn and the clover seeded early in the spring. Further south, both clover and the grasses may be seeded in the autumn. The four staple crops above mentioned may be arranged into several rotations with manure and fertilizers applied as suggested in the following tabulation.

METHOD OF FERTILIZING CROP ROTATIONS.*

3 Years.	4 Years.	5 Years.	7 Years.	Per Acre.
..	..	..	1	Corn: 6 to 10 loads of manure and 25 lbs. of phosphoric acid.
1	1	1	2	Corn: 6 to 10 loads of manure and 25 lbs. of phosphoric acid.
..	2	2	3	Oats: no fertilizer.
2	3	3	4	Wheat: 50 lbs. each of phosphoric acid and potash.
3	4	4	5	Clover and timothy: no fertilizer.
..	..	5	6	Timothy: 25 lbs. each of nitrogen, phosphoric acid and potash.
..	..	..	7	Timothy: 25 lbs. each of nitrogen, phosphoric acid and potash.

* Roughly speaking, 25 pounds each of nitrogen, phosphoric acid and potash may be obtained by buying 150 pounds nitrate of soda, 175 pounds of acid phosphate and 50 pounds of muriate of potash. The cost of the ingredients may be estimated from the following prices per pound, which will vary according to circumstances: nitrogen, 18 cents; phosphoric acid, 4 cents; and potash, 5 cents.

In the trucking regions of New Jersey, Delaware, Maryland and Virginia, two crops may frequently be secured in one season. Over much of this region tomatoes may be set as late as June 1st. This gives opportunity to grow a quick-maturing crop before the land is needed for tomatoes. If hay is needed crimson clover may be seeded in the fall and cut for hay the next spring, before the land is needed for tomatoes. Where canneries are available, early peas may be harvested before time to set tomatoes. This gives two crops in one season, both of which provide for the operation of the cannery and prolong its season of activity. Crimson clover may be seeded in the tomatoes at the last cultivation, and growth turned under the following spring for the benefit of a succeeding crop.

In this district a two-year rotation in which four crops are grown is found to be quite successful. Two of these are cash crops and two are renovating crops. The cash crops are corn and either potatoes or tomatoes. The renovating crops are crimson clover or soy beans or winter rye mixed with winter vetch. This makes the purchase of nitrogen in fertilizers unnecessary. Acid phosphate and potash are applied in moderate quantities and generally to the cash crops only. This system, without any manure and with the occasional use of lime, maintains the fertility of the soil.

In portions of Ohio and Indiana a three-year rotation of corn, wheat and clover is common. One strong point in this rotation is that one plowing answers for three crops. When the clover sod is plowed for corn in the spring the ground breaks up easily and makes an ideal seed-bed for corn. The cultivation given the corn provides a good seed-bed for wheat with no other preparation than thorough disking and harrowing of the corn stubble. This, of course, necessitates a removal of the corn stalks sufficiently early to seed wheat. It is not applicable where the growing season is too short. This rotation not only economizes in labor as above suggested, but makes a good distribution of labor. Furthermore, it provides for rather continuous occupation of the soil. If the sod devoted to corn is not plowed until spring and corn is followed by fall seeding of wheat in which grass and clover is seeded, the soil will be subject to erosion only during the time it is in corn. Erosion in this case may take place in times of heavy rains and on rolling land, by the water running down the furrows between the corn rows. This may generally be overcome by having the rows and cultivation at right angles to the slope.

This is a fairly good rotation for the stockman and dairy farmer. Corn furnishes the material for the silo, while clover hay supplies the protein in which corn is deficient, thus giving a well-balanced ration. The wheat straw makes good bedding, while the wheat may be either sold or exchanged for concentrates. On farms having no permanent pasture the clover and timothy may be left for another year, cut once and pastured afterwards, or, if necessary, it may be pastured throughout the fourth year. If used for this purpose, both timothy and alsike clover should be seeded with the red clover.

The following five- and six-year rotations have been found successful

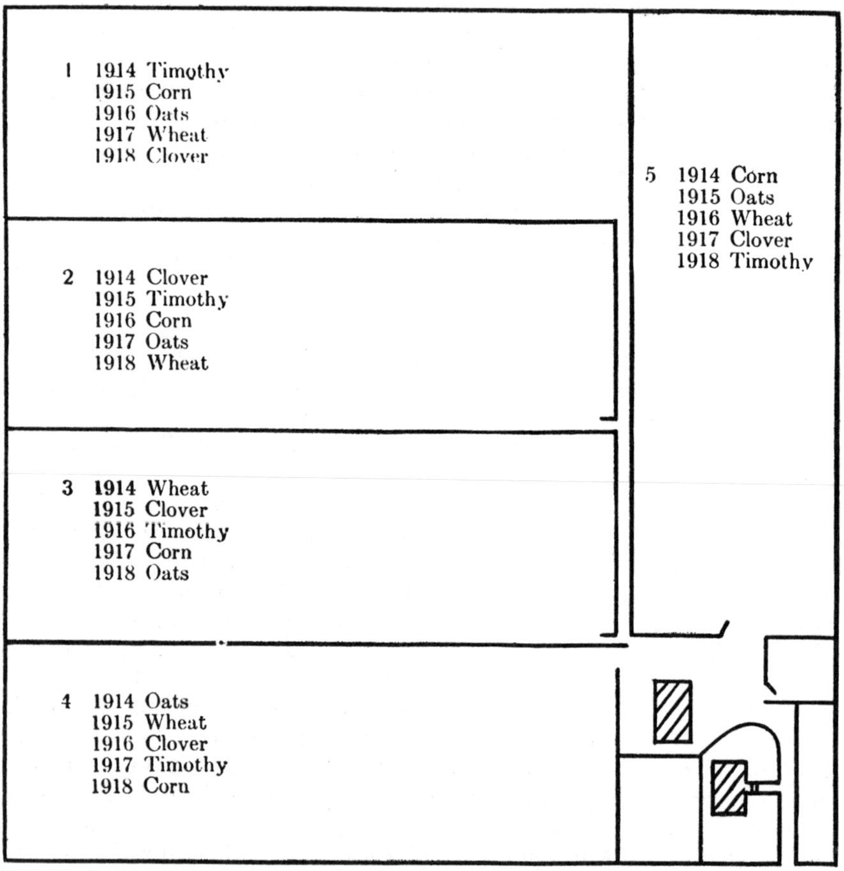

A FIVE-YEAR ROTATION.

Field.	1914.	1915.	1916.	1917.	1918.
1—25 A.	Timothy	Corn	Oats	Wheat	Clover
2— "	Clover	Timothy	Corn	Oats	Wheat
3— "	Wheat	Clover	Timothy	Corn	Oats
4— "	Oats	Wheat	Clover	Timothy	Corn
5— "	Corn	Oats	Wheat	Clover	Timothy

in the Great Plains area: (1) corn; wheat; brome-grass; brome-grass; oats, barley or emmer; (2) corn; wheat; brome-grass; brome-grass; brome-grass; oats, barley or emmer. In these rotations the wheat may be either winter or spring, and, furthermore, wheat may be substituted for any of the last-mentioned crops in either of the rotations.

Space will not permit the enumeration of all the rotations that are possible. With a clear understanding of the underlying principles and a knowledge concerning the utilization and market value of the crops to be grown, any farmer may plan crop rotations best suited to his farm.

Methods of Planning and Recording Rotations.—It is a principle that there should be as many fields as there are years and crops in the rotation, unless two crops can be harvested from the land in one year. It is also advisable that the fields be as nearly of equal size and productivity as possible. This provides for uniformity in distribution of work from year to year, as well as in the utilization of the products. Where livestock dominates the type of farming, it will often be found advisable to adopt two rotations, one known as the major and the other as the minor rotation. The former will include the staple crops grown both for feed and market, while the latter provide soiling and annual pasture crops. In such a scheme the minor rotation should be located near the farmstead where the small fields will be easily accessible. The tabulation on preceding page shows how a five-field rotation may be planned, and serves as a record of what has been and what will be in every field in any particular year.

REFERENCES

"Field Crops." Wilson and Warburton.
"Soils and Fertilizers." Snyder, pages 131–159.
Minnesota Expt. Station Bulletin 109. "Rotation of Crops."
Ohio Experiment Station Bulletin 182. "Maintenance of Fertility."
Rhode Island Expt. Station Bulletin 135. "Crop Rotations."

THE HEIGHT OF STALKS AND POSITIONS OF EARS MAY BE GREATLY CHANGED BY SELECTION OF SEED FOR THESE CHARACTERS.[1]

[1] Courtesy of Ohio Agricultural Experiment Station, Bulletin 282, "Corn Experiments."

CHAPTER 11

CORN (ZEA MAIZE)

The average acre of corn produces more food value than an equal area of any other staple crop except potatoes. Corn has a longer season of growth than most other staple crops, and, consequently, it more fully utilizes the plant food that is made available by processes going on in the soil when reasonably warm and moist. It is adapted to a wide range of soil conditions. It fits well into the crop rotations without seriously competing with other crops for labor. It has a wide range of uses. The tillage which the crop receives leaves the soil in excellent condition for the crops which follow.

Classification of Corn.—There are six types of corn: dent, flint, sweet, pop, soft and pod. The first four only are of importance in America. Fully 90 per cent of the corn grown in North America is of the dent type. There are several hundred varieties of dent corn and a score or more varieties of flint corn. The types are classified according to color and size. Dent corn is divided into three classes with reference to size and time of maturity, namely: early, medium and late maturing varieties. It is also divided according to color into yellow dent, white dent, white cap yellow dent and mixed dent varieties.

Varieties of Corn.—Of the several hundred varieties of dent corn, comparatively few are worthy of cultivation in any particular locality; and yet one often finds many varieties within a restricted area. Where soil conditions are uniform over several counties, one or two varieties may be found best suited to the whole of the area.

Corn is a very minor crop in Canada, the most of it being grown in the Province of Ontario. Flint is the prevailing type. In the northeastern part of the United States, including New England, New York, Pennsylvania and New Jersey, varieties of flint corn are extensively grown on the higher elevations and in the northernmost latitudes. Among the best known varieties of this class may be mentioned Longfellow, King Phillip, Smut Nose, Stickney's Yellow, Taylor's Improved Flint and Davis' Eight Rowed Flint. The prevailing varieties of dent corn in this section are Pride of the North, Early Huron Dent, Funk's 90 Day, Leaming and numerous strains of white cap dent, seldom having local names.

In the typical corn belt of Ohio, Indiana, Illinois, Iowa, Missouri and eastern Kansas and Nebraska, the leading varieties are Reed's Yellow Dent, Funk's Yellow Dent, Leaming, Reilley's Favorite, Clarage, Hogue's Yellow Dent, Hildreth's Yellow Dent, Hiawatha Yellow Dent, Boone

County White, Johnson County White, Silver Mine, St. Charles White and Kansas Sunflower.

In the Southern states we have among the large-eared varieties: Huffman, Excelsior, Chisholm, McMacnin's Gourdseed, St. Charles White, Boone County White, Rockdale, Singleton and Ferguson's Yellow Dent. Among the two-eared varieties may be mentioned Lewis' Prolific, Hickory King and Neal's Paymaster. Prolific varieties, producing two or more ears to a stalk, are Cocke's, Albemarle, Whatley's, Mosby's, Hasting's, Marlborough and Batts'.

In the northern portion of the corn belt, including the states of Michigan, Wisconsin, Minnesota, the Dakotas and the northern portions of Illinois and Iowa, the most common varieties are Silver King, Pride of the North, Wisconsin No. 7, Murdock, Wimple's Yellow Dent, Pickett's Yellow Dent and Golden Eagle.

The best variety for any locality can be determined only by local variety tests. Such tests have been conducted in many counties through the effort of the local organizations in co-operation with the state experiment stations. The results for such tests for sixteen counties in Iowa for the year 1911 are given in the following table:

VARIETY TEST, 1911.

Average of Sixteen Counties in Iowa.

	Number of Samples.	Yield per Acre, bushels.	Standing, October, per cent.	Strong, per cent.	Weak, per cent.	Dead, per cent.	Barren, per cent.
Farmer's variety test.....	966	54.3	78.0	78.1	14.6	7.3	5.2
One-tenth highest yielding	97	62.0	81.5	80.5	14.5	5.0	4.4
One-tenth lowest yielding.	97	44.5	71.0	73.5	15.0	11.5	6.1
Imported seed...........	128	53.0	81.5	67.0	27.0	6.0	5.9
Seed-house seed..........	190	49.5	72.0	61.5	26.5	12.0	4.6

INDIVIDUAL EAR TEST, 1911.

Average of Sixteen Counties in Iowa.

	Number of Samples.	Yield per Acre, bushels.	Standing, October, per cent.	Strong, per cent.	Weak, per cent.	Dead, per cent.	Barren, per cent.
Individual ears...........	1,440	53.5	78.5	83.5	11.5	5.0	5.7
One-fourth highest yielding	360	62.0	83.0	85.5	11.5	3.0	4.5
One-fourth lowest yielding	360	43.5	71.5	77.5	11.5	11.0	7.6

The large number of samples tested and the average results secured make conclusions relative to the differences found in yield and other qualities rather definite. It will be noted that one-tenth of the samples giving highest yields averaged 62 bushels per acre, while one-tenth of the samples giving lowest yields average 44.5 bushels per acre, or only about two-thirds as much as the best yielding samples. Note also that over 100 samples of imported seed averaged less per acre than did nearly 1000 samples of home-grown seed. Likewise, the 360 ear-to-row tests giving the highest yields were no better than the best one-tenth of the larger samples tested. One-fourth of the ear-to-row samples giving the lowest yield averaged a little more than two-thirds as much as the one-fourth giving the highest yields. The results show wide differences and emphasize the importance of the farmer selecting for his soil and locality the variety that will do best. Such selection will evidently make a great difference in the total yield of corn on a given acreage.

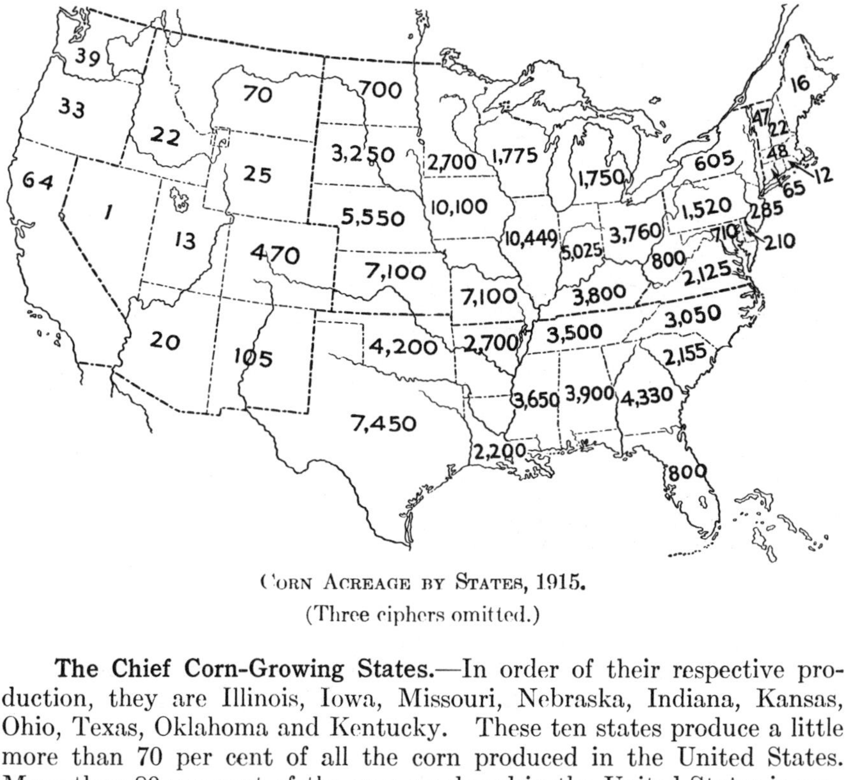

CORN ACREAGE BY STATES, 1915.
(Three ciphers omitted.)

The Chief Corn-Growing States.—In order of their respective production, they are Illinois, Iowa, Missouri, Nebraska, Indiana, Kansas, Ohio, Texas, Oklahoma and Kentucky. These ten states produce a little more than 70 per cent of all the corn produced in the United States. More than 80 per cent of the corn produced in the United States is consumed within the counties in which it is grown. The great use of corn is as a feed for livestock. There are a few counties, especially in the State of Illinois, where a considerable portion of the corn is marketed and goes outside of the counties in which it is produced.

North America produces three-quarters of the world's corn, nearly all of which is produced within the borders of the United States. Of the remaining one-quarter of the world's production, Europe produces about two-thirds and South America and Australia the remainder.

Soil and Climatic Adaptation.—Corn is best adapted to well-drained soils that are deep, loamy and warm. Large yields demand a high-water capacity of the soil and this is materially increased by deep drainage, deep plowing and organic matter. Corn requires a growing season ranging from 100 to 170 days, through which period the temperature should be high and accompanied by warm rains. An abundance of rainfall properly distributed is essential. In the typical corn belt the rainfall during July and August is most important, and the yield of corn is determined to a considerable extent by the rain during these two months.

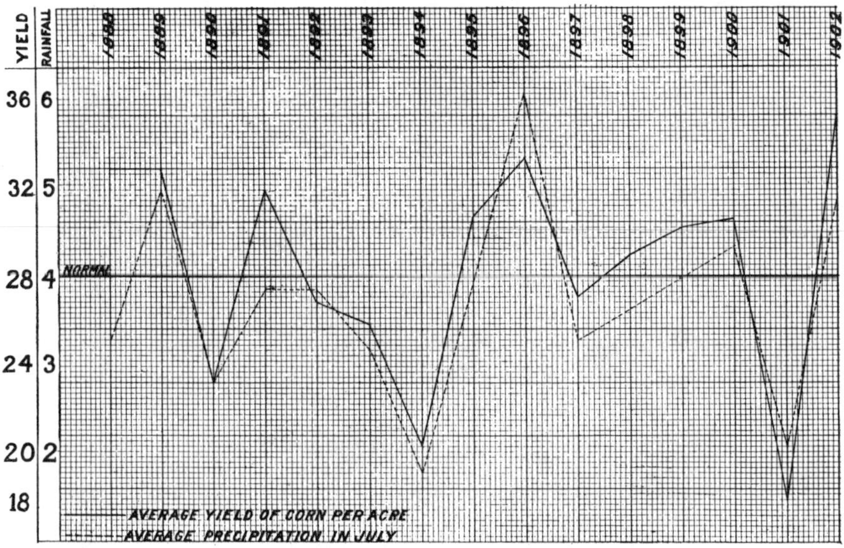

Chart Showing How Closely Corn Yield Follows Amount of Rainfall.

The accompanying chart shows the average yield of corn for a period of fifteen years, together with the July precipitation for the same years. There is a fairly close correlation between July rainfall and the average yield of corn.

It is not profitable to grow corn on very poor land. The nature of the corn plant is such that it will not produce grain unless the soil is sufficiently rich to afford considerable growth of stalk. In general, the richer the soil the heavier will be the yield of grain. Some other crops will produce fair yields on soil too poor to produce corn.

Crop Rotation for Corn.—Corn cannot be grown continuously on the same soil without diminished yields. A rotation of crops is, therefore, essential. In this rotation should occur at least one leguminous crop.

East of the Mississippi River and north of Mason and Dixon's Line, common red clover is best suited for this purpose. Alfalfa, crimson clover and alsike clover may be substituted for it under certain conditions. Over a considerable portion of this region the most usual rotation is corn, oats, wheat, and clover and timothy. This provides for a rotation ranging from four to seven years, depending on the length of time the land is left in grass and whether or not corn is grown more than one year in the rotation. In this rotation the corn should follow the sod on which may be scattered the manure prior to plowing. No other crop is better adapted to utilize the available nitrogen and mineral constituents that are slowly brought into a state of availability through decomposition of the roots, stubble and manure.

On fertile soils in a high state of cultivation corn may be grown two years in succession. This will require sufficient manure to apply on the corn land two years in succession, or will demand an application of commercial fertilizers for the second year's crop.

In the South the crops associated with corn in the rotation are quite different. In most cases cotton is the chief money crop; cowpeas and soy beans are the chief legumes; and winter oats is the principal small grain. The rotation frequently consists of cotton followed by cotton, with cowpeas planted between the cotton rows. The third year the land is planted in corn and seeded to winter oats after the corn has been removed. After the oats are harvested in the fourth year the land is broadcasted with cowpeas, and these harvested for hay. This rotation has proven successful in many parts of the cotton belt.

Many of the experiment stations have tested different rotations. The following tabulation gives the average results with corn in two rotations covering a period of more than twenty years at the Ohio Experiment Station:

CONTINUOUS *vs.* ROTATION CORN. TWENTY YEARS' WORK.

System.	Treatment.	Application per Acre.		Average Yield per Acre, bushels.				Average Yield for 20 Years.
		Per Crop.	Per 5 Years.	1st Period.	2d Period.	3d Period.	4th Period.	
Continuous....	None........	....	....	26.26	16.76	10.43	8.44	15.47
Rotation*.....	None........	....	....	31.89	30.82	31.04	20.31	28.95
Continuous....	Manure.....	5 tons	25 tons	43.13	40.11	34.62	30.22	37.02
Rotation*.....	Manure.....	8 tons	16 tons	40.73	49.52	59.75	55.83	51.81
Continuous....	Com. fert....	250 lbs.	1250 lbs.	38.86	39.09	28.00	26.83	33.19
Rotation*.....	Com. fert....	320 lbs.	985 lbs.	35.78	49.54	53.91	44.10	46.49
Rotation†.....	Manure.....	8 tons once in 3 years on corn.						60.20‡
Rotation†.....	None.......	Average of 8 unfertilized plots.						35.19‡

* Five-year rotation. † Three-year rotation. ‡ Average for 17 years.

It will be noted that where corn was grown continuously the yields have declined regardless of the character of the manure or fertilizer applied, whereas corn grown in a rotation has increased decidedly in yield when either manure or fertilizers have been used.

Plowing for Corn.—Plowing for corn may be done either in the fall, winter or spring. In many sections of the country fall plowing gives better results than spring plowing. The difference, however, is not sufficient to justify the advice that fall plowing should be universal. Every acre that is plowed in the fall or winter facilitates getting crops in the ground at the proper season in the spring. Deep plowing for corn deposits the trash and manure to a greater depth and induces the roots to go deeper into the soil, thus coming into contact with more plant food and soil moisture from which to draw nourishment. Deep plowing enables the soil to absorb a larger proportion of the rainfall, thus increasing its capacity for water. The further preparation of the seed-bed by disking and harrowing should leave it in a loose, friable condition to a considerable depth. Such a seed-bed is in marked contrast to the compact and finely pulverized one that is essential to wheat.

Manures and Fertilizers for Corn.—The amount and character of fertilizer for corn varies greatly in different localities, depending on the character of soil, length of time it has been in cultivation, and the rotation of crops. No definite formula is applicable to any very large territory. As a rule, no crop makes better use of barnyard manure than corn. Six to ten tons of stable manure to an acre of grass sod is generally sufficient. In growing corn, all of the nitrogen needed should be secured from the manure and leguminous crops that enter into the rotation. On soils not in a high state of fertility, the manure may be supplemented by about 200 pounds per acre of acid phosphate. In portions of Indiana, Illinois and Iowa, rock phosphate may be advantageously substituted for acid phosphate. On sandy soils and on swampy soils some potash may be advantageously used.

In the absence of barnyard manure good corn crops may be secured by the liberal use of a complete fertilizer in which phosphoric acid is the dominant ingredient. The amount of such fertilizer and its exact composition will depend on the character and condition of the soil in question. The average composition of such a fertilizer would be from 2 to 3 per cent of nitrogen, 7 to 10 per cent of phosphoric acid and 3 to 6 per cent of potash. The amount to use will range from 100 to 500 pounds per acre, depending on location. The character of fertilizer and the amount required can best be ascertained by actual test. In general, applications of less than 200 pounds may be applied through the fertilizer attachment to the corn planter. Where large amounts are used, it is best to distribute it throughout the soil before planting the corn.

Experiments that have been in progress for twenty years at the Ohio Experiment Station emphasize the importance of phosphorus in corn

TIME OF PLANTING, APRIL 29TH.[1]

production. A series of plats which received nothing save 320 pounds of acid phosphate per acre during each five-year rotation showed an increase in the yield of the several crops valued at $16.52 per acre. The acid phosphate cost $2.24, thus leaving a net gain of $14.28.

The addition of phosphorus to manure also increased the yield very materially.

Time and Method of Planting.—The time of planting corn varies with the location and character of season. It is never advisable to plant until the soil is sufficiently warm to cause a prompt germination of the

TIME OF PLANTING, MAY 7TH.[1]

[1] Courtesy of Ohio Agricultural Experiment Station, Bulletin 282, "Corn Experiments."

Time of Planting, May 16th.[1]

seed. The best of seed will often rot in a cold, wet seed-bed. In the United States the corn planting season from the Gulf northward ranges from the 15th of February until June 1st, a period of three and one-half months. In the heart of the typical corn belt corn is generally planted between the 1st and 10th of May, while in the northernmost limit of successful corn production, the planting season ranges from the 15th to 31st of May. In any locality the best time to plant will not be far from the time when the leaves of the oak trees are the size of a squirrel's ear. If seasonal conditions retard the work and necessitate planting two weeks

Time of Planting, May 26th.[1]

[1] Courtesy of Ohio Agricultural Experiment Station, Bulletin 282, "Corn Experiments.

later than the best time, it will be wise throughout most of the typical corn belt, and especially in the northernmost districts, to resort to varieties of corn of earlier maturity than those generally grown in the locality. In the Southern states the season is so long that there is a much wider range in the planting period. A uniform stand of vigorous plants is most easily secured by deferring planting until the soil is in the proper moisture and temperature condition.

Several of the state experiment stations have conducted tests extending over a number of years relative to the best time to plant corn. As an average of six years' work at the Ohio Experiment Station there was little difference in yield in planting any time between the 1st and 20th of May. For dates much later than the 20th there was a marked reduction in yield. Planting in the last week in April was nearly as good as

Time of Planting, June 6th.[1]

planting between the 1st and 20th of May. It is better to plant too early than to plant too late. Failure in case of early planting may be corrected by replanting, but there is no remedial measure for a planting that is made too late.

Rate of Planting.—A full stand of corn is essential. The number of plants per acre will vary with the fertility of the soil, the kind of corn and the purpose for which it is grown. Fertile soils will support more plants per acre than poor ones. Small varieties may be more thickly planted than large ones, and an abundant moisture supply in the soil will mature more plants than when dry. When planted for grain, 10,000 to 12,000 plants per acre are probably best throughout the greater portion of the corn belt. In the South, on thinner soils, fewer plants are often desirable. If grown largely for fodder or ensilage, corn may be planted one-quarter thicker than when grown for grain.

[1] Courtesy of Ohio Agricultural Experiment Station, Bulletin 282, "Corn Experiments."

Numerous experiments indicate that there is little difference within a reasonable range whether corn is planted in hills or drills. When planted in checks three kernels per hill, 3 feet 8 inches apart, an acre will contain 9720 plants. When planted in drills with the rows 3 feet 8 inches apart and one plant every 14 inches in the rows, an acre will contain 10,180 plants. Drilling is somewhat easier and safer on small, irregular fields and on land that is of uneven topography, and is preferable on most lands that are reasonably free of weeds. On badly weed-infested land checking the corn is recommended, because of the better facilities offered for cultivation and weed extermination.

On the better lands in the corn belt there has been a tendency in recent years to lessen the distance between hills, and in many districts 40 inches is now the common planting distance.

At the Ohio Experiment Station the average annual yield per acre for a period of ten years when corn was planted at the rate of 1, 2, 3, 4 and 5 kernels per hill, with hills 42 inches apart, the largest yield was secured from 4 kernels. The yields were as follows: 1 kernel, 31.7 bushels; 2 kernels, 50.8 bushels; 3 kernels, 60.8 bushels; 4 kernels, 64.9 bushels, and 5 kernels, 63 bushels per acre. The yield of stover was largest in case of 5 kernels per hill. The reduced size of ears and the increased labor in husking are such as to indicate 3 kernels per hill as the best rate of planting when grown for grain.

In regions of abundant rainfall corn is planted on the level, but in regions of low rainfall it is frequently planted in furrows by what is known as listing. This encourages a deeper rooting of the plants, which protects them from severe droughts.

Depth of Planting.—The depth at which to plant corn will vary with the character and condition of the soil and the nature of the season. In loose, loamy soils the depth may safely be 3 inches, and in the absence of sufficient moisture near the surface 4 inches in depth may be justified. On wet, heavy soils $1\frac{1}{2}$ inches to 2 inches will be better than to plant deeper. No matter at what depth corn is planted, the permanent roots start at a point about one inch beneath the surface of the soil. The depth of rooting is not influenced by the depth of planting, unless the depth is less than one inch.

Preparation of Seed for Planting.—Before shelling corn for planting it is important to remove all irregular kernels from the butts and tips of ears. Such kernels will not pass through the corn-planter with uniformity. Before being shelled the ears should be assorted into two or three lots, according to the size of kernels, and the shelled corn from each lot kept separate so that the planter plates may be adjusted to each size. The same results may be secured by the use of a seed-corn grader, of which there are several kinds on the market.

The planter should be carefully adjusted to each lot of seed. A poorly adjusted machine may offset the advantages derived from the carefully selected and graded seed. An actual count of the number of missing hills or plants on an acre would prove to the grower his loss through imperfect planting. Extensive investigations over large areas have shown that in certain years farmers secured not more than three-quarters of the full stand. If 75 per cent of a full stand produces 40 bushels to the acre, what will 95 per cent of a full stand produce?

Cultivation of Corn.—It is a trite saying that the cultivation of corn should begin before it is planted. This means that the final preparation of the seed-bed should take place just before planting, in order that all weeds that have just begun to grow will be destroyed. In the absence of such preparation weeds that have started will make so much growth before the corn comes up that it will make the first cultivation difficult. Small corn may be harrowed with a slant-toothed smoothing harrow without injury. A thorough harrowing at such a time will destroy many weeds that are beginning to grow, and is equally as effective as one good cultivation, and much more quickly done.

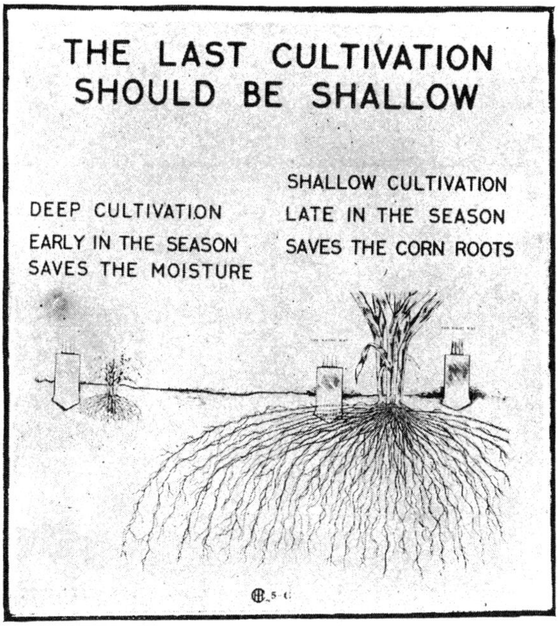

The Right and Wrong Way of Cultivating Corn.[1]

The chief objects of cultivation are: (1) to destroy weeds, (2) conserve moisture, (3) aerate the soil, and (4) increase the absorption of rainfall by keeping the surface loose. Under most conditions level and shallow cultivation is superior to deep cultivation and the ridging of the soil. Deep cultivation cuts many of the corn roots, thus reducing the ability of the plants to secure both plant food and moisture. In general, the first cultivation may be fairly deep, thus inducing a deeper rooting of the corn plants, after which shallower cultivation should take place which will interfere but little with the roots. One hundred and sixteen tests at thirteen experiment stations relative to the depth of cultivation for

[1] Courtesy of The International Harvester Company, Agricultural Extension Department. From pamphlet "Corn is King."

corn show a difference of more than 15 per cent in yield in favor of shallow cultivation. Sixty-one tests of deep cultivation gave an average yield of 64.9 bushels per acre, while 55 tests of shallow cultivation gave an average yield of 74.7 bushels, a difference of nearly 10 bushels per acre. One to two inches is considered shallow cultivation and four to five inches deep cultivation.

The frequency of cultivation will depend chiefly on the surface condition of the soil and the presence of weeds. In the absence of weeds and with the surface soil in a loose condition, little is to be gained by cultivation.

Methods of Harvesting.—Throughout the typical corn belt a large proportion of the corn is harvested from the standing stalks in the field, and the stalks are pastured or allowed to go to waste. This method fails to fully utilize the by-products of corn production, and is wasteful in

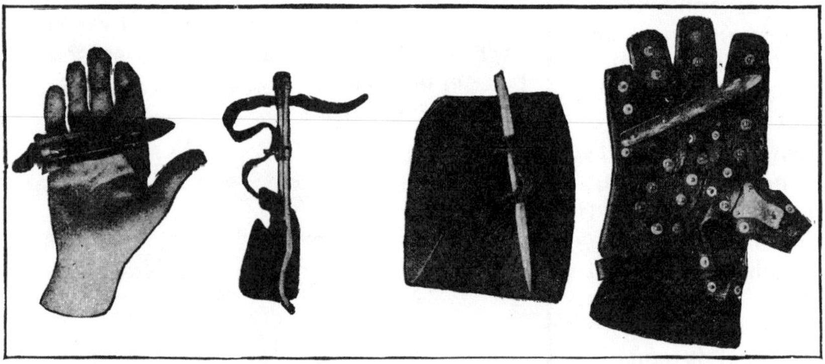

SEVERAL FORMS OF HUSKING PEGS.[1]

the extreme. In the eastern part of the United States the whole plant is generally harvested and utilized. When corn is grown for feeding dairy cows or steers the fullest utilization of the entire product is attained by storing in the silo. For this purpose it should be cut when the kernels have begun to glaze and the husks and lower leaves are turning brown. When not to be used for silage, corn should be put in shocks at a somewhat more advanced stage of maturity. Three to four hundred stalks make a shock sufficiently large to stand well and cure properly. The corn should be husked in three to six weeks after shocking, the ears stored in a well-ventilated crib, and the stover reshocked. Care should be exercised to so stand and slant the stover that the shocks will stand. They should be securely tied about two feet from the tops with strong binder twine. It is a waste of good material to allow the shocks to stand in the field until March or April.

[1] From Farmers' Bulletin 313, U. S. Dept. of Agriculture.

It is wise to feed stover during the winter period. Its feeding value may be increased by shredding. This encourages livestock to consume a larger proportion of the stalks. Fifty per cent of the feeding value of the corn stover lies in the portion of the stalk below the ear. When this is neither cut nor shredded very little of it is eaten by livestock. Shredding or cutting better fits the refuse for bedding purposes and facilitates the handling of the manure in which the refuse is finally deposited.

In storing cut or shredded fodder one should be certain that it does not contain too much moisture. It should be reasonably dry when stored in large bulk in order to prevent heating and spoiling. It is well, therefore, to shred when weather conditions are fairly dry, and not until the corn stover has become thoroughly cured.

SHRINKAGE OF CORN IN CRIB BY MONTHS. AVERAGE 8 YEARS, IOWA EXPERIMENT STATION.

Month.	Total Shrinkage to Date, per cent.	Average for the Month, per cent.
November	5.2	5.2
December	6.9	1.7
January	7.5	.6
February	7.8	.3
March	9.7	1.9
April	12.8	3.1
May	14.7	1.9
June	16.3	1.6
July	17.3	1.0
August	17.8	.5
September	18.2	.4
October	18.2	.0

The cheapest method of harvesting corn is to pasture with hogs. This is known as hogging down corn. The results of a four years' test at the Missouri Experiment Station showed that hogging down corn gave a return of 324.5 pounds of pork per acre, which, at 6 cents per pound, was valued at $19.48. The average number of hogs per acre was 14, and the number of days kept in the field was 35. This was on poor land and with corn yielding 25 to 30 bushels per acre.

Storing Corn.—The grain of corn is best stored for a time on the ear in a well-ventilated crib or building. Corn cribs of slatted sides with openings just small enough to prevent ears passing through are almost universally used for this purpose. They should be covered with roofs projecting some distance beyond the sides, and turn water without leaking. Cribs should be on elevated foundations, preferably of masonry or concrete. For the ideal crib, see Chapter 57. All precaution must be taken to prevent serious loss by rats and mice. Corn should not be put in the crib until reasonably well cured. If very wet when cribbed it is likely to mould and decay. Ear corn at husking time will contain 15 to 40 per cent of moisture, depending on conditions. After standing for six months or more in the crib, the moisture, under normal conditions, will range from 10 to 12 per cent. After this time shrinkage from loss of moisture will be slight.

Shrinkage of Corn.—A knowledge of the average shrinkage of corn is important in connection with future prices, and should be taken into consideration by the farmer in connection with the holding of corn for a future market. The table on preceding page shows the average shrinkage of corn at the Iowa Experiment Station as determined for eight successive years.

Market Grades of Corn.—According to the act of Congress of June 30, 1906, and March 4, 1913, the Secretary of Agriculture has fixed the following definite grades of grain, which went into effect on July 1, 1914:

STANDARD GRADES OF CORN AND SPECIFICATIONS FOR SAME.

Grade and Classification: White, Yellow and Mixed Corn.	Moisture.	Maximum Percentage of Damaged Corn.	Maximum Percentage of Foreign Material, Including Dirt, Cob, Other Grains, Finely Broken Corn, etc.	Maximum Percentage of "Cracked" Corn, not Including Finely Broken Corn. (See General Rule 9.)
No. 1	14.0	*...	1	2
No. 2	15.5	*...	1	3
No. 3	17.5	*...	2	4
No. 4	19.5	†0.5	2	4
No. 5	21.5	†1.	3	5
No. 6	23.0	†3.	5	7

* Exclusive of heat-damaged or mahogany kernels.
† May include heat-damaged or mahogany kernels not to exceed the percentage indicated.
"Sample"—See General Rule No. 6 for sample grade.

GENERAL RULES

1. The corn in grades No. 1 to No. 5, inclusive, must be sweet.
2. White corn, all grades, shall be at least 98 per cent white.
3. Yellow corn, all grades, shall be at least 95 per cent yellow.
4. Mixed corn, all grades, shall include corn of various colors not coming within the limits for color as provided for under white or yellow corn.
5. In addition to the various limits indicated, No. 6 corn may be musty, sour, and may also include corn of inferior quality, such as immature and badly blistered.
6. All corn that does not meet the requirements of either of the six numerical grades, by reason of an excessive percentage of moisture, damaged kernels, foreign matter, or "cracked corn," corn that is hot, heat-damaged, fire-burnt, infested with live weevil, or otherwise of distinctly low quality, shall be classed as sample grade.
7. In No. 6 and sample grade, reasons for so grading shall be stated on the inspector's certificate.
8. Finely broken corn shall include all broken particles of corn that will pass through a perforated metal sieve with round holes $\frac{9}{64}$ of an inch in diameter.

9. "Cracked" corn shall include all coarsely broken pieces of kernels that will pass through a perforated metal sieve with round holes ¼ of an inch in diameter, except that the finely broken corn as provided under Rule No. 8 shall not be considered as "cracked" corn.
10. It is understood that the damaged corn, the foreign material, including pieces of cob, dirt, finely broken corn, other grains, etc., and the coarsely broken or "cracked" corn, as provided for under the various grades shall be such as occur naturally in corn when handled under good commercial conditions.
11. Moisture percentages, as provided for in these grade specifications, shall conform to results obtained by the standard method and tester as described in Circular No. 72, Bureau of Plant Industry, United States Department of Agriculture.

Composition and Feeding Value of Corn—The following is a compilation of American analyses of the grain of the three principal types of corn and the stalks of dent corn, under three conditions:

COMPOSITION OF CORN (MAIZE).

	Grain.				Silage.	Fodder.	Stover.
	All Varieties.	Dent.	Flint.	Sweet.	Fresh.	Field Cured.	Field Cured.
Number of analyses	208.	86.	68.	26.	99.	35.	60.
Water	10.9	10.6	11.3	8.8	79.1	42.2	40.1
Ash	1.5	1.5	1.4	1.9	1.4	2.7	3.4
Protein (Nitrogen x 6.25)	10.5	10.3	10.5	11.6	1.7	4.5	3.8
Crude fiber	2.1	2.2	1.7	2.8	6.0	14.3	19.7
Nitrogen-free extract	69.6	70.4	70.1	66.8	11.1	34.7	31.9
Fat	5.4	5.0	5.0	8.1	0.8	1.6	1.1

The following tabulation gives the farm value and feeding value of corn per acre as compared with oats, wheat and hay, when grown in a four years' rotation on the limestone soil at the Pennsylvania Experiment Station:

THE AVERAGE ANNUAL YIELD DURING 25 YEARS OF 24 TREATMENTS ON 36 PLATS ON EACH OF 4 TIERS AT THE PENNSYLVANIA STATION.

	Average Yield per Acre.		Price per 100 pounds.	Farm Value per Acre.	Digestible Protein, pounds.	Energy Value, therms per Acre.
	Pounds.	Bushels.				
Corn, ears	3,534	50.5	$0.75	$26.51	160	3,198
Corn, stover	2,528	....	.125	3.16	40	671
Oats, grain	1,227	38.1	1.00	12.27	102	813
Oats, straw	1,772	....	.125	2.22	19	370
Wheat, grain	1,192	19.9	1.33	15.85	106	985
Wheat, straw	2,099	....	.125	2.62	8	348
Timothy and clover hay	3,609	....	.50	18.05	135	1,232

These figures may be condensed into a table that will bring out the comparison in a more striking manner, as shown below:

Comparison of Digestible Protein, Energy Value and Farm Value per Acre of 4 Crops when Grown in Rotation During 25 Years (1882–1906).*

	Digestible Protein, pounds.	Energy Value, therms.	Farm Value.
Corn	206	3,869	$29.67
Oats	121	1,189	14.49
Wheat	114	1,333	18.47
Hay	135	1,232	18.50

CORN IMPROVEMENT

No crop is more easily and rapidly improved by selection and breeding than corn. No work on the farm will come so near producing something for nothing as time intelligently spent in improving this crop. It is just as important to use well-bred seed-corn as it is to breed from an animal of good pedigree. The same principles apply in the breeding of both plants and animals. Well-bred seed-corn has often produced from five to twenty bushels per acre more than seed which has received no special attention when grown under identical conditions. A bushel of seed-corn will plant six acres; 10 bushels increase on each of six acres equals 60 bushels; 60 bushels at 60 cents per bushel equals $36 the value, of a bushel of good seed.

Securing Seed.—Seed-corn should be purchased in the ear so the buyer can see if it is as represented in regard to type, size and uniformity. It should have been grown on soil and under climatic conditions very similar to those surrounding the purchaser. Do not send far away for seed-corn. Many farmers have done so and have generally been disappointed.

Selecting Seed.—Selection should be made in the field where both plant and ear can be seen. Seed plants should be under normal conditions relative to soil and stand. Good plants should be of moderate height. Short nodes or joints are preferable to long ones, for each node bears a leaf. The more the leaf surface, the greater the power of the plant to manufacture the elements of the air and soil into corn. The leaves are the most palatable, digestible and nutritious part of the forage. The plants should be free from smut, rust and any other fungous diseases.

The ears for a medium maturing variety of dent corn should be attached to the stalk at a convenient height of about four feet, and by a shank of moderate length and thickness. For very early varieties the ears may be a little lower and for large late maturing ones, there will be no objection to having the ears five feet above the ground. When the shank

* Refer to Bulletin No. 116, Agricultural Experiment Station, The Pennsylvania State College.

is too long it allows the ear to pull the stalk over, and when too short the ear is too erect and may be damaged at the tip by allowing water to enter the husks. The husks should be moderate in amount and sufficiently long to cover the tip of the ear and protect the kernels from insects, birds and damage by rain.

The size of the ear will vary in different districts, but for a medium maturing variety a good seed ear should be 8 to 10 inches long. The circumference two-fifths the distance from the butt should equal three-fourths but not exceed four-fifths of the length. The form should be cylindrical or but slightly tapering from butt to tip. The tip and butt should be well filled with kernels and the rows, 16 to 20 in number, should

HIGH AND LOW EARS.[1]

be straight and carry out well to the butt and tip with kernels of regular and uniform shape.

The depth of kernels should equal one-half the diameter of the cob. Kernels five-eighth inch long, three-eighth inch wide and one-sixth inch thick are a good size. The tips should be strong and full, for such indicates good vitality. The embryo or germ should be large and extend well up toward the crown. Large embryos produce vigorous plants and indicate high fat and protein content and consequently high feeding value.

Care of Seed.—Seed-corn should be well cared for by storing in a dry and well-ventilated room and out of reach of rats and mice. Corn, thoroughly dried, will stand a very low temperature without injury, but

[1] Courtesy of Ohio Agricultural Experiment Station, Bulletin 282, "Corn Experiments."

if not well dried, a temperature not far below freezing will injure it and destroy its vitality or germinating power and make it worthless for seed.

Germination Test.—The importance of securing a perfect stand of strong plants in the cornfield cannot be overestimated. Aside from field conditions favorable to germination and the proper placing of the corn in its seed-bed, there are two dominant factors on which perfection of stand depends: first, the vitality of the seed; second, requisite number of kernels in each hill or regular and uniform spacing if planted in drills.

A vitality or germination test of seed-corn should always be made. It should be made several weeks before corn is required for planting, so that there may be time to secure a new supply in case the seed has been injured. There are several simple methods of making such tests, but in all cases every ear should be tested.

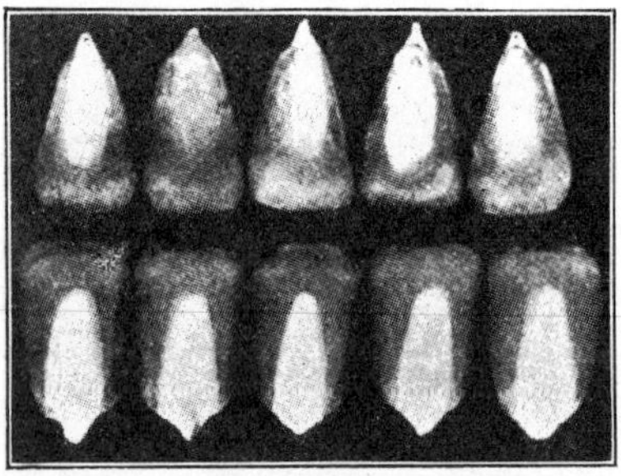

Good and Poor Types of Kernels.[1]

The top kernels came from an ear with too much space at cob, indicating low yield, poor feeding value, immaturity. Compare them with the kernels in the bottom row.

Germinating Box.—A box about 18 inches square and 3 inches deep, two-thirds full of clean sawdust or sand, is most convenient for germinating corn. The material should be thoroughly moistened and smoothed to a level in the box. Lay the ears of corn on the barn floor, tips to tips in double rows. Number every tenth ear with a small paper tag stuck between the rows. Remove from various parts around the ear, and from butt to tip, five grains from each ear. Now cover the sawdust in the box with a piece of white cloth marked off into squares $1\frac{1}{2}$ inches on a side with a lead pencil, preferably an indelible pencil, and numbered consecutively. In the squares, place the five grains from each ear separately, exercising care that the grains from each ear are placed in the square with the number corresponding. Cover the grains thus placed with another cloth of close weave or a fold of the one under the corn, to prevent the sprouts from coming through, and spread over all a piece of burlap or a gunny sack well soaked in water. The requisites for germination are air, warmth and moisture. The temperature of the living room or kitchen is about right, providing it does not fall below 55 degrees at night. If the temperature is favorable germination will have taken place in four to six days. Any ear failing to give five kernels vigorously germinated should be rejected. A handy man, working systematically, can test five or six bushels of corn in a day. It is work that should never be neglected

[1] Courtesy of International Harvester Company, Agricultural Extension Department.

and will pay for the labor involved many times over in a better stand and resulting larger yields of corn.

Improvement by Selection and Breeding.—The ear row method is the most satisfactory way of improving corn along any line. This method is based on the principle that like begets like, but fortunately this principle is not rigid. It is the variation in the progeny of any parent plant that enables us, through selection, to improve the variety, and it is the tendency for like to produce a larger percentage of progeny, differing but slightly from the parent that enables us to make progress in plant improvement.

Corn improvement by selection is easy, because the plant is large and its characteristics plainly visible; because the variations are sufficiently marked and frequent to enable man to select individuals with

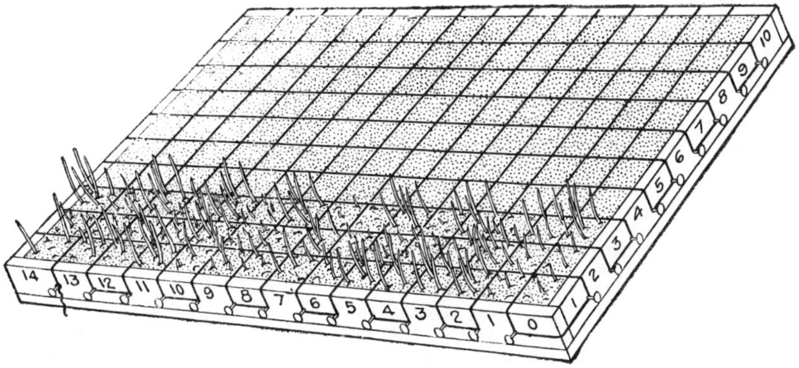

A Good Germination Box Seven Days After Planting.[1]

The box is filled with wet sand and marked into checks by means of cord stretched across the top at even intervals.

desirable characteristics, and also because of the large number of plants that can be secured from the individual and the consequent rapidity of multiplication.

Corn breeding is somewhat difficult because of the natural cross-fertilization and the impracticability of keeping the breed pure, and also because close and self-fertilization are difficulties that must be guarded against. None but the choicest ears selected for desirable qualities of both ear and plant should be used in the breeding plat, and any ears that do not show a high standard in the germination test should be rejected.

The selected ears should next be tested for yield and prepotency. The ears should be numbered and a portion of each planted in a separate row of a test-plat having uniform fertility. The rows should be sufficiently long to contain about 200 plants. This will require about one-fourth of the kernels of each ear. The rows should bear the same numbers as ears

[1] From Farmers' Bulletin 409, U. S. Dept. of Agriculture.

from which planted. The remaining portion of ears, with numbers securely fastened, should be saved for next year's multiplying plat. When corn is up, it should be thinned to a uniform stand for all rows. It should be frequently observed during growing season for rows that develop desirable characters. At harvest time each row should be husked separately and the corn weighed. The remnants of seed ears, from which a limited number of the highest yielding rows of best type were planted, should be shelled together and planted the following year in a multiplying plat which should supply seed for the general crop. From the multiplying plat should be selected choice ears for another test as above described. This method repeated each year makes progress in corn improvement.

REFERENCES

"Corn Crops." Montgomery.
"Book of Corn." Myrick.
"Manual of Corn Judging." Shamel.
"Study of Corn." Shoesmith.
Kansas Expt. Station Bulletin 205. "Growing Corn in Kansas."
North Dakota Expt. Station Circular 8. "Home Grown Seed Corn."
Pennsylvania Expt. Station Bulletin 116. "Corn Growing in the East."
U. S. Dept. of Agriculture Bulletin 307. "Tests of Corn Varieties on the Great Plains."
U. S. Dept. of Agriculture Bulletin 168. "Grades for Commercial Corn."
Farmers' Bulletins, U. S. Dept. of Agriculture:
313. "Corn—Harvesting and Storing."
317. "Increasing Productiveness;" "Shrinkage of Corn in Cribs."
400. "A More Profitable Corn Planting Method."
415. "Seed Corn."
414. "Corn Cultivation."
537. "How to Grow an Acre of Corn."
546. "How to Manage a Corn Crop in Kentucky and West Virginia."
553. "Pop Corn for the Home."
554. "Pop Corn for the Market."

CHAPTER 12

Wheat (Winter and Spring)

By W. H. Darst

Assistant Professor of Agronomy, Pennsylvania State College

The crop that furnishes the bread material of a country comes a little closer to the lives of the people than any other. In nearly all countries of the world wheat holds the first place as a bread crop, and for that reason deserves most careful attention.

The United States, with its rapidly increasing population, especially in the cities, and its constantly increasing demand for breadstuffs, may very soon find it necessary to import wheat. Under existing conditions the price of wheat must increase rather than decrease, and there will be more and more inducement for the farmer to increase his production.

The world's annual production of wheat for the last three years (1912-14) has been approximately 3,882,255,000 bushels. The six leading countries in production and in average acre yield are as follows:

Average Annual Production, 1912–1914.		Average Acre Yield, 1904–1913.	
Country.	Bushels.	Country.	Bushels.
United States	794,889,000	United Kingdom	32.8
European Russia	686,512,000	Germany	30.7
British India	349,273,000	France	20.1
France	325,650,000	Austria-Hungary	19.1
Austria-Hungary	226,732,000	United States	14.3
Canada	205,718,000	European Russia	10.0

It is an interesting fact that the two largest producing countries have the lowest acre yields. At one time these European countries had average yields very similar to our own. By years of systematic application of best known methods of production, the yields of these countries have increased enormously.

The climatic and soil conditions of some European countries are more favorable to the production of wheat than those in the United States. In European countries, also, the labor proposition is not so serious as it is in this country; consequently, they can afford to spend more time on their wheat crop.

Wheat Production in United States.—About one-half the wheat crop of the United States is produced in the North Central states west of the Mississippi River. This section includes the states of Kansas, Nebraska, North and South Dakota, Minnesota and Iowa. Hard winter wheat and hard spring wheat (including Durham) are grown in this section.

About one-sixth of the crop is produced in the North Central states east of the Mississippi River. The wheat in this section is known as the soft or red winter wheat.

About one-sixth of the wheat crop of the United States is grown in the far West. This includes the irrigated districts of the Rockies and the Pacific Coast wheat districts. White and red spring, and some winter wheat, are grown in this section.

All other states not in the general districts mentioned produce approximately 100,000,000 bushels annually.

Climatic and Soil Adaptation.—Wheat has a very wide climatic adaptation, which makes it a staple crop in many countries of the world. Wheat is best adapted, however, to regions having cold winters, especially cool weather during the first of the growing season. Cool weather during early growth causes wheat to stool more abundantly, which generally results in a larger yield. This applies to spring wheat as well as to winter wheat.

Climatic conditions, viz: rainfall, temperature, sunshine and humidity, influence the milling quality of wheat to a greater degree than does the type or fertility of the soil. The map, roughly dividing the United States into wheat districts, shows that climatic conditions existing in any section determine to a large extent the milling quality of the wheat.

In the hard spring and hard winter wheat districts, the season is comparatively hot and dry during the fruiting period, forcing early ripening of the wheat. This results in a hard, flinty kernel, high in protein and of good milling quality. The fruiting period being shortened, the wheat does not have the opportunity to store as large amounts of starch in the grain as it would under more favorable climatic conditions.

Where the fruiting season is longer and more favorable, as in the red winter wheat district and along the Pacific Coast, more starch is stored in the grain, which results in a starchy, light-colored wheat having lower milling quality.

A proper soil for wheat is important in that it determines the yield rather than milling quality. A large portion of the wheat in the United States is grown on the so-called "glacial drift" soils. These soils vary greatly in texture and structure, humus and plant food. The clay or clay loam uplands are usually better adapted to wheat than the low-lying dark-colored loamy soils. Dark-colored soils, rich in humus, are better adapted to corn. Wheat grown on such soil is apt to winter-kill and heave badly. The wheat grows tall and rank and may not fill out properly.

Rotations.—In parts of the Great Plains region, wheat is grown in continuous culture with fair returns, because the farming operations are so extensive. Rotations, therefore, are not profitable as yet. Eventually these large farms will be made into smaller ones, and it will be necessary to properly rotate the crops for profitable yields.

Continuous culture of wheat not only reduces the fertility of the soil, but multiplies the insects and fungous diseases injurious to wheat. Rotations are greatly modified in different localities by the crop-producing power of the soil and by the crops produced. Wheat is frequently grown in a rotation in order to obtain a stand of grass. The value of

EFFECT OF TIME OF PREPARING SEED BED. YIELD OF BAGGED WHEAT.[1]

rotations from the economic standpoint has been discussed in a previous chapter.

Preparation of the Seed-Bed.—The method used in preparing a seed-bed for wheat is determined by the rotation and kind of wheat grown. In winter wheat sections wheat may follow corn, oats, potatoes or tobacco. Wheat requires a firm, fine and moist, seed-bed, whether it be sown in the fall or spring. When wheat follows corn, potatoes or tobacco, the ground should be thoroughly plowed for these crops in the spring of the year, and the crop grown should receive thorough and regular cultivation as long as possible. After the crop is harvested double disking should put the ground in ideal shape for the seeding of wheat.

When winter wheat follows oats the stubble should be plowed as early as possible. The early breaking of oat stubble gives more time

[1] Courtesy of Kansas Agricultural Experiment Station.

for the preparation of the seed-bed, the firming of the soil and the conserving of moisture.

If plowing is done late in the season, each day's work should be harrowed as soon as finished. Plowed ground that is allowed to remain a few days before working is likely to become very dry and cloddy. A well prepared seed-bed insures quick germination, a good root system and results in less pulling and winter killing.

The following table taken from Bulletin No. 185 of the Kansas Experiment Station, shows that yield of wheat is greatly influenced by both the time and method of preparing the seed-bed:

METHODS OF PREPARING LAND FOR WHEAT. CROPPED TO WHEAT CONTINUOUSLY.

Method of Preparation.	Average 3 Years, 1911–1913.		
	Yield per Acre, bushels.	Cost per Acre for Preparation.	Value of Crop, Less Cost of Preparation.
Disked, not plowed	6.63	$2.07	$3.64
Plowed Sept. 15, 3 inches deep	13.24	2.83	8.35
Plowed Sept. 15, 7 inches deep	14.15	3.33	8.60
Plowed Aug. 15, 7 inches deep	22.19	4.00	16.34
Plowed Aug. 15, 7 inches deep. Not worked until Sept. 15	20.48	3.33	13.65
Plowed July 15, 3 inches deep	20.77	4.85	12.25
Plowed July 15, 7 inches deep	27.11	5.35	16.87
Double disked July 15. Plowed Sept. 15	19.71	3.93	12.37
Double disked July 15. Plowed Aug. 15, 7 inches deep	23.40	4.93	14.30
Listed July 15, 5 inches deep. Ridges split Aug. 15	22.90	3.92	14.73
Listed July 15, 5 inches deep. Worked down	22.77	4.05	14.53

Early preparation of the seed-bed gave a profitable increase in yields. Early disking of the stubble, and plowing later, also gave very good returns. The possible objection to early plowing (July 15th to August 15th) is the lack of labor and teams at this time. In this case the stubble may be disked early and plowed later when work is less pressing. Disking a stubble before plowing tends: (1) to conserve moisture, (2) to kill weeds, (3) to lessen the draft and cost of plowing the land, (4) to pulverize that portion of the seed-bed that eventually will be turned under, and (5) to aid in destroying the Hessian fly.

In the semi-arid districts of the United States the lister is often used in preparing the seed-bed for wheat. The lister leaves the bottom of the furrow in ridges, however, and should not be used year after year in the preparation of the soil.

Fertilizers for Wheat.—A detailed discussion of fertilizers has been given in a previous chapter. Two methods of supplying plant food to the wheat crop are: (1) by the application of barnyard manure, and (2) by the use of commercial fertilizers.

Where clover or grass is followed by corn in a rotation, better returns are obtained from manure when placed on the sod and plowed under for corn. For soils low in plant-food and humus, manure may be applied profitably to the wheat crop. Unless the ground is too rolling the manure

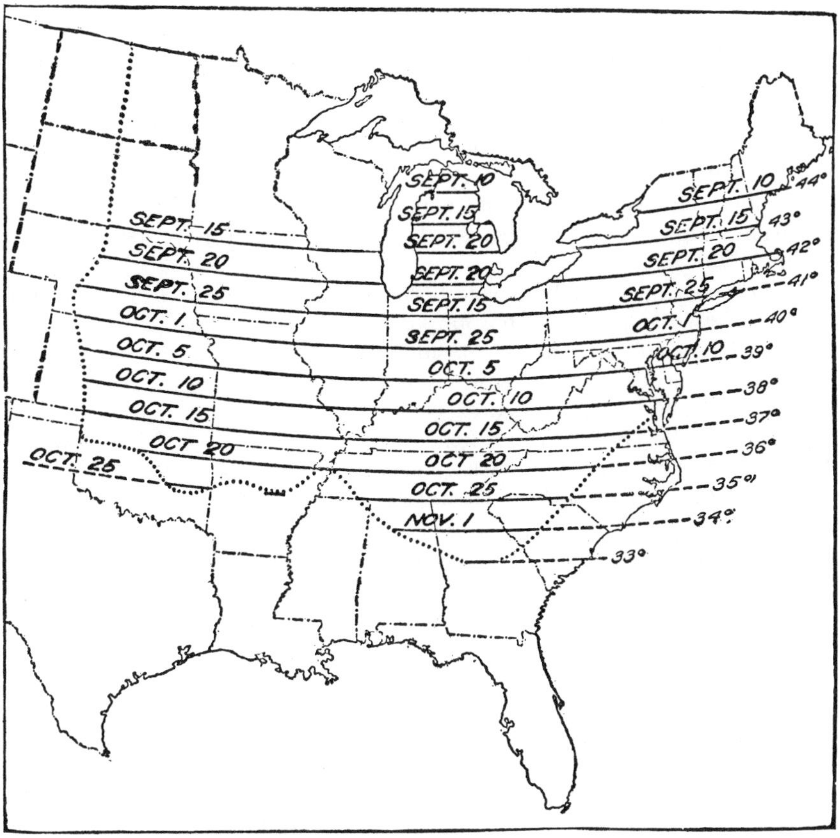

Approximate Date of Seeding Winter Wheat.[1]

should be applied to the wheat as a top dressing before seeding rather than plowed under, or it may be applied after seeding. Soluble plant-food from the manure will leach down into the soil and the strawy remains will act as a mulch during the winter.

The needs of the soil upon which the wheat crop is to be grown will determine the proportion of different plant-food elements to be used. The intelligent use of fertilizers for wheat calls for a knowledge of the

[1] Courtesy of U. S. Dept. of Agriculture.

needs of the soil. This may be ascertained partly by knowing the previous treatment of the soil and by studying the appearance of the crops now growing upon it. More definite information may be secured by the use of different fertilizing elements on small plats conducted as a test during one or more years for the purpose of ascertaining the needs of the soil.

Phosphorus is the element most often needed on ordinary wheat soils of most northern states, and is the one that usually gives the greatest increase in yields. In many localities the yield may be further increased by the addition of small to moderate amounts of potash. In many cases, some nitrogen will produce still further increase. However, it is poor policy to pay 18 cents a pound for nitrogen that can be produced more cheaply on the farm by the use of various leguminous crops in the rotation.

Time of Seeding.—The time to seed wheat in a given section will be determined largely by previous experience. The latitude, season, soil conditions and insect enemies all help determine the proper time for seeding.

The chart on preceding page prepared by the United States Department of Agriculture gives the approximate date of seeding winter wheat, where the Hessian fly must be considered as a factor.

Spring wheat should be sown as early as the ground can be prepared properly. Early seeding insures cool weather during the early growth and permits the crop to ripen before the severe storms of late summer. Wheat is generally seeded with a grain drill, although broadcasting is still practiced is some parts of the far West.

Rate of Seeding.—The rate of seeding varies greatly in different wheat districts of the United States. East of the Mississippi River two bushels of well-cleaned seed will generally give the best results. Results by the Ohio Experiment Station, located near the center of the humid region, teach a valuable lesson on this point.

THICK AND THIN SEEDING OF WHEAT. TEN DIFFERENT VARIETIES USED. SIXTEEN-YEAR AVERAGE.*

3 pecks per acre	20.26	bushels per acre
4 "	21.64	"
5 "	22.97	"
6 "	24.11	"
7 "	24.36	"
8 "	25.01	"
9 "	25.46	"
10 "	25.43	"

In the dry farming area of the West the amount of seed required ranges from two to three pecks in the driest sections to six or eight pecks in the more humid sections. The rate of seeding for any section should be determined by actual tests.

Wheat should not be covered too deeply. The depth of seeding will depend on the type of soil and the preparation of the seed-bed. The

*Taken from records of the Ohio Experiment Station.

usual depth of drilling is from two to three inches. To secure ideal condition for germination the seed should be placed in the drill furrow on firm, damp soil, which will supply moisture for rapid germination and the development of roots.

Grain Drills.—For general use a good single-disk drill does very good work. On stony, trashy land it does better work than double-disk or shoe drills. In the absence of trash and on a well-prepared seed-bed, the shoe drill is more readily regulated to a uniform depth of seeding. The press drills are preferred for use in light, droughty soils and drier climates.

Winter Killing.—Winter killing of wheat is a source of great loss throughout the winter wheat districts of the United States. Winter killing may be due to: (1) alternate freezing and thawing of wet soils, which gradually lifts the plants, exposing and breaking the roots; (2) weak plants, resulting from late sowing, lack of moisture or freezing in a dry, open winter; (3) smothering of the plants under a heavy covering of ice and sleet. A heavy growth of early seeded wheat is more apt to smother than that sown later. When unfavorable weather conditions exist, very little can be done to prevent winter killing. However, preventive measures such as the following are advised: (1) Grow a hardy variety of wheat; (2) drain wet spots in the wheat field; (3) thoroughly prepare the seed-bed; (4) sow seed early enough to secure strong, vigorous plants; (5) roll wheat that is pulled by freezing and thawing. Rolling early in the spring firms the soil about the roots and benefits the wheat if the pulling has not progressed too far.

Wheat Districts.—The United States may be divided into five wheat districts according to the color and composition of the grain. These districts are not sharply defined, but a brief outline of them should give the reader a better idea of the kind of wheat grown, the leading varieties and the milling qualities of the wheat in the different parts of the United States.

District No. 1.—All wheat east of the Mississippi River is known as Red Winter, or soft winter wheat. It varies in color from white to red and amber. The quality of this wheat varies from medium in the northern part to poor in the southern part of the district. The leading varieties in the northern portion are Fulcaster, Pool, Dawson's Golden Chaff, Gypsy, Harvest King, Fultz, Rudy and Michigan Amber. In the southern portion the leading varieties are Fulcaster, Pool, Purple Straw, Bluestone and Mediterranean.

District No. 2.—The hard spring wheat, including Durham, is located in the Dakotas, Minnesota and parts of Nebraska, Iowa and Wisconsin. The wheat in this district is small and shriveled in kernel, hard and dark in color. The milling quality of hard spring wheat is excellent. The principal varieties are Bluestem, Velvet Chaff, Fife and Durham (Kurbanka and Arnautha).

District No. 3.—The hard winter district includes Kansas, Nebraska, Oklahoma, Iowa and Missouri. The wheat in this district is red to amber in color. The grain is hard and flinty, but larger and plumper than the hard spring. The milling quality of the wheat is excellent, although the quality of the gluten is not as high as in the hard spring wheat. The principal varieties grown are the Turkey and Kharkof.

District No. 4.—White soft or Pacific Coast wheat, grown mostly in California, is soft and starchy, and yellow to red in color. The milling quality varies from fair to poor. For bread purposes this wheat must be blended with the hard wheats. The wheat in this district is classed as soft winter on the market. The leading varieties are White Australian, Sonora, Club, King's Early and Early Baart.

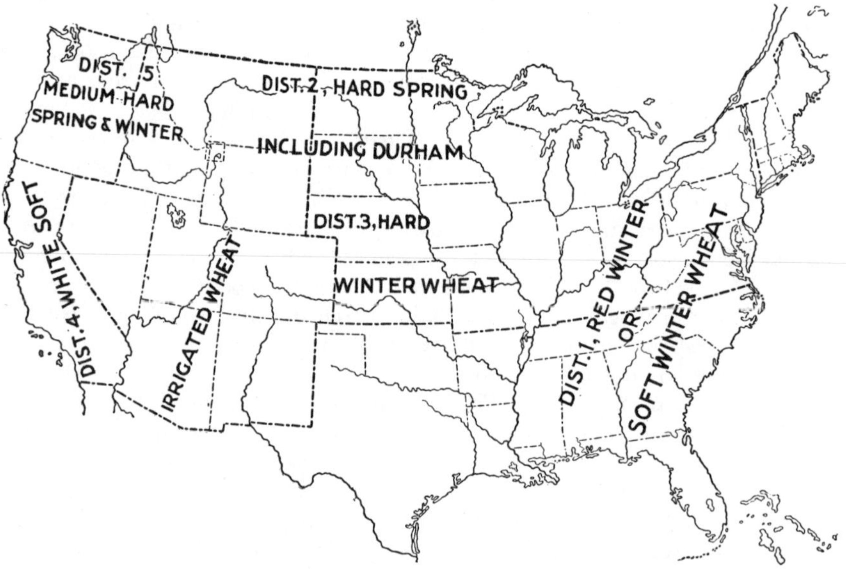

Wheat Districts of the United States.

District No. 5.—The medium hard spring and winter wheat is grown in the extreme Northwest, Washington, Oregon and Idaho. The wheat in this section is medium in quality, much of it having a bleached, dull appearance. The principal varieties of winter wheat are Forty Fold, Red Russian and Jones' Winter Fife. Bluestem is the leading spring variety.

Wheat Improvement.—Every wheat grower should ascertain, by test or otherwise, the variety best suited to his conditions. The variety tests at the nearest experiment station will generally indicate the best varieties for similar conditions. For a community located on soil different from that of the nearest experiment station, an ideal plan is to organize a community seed association. A variety test of wheat should be conducted on some central farm. After the best variety is determined, the farmers of that community will find it advantageous if all grow the same variety of wheat. The advantages of such a plan will be: (1) larger yields for all; (2) better and more uniform quality, resulting in higher prices; (3) the production of pure seed, true to name; and (4) the providing of a better opportunity to improve the variety. When farmers of a community are all interested in one variety of wheat, they will naturally be interested in its improvement.

The so-called "mass selection" will be found both practical and profitable in improving a variety of wheat. The procedure is as follows: A field of good wheat is examined at harvest time and enough of the choicest heads are selected to make a bushel or more of seed. This is threshed by hand and carefully stored until seeding time. This selected seed should be sown in a marked portion of the general wheat field. At harvest time choice heads are again hand selected from this special plat. The remaining wheat is harvested for seed to be multiplied for the general field wheat. By continuing this process of selection each year there will be a tendency to improve the variety continually, or at least to eliminate all danger of the wheat running out.

A PROFITABLE YIELD OF WHEAT.[1]

[1] Courtesy of Penn State Farmer, State College, Pa.

Harvesting.—Wheat is generally harvested as soon as ripe. The straw should be yellow in color and the grain in hard dough, before the wheat may be safely harvested. In the wheat-growing section along the Pacific Coast the wheat is allowed to stand a week or two after it is ripe, and is then harvested with a combined harvester and thresher.

Wheat should be shocked the same day it is cut. Considerable starch is transferred from the leaves and stems to the grain after the wheat is harvested. Immediate shocking of the grain prevents rapid drying and aids this action.

When not to be threshed from the shock, wheat should be hauled in and stacked or stored in the barn as soon as possible (a week or ten days). Hot sun bleaches wheat rapidly; rainy weather often damages and sometimes destroys the crop in the shock. In the eastern United States threshing generally takes place in the barn in the late fall. In the corn belt section and Great Plains region most of the wheat is threshed out of the shock or in the field by a combine.

Threshed wheat should be stored in tight, clean granaries. When it is to remain in storage for some time the granary should be cleaned thoroughly to make sure of the removal of grain moths, weevils and fungous diseases. If the granary is constructed so as to keep out vermin and insects, there is practically no loss of weight in storage.

Cost of Producing Wheat.—The fixed charges of growing an acre of wheat are about the same, whether the yield is 15 bushels or 30 bushels per acre. A rough estimate of the cost of growing wheat in the United States is between $10 and $12 per acre. The United States Department of Agriculture has secured from many farmers itemized estimates of the cost of producing wheat in all of the states. Those for a few of the widely separated states are as follows:

	Pennsylvania.	South Carolina.	North Dakota.	Illinois.	Kansas.
Plowing	$3.80	$1.46	$1.95	$2.01	$1.81
Seed	1.94	1.36	1.31	1.50	1.22
Planting	.60	.89	.44	.35	.41
Harvesting	1.79	1.23	1.03	1.19	1.49
Threshing	1.60	1.33	1.60	1.46	1.44
Rent	3.50	3.03	2.22	5.33	3.41
Fertilizer	2.83	2.66	.06	.27	.06
Miscellaneous	.62	.35	.38	.43	.45
Total	$16.68	$12.31	$8.99	$12.54	$10.29
Cost per bushel	.84	.96	.62	.64	.63
Net profit per acre	3.42	3.85	4.87	6.41	5.66
Number of reports	131	40	177	256	309

The estimated cost in Kansas was based on the reports of 309 farmers who, during the year 1909, secured an average yield of 16.3 bushels per acre. This is representative of Districts 2 and 3 that produce one-half of the wheat grown in the United States. The average acre yield in the United States is 14.8 bushels. It will be seen that there is little profit in raising less than 15 bushels to the acre.

Enemies of Wheat: Weeds, Insects and Fungous Diseases.—Weeds, common in wheat fields, are not, as a rule, difficult to eradicate. Weeds damage wheat by reducing the yield and by injuring the milling quality of the grain. The weeds most objectionable in wheat are garlic, cockle, cheat or chess, wild oats and wild mustard. These are usually controlled by proper cleaning of the seed wheat, by carefully preparing the seed-bed and by a suitable rotation of crops.

Insects.—The Hessian fly and chinch bug are probably the most destructive of wheat insects. The methods of control are preventive for the most part. The burning over of stubble land any time from harvest to the middle of August will destroy many of the Hessian flies and chinch bugs. The planting of trap crops also will aid in reducing Hessian fly trouble. A strip of wheat sown early in August will induce the fly to lay eggs. This wheat should be carefully plowed down after the first frost, so as to destroy the fly. Often an early strip of wheat may be plowed down in time for proper preparation and reseeding.

A stinging frost will kill the adult Hessian fly. If the season is not too backward it is well to delay seeding of wheat until this time. However, wheat should be seeded early enough to become rooted before winter sets in.

A patch of millet sown early in the spring will attract many of the chinch bugs, thus keeping them out of the wheat and corn.

The common insects of the granary are the granary weevil (*Colandra granaria*) and the Angoumois moth (*Sitotroga cerealella*). Both these insects multiply rapidly and should be attended to at once.

Used granaries should always be cleaned thoroughly before the new wheat is stored. Granaries should be repaired when needed so as to make the sides and floor as tight as possible.

Fumigation should be resorted to when insects first appear. Carbon bisulphide is a very effective chemical to use in a good tight granary. One and one-half pints to one ton of grain, or 1000 cu. ft. of space, is the recommended amount to use. The liquid should be poured into shallow pans and placed over the wheat. For the best results fumigation should be repeated in two weeks' time. Hydrocyanic acid gas is used in elevators and mills, but would be very dangerous in the ordinary barn where live-stock is housed.

Fungous Diseases.—Rust and smut are perhaps the most destructive among wheat diseases. There is no known remedy for rust other than the growing and breeding of rust-resistant varieties of wheat. Stinking smut may destroy as much as 10 per cent of the total wheat crop of the United States. It does not change the general appearance of the wheat head while in the field, but develops within the kernel as the wheat ripens. At threshing time the infected kernels may be broken, exposing a black, stinking, greasy mass of smut spores. The handling of smutty wheat aids in the infecting of all sound wheat that comes in contact with it. The smut spores adhere to the outside of the kernel until it is planted. The fungus grows within the wheat plant and finally takes possession of the newly formed berry. Stinking smut can be controlled by the formaldehyde treatment.

Treatment.—One pint of 40 per cent formaldehyde is added to 40 gallons of water. This is sufficient to treat 40 bushels of wheat. The wheat should be spread on a good tight floor and sprinkled with the solution. The wheat should then be shoveled over until the grain is well moistened, after which it should be shoveled into a pile or ridge and covered with canvas for several hours. The wheat should then be spread out on the floor to dry. The kernels will absorb water and become larger. If seeding takes place before the wheat is thoroughly dry, one-fifth to one-fourth more seed to the acre is sown than when untreated seed is used.

Loose smut is less injurious to wheat then the hidden or stinking smut, but is more difficult to treat and control. It destroys the head in the field, leaving the bare rachis as evidence of its presence. The mature spores are scattered by the wind. If they gain entrance to the growing berry in the head, they germinate and send mycelium into its tissues to await the time when the wheat is sown in the ground. The formalin treatment, which simply acts on the outside of the berry, is ineffective.

The hot-water treatment is recommended for the loose smut of wheat. This treatment requires careful and painstaking work, and is not practical for large quantities of seed. A small quantity of seed should be treated and sown in a separate plot to be used for seed purposes the following year, thus eliminating the smut.

Treatment.—The equipment required for the hot-water treatment is as follows: 3 large kettles, 1 tub, several wire baskets holding about 1 peck of grain, and 1 good thermometer. The seed wheat should be soaked several hours in cold water placed in tub. The water in kettle No. 1 is heated to 127° F., and in kettle No. 2 to 130° F. This can be done by heating water in the extra kettle and regulating to the required temperature the water in kettles No. 1 and 2. A wire basket should be filled with wheat from the tub of cold water, allowed to drain, and immersed in kettle No. 1 for two minutes. It should then be taken out and immersed in kettle No. 2 for ten minutes, after which the wheat should be spread out to dry. This treatment frequently kills a small percentage of the kernels, the amount of which should be determined so as to regulate the proper rate of seeding. A germination test is therefore advised before seeding.

REFERENCES

"Book of Wheat." Donalinger.
"Wheat." Ten Eyck.
Farmers' Bulletins, U. S. Dept. of Agriculture:
320. "Quality in Wheat."
534. "Durum Wheat."
596. "Winter Wheat Culture in Eastern States."
616. "Winter Wheat Varieties for Eastern States."
678. "Growing Hard Spring Wheat."
680. "Varieties of Hard Spring Wheat."

CHAPTER 13

OATS, BARLEY AND RYE

OATS

As a farm crop in North America, oats rank fifth in value. It has a short season of growth, is easily raised by extensive methods and brings quick returns. It is, therefore, a popular crop, especially with the tenant farmer. The yield and cash value per acre is low compared with the best oat-producing countries of Europe, and some question the advisability of continuing its cultivation so extensively in this country.

Oats fit into the general crop rotation and follow corn better than most other crops. In the North Central states it is extensively used as a crop in which to seed the clovers and grasses. It makes a desirable feed for all classes of livestock except swine, and is highly prized for horses. The straw is valuable as roughage and as an absorbent in stables and has considerable fertilizing value.

The average acreage, yield, production and value of oats in the United States for ten years ending 1914 is given in the following table:

Average Annual Acreage, Production and Farm Value and Mean Acre Yield of Oats in the Ten States of Largest Production for the Ten Years from 1905 to 1914.

	Area, acres.	Mean Yield per Acre, bushels.	Production, bushels.	Farm Value, December 1.
Iowa	4,581,000	31.9	146,618,000	$48,182,000
Illinois	4,160,000	31.2	130,096,000	46,920,000
Minnesota	2,697,000	30.8	84,739,000	27,526,000
Wisconsin	2,337,000	32.5	73,386,000	29,202,000
Nebraska	2,373,000	25.3	59,384,000	19,938,000
Ohio	1,636,000	32.4	53,581,000	20,881,000
Indiana	1,719,000	29.0	49,887,000	18,018,000
North Dakota	1,737,000	27.7	48,233,000	15,233,000
Michigan	1,424,000	30.8	43,704,000	17,327,000
New York	1,268,000	31.5	39,973,000	18,761,000

Soil and Climatic Adaptation.—In the production of oats, favorable climate and cultural conditions are more important than the character and fertility of the soil. They do best in a cool, moist climate. In North America oats succeed best in Canada and those states of the Union lying next to the Canadian border. The acreage of spring oats below 38 degrees north latitude is very small. Oats require an abundance of water and loam, and clay loam soils are generally best adapted to them.

Classes and Varieties.—Oats are divided into spring and winter oats. By far the larger proportion in North America belongs to the former class. Spring oats are divided into two classes, namely, those having open panicles and those with closed panicles. By far the larger number of varieties falls into the first class. They are further classified by color into white, yellow, black, red and shades of black and red. They are also divided according to time of maturity into early, medium and late varieties. The time for maturity ranges from 90 days to 140 days. In the Central states in favorable seasons early oats should ripen in 90 days from time of seeding.

The accompanying map shows the three oat districts of the United States.

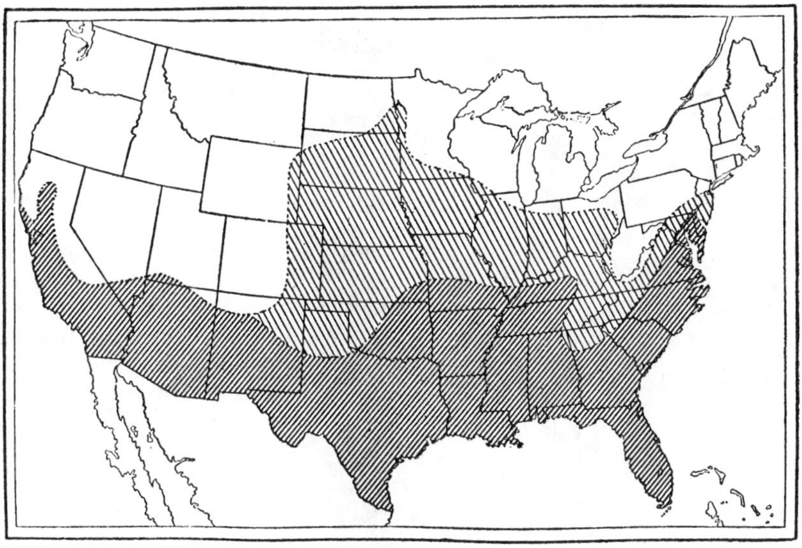

Map of the United States, Showing Approximately the Areas to which Certain Types of Oats are Adapted.[1]

In the unshaded portion rather late maturing, large-grained white oats are usually best; in the lightly shaded portion early, small-grained, yellow varieties are most important; while in the heavily shaded portion brownish-red or gray varieties, which in the warmer sections are sown in the fall, are most certain to succeed.

In the northern district the medium-maturing and late-maturing varieties generally give best results. The leading varieties in this district are American Banner, Big 4, Clydesdale, Lincoln, Probstier, Siberian, Silver Mine, Swedish Select, Tartarian, Wide Awake and White Russian.

In the central region the principal varieties are Big 4, Burt, Clydesdale, Kherson, Lincoln, Red Rust Proof, 60-Day, Silver Mine, Siberian and Swedish Select. In the southern district the chief varieties are Burt

[1] Courtesy of U. S. Dept. of Agriculture. Farmers' Bulletin 424.

and Red Rust Proof, together with Winter Turf, which is a strictly winter variety. Burt and Red Rust Proof may be seeded either in the winter or spring.

Seed Oats and Their Preparation for Seeding.—It is important to seed only varieties that are adapted to the conditions that prevail, giving particular attention to time of maturity as related to the prevailing climatic conditions during the oat-growing period. Seed oats should be thoroughly cleaned by the use of a good fanning mill before seeding. The screens of the mill and the blast of air should be such as to remove all foreign seed, hulls, trash and light and small oats. Frequently one-quarter or one-third of the oats may be removed in this way. Such thorough cleaning makes for a uniform stand of vigorous plant in the field.

Two Types of Oat Heads.[1]
Spreading, or panicled, oats (on the left); side, or horsemane, oats (on the right).

If there is any trouble from smut, seed should be treated with formaldehyde; one pound of 40 per cent formaldehyde to 45 gallons of water. This is sufficient for treating about 45 bushels of oats. The solution must be brought in contact with every berry in order to be thoroughly effective. The oats may be

[1] From Farmers' Bulletin 424, U. S. Dept. of Agriculture.

spread out in a thin layer on a clean floor and the solution applied with a sprinkling can. Several thin layers of oats may be placed one on top of another, and each sprinkled in this way, after which the whole pile should be thoroughly stirred, shoveled into a compact heap, covered with a wet blanket and allowed to remain for twelve hours. The blanket should then be removed and the oats spread out and occasionally stirred until thoroughly dry.

Preparation of the Seed-Bed.—A large portion of the oats grown in the corn belt are seeded on corn ground without any preparation. The ground is disked and harrowed, or sometimes cultivated once or twice after seeding the oats. It is much better to double disk and harrow once before seeding. The better preparation in this way will usually more than pay for the increased expense. In some localities shallow plowing for oats may prove to be the best method of preparing the seed-bed. When seeded on corn land the stalks should be broken down. This is most easily accomplished by dragging a heavy pole or iron rail broadside across the field on a frosty morning when the ground is frozen. A mellow, loose surface soil with a firm subsoil is best for oats. This character of seed-bed is secured on corn land by the methods above described.

Fertilizers and Manures for Oats.—Over most of the spring oat region oats are grown without the direct application of either manure or fertilizers. When soils call for manure or fertilizers it is best to apply them to the crop preceding oats. In this way the oats receive only the residual effect, but this generally meets the needs of the crop. This avoids the danger of too rank a growth of straw that is likely to cause oats to lodge. Oats that lodge badly are not only difficult to harvest, but generally cause a failure of grass and clover seeded with them and give rise to a reduced yield of grain. On soil that is in a low state of fertility, or which receives no manure or fertilizer for the preceding crop, rather light applications of either manure or a complete fertilizer may be applied for oats with profit. Experiments show that phosphorus is the most important ingredient to be applied. Some nitrogen, preferably in an immediately available form, is generally advisable. Nitrate of soda at the rate of 75 to 100 pounds per acre will generally fully meet the needs for nitrogen.

The fertility removed by oats is given in Table VII in the appendix.

Time, Rate and Manner of Seeding.—The time of seeding will vary with the season and locality, but generally should be as early in the spring as soil conditions will permit the preparation of the seed-bed. Throughout a considerable part of the oat region, oats are seeded during April. Those seeded during the first half of this month are found to give larger yields than those seeded during the last half. In the southern part of the district, seeding in March usually gives good results, and in the Southern states seeding may take place much earlier. Oats do best if they can make the major portion of their growth during the cool part of the season. They are often injured by a short hot spell as they near maturity. Frosts or even hard freezes after they are seeded seldom do injury, although prolonged wet weather immediately following seeding may cause the seed to rot in the soil and reduce the stand. A few farmers in the Northern states are now seeding ordinary spring oats in the late fall or early winter, so timing the seeding that the oats will not germinate until spring. This method is still in the experimental stage, and farmers should try it only on a limited scale and in an experimental way until it is demonstrated to be satisfactory.

The rate of seeding depends on the character and condition of the soil, the fertility of the soil, the quality of the seed, the size of the grains and the manner of seeding. Fertile soils require less seed than poor ones, because the plants tiller more. Abundance of seed should be used on weedy land, and seed broadcasted should be used more freely than when it is drilled. It will require more seed of the large-grained than of the small-grained oats.

In general, the rate of seeding ranges from 8 to 12 pecks per acre, the smaller amount being used when drilled and the larger amount when broadcasted. At several state experiment stations drilled oats have yielded three to five bushels per acre more than oats broadcasted under identical conditions. Where satisfactory results have not been secured by drilling oats, it has usually been attributable to covering the seed too deeply. Under average conditions oats should be covered from 1 to $1\frac{1}{2}$ inches in depth, although in very loose soils or in a dry seed-bed, deeper covering will be satisfactory.

Oats as a Nurse Crop.—Oats are frequently used as a nurse crop for clovers and grasses. When used in this way early varieties and rather thin seeding is advisable. This encourages a good catch and stand of the clovers and grasses.

Harvesting, Shocking and Threshing.—Oats should be harvested when the grain is in the hard dough stage. If allowed to become fully ripe, the grain shatters badly in the process of harvesting, thus causing considerable loss. When cut early the straw will have a higher feeding value than when allowed to fully mature. Any shrinkage in grain that may take place as a result of cutting early will be more than offset by the increased value of straw. In regions where the straw is not utilized (and there should be no such regions) the harvesting should be carefully timed in order to secure the largest possible yield and yet avoid loss of grain.

When cut rather green or when the straw is damp, or when the oats are foul with weeds and grass, the harvester should be set for small bundles. When harvested in a thoroughly ripe condition the bundles may be placed into shocks immediately. When damp or green it is generally best to allow the bundles to lie several hours before shocking. When the oats are in a good, dry condition round shocks with a cap sheaf are to be preferred. Twelve bundles to each shock, exclusive of the cap sheaf, is the best number. The cap sheaf should be broken near the band and the heads placed toward the direction of the prevailing wind. Where wind storms are very prevalent at this time of the year it is best not to use cap sheaves. When oats are green or damp, long shocks, made by standing the sheaves in pairs and extending north and south, are to be preferred. It pays to have the grain properly shocked, even though it is to be threshed in a short time and directly from the field.

If the grain is to be stacked, stacks should be well built. Rails or old straw should be used for the foundation to prevent damage to the first layer of sheaves. Stacks may be either round or long. The butts should

A Field of Good Oats being Harvested with a Modern Self-Binder.

always be laid toward the outside of the stack, and the outside layer should always slope downward so that the stack will turn rain. The greatest diameter of the stack at the time of construction should be five or six feet above the ground. This form in settling accentuates the sloping of the outside sheaves in the upper portion of the stack. Whether oats are to be threshed from the field or stack will be determined largely by the threshing custom of the locality. Where the custom of threshing from the field prevails, it will be difficult to get stacked oats threshed until field threshing is completed. Stacking entails some additional work, but generally improves the quality of the oats. Oats in the shock are often badly damaged and suffer great loss from rains. This is largely obviated by stacking as soon as in proper condition.

In threshing, the concaves of the machine should be so adjusted that all grain will be separated from the straw, but the adjustment should be such as not to cause serious hulling of the berries or undue cutting of the straw. Oats are easily threshed when in a dry condition. The straw at threshing time should be either carefully stacked or run directly into a hay-loft or storage-shed, depending on facilities.

Storing and Marketing.—The threshed grain should be dry when put in bins and should be kept dry by adequate protection from rains or absorption of moisture from any source. Mustiness lowers the feeding value and endangers the health of animals. It also lowers the market value of the grain. Where grain weevils and other insects seriously affect stored grain, tight bins which can be fumigated are advisable. Under favorable conditions oats may be stored for a considerable time with very little shrinkage and loss. The highest market price generally prevails during the early part of the year and just prior to the oat harvest.

Composition and Feeding Value.—A large portion of the oats grown in America are fed to livestock. Limited quantities are used for the manufacture of prepared cereals. Oats are high in protein and are well adapted for work horses and growing animals. They are especially desirable because of the hulls which they contain, and which dilute the concentrate to about the right extent for healthy digestion. They are generally fed whole, although not infrequently are they chopped and mixed with other grains. An average of thirty analyses of oats gives 13.3 per cent protein, 5.6 per cent fat and 67.1 per cent carbohydrates, as compared with 11.8, 6.1 and 78.1 per cent for those respective items in corn.

Value of Oats for Hay and Soiling Purposes.—If cut when the grain is in the milk, oats make a palatable and nutritious hay, especially well suited for horses. Oats seeded with Canada peas make a good hay for milch cows and other cattle. This mixture is also well suited for soiling purposes and provides an early soiling crop. By seeding at different times the season of available soiling crops from this source may be considerably prolonged. A common rate of seeding this mixture is 1 bushel of peas to $1\frac{1}{2}$ bushels of oats. This mixture also makes good pasture for stock of all kinds.

Oat Straw and its Utilization.—Oat straw has a higher feeding value and is more palatable than straw from the other grains. It is quite generally used for feeding horses during the winter, and as a maintenance roughage for cattle and sheep. Its feeding value and palatability are best when the grain is harvested fairly early and the straw is secured without damage by rains. It pays to store it carefully and utilize it for feed as fully as possible. The refuse portion makes a valuable bedding and the straw has a fertilizer value of about $3 per ton.

Cost of Producing Oats.—The Bureau of Statistics of the United States Department of Agriculture secured estimates from about 5000 farmers in all parts of the country on the cost of producing oats in 1909. The estimates show an average cost of $10.91 an acre, or 31 cents a bushel. On the same farms for that year the average value of the oat crop was $14.08 an acre, or 40 cents a bushel. The average net return from grain was estimated at $3.17 an acre, to which was added the value of by-product to the amount of $1.42, making an average total profit of $4.59 per acre.

Oat Improvement.—The improvement of this crop has received much less attention from plant breeders and farmers than has corn and wheat. There are, however, many varieties of oats, most of which have originated through selection and breeding. It is important for the farmer to secure a variety well suited to his local conditions, and to improve that variety by thorough cleaning and grading of seed. There are opportunities, however, for improvement by selecting exceptional stools of oats and threshing these by hand and planting each in a separate row. These should be harvested separately and the best ones retained, threshed and used for seeding longer rows the following year. In this way new strains are frequently secured that are superior to the general crop.

BARLEY

The world's production of barley is about 1,500,000,000 bushels, of which North America produces one-seventh. Of this the United States produces 166,000,000 and Canada 48,000,000 bushels. In the United States, California, Minnesota, Wisconsin, North and South Dakota lead in barley production. These five states produce 73 per cent of all the barley grown in the United States.

Soil and Climatic Adaptation.—Barley is adapted to a wide range of climatic conditions, but it does best in the North Temperate Zone. It is somewhat more exacting in its soil requirements than either wheat or oats. It does best on a well-drained loam that is well supplied with organic matter. It is quite resistant on alkali soils, and is, therefore, adapted to such soils in the irrigated districts.

Classes and Varieties.—Barley is divided into two-rowed and six-rowed forms, depending on the character of the spike or head. In the United States the six-rowed form predominates. Manchuria and Oderbrucken are the leading varieties of this type. It is also divided into spring and winter, and bearded and beardless types. The bearded spring varieties prevail.

Preparation of Land and Seeding.—Barley demands a well-prepared seed-bed, and should be seeded in the spring as soon as all danger of freezing is past. Best results are secured by drilling at the rate of six to eight pecks per acre. Broadcasting the seed usually gives much lower yields than drilling.

Harvesting and Use.—Barley is harvested in the same manner as oats. It should be shocked in round shocks with cap sheaves, and in threshing the cap sheaves are usually threshed separately in order to secure as large a proportion as possible of unstained grain. Barley that is discolored by rains commands a much lower price than bright, unstained grain.

More than half of the barley produced in North America finds its way into the market, and much of it is used in the manufacture of malt. Malt is largely used in the production of beer and other malt liquors. Barley for this purpose should be clean and bright in color, and should

A FIELD OF WINTER BARLEY SEEDED AFTER CORN, ANNE ARUNDEL COUNTY, MD.[1]

be free from foreign seeds and broken grains, and possess a high germinating power.

Use of By-Products.—Straw from barley is less palatable than that of oats or beardless wheat, and is also somewhat less nutritious. It makes excellent bedding, although the beards are more or less irritating to both man and beast.

RYE

Rye is of minor importance both in the United States and Canada. Pennsylvania, Wisconsin, Michigan, Minnesota and New York produce 64 per cent of that grown in the United States, while Ontario produces the most in Canada.

[1] From Farmers' Bulletin 518, U. S. Dept. of Agriculture

Adaptation and Culture.—Rye will grow on rather poor soil, and is most extensively grown in districts in the temperate zone where the soils are low in fertility. It is more hardy than wheat, and this is one of the principal reasons for growing it. The time of seeding and cultural methods are the same as those for wheat, although there is a somewhat wider range in the time of seeding. It may be seeded late in the summer and pastured so as to prevent heading during the autumn. It is quite extensively used as a cover crop and for green manure. Its hardiness and adaptation on poor soils make it especially valuable for these purposes in the temperate zone.

Rye is frequently broadcasted, although it gives better results when seeded with a drill. A well-prepared seed-bed is essential to a good stand of plants. Five to six pecks of seed per acre are required.

Uses of Rye.—Rye is frequently used as a soiling crop and occasionally cut for hay. When used for hay, it should be cut just before the heads are out. If not cut early, the straw hardens and makes a tough, unpalatable hay. A large part of the grain of rye in America is used in the manufacture of alcohol and alcoholic beverages. The grain is excellent for feeding stock, but it gives best results when used in small quantities and combined with other grains. It is best suited for hogs, horses and poultry. The grain, being very hard, generally gives best results when coarsely ground.

REFERENCES

"Small Grains." Carleton.
"Field Crops." Livingston.
Farmers' Bulletins, U. S. Dept. of Agriculture:
395. "Sixty-day and Kherson Oats."
420. "Oats: Distribution and Uses."
424. "Oats: Growing the Crop."
427. "Barley Culture in the Southern States."
436. "Winter Oats for the South."
443. "Barley: Growing the Crop."
518. "Winter Barley."

CHAPTER 14

BUCKWHEAT, RICE, FLAX, EMMER, KAFFIR CORN AND SUNFLOWER

BUCKWHEAT

Buckwheat is a minor crop in most parts of America. It can be considered a staple crop only in New York and Pennsylvania. For a number of years its acreage has remained about stationary. The entire area devoted to it in the United States is about 800,000 acres. New York and Pennsylvania produce about 77 per cent of the total production.

It is often spoken of as the "lazy man's crop." It lends itself well to the farmer who lacks capital. It brings quick returns and finds a ready market at fair prices. It is the only grain for which a farmer can buy fertilizer on a ninety-day note and pay for it out of the crop.

Soil and Climatic Adaptation.—Buckwheat does best in a moist, cool climatc and at high altitudes. High temperatures during the period of seed formation, accompanied by hot sunshine followed by showers, is generally disastrous to the crop. Buckwheat will mature a crop of grain in eight to ten weeks under favorable conditions.

Buckwheat is adapted to a wide range of soils, but does best on well-drained soils that are rather light in texture. It succeeds on poor soils and is most extensively grown in those regions where the soils are of rather low fertility.

Varieties.—The varieties common to the United States are Japanese, Silver Hull and Common Grey. The Silver Hull is slightly smaller than the Common Grey. The seed is also smaller, plumper and lighter in color than the Japanese. If there is no objection to mixing varieties it is thought larger yields can be secured by mixing the large and small growing varieties, which affords a better distribution of the seed heads in the field.

Preparation of Soil and Seeding.—Early plowing of the land in order to permit harrowing at intervals of two weeks and a thorough settling of the soil before seeding time, is advised. If early plowing is not possible, greater attention should be given to a thorough fitting of the seed-bed immediately following plowing.

The amount of seed per acre varies from three to five pecks, depending on manner of seeding, character of seed and condition of seed-bed. It may be seeded either with the grain drill or broadcasted and harrowed in. When drilled a smaller amount of seed will prove satisfactory, but the distribution of plants secured by broadcasting is preferable to that secured by drilling, unless the drill hoes are close together. The later buckwheat is sown so as to get ripe before frost, the better the yield will be. It is seldom advisable to seed earlier than the last week in June, and in some localities it may be seeded as late as the second week in July.

Fertilizers and Rotations.—Buckwheat seeded on poor land responds well to a moderate dressing of low-grade fertilizer. On heavy soils where it is desired to grow potatoes, buckwheat is recommended as a good crop to precede potatoes. The following rotation is recommended for such soils: clover, buckwheat, potatoes, oats or wheat seeded with clover. With this arrangement the first crop of clover is harvested early and the land immediately plowed and seeded to buckwheat. This gives two crops during the season preceding potatoes, and leaves the land in excellent condition for potatoes.

Harvesting and Threshing.—The harvesting of buckwheat should be delayed until the approach of cold weather, because the plants continue to bloom and produce seed until killed by frost. The self-rake reaper is well adapted to cutting buckwheat. The machine used should leave the buckwheat in compact gavels with as little shattering as possible. The self-binder is sometimes used, being set to deliver small bundles loosely bound. However it may be harvested, it should be set upright in the field so as to prevent the grain lying on the ground. It is customary to haul the grain directly from the field to the threshing machine, as it is likely to mould when placed in stacks.

In threshing by machinery, neither the crop nor the day need be especially dry. The spiked concave of the thresher is generally replaced with a smooth one or a suitable plank. This avoids serious cracking of the grain and unnecessary breaking of the straw.

Buckwheat weighs 48 pounds to the bushel, and 35 bushels per acre is considered a good yield, while 25 bushels is satisfactory. The average yield of buckwheat in the United States is 18 to 19 bushels per acre.

Uses of Buckwheat.—Buckwheat is used chiefly in the manufacture of pancake flour. In some sections, and especially when the market price is low, it is used quite extensively for feeding livestock. It is an excellent poultry feed. The straw, being coarse and stiff, is of little value except for bedding or to make manure.

In some localities buckwheat is used as a green manuring crop. It serves well for this purpose because it grows quickly, may occupy the land after an early crop is removed, and leaves the soil in a loose condition. The seed being comparatively inexpensive and requiring only a moderate amount, makes it inexpensive from the standpoint of seeding. It is frequently used as a catch crop, being seeded in fields where other crops fail from whatever cause.

Buckwheat is an excellent bee feed. It blossoms for a considerable period of time and affords an abundance of nectar which makes honey of good quality.

RICE

Rice is unique in its culture, because it depends upon irrigation. It is one of the oldest cereals, and is also one of the greatest food crops, being a staple article of diet for millions of people in India, China and Japan. The world's annual production is approximately 175,000,000,000 pounds of cleaned rice, the greater portion of which is grown in India, China and Japan. As an article of food in the United States it is of minor importance, and yet the production in this country falls short of the consumption by about 200,000,000 pounds annually.

Soil and Climatic Adaptation.—Rice is adapted to a moist, warm climate, and its production in the United States is confined to the South Atlantic and Gulf Coast states. The bulk of the crop is now produced in Texas, Louisiana and Arkansas. Prior to 1890 it was produced mostly in the Carolinas and Georgia.

Since the lowland forms which constitute the principal source of the crop require irrigation, it demands a level soil with a compact subsoil that will prevent rapid downward movement of water. Such soils are found along the bottom lands of the rivers and on the level prairies of Texas and Louisiana.

Preparation of Land and Seeding.—The land is usually plowed in the spring and disked and harrowed to provide a good seed-bed. The rice is seeded at the rate of one or two bushels per acre with a seed drill, usually from April 15th to May 15th. Unless water is needed to germinate the seed the land is not flooded until the plants are six to eight inches high. If the soil is too dry the land may be flooded immediately after seeding for a few days to sprout the seed, after which the water is removed until the plants are six to eight inches high.

Weeds are often a serious menace to rice culture. Such weeds may be brought on rice fields in the irrigation water or may find their way there in the seed rice. Red rice is a serious pest, and seed should not be used in which it occurs. The presence of red rice in milled rice lowers its grade and reduces its price. Red rice, being stronger, hardier and more persistent than white rice, soon gets a foothold in the fields unless precautions are taken to prevent it.

Fertilizers are seldom used in the production of rice, because the practice of irrigation brings to the land some fertility in the water. This is especially true when the water is not clear. Furthermore, rice lands, being either river bottom land or prairie land, are generally very fertile. In the course of time, however, if rice is grown continuously, fertilizers will be needed.

Flooding or Irrigation.—Water is let into the rice field to a depth of three to six inches, and is maintained at this depth until the crop is nearly mature. Water of a rather high and uniform temperature is preferred. Cold water from mountain streams is undesirable. The water is constantly renewed to prevent it from becoming stagnant. This necessitates a slow movement of water across the rice field, and for this reason it is not advisable to have the fields too large. Irrigation necessitates the land being practically level and surrounded by dikes.

There should be good facilities for draining, since land must be in good condition when prepared for seeding and should be fairly dry at the time of harvesting.

Harvesting and Threshing.—It requires from four to six months to mature a crop of rice and the date of harvesting in the United States varies from August to October, depending on time of seeding, character of season and variety of rice. The crop should be harvested when the grain is in the stiff dough stage and the straw somewhat green. The ordinary grain binder is used for harvesting the crop, and the methods of shocking, stacking and threshing are very similar to those used in wheat production.

Yields and Value.—Rough rice weighs 45 pounds to the bushel. It is generally put into barrels of 162 pounds each, and the yield is spoken of in barrels, and ranges from 8 to 30 barrels per acre; 12 barrels is considered a good yield. The hulls or chaff constitute 12 to 25 per cent of the weight of the rice, depending on variety and condition. In 1910 the total crop in the United States was valued at $16,000,000, or about $20 per acre. The rice is prepared in mills which remove the husk and cuticle and polish the surface of the grain. In this condition it is placed upon the market.

FLAX

Flax is grown in Canada and in a few of the Northern states. Nearly nine-tenths of the flax of the United States is grown in North and South Dakota and in Minnesota.

Soil and Climate Adaptation.—Flax grows best in a cool climate and on soils that are not too heavy. Sandy loams are better adapted to the crop than clay loams or heavy clays. It is extensively grown on virgin prairie soil, and is well adapted for seeding on the rather tough prairie sod when plowed for the first time. The roots of flax develop extensively near the surface of the soil. It is often considered an exhaustive crop, but the actual removal of plant-food constitutents is less than in most other farm crops Its shallow, sparse root system and the small amount of stubble usually left in the field probably explain why it is considered exhaustive.

Preparation of Land and Seeding.—Where grown on virgin prairie land, the sod should be broken about four inches deep and completely inverted in order to make a smooth surface for seeding the flax. On newly plowed land flax is seeded broadcast at the rate of one-half bushel per acre, and covered by harrowing. It is thought better to fall-break sod, and to provide a better prepared seed-bed the following spring by thorough disking and harrowing. In this process the sod should not be loosened from its place, and the roller is frequently used to compact the seed-bed and keep it smooth and also level to facilitate the covering of the seed at a uniform depth.

Where flax is grown on old land it follows corn to good advantage, and the seed-bed may be prepared by disking and harrowing in a manner similar to preparing the land for oats. In recent years a seed drill has been used for seeding flax with good results. The seed should be covered from one-half inch to an inch deep.

Thin seeding encourages the branching of the plants and within reasonable limits encourages large yields of seed. On land foul with weeds it is better, however, to seed somewhat thicker to prevent weed development.

A FIELD OF FLAX IN BLOOM.[1]

When flax is grown chiefly for the fiber one and a half to two bushels of seed per acre are used.

Harvesting and Threshing.—Flax may be harvested either with the self-rake reaper or self-binder. When harvested with the reaper the gavels should be rolled and set upright. The heads become entangled in such a way as to hold the rolled gavels together. The straw is frequently so short that it is necessary to cut as close to the ground as possible, and this calls for a level seed-bed that will facilitate close cutting with machinery. When cut with the binder the bundles should be set in small, loose shocks to facilitate drying. The highest quality of seed for market demands threshing from the shock as soon as it can be safely done.

Threshing is done with the ordinary threshing machine and necessi-

[1] Courtesy of Webb Publishing Company, St. Paul, Minn. From "Field Crops," by Wilson and Warburton.

tates having the concaves set fairly close in order to separate all the seed from the straw. The seed is small and flat and is but little broken in the process of threshing.

The threshed seed is generally placed in strong, closely woven bags and securely tied. The seed, being small, flat and exceedingly smooth, will run almost like water, and requires exceedingly tight bins for its storage and very tight wagon boxes in case it is to be hauled unbagged.

Yield and Value of Crop.—The yield of flax seed ranges from 8 to 20 bushels per acre. Since most of the flax is produced by extensive methods and on new land, the average yield for the United States is about 9 bushels. The price generally ranges from $1 to $1.50 per bushel. During the last few years a scarcity of flax has caused a somewhat higher price. A bushel of flax will produce about twenty pounds of crude linseed oil, and the oil cake after the removal of the oil is worth from 1 to 1½ cents per pound. The average annual production in the United States for ten years ending 1911 was about 24,000,000 bushels, valued at approximately $28,000,000.

Utilization.—Flax is grown chiefly for its seed, from which is made linseed oil, extensively used in the manufacture of paints. The meal, after the extraction of the oil, finds a ready sale as a nitrogenous stock food, and is extensively used as a concentrate for dairy cows.

The straw is utilized in only a limited way. It makes fair roughage for stock, although not as valuable as oat straw. In some localities the straw is used in the manufacture of tow, which is used in making rough cordage and twine.

In the old world the plant is extensively used for the manufacture of fiber. This necessitates pulling the plants by hand and requires special facilities for treating the straw and separating the fiber. Labor is too expensive in this country to enable American flax to compete with that of the old world in this respect. Ground flax seed in small amounts is a splendid feed for all kinds of stock. It acts as a tonic and has a good effect upon the digestive system.

Diseases of Flax.—Flax is so seriously troubled with a disease known as flax wilt that it necessitates the use of treated seed selected from wilt-resistant plants. The formalin treatment described for wheat serves equally well for the treatment of flax seed. Flax seed will require only about one-half gallon of the solution to each bushel of seed. It should be thoroughly stirred after sprinkling, covered with canvas treated with formalin, and allowed to remain two or three hours and then stirred and dried. After thoroughly dry it may be placed in bags which have been treated with formalin to prevent the presence of wilt spores.

Since this disease may live in the soil for several years in the absence of flax, it is necessary to practice long rotations in which flax will not be grown more frequently than once in five to seven years.

KAFFIR CORN

Kaffir corn is a non-saccharine sorghum. The sorghums are generally divided into three classes: (1) those cultivated chiefly for grain, of which Kaffir, milo and dura are the best types; (2) those cultivated for the manufacture of brooms; and (3) those grown chiefly for the production of syrup.

Regions of Production.—Kaffir corn, milo and dura are grown chiefly between the 98th meridian and the Rocky Mountains, and south of 40 degrees north latitude. This crop is drought resistant and adapted especially to the dry conditions of the Great Plains region.

HEADS OF FOUR VARIETIES OF KAFFIR.[1]

A—White Kaffir; B—Guinea Kaffir (Guinea corn of the West Indies); C—Blackhull Kaffir; D—Red Kaffir. (About one-fifth natural size.)

Value and Uses.—Kaffir corn is used chiefly as a source of stock food. The grain is similar in composition to ordinary corn, and has about the same feeding value. In composition there is very little difference between the stover of corn and Kaffir corn. Any surplus of the grain finds a ready market, and is in much demand for poultry feed. The grain may be fed either whole or crushed. It is somewhat softer than the grain of corn and the kernels, being smaller, can be used for poultry without crushing. It makes excellent feed for horses, cattle and swine.

Varieties.—There are many varieties in each of the three classes of non-saccharine sorghums. The Kaffir corn proper has erect, compact seed heads and the foliage is more leafy than that of milo. The seed heads of the latter are usually pendant, the stalks are less leafy and the plant is generally earlier in maturity. It is, therefore, adapted to the northern portion of the Kaffir corn region, and to those localities where seed production is most important.

[1] From Farmers' Bulletin 686, U. S. Dept. of Agriculture.

Production and Harvesting.—The preparation of the land, the planting and the cultivation of Kaffir corn are similar to those required for corn under the same conditions. The seed should be drilled in rows sufficiently far apart to facilitate cultivation with two-horse cultivators, usually 3½ feet apart. The seed is drilled at such a rate that the plants in the row will stand from 4 to 6 inches apart. For small growing varieties plants may be closer than in case of the larger varieties. Planting should not take place until the soil is quite warm. It is usually best to plant about ten days later than the best time for planting field corn. It is advisable to have a well-prepared seed-bed free from weeds. The plants as they first appear are small and make slow growth.

The crop may be harvested by cutting the whole plant and placing in small shocks, or the seed heads may be removed and stored in narrow, well-ventilated cribs. After removing the seed heads the stalks may be cut and shocked or they may be pastured as they stand in the field. In some localities the whole plant is cut and put in the silo in the same manner as making ensilage of field corn. The yield of grain is fully as large as that of field corn grain under similar conditions, and the drought-resistance of the crop makes it more certain than corn. Fifty bushels per acre is considered a good yield. The seed is separated from the head by means of a threshing machine. The weight of threshed grain per bushel is 56 pounds.

EMMER.[1]
A good substitute for oats and barley.

EMMER

Emmer, also known as spelt, is closely related to wheat, but is distinguished from it by the grain, which remains enclosed in the glumes when threshed. There are both spring and winter varieties. The spring varieties are most extensively grown in the northern portion of the Great Plains region. The crop is characterized by its ability to make a satisfactory growth on almost any kind of soil. All of the varieties are drought resistant, and the winter varieties are fairly hardy. It is not attacked by rusts and smuts to the same extent as wheat and oats.

[1] From Farmers' Bulletin 466, U. S. Dept. of Agriculture.

It stands up well in the field and is little damaged by wet weather at harvest time.

The methods used in the seeding of other spring grains will apply to emmer. The seed should be drilled at the rate of about two bushels per acre. It is important to sow early. The grain will stand a great deal of spring frosts.

Emmer is well adapted to the feeding of stock, and will easily take the place of oats, barley or rye.

A comparative test of emmer as compared with other spring grains covering a period of eight years at the North Dakota Experiment Station shows comparatively little difference in the yield of grain from the several crops. Oats led with 1969 pounds per acre, while emmer was second with 1945 pounds to the acre. The lowest yield, 1711 pounds per acre, was from wheat.

While this crop is especially adapted to the semi-arid conditions of the Northwest, it is suggested that it might prove a profitable substitute for oats in those portions of the Central, Southern and Eastern states where oats prove unsatisfactory.

SUNFLOWERS

Sunflowers are a native of America, and are widely but not extensively grown. The leaves and heads of the plant make good fodder for horses and cattle. The seeds are used for bird and poultry food and also for the manufacture of oil. Sunflowers succeed best on rather fertile soil and with warm climatic conditions. The requirements are similar to those for corn. The seed should be planted in drills sufficiently far apart for cultivation, and should be thinned to one plant every 12 to 14 inches in the row.

When the heads form, it is advisable to remove all but two or three on each plant.

The heads should be harvested before the seed is fully ripe. This prevents loss of seed by shattering and damage by birds. The heads should be spread out on a barn floor or other suitable place until dry. They may then be stored in bulk. Where used on the farm for poultry, there is no need for threshing the seed. The cost of growing sunflowers is much the same as for corn. The harvesting, however, is much more expensive, and until suitable methods for harvesting and threshing and storing are devised, the crop is not likely to be extensively grown.

Yields ranging from 1000 to 2250 pounds of seed per acre are reported. The seed weighs 30 pounds per bushel.

REFERENCES

"Manual of Flax Culture."
North Dakota Expt. Station Circular 6. "Flax."
North Dakota Expt. Station Circular 7. "Flax for Seed and Oil."
Farmers' Bulletins, U. S. Dept. of Agriculture:
274. "Flax Culture."
322. "Milo as a Dry Land Grain Crop."
417. "Rice Culture."
448. "Better Grain. Sorghum Crops."
466. "Winter Emmer."
552. "Kaffir Corn as a Grain Crop."
669. "Fiber Flax."
688. "The Culture of Rice in California.

CHAPTER 15

Meadow and Pasture Grasses

Meadow and pasture grasses constitute an important and desirable part of the roughage for most classes of livestock. Livestock is indispensable as a part of good agriculture. An old Flemish proverb says, "No grass, no cattle; no cattle, no manure; no manure, no crops." The history of agriculture of many countries shows that where the production of grasses has been neglected, agriculture has declined. England neglected the grass crops and her yield of wheat fell to less than fifteen bushels per acre. She then turned her attention to grasses and the yield increased to over thirty bushels per acre. Of her 28,000,000 acres of tilled land, over one-half are now in permanent pastures. For the past forty-five years permanent pastures of England have increased at about one per cent annually. This should convince the American farmer that in order to grow grain profitably crops must be rotated, and in this rotation grass should find a prominent place. Some far-sighted farmers in North America saw this many years ago, and in the corn belt those who have grown grass are today husking sixty bushels of corn per acre, while those who did not must be content with about thirty bushels.

Importance and Value of Grasses.—According to the last census the hay crop of the United States was 61,000,000 tons, valued at $750,000,000. This does not include the annual hay and forage crops and various kinds of by-products, such as straw and corn stover. This amount of hay will sustain the livestock of the United States about one-fourth of the year, and must be supplemented by about 200,000,000 tons of other forms of feed. Considerable of this comes from the pastures, for which we have no definite statistics. The combined value of hay and pasture grasses far exceeds that of any other crop excepting corn.

Regions of Production.—The perennial hay and pasture grasses succeed best in the northeastern one-fourth of the United States and in southeastern Canada. This grass region extends south to the Potomac and Ohio rivers and to the southern border of Missouri and Kansas, and is limited on the west by about the 96th meridian. The region is characterized by a cool, moist climate and moderate to abundant rainfall.

Principal Grasses of North America.—There are several hundred species of grasses, but of these there are less than one dozen that are of economic importance in North America. Those of greatest importance in the order mentioned are timothy, blue grass, redtop, Bermuda grass, orchard grass, smooth brome grass and Johnson grass. There are a number of others that are grown on a very limited scale, among which may be mentioned tall oat grass, meadow fescue, tall fescue, English rye grass, Italian rye grass, sheep's fescue, red fescue, Sudan grass and sweet vernal grass.

Valuable Characteristics.—To be valuable under cultivation grasses should give satisfactory yields, possess good feeding value, be capable of easy reproduction and be reasonably aggressive. To these might be added, habit of seeding freely so that seed can be cheaply harvested, together with hardiness or ability to withstand adverse climatic conditions.

Choice of Grasses.—The kind of grass to grow will depend on what one wishes to do with it. For pastures a mixture or variety of grasses is

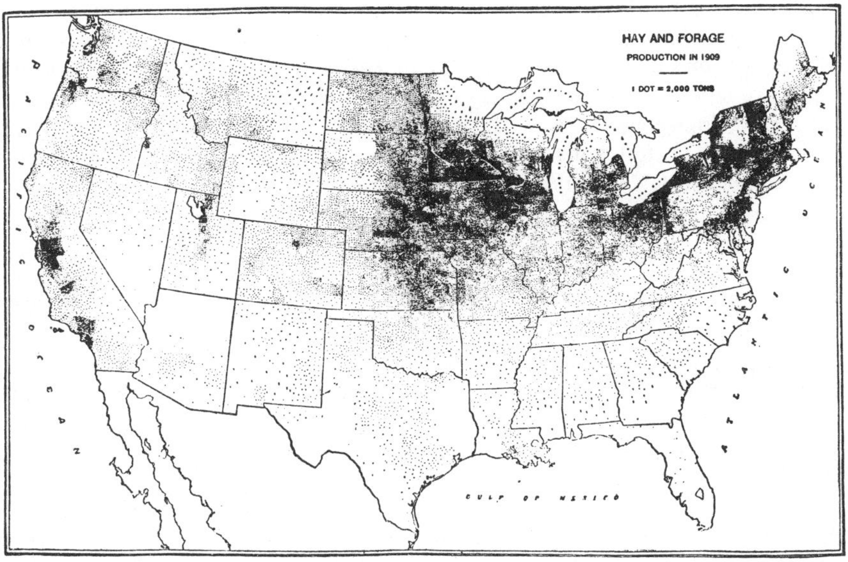

MAP SHOWING REGION OF GRASS PRODUCTION IN THE UNITED STATES.[1]

desirable for a number of reasons. In the first place, a variety of grasses lends variety to the forage for the pastured animals, and induces them to partake of more food and consequently make more growth. A variety often prolongs the season of pasturage, some grasses making their growth in the early and late portions of the growing season when weather conditions are cool, and others growing more freely in the warmer portion of the season. Variety also increases the total yield because of the variation in habits of growth of both roots and foliage.

When grown chiefly for hay, the yield, quality and palatability of the crop secured are important. The cost of establishing, both in direct outlay for seed and in the preparation of the seed-bed, should be considered,

[1] Courtesy of The Macmillan Company, N. Y. From "Forage Plants and Their Culture," by Piper.

as should also the duration of the grasses and the length of time required to come to perfection. Consideration should also be given to time of maturity as related to favorable or unfavorable weather. Abundant sunshine and freedom from rains facilitate making hay of good quality. Where two or more grasses are grown, those should be selected that will mature at approximately the same date.

Seed and Seeding.—There is no crop in which more seed is wasted than the grasses. Of the seed sown a relatively small percentage develops plants, and probably there is no crop in which failure to secure a satisfactory stand of plants is more common. This is due to a number of factors, among which may be mentioned the poor preparation of the seed-bed, the faulty covering of the seed and the adverse conditions that frequently follow seeding, thus causing a large percentage of the small plants to perish. The poor quality of the seed used is also a factor and one that can be largely avoided by the purchase of only first-class seed. As a rule, first-class seeds, although costing more than poor ones, are the cheapest. The following table gives the rate of seeding, the cost of seed per pound and the calculated cost per acre:

COST OF SEED PER ACRE, USING AVERAGE AMOUNT.*

Plant.	Rate of Seeding, pounds.	Cost of Seed per Pound.	Cost of Seed per Acre.
Timothy	15	$0.06½	$0.975
Orchard grass	20	.15	3.00
Redtop	10	.10	1.00
Brome grass	20	.10	2.00
Kentucky blue grass	25	.14	3.50
Italian rye grass	30	.05	1.50
Perennial rye grass	30	.05	1.50
Tall oat grass	30	.14	4.20
Tall fescue	20	.18	3.60
Meadow fescue	20	.11	2.20
Red clover	12	.17	2.04
Alsike clover	8	.20	1.60
Alfalfa	20	.15	3.00
Sweet clover	20	.20	4.00

Since failure to secure a satisfactory stand of grass is so common, farmers are advised not only to use every precaution in the preparation of seed-bed and time and manner of seeding, but also to use an abundance of good seed. As land values increase and the price of product becomes higher, the necessity for these precautions becomes greater. The extra expense for liberal seeding will pay abundantly in the vast majority of cases. The ideal seed-bed is moist and finely pulverized. The slant-toothed harrow is the best implement for making the final preparation.

* The prices given were New York wholesale prices in January, 1914, as given in "Forage Plants and Their Culture," by Piper. Rate of seeding for red, alsike and sweet clover changed.

Harvesting.—The time of harvesting grasses for hay will be determined: (1) by the weather conditions that prevail at the period of maturity, (2) the injury to the succeeding crop as determined by time of cutting, (3) the total yield as determined by stage of maturity, (4) the amount of digestible nutrients secured, and (5) the digestibility and palatability of the product. These factors will vary somewhat with different species of grasses and with the character of animals to which they are to be fed. In general, hay cutting should take place from the period of bloom until seeds are in the dough stage. The total pounds of dry matter will generally increase up to fair maturity. Palatability will be lessened and digestibility diminished if harvesting is too long delayed. If a large acreage is to be handled and weather conditions are uncertain, the harvest period is likely to be prolonged. It is, therefore, well to begin

THE SIDE DELIVERY RAKE.[1]

harvesting rather early in order that the harvest may be completed before the grass becomes too mature.

The market demands a product of timothy hay that is fairly mature when harvested. Such hay is more easily cured and less likely to contain dust and moulds. The large part of timothy that is placed upon the market is used for feeding horses, and feeders object to dusty and mouldy hay.

The quality of hay is determined to a large extent by the manner in which it is handled and cured. This in turn depends to no small degree upon weather conditions. Warm weather, accompanied by plenty of sunshine and a fairly dry atmosphere, is favorable to hay making. If the grass is fairly mature it may be cut late in the afternoon or early in the morning, and placed in the windrow or shock during the evening. Where hay is produced extensively, it is advisable to use up-to-date

[1] Courtesy of The International Harvester Company, Chicago.

mowing machines, side-delivery hayrakes, tedders and convenient and automatic forks for conveying the hay from wagons to mows or stacks. With such an equipment the hay is secured with the minimum of labor and the least possible handling and consequent loss of the leaves and finer portions.

It is maintained, however, that hay of better quality is obtained by curing it in the field in the shock. Cocking hay so that it will not be unduly exposed to rain entails additional labor. Canvas covers are advised if weather conditions are uncertain.

Hay placed in the mow or stack before thoroughly dry goes through a sweating process. A certain degree of sweating is deemed desirable, but should not proceed sufficiently far to develop moulds or cause discoloration. The amount of sweating is dependent on the moisture in the hay. The amount of moisture in hay as it is hauled from the field varies greatly, but ordinarily will not exceed more than 25 to 28 per cent; 20 to 25 per cent of moisture is favorable to a good quality of hay, and is better than to have it too dry or too moist when stored. Numerous determinations of the shrinkage of hay in stack or mow show a loss in a period of six months ranging from as low as 3 per cent to over 30 per cent. This loss is due chiefly to the loss of moisture from the hay. Where the sweating is intense and the temperature runs high, there will also be some loss of organic matter.

COMBINED SWEEP RAKE AND STACKER.[1]

In stacking hay great care should be exercised in the construction of the stacks in order that they shed water. The stacks should be built of good form, and the central portion should be more thoroughly compacted than the outsides. Where hay is valuable, it pays to cover the stacks with good canvas covers or to provide a roof of boards. The stack should be protected from the earth by a foundation of rails or by a thick layer of straw.

Hay is marketed both baled and unbaled. It is graded according to its quality and freedom from weeds and grasses other than that of the name under which sold. Market grades can be secured from grain dealers' associations, and are generally given in market quotations.

[1] Courtesy of The International Harvester Company, Chicago.

TIMOTHY

Timothy is the most important and the most extensively grown of any of the meadow grasses in North America. It is the standard grass for hay purposes and finds a ready sale in all of the hay markets.

Soil and Climatic Adaptation.—Timothy is a northern grass and seldom does well in North America south of latitude 36 degrees, excepting in high elevations. Cool, moist weather during the early part of the growing season is favorable to good yields of hay. It is best adapted to loam and clay loam soils. It is not adapted to swampy soil conditions, neither does it succeed on sandy or gravelly soils. It is not drought

A Field of Good Grass (Timothy), College Farm, Pa.
Yield, five tons per acre field-cured hay.

resistant, and does best on moist, well-drained soils. It calls for a fair degree of soil fertility and does not do well on acid soils.

Advantages of Timothy.—The importance of timothy lies chiefly in its ability to produce good yields of hay that find a ready market at a fair price. The plants seldom lodge and are easily cut and cured, and the period during which it may be cut is longer than that for most grasses. It seeds abundantly, and seed of a high degree of purity and of good germination can be secured at a low cost. It fits well into the crop rotations, and is adapted to seeding with small grains, such as wheat, oats, rye and barley, either in the autumn or in the spring.

Seed and Seeding.—The low price of timothy seed and its appearance make it difficult of adulteration. No grass seed on the market so nearly approaches absolute purity as timothy seed; consequently, the standard of purity is placed at 99 per cent, and that of germination at 98 per cent. Timothy seed contains about 1,200,000 seeds to the pound, and weighs 42 to 48 pounds per bushel. The legal weight is 45 pounds. Four pounds of timothy seed furnish 100 seeds to the square foot on an acre. If every seed produced a plant there would be a great many more plants than are required to make a satisfactory hay crop. The seeds, however, are so small, and the conditions for germination and early growth often so unfavorable, that 12 pounds per acre are usually required. Tests at several of the experiment stations with different rates of seeding show that the largest yield of hay has been secured by using amounts somewhat in excess of 15 pounds per acre.

THE HAY LOADER IN OPERATION.[1]

Under favorable temperature and moisture conditions the seed germinates in five to six days. Although a large percentage of seed three or four years old will grow, it is safest to use seed that is not more than one year old. New seed is sometimes adulterated with old seed. Old seed can generally be detected by its lack of luster, but a germination test to determine the quality of the seed is advised.

The seed is sown broadcast and where seeded with a nurse crop is generally applied by means of the grass seed attachment to the grain drill. There are two methods of distributing the seed by this attachment. In some cases the grass seed distributors are turned in front of the drill hoes. This provides for considerable covering of the timothy seed, and

[1] Courtesy of The International Harvester Company, Chicago.

is applicable only when the soil is of a sandy nature, or in excellent physical condition. Otherwise, it is generally best to distribute the seed behind the drill hoes, and allow it to become covered by the action of rain.

The wheelbarrow seeder is also used, and where the seeding by the above-mentioned method cannot be entrusted to thoroughly competent labor, it is better to use the wheelbarrow seeder. In this way the operator has only the seeding of grass to look after and will do a better job than is likely to be done when the seeding is combined with the distribution of grain and fertilizers all in one operation.

Rows of Timothy, Each Propagated by Slips from the Original Seedlings.[1]
Each row represents a distinct type. Note the variation in size and vigor.

When winter grains are grown, most of the timothy is seeded with them in the fall. When seeded in this way it makes but little growth the succeeding year, and no hay crop is secured. The second year a full crop of hay is secured. In some localities timothy is seeded alone in the fall. This method is applicable in the southern portion of the timothy region. It involves more labor, but results in a full crop of hay during the following season.

Where spring-sown grains prevail, timothy is more frequently seeded with them in the spring. With this method, no crop is secured the first

[1] Farmers' Bulletin, 514, U. S. Dept. of Agriculture.

season. In the southern portion of the timothy belt spring seeding without a nurse crop is practiced to a limited extent. Such seeding is successful only on land that is free from weeds and annual grasses. Under such conditions a light cutting of hay is secured during the first year.

Timothy may be seeded on wheat that has been severely winter killed. If seeded early and the wheat is not harvested too early, both wheat and timothy may be cut for seed at one and the same operation. By using a fanning mill with proper sieves the wheat and timothy seed are easily separated after threshing.

Fertilizers and Manures.—Timothy responds abundantly to light top dressings of manure. The manure should be applied with a manure spreader, and best results will be secured when used at the rate of six to ten loads per acre. It may be applied any time during the autumn or winter. In the absence of manure, a top dressing with a complete fertilizer early in the spring just as the grass begins to start is very beneficial. In several of the states 350 pounds per acre containing about seven per cent of each of the three constituents have given excellent results.

Tests at several of the experiment stations relative to the position of the roots of timothy in the soil show that 85 to 90 per cent of the roots are found in the first six inches of soil. In one case 63 per cent occurred in the upper two inches of soil. This is important in connection with the top dressing of timothy and shows that such top dressing is very close to the great bulk of the active roots of the crop.

Mixing Timothy with Other Grasses and Clovers.—If the hay product is to be fed on the farm, it is advisable to seed clover with timothy. In this practice the amount of timothy seed is reduced to eight or ten pounds per acre, and may be seeded either in the fall or spring, depending on local practice. In the northern part of the timothy region the clover can be safely seeded only in the spring. Six to ten pounds of clover seed per acre will be required, depending on soil conditions and the kind of clover. The first crop of hay will be largely clover, the second chiefly timothy.

Where meadow land is to be used for hay during the first year or two and afterwards devoted to pasture, it is well to include redtop, blue grass and some other grasses and clovers with it. It is also thought wise on very wet lands or on sour soil to include some redtop with the timothy for hay purposes.

Harvesting.—Many experiments relative to the time of harvesting show that the best results are secured only when cut between the time of full bloom and the soft dough stage of the seed. Since timothy is shallow rooted and much of its vitality depends on the thickened bulb-like base of the stem, it is desirable not to cut too closely. Close cutting, or pasturing closely with stock after cutting, injures the subsequent crops by exposure of the bulbs and by injury from tramping. Only when the aftermath is abundant should pasturing be allowed. In no case is it deemed desirable to pasture with sheep, since they are apt to nip off the crown of the plant and thus destroy it.

Pasturing.—Timothy is distinctly a grass for hay rather than for pasturing. It may be used in pasture mixtures to give early grazing, and will give way to the more permanent grasses which are slower in becoming established. It is a common practice to cut timothy for hay purposes for one or more years and then pasture during the year just preceding the devotion of the land to another crop.

FIELD OF TIMOTHY PLANTS GROWN FOR SELECTION, SHOWING VARIATION IN SIZE AND FORM OF INDIVIDUAL PLANTS.[1]

Slips and seeds from choice plants are used for propagating new strains.

Seed Production.—Timothy generally produces between five and twelve bushels of seed per acre. It is most conveniently cut with the self-binder, and is threshed with the ordinary threshing machine, using special sieves to clean and separate the seed. Loss from shattering will be severe if allowed to become over-ripe. If cut promptly the straw has considerable feeding value. The principal seed-producing states are Illinois, Iowa, Minnesota, South Dakota, Kansas and Ohio.

Composition and Feeding Value.—Timothy hay contains about 6 per cent of protein, 45 per cent of carbohydrates, 2.5 per cent of fat and 29 per cent of crude fiber. About one-half of this is digestible.

[1]Farmers' Bulletin, 514, U. S. Dept. of Agriculture.

Improvement of Timothy.—Although timothy has been an important crop and large quantities of seed are bought and sold, as yet no varieties have been developed. Timothy plants show marked variation in size, vigor, character of foliage and resistance to drought. Improvement of the crop for special purposes can be made by the selection and propagation of desirable plants. Several of the experiment stations have made progress along this line and have already developed strains of timothy that have

VARIATIONS IN TIMOTHY.[1]

outyielded that secured from commercial seed by as much as one ton per acre.

Marketing the Hay.—The bulk of timothy hay is placed upon the market in bales of about 100 pounds each. The market calls for bright, clean timothy hay, free from weeds and various grasses. When mixed with clover, redtop or other grasses, quotations will be somewhat lower than for pure timothy.

BLUE GRASS

There are two chief species of blue grass in North America, namely, Kentucky blue grass and Canada blue grass. These grasses spread by

[1] Courtesy of The Macmillan Company, N. Y. From "Plant Breeding," by Bailey.

means of seed and also by underground root stocks. They give rise to an even and continuous turf, and are especially adapted for pasture purposes. They are aggressive grasses and tend to take possession of the land and crowd out weeds and other grasses. The Kentucky blue grass is superior in both quality and yield. Its climatic adaptation is essentially the same as that for Canada blue grass, and ranges from Virginia northward into Canada, and westward to the central part of Kansas and Nebraska. It reaches its highest development in the region of limestone soils. Parts of Kentucky, Missouri, Virginia and Tennessee are noted for their blue grass regions. It also succeeds well on both the timber and prairie soils of Ohio, Indiana, Illinois and Iowa.

Soil and Climatic Adaptation.—These two prominent pasture grasses are adapted to a cool, moist climate having thirty inches of rainfall and upward. They are exceedingly resistant to cold, never freezing out in even the most severe winters. These grasses prefer well-drained loams or clay loams. They are not adapted to loose, sandy soils. The Kentucky blue grass calls for a fair to good degree of fertility, and where these two grasses are seeded together on such soil, the Kentucky blue grass will soon take full possession. The Canada blue grass has the ability to grow on poor soils, although it will produce only small crops and poor pasturage under such conditions. On poor soils the Canada blue grass will take possession finally to the exclusion of Kentucky blue grass.

Although these two grasses will make hay of fair quality, the yield is so low that they are not adapted to hay purposes.

Importance of Blue Grass.—As pasture grasses these are unexcelled for the temperate portions of North America where the rainfall is fairly abundant. They are not only valuable as summer pasture, but as winter pasture for horses and sheep, have no equal. When desired for winter pasture they should not be closely pastured during the summer. Winter pasture from these grasses can often be provided by turning the stock into the fields from which the spring crops have been harvested and on to meadow land during the late summer and autumn. This permits the blue grass to make good growth for winter pasture. Even when covered with snow, horses and sheep will paw off the snow and pasture on the grass.

Severe drought during the summer may completely suspend the growth of blue grass and cause it to appear dead. No matter how long the period of drought, rains will quickly revive the grass and it will resume its normal growth and condition. It will stand a great abundance of tramping without serious injury. The writer has seen calves retained in hurdle pens during wet weather on blue grass until the surface would be thoroughly puddled and no grass visible. A few weeks after removing the pens the grass would be in as thrifty a condition as ever.

Methods of Establishing.—Blue grass seed weighs from fourteen to twenty-eight pounds per bushel, the legal weight being fourteen pounds. The weight is determined chiefly by the presence or absence of the glumes or hulls that enclose the seed proper. Blue grass seed is frequently of low vitality, due to faulty methods of harvesting and curing. It is always well to test the seed before seeding as a guide to the amount of seed desirable to use. Blue grass is very slow in becoming thoroughly established, and good pastures can seldom be secured in less than two years from time of seeding, and in some cases more time is required. It is generally advisable to seed with a mixture of grasses and clovers, some of which will give prompt pasture. Timothy, orchard grass, and red and alsike clover are, therefore, frequently used. These ultimately give way to the blue grass. Virgin grass land and meadow land are frequently converted into blue grass pastures by seeding blue grass, which gradually spreads and takes possession. When used for lawn purposes, the rate of seeding should be three to four bushels per acre. As little as eight to ten pounds per acre may be used when seeded with other grasses and when plenty of time is allowed for becoming well established. Ordinarily, twenty to twenty-five pounds of blue grass should be used when it is the chief grass for the pasture.

It is difficult to distinguish between seed of Kentucky blue grass and Canada blue grass. The latter is sometimes used to adulterate the former, since it generally is less costly.

Pasture Maintenance.—Blue grass, because of its numerous underground root stocks, tends to form a sod-bound turf. This condition may be obviated by seeding blue grass pastures with red or alsike clover every three or four years. This can be done by using a disk drill early in the spring. The use of the disk will also help to overcome sod-binding. The presence of the clover will enhance the pasture for the time being, and especially during the dry period when the blue grass will remain dormant. The clover roots tend to loosen up the ground and supply nitrogen to the blue grass. White clover is advantageous when seeded with blue grass. It re-seeds itself and becomes permanent so long as soil conditions are favorable. Under favorable conditions and with proper treatment, blue grass pastures improve with age, at least for several years. There are many instances of such pastures having been undisturbed for thirty or forty years.

REDTOP

Redtop is a native grass of North America, and grows naturally in cold, wet soils. It is a perennial provided with long, creeping underground root stems, and spreads both by means of these and seeds. It forms a continuous and fairly even turf, and is, therefore, well adapted for pasture purposes. It has a wider range of adaptation, both from the soil and climatic standpoint, than any other cultivated grass. It is resistant to cold and withstands summer heat much better than timothy. It does not show much preference for type of soil, but does best on loams and clay loams. It is exceedingly tolerant of soil acidity. It is also fairly drought resistant and succeeds better than most grasses on poor, sandy soils.

Importance of Redtop.—Redtop is the third or fourth most important perennial grass in America. It is adapted to both pasture and hay purposes, although it is not equal to timothy as a hay producer nor to Kentucky blue grass for pasture purposes. As a pasture grass it is not so palatable as Kentucky blue grass.

Culture.—Like Kentucky blue grass, redtop is aggressive and frequently takes full possession of the land. It is seldom seeded alone, usually being included in mixtures. The rate of seeding depends on the quality of the seed and the nature of the mixture in which seeded. With re-cleaned seed, twelve to fifteen pounds per acre are sufficient when seeded alone. Much smaller amounts will meet the requirements in mixtures. The time and manner of seeding are similar to those for timothy.

Yields and Uses.—Redtop has been tested at a number of state experiment stations and yields of hay ranging from 3000 to 5600 pounds per acre are reported. In order to be of good quality redtop should be cut early. If allowed to become fairly mature it makes hay that is fibrous and unpalatable. Numerous analyses show that redtop hay contains more nutrients than timothy hay.

ORCHARD GRASS

Orchard grass, a native of Europe, is grown quite generally throughout the United States, except in the semi-arid sections and the extreme south. It is a rather deep-rooted, coarse grass which grows in tufts or bunches and is without creeping root stocks. It does best in a temperate climate, but will stand more heat than timothy, and is less resistant to cold. In the United States it is cultivated more abundantly southward than northward. It begins growth earlier than most grasses, and often produces a second cutting of hay.

Importance.—Orchard grass ranks fourth or fifth in importance among the perennial cultivated hay grasses in North America. It is most extensively grown in Maryland, Virginia, West Virginia, North Carolina, Kentucky, southern Indiana, Iowa and Oregon.

Culture.—The seed of orchard grass weighs from fourteen to twenty-two pounds per bushel, and when seeded alone requires about thirty-five pounds per acre. Germination of the seed is complete in about fourteen days. It may be seeded either in the fall or very early spring. When seeded in the fall, early seeding is desirable to prevent winter killing. The seed, being of an exceedingly chaffy character, does not feed well through a seed drill, and is generally sown by hand or with the wheelbarrow or other types of seeders.

Ordinarily, the grass does not form seed the first season. It is long-lived, and individual plants are known to live eight years, and will probably live longer.

Yields and Uses.—Whether seeded in fall or spring, the first year's growth rarely gives a hay crop, but it may be utilized for pasture. When used for hay it should be cut as soon as in full bloom. The stems become woody if it stands longer. It is usually about three weeks earlier than timothy and is advantageous on lands infested with ox-eye daisy, flea-bane and other weeds that do not ripen seed before time of harvesting it. It yields about as well as timothy, and yields reported from several experiment stations range from three-quarters of a ton to two and one-half tons per acre, the average being 1.4 tons.

It is considered valuable as a soil binder and serves to prevent soil erosin on land subject to washing.

It is recommended as a constituent of mixed pastures. It is valuable in this respect because of its early growth and its ability to grow during cool weather. It succeeds best under heavy grazing, and is admirably adapted for shady pastures and in orchards that are to be grazed.

Brome grass.—Brome grass is of comparatively recent introduction. It is a long-lived perennial, spreading both by seeds and root stocks. It forms heavy clumps, frequently twelve inches in diameter, but when seeded abundantly these join and form a compact sod. It is quite deep rooted and is adapted to a wide range of climatic conditions, both from the standpoint of temperature and rainfall. It is especially important, both as a hay and pasture grass, for the Great Plains region and the Pacific Northwest.

The method of seeding is similar to that for timothy. It is especially valued for hay during the first two years after seeding. There is then a tendency to become sodbound, after which it serves better for pasture. It is both palatable and nutritious, whether used as hay or for pasture.

Tall Oat Grass.—This grass has a climatic adaptation very similar to orchard grass. It is fairly drought resistant and does poorly on wet land. It does best on rather loose, deep loams, and succeeds well on calcareous soils; also does well on sandy and gravelly soil, but is not adapted to poor land. It is a perennial and is strictly a bunch grass.

When used for hay it should be cut promptly while in bloom. After this period the stems rapidly become woody. It yields well, but is of low quality, the hay being somewhat bitter in taste. For this reason it is generally best grown in mixtures.

The Fescues.—There are a number of fescues, among which may be mentioned meadow fescue, tall fescue, reed fescue, sheep's fescue and red fescue. None of these are of much importance in American agriculture. They have about the same range of adaptation as timothy.

Sheep's fescue is a fine-textured, small-growing species adapted for lawn grass mixtures. Sheep eat it quite freely, but cattle avoid it if other grasses are available.

Red fescue makes a dense growth under favorable conditions and may attain a height of two feet or more. It makes fair yields of hay, but is not equal to many of the better species for this purpose.

Rye Grasses.—Perennial rye grass is a short-lived, rapid-growing perennial, living usually only two years on poor land, but somewhat longer under favorable conditions. It is seldom employed except in lawn mixtures.

Italian rye grass is adapted to moist regions with mild winters. It succeeds best on loam and sandy loam soils. It is adapted for hay purposes and may be cut several times during the season.

Sudan Grass.—A tall annual grass resembling Johnson grass, but spreads only by seeds. It has been recently introduced and seems to be best adapted to the semi-arid belt. It has been tried in an experimental way in many of the states and has generally made a good growth.

SUDAN GRASS, A NEW ACQUISITION.[1]

Bermuda Grass.—Bermuda grass is a perennial with numerous branched leafy stems, which, under favorable conditions, attain a height of twelve to eighteen inches. Ordinarily, it is not so tall. This grass occurs chiefly in the southern part of the United States, but extends as far north as Pennsylvania and Kansas. It is especially adapted to the cotton belt, and is to the South what blue grass is to the North. While it is more particularly adapted as a pasture grass, it is also quite extensively used as hay. It will grow on all types of soil, but does best on rich, moist bottom lands that are well drained. It is also used as a lawn grass. Bermuda grass does not seed at all freely and most of the seed is imported. It is most easily propagated by cutting the culms into short pieces, scattering them on the

[1]Courtesy of The Macmillan Company, N. Y. From "Forage Plants and Their Culture," by Piper.

field to be seeded and covering them with disk, harrow or other suitable implements. These fragments of grass take root and spread rapidly by means of numerous root stocks or creeping stems.

Bermuda grass meadows and pastures frequently become sod-bound and fall off in yield. This condition may be alleviated by disking or by plowing and harrowing. After such treatment the growth will become much more vigorous.

Johnson Grass.—It is a coarse, large-growing species adapted to the whole of the cotton belt. It grows well in the summer as far north as 37 degrees north latitude, but usually will not withstand winters in such latitude. It spreads both by seeds and rhizomes, and when once established it is difficult to eradicate. It is utilized for both hay and pasture. Two or three crops per season are frequently harvested.

Para Grass.—This is a rank-growing tropical species adapted to moist loams or clay loams. In the United States it is adapted only to Florida, and the Gulf Coast to southern Texas. This grass is easily propagated by cuttings of the long, prostrate runners in much the same way that Bermuda grass is propagated. It is of value both for pasture and for feeding in the fresh state. It is seldom used for making hay.

Guinea Grass.—This is a long-lived perennial with short, creeping, root stocks. It generally grows in immense tufts, sometimes as much as four feet in diameter. The culms are large, erect, tall and numerous. It is adapted to tropical conditions, but may be grown in Florida and along the Gulf Coast of North America. Both this and the preceding grass may be cut several times each year. Under strictly tropical conditions, cuttings are frequently made every six or seven weeks.

REFERENCES

"A Textbook on Grasses." Hitchcock.
"Forage Plants and Their Culture." Piper.
"Forage and Fiber Crops in America." Hunt.
"Grasses and How to Grow Them." Shaw.
Farmers' Bulletins, U. S. Dept. of Agriculture:
361. "Meadow Fescue (Its Culture and Uses)."
362. "Conditions Affecting Value of Market Hay."
402. "Canada Blue Grass (Its Culture and Use)."
502. "Timothy Production on Irrigated Land."
508. "Market Hay."
509. "Forage Crops for the Cotton Region."

CHAPTER 16

THE CLOVERS

Clovers are important on account of their high protein content and nutritive ratio. They are especially valuable as forage for all classes of livestock. Clovers enrich the soil in nitrogen and organic matter, and improve its physical condition through the deep penetration of roots. For years farmers have paid out large sums in the purchase of nitrogen for the soil and protein for livestock. This can be largely avoided by growing an abundance of leguminous crops on the farm.

Characteristics of Clovers.—The true clovers are herbaceous leafy plants having three palmately arranged leaves. The larger growing species have deep roots on which occur nodules containing certain species of bacteria. These bacteria enable the plants to secure nitrogen from the air and use it in their development. For this reason legumes are richer in protein than other classes of plants. Of the total nitrogen in the plants about two-thirds are in the tops and one-third in the roots.

Uses of Clovers.—As a rule from one-half to two-thirds of the roughage in the ration for milk cows and young stock should consist of legumes, among which the clovers as hay are most convenient to use and most economical. The larger growing clovers are also quite extensively used for soiling purposes, and in some cases have been used for ensilage. The clovers are also among the most important crops for green manuring and as cover crops.

Inoculation.—Since all of the legumes contain bacteria in the nodules on their roots, it is best to inoculate many of the legumes when grown for the first time in any locality. In most of the clover region soils are already inoculated for the clovers. If inoculation is advisable, it may be effected either by soil transferred or by the use of artificial cultures. In this connection it should be borne in mind that as a rule each legume has a particular species of bacteria. Three to four hundred pounds of soil transferred from a well-established field of any species of clover to a new field will effect satisfactory inoculation of the latter. The soil should be taken from the zone of most abundant root activity, thoroughly distributed on the new field and at once mixed with the soil by disking or harrowing.

Artificial cultures have now been perfected and can be purchased at reasonable prices from many manufacturing firms. The culture is generally applied directly to the seed just before it is sown.

Composition and Feeding Value.—The composition of several species of clovers in the green state and in forms of preservation will be found in Table VI in the Appendix. Clovers, whether used for ensilage, soiling, hay or pasture, all possess high feeding value, and are especially desirable for the production of milk, butter and the growth of young animals. They are among the most highly nutritious forage plants, and should supplant as far as possible the expensive concentrates such as bran, oil meal, cottonseed meal, etc.

Harvesting Methods.—The purpose for which the product is used will determine the method of harvesting. When used for soiling, it is advisable to cut clover each day in quantities sufficient to meet the day's ration. If used for silage it should be cut when fairly mature, and go directly to the silo with but little loss of moisture.

When clovers are cut for hay, both the quality and quantity of feed should be considered. If the acreage to be harvested is large it will be

A Clover Field in Blossom.[1]

advisable to commence early in order to complete the work before the crop becomes too mature. The more uncertain the weather, the earlier the process should begin.

The best quality of hay is secured by a comparatively slow process of curing. In this process the moisture should leave the plants almost entirely through the leaves. Clover cut in the middle of a hot, dry day when the ground is dry and the sunshine bright, will dry so rapidly that the leaves soon lose their structure, become brittle and cease to give off moisture. Although there may still be much moisture in the stems of the clover, the leaves will break and be largely lost in the handling of the hay. These leaves are high in feeding value. It is wise, therefore, to cut in the evening and to place the hay in the windrow before the leaves become sufficiently dry to break and shatter. The best quality of hay is secured by placing in shocks before thoroughly cured and allowing curing to be completed slowly within

[1] Courtesy of Hoard's Dairyman.

the shock. This entails much additional work, and if weather conditions are favorable a good quality of hay may be secured without resorting to shocking.

Clover hay may go into the mow or stack with 25 to 30 per cent of moisture without injury. Good judgment and prompt and systematic work on the part of the haymaker are necessary to secure the best results.

The hay tedder and side-delivery rake are important adjuncts to securing a good quality of clover hay, and may be considered necessities where the acreage is sufficiently large to justify their purchase.

RED CLOVER

Red clover is a native of western Europe, and has long been cultivated in North America. It is now the most important leguminous crop in the Northern and North Central states and eastern Canada. While red clover is grown to some extent in every state and province of the United States and Canada, it is most extensively grown in those states lying north of the Ohio River and east of the Missouri River. Kansas and Nebraska, however, produce a large acreage. The accompanying map shows the distribution of red clover, grown alone and with timothy, by states and provinces for the United States and Canada.

Soil and Climatic Adaptation.—Red clover is quite resistant to cold and endures winters well in Nova Scotia, Maine and Minnesota. Northern grown seed is, therefore, generally preferable for seeding in cold latitudes. It does not do well in an extremely warm climate, and in the South succeeds only when planted in the fall, and usually survives only one year. A moderate to abundant rainfall is desirable.

It is adapted to quite a wide range of soils, but makes its best growth on fertile, well-drained soil well supplied with lime and organic matter and reasonably free from weeds. Any soil that will grow corn successfully is well adapted to red clover. It does not do well on poorly drained land. On such soil alsike clover succeeds better.

Endurance of Red Clover.—Red clover is generally considered a biennial, the plants dying at the end of their second year. Some plants, however, will live over for a third year and a few frequently die at the close of their first year. The time of seeding and the treatment during the first year doubtless influence the life of clover plants. It is a common belief that if clover blooms abundantly toward the close of the first year many of the plants will fail to continue their growth the following year. For this reason clipping or light pasturing is advised.

Clover on wet soil may be killed in severe winters by repeated freezing and thawing. The plants will be so nearly pulled out of the soil that they perish in the spring for want of moisture and plant food. If the ground is deeply frozen and the surface only thaws and freezes the taproots are broken. This difficulty is best overcome by a thorough drainage of the soil and by providing a surface mulch.

Securing Clover Seed.—The intelligent selection of clover seed calls for knowledge relative to the characteristics of both good and poor seed. Good seed is plump and has a bright luster, and is generally violet to bright yellow in color. The proportion of violet to yellow varies considerably in different lots of seed. Good seed should be free from noxious weed-seeds and adulterants of any kind. The standard of purity should not be below 98 per cent and the germination should be about 98 per cent. Frequently some of the clover seeds will be so hard that they will not germinate promptly. The hardness of the coat prevents absorption of moisture. The percentage of hard seeds is largest in new seed.

Home-grown seed possesses several advantages: (1) it is likely to be adapted to local climatic and soil conditions; (2) its use avoids the intro-

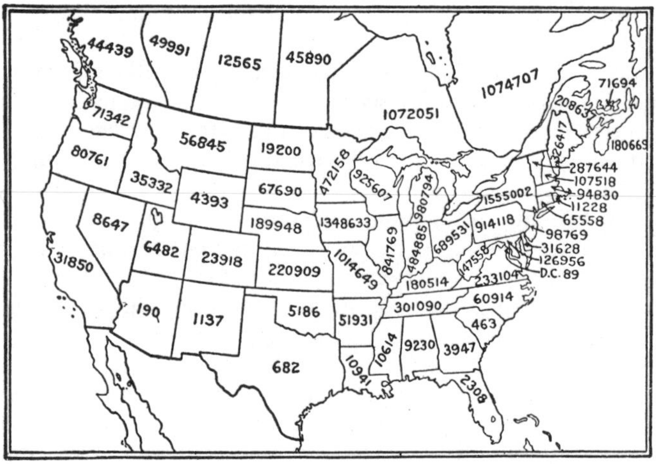

Map Showing the Acreage of Red Clover in the United States, 1909, and Canada, 1910.

duction of obnoxious weeds foreign to the neighborhood. Among the most obnoxious weeds are clover dodder, buckhorn, Canada thistle and dock. Most weed-seeds may be removed by the use of suitable screens. The longevity of clover seed is three years. The deterioration in vitality depends largely upon the conditions of storage. Continuous warm, moist conditions cause deterioration and make it inadvisable to use seed more than two years old. A considerable percentage of the seed as determined by numerous tests will retain its vitality for quite a number of years, and the hard seeds have been known to germinate after fifteen or twenty years.

Seed of mammoth clover is so much like that of red clover that it is difficult to distinguish between them. Ordinarily, mammoth clover seed is a little larger than that of red.

Preparation of Seed-Bed.—Red clover is usually seeded in the winter or spring, in which case no special preparation of the seed-bed is necessary. When seeded in this way natural covering results from the freezing and thawing of the ground and the beating of rains. If conditions for spring seeding with wheat necessitate seeding rather late, it is best to harrow the wheat, thus covering the clover seed.

When seeded with spring grain the preparation for the grain is generally sufficient for the clover. It will pay, however, to provide a well-prepared seed-bed that will fully meet the needs of clover seed, even though equally thorough preparation is not necessary for the spring grain. A fair degree of compactness and a thorough covering of the seed are desirable.

Time, Manner, Rate and Depth of Seeding.—In all regions of moderate to severe winters, winter or spring seeding is advisable, except when clover may be seeded in midsummer without a nurse crop. Further south, fall seeding may be practiced without winter injury to the young clover plants.

While clover seed is generally broadcasted, recent tests show that better results can be secured with less seed by using a grass seed drill. Such implements are now available and are so constructed as to drill the rows at intervals of four inches. Their adjustment permits of a shallow covering of the seed. The rate of seeding when clover is grown alone should be ten to twelve pounds of good seed per acre if broadcasted and a somewhat smaller amount when drilled. When seeded in mixtures the amount may be reduced, depending on the character of the grass seed mixture. Clover seed should be covered from one-half to two inches in depth. On very loose, dry soils it may be covered as much as three inches deep with fairly good results.

Failure to secure a satisfactory stand of clover frequently results from various causes. The condition of newly seeded clover fields immediately after the nurse crop is harvested should be observed. If there are indications of insufficient plants for a satisfactory stand, it is generally advisable to re-seed at once. This re-seeding may take place over those portions of the field where the stand is poor, or may cover the entire field as conditions require. A disk should be used to loosen the soil before seeding, and after seeding it should be harrowed. Disking may injure some of the clover present, but not seriously.

Good results are also secured by seeding in August without a nurse crop. Such seeding takes place after the wheat or oat harvest and provides for a full clover crop the following year. The chief objection to this method is the extra labor of preparing the seed-bed and seeding.

Nurse Crops for Clover.—Where clover grows without difficulty, it is common practice to seed with some nurse crop. In sections where winter wheat is grown, this crop is a favored nurse crop for clover. Winter wheat is seldom seeded before the latter part of September and this does not give sufficient time for clover to make enough growth to protect itself during the winter. As a result the clovers north of latitude 36 should be seeded in the late winter or early spring in the growing wheat. Of the spring-seeded grains, barley and oats are the best nurse crops for clover. These should not be seeded very thick, otherwise the clover may be smothered. The nurse crop should be cut sufficiently high to leave a stubble that will protect the young clover as much as possible.

Fertilizers for Clover.—As a rule, no fertilizers or manures are applied directly for the benefit of the clover. The residual effect of that applied to the crop preceding the clover is generally sufficient. This is especially true when seeded with winter wheat. On soils of low fertility, especially when there is little organic matter present, top dressing with manure previous to the time of seeding is very beneficial to the clover. No nitrogen is needed when commercial fertilizer is used. Moderate amounts of phosphorus and potash applied broadcast will meet the needs.

After-Treatment of Clover.—Clover seeded with a grain crop seldom requires any special treatment during the first year. Under favorable conditions it may make sufficient growth after the harvest of the grain to produce a cutting of hay. This is thought by some to be injurious to the following year's clover crop. It is, therefore, advised to clip the clover before it comes extensively into bloom, and allow the clipping to lie on the field. If so abundant as to smother the plants, it may be removed. Clipping is also advisable to prevent the ripening of the seeds of obnoxious weeds and grasses that are always present to some extent. The clipping should be so timed as to prevent the seeding of the largest possible number of such plants. If too early, seeds may develop after the clipping, and if too late some of the seeds may have already matured. The ordinary mowing machine with the bar set rather high is well suited for this purpose.

Light pasturing may be practiced instead of clipping. Pasturing with sheep is best, since sheep are fond of many of the weeds and grasses, and will eat the seeds in great abundance.

Since red clover lives only two years, the first crop during the second year is generally cut for hay and the aftermath is either used for a seed crop, is pastured or plowed under for the benefit of the soil. If the second crop is to be used for seed it is wise to cut the first crop early. This encourages a better development of the second crop and increases seed production. The first crop should be cut just as it is coming into bloom. If the clover is to remain for the third year, seed must be allowed to mature during the late summer of the second season, with a view of having the clover re-seed itself naturally. This is not a very satisfactory method, however, because the seed heads generally fall to the ground and give rise to an uneven distribution of the seed. This, however, may be obviated by thoroughly harrowing the field after the seed heads are mostly on the ground. The harrowing breaks up the heads and distributes the seed. It should be so timed as to avoid destruction of clover plants when just starting.

Harvesting Clover.—Red clover, harvested for hay, should be cut when one-third of the blossoms have begun to turn brown. At this time the plants will contain about all the nutrients they ever will have, and the product will cure readily and make a palatable, digestible hay. After this period the lower leaves begin to fall rather rapidly and the clover is apt to lodge so that loss takes place.

When used for soiling purposes, cutting may begin as soon as the first blossoms appear, and continue until the crop is fairly mature. When used for silage, the plants should be fully as mature as when cut for hay. If cut too green it makes a sloppy, sour silage of poor quality. When used for silage, clover gives best results when mixed with non-leguminous crops. The second cutting of clover can frequently be used to mix with corn in the making of silage.

The least expensive way of harvesting is to pasture. While red clover is not especially well adapted to pasture purposes, it makes a good quality of pasture, and especially when mixed with grasses. It is especially suited to cattle, sheep and swine. Sheep and cattle are sometimes subject to bloating when allowed to feed on red clover when it is especially succulent or when wet with dew or rain. Such trouble occurs only when the animals are unaccustomed to it and when they feed too heavily.

Clover Seed Production.—Red clover seed may be successfully produced in practically all areas adapted to the production of clover hay. It differs in this respect from alfalfa.

Seed production is encouraged by retarding somewhat the vegetative growth. Conditions that will produce a medium growth of plant usually induce the best setting of seed. Good seed crops are seldom secured from a rank growth of clover. Under such conditions the heads are few and are not well filled. The probable yield of seed and advisability of saving the crop for that purpose can be determined by a careful examination of a number of seed heads. If the seed heads are fairly abundant and contain an average of twenty-five to thirty seeds each, it indicates a yield of one to two bushels per acre, and justifies saving for seed purposes. If the average number of seeds is not more than twenty it will generally not pay to cut for seed. This determination must be made fairly early in order to cut the crop for hay before it becomes too mature in case it will not pay to save for seed.

It is a common belief that seed production calls for a pollination of the flowers by insects. The ordinary honey bee cannot reach the nectar of the average clover blossom, and is, therefore, not instrumental in the fertilization of the flowers. Bumble bees, however, are supposed to be the most effective agents in this process. There are probably numerous very small insects that also produce pollination. However this may be, the second crop is the one that gives best results for seed purposes. At that time insects are more numerous, weather conditions are drier and the plants tend to produce seed more abundantly than earlier in the year. Occasionally the first crop will produce plenty of seed. The seed crop should be cut when the largest number of heads can be secured. If cut too early, the late blossoms will have no seeds or will have poorly developed seeds. If cut too late, the early blossoms will have shattered off.

The old-fashioned self-rake reaper is best adapted to cutting the seed crop. It leaves the cut clover in bunches of convenient size, sufficiently far from the standing clover for the team and machine to pass for the next swath. These bunches of cut clover do not need to be disturbed until they are ready to be hauled to the threshing machine. In the absence of the self-rake reaper, a mowing machine with a buncher may be substituted. If the buncher leaves the clover in the path of the team and machine, a man should follow the machine with a barley fork and move the bunches. Serious shattering in the cutting process may be avoided by harvesting the crop in the evening or early in the morning, or on damp days.

The clover is generally threshed with a clover huller. This machine should contain two cylinders. Concaves must be set rather close in order to remove all of the clover seed from the hulls. The seed being valuable, it is advised to spread canvas beneath the machine to save the clover seed which shatters out in the threshing process. Where threshing is done on a barn floor canvas will not be required.

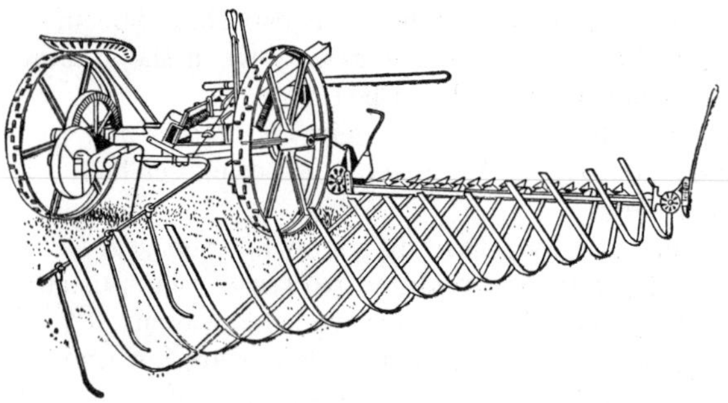

A CLOVER BUNCHER ATTACHED TO A MOWING MACHINE.[1]

The seed should be thoroughly cleaned before being placed upon the market or used for seed purposes. Nearly all foreign matter and weed seeds can be removed by use of a suitable fanning mill. Occasionally there are seeds present of about the same size and weight as clover seeds, and these will be difficult to remove. Buckhorn seed is difficult to remove in this way. It is a very troublesome weed in meadows and the following process of removing it from clover seed is recommended. Thoroughly wet the clover seed with water at about room temperature, and allow to stand in the water for five minutes, or as much as eight minutes if the temperature of the water is low. The water is then drained off and the moist seed thoroughly mixed with sawdust; about four parts of sawdust to one part of seed by measure will be required. Two or three minutes of thorough mixing will cause the sawdust to absorb the free surface moisture from the seed. The buckhorn seeds become mucilaginous and the sawdust adheres to them. The mixture is now run through two screens, preferably in a fanning mill. The upper one should be perforated with round holes

[1] Courtesy of U. S. Dept. of Agriculture. From Farmers' Bulletin 495.

one-fifteenth of an inch in diameter. The lower should be a No. 22 mesh wire screen. The buckhorn seeds with sawdust adhering will pass over the surface of the upper screen and be removed. The clover seed will pass through the openings and be retained by the lower screen, passing off at the edge, where it may be collected. The sawdust should be fine and will pass through the lower screen. In this way the separation is complete.

Red Clover Troubles.—The principal enemies of red clover are insects, fungous diseases and weeds. Much is heard concerning clover sickness, but little is known relative to the nature of the malady. Failure to grow

Red Clover on Limed and Unlimed Land.[1]

continuous crops of clover may be due to any one of several causes. Soil acidity is probably the most common cause of clover failure. This, as previously stated, is overcome by the use of lime. One of the most common diseases of clover is anthracnose. In some sections nematodes have also been responsible for clover failure. These difficulties will be mentioned under special chapters covering plant diseases, insect enemies and weeds.

Alsike Clover.—Is a perennial plant intermediate between red and white clover in size and appearance. It is adapted to ground that is too wet for red clover, and is also more tolerant of acidity.

[1] Courtesy of The Macmillan Company, N. Y. From "Crops and Methods for Soil Improvement," by Agee.

As a hay crop it will not yield as much as red clover, although it makes hay of finer textures and retains the leaves better. It is hardier than red clover, but lodges worse. The foliage is slightly bitter and not relished as well by cattle. For this reason it is better to mix it with red clover or with grasses. It matures about two weeks earlier than red clover; consequently, does not fit into mixtures as well as the red. It may be grown with early-maturing grasses such as orchard grass and redtop. Alsike clover and red-top make an admirable mixture for wet, sour soil, and may be used both for hay and pasture purposes.

It seeds abundantly and the seed, though much smaller than that of red clover, commands about the same price. Six to eight pounds per acre when seeded alone are sufficient. Smaller amounts may be used in mixtures.

White Clover.—White clover is a low-growing perennial, having abundant solid, creeping stems. It is well adapted to moist soils in nearly all of the temperate zone. It is especially well adapted for pasture purposes and is frequently used with blue grass both in pastures and lawns. It seeds abundantly, often producing from two to six bushels per acre, the price ranging a little above that for common red clover. It has long been valued as a honey plant. The blossoms when excluded from insects usually set no seeds.

Ladino Clover.—This clover is similar to white clover, but much larger. It has but recently been introduced into North America. It furnishes good yields of excellent pasturage and under favorable weather conditions attains sufficient size to be harvested for hay.

Crimson Clover.—This is a winter annual adapted only to regions of mild winters. It is extensively used as a green manuring and cover crop. It may be seeded from May to August, either alone or in other crops such as standing corn. It makes hay of a good quality if cut just as it comes in flower. The plant is somewhat hairy and the seed heads are abundantly supplied with long hairs. If the heads become rather mature, the hairs harden and cause serious trouble when fed to livestock. It is never advisable to feed straw from crimson clover to horses. Either the mature hay or straw causes hair balls in the stomach and intestines that frequently result in the death of the animals. Both the hay and seed crops are handled in about the same way as red clover.

Sweet Clover.—This plant has come into prominence in recent years, and has been extensively discussed in the agricultural press. Recent careful inquiries and investigations indicate that it is destined to become an important legume, both as a forage crop and for soil improvement.

There are several species of sweet clover, but the white sweet clover (*Melilotus alba*) is the most valuable under most conditions. It is adapted to a wide range of both soil and climatic conditions. It is exceedingly hardy and makes fair growth under adverse conditions. It is a biennial. It is often spoken of as a roadside weed, and occurs along roadways in many parts of nearly every state in the Union and the provinces of Canada. It seeds abundantly, the seed being similar to that of alfalfa. The plant also closely resembles alfalfa in its early stages of growth, although the blossoms and seed heads are quite different.

It is deep rooted and the tops often attain a height of four to five feet. The composition of sweet clover is nearly the same as that of alfalfa. It is high in digestible protein and very nutritious as feed. Because of a peculiar odor and taste, animals seldom eat it at first. They soon acquire a taste for it and eat it with avidity and thrive on it.

Sweet clover is especially valuable for soil improvement. Its greatest benefit will result by plowing it under the second season before it blooms. The seed should be sown at the rate of fifteen to twenty pounds of hulled seed, or at the rate of twenty-five to thirty pounds when hulls are present.

PASTURING SWEET CLOVER IN KANSAS.[1]

It may be seeded either in August or early in the spring. The methods of seeding are similar to those for red clover.

Lespedeza or Japan Clover.—This is a small-growing summer annual, attaining a height of six to eighteen inches, depending on soil conditions. It is adapted especially to the cotton belt. It is to the South what white clover is in the North. It is especially adapted for grazing purposes, and a mixture of Bermuda grass and Lespedeza makes a good pasture for many parts of the South. It begins growth in the middle spring and reaches maturity in September or October. It may be distinguished from the yellow-flowered hop clovers which it closely resembles by its purple blossoms, which do not appear until August or later, while the hop clovers bloom early. It seeds freely and perpetuates itself from year to year by self-seeding.

Bur Clover.—This is a rather small-growing clover indigenous to Texas and California, and is closely related to alfalfa. It is of very little value for hay, and will give only one cutting. It serves best for winter and early spring grazing. It is especially valuable because it affords

[1] Courtesy of Kansas Agricultural Experiment Station.

grazing in the South for about two months before Bermuda grass and other summer grasses are available for this purpose. It makes a good combination with Bermuda grass for an all-year-round pasture.

Hop Clover.—Common in the Southern and Eastern states on sandy soils and along roadsides. Periodically this plant comes into notice, due probably to favorable seasonal conditions inducing an abundant growth of unusual size. It often attracts the attention of farmers to such an extent that they become interested in its economic possibilities and send samples to their experiment station for information and advice.

There are several species of hop clover and the talier one doubtless is worthy of cultivation for pasture purposes and for soil improvement, especially on run-down soils that are best suited for pasture purposes.

REFERENCES

"Clovers and How to Grow Them." Shaw.
Kentucky Expt. Station Circular 8. "Clover Sickness."
Farmers' Bulletins, U. S. Dept. of Agriculture:
323. "Clover Farming on Northern Jack-Pine Lands."
441. "Lespedeza or Japan Clover."
455. "Red Clover."
550. "Crimson Clover: Growing the Crop."
676. "Hard Clover Seed and Its Treatment in Hulling."
693. "Bur Clover."

CHAPTER 17

ALFALFA

Alfalfa is one of the oldest forage crops. Its history has been closely related to that of man throughout past ages. It was highly esteemed by the ancient Persians as the most important of forage crops, and followed their invasion by Xerxes into Greece, 490 B. C. During the early centuries of the Christian era it spread throughout the countries of Europe, and it was brought to North America by the early colonists. It was introduced into the Eastern colonies under the name of Luzerne. It found its way into

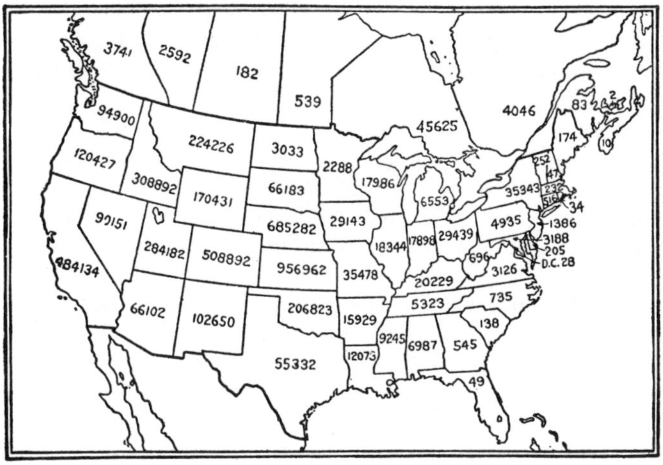

MAP OF THE UNITED STATES AND CANADA SHOWING ACREAGE OF ALFALFA. FIGURES = ACRES.[1]

California and other Western states probably by way of South America, and brought with it the Spanish name of alfalfa.

Alfalfa is characterized by its deep root system, on which are found nodules similar to those described under the clovers. The bacteria in these nodules enable the alfalfa to secure nitrogen directly from the air. Alfalfa plants are propagated only by seeds. They do not spread, as do some of the clovers and many of the grasses, by creeping stems and underground root stocks. Alfalfa is a perennial and under favorable conditions lives many years and attains a large size. The crowns of the plant become

[1] Courtesy of The Macmillan Company, N. Y. From "Forage Plants and Their Culture," by Piper.

much branched and old plants frequently give rise to as many as 200 stems.

Distribution of Alfalfa.—This plant is grown as a crop in every state in the Union and most of the provinces of Canada. The map on the preceding page gives the acres by states and provinces.

Soil and Climatic Adaptation.—Alfalfa is best adapted to a warm, dry climate. In North America it is most extensively and successfully grown under the semi-arid conditions that prevail in the western half of the United States. More than two-thirds of the hay grown in New Mexico is alfalfa. Over one-half of that grown in Colorado is alfalfa. Kansas produces more than Nebraska, and Nebraska more than the Dakotas. The relative production of alfalfa as compared with other forms of hay and forage decreases as we proceed northward and eastward. Alfalfa has been cultivated so long that strains and varieties have been developed that are adapted to a wide range of climate. For this reason it is extensively grown in North America. It will withstand great heat and extreme cold in the arid regions, but is more sensitive to such extremes in the humid regions, and rarely succeeds in tropical or sub-tropical regions where the humidity is high and rainfall abundant.

It is adapted to a wide range of soils, growing well on loose, sandy soils as well as upon heavy clays. It succeeds best on soils of medium texture that are capable of deep penetration by roots and well supplied with mineral plant foods. It will not succeed on soils closely underlaid with hardpan, impervious rock or standing water. Neither will it thrive on sour soils.

Essentials for Success.—In the western half of the United States there are probably few localities where alfalfa will not succeed with the most ordinary treatment. Its growth would be curtailed or possibly prevented by the presence of too much alkali or by over-irrigation. In the eastern half of the United States where conditions are less favorable, there are certain essentials necessary to the success of this crop that must be carefully considered. These are good drainage, freedom from weeds, absence of acidity or presence of plenty of lime, a fair degree of organic matter in the soil, thorough preparation of seed-bed, most favorable time of seeding, inoculation of seed or seed-bed and the use of plenty of good seed. Added to this will be the after treatment, such as time of cutting, care in pasturing, cultivation and mulching.

The treatment essential to success being so diverse and exacting, farmers contemplating growing alfalfa are advised to first undertake it on a small scale. There are a number of advantages in doing this, such as the practical experience gained and the providing of inoculated soil.

Varieties of Alfalfa.—Thus far varieties and regional strains of alfalfa have been relatively unimportant. At least 95 per cent of the alfalfa in North America may be called ordinary alfalfa. A number of strains have been introduced from time to time, some of which are superior for hardiness. Aside from the common or ordinary alfalfa, Turkestan, Arabian, Peruvian and Grimm are of some importance. The common or ordinary alfalfa is that generally grown in North America, Europe, Argentine and Australia.

Turkestan alfalfa closely resembles ordinary alfalfa, and neither plant nor seed can be easily distinguished from it. It is thought to be a little more drought and cold resistant than ordinary alfalfa, but is inferior to the ordinary alfalfa for the eastern half of the United States.

Arabian alfalfa may be recognized by its hairiness, large leaflets, rapid growth and short life. It begins growth and continues to grow at a somewhat lower temperature than common alfalfa.

Peruvian alfalfa may be recognized by its somewhat bluish appearance, coarse, erect stems and large leaflets.

Grimm alfalfa, brought to this country from Germany, has been cultivated here for a long time, and through elimination of the less hardy plants has become adapted to severe climatic conditions. It is, therefore, recommended for the Northern states. It is claimed to resist severe pasturing better than ordinary alfalfa, and is thought to be somewhat more drought resistant. The seed is higher priced than that of the ordinary alfalfa.

Sources of Seed.—Best results are usually secured by the use of locally grown seed. In the eastern half of the country, very little seed is produced, and imported seed must be relied upon. It is, therefore, advisable to secure seed from approximately the same latitude or preferably somewhat north of the latitude in which it is to be used. Nebraska-grown seed is good for Illinois, Indiana and Ohio. Kansas-grown seed is generally a little cheaper and will be good for Missouri and southern Illinois. Dakota seed will be higher priced, but should be used in Wisconsin, Minnesota and Michigan.

Alfalfa seed varies in purity, germination and price. It is, therefore, wise to secure samples from several sources before purchasing. These should be examined for impurities and tested for germination as a basis for calculating which will be the cheapest. None but first-class seed, free from noxious weed seeds and showing good germination, should be used.

A pound of alfalfa contains about 220,000 seeds. If evenly sown on an acre these would average over five seeds to the square foot. Alfalfa fields one year old rarely contain more than twenty plants to the square foot, and older fields usually have less than ten. It is evident from this that a large percentage of the seeds sown fail to produce plants. It is very important that a full stand be secured on all parts of the field. Vacant spots give an opportunity for grass and weeds to start, and these encroach upon the alfalfa.

The percentage of hard seeds in some lots runs very high and necessitates treating the seed to increase its germination. Hard seeds are treated with a mechanical device through which they are passed with much force, and the hard coats are weakened by striking against a hard, rough surface.

Alfalfa seed two years old may generally be used with safety. Old seed can be detected by its having a much darker color and less luster than fresh seed. Good seed will usually germinate in less than ten days. One hundred seeds placed between blotters or in a flannel cloth between two dinner plates will make a satisfactory test.

Need for Fertilizers and Lime.—In the western half of North America commercial fertilizers and lime are seldom needed for alfalfa, but in the eastern half these are frequently of great importance. Large crops of alfalfa remove from the soil considerable quantities of lime and the essential mineral plant foods. For this reason, large crops cannot be maintained except on fertile soils or soils that are well supplied with plant food and lime; 400 or 500 pounds of a fertilizer containing about 10 per cent of phosphoric acid and 6 to 8 per cent of potash should be applied at the time of seeding. If the field is continued in alfalfa for several years it should be top dressed with manure or commercial fertilizer every year or two. There is no danger of getting the soil too rich for alfalfa. Manure should be used that is as free from weed and grass seeds as possible. Their introduction into the alfalfa should be guarded against, and the alfalfa cultivated for weed destruction if necessary.

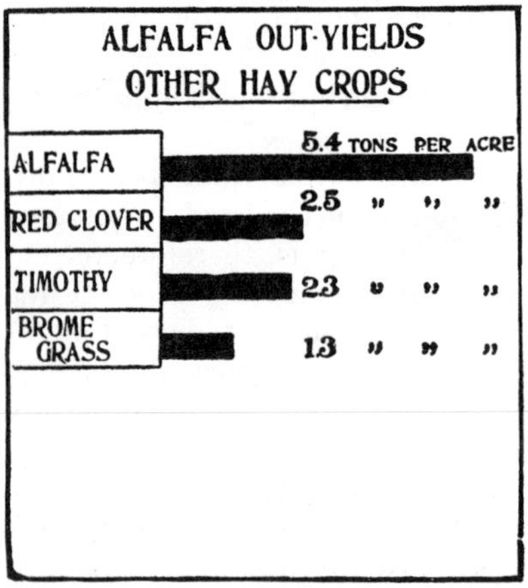

Alfalfa Out-yields other Hay Crops.

Alfalfa has but little tolerance for soil acidity. It removes much lime from the soil and grows best on soils well supplied with lime. Soils should be tested for acidity before seeding to alfalfa, and a liberal supply of lime provided wherever there is any indication of its need. It is immaterial in what form this is applied. The finely pulverized raw limestone is fully as effective as equivalent amounts in any other forms.

Preparation of Seed-Bed.—Alfalfa demands a finely pulverized, moist, fairly compact seed-bed, free of weeds. This can generally be best provided by devoting the land during the preceding year to an inter-tilled crop, such as corn, potatoes or tomatoes. The preceding crop, if liberally manured, will obviate the necessity of applying manure directly for the benefit of alfalfa. This has the advantage of permitting weed and grass seeds in the manure to germinate and be destroyed. The residual effect of the manure will be sufficient to start the alfalfa. The best seed-bed can be secured by plowing late in the spring and disking or harrowing at intervals of ten days or two weeks until the first half of August. Such treatment pulverizes the soil, compacts it, conserves soil moisture and destroys weeds. It provides an ideal seed-bed on which alfalfa may be seeded.

Time, Rate, Depth and Manner of Seeding.—Alfalfa may be seeded either in the spring or late summer. In the western half of the United States spring seeding predominates. In the eastern half, summer seeding is more certain. Seeding either very early in the spring or too late in the season should be avoided. A satisfactory stand is more certain when the seeding is made on soil that is sufficiently warm to produce prompt germination of the seed and rapid growth of the young plants. At 40 degrees north latitude spring seeding may be made during the last part of April or early May. Northward or at considerable elevations the date should be a little later, while southward or at low elevations it may be a little earlier. For latitude 40 degrees north, late summer seeding should generally be during the first half of August, northward it may be a little earlier, and southward considerably later depending on latitude. In any event there should be sufficient time for the alfalfa to become well established and make considerable growth before winter sets in.

The rate of seeding varies greatly, but in the eastern half of the United States and Canada twenty to thirty pounds of seed per acre is advised. In the western half of the United States seeding generally ranges from ten to twenty pounds per acre. Where grown under the dry land system of farming, five to ten pounds of seed per acre often gives satisfactory results.

The seed should be covered anywhere from one-half inch to two inches in depth, depending on character of soil and presence of moisture. The manner of seeding must be determined by local conditions and available machinery. Alfalfa drills are advised when they are available. The most of the seed, however, is sown broadcast and covered with the harrow. Summer seeding is made without a nurse crop and spring seeding generally with a nurse crop. The principal nurse crops are winter wheat, rye, spring oats and barley. Barley is considered preferable to oats, and winter rye seeded in the spring is considered best of all. The nurse crops should be seeded rather thinly in order to encourage the growth of the alfalfa.

Inoculation.—West of the Missouri River the soil seldom needs inoculation for the successful growth of alfalfa. East of that, however, inoculation is generally necessary. Wherever sweet clover is not a common weed and wherever alfalfa has never been grown, it is always advisable to inoculate this crop.

There are two general methods of inoculation: (1) by soil transfer, (2) by artificial cultures. Inoculation by soil transfer is simple, easy and, with reasonable precautions, generally successful. It consists in securing from a well-established field or from a field where sweet clover grows, soil from that portion of the root zone where nodules are most abundant. This is transferred to the new field and spread broadcast at the rate of 300 to 500 pounds per acre, and thoroughly mixed with the soil by disking or harrowing. The inoculated soil should be spread on a cloudy day, or in the morning or evening, and the field thoroughly disked and harrowed at once.

When soil must be secured from a long distance the freight charges, cartage, bags, etc., may make it costly, in which case smaller amounts may be used and more time allowed for the inoculation to develop. Certain precautions are advised relative to the introduction of noxious weeds in this way. If alfalfa is to be grown rather extensively, it is economical to first seed a narrow strip of alfalfa through the center of the field and thoroughly

A Standing Field of Alfalfa.[1]

inoculate it. At the end of one year this will serve as a source of inoculation for the entire field, and a suitable drill or fertilizer distributor may be used going back and forth across the field at right angles to this strip, and filling the distributor from the soil of the strip each time the machine passes.

Artificial cultures have recently been developed and may be secured from a number of sources. By carefully following directions, they are generally successful. The artificial cultures are applied, according to instructions, directly to the seed so that the bacteria are introduced into the soil on the seed and immediately where the young plants start growth. In this way the minimum number of bacteria accomplish the maximum result.

[1] Courtesy of The Pennsylvania Farmer.

After-Treatment.—The after-treatment of alfalfa is more important than in case of the clovers and grasses. Clipping the alfalfa at the close of its first season has been quite generally recommended, but is a doubtful practice so far as direct benefit to the alfalfa is concerned. If, however, weeds and grasses are abundant, or if the alfalfa was seeded early and is blooming rather freely, clipping in the fall is advised. The clipping should be so timed as to prevent maturing of weed seeds. The alfalfa should be clipped rather high and the clippings left on the field for winter protection.

Winter killing of alfalfa is most severe during the first winter and in severe climates or on soils subject to heaving. Winter protection by mulching or otherwise is advised. The more hardy varieties of alfalfa will stand a temperature twenty to thirty degrees below zero if the soil is reasonably dry. The chief trouble occurs as a result of the plants being heaved out of the soil by repeated freezing and thawing, generally toward the close of the winter.

Disking and harrowing alfalfa fields have been frequently recommended for the purpose of killing weeds and grass, for loosening the soil and for splitting the crowns of the alfalfa plants. The improvement of soil and destruction of weeds is justifiable, but injury to the alfalfa plants should always be avoided. Under favorable conditions considerable injury may not prove serious, but in the eastern part of the country, injury to the crowns of the plants results in decay of the roots and shortens their life. The ordinary disk is, therefore, not recommended. Suitable harrows and the spike-toothed alfalfa disk harrow may be used to good advantage. The spring-toothed harrow with the teeth brought to a sharp point is recommended. There is enough spring in the teeth so that they will pass around the crowns of the alfalfa plants without serious injury, and at the same time will uproot small weeds and grasses.

Cultivation should take place just after cutting, and is generally not necessary during the first year of the alfalfa.

Making Alfalfa Hay.—The time of cutting alfalfa should be carefully regulated in order not to injure it. If cut too early the second crop is slow in starting and the exposed crowns of the plants may be injured by hot, dry weather. Neither is it advisable to delay the cutting, for this will result in clipping off the new shoots that produce the new crop. Alfalfa should be cut for hay when the small shoots starting from the crown and which produce the next crop are one-half inch to one and one-half inches in length. At this time about one-tenth of the blossoms will usually be out. In the eastern part of North America leaf spot is quite common and spreads rapidly through the field as the plants approach the hay-making stage. If this trouble is very prevalent the leaves fall rapidly and harvesting should be hastened somewhat to prevent loss. A fair degree of maturity of the alfalfa makes the curing of hay easier than if cut when too succulent. In the western half of the United States there is very little difficulty in this respect. Weather conditions are more favorable and hay of good quality can be made with the minimum amount of labor. In the eastern half of the country rains are prevalent, especially at the time of the first cutting. This calls for special precautions and often necessitates extra labor and the use of canvas covers to secure hay without serious injury.

It is advised to cut in the evening and early morning, and follow the mower with the tedder before any of the leaves become dry. The second teddering at right angles to the first is advised if the alfalfa is heavy. With favorable weather it may be possible to put the alfalfa in the windrow toward evening of the first day. One more day's exposure in the windrow under favorable conditions will generally cure it sufficiently to go directly to stack or mow. This reduces handling to the minimum and prevents loss by shattering.

Curing Alfalfa Hay in Shocks.[1]

If weather conditions are threatening, it will be best to put into moderate-sized shocks at the close of the first day, and cover with canvas to protect from rains. It requires from three to seven days to cure in the shock, depending on weather conditions.

A little more than two-fifths of alfalfa hay is leaves and about three-fifths stems. The leaves, however, contain fully three-fifths of the protein. It is, therefore, advisable to save the leaves as fully as possible. Do not rake or tedder alfalfa in the middle of the day if dry. This is sure to shatter the leaves and cause serious loss.

Number of Cuttings and Yield.—Alfalfa is a remarkable hay and forage plant because of its long life and the frequency with which it may be cut every year. The number of cuttings varies with the locality and ranges from two or three cuttings in the provinces of Canada and the northern tier of states to as many as ten or eleven cuttings annually in the Imperial Valley in California. In the warmer portions of Texas seven or eight cuttings are not uncommon. In most parts of the country, a second crop may be harvested within from thirty to forty days after the first cutting. In warm regions where the growing season is long, cuttings during this season may be made about every five weeks.

[1] Courtesy of The Pennsylvania Farmer.

The yield is generally largest for the first cutting of the season and declines slightly for subsequent cuttings. Much, however, will depend upon rainfall and available moisture which influences the growth.

Alfalfa yields about twice as much as red clover and, being richer in protein, produces about three times as much protein per acre.

Other Uses of Alfalfa.—Alfalfa makes an excellent soiling crop and produces a succulent nitrogenous roughage, especially desirable for dairy cows. Since it may be cut three or more times each season it may be quite extensively used for this purpose. It, therefore, takes a very important place in a soiling system wherever it can be satisfactorily grown.

The last cutting of alfalfa comes at about the right time to combine with corn for the making of ensilage. One load of alfalfa to every three or four loads of corn makes an excellent combination. Alfalfa is sometimes made into silage by itself, but makes a rather sour, slimy product.

While alfalfa is not a pasture plant and is easily injured by pasturing, it may be used especially for young stock and for swine. It makes a most excellent pasture for the latter, and where it is to be used for this purpose will carry about forty pigs and their dams per acre without being injured. It is generally thought advisable to divide the field into two or three parts, pasturing one part for a period, and then turning into another part. Frequently some hay may be harvested in addition to pasturing.

Comparison of Hogs Fed on Corn and on Alfalfa.[1]

Alfalfa makes a range for poultry and may also be fed to poultry and swine in the form of hay.

Composition and Feeding Value.—The composition of alfalfa is given in Table VI in the Appendix. The nutritive ratio of alfalfa hay is about 1 to 4. Extensive experiments at a number of experiment stations

[1] Courtesy of The International Harvester Company, Agricultural Extension Department. From pamphlet "Livestock on Every Farm."

have clearly demonstrated the high feeding value of alfalfa. Experiments with forty cows covering a period of two years at the New Jersey Experiment Station clearly demonstrated that eleven pounds of alfalfa hay were equal in feeding value to eight pounds of wheat bran. Plenty of alfalfa as roughage materially reduces the bills for the purchase of protein in costly concentrates.

The hay is exceedingly palatable and highly digestible and is eaten with avidity by all classes of livestock. When fed to horses the ration should be limited. Horses, if allowed to eat their fill, generally consume nearly twice as much as is necessary to provide the required protein of their ration. This results in unnecessary waste of feed. Alfalfa hay and corn make a good combination, since the corn tends to properly balance the ration.

Considerable alfalfa hay is made into alfalfa meal for shipment to the eastern markets and is quite extensively used in rations for dairy cattle and also for poultry.

Irrigation of Alfalfa.—Alfalfa is exceptionally well adapted to irrigation and a large portion of that grown in North America is irrigated. The amount of water to use will be determined chiefly by the character of the soil and rainfall of the region. It is a good practice to irrigate rather liberally and at rather remote intervals. Alfalfa is so deep-rooted that the soil should be thoroughly wet to the depth of three feet or more. Ordinarily, one good irrigation should produce a full cutting of alfalfa. It is, therefore, customary to irrigate the fields immediately after the hay is removed and this irrigation should be sufficient to last until the next cutting. With this system certain precautions are called for such as to prevent the scalding of the young and tender shoots that are just starting to grow at this time. Where fields are deeply and rapidly flooded with water carrying much sediment, a deposition on the young shoots frequently causes injury. It is advisable to irrigate carefully, providing for slow movement of the water across the fields without attaining any considerable depth at any point. Over-irrigation is to be avoided, since it not only wastes water, but often causes a rise in the ground-water table and brings alkali salts to the surface of the soil.

Winter irrigation is practiced in some localities where the winters are mild and where water is abundant at this time of the year. The principal object is to conserve water which would otherwise go to waste. This is especially desirable where water is scarce in summer. Such winter irrigation will often result in one good crop that could otherwise not be secured.

Seed Production.—The production of alfalfa seed in North America is confined chiefly to the semi-arid regions. East of the Missouri River the production of seed is small, except when drought prevails. It is estimated that about one-half of the seed used in North America is produced on irrigated lands in regions of dry summers. There is also a considerable amount produced on unirrigated semi-arid lands, and such seed is considered preferable for dry farming purposes. When produced on unirrigated lands alfalfa is seeded very thinly. In some cases it is seeded in rows sufficiently far apart to permit of cultivation. Isolated plants that can branch abundantly and receive plenty of sunlight, seed more abundantly than when they are close together. When grown under irrigation, irrigation water is withheld during the period of seed formation. The presence of rains or the application of water stimulates the vegetative growth and reduces seed production. Usually the second crop is utilized for seed production, although in the extreme Northern states the first crop is necessarily used. There are various conditions that influence the yield of seed, such as thickness of stand, moisture supply, conditions favorable to pollination, etc. Yields of as much as twenty bushels per acre have been reported, but eight bushels are considered a good yield. Two to five bushels probably represent the average crop.

A Well-set Cluster of Alfalfa Pods.[1]

Little is known relative to seed production east of the Missouri River, although numerous observations have shown that plants frequently seed quite abundantly. In the corn belt it is quite possible that certain crops could be used for seed to good advantage. The probable yield of seed is indicated if the crop has been in bloom for some time and considerable seed is set before new shoots appear. If dry weather prevails when these conditions are evident there is a fair chance of a crop of seed.

The hope of securing varieties adapted to eastern conditions lies in the possibility of seed production in the various localities.

The method of harvesting the seed of alfalfa is essentially the same as that for red clover.

REFERENCES

"The Book of Alfalfa." Coburn.
"Alfalfa in America." Wing.
"Clovers and How to Grow Them," pages 118–193, Shaw.
Missouri Extension Service Circular 6. "Growing Alfalfa in Missouri."
Delaware Expt. Station Bulletin 110. "Alfalfa."
Wisconsin Expt. Station Bulletin 259. "Alfalfa Growing in Wisconsin."
U. S. Dept. of Agriculture Bulletin 75. "Alfalfa Seed Production."
Farmers' Bulletins, U. S. Dept. of Agriculture:
315. "Legume Inoculation."
339. "Alfalfa."
495. "Alfalfa Seed Production."

[1] Courtesy of U. S. Dept of Agriculture. From Farmers' Bulletin 495.

CHAPTER 18

MEADOWS AND PASTURES

Success with livestock is conditioned on the production of good grass. This may be in the form of meadows or pastures, but a combination of the two is generally desirable. In latitudes of long winters the importance of meadows may predominate, whereas in regions of short winters, pastures may be the more important. With minor exceptions, meadows and pastures are the most economical source of the farm income.

As a rule, the highest type of general agriculture includes the rearing of farm animals. They may be considered machines for the manufacturing of the roughage produced on the farm into more concentrated and valuable products, such as meat, milk, butter, wool, etc. These require more skill on the part of the farmer and give to him continuous employment.

Extent, Value and Importance.—It is estimated that about thirty per cent of the improved land in the United States is pasture land. The largest area of land used for grazing is embodied in the extensive ranges lying in the western half of the United States. To this range land and the permanent pastures on farms may be added large deforested areas that are capable of producing pasture. The value of the products per acre from the grazed land is exceedingly low, but since the area is so large, the aggregate return is great. The return per acre from meadow land is also comparatively low, but much larger than that from pasture lands. No statistics are available by which to estimate the returns from pasture lands, although there are fairly accurate statistics for the meadows, as indicated in the chapter on "Meadow and Pasture Grasses."

Essential Qualities of Meadows and Pastures.—The essential qualities of meadow grasses are given in the chapter under that name. It is not so essential that meadows become permanent, except in case of wet land or land too rough or stony to be cultivated, and which for any reason cannot be pastured.

It is generally important, however, that pastures be made as permanent as possible. This calls for a mixture of grasses that are either very long lived or that are capable of reproduction under pasture conditions. A good pasture should start growth early in the season and continue to produce until late in the fall. The grasses should be palatable, nutritious and present variety and give abundant growth. They should also form a continuous, compact turf that will withstand much tramping by animals. A variety of grasses that will provide for growth under both moist and dry soil conditions is also advantageous. The deep-rooted grasses and clovers can, therefore, be advantageously included with the shallow-rooted ones such as blue grass and white clover. The latter are more substantial, both in quality of grazing and in the character and durability of turf which they form.

Advantages of Meadows and Pastures.—Where land is moderate to low in price and labor is costly, no feed will produce results with cattle and sheep as economically as good pasture. While a given area in meadow will produce three times as much weight in hay as it will in pasture, yet there is about three times as much protein in a given weight of dry material in pasture grass as there is in the same material in hay. The increased energy value of the hay over that of an equal area of pasture will generally be offset by the increased labor required in harvesting and feeding the hay. Meadows require on an average one unit of man and horse labor per acre annually. This consists of ten hours work per year. The cultivated crops

LIVE STOCK ON PASTURE.

require from two to as high as fifteen or sometimes twenty units of labor per acre.

Pastures, on the other hand, require no labor unless it be for the purpose of applying manure or fertilizers, or for improvement by re-seeding or cultivating. It is from the standpoint of labor that meadows and pastures are especially economical. When land values become exceptionally high, farmers may be justified in reducing the acreage of pasture and resorting to cultivated crops as a source of feed for livestock. This is an economical problem that must be determined by local conditions.

Meadows and pastures make use of land which cannot be economically used for cultivated crops. This is especially true in the case of woodland pastures or pastures along streams that are irregular and subject to overflow. Stony portions of farms are often utilized as meadows or pastures. Irregular corners, cut off by roads or streams, are more economically devoted to hay than to a cultivated crop requiring tillage.

Soil and Climatic Requirements.—Most of the grasses and clovers succeed best in moist, cool climates and on soils that range from medium to heavy in texture. On the other hand, there are a few grasses and clovers that succeed in regions of continuous high temperature. There are, however, no regions in the world within the tropics that are especially prominent for the production of meadows and pastures. These attain their greatest perfection in temperate climates with abundant and well-distributed rainfall. England and Scotland represent the ideal conditions for meadows and pastures. The range in variety of grasses and clovers makes possible meadows and pastures which are more or less successful in all parts of North America. Of course, there are considerable areas of sandy soils, especially in the warmer sections, that are impracticable of utilization in this way.

Formation of Meadows and Pastures.—Since meadows and pastures are to remain for a considerable period of time, the necessity of thorough preparation for their establishment is more imperative than in case of annual crops. The successful orchardist goes to much expense in the preparation of land and the setting of trees for the orchard, realizing that orcharding is a long-time proposition. The same policy is applicable in case of permanent pastures or meadows. The shorter the period of time that a meadow is to remain as such, the less will be the expense justified in its establishment.

The first consideration is the adaptation of the land for meadow or pasture purposes. The value of the land and the possibility of its utilization for other purposes should be considered. Consideration must also be given to the variety and character of grasses adapted to the soil and climate and that will meet the requirement of the livestock to be pastured. No definite formula can be given, since conditions vary greatly.

Preparation of Soil.—The preparation of the soil for either meadows or pastures should begin at least a year in advance of the time of seeding. There are two things essential to the establishment of grasses and clovers, namely, absence from weeds and a good physical condition of the soil. This may be provided by growing an inter-tilled crop which is given thorough cultivation during the year preceding the seeding of grass.

Organic matter in the soil is decidedly helpful for both grasses and clovers, but not essential. In plowing for seeding grasses and clovers, manure and organic matter should not be turned under too deeply, but should be left as near the surface as possible. A thorough preparation of the seed-bed is essential for both meadows and pastures. For meadows, the soil should not only be thoroughly pulverized and made moist and compact, but should also be level to facilitate cutting at a uniform height. The presence of hummocks or depressions in a meadow means that some of the plants will be cut close to the crowns and others cut far above.

A moist, compact, finely-pulverized seed-bed is essential in pastures, but it need not be necessarily level, since animals can graze with as much satisfaction on uneven land.

When seeding is to be made in August it is well to plow the land in the spring. An occasional disking or harrowing during the summer will destroy the weeds, conserve moisture and provide a pulverized seed-bed for the grass. It is advisable to plow land for spring seeding of grass and clover the preceding year, or at least several weeks in advance of seeding time, in order that it may become thoroughly settled before seeding.

Meadow and Pasture Seed Mixtures.—From the standpoint of both variety and total yield, mixtures give best results in both meadows and pastures. Experiments at several experiment stations report yields for mixtures of two or more grasses and clovers that exceed the yield of any of the varieties entering into the mixture when seeded alone under identical conditions. There are a few exceptions, namely, that of alfalfa which is cut several times a year, and which generally gives best results when grown alone. The same has been found true with Italian rye grass.

Mixtures yield better than pure cultures because: (1) the requirements of the different plants entering into the mixtures are dissimilar and do not make them direct competitors for plant food; (2) the root habits being different, their distribution through the soil is more thorough; (3) the average yearly return is more nearly even through a long period of time; (4) variation in light requirements of different plants enable some species to do well in the shade of taller ones, thus increasing the quantity of herbage; and (5) legumes mixed with grasses increase the nitrogen supply for the latter.

As before indicated, mixtures for mowing purposes should contain only plants that mature near the same time. This will generally confine the mixture to two or three species, although occasionally a larger number may be advantageously used. Timothy and red clover constitute the mixture most common and practical over a large region of the hay-producing district of North America. Redtop and alsike clover are frequently included, especially where soils are wet and inclined to be sour. Alsike clover and redtop are occasionally used without the timothy and red clover. Orchard grass and alsike clover work well together, both as to character of growth and time of maturity.

In pasture mixtures there is opportunity for a much greater variety and wider range as to time of maturity in the plants used. In North America, however, mixtures made up of a great number of clovers and grasses are rather unusual, although these seem to be the rule in pasture mixtures of England and Scotland.

Soil and climatic conditions are so diverse that it is impossible to enumerate all the mixtures suited to different conditions and localities for any extensive region or for different purposes. Prominence should be given, however, to those grasses that are best adapted to local conditions and best meet the needs. One or more species that will make quick growth and give early pasture should be included in such a mixture. The following general suggestions are offered:

In regions adapted to Kentucky blue grass, add white clover, red clover and timothy.

On wet soils adapted to redtop, add white clover and alsike clover.

On poor upland soils use redtop, Canada blue grass and white clover. Under certain conditions brome grass may be included.

Where Bermuda grass thrives best, add Lespedeza clover, bur clover and Italian rye grass.

In addition to the grasses mentioned, orchard grass is desirable, because it furnishes early pasture.

If there is any doubt relative to the purity of the grass and clover seeds to be used, a sample should be submitted to the state experiment station for examination and test. One familiar with grass and clover seeds may make his own inspection by the use of a hand lens, and may also make his own germination test by the use of white blotting paper moistened and placed in an ordinary dinner plate covered with another to retain moisture. One or two hundred seeds placed between the blotters and kept at favorable temperature will enable one to determine the percentage of germination. Careful inspection every day or two should be made to keep the blotters continuously moist.

Seeding Grasses and Clovers.—A full crop of grass, whether for a meadow or pasture, necessitates a full stand of plants. The first essential to this is the requisite number of viable seeds, well distributed on every part of the field. There are many factors that influence the stand besides the rate of seeding.

Rates of seeding for the different grasses and clovers when used alone are given in the chapter on "Grasses and Clovers." A few species only enter into the average meadow mixture. As a rule, the ratio of the amount of seed for the different species entering into a meadow mixture will be a little larger than the amount when seeded alone. For example, timothy seeded at the rate of 15 pounds alone and red clover at the rate of 12 pounds, when seeded together would require on an average of about 9 pounds of timothy and 7 pounds of clover, making a total of 16 pounds as compared with one-half of the sum of the two individual rates, which would be $13\frac{1}{2}$.

The depth of seeding has already been discussed under several of the species of grasses and clovers. The depth in case of mixtures should be regulated with even more accuracy than in seeding one species only. It should meet as accurately as possible the needs of the leading grasses and clovers in the mixture. In special cases it may be found advantageous to drill the clovers and broadcast the smaller grass seeds, such as timothy, redtop and blue grass. The depth is also controlled largely by character of soil and weather conditions. In midsummer, when the soil is dry and the temperature high, seeds should be covered rather deeply. In the cool, moist portion of the year, very shallow covering is better. In no case can grass and clover seeds be covered more than two inches without suffering much loss. With the smaller grass seeds, one-half inch to an inch is generally sufficient.

The time of seeding is subject to considerable latitude, but there are two seasons of the year that generally give best results. These are very early in the spring or rather late in the summer. These two seasons will be subject to some modification, depending upon weather conditions. It is wise to seed when the soil is in a good moisture condition so as to insure quick germination. As a rule, it is not advisable in case of summer seeding to seed just before a heavy rain. Such a rain compacts the soil and the hot weather that is likely to follow will form a crust that the small plants cannot penetrate. Seeds deposited in a dry soil may be germinated by a light shower followed by dry weather that will cause the small plants to perish.

Grasses seeded in summer may be broadcasted on a well-prepared seed-bed immediately following the harrow. One additional harrowing will sift the seeds down into the soil and effect a satisfactory covering. If the soil is dry the first harrowing may be followed by the plank drag. This will mash the small clods, compact the soil, bring the moisture nearer the surface and germinate the seed.

The manner of seeding depends largely on seasonal condition of the soil and character of grass-seed mixture. Grasses and clovers are generally sown broadcast. There are a number of forms of seeders. The grass-seed attachment to the grain drill predominates where fall seeding with wheat occurs. It is also extensively used where the drill is used for spring seeding of oats. The wheelbarrow seeder and the hand seeder are extensively used when seeded alone or on grain fields where drills are not employed. Slant-toothed spike harrows are most generally used for covering the seed when broadcasted in this way. Brush harrows are sometimes used when the seed is very small and the seed-bed very mellow. This avoids covering too deeply. In any case, implements should be used that do not tend to drag trash or soil and result in bunching the seed. Much seeding is done in the winter and very early spring which calls for no covering. In this case the seed is covered by the freezing and thawing of the soil and by rains and winds.

Late fall and early spring seeding usually takes place with a nurse crop. In this way the cost of seed-bed preparation is charged chiefly to the grain. This is the cheapest possible way of seeding grass other than that of sowing it in the spring to be covered by the freezing and thawing and rains. The nature of the nurse crop is important. Moderately thin seeding and the use of early varieties generally favor a good catch of grass.

Seeding without a nurse crop calls for especially well-prepared seed-bed and freedom from weeds. Such seeding generally does best in the late summer.

Treatment of Meadows and Pastures.—Of all the farm crops, the meadows and pastures are probably the most neglected. Meadows usually receive more care and attention than pastures. The treatment accorded meadows will consist chiefly: (1) in the application of manures

HAY-MAKING SCENE—MILLET.[1]

[1] Courtesy of Virginia-Carolina Chemical Company, Richmond, Va. From V.- C. Fertilizer Crop Books.

and fertilizers, (2) re-seeding of the grasses and clovers in case of failure, (3) cultivation to maintain a good physical condition of the soil, and (4) cutting of weeds when they become serious.

The cultivation given to meadows, while rather unusual, will consist mainly in disking and harrowing. These operations will frequently be demanded wherever re-seeding is required and may be used for the destruction of weeds and the loosening of the soil. There are now on the market certain forms of spiked disks designed especially for this type of work.

Cultivation is even more applicable to pastures than it is to meadows. Pastures are more permanent, or at least remain for a long series of years without being disturbed. Certain grasses frequently become sod-bound. As a result of close grazing, weeds also frequently become numerous. The tramping of the animals tends to compact the soil. Cultivation is beneficial for all of these difficulties. Harrowing spreads the droppings of the animals and affords a more effective distribution of the manure for the benefit of the grass.

Meadows should not be maintained for too long a period. Better results have been secured by plowing and re-seeding than to continue too long in consecutive crops of grass. In pastures the situation is much different. There are records of pastures forty and fifty years in grass without being disturbed. This applies, however, to those regions in which the soils and climate are especially adapted to the typical pasture grasses and clovers, such for example as Kentucky and Canada blue grass and white clover. Where pastures are prone to run out in a few years, it is better as a rule to re-seed. This, of course, applies only to lands that are capable of cultivation and devotion to other crops.

Care of Meadows and Pastures.—The life of a meadow and the maintenance of its productivity may be prolonged by exercising certain precautions in connection with its care and the harvesting of the crops. It is unwise to pasture animals or to haul manure onto a field when the soil is too wet. The more permanent the nature of the meadow the greater should be the care exercised. Meadows should go into the winter well protected by either sufficient second growth or proper mulching with manure. It is, therefore, unwise to closely pasture the aftermath of meadows late in the season. In favorable years a moderate amount of pasturing will not be undesirable. If weeds occur in considerable numbers, late summer or fall clipping to prevent seeding is advised.

Pastures should not be grazed too early in the spring. It is undesirable: (1) from the standpoint of not giving the grass a sufficient start, and (2) through injury by tramping and compacting the soil when it is wet. It is also unwise to pasture closely too late in the fall, since pastures, like meadows, should have winter protection. It is never wise to pasture too closely at any time of the year. Close pasturing reduces the vitality of the plants and their subsequent producing capacity. The packing of the soil by animals under favorable conditions will be overcome in temperate climates by the freezing and thawing during the winter.

In grasses the growth takes place at the base of the leaves and lower portions of the internodes, so that grazing does not destroy the plants unless the plants or portions thereof are injured below the point of growth.

The grazing capacity of a pasture will be determined by the care given to it and the manner in which it is grazed. Its grazing capacity should be fully utilized, and it is believed that the pasture will be maintained fully as well, and sometimes better, in this way than when not fully grazed. In pastures that are not fully utilized many weeds occur that go to seed and result in weedy pastures within a few years. No animals are better for destroying weeds than sheep, although all classes of livestock will eat most kinds of weeds when there is a shortage of grasses. There are few experiments in America on pastures and pasturing.

Improvement of Meadows and Pastures.—"An ounce of prevention is worth a pound of cure" applies especially to meadows and pastures. This is pretty thoroughly covered in the treatment and care of meadows and pastures discussed in the preceding topics. Brush pastures may be improved by removing the brush by clearing, by firing or by pasturing with goats. The latter is perhaps the most economical method, provided goats can be secured and disposed of without loss. This not only cleans the pastures, but utilizes the removed product in the form of brush, weeds, etc.

Wet pastures may be improved by underdrainage. This not only encourages the growth of the more nutritious and better grasses and clovers, but protects the pasture against injury through tramping by animals when too wet. The expense of drainage for pasture land must not be too great.

Manuring, Fertilizing and Liming.—Sour soils should be liberally limed when prepared for meadows or pastures. Meadows that are to be continued for several years may be top-dressed with lime to good advantage, and pastures may be top-dressed at intervals of six to ten years. The benefits from liming will be determined chiefly by the acidity of the soil and the proportion of clovers that enter into the meadow and pasture mixtures.

Barnyard and stable manure is advantageously used in the establishment of meadows and pastures. It is often advisable to apply the manure to the crop preceding the one in which the grass is seeded. On the other hand, meadows that are to remain for several years may be advantageously top-dressed with light applications of manure, greatly to the benefit of the grass. Such top-dressing has been found profitable wherever manure is available, or may be purchased at low cost. The better sod resulting is also beneficial to the crops which are to follow the meadow.

It is unusual to apply manure to pastures once established, since the droppings of the animals, if properly distributed, go far towards meeting the needs of the soil. In all probability the manure can be more advantageously used on the meadows and other crops.

Experiments at several of the state experiment stations have demonstrated that moderate amounts of complete commercial fertilizers can be economically used on meadows. The more perfect the stand of grass, the larger the increased yields resulting from such treatment. While the composition of the fertilizer will differ somewhat for different soils and grasses, that for the grasses proper should contain about equal percentages of the

GOOD PASTURE LAND.[1]

three fertilizing constituents. Nitrogen is essential in increasing vegetative growth. A home-made mixture consisting of 150 pounds per acre each of nitrate of soda and acid phosphate, and 50 pounds of muriate of potash, is recommended. This should be applied broadcast very early in the spring just as the grass is beginning to start.

Since nitrogen is so expensive, clovers should be used in both meadows and pastures for the benefit of the grasses. They also increase the protein content of both the hay and grazed product.

Utilizing Aftermath.—The amount of aftermath or second growth on meadows depends on the nature of the grasses, the time of cutting the first

[1]Courtesy of The Macmillan Company, N. Y.

crop and the weather conditions which prevail. With early cutting of the first crop and favorable subsequent weather conditions, the second crop may be as large and well worth harvesting for hay. Certain precautions in this connection are necessary, namely, not cutting so late as to prevent further growth for winter protection. There is no objection to pasturing the aftermath if not pastured too closely and if the character of grasses is such as not to be seriously injured by the tramping of animals. The future life and use of the pasture will be a factor in this connection.

Capacity of Pastures.—The capacity of pastures varies all the way from fifty acres to the animal unit in case of the range pastures of the West to one acre per animal unit on first-class pastures in humid regions. The capacity is also measured by the length of grazing season, and this is dependent chiefly upon latitude and elevation. It is also influenced by the nature of the pasture grasses, some prolonging their growth into the cooler portion of the year. Experiments show that more product is secured as hay than can be secured when the same grasses are pastured. This has been determined by comparing the relative yield of cuttings at short intervals with cutting once at maturity. Such experiments have given nearly three times as much dry matter in the form of hay as was secured in frequent cuttings. The protein content of the new growth was much higher and aggregated nearly as much in frequent cuttings as in the matured product.

Pasture experiments in Missouri showed average daily gains of 1.65 and 1.85 pounds for yearlings and two-year-old steers respectively during the summer season. At the usual charge for pasturage in that state, the estimated cost per hundred pounds of live weight was $1.60 and $1.90 respectively. Pasture experiments in Virginia covering several years gave gains in live weight of 150 pounds per acre annually. This was on average blue grass pasture in that state. The average pasture in the humid region should produce 150 pounds live weight in cattle per acre annually.

Composition and Palatability of Pasture Grass and Hay.—The composition of various kinds of grasses and hay is given in Table VI in the Appendix. The composition of grass mixtures will be determined by the relative portions of the species entering into it, and also by the stage of growth when harvested, and the conditions under which grown. Nitrogenous fertilizers have been found to somewhat increase the protein content of the grasses.

The palatability and digestibility of grasses as grazed are doubtless much greater than those for mature hay. The labor required for harvesting the hay is also saved.

Temporary Pastures.—Temporary pastures are generally provided to meet early needs and are designed for short periods. They consist of annual plants, of which there are many species. These will be determined by soil and climatic adaptation and the character of animals to be grazed. Oats, sorghum and red clover make a good combination. Oats make rapid growth during the early part of the season, while sorghum grows more rapidly with the approach of warm weather. As these two crops are becoming exhausted, the clover takes their place. This mixture is suited to spring seeding and can be pastured from the latter part of June to the close of the season. Another mixture consists of spring wheat, barley and oats, using about one-third of the usual sowing of each. These may be pastured as soon as they attain sufficient size to afford a good supply of pasturage. Another mixture frequently used consists of rye, winter wheat and winter vetch sown in the fall. This will afford pasture in the spring earlier than the spring-sown grains, and if seeded fairly early may furnish some winter pasture. In pasturing the annual crops, waste by tramping may be prevented by restricting the area grazed by means of hurdles or temporary fences. Such pastures require knowledge relative to the date crops must be sown to afford pasture when needed. In this respect it resembles the provision for soiling crops which are to be cut and fed from day to day.

REFERENCES

"Meadows and Pastures." Wing.
"Forage and Fiber Crops." Hunt, pages 1–274.
"Farm Grasses." Spillman.
Pennsylvania Expt. Station Bulletin 101. "Meadows and Pastures."

CHAPTER 19

MISCELLANEOUS ANNUAL HAY AND FORAGE CROPS

Of the miscellaneous annual hay and forage crops the legumes take first place. They are important both from the standpoint of high feeding value and of the benefit derived from them by the soil. In regions adapted to alfalfa or the clovers, annual legumes find a minor place, chiefly as substitutes when for any reason the clovers fail.

Cowpeas and soy beans are by far the most important annual legumes. The former are especially adapted to the cotton belt, while the latter may be grown wherever corn is successfully raised. For northern latitudes, Canada field peas and winter vetch are hardy and promising.

Of the non-legumes, the millets and sorghums rank first as annual hay and forage crops.

COWPEAS

The cowpea is a warm-weather crop, and is the best annual legume for the entire cotton belt. It is suited for the production of both hay and seed. It is seldom grown above 40 degrees north latitude, and in the northern limits of its production only early-maturing varieties should be used. There are more than sixty varieties of cowpeas, differing greatly in size, character of growth, color of seeds and time of maturity. Only a few of them are extensively grown.

Varieties.—Whippoorwill is the best known and most extensively grown variety. It is of medium maturity and well adapted for making hay. It may be recognized by seed which has a mottled chocolate on a buff or reddish ground color. It makes a vigorous growth, quite erect and produces a large amount of vine. It can be handled readily by machinery.

Iron is also a well-known variety, and is especially valuable because it is practically immune to root knot and wilt, diseases which cause much trouble with cowpeas in many parts of the cotton belt.

New Era is one of the earliest of the cowpea varieties and is adapted to the southern portion of the corn belt. Its habit of growth is erect with few prostrate branches, thus making it easy to cut with machinery. It produces a heavy crop of small seed, characterized by innumerable minute blue specks on a gray ground color. Because of the small seed, less quantity is required for seeding.

Unknown or Wonderful is one of the most vigorous and largest growing varieties and is late in maturing. It is quite erect and is handled readily by machinery, either for hay or grain production. The seed is large and of a light clay color. It is not adapted north of North Carolina and Tennessee, except in a few localities at the lower altitudes.

Clay is the most variable of any of the varieties, and the name is given commercially to any cowpeas having buff-colored seeds, except the Iron. For this reason there are doubtless many varieties that masquerade under this name. This variety is vigorous, but of a trailing habit. It fruits sparingly and is consequently rather unpopular either for seed or hay purposes. It is especially valuable for pasturing and for soil improvement.

Groit is very similar to New Era, but makes a slightly larger growth and fruits more heavily.

Black is a variety characterized by its large black seeds that do not decay rapidly after ripening, even after lying on the warm, moist earth.

FIELD OF IRON COWPEAS PLANTED IN ONE-FIFTH-ROD ROWS AND CULTIVATED THREE TIMES.[1]

It is especially adapted to the sandy, coastal plain soils of Virginia and North Carolina. It is also popular in the sugar-cane section of Louisiana.

Time, Manner, Rate and Depth of Seeding.—Cowpeas should not be seeded until the soil is thoroughly warm. In most localities the date of seeding will be one or two weeks later than the best time for planting corn. The plants are tender and are injured by the slightest frost.

In the cotton belt, the time of seeding should be regulated so that when harvested for hay, the proper stage of maturity will occur when the weather conditions are favorable for hay making. This will usually be sometime in September.

The seed-bed for cowpeas should be prepared the same as for corn. The planting may be in drills or by broadcasting. When grown for seed it is generally best to plant in drills not less than thirty inches apart and

[1] From Farmers' Bulletin 318, U. S. Dept. of Agriculture.

cultivate the same as for corn. Good results, however, have been secured by seeding with the ordinary grain drill, which, of course, permits of no cultivation. When seed is costly, the saving of seed by drilling in rows thirty inches or more apart may offset the labor of cultivation. When grown chiefly for hay, broadcasting or drilling in rows close together is best.

The rate of seeding varies from one to eight pecks per acre, depending on the manner of seeding, the character of seed and the purpose for which grown. When seeded with the wheat drill, with all of the holes open, one bushel of seed per acre will give good results for hay and still provide for fair yields of seed. Small seed requires less in planting than large, and less seed is required for seed production than when grown for forage.

The depth of seeding will depend on the character and condition of the soil. It may vary from one to four inches. The looser the soil or the drier the seed-bed, the deeper should be the planting. The cowpea is really a bean and, like all beans, should not be planted too deeply.

Cowpea seed usually costs from $2 to $3 per bushel.

Seeding with Other Crops.—There are two principal advantages in seeding cowpeas with other crops, namely, the production of a better balanced ration when used as forage, and the increased facility with which the crop may be harvested and cured when supported by upright growing plants.

The best crops to seed with cowpeas are corn, sorghum and millet. These are all similar to the cowpea in soil and climatic requirements. It is never wise to seed cowpeas with oats, as the one requires warm weather and the other cool weather for best results.

The upright growing varieties of cowpeas may be grown with corn, preferably by planting both corn and peas in rows at the same time. By selecting the proper variety with reference to habit of growth and time of maturity, the cowpeas may be harvested at the same time with a corn harvester and used for making ensilage.

In the southern portion of the corn belt and in the cotton belt cowpeas are frequently drilled between the corn rows after the last cultivation. The pods are gathered for the peas and the vines turned under for the benefit of the soil. When planted with corn, the cowpeas should be four or five inches apart in the row and the corn about twelve inches apart. Best results are secured by using a cowpea attachment to the corn planter.

When grown for hay, seeding with sorghum or millet gives best results. Sorghum is generally preferable to millet, because it has a somewhat longer growing season and makes a more palatable hay. Best results are secured by mixing the seed at the rate of two bushels of peas to one bushel of sorghum and seeding with a wheat drill at the rate of one and one-half bushels per acre. The large varieties of millet may be used with the early maturing varieties of cowpeas.

Fertilizers, Tillage and Rotation.—Cowpeas respond to moderate applications of phosphorus and potash, but do not need nitrogen.

When planted in drills sufficiently far apart to enable cultivation, cowpeas do best when given frequent, shallow and level cultivation. The earth should not be thrown on the foliage and tillage should cease as soon as the vines begin to run.

Cowpeas are adapted to short rotations. They may frequently follow an early-maturing crop, such as wheat, oats and early potatoes, thus providing two crops from the land in one season. A rotation of wheat or oats and cowpeas is giving excellent results in portions of Tennessee, Arkansas and Missouri.

Time and Method of Harvesting.—For hay purposes cowpeas should be cut when the first pods begin to ripen. A large growth of vines is somewhat difficult to cure. The cut vines should lie in the swath for one day. They should then be placed in windrows where they may remain until fully cured. If weather conditions are not most favorable the vines, after remaining one or two days in the windrow, should be put into tall, narrow cocks and left to cure for a week or more. If rains threaten, canvas covers are advised.

The leaves are the most palatable and nutritious portion of the forage, and every effort should be made to prevent their loss. When so dry that no moisture appears on the stems when tightly twisted in the hands, the hay may be put into stack or mow.

Harvesting for seed is most cheaply done by machinery. The crop should be cut with the mowing machine or self-rake reaper when half or more of the pods are ripe. When thoroughly dry they may be threshed with the ordinary threshing machine by removing the concaves and running the cylinder at a low speed to prevent breaking the peas. Better results are secured by using a regular cowpea threshing machine.

Feeding Value and Utilization.—Well-cured cowpea hay is superior to red clover and nearly equal to alfalfa hay. It is very high in digestible protein. Experiments relative to its feeding value show that one and one-quarter tons of chopped cowpea hay is equal to one ton of wheat bran. It is a satisfactory roughage for work stock and for beef and milk production.

SOY BEANS

Soy beans are adapted to the same soil and climatic conditions as corn. They are most important in the region lying between the best clover and cowpea regions. This is represented by Delaware, Maryland, West Virginia, Virginia, Tennessee and the southern portion of the corn belt. They do well on soils too poor for good corn production, but are not so well adapted to poor soils as the cowpea. They stand drought well.

Varieties.—There are several hundred varieties of soy beans, but only about fifteen are handled by seedsmen. The most important of these are described in the accompanying tabulation. The selection of a variety should be based upon time of maturity as related to the length of season for growth and the purpose for which grown. For seed production, good seed producers should be selected, and for hay and ensilage the leafy varieties are better.

Time, Method, Rate and Depth of Seeding.—The seed-bed for soy beans should be prepared the same as for corn, and the seed may be sown broadcast or drilled, according to the purpose for which grown. On land that is not weedy the seed may be drilled solid with a grain drill. About one bushel of soy beans should be used per acre and they should be covered with one to two inches of soil. If land is weedy or if crop is grown for seed the corn planter may be used, the rows narrowed to three feet if possible

LEADING VARIETIES OF SOY BEANS AND THEIR CHARACTERISTICS.

VARIETY.	COLOR OF SEED.	NUMBER OF SEEDS PER LB.	TIME OF MATURITY.	PURPOSE TO WHICH ADAPTED.	HABITS OF GROWTH.
Mammoth.	Yellow.	2100	Late, 120 to 150 days.	Roughage and grain for entire South.	Large and bushy; 3 to 5 feet high. Will not mature seed north of Virginia and Kentucky.
Hollybrook.	Yellow.	2100	Medium, 110 to 130 days.	Principally for seed. South.	Three feet or less; coarse; poor for hay. Not so valuable as Mammoth.
Haberlandt.	Yellow.	2400	Medium-early, 100 to 120 days.	Principally for seed. South.	Stocky; seldom more than 30 inches tall.
Medium Yellow or Mongol.	Yellow, with pale hilum.	3500	Medium-early, 100 to 120 days.	Forage.	Erect; bushy; 2½ to 3 feet.
Guelph or or Medium Green.	Green.	2600	Early, 90 to 100 days.	Principally for seed. North.	Coarse; not satisfactory for hay; stout and bushy; 1½ to 2 feet. Seed shatters easily.
Ito San.	Yellow, with pale hilum.	3200	Early, 90 to 110 days.	Hay and seed. North.	Bushy, with slender stems; 2 to 2½ feet. Much grown in North.
Wilson.	Black, yellow germ.	2400	Medium-early, 100 to 120 days.	Hay and seed.	Tall, slender; 3 to 4 feet. Excellent for hay.
Peking.	Black, yellow germ.	C300	Medium, 110 to 130 days.	Hay and seed.	Bushy with slender, leafy stems; 2½ to 3 feet. Shatters very little.
Sable.	Black.			Hay and silage.	

and the seed drilled two inches apart in the row. This should require not more than one-half bushel per acre. The drill will accomplish the same result if every fifth drill hoe is used and the planting is made in rows for cultivation.

Seeding should not take place until danger of frost is past. In the Central states it is safe to seed as late as July 1st, and further south seeding may take place later. Soy beans are adapted to seeding with corn to be used as ensilage, in which case varieties should be used that mature about the same time as the corn with which planted. This mixture is also well adapted for hogs and they may be turned into the field as soon as the corn reaches the roasting-ear stage.

Inoculation, Tillage and Fertilizers.—On land which has not before grown soy beans it is advisable to inoculate, either by soil transfer or by artificial cultures. When sown in rows, inoculated soil may be put into the fertilizing box and distributed with the beans at time of planting. This reduces the amount of soil required and gives perfect inoculation. The precautions pertaining to inoculated soils and artificial cultures are the same as those given for alfalfa.

The fertilizers for soy beans are the same as for cowpeas.

When planted in rows far enough apart to permit of cultivation, cultivation should begin early and be sufficiently frequent to keep down all weeds and maintain a soil mulch. Soil should not be thrown on the plants when they are wet. Cultivation should cease when the plants come into bloom.

Time and Method of Harvesting.—Beans grown for hay may be cut with the mowing machine and cured in the same manner as cowpeas. For this purpose it is best to cut when the leaves first begin to turn yellow and the best developed pods begin to ripen. When harvested for seed it is best to wait until the leaves have fallen and at least half of the pods have turned brown. If much value is attached to the straw, harvesting for seed may take place a little earlier. The method of threshing is the same as that for cowpeas.

When grown with corn for silage purposes, the beans should be a little more mature than when harvested for hay.

Composition, Feeding Value and Utilization.—Well-cured soy bean hay is superior to clover hay and equal to alfalfa. It is more palatable than cowpea hay. Whether used for hay, grain, straw or ensilage, it is very valuable as a feed for nearly all kinds of livestock. It is especially valuable in all kinds of rations where high protein content is desired. The whole plant is high in protein and the beans are very high in both protein and fat.

Vetches.—The hairy vetch is a winter annual and is important as a forage and soil improvement crop in the United States and Canada. It belongs to the same family of plants as cowpeas and soy beans. It is best adapted to a cool, moist climate and succeeds best in the northern half of the United States and southern portion of Canada. Although it may be seeded any time during the summer, it does best when seeded in the late summer or autumn. It generally blossoms in May and matures seeds in June or July.

It is valuable as a winter cover crop. The plant has a reclining habit. It is, therefore, best to seed rye and vetch together. About twenty-five pounds of vetch and one-half bushel of rye per acre makes a suitable mixture. The crop may be turned under early in the spring for the benefit of the soil, or pastured or cut green for soiling purposes, or made into hay.

Canada Field Peas.—This term is used for field peas regardless of their variety. The plant is adapted to a cool, moist climate and succeeds best when seeded early in the spring. When used for haying or soiling purposes, it is best to seed it with oats. The oats support the peas and facilitate the harvesting of the crop.

The amount of seed to use will vary with the size of the pea and the character of the soil. It will vary from two bushels per acre in case of small

Hairy Vetch and Rye Growing Together.[1]

seed to three and one-half bushels of the large seed. When seeded with oats, two bushels of peas and one bushel of oats per acre is about the right proportion.

On light soils peas may be sown broadcast and plowed under to a depth of three to four inches. Peas should not be buried so deeply on stiff clays. Best results will be secured by drilling the seed with a grain drill. Some of the peas will be broken in passing through the drill, but the loss

[1] From Farmers' Bulletin 515, U. S. Dept. of Agriculture.

will not be serious. When oats and peas are drilled together, it is best to drill the peas first, after which the oats may be drilled at right angles to the peas and not so deeply. Since the oats come up more promptly than the peas, some advocate deferring drilling the oats until three or four days after drilling the peas.

Harvesting.—Peas are ordinarily cut with a mowing machine when the first pods are full grown but not yet filled. At this time they make an excellent quality of hay. They are cured in the same manner as clover or timothy. Care should be taken to prevent loss of leaves by shattering and injury from rain.

Other Annual Legumes.—The Velvet Bean is a rank growing vine requiring seven to eight months to mature seeds, and is especially adapted as a cover crop in Florida and along the Gulf Coast.

The Beggar Weed is also well adapted to the extreme South and is utilized both as forage and for cover crop purposes. It is adapted to light, sandy soils, and when seeded thickly, can be converted into hay or silage. It grows six to ten feet high and is relished by all kinds of livestock.

Sorghum.—The non-saccharine sorghums were discussed under the head of Kaffir corn. The sweet sorghums, of which there are a number of varieties, are utilized for forage purposes as well as for the manufacture of molasses. The sweet sorghums are not so drought resistant as the non-saccharine sorghums, and a small acreage may be advantageously grown on many livestock farms east of the semi-arid region.

The season of growth is similar to that of corn and the plant demands the same kind of soil and methods of treatment. When used for hay, it should be seeded thickly either by broadcasting or by drilling with a wheat drill, using 70 to 100 pounds of seed per acre.

The Early Amber is considered the best variety for general purposes.

Sorghum should be cut for hay when the seeds turn black. It may be cut with a mowing machine the same as any hay crop. Best results are secured by putting it into large shocks and allowing it to remain until thoroughly cured. If cut too early or stacked before the weather becomes quite cool, it is likely to sour and make a poor quality of hay.

Millet.—There are three common varieties of millet: German, Hungarian and common millet. The common millet is drought resistant and grows well on rather poor soil. It matures in from two to three months. It makes a good quality of hay and can be fed with less loss than the coarser varieties.

The German variety is the largest and latest maturing variety. It will outyield common millet, but is not so drought resistant.

Hungarian millet is about midway between the common and German millet as regards time of maturity, drought resistance and yield. Its tendency to produce a volunteer growth has brought it somewhat into disfavor.

The millets may be seeded any time after the soil is thoroughly warm. In latitude 40 degrees north, German millet should be seeded the last week in May or the first week in June. Hungarian millet may be seeded two or three weeks later, while common millet will frequently produce a crop when seeded as late as the middle of July.

Millet is used chiefly as a catch crop for hay. It is well adapted for this purpose and may be substituted where a catch of clover or timothy fails. It is also excellent to fill in where areas of corn have failed.

The preparation of the seed-bed should begin as early in the spring as conditions will permit. This gives an opportunity to rid the soil of weeds by occasional harrowing prior to seeding. Millet is seeded broadcast at the rate of one peck per acre when grown for seed, and one-half bushel per

Millet Makes an Excellent Catch Crop and is Profitable Either for Hay Purposes or for Seed Production.

acre when grown for hay. Three pecks of seed is advised by some for hay. This results in smaller plants with a finer quality of hay.

Where extensively grown for seed, millet should be harvested with the self-binder when the seed is in the stiff dough stage. The after-treatment is similar to that for wheat and oats. The best quality of hay is secured by cutting before the seeds begin to ripen. The seeds act as a diuretic to animals and it is not safe to feed too much of it to horses. Hay that is to be used for horses should be harvested before seeds form.

Rape.—Rape belongs to the same family of plants as cabbage and turnips. There are two varieties, annual and biennial. The latter bears seed in the second year. The best known variety of biennial is the Dwarf Essex. This gives best results for soiling and pasture purposes. Cattle and sheep are fond of rape. It is especially fine for hog pasture.

Catch Crops for Pasture and Hay.

Crop.	Time of Seeding.	Variety.	Amount of Seed per Acre.	Cost per Bushel.	Cost per Acre.	Method of Seeding.	Time to Harvest.	Yield.	Value.
Field Peas and Oats.	Same time as spring small grain.	Golden Vine or Canadian Beauty field peas; any late oats.	1½ bushels oats, 1½ bushels field peas.	Field peas $2.75, oats 60 cents.	$5.00	Drill in together 3 inches deep.	Oats in milk and pea pods forming.	2 tons.	Almost equal to clover hay.
Succotash.	Same time as small grain.	Any mixture of small grains.	2½ bushels.	$1.00	$2.50	Drill or broadcast.	Use for pasture if needed, or cut in the milk for hay.	Good pasture to middle of July or 2 tons of hay.	Not quite equal to field peas and oat hay, but equal to timothy.
Rape.	From early spring to August 1st.	Dwarf Essex.	5 pounds.	$4.20	$0.40	Drill or broadcast alone or harrow in on small grain.	Pasture 8 or 10 weeks after seeding, 12 or 15 inches high; not much good during hot weather.	Large yield of good pasture.	Furnishes splendid pasture, especially for hogs and sheep, early in season and late.
Millet.	Late May, June or early July.	Common (early), Hungarian (medium), German (late).	3 pecks.	$1.75	$1.30	Drill or broadcast.	Just after blossoming, but before the seeds form.	3 tons.	A fair quality of hay, but not nearly equal to clover; dangerous for horses.
Sorghum.	Soon after corn planting or same time as millet.	Amber.	80 to 100 pounds broadcast; 40 pounds of germinable seed if drilled.	$1.50	$3.00	Drill or broadcast.	Early fall just before frost and when the heads are in dough.	4 tons.	Palatable hay, much the same as corn fodder; deteriorates in spring.
Cowpeas.	Same time as sorghum.	New Era (early), Whippoorwill (medium), Black (late), Clay (late).	1¼ bushels, less if in rows.	$2.75	$3.40	Drill thickly, broadcast or sow in rows three feet apart.	When pods are partly ripe and lower leaves turning.	2 tons.	Fully equal or superior to clover; good late summer pasture.
Soy Beans.	Same as cowpeas or a little earlier.	Ito San, medium brown, medium yellow, medium green.	Same as cowpeas.	$3.00	$3.75	Same as cowpeas.	Same time as cowpeas.	2 tons.	About same as cowpeas.
Sorghum and Cowpeas.	Late May, June or early July.	Any of the varieties mentioned in foregoing.	30 pounds sorghum, 1 bushel peas.		$3.55	Drill or broadcast.	When sorghum is in dough stage.	3 tons.	Not quite equal to clover hay.

This plant is best adapted to cool, moist climates and does best in the Northern states and Canada. South of latitude 38 degrees it is best to sow it in the fall. This allows it to make most of its growth during the cooler part of the year. North of this, rape should be seeded in the spring so that it may make most of its growth before hot weather.

Three to six pounds of seed per acre are required. It may be either broadcasted or seeded with a drill on a well prepared seed-bed.

Rape is usually ready to pasture in six or eight weeks after seeding. If not pastured too closely, it continues to grow until freezing weather.

MAKING HOGS OF THEMSELVES.[1]

Rape makes an excellent late fall and early spring pasture for growing hogs.

Care must be taken in pasturing cattle and sheep in rape. They should be allowed on the rape only a short period at a time, until they become accustomed to it. Very bad cases of bloat may result if this caution is unheeded.

The preceding tabulation taken from "Wallace's Farmer" summarizes the requirements for catch crops when used for pasture and hay. It gives the approximate requirements for average corn-belt conditions, but is subject to modifications as regards time of seeding and amount of seed, depending on climatic conditions.

[1] Courtesy of Dept. of Animal Husbandry, Pennsylvania State College.

REFERENCES

"Soiling Crops and the Silo." Shaw.
"Forage Crops for the South." Tracy.
"Forage Crops." Voorhees.
"Forage Plants and Their Culture." Piper.
Michigan Expt. Station Circular 27. "Hairy Vetch."
Mississippi Expt. Station Bulletin 172. "Forage Crops."
Farmers' Bulletins, U. S. Dept. of Agriculture:

458. "Best Two Sweet Sorghums for Forage."
515. "Vetches."
529. "Vetch Growing in the South Atlantic States."
599. "Pasture and Grain Crops for Hogs in the Pacific Northwest.
605. "Soudan Grass as a Forage Crop."
677. "Growing Hay in the South for Market."
686. "Uses of Sorghum Grain."
690. "The Field Pea as a Forage Crop."

CHAPTER 20

Annual Legumes, Grown Principally for Seeds

The annual legumes most grown in North America for seed are the white or navy bean, the common pea and the peanut. They are used extensively as food for man. In addition to these, cowpeas and soy beans are grown for seed, some of which is used for human food, some for stock food, but still more for seeding purposes.

The production of crimson clover, vetch and castor bean for their seed is of minor importance in North America.

Field Bean.—Is extensively grown under field conditions for the production of dried beans. These become the baked beans of New England fame. According to the census of 1910 the production in the United States was 11,250,000 bushels of 60 pounds from 803,000 acres. Michigan, California and New York lead in bean production. During the same year Canada grew about 1,000,000 bushels from 50,000 acres.

Field beans do best in a cool, moist climate. They are not adapted to conditions south of 40 degrees north latitude. Field beans are adapted to loamy soils of a calcareous nature, but may be grown fairly well on clay loams and silt loams when well supplied with organic matter. The underdrainage must be good and cultural methods such as will produce a fine, mellow seed-bed.

Time, Rate, Manner and Depth of Seeding.—Beans are tender plants and seeding, therefore, should be deferred until danger from frost is past. This makes it convenient to plant them immediately after planting corn.

They give best results when planted in rows far enough apart to permit horse cultivation. The beans may be drilled or planted in hills. Drilling usually gives best results, distributing the seed from three to six inches apart in the row. With rows thirty inches apart about one-half bushel of seed per acre will be required.

Great care must be taken not to plant too deeply. The habit of growth is such that the plant cannot reach the surface if planted deeply. An inch and one-half to two inches is the maximum depth on any except sandy soils. On sandy soils they may be three inches deep.

The beans should be thoroughly and frequently cultivated during their early stages of growth to destroy weeds and conserve soil moisture. They should not be cultivated when dew is on the plants. This precaution must be taken to guard against certain diseases, the spores of which may be in the soil. Disturbing the plants while they are wet tends to scatter the spores and spread the disease.

Harvesting.—The ripe beans are harvested with a bean harvester. This implement cuts two rows at a time, leaving the vines in a single windrow. If the vines are practically dead when harvested they may be placed at once in small piles, and later built into large cocks around poles five feet or more in height.

Threshing and Cleaning.—Beans grown commercially are threshed with a machine especially adapted to the purpose. It is operated in a manner similar to the ordinary threshing machine. If only a few beans

HARVESTING FIELD BEANS WITH A HARVESTER.[1]

are grown an ordinary threshing machine may be used. All except four teeth should be removed from the concaves and the speed of the machine should be such as not to break the beans. Most satisfactory results will be secured by having all the beans uniformly dry.

Beans fresh from the thresher generally contain fragments of straw, stones and particles of earth which must be removed before being placed upon the market. This calls for the use of a special cleaning machine, which removes most of the foreign matter. After this the remaining broken and discolored seeds must be removed by hand.

[1] Courtesy of U. S. Dept. of Agriculture, Bulletin 89.

Yield.—Variations in weights of measured bushels range from fifty-seven to sixty-five pounds. The standard weight is sixty pounds. Beans yield all the way from five to thirty-five bushels per acre. There is usually no profit in a ten-bushel crop. According to the last census the average yield per acre was fourteen bushels.

Field Peas.—The Canada field peas, described in the preceding chapter, are extensively grown in Canada and a few of the Northern states for the dried peas. These are adapted to a wide range of uses as feed for livestock. They also furnish the supply of seed for all localities where the crop is grown for forage purposes.

Peas are very high in protein and are especially adapted as feed for young stock and for the production of milk and butter. When given with oats and bran to cows in milk, they may constitute from one-third to one-half of the concentrates fed.

When harvested for seed, the vines are cut with a mowing machine to which special guards are attached for lifting them from the ground. There is also a device attached to the rear of the cutting bar, which leaves the vines in a swath far enough from the standing peas to enable the team and machine to work without tramping the peas. It is customary to cut when two-thirds of the pods are yellow.

When dry the peas should be stacked under cover or threshed immediately with a pea huller or with an ordinary threshing machine in the same manner as described for field beans.

The legal weight of field peas is sixty pounds to the bushel. They are quite prolific and under favorable conditions will yield forty bushels to the acre. At Guelph, Ontario, eight varieties during eleven years gave an average yield of 31.5 bushels per acre. Four varieties at Ottawa averaged 34.4 bushels for five years, while six varieties grown for five years in three other localities averaged 40, 41 and 41.2 bushels respectively per acre.

The most suitable varieties to grow depend somewhat on soil and climatic conditions. Three good all-around varieties are Prussian Green, Canadian Beauty and White Marrowfat.

Cowpeas.—The seed of cowpeas has been very little used as feed, because the price has been too high to justify its use in this way. The introduction of suitable harvesting and threshing machinery should make it possible to produce grain of the more prolific varieties at prices that will put it in reach for feeding purposes. At present practically all of cowpea seed is used for seeding purposes, the price ranging from $2 to $4 per bushel.

The dried shelled peas contain 26 per cent of protein, 1.5 per cent of fat and 63 per cent of nitrogen free extract. A comparatively low rainfall is favorable to seed production. Continuous wet weather causes a development of vines at the expense of seed. At one of the southern experiment stations during a series of five years, the yield of peas with a yearly rainfall of 62 inches was only 12 bushels per acre, whereas, with only 22 inches of rainfall, the yield was 28 bushels per acre. The yield of hay in both cases was practically the same.

The methods of seeding and harvesting for seed production are treated in the foregoing chapter.

Soy Beans.—The growing of soy beans for grain to be used as feed is profitable if the yield is sixteen bushels or more per acre. The seed is very rich in oil and protein and occupies the same place in concentrates as cottonseed meal and oil meal. The seed should be ground before being fed. Some of the varieties with highest fat content are being utilized for the manufacture of oil. This is used as a substitute for linseed oil in the manufacture of paints. The best varieties under proper cultivation yield from

SOY BEANS, BRADFORD COUNTY, PENNSYLVANIA.[1]

This annual legume is excellent for both forage and seed production. May be grown nearly as far north as dent corn.

thirty to forty bushels of seed to the acre Hollybrook, Mammoth and Haberlandt are three especially good varieties for seed production. Tall varieties that bear pods some distance from the ground are most desirable and most easily harvested.

The methods for harvesting and threshing are given in the preceding chapter. The threshed beans should be thoroughly dried when stored. Otherwise they are likely to heat and spoil. They should be carefully watched when first stored and at once spread out to dry if there are signs of heating.

Soy bean seed is especially exempt from weevils.

[1] Courtesy of Department of Agricultural Extension, Pennsylvania State College.

Castor Bean.—There are two classes of castor beans, one a perennial, bushy plant with large seeds; the other a small seeded variety which yields oil of superior quality. The plant grows within a wide range of climate, from the tropics to the north temperate zone. In Florida it is a perennial plant growing from fifteen to thirty feet high. Further north, it becomes an annual, matures seed in a short season and grows only four or five feet high.

The castor bean thrives in sandy soils and its culture is simple. The seeds germinate with difficulty and it is advised to place them in hot water twenty-four hours before planting.

It is customary to plant them in hills two inches deep, eight to ten beans to a hill. They are afterwards thinned to one or two plants per hill. The rows should be five or six feet apart and the plants from two to three feet apart in the North, and from five to six feet apart in the South, where the plant grows more luxuriantly. They require about the same tillage as corn.

CRIMSON CLOVER, A GOOD WINTER COVER CROP WHERE WINTERS ARE MILD.

Well suited to the lighter soils in the Coastal Plain Region south of Philadelphia.

Planting should be done as early in the spring as possible, but must escape injury from frost.

As soon as the pods begin to open the fruit branches should be removed. This process must be repeated at least once a week as soon as seeds ripen. The branches are spread out to dry on the floor of a suitable building.

In the United States most of the castor beans are produced in Kansas, Oklahoma, California, Oregon and Wisconsin.

The chief use of the beans is for the manufacture of castor oil. This oil is one of the best lubricants for machinery and is used in the manufacture of many articles.

Vetch.—Common vetch and hairy vetch are the two most important varieties of vetches. Common vetch seed is produced in large quantities in the United States only in parts of Oregon. Hairy vetch has a wider range of growth, but is grown mostly for forage, most of the seed being imported from Russia. Both of these varieties seed freely wherever grown and the prevailing high price of the seed ($5 to $8 per bushel) should induce farmers to grow more of it for seed purposes. Yields ranging from twenty to twenty-five bushels per acre have been reported for common vetch, the average estimated yield being ten bushels. Hairy vetch is somewhat less prolific, but yields ranging from two and one-half to twenty-one bushels per acre have been reported by different experiment stations, the average yield being seven and one-half bushels.

The method of harvesting for seed is similar to that of cowpeas. It is threshed with the ordinary threshing machine.

Crimson Clover.—The chief demand for seed of crimson clover is for seeding purposes. The seed is larger than that of red clover, one pound containing 125,000 to 150,000. The weight is sixty pounds to the bushel. It yields better than red clover, averaging about six bushels to the acre. Most of the seed is produced in Delaware and nearby states.

Crimson clover should be harvested for seed as soon as perfectly ripe. The seeds shatter badly. For this reason it should be cut promptly, preferably in the morning or evening when the plants are damp. The mowing machine with a clover buncher or the self-rake reaper are best adapted for harvesting the crop. If the clover becomes wet the seeds sprout, causing serious loss. For this reason threshing should promptly follow the harvest.

Fresh seed is shiny and of a pinkish color. Seed two years old loses its bright color, becoming dark brown. It is then worthless for seeding purposes.

The cultural methods for crimson clover are given in the preceding chapter.

PEANUTS

During the last decade there has been a great increase in the production and use of peanuts in the United States. Their annual commercial value in the United States, according to the last census, was $18,272,000. The states leading in production are North Carolina, Virginia, Georgia and Florida, three-fourths of the marketable nuts being produced in these states. They are valued for forage as well as for a money crop, having a feeding value equal to that of clover hay. Peanut products, such as peanut butter, oil and meal, also have a market value. The peanut kernel has a high percentage of fat. After the oil has been extracted the meal is noted for its high percentage of protein. Being nitrogen gathering like other legumes, they are valued as a soil improvement crop.

In parts of the South where corn is not a successful crop, its place is being taken by the peanut, the entire plant being fed. It also enters usefully into the cropping system, on the cotton and tobacco lands of the Southern states. In parts of the South where the cotton-boll weevil is troublesome, peanuts are more advantageously cultivated than cotton.

Soil and Climatic Conditions.—A light, loamy, sandy soil is best suited to peanuts. A dark soil will produce the forage crop satisfactorily, but is apt to discolor the nuts for market purposes. Heavier soils may be used for forage purposes, but if grown for nuts, a loose soil is necessary, owing to the fact that the nuts must burrow into the soil in order to develop.

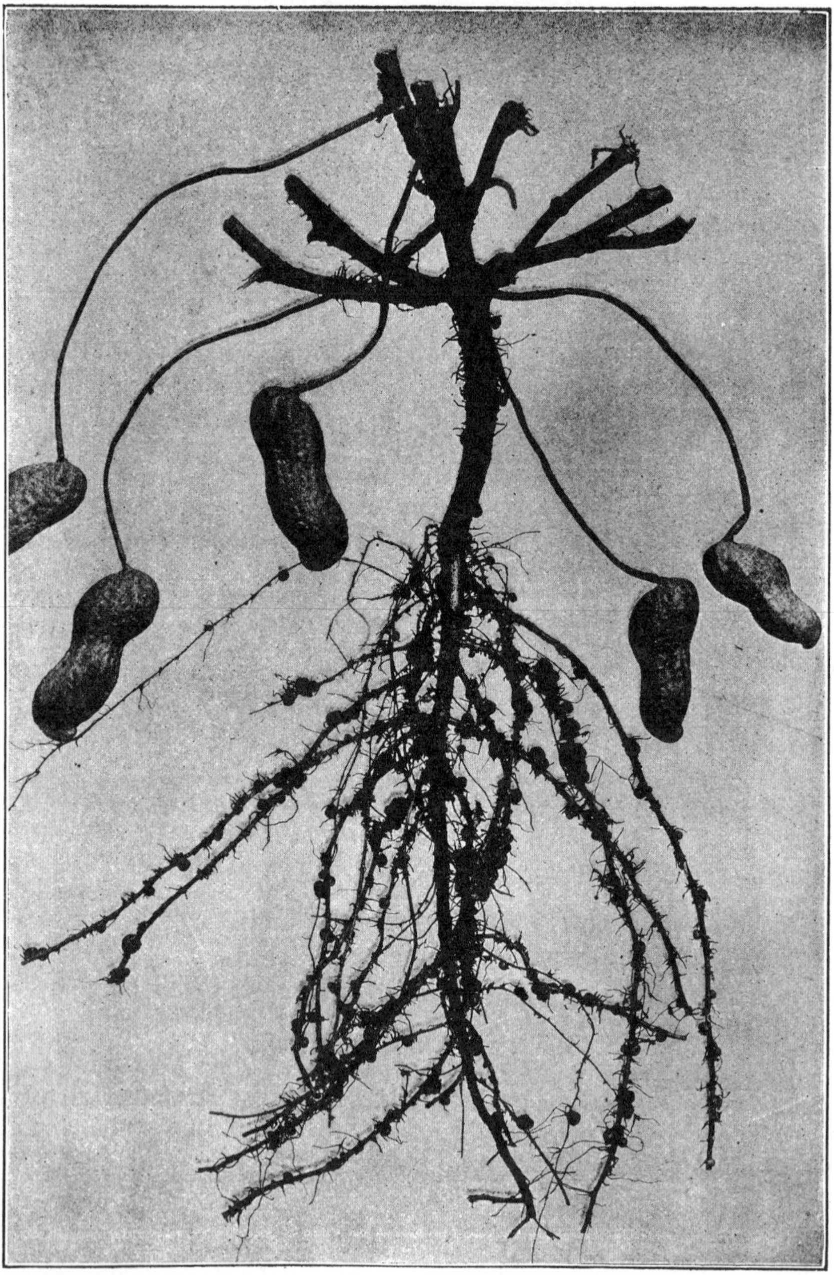

A PEANUT PLANT.[1]

[1] Farmers' Bulletin 431, U. S. Dept. of Agriculture.

A compact soil does not facilitate this very necessary process. The peanut is more susceptible to frost than the bean plant. It requires a long season without frost in order to develop nuts. The small Spanish peanuts require about 115 days to mature and the large varieties need a still longer period. For this reason they are most successfully grown in the frost-free regions, such as the South Atlantic and Gulf states and westward into California. When grown for forage, however, a wider range of climate is possible, peanuts being successfully grown as far north as Maryland and Delaware.

Fertilizers and Lime Required.—Soils that are adapted to peanuts will not require much commercial fertilizer, although the peanut responds readily to a moderate use of it. On river bottom lands no fertilizer will be needed, but in hillside regions applications of fertilizers and lime are advisable. Practically the same fertilizer that is suitable for potatoes is suitable for peanuts. The peanut responds well to the application of manure, but the manure should be applied to the crop preceding the peanuts. For this reason, peanuts should follow a cultivated crop if possible. This also aids materially in freeing the peanuts from weeds. Too much manure causes a heavy growth of tops to the detriment of the pods. If the forage is fed and returned to the land in the form of manure, the peanut is not an exhaustive crop, but if the entire crop is removed it soon robs the soil of fertility.

Peanuts also require an abundance of lime in the soil. Soils that show any indications of sourness should receive from 600 to 1000 pounds of lime (preferably fresh burned) to the acre. This treatment should be given at least every five years. The sorrel weed is an indication of a sour soil.

The fertilizer may be distributed in the row to be planted and thoroughly mixed with the soil. Lime should not be applied at the same time, but some time previous, either during the fall before or just after plowing.

Time, Rate, Depth and Manner of Planting.—Peanuts should be planted as soon as the ground is thoroughly warm and all danger from frost is over. This insures quick germination. The larger varieties must be planted somewhat earlier than the Spanish variety, as more time is needed to mature.

The soil is prepared much the same as that for potatoes. The peanuts are planted in furrows about three feet apart. The nuts may be dropped by hand or a one-horse peanut planter may be used. The running varieties should be planted from twelve to sixteen inches apart in the row but the bunch varieties somewhat closer, from nine to twelve inches apart. The richer the soil, the greater should be the distance between plants, in order to allow for growth.

Only one seed in a place is necessary, but in order to insure a good yield, two seeds are preferable. Two pecks of shelled peanuts are generally sufficient to plant an acre, while two bushels of the Spanish peanut in the pod are required.

Peanuts should be covered from three-quarters of an inch to two inches deep, depending upon character of soil. Light, sandy soils require a deeper planting, while on heavy soils from three-quarters to one and one-quarter inches is sufficient.

Seed Selection and Preparation.—Selecting a good grade of seed is just as important in peanut culture as it is with corn or any other crop. Seed should be selected only from mature plants and from those producing the largest number of pods. It must be properly cured and kept thoroughly dry during the winter. It is not safe to use seed older than the preceding crop.

Seed from the large pod varieties should always be shelled before planting. Shelled seed is surer and more rapid of germination than seed in the pod, and insures a better stand. Machine-planted seed must be shelled.

The small or Spanish varieties may be planted in the pod with but little disadvantage. Some growers make a practice of soaking the pods for a few hours before planting in order to soften them and hasten germination. Soaked seed must be planted at once, however, or it becomes useless. Shelled seed should not be soaked.

Preparing the large varieties for seed entails much work, as they must be shelled by hand. The smaller varieties, however, are usually shelled by machinery, although some loss is experienced by this process.

Varieties.—Peanuts are divided into large-podded and small-podded varieties, according to their size. The Virginia bunch and the Virginia runner are the two most grown large varieties. These varieties are the most used when roasted and sold for human consumption. They have about the same weight per bushel.

The Spanish peanut is much used for forage and for shelled purposes. Its range of growth is wider than that of the Virginia variety.

Other varieties are the African, the Tennessee Red and the Valencia. They are all small varieties.

Cultivation, Harvesting and Curing.—Peanuts should be cultivated in much the same manner as beans, corn or similar crops. Cultivation should begin as soon as the crop is up and continue until the vines spread over the ground. The soil should be kept loose and free from weeds. Peanut pods have the peculiar habit of burrowing in the ground when they begin to form. For this reason the dirt should be worked towards the vines in the last cultivation and the vines should not be disturbed after the process of burrowing begins.

The same implements may be used as for cultivating corn and beans. A one-horse weeder is the general form of cultivator used.

Harvesting should occur just before frost, as frost will injure the forage as well as the peanuts. Peanuts may be plowed from the ground with a common turning plow, but the use of a potato-digging machine is a much better method. The initial expense of such a machine is about $75, but it lasts many years and does the work much more efficiently than it can be done otherwise. If dug by plow the soil must be shaken from the roots by hand, whereas the machine shakes off the soil as it digs.

A few hours after harvesting the peanuts should be stacked about a pole. These poles should be driven firmly into the ground and pieces nailed at right angles to them just above the ground in order to keep the vines from the ground as much as possible. The stacks should be small and conical and stacked as loosely as possible so that air will pass through. It is not advisable to store peanuts in the barn until thoroughly cured. Then the forage part may be stored after the nuts are picked.

The nuts should not be picked from the vines until they are thoroughly dry and solid, else they will shrivel and become unfit for market purposes.

Harvesting and Curing Peanuts.[1]

On the other hand, picking should not be delayed too late in the season on account of ravages from crows and mice.

Hand-picked peanuts command the highest price, but owing to the dusty, irksome labor involved, picking machines are coming into general favor. There are two kinds on the market: one is a cylinder type used mostly for Spanish peanuts; the other machine drags the vines over a horizontal wire mesh, thus removing the nuts without breaking them.

Peanuts must be kept continually dry or they become discolored. After picking they are usually covered with dust and kept in a dry, well-ventilated place until stored in bags ready for market.

[1] Courtesy of U. S. Dept. of Agriculture, Farmers' Bulletin 431.

Preparing for Market.—Threshed peanuts contain much trash, necessitating a thorough cleaning before marketing. This can be done on a small scale by the grower, but if large quantities are involved, the process is more economically done in a cleaning factory, which is equipped with all necessary fanning and grading machinery.

Yields.—An average yield of peanuts is about thirty-four bushels an acre, although it is quite possible on fertile soil and by expert methods to increase this to sixty bushels an acre, with from one to two tons of forage. Peanut forage is worth from $8 to $10 per ton. Sixty bushels of nuts are worth from $40 to $60, according to quality. Estimating upon this basis, allowing an expenditure of from $12 to $25 per acre to grow the crop, the grower would realize a profit of from $36 to $45 per acre. This is a conservative estimate and, all conditions being favorable, might be much larger.

REFERENCES

"The Peanut." Jones.
"The Peanut and Its Culture." Roper.
"Peas and Pea Culture." Sevey.
Canadian Dept. of Agriculture Bulletin 232. "Field Beans."
Farmers' Bulletins, U. S. Dept. of Agriculture:
315. "Legume Inoculation."
318. "Cowpeas."
372. "Soy Beans."
431. "Peanuts."
561. "Bean Growing in Western States."
579. "Utilization of Crimson Clover."
646. "Crimson Clover Seed Production."

CHAPTER 21

Roots and Tubers for Forage

In the United States roots and tubers are grown principally as vegetables or for sugar production, but in Canada they are quite extensively grown for forage purposes. In such root crops as the beet, turnip, parsnip and carrot, the edible part is really an enlargement of the upper portion of the root and the lower portion of the stem merged together. Roots, such as cassava and chufa, are enlargements of the roots.

According to the last census Canada produced nearly 200,000 acres of root crops, while those grown in the United States for forage purposes

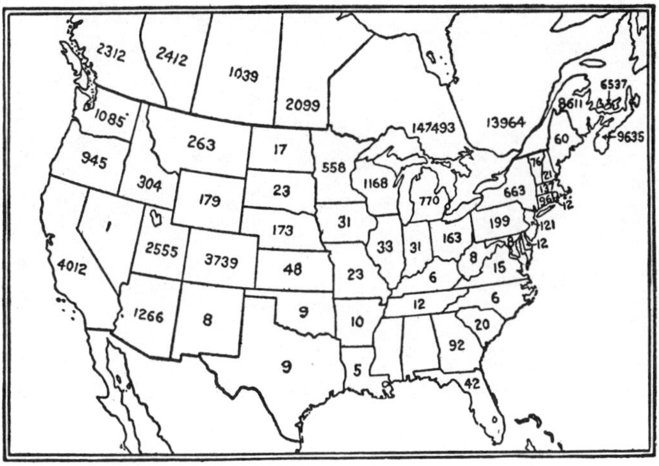

Root Crops, 1909-1910. Figures=Acres.[1]

aggregated only about 15,000 acres. Mangels, rutabagas, turnips, beets, carrots and cabbage are best adapted to cool, moist climates. Of these the rutabaga and turnip may be successfully grown further south than the others. The accompanying map gives the acreage of root crops in the United States and Canada by states and provinces according to the latest census figures.

Relation to Other Crops.—The economy in growing root crops for forage purposes depends chiefly on whether or not other succulent crops suited to feeding livestock can be more cheaply produced. It also depends on the relative yields of the different crops.

[1] Courtesy of The Macmillan Company, N. Y. From "Forage Plants and their Culture," by Piper.

The longer the winter period, the greater the need for succulent food for livestock during the stabling period. For this reason there is more need of such foods in the northern part of the United States and in Canada than farther south where the season for plant growth is longer. Many of the root crops are adapted to a short growing season where corn cannot be successfully grown.

Numerous experiments on the relative cost of producing corn and roots show that corn is the cheaper source of feed wherever it can be successfully

A LOAD OF MANGELS, NOTE SIZE AND CHARACTER OF ROOTS.[1]

grown. The root crops require more labor than corn in culture, harvesting and feeding. Less of the work can be done by labor-saving machinery. It is for this reason chiefly that they are the more expensive source of succulent food. Roots have the advantage in that they may be grown in small quantities for small numbers of livestock when it would not be practicable to have a silo. They also fit well into crop rotations and the tillage required by them leaves the soil in excellent condition for crops that follow.

Utilization and Feeding Value.—The root crops are best utilized for dairy cattle, especially during the winter period. The various roots differ considerably in their percentage of dry matter and feeding value. Sugar

[1] Courtesy of Webb Publishing Company, St. Paul, Minn. From "Field Crops," by Wilson and Warburton.

beets rank first, as they have about 20 per cent of dry matter, three-quarters of which is sugar. Mangels, rutabagas and turnips frequently have no more than 10 to 12 per cent of dry matter, not more than one-half of which is sugar.

Some of the flat-topped turnips that grow principally on the surface of the ground may be grown for pasturage and are readily eaten by sheep.

The dry matter in roots is slightly lower in feeding value, pound for pound, than that in cereals. It is about equal in digestibility to the dry matter in cereals.

The yield of some of the more important root crops, as grown at a number of experiment stations, is as follows: mangels, average yield in tons per acre during five years in five localities, 31; rutabagas, same localities and same number of years, 26.5 tons per acre; carrots, same localities and same number of years, 23.6 tons per acre; sugar beets, same localities, average five years in two of them and three years in other three, 20.6 tons per acre; turnips, three localities average of five years, 21.3 tons per acre.

The cultural methods of most of the crops, brief description of which will follow, are given in the chapter on "Vegetables and Their Requirements."

Sugar-Beets.—While sugar-beets have a high feeding value they are not extensively grown as forage because the yield is generally much less than can be secured from mangels and rutabagas. The by-products of the sugar factories in the form of beet pulp is quite extensively used as roughage for livestock. For cultural methods of beets see not only the chapter above referred to, but also the chapter on "Sugar Crops."

Mangels.—Mangels differ quite materially from sugar-beets in form, color and size. Sugar-beets grow mostly in the ground, are tapering in form, and both the skin and flesh are white. Mangels average four times as large, are more cylindrical in form, and a considerable portion of the root grows above ground. The flesh of the mangel is usually reddish or yellow, while the skin may be white, red, golden, purplish or even black. Mangels are planted in rows twenty-eight to thirty-six inches apart. The rate of seeding ranges from six to eight pounds of seed per acre. The seed should be covered about one inch deep and as soon as the plants are well established they should be thinned by use of a hoe to little groups of plants at intervals of twelve inches. These should be thinned later by hand to one plant to each place. They should be cultivated to destroy weeds and maintain a good soil mulch. They are generally harvested by plowing a furrow on one side of the row, and are pulled by hand. On account of their large size they require much cutting before being fed. They may be stored in root cellars or in pits, and call for a low, uniform temperature and fair ventilation during the storage period.

Turnips and Rutabagas.—There are a great variety of turnips. Rutabagas or Swedes are but a few of the large growing varieties that are especially adapted for forage purposes because of the large yields they give. From two to three pounds of turnip seed and four to five pounds of rutabaga seed per acre are required. The seed may be either drilled or broadcasted, although in case of rutabages drilling is decidedly preferable. The preparation of the ground, planting and method of tillage is very similar to that given mangels. As turnips make their growth in two or three months, they may be seeded late in the summer and yet mature before frost. Rutabagas require more time for maturity and should be sown in May or June.

Turnips do not keep well and should be fed in the fall and early winter. Rutabagas, on the other hand, keep through the winter without difficulty. The methods of storage are similar to those for rutabages.

Carrots.—This crop is less extensively used for forage purposes, chiefly because it yields less abundantly than rutabagas and mangels. It makes

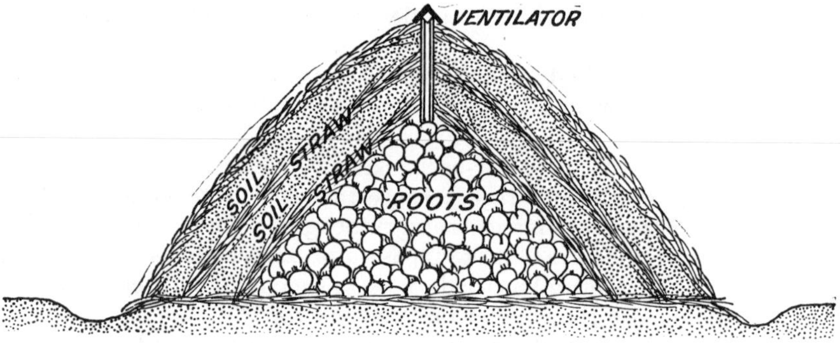

CROSS SECTION OF AN EASILY CONSTRUCTED PIT FOR ROOTS.[1]

an excellent quality of feed and calls for about the same soil conditions and cultural methods as the other root crops. The roots are much smaller and consequently there should be more of them in a given area. From four to six pounds of seed per acre are required. It should be drilled in rows, and the plants should ultimately stand two or three inches apart in the row.

Parsnip.—This crop requires a rich, fertile soil, and demands the same cultural methods as the carrot. The roots of the parsnip may be dug late in the fall and stored or allowed to remain where grown and dug as required for use. Whether they are allowed to remain in the field will be determined largely by winter conditions and the possibilities of digging them in the winter time. When used as human food, the freezing of the roots improves their flavor.

Cabbage.—While this crop is grown chiefly for human consumption, in some sections of the country it is extensively grown for forage purposes. The usual method of storing cabbage is to dig a trench about eighteen inches

[1] From Farmers' Bulletin 465, U. S. Dept. of Agriculture.

deep and three feet wide in which the cabbage is set with the heads close together and the roots bedded in soil. As cold weather approaches they are covered with straw and a few inches of earth. Slight freezing does not injure them, but they should not be subjected to alternate freezing and thawing. They should be well ventilated while in storage. Cabbage makes a good roughage for dairy cows and young stock.

Kale.—Thousand-Headed kale is the variety best adapted for forage purposes because of its large, rank growth and heavy yield. It somewhat resembles cabbage and makes a succulent forage which can be fed from October until April in regions where the winter is mild. It is best fed fresh or allowed to wilt, but should not be cut more than four or five days before feeding; neither should it be fed while frozen.

The methods of growing are similar to those for cabbage, the plants being grown in a seed-bed and transplanted in the field early in the spring.

Cabbage and any of the root crops that tend to give a peculiar taste to milk should always be fed soon after the milking period and never for several hours just prior to it. This precaution in feeding is said to obviate the disagreeable flavor which is frequently imparted to the milk.

Artichokes.—This crop, of which there are several varieties, belongs to the sunflower family, and both the tops and tubers are relished by livestock. They are cultivated much after the manner of potatoes, although planted somewhat farther apart. Yields of 200 to 500 bushels of tubers per acre have been reported.

Artichokes are valuable as forage, chiefly for hogs, which may be turned into the fields and allowed to harvest the crop themselves. The tubers keep in the ground all winter and usually enough of them are left by the hogs to produce a new crop for next year.

Cassava.—This plant is a native of the tropics and is adapted to Florida and the Gulf Coast portion of the states bordering on the Gulf of Mexico. It is a large growing, bushy plant attaining a height of four to ten feet and produces horizontal, fleshy roots or tubers three to five feet long and from one to two and one-half inches in diameter. While it will grow on quite a variety of soils, it can be economically produced only on loose, sandy soils which will enable the easy harvesting of the roots. On fertile soils and with good cultural methods, yields of five to ten tons per acre of roots are reported. The roots are very high in starch and sugar content and make an excellent food for all kinds of livestock. The crop is quite extensively used in the manufacture of starch.

Cassava is propagated by means of portions of the roots or stems which are stored in the dry during the winter. The roots or seed canes are cut into pieces of the desired length and planted in the spring after danger of frost is past. They are usually planted four feet apart each way and covered with a few inches of moist earth.

Chufa.—This is a sedge-like plant with creeping root stocks which produce great numbers of edible tubers. These are small, sweet and frequently used as human food or pasture for hogs. The yield varies greatly ranging from 50 to 300 bushels of tubers per acre. The plant is propagated by planting the tubers in the spring in rows sufficiently far apart to permit cultivation. The rate of planting is about the same as for potatoes.

Taro.—Thi plant, commonly grown for its edible roots in the tropics, is more familiar to persons in the United States as seen in the large-leaved, ornamental plant sometimes called "elephant's ear." The tubers are similar to potatoes in composition. It requires a long season for its growth and is adapted only to Florida and the lower portions of the Gulf states. The bulbs are from six to twelve inches long and three to four inches in diameter. It is grown chiefly for human food, but in semi-tropical districts may be used as a forage for livestock.

Youtia.—This plant closely resembles the taro and is similar in its requirements and uses. The yield of tubers under favorable conditions may be ten to fifteen tons per acre. They are harvested by pulling, supplemented by the use of the hoe. No doubt machinery such as is used for the harvesting of sugar-beets could be utilized for the harvesting of this crop and the one preceding.

REFERENCES

Farmers' Bulletins, U. S. Dept. of Agriculture:
309. "Root Crops." Pages 7 to 15.
465. "Methods of Storing Root Crops."

CHAPTER 22

THE POTATO

BY ALVA AGEE
Director, Agricultural Extension, Rutgers College, N. J.

The potato is one of the world's most important products for human food. The United States have been producing between 350,000,000 and 400,000,000 bushels, and Canada between 70,000,000 and 85,000,000 bushels annually. Adaptability to this crop gives high value to land near good markets, and good transportation facilities have made the crop profitable

THE POTATO CROP.[1]

in sections of the country that must ship their products long distances. The Southern states, growing their crop in the cool months of the spring, supply Northern markets during the summer, and in the fall scores of millions of bushels are sent southward from the Northern states. The crop is important not only for the reason that it produces a large amount of human food per acre, but on account of the reward it offers to the grower's skill. The limit to production per acre is unknown, but it is a conservative statement that the present average yield in this country could be doubled.

[1] From Farmers' Bulletin 365, U. S. Dept. of Agriculture.

The Soil.—The potato is a tuber developing below the surface of the ground and displacing soil particles as it grows. Therefore, a mellow soil is essential. The best potato lands are naturally loose, but somewhat heavy soils have been brought into profitable production by the free use of organic matter from sods and cover crops. A good potato soil is retentive of moisture, and rotted organic matter in it serves as the best insurance against drought. Some light, sandy soils of the seaboard states are put into productive condition by means of cover crops and manure which give them body and excellent physical condition. Soils naturally too compact for the potato may be made loose, friable and retentive of moisture by the same means.

Crop Rotation.—The history of potato production in other countries as well as our own teaches clearly that this crop should be grown in rotation with others and that when the crop rotation is shorter than four years there is great danger of ultimate failure. The practice of growing potatoes year after year on the same land, using a winter cover crop, or of using a rotation of two years only, may prevail for a number of years in a region peculiarly adapted to the crop, but it is only a matter of time until yields will be badly cut by disease and lack of vegetable matter in the soil. One excellent crop rotation is clover, corn, potatoes and grain, followed by clover. The manure is put on the field for corn, and both it and the sod are thoroughly rotted for the potato the following year. Another rotation of some reputation is clover, potatoes and wheat. The clover sod rots more readily than a grass sod and feeds the potato and at the same time keeps the soil mellow. A fresh-turned grass sod does not favor this crop. When it is necessary to follow grass with potatoes the sod should be broken in the fall, and if there is danger of undue leaching, a winter cover crop of rye or wheat should be grown.

Soil Preparation.—A deep soil holds moisture better than a shallow one, and our more productive potato lands have been made and are kept deep by proper plowing. A shallow soil should be deepened gradually, and the best part of the sod never should be thrown into the bottom of the furrow. A breaking-plow having a short, straight mold-board is to be preferred for all land that is at all deficient in humus, as it is essential that some organic matter be in the surface soil. The time of plowing is a local question. Wherever leaching is not to be feared and early planting is practiced, fall plowing is advised. When land is broken in the fall or very early in the spring, it is less subject to summer drought than late-plowed ground. We should bear in mind all the time that a supply of moisture is a big consideration and in the preparation of the ground that should be kept chiefly in view. The use of a heavily weighted, sharp, disk harrow on sod land before it is broken does much to hasten decay after the plowing and to insure prime physical condition. It is easy to do harm by tramping plowed land with horses in the spring, and disking before plowing reduces the amount of required preparation after the plowing.

The Seed.—The potato thrives in a relatively cold climate and loses vigor when grown during midsummer in warm latitudes. The best seed is obtained from our northernmost states, grown in midsummer, or from more southern states when grown in the cool months of autumn. As a rule, the northern seed is preferred, partly because it is in abundant supply.

Successful growers prefer potatoes of marketable size for planting. The tubers are enlarged underground stems, and their vitality may be measured by that of the vines which produce them. A small potato, known as a second, may have been set late by a vine of strong vitality which produced also a big crop of merchantable tubers. In that case the small potato makes fairly good seed, and would be just as desirable as a section of a large potato if it did not put out any more sprouts than the cut portion of a large tuber. On the other hand, many seconds are small because the vines producing them lacked in vitality. Experience has taught that growers depending upon seconds soon have a large percentage of plants that lack full productive power. Potato yields in the warmer latitudes of the Northern states are kept low by the use of home-grown seed which necessarily has had vitality impaired.

The amount of seed per acre depends somewhat upon variety, but relatively heavy seeding is profitable. The grower wants sufficient foliage to cover and shade the soil thoroughly, and ordinarily, that requires the use of thirteen or fourteen bushels of seed per acre. The seed piece should be a block of potato sufficiently large to average two eyes to the piece. The size of the seed piece is important in insuring a good stand, and the cutting should be related more to size of the piece than to number of eyes. In some instances there will be only one bud which may produce two or three good stalks, and in other cases a seed piece of right size may have three eyes. Close cutting and any skimping of the amount of seed result in loss under ordinary conditions, however successful they may be in a very fine and fertile soil having the right amount of moisture immediately after planting.

Fertilization.—Large areas of sandy loams are planted with potatoes because they have right physical condition and partly because they mature a crop early in the season. Sandy soils are badly deficient in potash, and it has come about that most growers think of the potato as a plant requiring unusually heavy applications of potash. Manufacturers of fertilizers have fostered this idea, but the results of careful experiments have shown within recent years that phosphoric acid should be the controlling element in the potato fertilizer, just as it is in the fertilizer for corn and most other staple crops. In normal soils of great natural strength no commercial fertilizer may be used, but when need first develops, phosphoric acid is the requirement. This occurs even where clover and stable manure are freely used. Commercial growers, as a rule, make no use of stable manure direct to potatoes, as it furnishes ideal conditions for the development of disease, and especially of the scab. In the case of naturally fertile land the manure applied for corn and the legumes in the rotation may furnish the most of

THE CONDITION OF SEED POTATOES DEPENDS ON CHARACTER OF STORAGE.
1—Stored in cool place. 2—Stored in warm place, tubers shrunken and vitality impaired.

the needed nitrogen, and the decay of the vegetable matter may free all of the potash required, but we now have relatively small areas in which phosphorus does not add materially to crop yields. As potato production continues, a need of nitrogen develops, and as has been said, potash is a requirement for most sandy soils. A lack of fertility may be met by use of a fertilizer containing 3 per cent of nitrogen, 10 per cent of phosphoric acid and 5 or 6 per cent potash excepting, naturally, areas where the percentage of nitrogen must be increased. The amount of fertilizer used per acre varies greatly. Some growers in the seaboard states apply one ton of a high-grade complete fertilizer per acre, and many growers on naturally

A POTATO PLANTER.[1]

fertile soil in the Central states use none at all. It is a common practice to apply all of the fertilizer in the row, and when the amount is in excess of 1000 pounds per acre, there is danger of injury to the plants as they start growth

Lime is not applied to land immediately before potatoes are planted, as it favors the development of potato scab. Acid soils are more free from this disease than alkaline ones, but clover demands lime and is needed in a rotation with potatoes. The best practice is to use finely pulverized limestone rather than burned lime and to make the application immediately after the potato crop in the rotation.

The Planting.—As the potato thrives best in cold latitudes the planting should be made as early as possible in the spring in the Southern states and

[1] From Farmers' Bulletin 365, U. S. Dept. of Agriculture.

the southern tier of the Northern states. The only exception is in the case of midsummer planting with the aim of securing a crop in the fall. Farther north the planting may be later in the spring, although the tendency in recent years has been away from June planting.

The depth of planting depends upon the character of the soil and the variety. Where an early crop is wanted, the planting is shallow, but for a main crop in loose soils the depth should be at least three inches below the dead level of the surface.

A planter does more satisfactory work than can be done by hand, dropping the seed in a more direct line. The width between rows may vary from thirty to thirty-five inches and the distance between the seed pieces in the row should be sufficient to require about fourteen bushels of seed per acre. This is a surer rule than any fixed number of inches, as much depends upon the cutting.

Cultivation.—A soil that is sufficiently retentive of moisture for the potato usually inclines to become more compact than is desired. The preparation of the soil and the planting compacts some of the ground beneath the surface. A few days after the planting is finished it is good practice to give a very deep and close cultivation, the shovels being guided by the furrows made in covering. Later the weeder or harrow should be used to level the ground and kill all weeds so that the potato plants will come up in a fresh, clean soil. Close and deep tillage should be given when all the plants are above ground, and later the cultivation should be more shallow so that the roots of the plants will not be unduly disturbed. Level culture enables the grower to keep the maximum amount of moisture in the soil, but dependence upon mechanical diggers has led practical growers to ridge the rows and, when the growing is on a large scale, this is the only practical method of controlling grass and weeds. Cultivation should continue until the vines fill the middles, and the last cultivation should be given by a light one-horse cultivator that will slip under fallen vines. The early cultivation should keep the soil loose and later cultivation should keep the surface well mulched with loose earth and should prevent any growth of weeds.

Diseases.—The number of virulent potato diseases is increasing in this country, and the grower should study the latest bulletins from his state experiment station. He will be informed regarding the formalin treatment for the seed before planting, that gives control of some diseases. All potato seed should be given this treatment, which consists of soaking the seed for two hours in a solution of formaldehyde made by diluting one pint of 40 per cent formaldehyde in 30 gallons of water. This should be done before the seed is cut and under no circumstances should scabby seed be planted without this treatment.

Close examination of the seed pieces when cutting is an aid. Mechanical cutters are not advised and partly for this reason. All tubers showing discoloration of any sort should be rejected.

Spraying with Bordeaux mixture increases the yield of potatoes through stimulation, and is profitable, except in case of very highly fertilized soil, even when no blight prevails. The early blight which is prevalent in the southern tier of our Northern states is not well controlled by spraying, but in cooler latitudes where the late blight prevails spraying should never be omitted. Directions for making the Bordeaux mixture and applying it are furnished by the experiment stations. The only point to be emphasized here is that the spraying should be thoroughly done, insuring a perfect coating of the plants, and that is possible only by use of strong pressure and two nozzles to the row when the plants have reached some size.

Insect Pests.—For white grubs and wire-worms, which may render a potato crop unmarketable, there is no remedy. There is no soil treatment that will kill these pests. The grower should know the life history of these insects and plan his rotation as far as possible for their control. Examination will show whether a soil is infested or not at planting time, and potatoes should not be planted where serious injury is sure to come.

The potato beetle is easily controlled by use of arsenical poisons and these should be on the plants when the larvæ of the potato beetle are hatching. Two pounds of Paris green or four pounds of arsenate of lead in fifty gallons of Bordeaux will prevent injury by this insect.

The flea-beetle does great injury not only by impairing the vitality of the plant, but by opening the way for disease attacks. Control is very difficult. The Bordeaux mixture repels for twenty-four to forty-eight hours and to that extent is a help.

Harvesting the Crop.—An early crop of potatoes when dug for market in hot weather must have careful handling. All cut and bruised tubers should be discarded. If there is reason for not marketing promptly, the crop is safer in the ground than out of it, although excessive wet weather may cause rot. Later varieties, dug usually in the fall when nights are cool, will bear placing in large bulk.

The best diggers elevate the soil of the row with the tubers and, having sifted the soil back, drop the tubers on top of the fresh surface. Such diggers are relatively expensive and small growers use low-priced diggers that do fairly good work. When good seed is planted in highly fertilized soil the percentage of seconds may be so small that little grading is required, but it never pays to send to market any tubers below merchantable grade.

REFERENCES

"The Potato." Fraser.
"Potatoes for Profit." Van Ornam.
"The Potato." Grubb and Gilford.
"Potatoes: How to Grow and Show Them." Pink.
South Dakota Expt. Station Bulletin 155. "Selection of Seed Potatoes."
Farmers' Bulletins, U. S. Dept. of Agriculture:
365. "Farm Management in Northern Potato Sections."
386. "Potato Culture on Irrigated Farms of the West."
407. "The Potato as a Truck Crop."
533. "Good Seed Potatoes and How to Produce Them."

CHAPTER 23

SUGAR CROPS (CANE, BEET AND MAPLE SUGAR, AND SORGHUM)

By W. H. Darst
Assistant Professor of Agronomy, The Pennsylvania State College

The world's sugar supply is manufactured from two plants, namely, the sugar-beet (*Beta vulgaris*) and the sugar-cane (*Saccharum officinarum*). The amount of sugar secured from the maple tree is insignificant.

SUGAR-BEETS

The development of the sugar-beet industry dates back to March 18, 1811, when the French Emperor dictated a note to his Minister of the Interior, instructing him to see that 90,000 acres of beets were planted. He then appropriated 1,000,000 francs with which to establish schools of instruction, and to be given in bonuses to those who erected factories. Even though sugar-beet was an unknown crop, the farmers were compelled to grow them. At the end of two years France was producing 7,700,000 pounds of sugar. By 1836 the production of sugar in France amounted to 40,000 tons. At this time Germany observed that sugar-beets in France had revolutionized French agriculture. By growing beets in the rotation the yield of all the cereals was increased to an even greater extent than where turnips were grown, as in England. Up to this time Germany had not been able to induce her farmers to grow beets of their own accord. Germany then adopted the French plan of governmental aid to establish the industry. Other European countries soon followed the same plan, with the result that today one-half of the world's supply of sugar is derived from European sugar-beets.

The following table gives the total world's production of beet and cane sugar compared:

World's Production.	Short Tons.		
	1911–12.	1912–13.	1913–14.
Cane sugar	10,253,000	10,699,000	11,118,000
Beet sugar	7,072,000	8,365,000	9,765,000
Total production	17,325,000	19,064,000	20,883,000

The countries leading in the production of both beet and cane sugar in 1914 are as follows:

Beet Sugar.		Cane Sugar.	
Country.	Short Tons.	Country.	Short Tons.
Germany	2,886,000	Cuba	2,909,000
Russia	2,031,000	British India	2,534,000
Austria-Hungary	1,858,000	Java	1,591,000
France	861,000	Hawaii	612,000
United States	733,000	Porto Rico	364,000
Italy	337,000	United States (Louisiana and Texas)	300,000

The development of the sugar-beet industry in the United States is of comparatively recent date. It was not until 1906 that the production of

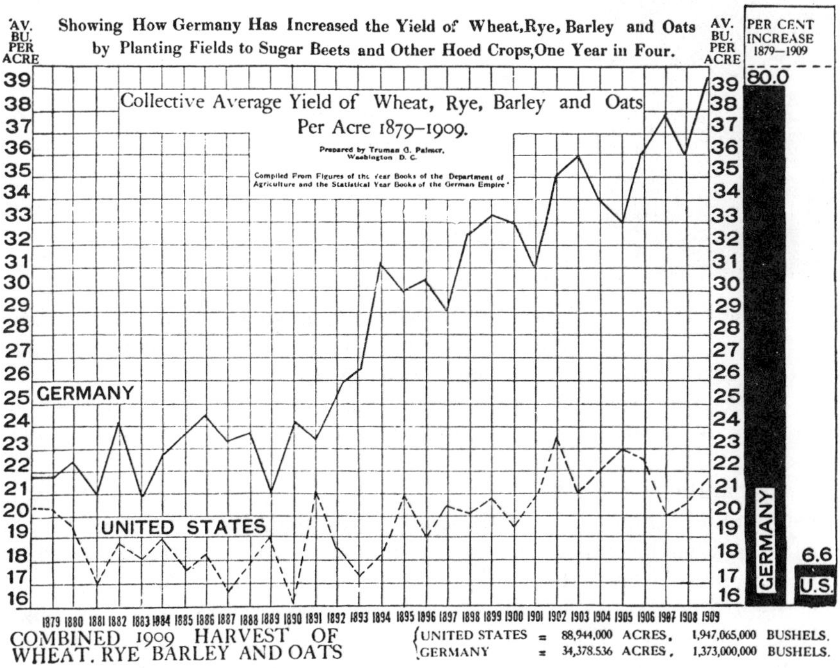

Agricultural Progress in the United States and Germany.

sugar from beets exceeded that from sugar-cane. At present the production of beet sugar has more than doubled that of cane sugar in the United States. (See above table.)

The leading states in the production of beet sugar, in the order of their production, are: Colorado, California, Michigan, Utah, Idaho and Ohio.

In the past and even at present, many farmers think beet culture injures the soil. This, with the high cost of extracting the sugar from the beet, has made progress in beet culture in this country very slow.

Results obtained in Germany and other European countries, when beets are introduced into the rotation, suggest that the farmers of the United States, having the proper conditions for production, would do well to introduce them into their rotations. European farmers do not find the beet crop in itself highly profitable, but the extra cultivation and fertilization necessary to grow them, has greatly increased the yields of all other crops, in the rotation, especially the cereals.

Sugar-Beet.[1]

For the most part the profit is made indirectly from the beet crop. The preceding chart from the loose-leaf service of the United States Sugar Manufacturers' Association compares the average yields of cereals in Germany, a beet-raising country, with those of the same crops in the United States, where very few beets or roots are grown.

Adaptation.—The soil and climatic conditions are very important factors in growing beets with high sugar content. They are not as widely adapted as other farm crops commonly grown in this country. Plenty of moisture and sunshine, particularly during early growth, are essential to the production of beets with high sugar content. Ideal conditions are found most commonly in the irrigated districts of the Rocky Mountains and the Pacific Coast, although many Northern states have favorable conditions for sugar-beet growing.

Sugar-beets require deep, well-drained soils. They do best on rich loam or sandy loam and are not adapted to clays, muck or peaty soils.

Preparation of Land.—The root of the sugar-beet grows entirely or mostly underground, the smaller roots often reaching a depth of four to six feet. For this reason, a deep soil and a deeply prepared seed-bed are necessary. Beet ground should be plowed eight to twelve inches deep, and where possible a subsoiler may be used with good results. Fall plowing is advised where conditions will permit. It is very important that the seed-bed be well prepared. The land should be worked often enough to secure a fine, firm, moist seed-bed. It is necessary to obtain a soil free from weeds or weed-seeds. Beets grow slowly at first, and if weeds are allowed to start, considerable hand labor will be required to eradicate them. Beets should never be grown in continuous culture. The rotation will depend on the

[1] Courtesy of California Agricultural College.

crops common to the region where grown. A three, four or five-year rotation, including a legume crop, should be used when growing beets.

Fertilization.—Barnyard manure and high-grade fertilizer are used with profit on beets. The manure should be well rotted when applied, so as to lessen the chances of weed-seed. High-grade fertilizers, selected to meet the needs of the soil, should be used.

In European countries beets are fertilized very heavily. This produces a large tonnage of beets and the residual effect of the fertilizer is taken up by the crops that follow. (See chapter on "Fertilizers.")

Seeding and Cultivation.—The beet plant produces seed in balls or capsules containing one to five seeds. It is impossible, therefore, to regulate the rate of seeding so as to get a satisfactory distribution of plants in the row. The seed is drilled rather thickly, and when the plants are large enough, they are thinned to the required distance in the row. The seed is ordinarily sown with a beet drill, which sows several rows at a time. The distance between rows varies from twenty to twenty-eight inches. To insure a full stand of plants, about twenty pounds of seed are sown to the acre. In irrigated sections, beets are often sown in double rows one foot apart and twenty-four to twenty-eight inches between each pair of rows. Beet-seed should be sown early in May or after the ground warms up. Cultivation should begin as soon as the rows can be followed and continued at intervals of six to ten days, until the tops nearly meet between the rows. A special beet cultivator is used that will cultivate several rows at a time.

The thinning of the plants should be done about the time the fifth leaf is formed. Thinning is done by first blocking or bunching with a hoe. This consists of cutting out the plants in the row, leaving small bunches eight to ten inches apart. After blocking, further thinning is necessary, leaving but one plant in each bunch. The blocking and thinning, hoeing, pulling and topping of the beets are done by hand labor. On the larger beet farms this work is generally done by foreigners under contract.

Harvesting.—Beets should be harvested before danger of frost in the fall; if not worked up immediately, the roots should be protected from freezing. Harvesting consists of lifting, pulling, topping, piling and hauling away the roots. Lifting is done by a special implement that loosens the roots in the soil. The pulling, topping and piling are done by hand. In topping, the leaves are sometimes simply twisted off. A much better method of topping, from the standpoint of the manufacturer, is to remove the tops with a sharp knife at the lowest leaf scar on the root. The part of the beet that grows above ground is not desirable. The sugar content of this part is low, and there is a high percentage of minerals that may crystallize the sugar at the wrong time in the process of manufacture.

Seed Production.—The sugar-beet is a biennial, producing seed the second year. Almost all of the seed used in this country is imported. When grown for seed, only beets with high sugar content should be saved. This selection is based on the percentage of sugar as determined in a small sample taken out of the side of the root with a trier. The hole made by the trier is filled with charcoal or clay to prevent rotting. The selected beets are stored over winter in sand, in a dry cellar or pit. The next spring these roots are planted in rows to produce seed. From three to five plants will produce a pound of seed.

Manufacture of Beet Sugar.—At the factory the beets are washed in sluiceways, then sliced into long strips called "cosettes" The juice is

A Good Stand and Vigorous Growth of Sugar-Beets.[1]

removed by applying hot water to the sliced beets, leaving a product known as beet pulp. This juice is purified by adding small quantities of lime. The lime combines with the foreign matter and is filtered out. The purified juice is then placed in vacuum pans and boiled until the sugar crystallizes. The sugar is removed by placing the product in a large centrifugal machine, lined with fine sieves. The whirling motion drives off the molasses through the sieves, and the sugar is retained. The sugar is then dried and is ready

[1] U. S. Dept. of Agriculture, P. I. Bulletin 238.

for market. The molasses, to which is added a little fresh juice, is again boiled in vacuum pans until the remaining sugar crystallizes. The sugar is separated out as before, the product being known as second sugar. The molasses, after the second boiling, is sold as stock feed.

By-Products of Beet Farming.—Beet tops left on the field after harvesting may be cured as forage to be fed to livestock. If not fed, they should be spread evenly over the ground and plowed under as a fertilizer.

Beet pulp, a by-product of the sugar factory, is an excellent substitute for corn silage. Wet beet pulp contains about 90 per cent of water and 10 per cent of solids, which compares favorably with mangels as a feed. Many factories dry the pulp. Dried pulp makes a better feed, in that it remains in better condition for a longer time and is worth about eight times as much as the wet pulp.

Beet molasses, another by-product, is not palatable when fed alone; but when mixed with dried pulp, chopped hay or straw, has considerable feeding value.

CANE SUGAR

Sugar-cane has been cultivated for many centuries in the tropical and semi-tropical portions of the world. According to the best authorities, sugar-cane appears to have originated in India. From there it was taken to China and other parts of the Old World, where it has been extensively cultivated from time immemorial. After the discovery of the New World sugar-cane was introduced first in San Domingo, then into Mexico, Martinique, Guadaloupe, Cuba, the Guianas and the warmer states of South America.

The State of Louisiana produces almost all of the cane sugar produced in the United States. Texas and Florida produce some. Sugar-cane was first introduced into Louisiana in 1751, but sugar was not manufactured from it until about 1792.

Description and Mode of Reproduction.—Sugar-cane is a perennial grass, growing from eight to fifteen feet tall. The stalks are thick and heavy, being filled with a sweet, juicy pith. The flowers are borne in silky-like panicles. Seed is never formed in this country, and is not abundantly produced in Egypt or India. Cane in its wild and native state reproduces vegetatively more often than by seeds.

The stalk of cane is divided into joints or nodes and internodes. At each joint is a bud which under proper conditions develops into a stalk. Around each bud, on the stalk, are semi-transparent dots which develop into roots that feed the bud when planted.

Soils.—Sugar-cane requires a large quantity of water during the growing season; consequently, it grows best on soils well supplied with humus and having a high water-holding capacity. Well-drained alluvial bottoms and muck soils are very good soils for sugar-cane. The more fertile clay uplands produce cane higher in sugar, but do not supply the required amount of water for large yields.

Sugar-cane is adapted to tropical or semi-tropical latitudes, the two predominating essentials to growth being warmth and moisture. A mean annual temperature of 70° F. and a minimum annual rainfall of about 60 inches are essential to the successful growth of sugar-cane. One of the difficulties in growing sugar-cane is in the control of water. In Louisiana as much as five to seven inches of water may fall during one rain. The problem, then, is to get rid of the excess water before it damages the crop. Good tile drainage is necessary on most of these sugar plantations. If for any reasons, tile drainage is not possible, it is then necessary to depend on surface drainage.

There are times when irrigation is necessary. The ideal sugar-cane plantation should be equipped with underdrainage as well as irrigation ditches. In Louisiana, scarcely a year passes that irrigation water cannot be used at some time. Irrigation may be used to help prepare the seed-bed, as well as to supply water when needed for the growing crop.

Varieties of Cane.—Many varieties of cultivated cane are grown in this country. These have been and are being introduced from various parts of the world. The Louisiana Agricultural Experiment Station has arranged the varieties into groups and then under classes as follows:

Class one—white, green and yellow canes.

Class two—striped canes.

Class three—solid colors other than class one.

In the Louisiana Bulletin No. 129, the variety known as D.74, a light-colored cane, is recommended very highly. It is very high in sugar and outyields by 20 per cent the green or ribbon canes.

Rotation and Preparation of the Land.—It is not desirable to grow sugar-cane continuously. A common rotation is two years cane and one of corn and cowpeas. The cowpeas are sown in the corn to be plowed down for the benefit of the cane crop which follows. The plowing is generally done in the fall of the year. The land must be plowed very deep, the deeper the better, up to twenty to twenty-four inches. Traction plows are quite generally used, as the work is too heavy for horses. On small plantations, heavy mules and disk plows are used to break the soil.

After the land is plowed it is bedded with a two-horse mold-board plow. This gives surface drainage between each two rows of cane. When ready to plant, the rows are opened with a double mold-board plow. Two or more running stalks are deposited in this furrow and covered by a disk cultivator.

It has been demonstrated in Louisiana that fall planting gives best results when winter freezing is not too severe and when the seed-bed is properly prepared and drained. Planting may take place any time from the middle of September to the first of April.

Fertilizers.—Cane is a rank-growing plant and demands the liberal use of fertilizers. Since most of the potash and phosphoric acid removed by the crop is returned in the ash and the waste from sugar factories, as explained later, nitrogen is the only element of fertility that need be purchased in large quantities. The humus of the soil must be kept up by the application of barnyard manure and by plowing down legumes. When nitrogen is used as a fertilizer it should be applied in the organic form. The nitrogen in cottonseed meal becomes available more slowly than in nitrate of soda, hence this carrier is better adapted to the long-growing season required for sugar-cane.

Cultivation.—Sugar-cane is cultivated frequently to keep down weeds and to insure rapid growth by conserving the moisture. Considerable hand hoeing is necessary as the cane rows can be cultivated only one way. The disk is a favorable type of cultivator; however, the tooth or shovel types are also used.

Harvesting.—The sugar in the plant increases up to a certain stage of ripeness. While the maximum amount of sugar can be determined only by chemical means, the grower learns to determine the proper stage quite accurately by the appearance of the stalks and inflorescence or flower cluster. For economy of production, it is desirable to continue the grinding of cane over as long a period as possible. The season may be extended by planting at different times and by using varieties that vary in time of maturity on different types of soil.

In Louisiana the harvesting begins the first of November. The cane is cut by hand and is a very slow process. The plant is first stripped with the back of the cane knife, then topped and cut close to the ground. The stalks are thrown in piles for loading. As the canes begin to lose sugar rapidly in twenty-four hours after cutting, they are usually hauled immediately to the mill.

Cane Sugar Manufacture.—At the factory the stalks are first shredded. The juice is then pressed out by running this shredded material through three sets of heavy steel rollers. After passing through the first set of rollers, the pressed material is sprayed with hot juice, then passed through the second set of rollers. In turn, this material is sprayed with hot water and again pressed. In this way from 90 to 95 per cent of the juice is removed. The pressed material is used as fuel and is converted into the heat and power necessary to operate the mill.

The juice is heated and purified by adding milk of lime. The lime combines with the impurities and is filtered out. The purified juice is then concentrated by boiling in vacuum pans and is finally crystallized.

The principal by-products of the sugar-cane factories are the impurities combined with lime, the different grades of syrup and molasses and the ashes from the pressed cane.

Since the impurities taken out in combination with lime contain a large part of the phosphorus and potash removed by the crop, this product with the ashes is returned to the soil as a fertilizer.

A Field of Sugar-Cane.[1]

[1] Courtesy of Virginia-Carolina Chemical Company, Richmond, Va. From V.-C. Fertilizer Crop Books.

MAPLE SUGAR

The making of maple sugar, like every other farming industry, has changed greatly within the last fifty years. In this country maple sugar has become more and more a luxury, and less a necessity, owing to the low price of cane and beet sugar.

The maple sugar production of the United States during the year 1909 was 14,060,206 pounds, valued at $1,380,492. The following states lead in the production of maple sugar: Vermont, New York, Pennsylvania and New Hampshire.

Sugar is made from the saps of several varieties of maple trees. The two most important are the Rock Maple (*Acer saccharinum*) and the Red Maple (*Acer rubrum*). Ideal sugar weather occurs in the late winter or early spring when the days are warm and sunny and the nights cold and frosty. This weather starts a rapid flow of sap in the tree. The tree is then tapped and the sap collected in covered buckets made for the purpose. The sap as it comes from the tree is colorless and contains on the average about three per cent of sugar.

Sugar Making.—In the process of sugar making, the sap is first boiled down in evaporators; then boiled to a much greater density in concentrating pans.

In making maple syrup the sap is boiled until the temperature reaches about 219° F.; in making sugar, the temperature must reach 234° to 245° F. The boiling of maple sap for syrup must be done over a hot fire. Boiling over is prevented by adding cream or skim milk from time to time. While the thermometer is used to determine the amount of boiling necessary, an experienced individual can tell simply by the way the syrup boils.

The brown syrupy fluid is then cooled, during which it must be stirred vigorously until graining begins. The soft mass is then poured into molds.

SORGHUM

Sugar from sorghum has never been manufactured on a commercial scale, although it has been made in small quantities and in an experimental way. The difficulty in making sugar from sorghum lies chiefly in the fact that there is only a very short period in the life of the plant when it is possible to crystallize sugar from its juices. The period is so short and the possibilities of detecting the right period are so difficult that it makes sugar making from this plant impracticable.

The plant is quite extensively used, however, in the manufacture of molasses or syrup. It is best known as sorghum molasses, and is used for cooking purposes more extensively than for the table.

The requirements and cultural methods for sorghum are given in the chapter on "Annual Forage Plants." When used for molasses the crop should be planted in drills and given thorough cultivation. The plants should be about six inches apart in the row.

There are many varieties of sorghum, but the Early Amber is the only early variety given any particular preference. There is much uncertainty as to the quality of molasses that will be secured, and it does not seem to depend either upon the variety used or the method of making. Experiments indicate that there are frequently impurities in the juice which interfere with the making of a good quality of molasses.

In general, the best quality of molasses is secured in the northern region of production and in seasons of comparatively low rainfall and abundant sunshine. It is essential that the canes be harvested at the right stage of maturity and that there be uniformity in maturity. Carelessness in the selection of seed and the manner of planting often give rise to canes varying greatly in maturity at harvest time. It is very important to have all the canes about the same height and of the same maturity. This facilitates the removal of the seed heads and is more likely to produce good molasses.

REFERENCES

"American Sugar Industry." Myrick.
"Cane and Beet Sugar Industry." Martineau.
Utah Expt. Station Bulletin 136. "Production of Sugar Beet Seed."
U. S. Dept. of Agriculture, Bulletin 238. "Sugar Beets: Preventable Losses in Culture."
U. S. Dept. of Agriculture, Bureau of Plant Industry, Bulletin 260. "American Beet Sugar Industry."
Farmers' Bulletins, U. S. Dept. of Agriculture:
516. "Sugar Beet Growing Under Humid Conditions."
567. "Sugar Beet Growing Under Irrigation."
568. "Production of Maple Syrup and Sugar."

CHAPTER 24

COTTON PRODUCTION

BY PROF. E. F. CAUTHEN

Associate in Agriculture, Alabama Agricultural Experiment Station

Cotton, the second most valuable crop produced in the United States and the first most valuable export, is grown in that part of the country lying south of 36 degrees north latitude and east of western Texas. This section is known as the "Cotton Belt." The climate and soil are peculiarly adapted to its growth. The warm, moist spring and hot, humid summer favor the growth of the plant and its fruit; the dry, warm autumn matures and opens the bolls and permits the picking of the cotton.

Species.—The genus (*Gossypium hirsutum*) includes the common long and short staple varieties grown in the United States. The length of lint varies from one-half inch to one and a half inches.

Sea Island cotton (*Gossypium barbadense*) grows on the narrow Sea Islands along the coast of South Carolina and in some of the interior counties of south Georgia and north-central Florida. It makes the longest, finest and most valuable of all cotton fibers. Sea Island cotton may be distinguished from the ordinary upland cotton by: (1) its long, slender bolls bearing usually three locks, (2) deeply lobed leaves, (3) yellowish flowers with a red spot on each petal, and (4) many black seeds almost necked, with long slender, silky fiber. Its fiber may be two inches long, and is separated from the seed by the roller-gin, which does not cut the fiber from the seed, but pushes the seed out of the fiber. This cotton is used in the manufacture of fine fabric and laces and in the finer grades of spool cotton thread.

Characteristics of the Plant.—Cotton is a tap-root plant. In loose soils this root penetrates to considerable depth, even into the subsoil. When the subsoil is hard, poorly drained or near the surface, the tap-root is forced aside and the plant becomes dwarfed. Most lateral roots branch from the tap-root near the surface and feed shallow, hence the need of shallow cultivation.

On fertile soil cotton may grow five or six feet high. From its nodes spring two kinds of branches, vegetative and fruit-bearing. The lowest branches or vegetative ones are often called base limbs; they may bear short fruit-limbs. As the top of the plant is approached, the branches shorten, giving it a conical shape. The bolls of cotton are borne only on fruit-limbs.

Some varieties, like Russell and Triumph, produce bolls from one and one-half to two inches in diameter, and require from 60 to 70 to make one pound of seed cotton; others, like King and Toole, having smaller bolls, require from 100 to 120 to make a pound.

Some varieties are much more easily picked than others. If the parts of the boll open wide, the locks of cotton are easily picked out by hand or blown out by wind or beaten out by rain; but if the parts of the boll do not open wide, the locks may cling to the burs and suffer less damage from wind and rain.

Cotton fiber varies in length from three-quarters of an inch in the upland varieties to two inches in Sea Island cotton, and may be likened to a long, slender, flattened tube with two-thirds of its length slightly curled. It is this curled condition of a fiber that makes it valuable, for without it the fiber could not be spun into thread.

A Good Cotton Plant Showing Good Base Limbs; Variety, Cook.

Seed. — The number of seed in a boll varies from twenty-five to fifty. The size of seed in some varieties is larger than in others. Some varieties have green seed, some gray and still others have blackish or necked seed. In the upland varieties most seed are covered with a short fuzz. A bushel of seed weighs $33\frac{3}{4}$ pounds.

Varieties of Upland Cotton Grouped. — The cotton plant is a native of the tropics; but under the influence of man, its growth has been extended far into the temperate zones and its habit changed from a biennial to an annual. Climate, soil, selection and cultivation have wrought many changes in the plant. The true and so-called varieties now number several hundred.

To facilitate the study of so many varieties, a system of grouping, worked out by the Alabama Experiment Station, is followed. According to form of plant, size of boll, time of maturing and other characteristics, they are classified into six groups: cluster, semi-cluster, Peterkin, King, big-boll and long-staple upland. There is no striking demarcation between any two groups, but a gradual blending of the characters of one into the next group.

Cluster Group.—The distinguishing characteristics of the cluster group are the one or two long base limbs near the ground and above them the many short fruit-limbs that bear the bolls in clusters of two or three. The plants are usually tall, slender and bend over under the weight of the green bolls; the bolls of most varieties are small, pointed and difficult to pick.

The leading varieties of the cluster group are Jackson and Dillon. The Dillon variety is important where cotton wilt (*Neocosmospora vasinfecta*) exists, because of its considerable immunity to this disease.

Semi-Cluster Group.—This group somewhat resembles the cluster group, except that its fruit-limbs are longer and the bolls do not grow in clusters. Its varieties have medium to large bolls and large, white, fuzzy seed.

Two well-known varieties of this group are Hawkins and Poulnot. Bolls of both are medium size, slightly pointed and easily picked. One hundred pounds of seed cotton yields about thirty-four pounds of lint.

Peterkin Group.—The fruit and vegetative branches of the varieties of this group are long and nearly straight; its leaves are small and have rather sharp-pointed lobes; its bolls are medium to small in size; its seed is small and many of them are without much fuzz. A striking characteristic of the members of this group is the high percentage of lint that they yield—often as high as 40 per cent.

Some of the well-known varieties of this group are Peterkin, Toole, Layton and Dixie. Layton and Peterkin are very much alike, except that Layton does not have as many necked seed and is probably more uniform in type. Toole and a selection from it called Covington Toole, resemble both King and Peterkin groups. Toole has small bolls, is early and very productive. Some selections from Covington Toole are fairly immune to cotton wilt and are extensively grown in sections affected by this disease. Dixie is a variety that is being bred up by the United States Department of Agriculture to resist cotton wilt.

King Group.—This group embraces the earliest varieties. The plants do not grow large; the leaves and bolls are usually small. Its base limbs are often wanting, and its fruit limbs are usually long and crooked. A distinguishing mark of the group is the red spot on the inner side of the petals of many plants. Most varieties drop the locks of cotton on the ground when they are rained on or blown by hard wind.

The leading varieties are King, Simpkins, Bank Account, Broadwell, etc. On the northern border of the cotton belt these varieties are well adapted because of their earliness.

Big-Boll Group.—This group is marked by the size of its bolls. When seventy or less will yield a pound of seed cotton, the bolls are considered large and classed as a big-boll variety. Some varieties have long limbs; others have short ones, giving the plant a semi-cluster appearance. As a general rule, all big-boll varieties have rank stalks, large, heavy foliage and mature their fruit late.

Some of the widely grown big-boll varieties are Triumph, Cleveland, Truitt, Russell, etc. Triumph originated in Texas and is grown extensively there. It shows considerable storm resistance, has big bolls, is easy to pick and yields well under boll-weevil conditions. Cleveland has medium size bolls and is medium early, but it lacks storm resistance. Russell is late in maturing, has many large green seed and turns out a low percentage of lint.

Cook Improved is a leading variety whose bolls are scarcely large enough to belong to the big-boll group. The type of plant is variable. This variety yields a high percentage of lint, is early and easily picked and has stood at the top in yield of seed cotton in many experiments. However, it has two faults—a tendency to boll-rot (*Anthracnose*), and a lack of storm resistance.

Long-Staple Upland Group.—The chief characteristic of this group is the length of its fiber, which measures from $1\frac{1}{8}$ to $1\frac{1}{2}$ inches long. Most long staple varieties are late and, therefore, are not suited for that part of the country infested with boll-weevils. The percentage of lint is lower than the other upland varieties, but it commands a premium of three or four cents a pound. Some of the better known long-staple varieties are Webber, Griffin and Allen Long-Staple.

Desired Qualities of a Variety.—By careful selection, the type of plant or yield of seed cotton of any common variety may be greatly improved in a few years.

Some of the desirable qualities of a variety are:

(1) Large yield of lint.
(2) Medium to large size bolls that are easy to pick.
(3) Plants that are true to type and healthy.
(4) Medium earliness with some storm resistance.

Selection.—Field selection is the one method most frequently employed to improve a variety of cotton. It consists in sending a picker, who is familiar with the points to be improved, ahead of the other pickers to select the best plants and to pick the well-matured bolls on them. In this way a few hundred pounds of well-selected seed cotton is gathered and then carefully ginned. The next year the selected seeds are planted in a well-prepared and fertilized field away from the other varieties for a seed patch. From the seed patch selection is made in the same way as the year before in the field. By repeating this operation for several years a variety may be greatly improved. However, no variety will continue pure if the seeds are handled at the public gins in the usual careless way.

Soils Adapted to Cotton.—Cotton is grown on all types of soil from the light sandy to the heavy clays, from the badly eroded hills to the rich alluvial bottoms. However, in this wide range of soils are planted many acres that would yield a better income if they were planted in some other crop. It is the low yield of the poorly adapted acres that makes cotton an unprofitable crop on so many farms.

The type of soil influences the earliness of the cotton plants. As a general rule, cotton grown on light, sandy soil makes a rapid growth and matures the fruit early—a decided advantage where boll-weevils exist; while that on heavy clay soil may grow until frost stops it, if the season is favorable. Light soils are not naturally productive, but by the use of 500 to 1000 pounds of complete commercial fertilizer per acre, the yield is increased from one-third of a bale to one or two bales an acre.

Special Types of Soil.—Of the different types of soil, the heavier members of the Orangeburg series are the best adapted to cotton culture.

COTTON GROWN BY SINGLE STALK METHOD.[1]

They are marked by a reddish-brown to gray color and open structure soil with a friable, sandy-clay subsoil.

The Greenville series is very much like the Orangeburg in its adaptation to cotton.

The Norfolk soils are not so productive; but when there is an abundance of humus and a liberal supply of commercial fertilizer, they will produce a heavy early crop of cotton.

The Houston series east of the Mississippi and the Victorian west, with good cultivation and proper seasons, produce above an average crop. However, the cotton plants often suffer from rust.

In the Piedmont regions are located the Cecil soils. Where there is not a deficiency of humus, these soils are productive, but the plant grows

[1] From P. I. Bulletin 279, U. S. Dept. of Agriculture.

slowly in the spring and late in the fall—a condition favorable to boll-weevils.

Along the rivers and smaller streams are strips of alluvial land called bottoms. They are usually fertile, well watered and produce a rank growth of plants that do not make fruit in proportion to their size. On such land, hay or corn is a more profitable crop.

FERTILIZER AND CULTIVATION

Plant Food Removed by Cotton.—There is probably no cultivated crop that draws so lightly upon the fertility of the soil as cotton. The average crop per acre in the United States is slightly less than 600 pounds seed cotton yielding 200 pounds lint. This amount of lint removes from the land only.42 pound nitrogen, .15 pound phosphoric acid and 1.32 pounds potash. When both seed and lint are removed, the loss is 13 pounds nitrogen, 4.74 pounds phosphoric acid and 5.70 pounds potash. The roots, stems, leaves and burs contain about as much nitrogen and phosphoric acid, and about three times as much potash, as the seed cotton. These parts of the plants are seldom removed from the field.

Need of Humus.—In the cotton belt the amount of humus in the soil is small. The warm, moist conditions that prevail during a large part of the year favor rapid nitrification; and the heavy winter and spring rains rapidly leach out the soluble plant-food. As a general practice, cotton follows cotton year after year and receives clean cultivation and furnishes little organic matter to replenish the humus. There is needed on every farm some system of crop rotation in which one crop is plowed under to renew the humus.

Need of Nitrogen.—The small size of the cotton plants over large areas is evidence of the deficiency of nitrogen in the soil. In many fields the plants are large enough to make only two or three bolls. To make a profitable crop they should be two or three feet high, full of fruit and have a rich black color during the growing season. The only lands that do not need a supply of nitrogen are the rich bottoms or those that have received a heavy crop of clover or some other legume for soil improvement.

The chief sources of nitrogen in commercial fertilizer are cottonseed meal, which also furnishes some phosphoric acid and potash, nitrate of soda, tankage and calcium cyanamid. If quick results are desired, as in the case of a side application to a growing crop, some soluble form like nitrate of soda is used.

Need of Phosphoric Acid.—The need of phosphoric acid is almost universal. Most fertilizer experiments show an increased yield whenever it is used. The only soils that do not show an increased yield from its use are the rich alluvial lands and Houston and Victoria clays. A liberal application of acid phosphate on heavy clay soil often hastens the maturing of a crop of bolls that would not ripen and open before frost. When a crop of 200 or 300 pounds lint cotton is expected, it is usual to apply 150 or 200 pounds acid phosphate either before planting or as a side dressing.

Need of Potash.—Loose, sandy soils and the Houston clays show an increased yield when kainit or some other potash fertilizer is used; but most red clay and some silty soils do not seem to need artificial potash to make an average crop. The red clay soils, as a rule, have a great deal of potash, but it is slowly available.

When used alone, an excess of potash tends to delay the maturity of the fruit. When used in connection with other materials making a complete fertilizer, the tendency to lateness is obviated. Some soils subject to cotton rust are greatly improved by the use of 150 to 200 pounds kainit or 35 to 50 pounds of muriate of potash per acre.

Commercial Fertilizers Profitable.—Commercial fertilizers usually pay a good profit, when the season is favorable and they are intelligently used. Lands that formerly produced a half a bale of cotton, now by the use of $8 or $10 worth of high-grade commercial fertilizer adapted to the needs of the land, produce a bale per acre without much additional expense. There is a strong tendency all over the cotton belt to increase the amount of fertilizer and especially the amount of nitrogen. Many farmers are using 400 to 600 pounds of a formula that analyzes 5 per cent phosphoric acid, 4 per cent ammonia and 3 per cent potash for sandy soils and the same with less potash for the clay soils.

Three-Year Rotation Suggested.—The long practice of planting cotton continuously on the same land has destroyed nearly all the humus in the soil. To increase the humus and to maintain soil fertility in the cotton states, the following three-year rotation is recommended:

First year.—Cotton, following in the fall with crimson clover or some other winter cover crop.

Second year.—Corn with cowpeas sowed or drilled between the rows at the last cultivation.

Third year.—Oats or wheat followed by cowpeas sowed broadcast for hay or soil improvement.

Preparation of Land.—The only preparation a great deal of the cotton land receives before planting is one plowing, which consists in throwing up beds or ridges on which the seed is planted. Many farmers are beginning to recognize the need of better preparation and are plowing the land flat and then bedding it before planting.

Much of the plowing is done with a one-horse plow to a depth of four or five inches. However, the lands that are producing a bale of cotton to the acre are plowed with a team to a depth of six or eight inches. Subsoiling, as a special operation, is not recommended, but deeper plowing is proving beneficial in many parts of the cotton belt.

Time of Plowing.—Late fall or winter plowing is commendable for heavy soils and those that have a great deal of litter on them, if such lands are not subject to severe erosion. Light, sandy soils are liable to winter leaching if plowed early. All fall-plowed lands, especially if they are sandy or subject to erosion, should have some winter cover crop like crimson clover or grain so that their roots may take up the plant food as fast as it becomes available and prevent washing of the surface. In the early spring the cover crop is plowed under in the final preparation for planting. In a large measure the date of the first plowing should be governed by the labor on hand, the amount of litter and stiffness of soil.

Seed-Bed.—Land that was plowed broadcast in the winter or early spring is marked off in rows by a furrow that receives the fertilizer. Where cotton follows cotton without any previous plowing, as is too often the practice in a large part of the cotton belt, a furrow with a middle-buster is run in the row of old stalks or in the middle of the previous rows, and the

TURNING UNDER CRIMSON CLOVER FOR COTTON.

fertilizer is distributed in this open furrow with a one-horse machine that has a shovel-plow to mix soil and fertilizer together. By throwing over the fertilizer four or five furrows with a turning plow, a bed or ridge is formed four or five inches high and two feet wide. When no fertilizer is used, many farmers omit even the center furrow and "list" or bed without running the center furrow as a preparation for the row.

Planting.—Just before planting a drag or spring-tooth harrow is drawn across the beds or lengthwise to smooth them down and freshen the surface. On well-drained land some farmers are discarding the high beds and planting on a level surface. In the western part of the cotton belt, where the rainfall is below twenty-two inches, much planting is done in a water-furrow made with a two-horse lister.

In the southern part of the cotton belt, planting begins in March and is usually completed in the northern part of the cotton belt by the end of May.

Most of the crop is planted in April. Where boll weevils are present, planting should be made as soon as the danger from frost is past.

The seeds are sown or dropped in a shallow furrow and covered one or two inches deep in soil. If the soil is dry the seed should be planted deeper and the soil slightly packed on the seed. When the seed is drilled, one-half to one bushel of seed is required to plant an acre; when planted in hills, one or two pecks are required. If the land is rough, the planting should be thicker to secure a stand without replanting.

Tillage.—Prompt germination is desirable. If a rain packs the surface or a crust forms before the seed comes up, the surface should be stirred with a spike-tooth harrow or weeder to help the young plants to break through the crust. The harrow or weeder may be drawn across the rows after the plants come up to destroy small weeds and to cultivate the cotton plants. When the cotton begins to show its true leaves, it should be cultivated with a scrape or turner, which leaves the plants on a narrow ridge. The cotton is then thinned to one plant in a hill about one foot apart on poor land and about one and one-half to two feet apart on fertile land. Soon after thinning a little soil should be pushed up round the young plants. This may be done with a small scrape, sweep or spring-tooth cultivator.

Flat, shallow, frequent cultivation should be given the growing crop until about the first of August, when it may cease, unless the crop is very late.

HARVESTING AND MARKETING

Picking.—Cotton is picked by hand. A picker hangs a bag over his shoulder, picks the cotton out of the open bolls and drops it in his bag. He picks 150 to 200 pounds seed cotton a day and receives from forty to seventy-five cents per hundred pounds.

Picking begins in the latter part of August or early in September and ends about the first of December. When labor is scarce, the time of harvest may be prolonged until midwinter. Cotton should be picked out as fast as it opens to prevent damage from storms or rotting of fiber.

Picking is an expensive operation because it has to be done by hand. However, it does not require much skill and much of it is done by the cheapest of labor—women and children. Many cotton picking machines have been invented, but none of them have proven successful. They damage the plant and gather much trash with the cotton.

Cotton should not be picked when it is wet, nor should locks fallen on the ground and badly stained be picked up and mixed with the white cotton. The damaged cotton should be placed in a separate bale. If cotton is picked when it is slightly wet, it should be dried before ginning, as damp cotton cannot be ginned without injury to the fiber.

Ginning.—When 1200 or 1500 pounds of seed cotton have been picked, it is usually hauled to a public ginnery. A suction pipe draws the seed cotton into a screen where a great deal of the dirt and trash is blown out, and then drops it into a feeder. The feeder picks up locks or small wads of cotton and drops them into the gin-breast, where they form a revolving roll of seed cotton. On the under side of this roll are many small circular saws rapidly revolving in opposite directions and cutting the lint off the seed. A rapidly revolving brush takes the lint off the saws and drives it into a condenser. The lint is then dropped into a large box and packed into a bale of cotton, which is now ready for the market or warehouse.

Cotton Seed.—The seed is usually sold to a cottonseed-oil mill. The short lint or fuzz is cut off the seed and is called "linters." The seed is then run through a mill that takes off the hulls, which are used for cattle food; the kernels, or meats as they are called, are ground and cooked, after

A Field of Cotton.

which they are put in a powerful press that removes the crude oil and leaves a hard yellow cake.

The crude oil is refined and from it are obtained: (1) "summer white oil," which is used in the manufacture of a compound of lard; (2) stearin, used in making solid oils, etc.; and (3) a residue that is used in making soap. On the dry western stock ranches, a great deal of the yellow cake is fed to cattle and sheep in the winter; the cake is ground, forming what is known as cottonseed meal, and is used as stock feed and commercial fertilizer. Recent experiments show that specially prepared meal mixed with wheat flour makes an excellent nutritious bread.

Not many decades ago, cottonseed was a waste product on the farm, but now the commercial value of the seed equals one-seventh the value of the lint.

On an average 1500 pounds of seed cotton make a 500-pound bale and 1000 pounds of seed. When the seed passes through an oil-mill, it produces about 150 pounds crude oil, 337 pounds meal, 500 pounds hulls and 13 pounds linters.

Storing.—After the cotton is ginned, the bales may be marketed at once, or stored on the farm or in a public warehouse. The bales of cotton are often left lying about the ginhouse or homes, exposed to the weather. As a result of the weather, their covering becomes badly damaged and the lint tinged with a bluish color, and the buyer "docks" them to cover the damage.

Bales of cotton should be stored under a shed on timber to prevent their touching the damp ground and absorbing moisture. In many markets are large public warehouses where cotton can be weighed, stored and insured at a small cost per bale.

Before selling a bale, a sample of lint is drawn from each covered side and placed together as a sample of the bale. The buyer judges its grade and makes a bid. The price is based on the grade and the demand for that grade of cotton in the markets. Most farmers do not know the grade of their cotton, as it takes expert knowledge to classify cotton correctly. They accept the highest price bid on the cotton as the top of the market for that grade. Where a large number of bales are offered in the market, often an expert grader is employed to classify the cotton, which method usually gives satisfaction to seller and buyer.

When a foreign or domestic mill wishes a quantity of a given grade, an order is placed with an agent, and this agent goes to the warehouses or dealers and buys the grades desired. If the bales have to be shipped far, they are sent to the compress, where the size is greatly reduced by a powerful press and thereby the cost of transportation is reduced.

Grades of Cotton.—The grades of cotton depend mainly on (1) color of fiber, (2) amount of trash, and (3) quality of ginning. A high grade requires that the fiber be white, with a slightly creamy tinge, strong and free from trash or dirt. When the cotton shows a yellowish or bluish tinge, the fiber usually is not strong; immaturity or exposure to the weather are the usual causes for this condition. To get a high grade, the farmer should pick the cotton from only the fully opened and matured bolls, and pick it free from trash and dirt.

There are seven primary grades in the commercial classification of lint cotton. They are named in the order of value: (1) "fair," (2) "middling fair," (3) "good middling," (4) "middling," (5) "low middling," (6) "good ordinary," (7) "ordinary." The half grades, which lie between the primary grades are named by prefixing the word "strict" to the name of the next lower grade, as "strict good middling," which is a half grade better than "good middling." The telegraphic dispatches from the cotton exchanges quote prices on "middling," and the prices of better and lower grades are calculated on the basis of "middling."

The larger part of the cotton crop of the United States falls under the following grades: strict good middling, good middling, strict middling and middling. Storms and early frost increase the quantity in the lower grades.

The diseases and insect enemies of cotton are discussed in Part VIII of this book.

REFERENCES

"From Cotton Field to Cotton Mill." Thompson.
"Hemp." Boyce.
"Cotton." Burkett and Poe.
"Southern Field Crops." Duggar.
U. S. Dept. of Agriculture:
Bulletin 38. "Egyptian Cotton Seed Selection."
Bulletin 279. "Single Stalk Cotton Culture at San Antonio."
U. S. Dept. of Agriculture, Bureau of Plant Industry:
Circular 26. "Egyptian Cotton in Southwestern U. S."
Circular 57. "Cultivation of Hemp in U. S."
Circular 123. "Production of Long-Staple Cotton."
Farmers' Bulletins, U. S. Dept. of Agriculture:
302. "Sea Island Cotton."
314. "Method of Breeding Early Cotton to Escape Boll-Weevil."
326. "Building Up a Run-Down Cotton Plantation."
364. "A Profitable Cotton Farm."
501. "Cotton Improvement Under Weevil Conditions."
577. "Growing Egyptian Cotton in Arizona."
591. "Classification and Grading of Cotton."
601. "A New System of Cotton Culture and Its Application."

CHAPTER 25

TOBACCO

By George T. McNess
Tobacco Expert, Texas Experiment Station

Types and Their Commercial Uses.—The commercial tobaccos of North America are divided into three principal types, known as cigar leaf, manufacturing and export. These types are again subdivided according to their market grades and commercial use. The cigar type consists of three grades: wrapper, binder and filler leaf. The wrapper is a fine-textured leaf used for covering the outside of the cigar, and must have good appearance, length and width, be uniform in color and have fine veins. Cigar wrapper leaf is the highest priced tobacco produced in North America. The binder is that part of a cigar which holds the filler leaf or bunch together. This grade of tobacco must have fair size and possess good burning qualities. It is generally selected from the poorer grades of wrapper leaf. The filler is that part which constitutes the bulk of the cigar, and varies in quality according to the kind of tobacco used for this purpose. Filler tobacco should possess good aroma, taste and perfect combustion.

There are quite a number of tobaccos used for cigar purposes, each having distinctive characteristics and grown in different parts of the country. The kind of seed used, the influence of climate, soil conditions and methods of culture and curing determine the ultimate use of the leaf.

The tobaccos used in the manufacturing of cigars are: the Havana Seed, Broadleaf, Cuban Seed, Florida Sumatra, Georgia Sumatra, Texas Hybrid, Wisconsin Seed, Pennsylvania Seed, Zimmer Spanish, Gebhardt and Little Dutch. Several types of tobacco are used in the manufacture of smoking and chewing tobaccos, the principal type used in this country being the White Burley, which is grown in Kentucky and parts of Ohio. Cigarette tobacco is manufactured from the bright flue-cured leaf of the Carolinas and southern Virginia. About 60 per cent of the crop is used for home consumption. The heavy or fire-cured tobaccos are mostly exported to Europe, although some of the finer grades are used for plug wrappers.

Principal Tobacco Districts.—The finest cigar tobaccos are grown in the New England states of Connecticut and Massachusetts, and in the South in Florida, Georgia and Texas. These states produce the fine grade cigar wrapper leaf. In the New England states it is grown under cloth shades, while in the Southern states a slat shade is used. These shade-grown tobaccos rival the fine tobaccos imported from Sumatra and Cuba both in quality of burn and taste and in wrapping capacity. The binder tobaccos are produced in the states of Connecticut and Wisconsin; while filler leaf of the various types comes from the Miami Valley of Ohio, and from Pennsylvania, Florida, Texas, Georgia and Connecticut.

The manufacturing tobaccos, air, sun, flue and fire-cured, are grown in Kentucky, Ohio, Virginia, Tennessee and North and South Carolina. Maryland also produces a fine grade of pipe tobacco, but most of this tobacco is exported to England and France. Nearly all of the fire-cured

FIELD OF VIRGINIA HEAVY TOBACCO.

tobaccos produced in the above states are exported to the various parts of the world.

Tobacco Soils.—It might be well to mention briefly a few of the principal soils upon which tobacco is grown. The heavy tobaccos of Virginia are grown in the Piedmont District on soil known as the Cecil clay or Cecil clay loam. This soil is a heavy, red clay soil and produces a heavy-bodied dark-colored tobacco. This type of soil is also found in the tobacco districts of Tennessee and part of Kentucky. The soil of the Carolinas is a very light-gray, sandy soil and belongs to the Norfolk series of soils as classified by the U. S. Bureau of Soils. This soil produces a light-colored, thin-textured leaf which is used in the manufacture of cigarettes and granulated tobaccos. The soil upon which the burly tobacco is grown is also a light soil, as is also the tobacco soil of Maryland. The tobacco soil of Connecticut and Massachusetts is a light, gravelly soil belonging to the Hartford series of soils, and when well fertilized produces a fine quality of tobacco.

The principal tobacco soils of the South Atlantic and Gulf states are light sand to sandy loam, underlaid by either a yellow or red sandy clay. These soils run from gray to red in color, and where they have the yellow clay subsoil they belong to the Norfolk series, while the red clay subsoil places them in the Orangeburg series. The Orangeburg soils are more productive than the Norfolk and the better grades of cigar leaf are produced upon the former soil. These southern soils are responsive to fertilization,

FIELD OF CIGAR LEAF TOBACCO.

and as high as one ton of commercial fertilizer is used to the acre by the best growers.

The soils of Ohio are of limestone formation, and produce a heavy-bodied cigar filler leaf having good aroma,but on account of the lime content are apt to flake. The Pennsylvania soils are a little heavy for the production of wrapper leaf, but the standard cigar filler leaf used in this country is produced upon these soils. The soils of Texas are the Orangeburg and Norfolk, which produce the same grade of tobacco as Florida and Georgia. They are found in the eastern portion of the state. For additional information on soils, see Chapter 1 on "Soil Classification and Crop Adaptation."

Preparation and Care of Seed-Bed.—The preparation of the seed-bed varies in the different tobacco districts, owing to some extent to the varied climatic conditions, financial condition of the grower and type of tobacco being grown. The most expensive and complete seed-beds are to be found in the New England states, while less pretentious ones are to be found in the Carolinas. The object, however, is the same, that is, to produce a supply of good, healthy, vigorous plants; for a failure of the seed-bed means a failure of the crop.

In Connecticut and Massachusetts the young plants are grown under glass in steam-heated beds, and the tobacco seed is sprouted before being sown in order to produce plants by the time danger of frost is over. It is only in the Northern states that it is necessary to go to this expense and trouble. In most of the heavy tobacco-growing districts, as well as in the

TOBACCO PLANT-BED, OR TOBACCO SEED-BED.

South Atlantic and Gulf states, the open seed-bed is used, the only covering being a thin cheesecloth to keep out the cold and conserve the heat and moisture in the bed.

In locating a good seed-bed for any type of tobacco the prospective grower should select a piece of ground near to water, having a southern exposure and protected on the north either by buildings or timber. The best plan is to select a piece of woodland near a small stream having the desired exposure. The timber should be cut off the land in the fall of the year, split into desired lengths and sizes and stacked to dry. January is the best time to burn a seed-bed, excepting in the Northern states. In these states this form of bed is not used. The first operation is to rake from the bed all leaves and trash, then lay across the bed skids of green pine poles, upon which the cut timber with a good supply of small brush is placed. This pile of wood and brush should extend clear across the bed, but not over the entire length. The fire should then be started and let burn until the soil directly under the fire has been burnt to a depth of three inches. It is then dragged on the skids and another section of the bed burnt. This operation is repeated until the entire bed is burnt. As soon as the ground has cooled off, the coals should be raked off the bed and the fine ashes spaded or plowed under.

The bed is now ready for the frame to be placed around it. In some states logs are used for this purpose, but one-inch planks twelve inches wide and any desired length, best serve the purpose. The most convenient size to make a seed-bed is six feet wide and fifty feet long, which will make 300 square feet of bed. In building the frame to go around the beds the planks should be set upon edge and where the ends meet they are nailed to a stake which has previously been driven in the ground (see preceding page). After the frame is complete a No. 9 wire should be stretched from the center of one end of the frame to the other, supported at intervals by stakes, the tops of which are about two inches higher than the top of the frame. When the cloth is stretched over the frame this will cause a peak or ridge to the cloth roof.

Prior to stretching the cloth over the frame, fertilizer should be applied to the bed. Best results have been obtained by using twenty-five pounds of cottonseed meal and ten pounds of acid phosphate to every fifty square yards of seed-bed. This should be thoroughly mixed with the soil, and should be applied several days before the tobacco seed is sown. This form of seed-bed is now used in nearly all of the tobacco districts of the United States with the exception of the New England states, where, on account of their severe climatic conditions and short growing season, glass frames and steam heat are used in order to obtain early seedlings.

In sowing a seed-bed it is very important to secure a uniform stand of seedlings and in order to have a stocky growth they must not stand too thick in the bed. On account of the small size of tobacco seeds, it is necessary to mix them with some foreign substance in order to facilitate uniform distribution in the bed. The best material to use for this purpose is fine-sifted dry ashes. One ounce of tobacco seed mixed with one gallon of sifted wood-ashes will plant three hundred square feet of bed. More than this amount of seed sown to three hundred square feet of bed will cause the plants to grow too thick; consequently, they will not have that desired stocky growth. The seed should not be raked in, but simply pressed into the surface of the soil either by a small roller or by a board placed upon the bed and pressure applied. As soon as the seeds have been pressed into the soil the bed should be watered and the cloth covering placed in position.

If the seed-bed has been well burnt and otherwise prepared very little attention will be needed except the daily watering, and this must not be neglected if a good germination is desired, for the grower must remember that the seed is upon the surface of the soil and that it takes moisture and heat to cause the seed to germinate. Tobacco seed germinates in from ten to fourteen days under normal conditions.

In the Southern states it may be necessary to weed the plant beds, and wherever weeds or grass appear in the bed they should immediately be pulled out. From six to seven weeks after sowing the seed the young plants will be ready to transplant to the field. The cloth cover should be removed for a few days prior to transplanting so as to harden the plants, and the beds should be well watered before the plants are pulled in order to lessen the injury to the roots. Plants should be taken from the plant-bed in the early morning and placed in a shady place until used.

Preparation of the Soil.—Tobacco requires a good seed-bed, therefore, the preparation of the soil is one of importance, and although the minor details of soil preparation may differ in the various tobacco districts, the ultimate object should be the same. Fields intended for tobacco culture should be plowed the previous fall to a depth of at least ten or twelve inches, and, if it is desirable, as in some localities,to apply stable manure,this should be applied at the rate of from fifteen to twenty loads to the acre, broadcasted over the field before plowing. Lime has been found beneficial upon some tobacco soils and should be applied after the land is plowed, and disked in during the preparation of the seed-bed.

The spring preparation of the soil depends largely upon the method to be used in transplanting the seedlings, either by machinery or by hand. In most of the Northern states, especially where cigar leaf tobacco is grown, machine setting is practiced, while in the Central Atlantic and Southern states most of the tobacco is transplanted by hand.

In the North where machinery is used the fertilizer is applied broadcast after the spring plowing and harrowed in by means of a disk harrow. Smoothing harrows, such as the Acme or Meeker, are then run several times over the fields, pulverizing the soil and leaving it in good condition for the planter.

In the Central Atlantic and Gulf Coast states most of the tobacco is transplanted by hand and the fields require entirely different treatment than where the machine is used. The field to be used for tobacco culture is bedded up during February, the beds varying from three to three and one-half feet apart for cigar tobaccos. The commercial fertilizer is applied in the drill and mixed with the soil by having a single-shovel plow furrow run in the drill, after which two furrows are made with a one-horse turning plow forming a list.

The field is left in this condition until the plants are large enough on the plant-bed to transplant to the field. At this time this list is leveled off either by a small harrow or with a log. Where the land has been listed for some time, it is good practice to re-list and then log off, as the small plants will take root much quicker in fresh-plowed mellow soil.

Fertilizers.—Tobacco responds to good fertilization and feeds heavily on nitrogen and potash. Larger amounts of commercial fertilizer are used in the production of cigar leaf tobacco than with tobacco used for other purposes. The principal source of nitrogen is from cottonseed meal, although where the heavy tobaccos are grown, castor pumace or ground blood is used to some extent. Potash is needed in the production of all tobaccos in order to improve the burning qualities of the leaf. Only sulphate or carbonate of potash should be used, as the salt contained in the muriate of potash is detrimental to the burning quality of the leaf. Phosphoric acid is also necessary in small amounts.

In the tobacco-growing regions of Florida, Georgia and Texas a vast amount of money is spent each season for commercial fertilizers. In addition to a liberal application of stable manure, as high as 2000 pounds of cottonseed meal, 400 pounds of sulphate of potash and 200 pounds of acid phosphate are used to the acre in the production of cigar wrapper leaf. Like amounts are used in the New England states. Smaller amounts are used in the production of heavy and export tobacco, and in such states as Virginia a crop rotation in which clover appears as one of the crops in the rotation, reduces the amount of commercial fertilizer, especially that which is used as a source of nitrogen.

Transplanting and Cultivation.—When the seedlings in the plant-bed have reached a height of from four to six inches, they are ready to be transplanted to the field. Great care is necessary in taking the seedling from the bed that the roots are not injured; therefore, it is necessary to water the bed well before pulling up the plants. Plants should be taken from the bed early in the morning and placed in a cool, shady place until they are to be used. If pulled during a rainy season there is no use in watering the bed and they can be used at once. Plants should be pulled one at a time with the finger and the thumb taking hold of the plant close to the ground. They should be shaken off or, if water is near, the soil washed from the roots, and then packed with the roots down in a basket or box.

Where a machine is used for transplanting, the field is usually left flat, having been previously harrowed so as to present a fresh surface. Two men are required to feed the machine and one to do the driving. There are several makes of transplanters, the most popular being the Beemis and the Tiger. These machines open the furrow, set the plants and place any amount of water desired around the roots. Tobacco transplanted by means of these machines appears to recover from the shock of being transplanted, and grows off much sooner and with more uniformity than when planted by hand. Another advantage of machine transplanting is that the transplanting can be done just as well, if not better, during dry weather as during wet, or when the soil is in favorable condition. These machines have been in use in the northern tobacco states for years, and they are gradually finding favor with the southern grower. The cheap negro labor of the South has been the principal cause of their restricted use, but as the price of labor has risen in the last few years, tobacco transplanters are now being used with success where formerly hand setting was practiced.

When hand setting of tobacco is practiced, the field is bedded instead of flat, the beds are marked off the distance required to plant the seedlings, and if the soil is at all dry, water is applied at these places. The plants are then dropped at each mark and a laborer sets them at these places with a dibble. Transplanting by hand should be done only when the soil is in a good moisture condition, or during cloudy or rainy weather. The distance at which the plants are set in the rows depends entirely on the type and commercial use of the tobacco. The heavy tobaccos of Virginia and Tennessee and the fiue-cured tobaccos of the Carolinas are usually checked at a distance of thirty-six inches, while cigar leaf tobaccos are set in the drill from twelve to eighteen inches, according to their type.

A PLANT READY TO SET IN FIELD.[1]

Tobaccos of all types require frequent and thorough cultivation. No weeds or grass should ever be allowed to grow in the field. Cultivation usually begins about eight days after transplanting, when the young plants should be hoed and given a reasonably deep plowing. This is the only time that a deep cultivation should be given. In the North, riding and walking cultivators are used, having an attachment of shallow running

[1] Courtesy of The Pennsylvania Farmer.

plows, while in the Southern states single stocks with sweeps are mostly used. Cultivation usually ceases when the plants have received their final topping. As soon as the seed-head appears it should be taken out along with about three or four leaves with cigar type tobaccos, while the heavy and export types are topped down to eight or ten remaining leaves, according to the growth of the plant and the style of leaf desired. The Maryland and Burley tobaccos have more leaves left on the plant after topping, but not as many as the cigar types. All types of tobacco will send out shoots or suckers after being topped, and these should be broken off, so that all the strength of the plant will go into the leaves on the main stalk.

Tobacco is subject to insect pests from the time it germinates in the plant-bed to the time it is harvested. The flea beetle which lives on the young plants in the bed can be controlled by using kerosene and wood ashes. In the field the bud worm, horn worm and grasshopper destroy the leaves. These can be controlled by the use of Paris green, either applied dry mixed with cornmeal or ashes for the bud worm and in a solution at the rate of one pound of Paris green to 100 gallons of water, for the horn worm. More detailed instructions for controlling these pests will be found in the chapter on "Insect Pests."

Methods of Harvesting.—Various methods are used in the different tobacco districts in harvesting tobacco. In the heavy and export districts the entire plant is cut. The stalk is first split down the middle about two-thirds its length; then cut off close to the ground. The plant is then hung across a stick about four feet in length holding from six to eight plants, according to their size. When a stick is filled it is placed upon a wagon and taken to the curing barn. In the Burley and Maryland tobacco districts the plant is simply cut close to the ground and speared upon the stick, the stalk not being cut as in the former method. This method of harvesting is also used in Ohio, Pennsylvania, Wisconsin and to some extent in the New England states with the binder and filler grade of cigar leaf tobacco.

For the cigar wrapper tobaccos of Florida, Georgia, Texas and the New England states, the leaves are picked off the growing plants as they ripen, beginning with the sand or bottom leaves. The leaves are placed in baskets and taken to the curing barn, where, by means of a needle, they are strung upon strings attached to sticks, each string holding about thirty-five leaves. The ends of the string are fastened to each end of the stick, which is then hung upon the tier poles in the barn where they remain until cured. The bright flue-cured tobaccos of North and South Carolina, also Virginia, are harvested by a similar method, differing in that the leaves are tied upon the string in pairs and sometimes in triplets instead of the individual leaves being strung upon the string by means of a needle. Cigar leaf tobacco, harvested by the priming or single-leaf method, will cure much quicker than when the whole plant is cut and will produce tobacco of more uniform color and finer texture; besides, there will be less waste of the bottom leaves and every leaf can be harvested at the desired stage of ripeness.

Barn Curing.—There are four methods of barn curing practiced: air curing, fire curing, flue curing and sun curing. All cigar leaf, Burley and Maryland smoking tobaccos are air cured. The tobacco, either primed or cut on the stalk, is hung upon the tier poles in the curing barn and there allowed to cure out by a gradual dying of the leaf tissues and evaporation of moisture. Favorable curing conditions exist when the tobacco will come and go "in kase" several times during the period of curing. Barns for air curing are provided with ventilators which can be opened or closed according to the climatic conditions and the stage of the cure. For the first few days the barn should be kept closed until the tobacco has wilted and taken on a yellow shade of color; then the ventilators should be opened so as to admit a free circulation of air until the tobacco assumes the brown color. During dry, windy weather the ventilators should be kept closed during the day and opened at night.

Fire-curing Barn.

The heavy and export tobaccos of Virginia, Tennessee and Kentucky are fire cured. As soon as the barn has been filled with green tobacco, small wood fires are started to wilt the leaf until the yellow color appears; then the amount of heat is gradually increased until the leaf turns brown. When the tobacco reaches this stage the heat is again increased to cure the midrib or stem of the leaf, after which the fires are allowed to die and the tobacco cool off. During the curing process the heat should never be allowed to fall until the final cure is obtained, as a fall of temperature during the curing process will injure the color and texture of the tobacco. Small log barns with tight walls are used for fire curing without any ventilation except the door.

The bright tobaccos of Virginia and North and South Carolina are cured by flues. The barns used in this process are similar to those used in the open fire process, except that the fires are built on the outside of the barn in brick fireplaces, having a metal flue running around the inside of the barn about two feet from the bottom. There are usually two fireplaces, the flues of each uniting at the opposite end of the barn, and merging into a single return flue coming out at the same end of the barn as the fireplaces. Some barns have the two flues independent of each other, in which case both flues return to the front end of the barn.

The tobacco, after being hung in the barn, is first given a low heat so as to wilt the leaf and produce the yellow color. The temperature is then rapidly increased so as to set the yellow color in the leaf and prevent the leaves turning brown. As soon as the color is set the temperature is again increased to cure the midrib, when the temperature is allowed to fall and the tobacco to cool off. Three days and nights is the usual time taken to cure a barn of tobacco by this process.

FLUE-CURING BARN, VIRGINIA.

The sun-curing process is similar to the air curing, the difference being that the tobacco is allowed to wilt on scaffolds before being placed in the barn and when climatic conditions are favorable it is also sunned before being placed in the barn. This process gives the tobacco a delicate flavor when smoked in the pipe.

Preparation for Market.—The first damp season after tobacco is cured is usually the best time to take it from the tier poles and prepare it for the market, especially with cigar leaf tobacco. At this time the tobacco is soft and pliable, and can be handled without injury to the leaf. The only grading done by the grower in preparing cigar leaf for the market is to separate the leaves into three groups, namely, sand or bottom leaves, middle leaves, and top leaves. Where the tobacco has been cured on the stalk the leaves are stripped off and tied into hands containing about forty leaves. Where the tobacco has been primed, or leaves picked off the stalk in the field, the cured leaves are simply bunched on the string and the string wrapped around the butt-end of the leaves. The tobacco is then packed in boxes and hauled to the packing house or kept in the barn until a buyer comes.

With the heavy, export and bright tobaccos, the grower usually grades the tobacco into the commercial classes as sand lugs, lugs and wrappers, the finer classification being performed by the buyer, who is usually a rehandler of tobacco.

These tobaccos are packed in hogsheads and remain for some time in the warehouses to undergo an ageing process which mellows the tobacco and brings out its best qualities. All cigar tobaccos have to go though a fermentation process, after which they are graded out according to color, texture and size. The tobaccos of Ohio, Wisconsin, Pennsylvania and certain grades of the New England tobaccos are packed in boxes, while the wrapper grades of Connecticut and Massachusetts are packed in mat bales weighing about 160 pounds. The cigar leaf tobaccos of the southern tobacco states are all packed in bales, either the Cuban or Sumatra style.

Methods of Selling.—All cigar tobaccos are sold by the grower in their unfermented condition to dealers in leaf tobacco, who either buy the tobacco from the curing barn, or upon delivery by the grower at the warehouse. In some cases the tobacco is grown upon contract at a stipulated price per pound for the various grades. All transactions are upon a cash basis upon delivery of the tobacco.

The heavy, export, manufacturing and bright tobaccos are sold at public auction, either in the hogshead or as loose tobacco. The place of sale is a public warehouse and all transactions are cash at the close of each auction. The sales are attended by buyers not only of this country, but of foreign governments where the regi system (government monopoly) exists; such countries as Spain, Italy, France and Japan having buyers attending these auctions.

Danville, Virginia, is the largest market for loose tobacco, especially the bright tobaccos; while Richmond, Lynchburg and Petersburg, Virginia, handle mostly dark, fire-cured tobaccos. Public auctions are held at various places in Tennessee, Kentucky, Ohio and the Carolinas.

REFERENCES

"Tobacco Leaf." Killebrew and Myrick.
"Tobacco: History, Culture and Varieties." Billings.
Kentucky Expt. Station Bulletin 129. "Cultivating, Curing and Marketing Tobacco."
Kentucky Expt. Station Bulletin 139. "Tobacco Improvement."
Ohio Expt. Station Circular 156. "Disinfecting Tobacco Beds from Root Rot Fungus."
Ohio Expt. Station Bulletin 239. "Breeding Cigar Filler."
Canadian Dept. of Agriculture Bulletins A2, A3, A4, A5, A6, A7. "Tobacco in Canada."
U. S. Dept. of Agriculture Bulletin 40. "Mosaic Disease of Tobacco."
Farmers' Bulletins, U. S. Dept. of Agriculture:
343. "Cultivation of Tobacco in Kentucky and Tennessee."
416. "Production of Cigar Leaf Tobacco in Pennsylvania."
523. "Tobacco Curing."
571. "Tobacco Culture."

CHAPTER 26

WEEDS AND THEIR ERADICATION

Weeds are the farmer's most active and persistent enemy. If he would keep them under control, he must wage a continual warfare against them. Seldom is there soil so poor that it will not grow weeds, and the richer the soil, the greater the weed crop. They seem to have been equipped by nature to hold their own in the struggle for existence, for they manage to thrive despite heat or cold, drought or flood.

Some may ask: Why do weeds exist? They undoubtedly have a place in nature's great plan. They are her agents in restoring fertility to the soil. If unmolested they will cover the soil as a blanket, first as weeds, then as brush and finally as a forest. In fact, in some parts of our country, land is farmed until crops are no longer profitable, and then abandoned. Weeds then take possession, and by returning nitrogen to the soil, they become restorative agents. Give nature time enough and she will restore any land to its normal fertility.

Damage Done by Weeds.—It is impossible to calculate the damage done yearly in the United States by weeds. Investigators roughly estimate it to be hundreds of millions of dollars.

Weeds Reduce Crop Yields.—Weeds are more rapid of growth and more tenacious of life than cultivated plants. They crowd out the rightful occupants of the soil, depriving them of air and sunshine. Being more vigorous, they absorb from the soil the plant food that should be used by the crops, thus reducing the yield. A ton of dried pigweed contains as much phosphoric acid, twice as much nitrogen and five times as much potash as a ton of ordinary manure.

Weeds also absorb moisture in greater quantities and more rapidly than crop plants. They are more drought resistant, for, having appropriated all the moisture to themselves, they continue to thrive while the plant beside them dies. Experiments prove that some weeds transpire 250 to 270 pounds of water to develop a pound of dry matter.

In addition to this, it is a well-established fact that weeds exert an injurious effect upon crop yields by giving off from their roots substances which are poisonous to crop plants.

Weeds cause a direct money loss by reducing land values. A would-be purchaser is not so apt to buy a farm where the fields are thickly covered with mustard, wild carrot or the ox-eye daisy. Naturally, the loss in value should be borne by the man who has allowed his land to be so abused.

Weeds increase the expense of harvesting the crop. A field overrun with weeds calls for extra labor and entails extra strain on the machinery.

It sometimes necessitates hand labor, which is most expensive. Also the cost of threshing and cleaning the grain is increased by the presence of weed seeds.

Market values are lessened by impure grain. Many crops are docked full half their value on account of noxious weed seeds. It is estimated that the State of Minnesota alone suffers a loss annually of over $2,000,000, because of weed seeds in the wheat.

But the loss does not stop here. Some weeds harbor and encourage harmful fungi and insects. For example, the very common clubroot of cabbage thrives on the various members of the wild mustard family. Stubble overgrown with weeds harbors cut worms, beetles and other insect pests. Mildew, smut and rust are often transferred from friendly weeds to the grain crop.

Furthermore, livestock and even human beings lose their lives as the result of eating poisonous berries or roots. The water hemlock or cowbane is fatal to sheep and cattle. The deadly loco-weed on the western plains has caused the death of many horses and cattle.

How Introduced and Spread.—Weeds are great travelers. They travel by means of the wind and water. They are carried by birds, beasts and human beings. They are disseminated by means of manure, feedstuffs, machinery and grain seeds.

Such weeds as the thistle, milkweed and the dandelion have downy plumes attached to each seed. The faintest breeze will carry them miles away, where they begin life anew. Members of the dock family have seeds supplied with wings which enable them to float upon the water as well as upon the breeze. Some weed seeds have sharp barbs and stickers by which they attach themselves to the hair of animals and to the clothing of human beings, and are thus carried into new localities.

When it is known how many thousand seeds one weed-plant produces, it can readily be seen how great a calamity it is to let a weed bloom and go to seed. One thistle head contains enough seeds to start several thousand plants the next year. It is estimated that one wild mustard plant produces 10,000 seeds, and one pigweed 115,000 seeds. If only a few of these seeds germinated, the situation would not be alarming, but the chances are that if allowed to seed a very high percentage of them will find opportunity to propagate their kind.

One of the most prolific sources of weed dissemination is in the buying and sowing of impure seeds. Especially is this true of clover and grass seeds. In an analysis of several samples of commercial seed at one of the experiment stations, one sample of red clover seed was found to contain 36,000 weed seeds to the pound. A pound sample of timothy seed contained 79,000 weed seeds.

Care should be taken to procure seeds only from uninfested districts. A farmer should have sufficient knowledge of seeds to enable him to detect impurities. It is a wise precaution to send first for samples of seed under consideration for purchase. If the farmer cannot determine with reasonable certainty as to their purity and germinating power, he should submit his sample to the experiment station of his state, that the weed seeds may be identified. These institutions gladly test samples of seeds for farmers free of charge.

Careful screening will overcome much of the difficulty with weed seeds.

Classification of Weeds.—It is not enough to know the name of a weed. In order to win in the struggle against a weed enemy, one must know its habits of life and its methods of propagation. There is no weed so vicious that it cannot be subdued or even entirely eradicated if its habits are understood.

Weeds may be divided into three classes according to their life cycle: annuals, biennials and perennials. Annuals complete their growth and ripen seed in one season, such as wild mustard and ragweed. These weeds must depend upon seed in order to grow again the following year. It can readily be seen that if the plant is destroyed before seeds form, the source of next year's crop is much lessened.

Unfortunately, some of these seeds are encased in an oily covering, enabling them to resist decay. Wild mustard seed, for instance, has such power of vitality that it has been known to germinate after having lain in the ground for many years.

Biennials are not so easily disposed of. They require two years in which to complete their growth. Some of them have long tap roots in which they store up plant food during the first year. This food is used to produce seed during the second summer. Burdock and wild carrot are common examples of this class. These weeds are seldom seen in cultivated fields, for the plow and cultivator are disastrous to their roots. If these weeds are cut off even with the ground, they branch out and become thicker than before. Cutting two or three inches below the surface so as to destroy the crown of the plant is effective.

Perennials grow year after year and produce seed indefinitely. Some of them reproduce by seed only, such as the ox-eye daisy and dandelion. Others have roots running under ground from which they send up new plants yearly. Common examples of this kind of weed are Canada thistle and bindweed. This class of weeds is the most difficult to eradicate, for wherever these roots are cut or bruised new stalks are sent forth and the difficulty increased. There is one time, however, during the growing season when these weeds are most effectively attacked; that is, when they are in full growth, but before seeds form. No plant can live long without a leaf system. If the plants are cut off and plowed under at this time, many of them will be eradicated.

Repeated and persistent attacks, however, on the part of the farmer will be necessary for ultimate success. The poorer the land becomes, the greater the number of biennial and perennial weeds. They seem to be best adapted to the poor conditions and will thrive where other crops fail.

Weed Habitats.—Bindweed, Canada thistle and horse nettle are entirely at home in any field, whether it be corn field, meadow or feed lot. However, it is a well-established fact that certain weeds seem to follow certain crops. For instance, corn fields are mostly infested by such weeds as foxtail, cockle-bur and butter-print. These can be overcome by persistent cultivation. Milkweeds and the large family of mustards, of which shepherds' purse and wild radish are members, seem to follow the small grains. The mustard family is easily overcome by cutting before it goes to seed. The milkweed, however, is more difficult to eradicate, as it spreads by means of underground roots. Meadows and pastures have a different

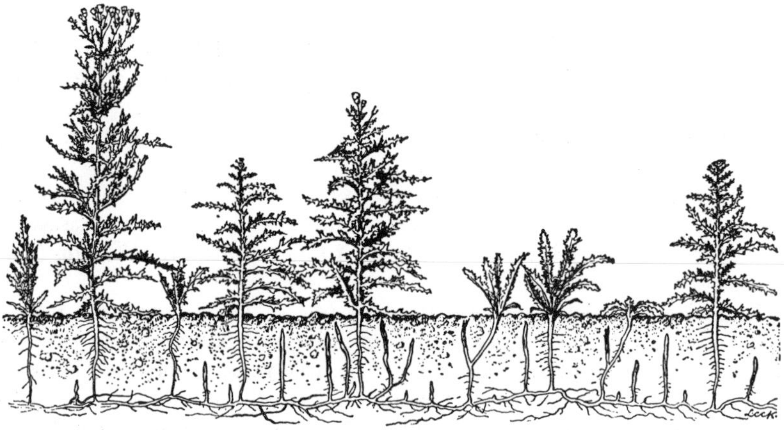

THE MANNER IN WHICH CANADA THISTLES SPREAD BY UNDERGROUND ROOTSTOCKS.[1]

When the rootstocks are brought to the surface by plowing at the right depth they may be raked up and destroyed.

type of weed from corn and small grains. Three of the worst meadow weeds are dock, ragweed and buckhorn. They spread by seed only and can be kept down by mowing before they go to seed. The ground must also be kept well seeded to grass or clover, for if bare spots appear, the weeds are quick to appropriate them. Sorrel is a pasture weed which is hard to eradicate, as it spreads by means of underground roots. It cannot compete with red clover, however, for a place in the meadows. For this reason it is usually found in meadows where the acidity of the soil does not encourage red clover. Plowing and sowing to cultivated crops is the best method of eradication. If the land will not admit of cultivated crops, common salt put on the plants will kill them and keep them from spreading.

Principles Governing Control.—The foregoing discussion suggests

[1] Courtesy of U. S. Dept. of Agriculture.

the necessity of working out a system of farm management that will afford weeds the least opportunity to gain a foothold. The problem is not how to rid a farm of weeds, but how to prevent weeds in the first place. This can be solved by a system of cropping which takes into consideration the needs of the different fields as regards weed eradication. This phase of the problem has been discussed under the chapter on "Rotations."

A few general principles for weed prevention and eradication are here given:

1. Cut all weeds before seeding, if possible.
2. Burn all weeds with mature seeds. Do not plow them under.
3. Practice crop rotations.
4. Sow clean seed.
5. Watch for new weeds in your locality. If you can not identify them, send them to your experiment station for identification.
6. See that the laws in your state dealing with control of weed plagues are enforced.

A few of the most common weeds are here considered.*

Canada Thistle.—The Canada thistle is a perennial of European origin, and is the most dreaded of all weed pests. It is a common weed of the northern half of the United States. The stems of Canada thistle grow from one to three feet tall; they are much smaller and smoother than other thistles. The leaves are very spiny and the margin has a ruffled appearance. The upper side of the leaves is smooth and bright green in color, while the lower side is downy or hairy. The flowers are rather small, about one-half inch in diameter and of a rose-purple color. The Canada thistle flowers from June to September, maturing the first seed by the middle of July. The seed is smooth and light-brown in color, measuring one-eighth of an inch in length. The seed is easily carried by the wind and is most commonly found in medium red and alsike clover seed.

Propagation.—The Canada thistle propagates by underground rootstocks as well as by seed. The underground rootstocks grow rather deeply in the soil and run parallel with the surface. They are the storehouse of the plants, and are capable of sending up young shoots for some time after the parent plant has been destroyed. This fact explains the persistency of the Canada thistle. As long as the plant is permitted to form green leaves, it will manufacture plant food, which is stored in the rootstocks. As long as plant food is present in the rootstocks they are capable of sending up new plants, and will continue to thrive as long as they grow leaves at the surface.

Control.—The Canada thistle occurs in all crops in the rotation, consequently no one method of control will be effective. The details of control had best be worked out for each particular condition.

To thoroughly subdue the pest it is necessary to starve out the root-

*Taken, with modifications, from The Pennsylvania Farmer, prepared by Professor Darst, of The Pennsylvania State College.

stocks by cutting off all green parts above ground. This requires destroying the plants in some manner every week at first, and then every two weeks until the rootstocks die of starvation.

In a small grain crop keep them cut with scythe or hoe, so as not to let them go to seed. After harvest the land should be plowed rather deeply and then worked down with the drag harrow. All roots harrowed out should be piled, dried and burned. The land should be disked regularly about every ten days, so as to destroy stray plants. In the late fall the land should be re-plowed, but not worked down, so as to expose the remaining roots to frost action. In the early spring the ground should be worked with a disk and a smother crop sown, such as oats and Canada field peas, millet or buckwheat.

THE CANADA THISTLE (*Circium arvense*).[1]
B—Seed enlarged.

Canada thistles occurring in pastures must be cut out below the ground every ten days until starved out. After cutting the plants, it is well to pour a little kerosene on the stem and roots. Often spraying with strong concentrated salt solution will be effective in pastures and waste places. The spraying should be done thoroughly and repeated when young shoots reappear. Spray materials should be applied under high pressure, and in a vapor spray, to be effective.

After all, sprays used as a substitute for the scythe and mower will not kill the roots below the ground.

When the thistle occurs in a cultivated crop, knives or sweeps should be used on the cultivator instead of the ordinary shovels. The sweeps will be more effective in cutting off the plants. The thistles that grow within the row should be kept cut out with the hoe.

Quack Grass.—Quack grass is perennial and propagates both by seeds and creeping underground rootstocks. The stems grow from one to two feet tall. The leaves are ashy green in color, rough on the upper side and smooth beneath. The plant flowers in June and seeds in July.

The plant sends out underground rootstocks which are jointed, each joint capable of budding a new plant. Quack grass grows an enormous root system, which soon crowds and smothers out other plants.

Control.—Quack grass may be subdued if no green leaves are allowed to develop. Since quack grass makes fair pasture, a good plan, where possible, is to pasture it close to the ground during the midsummer; then plow deeply in the early fall. The ground should be worked down immedi-

[1] Courtesy of The Pennsylvania Farmer.

ately with the spike-tooth harrow. This will drag out a great many of the roots. These should be dragged or raked to one end of the field, to be dried in piles and then burned. The ground should be plowed the second time, late in the fall. This should be done crossways of the first plowing. The ground should be worked down again with the harrow and as many of the rootstocks dragged out, piled and burned as possible.

Quack Grass (*Agropyron repens*).[1]

The next spring cultivation should begin as early as possible, the ground being worked every few days. Then a cultivated crop should be planted, preferably corn. The corn should be planted in hills so that cultivation can be given both ways of the field. The corn should be cultivated thoroughly and a close watch kept for any stray quack grass which may come up.

Green Foxtail (*Getaria viridis*).[1]

1—Concave side of seed.
3—Convex side of seed.
Both enlarged.

If there is any doubt as to whether the quack grass is completely killed, a mixture of hairy vetch and rye should be sown in the last cultivation of the corn. This cover crop should be plowed down the next spring and a heavy seeding of millet sown. The cover crop and the millet following the corn will surely smother out the last of the quack grass.

A thick covering of straw or manure a foot or so thick and well packed down, will smother out the grass. It will take from two to three months to smother out quack grass, as the roots remain alive for some time even though the tops above ground be dead.

Foxtail.—The green and yellow foxtails are very similar in appearance and in habit. They are both annuals and propagate by seed alone, seeding from August to October.

The yellow foxtail is a common weed all over the world, while the green foxtail is found mostly in North America. The seed of the foxtails are common impurities in many grass and legume seeds. Once in the

[1] Courtesy of The Pennsylvania Farmer.

ground, they retain their vitality for many years, germinating only when brought near the surface of the ground. The stems of the foxtails grow from one to four feet tall. The leaves are three to six inches long and are rather wide, flat and smooth. The seed heads are from two to four inches long. The seed is yellowish-brown in color, about one-tenth of an inch in length and ovoid in shape.

Control.—In grain fields the stubbles should be given surface cultivation or, if the soil is dry enough, burning over will destroy the seeds that have fallen on the ground.

In cultivated crops tillage should be continued very late, in order to prevent the development and distribution of seed from late-grown plants. Sheep may be turned in to graze down the aftermath of infested meadows.

A FIELD DODDER. B FLAX DODDER. C ALFALFA DODDER.[1]
Seeds enlarged. Blossom of Alfalfa Dodder.

Dodders.—There are several kinds of dodders; those found in this country are known as common dodder, field dodder, clover dodder and alfalfa dodder.

They are annuals and propagate by seed, and are very peculiar weeds because they live upon other plants. From their habit of growth they are known as parasites.

The seeds of the dodder germinate in the soil and the young plants soon perish unless they come in contact with a clover or alfalfa plant. Once in contact with a suitable host plant, the roots of the weed soon decay. The fine yellowish and reddish stems twine about the host and spread from plant to plant; a single dodder plant often growing on several different host plants at the same time.

This weed derives its nourishment by sending little suckers out into the stems of the plant on which it grows. Dodders appear lifeless to the casual observer as the leaves are reduced to mere scales.

The white or pink flowers occur in clusters along the slender stems. During the latter part of summer the cluster may contain flowers in bloom and the matured seed at the same time.

Small patches of dodder may result from a single seed. A patch can

[1] Courtesy of The Pennsylvania Farmer.

be distinguished at quite a distance on account of its distinct yellowish cast.

Dodder seed is very difficult to remove from clover and alfalfa seed.

Eradication.—The first step in the control of dodder is to sow clean seed. Clover and alfalfa seed should be carefully examined for the pest before it is sown.

When a field is badly infested the crop should be plowed under before the seeds form. Dodder seed, plowed under, may remain in the soil for seven or eight years and then germinate. After plowing, plant a cultivated crop for a year or two; as the weed is an annual, it yields readily to cultivation.

When dodder occurs in small patches it may be successfully eradicated by digging up the infested areas. To avoid scattering the seed, dry and burn the plants on the spot.

The dried plants may be burned by covering them with straw or shavings soaked with kerosene. After the first burning, stir the surface with a rake, then burn over for the second time, so as to destroy any seed that may have matured and fallen to the ground.

After a patch has been dug up, burned and re-seeded, it will be well to watch for stray plants which may come. If such plants appear, destroy them before flowering time.

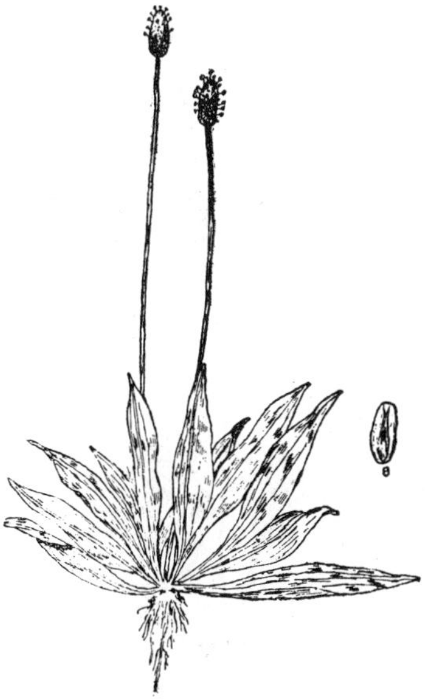

BUCKHORN OR NARROW-LEAVED PLANTAIN (*Plantago lanceolata*).[1]
B—Two times natural size of seeds.

When dodder seed is allowed to mature in clover or alfalfa hay it should not be removed from the field, but should be dried and burned.

Weed sprays are sometimes recommended for killing dodder in clover and alfalfa. The spraying should be done before or at blooming time in order to prevent the seed formation. A twenty per cent solution of iron sulphate is found effective on alfalfa fields. The spray will kill the parasite and apparently destroy the crop, but a new growth of alfalfa will spring from the roots as soon as the plant is relieved of the pest.

Buckhorn.—Buckhorn is a perennial of European origin. The weed propagates by seed, which matures any time from May to November. The plant has a short, thick root-stem which enables it to exist over winter. Buckhorn leaves are long and slender, borne in rosette-like clusters at the surface of the ground. The plant does not produce stems as other weeds

[1] Courtesy of The Pennsylvania Farmer.

commonly do, but sends up a stiff, slender flower stalk one to two feet long, at the end of which there is a short, compact head or spike. The figure shows the plant in bloom. After blooming, the heads elongate somewhat and turn a dark brown color. The seeds are shaped like a coffee grain, but very much smaller, measuring about one-tenth of an inch long. The seeds are a shiny amber to rich brown color. A dark-colored scar is present on the concave surface, while a narrow yellowish stripe is generally present on the back of the seed.

Control.—Buckhorn is a very common and persistent weed in meadows, pastures and lawns. It is without doubt the most common of all weed seed in clover and alfalfa seed. In fact, it is very difficult to buy seed, even from the best of seedsmen, that is entirely free from this weed. The first step in controlling the weed on the farm is to sow nothing but clean seed.

Common or Broad-leaved Plantain (*Plantago major*).[1]

B—Two times natural size of seeds.

Where meadows and pastures are badly infested they should be plowed and a cultivated crop grown for one or two years. By thorough cultivation and the use of hoes, all plants may be destroyed before going to seed. The seed remaining in the soil will be induced to germinate by frequent cultivation, thus making it possible to rid a field of the pest in one or two seasons.

Small areas in lawns may be controlled by cutting out with a spud or narrow-bladed hoe. This method is more effective during hot, dry weather. Buckhorn can be destroyed with carbolic acid without injuring or defacing the lawn. Stab each plant in the center, down to the fibrous cluster of roots, with a pointed stick and squirt into the opening a few drops of the acid with a common machine oil can.

Plantain.—The common plantain is a very persistent weed in lawns and yards. The plant is a perennial and propagates by seed. Plantain does not produce a true stem as most other plants. The leaves, which grow in a tuft near the ground are large, coarse and oval in shape. The weed flowers from May throughout the summer and ripe seed may be found by July.

The seeds are dark brown to black, slightly flattened, with acute edges. They are variable in shape, measuring one-twentieth of an inch in length.

Control.—Common plantain when occurring in fields will yield readily

[1] Courtesy of The Pennsylvania Farmer.

to thorough cultivation. The control is the same as recommended for buckhorn.

In lawns the most practical method of eradicating it is by hand digging. If the plant is cut off several inches below the ground during dry weather, it will give no more trouble.

Carbolic acid may be used in the same manner as recommended for buckhorn, where digging roughens the lawn.

Pigweed.—Pigweed is an annual and is commonly found growing in cultivated fields and waste places. While the weed itself is not hard to eradicate, yet it produces abundantly seeds which have long vitality. The seed has been known to survive in the ground for more than twenty years.

The pigweed has a long, fleshy, red taproot. The main stem is erect, stout, woody and slightly branched. The stem and branches are covered with stiff, short hairs.

Usually the plant will grow from one to four feet tall, but under more favorable conditions it will often reach six feet.

The leaves are long and ovate in shape, measuring from three to six inches in length. The small greenish flowers are crowded into thick, compact heads which are borne at the ends of the branches or in the axils of the leaves. The pigweed flowers from July to September and produces enormous quantities of small, shiny seeds. The seed is a jet black color, oval and flat in shape. It propagates by seeds only.

PIGWEED (*Amaranthus retroflexus*).[1]
C—Root.

Control.—Pigweed seed is commonly found in commercial seeds of different kinds. The first step in its eradication is to guard against buying seed containing this weed.

Thorough cultivation will suppress the weed. In case cultivation cannot be continued late into the summer the weed should be pulled or cut out with a hoe before going to seed. Plants which are pulled or cut while blooming should be destroyed, as they frequently mature seed after cut.

Pigweed may be destroyed in small grain crops by the use of the weeder or the spike-tooth harrow. By going over the grain field when the crop is but a few inches tall the small seedlings may be dug out without injuring the grain. If the weed makes its appearance later on in the growth of a small grain crop, it may be killed with an iron or copper sulphate spray.

[1] Courtesy of The Pennsylvania Farmer.

The spray to be effective must be applied before the grain begins to head and before the weeds bloom.

Lamb's-Quarters.—Lamb's-quarters is sometimes known as smooth pigweed or white goosefoot. This weed is a very common annual throughout the world. It is commonly found in cultivated fields, orchards and gardens.

Lamb's-quarters is distinguished by its upright grooved and many branched stem. The stems are often striped with purple. The plant is a rapid grower and attains a height of from two to four feet. The leaves are quite variable as to size and shape, the lower ones on the stock being comparatively large and irregular, while the upper ones are rather small and narrow.

Lamb's-Quarters, or Smooth Pigweed (*Chenopodium album*).[1]

A—Root.
B—Seed enlarged three times natural size.

The small greenish flowers are borne on the ends or in the axils of the branches. The entire plant presents a silvery gray or mealy appearance which distinguishes it from the true pigweed. The seed is about one-twentieth of an inch in diameter, lens-shaped and a dull black color. The seeds have long vitality, lying dormant in the soil for many years.

The control of lamb's-quarters is similar to that of pigweed. In hoed crops the weed is very persistent and cultivation should be continued until late in the season. In gardens and other small areas the weed should be pulled or chopped out while young.

Since the plants are very succulent while young, sheep may be used to pasture them where conditions will permit. Cultivation in the late summer or fall will germinate seed remaining in the soil which will eliminate the seed that may germinate the next year.

Wild Mustard or Charlock.—The cruciferæ or mustard family contains a large number of weeds, of which the wild mustard and tumbling mustard are the most troublesome. The plants of this family may be recognized by the shape of the flowers, which consist of four petals arranged like arms of a cross. This character was used as the basis for naming the family.

Wild mustard, because of its immense productiveness and the exceedingly long vitality of its seeds, is one of the most difficult weeds to dislodge.

It is an annual plant, which in its earlier stages of growth bears some resemblance to the radish or yellow-fleshed varieties of the turnip. It

[1] Courtesy of The Pennsylvania Farmer.

produces erect branching stems from one to three feet in height which are somewhat roughened by short stiff hairs.

The leaves are quite variable; the lower ones are slender-stalked and deeply pinnatifid, forming one large terminal lobe with two to four smaller lateral lobes.

The upper leaves are irregularly toothed, somewhat hairy and have very short petioles; the lobes are not very pronounced, while the terminal one is much narrower than the terminal lobe of the lower leaves.

The plant flowers from June to September and mature seeds may be found as early as August. The bright yellow, fragrant flowers which are about one-half inch in diameter are borne in elongated clusters at the end of the stem and branches.

The flowers begin to open at the bottom of the cluster, which lengthens as the season advances, and the pods form and empty so that there may be emptied pods below and forming pods above. One of the pods may contain from four to twelve seeds.

The round dark reddish-brown seeds are about one-sixteenth of an inch in diameter. They are a common impurity of grass and clover seeds.

Wild mustard grows in all kinds of grain crops that are sown in the spring and usually matures its seeds before the grain in which it grows is ripe. Where spring grains are chiefly grown the contest with this weed will be a difficult one.

WILD MUSTARD
(*Brassica arvensis*).[1]

A—Pod, natural size.
B—Blossom, one-third natural size.
C—Seed enlarged four times.

Wild mustard is distributed by different agencies. Some of the small seeds are carried from place to place by the birds, but usually the weed finds its way to new centers in grain seed. The threshing machine is also a potent means of carrying it from farm to farm.

It is further distributed over farms on which it grows by means of manures. It is also very frequently distributed by spring floods; when this is the case the farmer has a difficult job.

Control.—Wild mustard seed is a common impurity of small grain, clover and alfalfa seed. The first step in its eradication is to avoid sowing seed containing the pest. As the seeds are small, they are easily removed from wheat, oats and barley by screening. Wild mustard is most common in grain fields and generally disappears in grass and cultivated crops.

[1] Courtesy of The Pennsylvania Farmer.

A good short rotation will in time reduce the seed in the soil. In grain fields, young seedlings may be harrowed out when the grain is but a few inches high. Later on the mustard may be destroyed by the use of iron sulphate or copper sulphate spray. Iron sulphate is probably the most efficient and cheapest spray. The spraying should be done before the grain heads are out and when the mustard is just beginning to flower. The spraying at this time will kill practically all the mustard with little or no injury to the grain. Spraying for weeds should be done on a bright, still day when there are no immediate prospects of rain. For spraying, use a solution made by dissolving 100 pounds of iron sulphate in 52 gallons of water. This solution should be used at the rate of 50 gallons to the acre and put on at a high pressure of 100 or more pounds to the square inch. The spraying can be done at a cost of approximately $1.25 per acre.

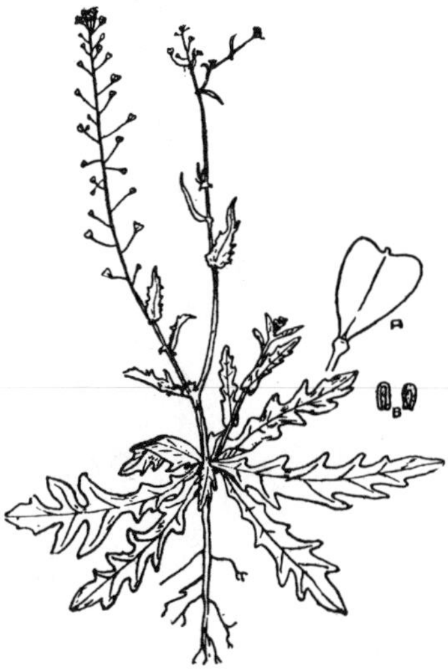

SHEPHERD'S-PURSE
(*Capsella bursapastoris*).[1]

A—Enlarged seed pod.
B—Enlarged seed.

Shepherd's-Purse.—Shepherd's-purse is sometimes known as Case weed, St. James' weed and Mother's Heart. "With the exception of the chickweed, it is probably the most common weed on earth," being found in all cultivated regions of the world.

Shepherd's-purse is very prolific and its seeds have long vitality. This weed often harbors the club-root fungus, so common on cabbage, cauliflower, turnips and radishes, and will infect soil where those plants are cultivated.

Shepherd's-purse is an annual, or winter annual. It is one of the first plants to make its appearance in the spring and it is not uncommon to find it making a good growth in March.

The seeds, as a rule, germinate in the fall and form a rosette of leaves, close to the ground. From this rosette the following spring, arise the stems which vary from three inches to two feet in height. The leaves growing close to the ground are rather long and deeply cut, while those on the stem are small and arrow shaped. The small white flowers are borne in elongated heads or racemes. On the flowering stalks will be found all graduations from a small bunch of flowers in bloom at the end, down to mature pods below.

Shepherd's-purse is propagated by seeds only, which germinate either in the fall or spring. This weed flowers and produces seed throughout the

[1] Courtesy of The Pennsylvania Farmer.

season. An average size plant will produce about 2000 seeds. The seeds are very small and covered with a transparent, mucilaginous material which preserves them in the soil for several seasons.

Control.—Shepherd's-purse may be controlled in the field or garden by thorough cultivation and use of the hoe.

In meadows or small grain fields the weed may be destroyed by spraying. Spray while the weed is still young, and if possible before it blooms. Seventy-five pounds of iron sulphate to fifty gallons of water is recommended as a good weed spray. The spray material should be applied under high pressure and in mist. It will take about fifty gallons to the acre.

If one intends to spray weeds in pastures or small grain crops on a large scale, it will pay to purchase a weed sprayer, which is better adapted for the work and will be more effective and economical than ordinary sprayers.

Weeds can be sprayed in grass fields, but not in alfalfa or clover fields, as the spray will kill the clover as well as the weeds.

Peppergrass.—Peppergrass is a native of this country; an annual, and propagates by seed. It seeds from June to October and will be found in small grains and clover fields. The seed is often found in timothy and red clover seed on the market.

Peppergrass grows from six inches to two feet tall and is much branched. The weed sometimes becomes a tumble weed because of its spreading growth.

The flowers are white and very small, borne on racemes or elongated heads. The seeds are formed in round but flattened pods. They are small, measuring about one-sixteenth of an inch in length.

Birds are very fond of the seed and dispose of large quantities.

Control.—The control is similar to that of shepherd's-purse. Care should be taken in plowing under plants that are nearly mature, since part of the seeds will germinate.

Badly infested land should be planted to a cultivated crop and thoroughly cultivated; thorough cultivation being all that is necessary to control the weed.

Cocklebur.—Cocklebur is known by several other common names, *i.e.*, "Clotbur," "Sheepsbur," "Buttonbur" and "Ditchbur." This weed is an annual and native of this country. The plant is coarse, rough and branched, growing from one to four feet tall. The stems are angled and often reddish, spotted with brown. The leaves are broad, bristly rough on both sides and placed alternately on the stems.

Cocklebur bears the male and female flowers at different places on the plant. The male flowers are borne above and near the end of the main stem, while the female flower clusters are borne below in the axils of the leaves.

The seeds of the cocklebur are borne in reddish-brown, two-peaked burs which are covered with stout hooked prickles. Each bur contains two seeds. It is claimed that one of the two seeds germinate the first year and the other the following year, thus insuring at least seed for two years.

Control.—Clean cultivation and the rotation of crops are recommended for this obnoxious weed. Infested corn fields should be put into a small grain crop, followed by clover or grass. The harvesting of these crops will kill or behead the weed before it has time to grow much or develop burs. Plants that have formed burs should be cut, raked and burned.

Field Bindweed or Wild Morning Glory.—It is most commonly found in grain fields, meadows and waste places. "It is a most obnoxious weed, spreading chiefly by means of its long, creeping, cord-like roots, which at any part of their length may bud new plants." Small bits of the roots may be broken off and carried quite a distance by a cultivator and produce new plants.

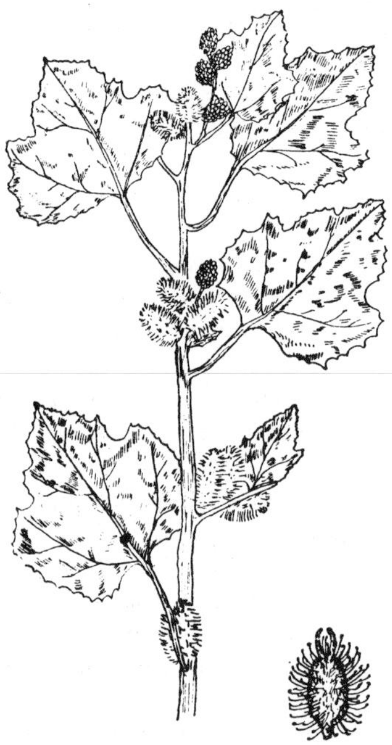

Cocklebur (*Xanthium canadense*).[1]
A—Bur two-thirds natural size.

"The stems are smooth, slightly angled, slender, one to three feet long, twining about and over any plants within reach," tending to smother them. The leaves are cordate or heart-shaped, one to one and one-half inches long, with backward pointing lobes at the base. The flowers are pink, sometimes nearly white, funnel-shaped, about one inch in diameter at the mouth of the tube.

This plant is propagated by seed and the underground fleshy root-stocks. It flowers from June throughout the summer and mature seed may be found in July.

The seeds, which are borne in globular, two-celled capsules, are dull, dark brown, about one-eighth inch long, rough, oval, with one side flat and the other rounded.

Control.—If the land is planted to crops that can be cultivated very often throughout the growing season, field bindweed can be completely eradicated in two years.

When this weed grows in pastures and waste places, its growth may be checked by allowing sheep to have access to the places where it grows. Three years of pasturing with a large number of sheep will greatly weaken this pest, if not kill it entirely.

Infested land should be plowed in the late summer after a crop has been removed and hogs which have not been ringed turned in for the purpose of turning out and eating the succulent roots, of which they are

[1] Courtesy of The Pennsylvania Farmer.

very fond. If hogs are again turned into the field as soon as possible in the spring and left until planting time the weed will be considerably weakened in vitality.

Hedge Bindweed.—This weed is very similar to field bindweed and about as hard to eradicate; its rootstocks are larger and not so difficult to remove from the soil. The trailing or twining stems are three to ten feet or more in length and have the same method of destroying other plants. The leaves are smooth, long, triangular and pointed at the end instead of rounded as the field bindweed. The base of the leaves forms pointed lobes. The funnel-shaped flowers are about two inches long, pink with white stripes or clear white. They are borne singly on slender flower stocks in the axils of the leaves. The seed capsules are globular and may contain four seeds, but often only three are fertile; the dark-brown, kidney-shaped seeds are angular and about one-eighth inch long. They retain vitality for several years.

Control.—The rootstalks should be starved persistently by the frequent cutting of the stems. The weed loves the mellow soil of a cornfield. If the land is planted to corn, this will mean that as soon as the corn is too large to be cultivated by a cultivator the field should be gone over with a hoe and all young shoots cut off as soon as they make their appearance. Keeping them cut prevents the pest from maturing seed and leaves to re-stock the underground storehouse with food for another year.

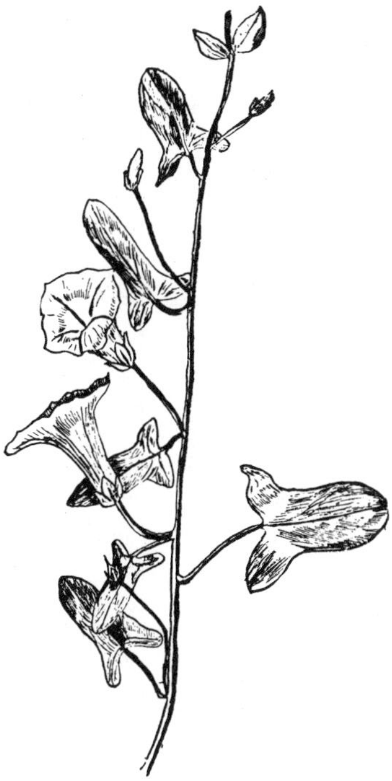

FIELD BINDWEED (*Convolvulus arvensis*).[1]
After F. S. Matthews in Manual of Weeds, by Ada E. Georgia.

Fifty Worst Weeds.—The following table gives an alphabetical list of the fifty worst weeds of the United States, with such information as will enable the reader: (1) to identify them; (2) to determine the nature and place of their greatest injuriousness; (3) to determine their duration or natural length of life, that is, whether annual, biennial or perennial; and (4) some methods of eradication. With this knowledge one will be able to attack much more intelligently any troublesome weed.

[1] Courtesy of The Pennsylvania Farmer.

Descriptive List of the Fifty Worst Weeds in the United States.*

(A—Annual. B—Biennial. P—Perennial.)

Common Name, Botanical Name, Duration of Life.	Color, Size and Arrangement of Flowers.	Sections where Injurious.	Method of Seed Distribution; Vegetative Propagation of the Perennials.	Place of Growth and Products Injured.	Methods of Eradication.
Bermuda grass, wire grass (*Capriola dactylon*), P.	Purple, $\frac{1}{12}$ inch, spikes.	Maryland to Missouri and southward.	Seeds sparingly, rootstocks.	Fields and lawns; hoed crops.	Plowing and planting cowpea, sorghum or millet to smother the wire grass.
Bindweed, field bindweed (*Convolvulus arvensis*), P.	White or pink, 1 inch, solitary.	Entire United States, especially California.	Grain and flax seeds, creeping roots.	Rich, moist soils; grain and hoed crops.	Cultivation: if very bad, close grazing with sheep.
Bindweed, wild morning glory (*Convolvulus sepium*), P.	White or rose, 2 inches, solitary.	Mississippi Valley region.	Grain and flax seeds, rootstocks.	Rich prairie and river bottoms; corn and small grains.	Starvation of rootstocks by close cutting.
Bitterweed, fennel, yellow dog fennel (*Helenium tenuifolium*), A.	Yellow, $\frac{3}{4}$ inch, head.	Virginia to Kansas and southward.	Wind, hay, animals.	Meadows and pastures; injures livestock and taints milk.	Close cutting or hand pulling before seeding time.
Broom sedge (*Andropogan virginicus*), P.	Green, $\frac{1}{4}$ inch, racemes.	Massachusetts to Michigan, Florida and Texas.	Wind, short rootstocks, plants in tufts.	Fields and waste places; pastures and meadows.	Prevent seeding, burn mature plants, cultivate infested ground.
Buffalo bur, sand bur (*Solanum rostratum*), A.	Yellow, $\frac{1}{2}$ inch, solitary.	Illinois and Colorado to Texas.	Plants rolled by wind, seeds in hay and by animals.	Fields; grain and hoed crops, wool.	Heavy seeding, close cultivating.
Bull nettle, horse nettle (*Solanum carolinense*), P.	Purple, 1 inch, solitary.	Entire United States.	Plants rolled by wind, running roots.	Everywhere; grain and hoed crops, pastures.	Alternate cultivating and heavy cropping.
Bur grass, sand bur (*Cenchrus carolinianus*), A.	Green, $\frac{1}{3}$ inch, bur.	Maine to Florida and westward to Colorado.	Animals, especially sheep.	Sandy land, pastures and waste places; pastures and wool.	Cultivation and burning.
Chess, cheat (*Bromus secalinus*), A.	Green, spikelets in panicles.	All grain sections.	Grain seed, especially wheat.	Everywhere; grain fields.	Clean seed, cultivation.
Chickweed, common chickweed (*Alsine media*), A.	White, $\frac{1}{8}$ inch, cymes.	Entire United States.	Grass and clover seed; animals; has a long seeding period.	Meadows, lawns; winter crops.	Cultivation in late fall and early spring.
Cocklebur, clotbur (*Xanthium americanum*), A.	Green, $\frac{1}{4}$ inch, head.	Entire United States.	Carried by animals.	Cultivated fields and waste places; hoed crops and wool.	Prevention of seeding, cultivation.
Crab-grass (*Syntherisma sanguinale*), A.	Green, spikes.	Entire United States, especially the South.	Clover and grass seed, hay; animals.	Cultivated fields, gardens, lawns; hoed crops.	Prevention of seeding, closer cultivation.

*Taken from Farmers' Bulletin No. 660, United States Department of Agriculture; with remedies added.

Descriptive List of the Fifty Worst Weeds in the United States (*Continued*).

Common Name, Botanical Name, Duration of Life.	Color, Size and Arrangement of Flowers.	Sections where Injurious.	Method of Seed Distribution; Vegetative Propagation of the Perennials.	Place of Growth and Products Injured.	Methods of Eradication.
Daisy, ox-eye daisy (*Chrysanthemum leucanthemum*), P.	White with yellow center, 1 inch, heads.	Maine to Virginia and Kentucky.	Clover seed, hay; woody, rather short rootstocks, but largely by seed.	Pastures, meadows, roadsides; hay, pasturage.	Prevention of seeding, close cultivation.
Dandelion (*Taraxacum officinale*), P.	Yellow, $1\frac{1}{4}$ inches, head.	Entire United States.	Wind; taproot, which spreads but little.	Lawns, meadows, waste places; hay and lawns.	On lawns, continued cutting below surface; in fields, cultivation.
Dock, yellow dock, sour dock (*Rumex crispus*), P.	Green, $\frac{1}{4}$ inch, panicle.	Entire United States.	Hay and straw, clover and grass seed; taproot, which spreads but little.	Hay, small grain and hoed crops.	Prevent seeding, continued cutting to destroy leaf system.
Dodder, alfalfa dodder, field dodder (*Cuscuta arvensis*), A.	Yellow, $\frac{1}{8}$ inch, clusters.	All clover and alfalfa regions.	Hay, clover and alfalfa seed.	Clover and alfalfa fields.	Clean seed, cultivation of crops other than clover or alfalfa.
Dogbane, Indian hemp (*Apocynum cannabium*), P.	Greenish white, $\frac{1}{4}$ inch, terminal clusters.	Upper Mississippi Valley.	Wind; creeping root.	Fields with sandy soil; pasture, grain and hoed crops.	In small spots, strong hot brine; in fields, plowing up roots during hot summer.
Fern, brake (*Pteridium aquilinum*), P.	No flowers.	Northwestern States and the Pacific Coast.	Spores scattered by wind; running roots.	Logged-off land. meadows and pastures.	Heavy seeding, cultivation.
Fleabane, horseweed (*Erigeron canadensis*), A.	White, $\frac{1}{4}$ inch, heads in cymes.	Entire United States.	Hay, grass and clover seeds.	Meadows, pastures and grain fields.	Prevent weeds seeding, close cultivation.
Foxtail, yellow foxtail, pigeon grass (*Chaetochloa glauca*), A.	Green, spikes.	Entire United States.	Animals, hay, grain and grass seeds.	Land cultivated in early part of season; young grass and clover seedlings.	Clean seed, cultivate land late in season.
Hawkweed, orange hawkweed, devil's paintbrush (*Hieracium aurantiacum*), P.	Orange, 1 inch, heads.	Maine to Ohio.	Wind, grass and clover seeds; runners similar to strawberry.	Untillable pastures and meadows.	Sheep pasturing, heavy cropping, cultivation.
Iron weed (*Vernonia noveboracensis*), P.	Purple, $\frac{1}{3}$ inch, heads.	Maine to Maryland and Iowa to Kansas.	Wind; short, thick rootstocks making plant grow in bunches.	Pastures and meadows.	Repeated close cutting to destroy leaf system, thus starving the roots.
Jimson weed (*Datura stramonium*), A.	Purple, 3 inches, solitary.	Maine to Minnesota and Texas.	Pods and plants blown by wind.	Pastures, barnyards and waste lands; seeds, flowers and leaves poisonous.	Prevention of seeding.
Johnson grass (*Holcus halepensis*), P.	Green, $\frac{1}{8}$ inch, panicle.	Virginia to Texas and California.	In hay, grain, and grass seed; running rootstocks.	All crops except hay.	Alternate cultivating and heavy cropping.

Descriptive List of the Fifty Worst Weeds in the United States (*Continued*).

Common Name, Botanical Name, Duration of Life.	Color, Size and Arrangement of Flowers.	Sections where Injurious.	Method of Seed Distribution; Vegetative Propagation of the Perennials.	Place of Growth and Products Injured.	Methods of Eradication
Lamb's-quarters, pigweed (*Chenopodium album*), A.	Green, very small, panicle.	Entire United States.	Grain and grass seed.	Grain fields and hoed crops.	Prevention of seeding.
Lettuce, prickly lettuce (*Lactuca virosa*), A.	Yellow, $\frac{1}{4}$ inch, heads in panicles.	Ohio to Iowa, Utah to California.	Wind.	Everywhere; all crops.	Prevent seeding, burn all mature seeds.
Milkweed, common milkweed (*Asclepias syriaca*), P.	Purple, $\frac{1}{2}$ inch umbels.	New York to Minnesota.	Wind; creeping roots.	All crops and in pastures.	Prevent seeding, cultivation and heavy cropping.
Morning-glory (*Ipomea hederacea*), A.	White, purple or blue, $1\frac{1}{2}$ inches, solitary.	New York to Missouri.	Corn stover, straw and wind.	Cultivated fields, especially corn and small grain.	Prevent seeding, thorough cultivation.
Mustard, wild mustard, charlock (*Brassica arvensis*), A.	Yellow, $\frac{1}{2}$ inch, racemes.	Maine to Washington.	Grain, grass, clover, and rape seeds.	Small grain fields and meadows; grains.	Prevent seeding, cultivation, hoed crops.
Nut-grass, coco (*Cyperus rotundus*), P.	Brown, $\frac{1}{16}$ inch, spikelets.	Maryland to Florida and Texas.	Wind, nursery stock, hay and grass seed; tubers.	All soils; hoed crops.	Alternate cultivation and smothering crops.
Penny cress, Frenchweed (*Thlapsi arvense*), A.	White, $\frac{1}{8}$ inch, racemes.	North Dakota and Minnesota.	Wind.	Grain fields and pastures; grain and dairy products.	Burning and thorough cultivation.
Pigweed, redroot, careless weed (*Amaranthus retroflexus*), A.	Green, quite small, spikes in panicles.	Entire United States.	In grain and grass seeds; plants blown by wind.	Plowed land; hoed crops.	Prevention of seeding, thorough cultivation.
Plantain, buckhorn, ribgrass (*Plantago lanceolata*), P.	White, $\frac{1}{16}$ inch, spike.	Entire United States.	Hay, clover and grass seed; spreads but slowly from a crown.	Everywhere; meadows, pastures and lawns.	Clean seed, cultivation.
Poison ivy, poison oak (*Rhus toxicodendron*), P.	Greenish white, $\frac{1}{8}$ inch, panicles.	Entire United States.	Does not spread fast by seeds; running rootstocks.	Moist, rich land along fences; poisonous by contact.	Cultivation, repeated grubbing.
Purslane, pusley (*Portulaca oleracea*), A.	Yellow, $\frac{1}{4}$ inch, solitary.	Entire United States.	Tillage implements; has a long seeding period.	Rich, cultivated land, especially gardens; hoed crops.	Close cultivation.
Quack-grass, witchgrass (*Agropyron repens*), P.	Green, spike.	Maine to Pennsylvania and Minnesota.	Seeds of grain and coarse grasses; creeping rootstocks.	All crops on the better soils; hoed crops.	Alternate cultivation, heavy cropping; close grazing.
Ragweed, smaller ragweed (*Ambrosia elatior*), A.	Yellow, $\frac{1}{4}$ inch, small heads on spike.	Entire United States.	Wind carrying matured plants; in grain and red clover seed.	Everywhere, especially grain stubble; hoed crops and young grass seeding.	Prevent seeding, burning.
Russian thistle, tumble weed (*Salsola pestifer*), A.	Purplish, $\frac{1}{4}$ inch, solitary.	Minnesota to Washington and southward.	Wind rolling matured plants.	Everywhere; small grain and hoed crops.	Cultivating, close grazing; burning developed seeds.

DESCRIPTIVE LIST OF THE FIFTY WORST WEEDS IN THE UNITED STATES (*Concluded*).

COMMON NAME, BOTANICAL NAME, DURATION OF LIFE.	COLOR, SIZE AND ARRANGEMENT OF FLOWERS.	SECTIONS WHERE INJURIOUS.	METHOD OF SEED DISTRIBUTION; VEGETATIVE PROPAGATION OF THE PERENNIALS.	PLACE OF GROWTH AND PRODUCTS INJURED.	METHODS OF ERADICATION.
St. John's-wort (*Hypericum perforatum*), P.	Yellow, $\frac{3}{4}$ inch, cymes.	Maine to North Carolina and Iowa.	In hay and grass seed; rootstock.	Meadows, pastures and waste places.	Frequent cutting to destroy leaf system.
Smartweed (*Polygonum pennsylvanicum*), A.	Light rose, $\frac{1}{16}$ inch, racemes.	Maine to Minnesota, Florida and Texas.	Wind carrying matured plants.	Moist, rich soils; hoed crops and young grass seedings.	Turn under badly infested meadows for hoed crops.
Smartweed, marsh smartweed, devil's shoestring (*Polygonum muhlenbergii*), A.	Rose color, $\frac{1}{16}$ inch, spikes.	Indiana to Iowa.	Wind and farm machinery; rootstocks.	Wet land, prairie and muck soils; hoed crops, hay, pasture.	Prevent seeding, cultivation.
Sorrel, sheep sorrel, horse sorrel (*Rumex acetosella*), P.	Red, $\frac{1}{8}$ inch, panicles.	Entire United States.	In clover seed; creeping roots.	Meadows and pastures.	Cultivation, smothering crops.
Sow thistle, perennial sow thistle, field sow thistle (*Sonchus arvensis*), P.	Yellow, $\frac{3}{4}$ inch, heads.	Maine to Minnesota.	Wind; running rootstocks.	Grain fields and hoed crops.	Thorough cultivation and smothering crops.
Squirreltail grass, squirrel grass, foxtail, wild barley (*Hordeum jubatum*), A.	Green, spike with long bristly glumes.	Minnesota to Texas and California.	Hay, animals, wind.	Meadows and pastures; barbed seeds produce sores on livestock.	Prevention of seeding, cultivation.
Thistle, Canada thistle (*Cirsium arvense*), P.	Purple, $\frac{3}{4}$ inch, heads.	Maine to Pennsylvania and Washington.	Wind, in hay and straw and in clover and grass seed; creeping roots.	All crops.	Alternate cultivating and heavy cropping.
Thistle, common thistle, bull thistle (*Cirsium lanceolatum*), B.	Reddish purple, 1 inch. heads.	Maine to Virginia and Washington.	Wind, in alfalfa, clover and grass seeds.	Pastures, meadows and winter wheat.	Prevent seeding, cut in the fall.
Wild carrot (*Daucus carota*), B.	White, very small, umbels.	Maine and Virginia to the Mississippi.	In foreign clover and alfalfa seed; carried by animals and wind.	Meadows and pastures.	Fall grubbing, cultivation.
Wild oats, (*Avena fatua*), A.	Green, panicles, similar to oats.	Wisconsin to Washington.	In seed oats.	Oat fields; lawns, injurious to stock.	Clean seeding, burning and pasturing.
Wild onion, garlic (*Allium vineale*), P.	Flowers rare, umbels, with bulblets.	Rhode Island to Georgia and west to Missouri.	Seeds rare; bulblets carried in wheat, underground bulbs.	Everywhere; wheat and dairy products.	Alternate cultivation and heavy cropping.
Winter cress, yellow rocket (*Barbarea vulgaris*), P.	Yellow, $\frac{1}{4}$ inch, racemes.	Maine to Virginia and westward.	In grain, clover and grass seeds.	Grain fields, pastures and meadows.	Prevent seeding, hoe grubbing beneath surface.

REFERENCES

"A Manual of Weeds." A. E. Georgia.
"Farm Weeds of Canada." Clark.
"Common Weeds of the Farm and Garden." Long and Percival.
"Weeds and How to Eradicate Them." Shaw.
North Dakota Expt. Station Bulletin 112. "Fertility and Weeds."
Canadian Dept of Agriculture Bulletin 188. "Weeds of Ontario."
U. S. Dept. of Agriculture, Bureau of Plant Industry, Bulletin 257. "Weed Factor in Corn Cultivation."
Farmers' Bulletins, U. S. Dept. of Agriculture:
306. "Dodder in Relation to Farm Seeds."
334. "Weed Seeds in Feeding-stuffs and Manure."
368. "Eradication of Bindweed or Wild Morning Glory."
464. "Eradication of Quack Grass."
545. "Controlling Canada Thistles."
610. "Wild Onion: Methods of Eradication."

BOOK III

HORTICULTURE, FORESTRY AND FLORICULTURE

CHAPTER 27

The Principles of Vegetable Gardening

By R. L. Watts

Dean and Director, School of Agriculture and Experiment Station, The Pennsylvania State College

A thorough knowledge of the underlying principles of vegetable gardening is exceedingly important, whether the vegetables are to be grown for the home table or for commercial purposes.

Soils and Locations.—Soils containing a considerable quantity of sand are best adapted to the growing of vegetables. Such soils are well drained, easily cultivated, and may be worked early in the spring. Sandy soils are warmer than clay soils, and for this reason crops mature earlier in them. They are especially desirable for crops requiring high temperatures, such as eggplants, peppers and melons. Any soil, however, which satisfactorily produces general farm crops, will, with proper treatment, grow good garden crops. The clay soils are avoided so far as possible by market gardeners and Southern truck growers. Southern or southeastern exposures are preferable for vegetable gardening because they are warmer and, therefore, conducive to earlier crops. Northern and western exposures are satisfactory for the later crops. Natural or artificial windbreaks are of advantage where there are cold exposures.

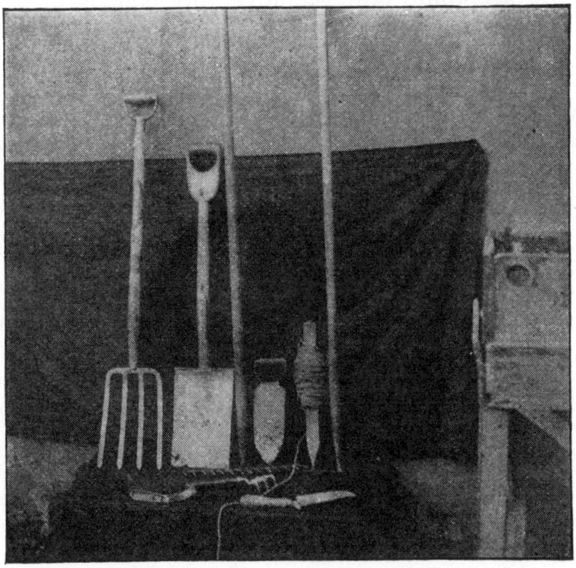

Necessary Garden Tools.[1]

[1] Courtesy of New York State College of Agriculture, Ithaca, N. Y. From Cornell Reading Courses, Vol. II.

Tillage and Tools.—The importance of thorough tillage in the production of vegetables cannot be over-emphasized. It counts for high yields as well as high quality. The conservation of soil moisture should be kept constantly in mind. Vegetables are composed largely of water and enormous quantities of it are required in their growth. Fall plowing is often advisable, especially in clay soils which are to be planted early the following spring. Early spring plowing, followed by immediate harrowing, is favorable to the retention of moisture.

The prudent garden maker will possess at least a small assortment of carefully selected modern tools or implements. Of the hand tools, the hand seed-drill and hand wheel-hoe are great time and energy savers and should be employed in all market gardens and in most home gardens. A variety of hand hoes and rakes should also be available.

Stable Manures.—All classes of vegetable growers recognize the value of stable manure. It not only supplies plant food, but adds humus to the soil, thus making it more retentive of moisture and more favorable to chemical and bacteriological changes which are essential to plant growth. Horse manure is most universally employed. Market gardeners nearly always compost it in large piles, which are kept moist and turned one or more times before the manure is used. If the piles have rectangular sides and are kept moist there will be practically no loss of fertility during the process of composting. From four to six weeks of composting will kill all weed-seeds and leave the manure in the finest state of texture. Cow manure is most excellent for all classes of vegetables, but it is slower in action than horse manure. Sheep and poultry manures are rich in nitrogen and their texture makes them particularly desirable for vegetable gardening.

Cover Crops.—In vegetable gardening it is absolutely essential to maintain the supply of vegetable matter in the soil. If stable manures are not available, cover crops must be produced for manurial purposes. The legumes, such as vetch, cowpeas, soy beans and the clovers, are most desirable, provided they can be grown satisfactorily, because they materially add to the supply of soil nitrogen. Rye, oats and buckwheat, however, can often be used to great advantage. The usual practice is to sow the seed of cover crops before the last cultivation of vegetables which mature and are harvested during the fall months.

Commercial Fertilizers.—Commercial vegetable growers are seldom able to obtain, at reasonable prices, all the stable manure that they need for the maximum production of crops. In many instances they rely wholly upon green crops for humus, and purchase commercial fertilizers to supply plant-food. There is the most varied practice with reference to the kinds and amounts of fertilizer applied for the various crops. The character of the fertilizer depends upon the crops to be grown, nature of soil, previous treatment and seasonal conditions. If stable manures have been used in liberal amounts, say twenty to forty tons to the acre, and for truck crops like cabbage and sweet corn, it is seldom necessary to use more than half a ton of fertilizer to the acre, containing four per cent of nitrogen and seven or eight per cent of each of the mineral elements—potash and phosphoric acid. As a rule, a complete fertilizer should be applied before the crops are planted, and thoroughly mixed with the soil by harrowing. If additional plant-food is needed after the crop is started, it may be applied along the rows. Nitrate of soda is largely used for this purpose. Applications may range from 100 to 250 pounds to the acre, and if desired may be applied at intervals of ten days to three weeks.

The Use of Lime.—The values of lime are discussed in Chapter 6. Vegetable growers recognize the necessity, more than ever, of keeping their soils in a neutral or slightly alkaline condition, so that liming at regular intervals is probably a necessity on most soils, and especially those which receive large annual applications of acid fertilizers. Serious troubles are likely to develop in such soils and it is desirable to take preventive measures by liming the land. This is the best known treatment of soils to prevent clubroot which infects cabbage and other members of this family.

Seeds and Seed Sowing.—The utmost care should be exercised to obtain seed of the highest quality. Numerous experiments show that there is marked variation in the strains of our most common varieties of cabbage, tomatoes, lettuce, onions and other classes of vegetables. A superior strain may mean a profit of fifty to one hundred dollars more to the acre than one which is inferior. The most reliable seedsmen should be patronized. It may even pay to grow seed at home or to buy from specialists who have developed strains of unusual merit.

A fine, moist seed-bed is essential to germination, whether the seeds are sown under glass or in the open ground. The surface of the ground should also be smooth, so that the seeds will be covered with a uniform depth of soil.

Transplanting.—Vegetable growers find that transplanting is often a great advantage if not a necessity. It makes it possible to care for thousands of seedlings on a very small area. For example, it is easily possible to start 10,000 cabbage plants under a 3 x 6 foot hotbed sash, while ten sash are necessary to protect that number of seedlings after they have been transplanted.

Vegetable plants should be transferred to their new quarters before they have become crowded and spindling. The time of sowing should be carefully planned so that this condition will be avoided.

Machine planters are largely used in field operations. If they are properly managed, they do the work fully as well as it can be done by hand. Whatever the method employed, the main essential is to bring a considerable quantity of fine, moist soil into close contact with the roots.

Starting Early Plants.—Soil to be used for starting early plants should be fine, free from stones and sticks and fairly rich. For cabbage or cauliflower, it should be taken from land that has not grown either those or other members of the cabbage family for seven or eight years in order to avoid clubroot. The soil should also contain considerable humus, and some sand is a great advantage. If composted, two parts of loam, one part of rotten manure and one part of sand will give good results. It is always desirable to prepare and store the soil in the fall, so that it will be ready for use when wanted in February or March.

Flats or shallow plant boxes are a great convenience in starting early plants. They may be made of new lumber or of empty store boxes. Chestnut and cedar are very durable woods for this purpose. The thin pine boards of boot and shoe boxes are easily made over. A common plan is to rip soap and tomato boxes into sections, using any kind of thin lumber

One of the Many Good Types of Seed Drills.[1]

for the bottom of the flats. Plant boxes need not have a depth of more than two inches, though deeper boxes require less attention in watering because they hold more soil, and, consequently, more water.

Seed sowing with such crops as cabbage and lettuce usually begins about the first of February in northern districts and earlier in the South. While the seed may be sown broadcast in flats or beds, the better plan is to sow in rows about two inches apart. This is ample space for all of the vegetables which are ordinarily started under glass. If ten to twelve good seeds are dropped to each inch of furrow, there should be a satisfactory stand of plants. The furrows should be about one-quarter inch deep for

[1] Courtesy of New York State College of Agriculture, Ithaca, N. Y. From Cornell Reading Courses, Vol. III.

seeds sown under glass, with the exception of celery, which should barely be covered. After the seeds are sown, the furrows may be closed quickly by drawing a small pot label or the thumb and index finger along the rows. The soil should be firmed with a block and thoroughly watered. If possible, the soil should be made so moist that no additional watering will be necessary until the plants are up.

Some fresh air should be admitted to the hotbed or greenhouse daily, unless the weather is unusually severe. This is essential to strong, stocky plants. High temperatures and excessive moisture, both in the soil and atmosphere, are conducive to the growth of weak, spindling plants which are liable to damp off, and if they do not die, are very tedious to transplant. A safe rule is not to water unless it is absolutely necessary and then to water thoroughly. If the flats are in hotbeds and the weather is severe, it will be necessary to protect the plants at night by means of mats those made of rye straw being the most satisfactory.

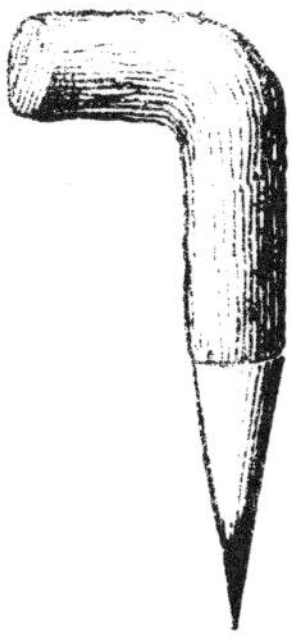

A DIBBLE.[1] (One-fifth actual size.) Very handy for making holes in which to set small plants.

In three to five weeks from sowing cabbage and many other vegetables the seedlings will be large enough to transplant. This operation may be performed any time after the rough or true leaves make their appearance. Soil such as has been described for seed sowing will be found satisfactory for this purpose. The flats or shallow plant boxes are also exceedingly useful receptacles in which to care for the plants until they are taken to the field. It is desirable to place about an inch of rotten manure in the bottom of the flat before filling it with soil. The soil should be moist enough to work well, and it is important to press it well along the sides and in the corners of the boxes.

Cabbage, lettuce and other plants are set from an inch and one-half to two inches apart. If they are to be kept in the flats for an unusual length of time, more space should be allowed. The work of transplanting may be done rapidly by the use of a small, pointed stick, often called a dibble. This simple tool is used to make the holes as well as to press the soil against the small roots of the plants. Sometimes transplanting boards, with holes bored in check rows, are used, and then a dibble is employed to punch all the holes before any plants are dropped. This method provides for a uniform number of plants to each box.

If the soil is just moist enough to work well during the transplanting, it will not be necessary to apply any water until the plants are well established. This is a great advantage, especially if the plants are placed in cold-frames and the weather is very severe.

After the plants have been in the cold-frame from several days to a

[1] Courtesy of New York State College of Agriculture, Ithaca, N. Y. From Cornell Reading Courses, Vol. II.

week, some fresh air may be admitted daily. Straw mats should be used at night. The mats will also be found useful in shading the plants in hotbeds and cold-frames.

Plants should be hardened as much as possible before they are taken from the frames to the field. This is accomplished by gradually subjecting them to fresh air and by not watering more than is absolutely necessary. The latter factor is more important than fresh air and low temperatures.

REFERENCES

"Farm Gardening." Corbett.
"Productive Vegetable Gardening." Lloyd.
"The Practical Garden Book." Hume and Bailey.
"Culinary Herbs." Kains.
Cornell Reading Course, Garden Series 1. "Vegetable Gardening."

CHAPTER 28

VEGETABLES AND THEIR CULTURE

BY R. L. WATTS

*Dean and Director, School of Agriculture and Experiment Station,
The Pennsylvania State College*

Asparagus.—The farmer's garden is not complete without this popular, hardy perennial, and it offers special inducements for cultivation near good local markets.

Numerous varieties are catalogued by our seedsmen and nurserymen but Palmetto is most largely grown. It is vigorous in growth, the shoots are large and the plants are regarded as more resistant to rust than other well-known varieties. Some meritorious new varieties are being developed at the Asparagus Experiment Station, Concord, Mass., and prospective growers should keep in touch with the work there and test for themselves the new sorts as soon as they are available.

It is universally conceded that asparagus thrives best in deep, rich, moist, sandy loams. Any soil, however, which will grow a satisfactory crop of corn will, with proper management, produce a good crop of asparagus. Thorough drainage is necessary. The character of the exposure is not important, though cuttings may be made earlier in the spring on southern slopes than on northern or western exposures.

It is exceedingly important to start with good stock. If a plot of even a few hundred plants is contemplated, it will pay to buy selected seed or roots from a specialist. The young plants are easily grown. A fertile plot should be chosen for the purpose and should be enriched still further by application of rotten stable manure at the rate of twenty-five tons to the acre. A good seed-bed should be prepared. Shallow furrows two feet apart should be made. The seeds should be dropped by hand at intervals of two or three inches and covered with an inch of soil. Radish seeds dropped eight to ten feet apart in the same drills with the asparagus will define the rows, since the radishes germinate within a few days, while the asparagus plants will not appear for about a month. It is quite an advantage to have the rows marked thus, so that the hand cultivator can be used before the asparagus plants are up.

Experiments at The Pennsylvania State College and elsewhere show that it pays to plant only the strongest roots or crowns. In other words, it is a good business proposition to grow two or three times as many roots as are actually needed for the plat in mind, and then plant only the largest. The selection or grading of the roots should be done late in the fall after the foliage has fallen. They may be kept in good condition until spring by packing in barrels with a little moist sand or sawdust. The barrels are then covered with straw and a few inches of soil added to protect the roots against severe freezing.

Bunching Asparagus Ready for Market.[1]
It pays to grade to a uniform size relative to color, length and size of bunches.

The grower should bear in mind the fact that the asparagus plantation should last at least ten years; therefore the soil should be prepared with the utmost care. The land should be heavily manured, plowed and then harrowed until it is thoroughly pulverized.

Planting distances vary greatly. If blanched or white shoots are to be grown, there should be at least five feet between rows, in order to provide plenty of soil for ridging. If green shoots—and they are gaining in popularity on American markets—are to be grown, four feet between rows will provide sufficient space for the use of horse cultivators, though many growers prefer a distance of four and one-half feet. Two feet between plants in the row is the most common distance, whether white or green shoots

[1] Courtesy of Department of Horticulture, Pennsylvania State College.

are to be grown. The crown of the asparagus comes nearer to the surface of the ground every year, because the new bulbs form somewhat higher than those from which the shoots of the previous season grew. This necessitates planting in trenches, which is also an advantage when the rows are to be ridged for the production of white shoots.

The roots should be planted just as early in the spring as the ground can be prepared. It is not desirable to make the trenches any deeper than the soil is ordinarily plowed. The fleshy roots are set on a tiny mound of soil, spread out and covered at first to a depth of not more than two inches. If conditions for growth are favorable, the new plants will soon appear and the trenches should be filled in gradually as the plants increase in height.

Thorough tillage is essential to the highest success. In new plantations it should begin early in the spring and continue until fall; old plantations should be cultivated as long as a horse and cultivator can be used without damaging the plants. It is especially important to disk the land as early as its condition will permit in the spring and again at the close of the cutting season, thus incorporating into the soil whatever manure has been applied.

No specific rule can be given for the fertilization of asparagus. One of the best methods is the application of ten tons or more of stable manure to the acre—late in the fall or early in the spring—supplemented with at least 1000 pounds of commercial fertilizer containing about 5 per cent of nitrogen, 8 per cent phosphoric acid and 6 per cent potash; half of the formula to be used early in the spring and half at the close of the cutting season. Intensive growers use at least a ton to the acre.

Some commercial growers cut approximately $50 worth of asparagus to the acre the second season from planting. If vigorous crowns have been used and a satisfactory growth obtained, this practice is not regarded as harmful to the plants. The cutting period of the third year should not last more than three or four weeks, but after the third year the usual practice is to cut until about the first of July.

The length of the shoots depends upon the demands of the market. Ordinarily, they are cut about nine inches in length and tied into bunches four and one-half inches in diameter. Two thousand bunches to the acre is a good yield, but this number is often exceeded. Twenty cents a bunch is a fair average price.

The asparagus beetle is the most destructive insect pest. In small plantations it is best controlled by setting coops of young chicks near the plot. Arsenate of lead is effective and is often used in young plantations and in old plantations after the cutting season. Air-slaked lime will also kill the larvæ.

Rust is practically the only disease to be feared. If it appears anywhere in the patch, the affected plants should be cut and burned before the foliage drops in the fall. Burning the tops is not regarded as a desirable practice in plantations which are free from rust.

Bean.—The bean occupies a most important place among the farm garden crops. All classes of beans, being legumes, possess high nutritive value and may often be served as substitutes for meats with satisfaction to the consumer.

Improved Golden Wax is an excellent wax-podded variety. Burpee Stringless is a leading dwarf green-podded bean. Goddard is a bush variety largely grown as a green-shelled bean. Lazy Wife is a superb green-podded pole bean. Early Leviathan is one of the best early lima beans and King of the Garden is valued as a late lima. White Marrow is one of the best varieties to grow for soup and baking.

The bean requires a well-drained soil. Sandy loams are preferred, but it is grown successfully in all types of soils. Applications of phosphoric acid are usually beneficial. Extensive plantings should not be made in the spring until the ground is thoroughly warm. Chances may be taken, however, in planting bush wax and green-podded varieties for the home table before conditions are ideal, and replanting can be made if the seeds decay or the plants are killed by frost.

The rows of bush beans should be far enough apart to be cultivated with a horse, and the seeds dropped two to three inches apart in the rows. An excellent plan is to drop four beans to the hill, the hills being eight or nine inches apart. Pole beans are usually planted in hills 4 x 4 feet apart.

Beet.—The beet may be grown in any good garden soil. The smoothest and finest roots are grown in sandy loams. Liberal applications of rotten stable manure are always beneficial. Excessive applications of nitrogen should be avoided. Potash and phosphoric acid are often used to advantage. Crosby Egyptian, Eclipse, Early Model and Egyptian are the leading early varieties. Edmond Blood turnip is good to follow early varieties.

Seed for the early crop is sown in the spring as soon as the ground can be prepared. The seed-bed should be fine and as level as possible. Drills should be made a foot apart for wheel-hoe cultivation. About ten seeds to the foot of furrow should be sown. Plants of early varieties should be thinned to about three inches apart in the row, and late sorts to five or six inches. Clean tillage is essential. An earlier crop may be obtained by starting the plants in hotbeds and greenhouses and transplanting them to the open ground after danger from hard frosts has passed by.

Brussels Sprouts.—This is a member of the cabbage family which is grown for fall consumption. The seed should be sown at the same time as for late cabbage and under the same conditions. Plants should be thinned to an inch apart. They should be transplanted early in July to rich, moist soil. Clean tillage should be given. Toward the end of summer, when the plants are well grown, the leaves should be cut off along the stalk, except a tuft at the top. This will induce the growth of large buds or "sprouts" in the axils of the removed leaf-stems. Brussels sprouts is regarded as a more delicate dish than cabbage.

Cabbage.—This is universally regarded as one of the most important farm garden crops. It fits in well with the general rotations practiced on American farms, and takes the place of potatoes after clover. It returns satisfactory profits wherever good markets are available.

Jersey Wakefield is the leading early variety. Charleston Wakefield, which is somewhat larger and a few days later, is also popular. Copenhagen Market is a round-headed early variety of special merit which has recently come into prominence. Early Summer, Succession and All Heart are very good midsummer varieties. Succession is also largely grown for late use, seed being sown later than for late varieties. Flat Dutch and Drumhead are well-known late sorts. Danish Ball Head is extensively grown for winter use. It possesses better keeping qualities than any other late variety. The heads are roundish and very solid.

FOUR STRAINS OF JERSEY WAKEFIELD CABBAGE.

Grown at The Pennsylvania State College, which show extreme variations in the germinating power of the seeds.

Cabbage requires a very rich soil for the best results. Stable manures are used extensively for this crop. Commercial fertilizers containing not less than four per cent of nitrogen and six to eight per cent of each of the mineral elements, are also applied at the rate of one-half ton to a ton to the acre.

Seed for the early crop should be sown in the hotbed or greenhouse about ten weeks in advance of planting in the field. In most northern sections the seed is sown about the first of February and the seedlings are transplanted to the cold-frame about the first of March. With proper frame management they will be well hardened and ready for the field April 10th or 15th.

The late crop is usually started in May. Danish Ball Head requires a full season and it is a mistake to sow too late, though local climatic conditions should be carefully considered.

Ordinarily, the best planting distance for Jersey Wakefield is 14 x 26, Charleston Wakefield 16 x 28, Succession 18 x 28, Danish Ball Head 18 x 30; and other late flat-headed varieties 24 x 36 inches. Close planting is conducive to small heads, and most of our markets prefer heads that are solid but not too heavy.

The early crop of the South is always marketed in crates of nearly one barrel capacity. Much of the crop in the North is sold by count, often by weight and frequently by the barrel. When the early crop is shipped in barrels it is important that they be well ventilated.

The late crop is stored in a great variety of ways. Although burying is troublesome, no other plan keeps the cabbage in better condition. The soil must be well drained. Windrows of cabbage, three heads side by side and two heads above, should be placed so as to drain the water away from the cabbage. The cabbage is then covered as nearly as possible with a plow and the work finished with hand shovels. Four or five inches of soil is sufficient covering and then enough manure is added to keep out frost. In central Pennsylvania, for example, four inches of soil and four inches of manure will keep the cabbage in perfect condition, provided the location is protected on the north and west from hard winds. There is no advantage in burying cabbage with the roots on. The best plan is to cut the stems with a sharp hatchet, leaving stubs four or five inches long for convenience in handling the crop.

A Plant Transferred with Plenty of Earth is not Checked in Growth.

Cabbage should be grown in a long-period rotation in order to avoid losses from clubroot, and the land should be kept well limed as a preventive measure against this most dreaded disease. The common green cabbage worm is best controlled by spraying with arsenate of lead.

Carrot.—The carrot is becoming more popular in America every year. It is easily grown in any rich soil, but attains its best development in sandy loams. By using early and late varieties and by making successive sowings, it is possible to have roots for sale and for the home table from June until late in the fall, and then the crop may be stored for winter use.

There are numerous varieties of carrots, but the best known early varieties are Early Short Scarlet and Early Scarlet Horn; for medium early, Model, Danvers Half Long Orange and Danvers Half Long Scarlet, Oxheart and Rubicon are popular. Long Orange is the leading late long-rooted variety.

For the early crop, seed should be sown as early in the spring as the ground can be prepared. It is customary to allow about a foot of space between rows for the early varieties and fifteen inches for the late. The early kinds may be thinned to stand two or three inches apart in the row,

STRAIN TESTS OF CABBAGE AT THE PENNSYLVANIA STATE COLLEGE, SHOWING FIRST CUTTING.

Note that only two heads of cabbage were marketable at the first cutting in Row 13.

while the late sorts should be four to six inches apart. Fairly liberal applications of phosphoric acid and potash are considered valuable for the carrot. It is easily kept until late winter by storing in pits or in cool cellars, where the roots should be covered with moist sand or soil.

Cauliflower.—Cauliflower is considered the most refined member of the cabbage family. The heads are more delicate in quality than cabbage, kale or even Brussels sprouts. It is also more difficult to grow than cabbage. This crop has two marked tendencies: first, not to form heads; and second, for the heads to "bolt" or "button" instead of forming hard, compact heads. The failure of this crop is very frequently attributed to the use of poor seed, and there is no question but that good seed is a most important factor in the growing of a satisfactory crop of cauliflower.

The early crop is started under glass and the plants are handled in the same way as cabbage. It is important, however, not to check the growth of the plants at any time, as this may cause "bolting" or "buttoning."

Seed for the late crop should be sown a trifle later than for cabbage. It is important to sow thinly so that every plant will have plenty of space for its full development.

Cauliflower should be planted in even richer soil than cabbage. It is especially desirable to use an abundance of rotten manure. Planting distances should be about the same as for early cabbage. When the heads are an inch or two in diameter, the leaves should be bent over them, or perhaps tied together over the heads, in order to protect the latter from rain and sunshine. The markets demand pure white heads.

Celery.—Celery occupies a most important place in American gardens, though it does not receive as much attention as it should. When the methods of culture are well understood, it may be grown with great ease, and no vegetable is more appreciated when it appears on the farmer's table. An immense quantity of celery which is grown in muck soils finds its way to our great markets. The crop is also well adapted to rich, sandy loams, but any soil which is properly fertilized will grow an excellent crop of celery. The two great essentials are a liberal supply of plant food and an abundance of moisture. Stable manure is universally regarded as the best fertilizer. It should be applied in a decayed condition and worked well into the soil as a top dressing rather than plowed under. Commercial fertilizers are also extensively used for this crop. As a rule the fertilizers employed by commercial growers contain four to six per cent of nitrogen and from eight to ten per cent of each of the mineral elements. A ton to the acre, mixed directly with the soil after plowing, is a very common application, and some growers use double this amount.

There are two general classes of celery: First, the so-called self-blanching, best represented by Golden Self-Blanching, which is more generally grown in this country than any other sort; and, second, the green varieties, such as Winter King, Winter Queen and Giant Pascal. The dwarf self-blanching varieties are most popular among commercial growers because they are easily blanched. Green winter varieties are better in quality than the self-blanching and are grown more largely for winter use.

Too much care cannot be exercised in purchasing celery seed. The grower should make certain that the stock is good, because many of the failures of celery growers are attributable to poor seed. The best seed of the self-blanching varieties is grown in France.

For the early crop, seed should be sown in hotbeds or greenhouses after the first of March. It is usually a mistake to sow earlier than this date, because the plants are likely to become crowded in the beds before planting time in the field, a condition which may check their growth and cause them to produce seed shoots instead of marketable stock. On account of the very small size of the seed, there is always likelihood of sowing too thickly. The plants should be thinned if that happens, and in a month or five weeks transplanted one and one-half to two inches apart in flats or beds. In the latter a constant supply of moisture should be maintained until the plants are set in the field.

Celery Under Irrigation, Skinner System.

Seed for the late crop should be sown in the spring as soon as the ground can be prepared.

When boards are to be used for blanching the early crop, it is customary to allow about two feet of space between rows and to space the plants three to five inches apart in the row. In most sections of the North, plants should not be set in the open ground before May 10th. The crop will stand considerable cold, but heavy frosts almost invariably check the growth and have a tendency to cause the production of seed stalks. The late crop should have more space and it is not uncommon to allow four to five feet between rows, the distance depending upon the method to be used in blanching.

The mulching system of celery culture makes the early crop much more certain. The plan includes a mulch of three to five inches of fresh horse manure placed between the rows immediately after the plants are set out. This conserves soil moisture, prevents weed growth, renders tillage unnecessary and supplies food to the plants after each rain.

Good Celery Well Prepared for Market.[1]

Boards are used almost entirely for blanching the early crop. They are placed along both sides of the rows and held in place by any convenient device at hand. From ten days to two weeks are required to blanch the crop. The boards may be used over and over again; with care they will last fifteen years.

The late crop is blanched by means of ridging with earth. This work should not begin until the cooler weather arrives in early September. The work of ridging proceeds until about the middle of October and commercial growers begin to store the crop soon after the first of November. Various methods of storage are in common use. One of the best is to dig trenches ten or twelve inches wide and not quite as deep as the height of the plants. The plants are placed close together in the trenches and covered with boards, which are nailed together in the form of a trough. The boards afford ample protection until freezing weather occurs and then additional covering is provided by placing

[1] Courtesy of Department of Horticulture, Pennsylvania State College.

manure or straw over them. The plants should be dry when stored and they should not be unnecessarily exposed to sun and hard, drying winds.

Cucumbers.—Most farmers are familiar with the ordinary method of growing cucumbers. If hotbeds are available, it is best to start a few hills under glass. This is a very simple operation. A good plan is to fill quart berry baskets with soil containing a large proportion of rotten manure; drop about eight seeds in each basket and after the plants are up thin them to two or three. See that the boxes are not lacking in moisture at any time. The seed should be sown not more than four weeks in advance of the time suitable for planting in the field. Overgrown plants are a disadvantage. It is very much better not to use plants more than a month old. Whether the seed is sown under glass or in the open ground, the soil should be made very rich by using plenty of rotten manure. Planting distances vary, but 5 x 5 feet will be found satisfactory when the ground is very fertile. There are several strains of White Spine which are popular for general planting. For picklers, Chicago Pickling, Boston Pickling and Fordhook Pickling are popular.

The striped cucumber beetle is one of the most serious enemies of this crop. The most thorough means of prevention is to cover the plants with mosquito netting or with wooden frames with netting over them. Air-slaked lime, sprinkled on the plants, is usually effective as a repellant. Tobacco dust may also be used.

Eggplant.—The eggplant is often overlooked in the planting of the farmer's garden. This crop thrives best in a warm climate and for this reason many of the northern gardeners do not attempt to cultivate it. It may be grown, however, in all parts of the North, especially if the plants are started under glass and planted in rich, moist soil. It is also important to select an early variety such as Early Long Purple. The fruit of this variety is not as large as that of New York Improved or Black Beauty, but it will be found quite satisfactory for the home table. A high temperature is required for starting the plants; therefore it is best not to sow too early. They should be transplanted into two-inch pots and later into three or four-inch pots, and then the gardener can transfer them to the field without checking their growth.

Horse Radish.—There should be at least a few plants of horse radish in every farm garden. It is easily grown in any moist, rich soil. The crop is easily propagated by root cuttings, which are made when the roots are dug for market or for the home table; that is, the small lateral roots are trimmed from the large ones and saved for planting. It is customary to cut the roots intended for propagation square at the upper end and slanting at the lower end so that you will know which end to plant up when they are set in the garden.

Kale.—This crop is quite successful in some parts of the South and is seen occasionally in northern districts. It requires the same cultural conditions as cabbage. The most prominent varieties are Imperial Long

Standing, Dwarf German, Dwarf Curled Scotch and Fall Green Curled Scotch. Sowing should be made about midsummer in order that the plants may attain full size before cold weather. The plants are thinned to stand eight to ten inches apart, according to variety.

Kohl-Rabi.—This vegetable is also called "turnip-rooted cabbage." It is easily grown in any rich soil. Plants may be started under glass, or the seed may be sown direct in the open ground and the plants thinned to about eight inches apart in the row. Green Vienna and Earliest Erfurt are the leading varieties. It is possible to have fresh roots in the garden from the middle of June until late fall, when they may be stored for winter use.

Leek.—This vegetable requires the same cultural conditions as onions. It is regarded as milder and more tender than the onion. The seed should be sown in the spring as soon as the ground can be worked. Market gardeners often transplant the seedlings in July, so that the crop will be ready to use in the fall. It is always an advantage to clip the tops at transplanting time.

Lettuce.—Most farmers are perfectly familiar with the methods which are ordinarily employed in growing lettuce. The usual practice is to sow the seed broadcast in small beds. A very much better plan is to sow in hotbeds or in a sunny window of the house and transplant the seedlings to the open ground after it is dry enough to work. This method will insure an earlier crop than is possible from sowing directly in the open ground. If it is desired to make sowing out of doors, the seed should be drilled in rows about a foot apart, and the plants thinned to stand from six to eight inches apart in the row. This will result in much finer heads than is possible by the broadcast method. There is a long list of varieties from which to select. Grand Rapids is grown largely under glass and is also suitable for culture out of doors. Wayahead is a comparatively new but most excellent head variety for out-door culture. Big Boston is one of the leading varieties for frame culture and for sandy and muck soils. All Heart and Sensation are also good varieties. Hanson, Iceberg and Brittle Ice are popular varieties of the "crisp-head" class.

Sandy soils are selected when an early crop is desired, though this vegetable may be grown with entire success on any soil properly fertilized.

Rotten stable manure is undoubtedly the best form of fertilizer. It may be used at the rate of twenty or more tons to the acre. Commercial fertilizers are also used extensively for the lettuce crop. The early crop may be started under glass as previously explained for cabbage.

Muskmelon.—The remarks made concerning the cucumber apply equally well to the muskmelon, although this vegetable requires better cultural conditions than the cucumber. By starting the plants under glass, practically every farmer could have a liberal supply of muskmelons. It requires more heat and a longer season than the cucumber, but plants which are well started by the time it is safe to plant them out of doors should mature a satisfactory crop, especially if the soil is well enriched with rotten manure. This vegetable will not thrive in any northern section unless the soil is well filled with organic matter. The planting distances for muskmelons should be more liberal than for cucumbers. Among the varieties which are popular throughout the country may be mentioned Rocky Ford, Paul Rose, Netted Gem, Hackensack, Osage, Emerald Gem, Eden Gem and Burrell Gem.

Onion.—No vegetable is found so universally in the farmer's garden as the onion. Indeed, it is rare that the onion is omitted from the home garden. A long list of varieties is available. Among the best yellow kinds

ONIONS UNDER SKINNER SYSTEM OF IRRIGATION.[1]

may be mentioned Danvers, Southport Yellow Globe and Strasburg. Weatherfield is the best known red onion and Southport Red Globe is a general favorite in many parts of the country. Silver Skin and Southport White Globe are popular white onions. The Egyptian (Perennial Tree Onion) is a valuable variety for fall planting in the North. Prizetaker is exceedingly valuable for starting under glass and transplanting in the open ground.

The onion thrives best in a moist, sandy loam, but may be grown with success in any rich soil. It is important to plant the seed in ground which is practically free from weed seeds. An excellent plan is to precede this vegetable with a crop like corn or cabbage which requires clean tillage.

[1] Courtesy of The Pennsylvania Farmer.

The soil may also be highly enriched the preceding year by the application of a large quantity of stable manure, and weed seeds should be completely destroyed by the time the onions are started. Commercial fertilizers are also largely employed for the onion. It is not uncommon to use a fertilizer containing four per cent nitrogen and six to eight per cent of phosphoric acid and potash, at the rate of a ton to the acre. The fertilizer should be well mixed with the soil before any planting is done.

The bulk of the mature bulbs which are sold on American markets is grown from seed sown in the open ground. The most common spacing between rows is a foot, and seed is sown sufficiently thick to give eight to twelve plants to a foot of furrow. Ordinarily four and one-half pounds of good seed to an acre will give the proper stand of plants. Seed more than a year old should never be used. The transplanting method, often referred to as the new onion culture, provides for sowing seed under glass and setting the plants in the open ground after danger of hard freezing has passed. Prizetaker is the leading variety for this purpose. The most common plan is to sow seed in January or February. After the tops attain a height of five inches they are clipped back every week to about four inches, and when planted in the field they are clipped to three or three and one-half inches. Bulbs of extra size for exhibition purposes may be obtained by starting the plants under glass in the fall, clipping the tops repeatedly, as explained, until they are set in the field, about May 10th in the North. Most farmers grow the bulk of their crop from sets which are planted as early in the spring as the ground can be worked. This is the most certain method of procuring a crop, though as large bulbs cannot be obtained as from the transplanting method. Clean tillage is absolutely essential to the success of a crop of onions, and this requires a certain amount of hand-weeding and hoeing between the plants. The mature bulbs for winter use are pulled after most of the tops have turned yellow and are partly dead. The bulbs are thrown together in windrows for a few days until partly dry and then placed in crates or bags which are hauled to sheds or well-ventilated buildings. Onions may be kept throughout the winter in a room where the temperature may be controlled, or allowed to freeze and then be covered with hay and kept in a frozen condition throughout the winter. The latter plan is very satisfactory and should be more generally used by farmers.

Bunching onions are most largely grown from sets, though many gardeners grow excellent green onions from seed sown in the open ground. The Prizetaker produces a particularly mild onion. Onion sets are grown by sowing the seed more thickly than is done for mature bulbs. There is no reason why every farmer should not grow his own sets. A good plan is to sow the seed very thickly in furrows which are about two inches wide and one-quarter inch deep. The plants come up so thickly that it is impossible for any of the bulbs to attain a large size. The sets are easily kept over winter in any dry room where alternate thawing and freeing does not occur.

Parsley.—There should be a supply of parsley in every farmer's garden. It is found useful as a flavoring for soups and other dishes and also for garnishing purposes. Seed may be sown under glass and the plants grown in the same way as explained for cabbage. The crop may also be started out of doors, the rows being spaced a foot or fifteen inches apart and the plants thinned to stand one foot apart in the row. Parsley thrives in any moist, fertile soil. Rapid growth may be encouraged by top-dressing with nitrate of soda.

Parsnip.—Parsnips are grown successfully in various types of soil. It is a long-season crop, hence should be sown as early as possible in the spring, and the soil should be made as rich as possible by the application of rotten manure. The deep, sandy loams are preferred. Guernsey, Hollow Crown and Early Short Round are leading varieties. The rows should be from fifteen to eighteen inches apart and the plants should stand from six to seven inches apart in the row. The roots may be sold or used on the home table any time after they have attained full size, but the usual custom is to leave most of them in the ground until spring, because freezing improves their flavor. It is rare that the roots are damaged by the hardest winter freezing.

Pea.—The pea is universally popular in American farm gardens. It is highly appreciated because it is very early and also very nutritious. This crop should be started just as soon as the ground is dry enough to work. It is not uncommon to make plantings the latter part of March. The early, smooth varieties are considered hardier than the wrinkled kinds. Alaska and Extra Early are well-known varieties of the smooth type. Gradus, Thomas Laxton and Nott's Excelsior are popular wrinkled kinds. Most farmers plant a few rows of medium or late varieties, such as Improved Stratagem and Telegraph. These varieties are excellent in quality and very prolific. The pea thrives best in cool, moist but well-drained soil. When very late plantings are made it is desirable to plant in trenches; cover at first with about two inches of soil and, after the plants are up, gradually fill in the trenches until the ground is level. This deep covering is favorable to moisture conditions and the ground is also cooler about the roots, which is an advantage. The dwarf varieties, such as Alaska and Extra Early, do not need support, while the late kinds should be supported by means of brush or wire trellises.

Pepper.—The pepper requires practically the same conditions as the tomato, although more heat is beneficial to its growth. For this reason the plant thrives best in warm, sandy soils. An abundance of decayed organic matter is a decided advantage in northern districts, which are not very favorable to peppers. Among the mild-fruited varieties may be mentioned Bull Nose, Chinese Giant and Ruby King. The Neapolitan is a very early variety that is popular throughout the North. Long Red Cayenne and True Red Chili are popular pungent-fruited varieties. The seed should be sown under glass about the time tomatoes are started. The plants of most varieties should have eighteen inches between them in the row and the rows should be far enough apart to use a horse cultivator.

Radish.—The radish is common to nearly all farm gardens. It does best in deep, rich, loamy soils. Though grown successfully when the seed is broadcast, it is better to sow in drills a foot apart and use enough seed to produce plants an inch or two apart; while late varieties should have two to five inches between plants in the row. Among the early varieties which are popular with home and commercial gardeners may be mentioned Earliest White, Round Red Forcing and Scarlet Frame. French Breakfast is a well-known radish, it is bright carmine above and clear white below. The first sowing should be made as soon as the ground can be worked and successive sowings should be made from week to week.

Rhubarb.—Rhubarb requires a deep, rich, moist soil. It is propagated commonly by roots. Annual applications of manure should be made in order to maintain the supply of organic matter in the soil and to furnish an abundant supply of plant food. Nitrate of soda may be used to advantage as a top dressing. It is ordinarily planted 3 x 4 or 4 x 4 feet apart. Victoria and Linnæus are leading varieties.

Salsify.—Salsify or "oyster plant" is not as generally grown as it should be in American gardens. This root crop requires the same cultural conditions as the parsnip. It is also a long-season crop and, therefore, the seed should be sown as early as possible in the spring. The roots are stored in the same manner during the winter as parsnips and will not suffer from freezing if left in the ground until spring.

Spinach.—Spinach is more largely grown in the South than in northern districts, although it should be a most important vegetable in all parts of the country. The usual practice is to sow the seed late in the fall, and the crop will be ready to harvest the following spring. In the North, the better plan is to sow very early in the spring. A successful method is to sow broadcast on the frozen ground and then cover the seed very lightly with fine, rotten manure. This vegetable requires a rich, constantly-moist soil to obtain the best results. Late plantings should be made in drills and the plants thinned to stand from five to six inches apart.

Squash.—The squash requires practically the same cultural conditions as cucumbers, but much more space is required. If the ground is a rich garden loam, the hills should be at least 10 x 10 feet apart, and more liberal spacing for the winter varieties will be an advantage in very rich soil. Summer squash need not be planted any farther apart than cucumbers, or even less space will meet their requirements. Early White Bush, Yellow Bush and Summer Crookneck are popular summer varieties. Hubbard, Warted Hubbard, Golden Hubbard and Boston Marrow are largely grown winter kinds. Squash must be stored in buildings where there is no freezing during the winter and a uniform temperature of 50° F. is most favorable to successful storage.

Sweet Corn.—Sweet corn requires the same conditions as field corn, if a good crop is expected. Among the early varieties which are popular and largely grown may be mentioned Fordhook and White Cob Cory. Golden Bantam matures somewhat later than these varieties and is superior in quality. Popular midsummer varieties are Cosmopolitan and Sweet Orange. Country Gentleman and Stowell Evergreen are the best known late varieties. Experiments made at various experiment stations show that it pays to select seed for sweet corn with as much care as for field corn. If space is available it pays to start one or two hundred hills in soil under glass by sowing seed two weeks before it is considered safe to set the plants in the open ground. This will make an early crop and insure a good stand of plants.

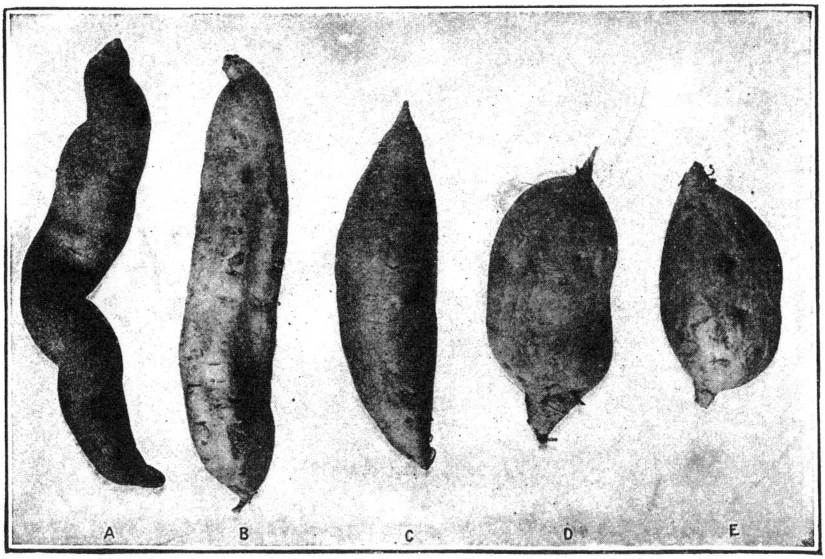

SOME COMMERCIAL TYPES OF SWEET POTATOES.[1]
A—Black Spanish. B—A Long, Cylindrical Type. C—Jersey Group. D—Red Bermuda. E—Southern Queen.

Sweet Potatoes.—The sweet potato is not universally grown in the farm gardens of the United States. It thrives only in warm soils and prefers one which is sandy in character. There are numerous varieties of sweet potatoes, some of the most popular being Big Stem Jersey, Yellow Jersey, Red Jersey, Southern Queen, Georgia Yam, Red Bermuda, Florida and Pierson. It is propagated by slips and these are obtained by bedding the tubers in fine soil with the proper amount of heat and moisture. The tubers soon send out sprouts and produce rooted plants which are set in the field after all danger of frost has passed. Field planting should not be

[1] From Farmers' Bulletin 324, U. S. Dept. of Agriculture.

Tomatoes Supported by Stakes.[1]

[1] Courtesy of Virginia-Carolina Chemical Company, Richmond, Va. From V.-C. Fertilizer Crop Books.

attempted until the ground is thoroughly warm. The tubers require the same conditions for storage as squash.

Tomato.—This is unquestionably one of the most important c ops of American home gardens as well as commercial plantations. It does well in a great variety of soil types. The sandy loams are preferred, though very heavy yields have been obtained in clay and silt soils. Earliana is the best known and most widely planted very early variety. It is fair in quality and very productive. Bonny Best matures soon after Earliana and is superior in some respects. Chalk Jewel and June Pink are also popular early varieties. Among the leading late varieties may be mentioned Stone and Matchless. Beauty and Trucker Favorite are desirable varieties of pink fruits. Good seed is highly essential to this crop and not a few of our commercial growers make careful selection from their own plantations. Seed for the early crop should be sown under glass not later than the 1st of March and, if extremely early tomatoes are desired, the 20th of February will not be too soon. The finest plants are obtained by first transplanting the plants one and one-half to two inches apart and then three or four inches apart, and finally into pots which vary in size from four to six inches. If the plants contain a blossom or two or perhaps a cluster of fruit when set in the field, a few ripe tomatoes should be available by the tenth of June and a large quantity should be available for market before the first of August. The plants should be hardened as well as possible before setting in the field, but no more water than is absolutely necessary should be applied. Such plants will stand a considerable amount of freezing in the field. Tomatoes of a superior quality may be obtained by training the vines to single stems. The usual practice is to make the rows about four feet apart and set the plants about fourteen to eighteen inches apart in the row. The plants are secured to stakes or wire trellises and the lateral branches are pinched out as fast as they appear. When a plant attains a height of four or five feet a trellis is always used and this causes the rapid development of fruit all along the stem. This method, however, should not be practiced unless there is plenty of labor to attend to the training.

Turnips.—Our farmers are familiar with the growing of turnips because they are produced not only for the home table but also to be stored during the winter for the farm stock. Roots most uniform in size are obtained by sowing in drills a foot to fifteen inches apart and thinning the plants to four to five inches apart. The roots are usually preserved during the winter by burying or covering with moist soil in pits. Some of the most popular varieties are White Milan, Red Purple Top, White Flat Dutch, Purple Top White Globe, White Egg and Yellow Globe.

Watermelon.—The watermelon requires the same cultural conditions as muskmelon. It should be planted in hills 8 x 10 feet to 10 x 10 feet apart. A bountiful supply of rotten manure should be used in the hills. Commercial fertilizers can also often be employed to advantage. Planting should not occur until there is no danger of frosts. Among the varieties which are popular may be mentioned Kleckley Sweet, Kolb Gem, Cuban Queen, Halbert Honey, Dixie and Sugar Stick. Cole and Fordhook are very hardy varieties desirable for planting in northern districts.

REFERENCES

"How to Grow Vegetables." French.
"Garden Farming." Corbett.
"Vegetable Gardening." Watts.
"Sweet Potato Culture." Fitts.
"Market Gardening." Yeaw.
Pennsylvania Expt. Station Bulletin 137. "Cabbage Experiments."
Canadian Dept. of Agriculture Bulletins:
- 196. "Tomatoes."
- 199. "Onions."
- 203. "Cabbage and Cauliflower."
- 231. "Vegetable Growing."

Farmers' Bulletins, U. S. Dept. of Agriculture:
- 324. "Sweet Potatoes."
- 354. "Onion Culture."
- 434. "The Home Production of Onion Seed and Sets."
- 433. "Cabbage."
- 520 and 548. "Storing and Marketing Sweet Potatoes."
- 642. "Tomato Growing in the South."

CHAPTER 29

The Farm Vegetable Garden

By Paul Work
Superintendent and Instructor, Department of Vegetable Gardening, Cornell University

In the rural sections the vegetable gardens adjoining the homes of the farmers show marked differences. In some sections almost no attempt is made to supply the home table with home-grown vegetables. In other districts the gardens are of good size, well planned and uniformly well cared for throughout the whole season. Those who devote no attention to the home garden little realize the advantages missed through this neglect. The diet of these families is usually not well balanced. Meats and cereals probably predominate and the elements which are supplied in vegetable food are lacking. These elements are not so much concerned in furnishing energy and building body material as they are in supplying the flavoring and mineral requirements. Moreover, man, as well as animals, requires a certain amount of more or less bulky feed. These factors in the diet are seemingly of minor importance, but are, nevertheless, absolutely essential. Just as no animal can thrive without a small amount of salt, so the absence of these elements from the table results in the weakening of the whole system and the undermining of the general health. The old-fashioned idea that one must necessarily be in poor physical condition when spring opens, is based upon the absence of vegetables and fruits from the old-fashioned winter diet. Nowadays, when canning is much more economically practiced and when the products of the garden and orchard are to be had during every month of the year, the old-time spring tonic is less in demand.

The value of the home garden must be further considered in its contribution to the joy of living and to the relish of a good table. Much can be said in praise of the endless array of delicacies which may be provided by the skilful housewife who is in league with the skilful gardener.

The economic value of the products which the home garden offers has been investigated by a number of experiment stations. At the Illinois Station it was shown that the average annual gross return from a half acre amounted to $105 through a period of five years, with an average cost, including all labor and materials, of $30. Some contend that the ordinary farmer cannot afford to devote the requisite amount of time and energy to the cultivation of the garden. It makes demands upon him which conflict with the demands of his fields and crops. It is true that a delay of a day in the planting of a field of oats may result in a very serious reduction in the yield. On the other hand, it is possible to so plan the work that both crops and garden receive the best of care. In fact, one is impressed with the correlation which exists between good farming and good gardening. It is largely a matter of management.

It is the aim of this article to offer some suggestions and general hints on garden making which may be of service to those trying to meet a given set of conditions. Rules are subject to numerous exceptions depending on conditions of soil, climate and exposure. These vary so widely that each must expect to work out his own salvation. An increasing measure of success from year to year is the reward to him who is willing to see and to think and to do the best that he knows, even though his knowledge in the beginning be exceedingly meagre.

Choosing a Site.—The gardener should carefully avoid the mistake of undertaking to cultivate a plot which is too large. A small area well kept and intensively managed will be much more satisfactory. It is safe to say that half an acre is the extreme for the ordinary family. Such a plot may be expected to yield an abundance for summer and autumn use, as well as for canning and storage. It is better to start with a garden too small than one too large.

On most farms, some choice as to location is possible. The garden should be near the buildings. It should be within easy reach of the housewife so that she may gather the products just as she is ready to use them. A distant garden seldom receives the care which is required. If the location is convenient, the hired men can make use of odds and ends of time which would otherwise be wasted. The cultivator which has finished its task a half hour before noon may loosen the crusted soil of many rows.

A southeasterly exposure is earlier and ordinarily offers protection from the severest winds. Roots of trees and shade of buildings should be carefully avoided.

If it is possible to choose from different types of soil, it is best to select a sandy loam. Heavy soil, the clays and clay loams, are lumpy when dry and are sticky and unworkable when wet. They cannot be cultivated early in the spring. On the other hand, the lightest sands ought to be avoided, in spite of the fact that they are loose and friable and may be tilled even when wet. They are not retentive of moisture or fertility. Of course, high fertility is of prime importance, but a soil may be improved in this respect more readily than in physical character. Freedom from weed-seed and disease must also be sought.

The Garden Plan.—Good planning is no less important in the garden than on the farm, although it is more often neglected here than in connection with the broader fields. During the winter the thoughtful gardener gathers about himself a supply of catalogues, a few good garden books and bulletins, together with paper, pencil and ruler. Furthermore, he refers to the concise but comprehensive notes which he has made during the previous season and which enable him to take advantage of points which would otherwise have escaped his mind. It is better to till a garden which is smaller by a few square yards and to keep an adequate record than to neglect this most important part of the gardener's task. The returns in later years will amply repay for the time and energy involved.

The first task is to decide what is wanted, making a list of crops, having in mind the likes and dislikes of the family. This should provide for an even distribution of products throughout the season and an adequate supply to be canned or stored for winter use. It should also take into consideration adaptation to climate, soil and space available.

In most cases the rows should be laid out lengthwise of the garden, and the spacing for all but the most intensive crops should be wide enough for

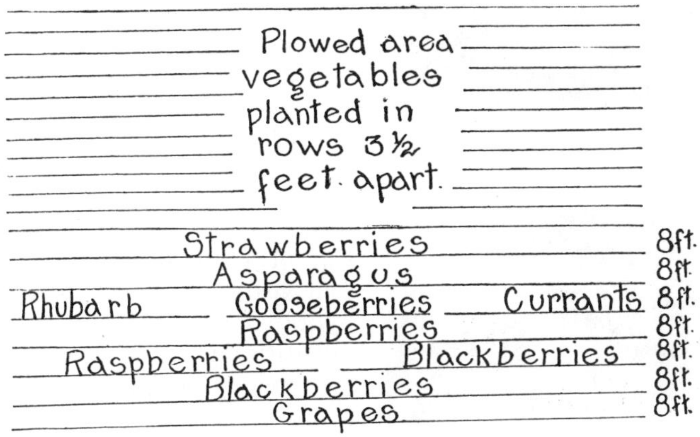

A Farm Garden Laid Out for Convenience in Working.[1]

horse cultivation. At the same time, the possibility of a much smaller garden to be tilled with wheel hoe and hand hoe may well be considered. The permanent crops such as asparagus and rhubarb ought to be placed at one side to avoid interference with tillage operations. In this same section of the garden the hotbeds and cold-frames may well be placed. Early crops should usually be kept together in order that the space made vacant by their removal may be more conveniently utilized. Attention should also be devoted to the symmetry, balance and neat appearance of the garden.

Fertility.—The first requirement for garden soil, as well as for farm soil, is good drainage. In case of surplus water, tile drains should be laid. Many soils which are not recognized as being especially wet are very materially benefited by drainage.

The farm gardener enjoys at least one great advantage over the city

[1] Courtesy of The Macmillan Company, N. Y. From "Farm Management," by Warren.

gardener. He has available an ample supply of stable manure. This material is the main reliance for the maintenance of fertility. Manure supplies nitrogen, phosphorus and potassium, the only chemical elements which are frequently lacking, and if the quantity applied is sufficient to maintain the humus content of the soil, there will be an abundance of these three elements. Manure that has been in the pile for several months is ordinarily preferred, but fresh material may be plowed under each fall with the assurance that it will be fairly well decayed and ready to aid the plants by spring.

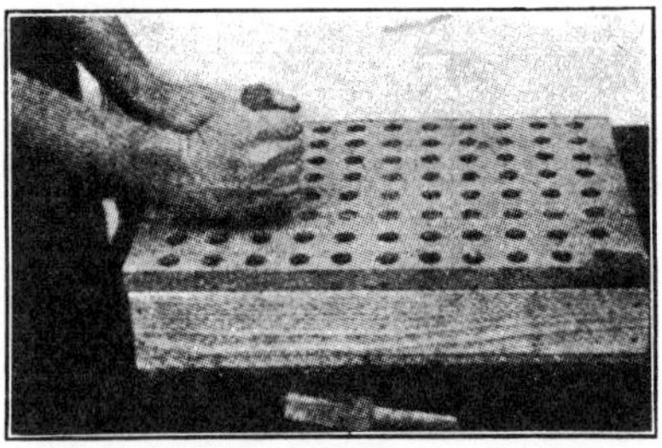

A—TRANSPLANTING BOARD AND DIBBLE IN USE.

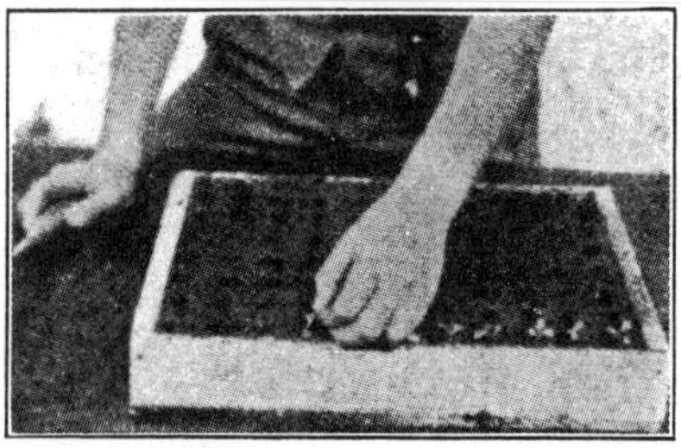

B—PLANTING THE SEEDLINGS. GROWING EARLY PLANTS.

In case the garden soil is of a refractory character, special treatment will be necessary. Heavy soils may be improved by hauling sand or gravel, by the addition of ashes, by the use of lime and especially by liberal applications of manure.

The lime content of the soil must in any case be maintained. Applications of air-slaked lime or finely pulverized limestone at the rate of a ton per acre every two to four years, are usually sufficient. If hydrated or quicklime be used the quantities may be reduced by approximately a quarter and a half respectively.

Tillage.—Every farmer realizes the danger incident to the plowing of the soil when it is wet. If such an error is harmful in ordinary farm practice it is doubly disastrous in the garden. Vegetable plants insist upon favorable growth conditions. In case the soil is shallow, it ought to be gradually deepened from year to year. The plowing should be done in the fall and the soil should be left in furrows to benefit from exposure to the frosts of the winter. In the spring it may be gone over with the disk harrow and worked down, making use of such other tools as are best adapted to the type of soil involved.

It is wise to prepare a few raised beds or ridges in the autumn for the earliest plantings. These will be ready to work much earlier in the spring, although they will dry out more rapidly in midsummer. Their direction should be such that they will gain full advantage of the warm southern sun.

Garden Seed.—Few problems connected with the garden are more bewildering than the choice of varieties to be planted. Each seedsman lands his own productions and impartial descriptions are seldom to be found. A variety well adapted to one locality may be utterly unsuited to another. Experience alone will enable one to meet this problem in a satisfactory way. Selections ought to be made and orders placed early in the season, in order to avoid disappointment and to allow time for testing. Many well-known seed houses are striving to supply good, clean, viable seed that is true to type. Packet seeds found in grocery stores may be more or less unreliable. Many local seed houses carry excellent stock, however.

To the gardener who is able to devote a bit of extra time to his plot, no hobby is more fascinating than the selection and saving of seeds from his own plants. There is always wide variation in excellence and these differences are inherited to a greater or less degree. By careful attention for a series of years, remarkable progress may be made in increasing the returns from a given area. Selections should be made on the basis of the individual plant rather than that of the individual fruit. The amateur plant-breeder should first clearly establish in his own mind a definite ideal. If he changes his ideal from year to year, no progress will be made. Considerable care should also be exercised in the harvesting, curing, labeling and storing of his seed crop.

Seed that will not start growth will certainly not produce a crop and such should be eliminated before the garden is planted by means of careful germination tests. A definite number of seeds may be counted out and planted in a small box of soil which should be placed under good growing conditions as regards moisture and temperature. Other tests which are less thorough may be made by the use of blotters, cloth, porous dishes and the like. These, however, indicate only whether seed will sprout or not. They do not afford knowledge as to whether the seed is able to establish in the soil a plant that is of sufficient vigor to grow independently of the supply of food material which is stored within the seed coats. The final test consists in growing the crop to maturity.

Growing Early Plants.—Every gardener is anxious to mature his crops at the earliest possible moment and to this end he is willing to employ special equipment and special methods. He selects the varieties which grow most rapidly and sows the seed long before outdoor planting is possible. He aims to have plants of such vigor and hardiness that they will make steady growth in spite of unfavorable conditions which they may encounter. These early plants enable him to more fully utilize the space of his garden, to care for both soil and plants more easily, to secure a better root system and in some cases larger yields.

Early plants may be started in window boxes in the house and may later be set directly in their permanent place, or the seed may be sown indoors and the seedlings transplanted to the cold-frame for hardening before they go to the garden. Whether in greenhouse or hotbed and cold-frame, the temperature should be relatively low, ventilation free, watering not too heavy and sunshine unimpeded. These conditions make for stockiness, hardiness, good root systems, vigor and freedom from disease. Conditions which are unfavorable in any of these respects encourage a soft and spindling growth and result in plants which do not yield as satisfactory results as plants from seed sown in the open.

Small greenhouses adjoining the farm home should be more common. The cost need not be heavy, as the construction may be exceedingly simple. The farmer may do the work himself at odd times. Old or second-hand pipe may be used for heating. The heating arrangement may be exceedingly simple, perhaps, using no pipe at all and merely setting a small stove in the middle of the house. In case it is not feasible to have a greenhouse, a hotbed may be used to excellent advantage. Cold-frames and hotbeds are described in the next chapter.

Seed Sowing.—Each vegetable has its own peculiarities as to time and manner of planting, and these peculiarities vary greatly with different climates and soils. Definite information upon these points, as well as upon many others, can be best secured from neighboring gardeners who have enjoyed long experience. The requirements for germination are moisture, warmth and air. Light is not necessary, although, of course, it is required immediately after seedlings break the ground. If the best results are to be obtained, the soil must be in excellent physical condition, especially for the smaller and more delicate seeds. These must also be sown a little more thickly than the more vigorous sorts, as is also the case when plantings are made very early in the spring when soil conditions are not strictly favorable and when damage by insects or diseases is feared. The skilful gardener should know his soil and his seed in order to sow just right both as regards thickness of sowing and depth of planting. Extreme thickness of sowing results in weak seedlings and requires much tedious work in thinning. Nevertheless, it is better for the novice to plant moderately heavily and to thus insure a good stand, even though some thinning is necessary after the plants have come up.

Particular attention must be devoted to the covering of the seed after it has been sown. The miniature plant enclosed within the seed coat depends upon the capillary movement of water in the soil for the moisture necessary for its growth. This movement is favored by thoroughly compacting the soil, and there is little danger of getting it too firm except in the case of heavy soils and of those which are rather moist at the time. Sowing in drills is preferred to broadcasting because it is easier to sow the seed at uniform depth. The seedlings help each other in breaking ground, and thinning and other work are more easily performed.

In the smaller gardens, seed is usually sown by hand. An envelope sealed at the side and cut squarely across the end is an excellent aid in this work. It is held the flat way and gently shaken with a movement lengthwise of the row, so manipulating it that the seed will drop evenly from the edge. Many gardeners, however, prefer to use the unaided fingers, working the seed over the second joint of the index finger by means of the thumb.

Mechanical drills are much more widely used in home gardens than ever before. Where fairly long rows are the rule, these implements are great time savers and in addition they may be depended upon to distribute the seed uniformly both as to thickness and depth. The drill requires as much skill for its adjustment as does the finger or envelope method. The scale on the machine which shows the approximate rate of sowing for the different seeds can be used only as a general guide, as there is wide variation in the size of seed of each vegetable.

Transplanting.—A seed consists of a miniature plant with its temporary food supply enclosed in such protecting covering as is necessary to insure safe removal to a situation far distant from the parent plant. This tiny plant is accordingly well adapted for a shift. However, the gardener in his eagerness for early fruition is not satisfied to let nature have her way. He must remove a plant which has discarded its protecting coat and which has already established its roots in the soil and begun to spread its branches in the air. This modification of nature's plan makes it necessary to exercise special precaution if he is to succeed. The soil should be in good physical condition and contain a reasonable amount of moisture. If possible, the work should be done on a cloudy day or in the evening so that the plant may recover from the shock before it is exposed to the unbroken rays of the sun. The little plants may be protected by special shading if it seems necessary. Care should be exercised to remove a good-sized ball of earth with the plant, thus establishing the foundling in its new place without serious disturbance of the roots. Plants should ordinarily be set just a little deeper than they stood in their previous place.

SOWING FROM SEED PACKAGE OR ENVELOPE.

It is not wise to set warm-blooded plants like tomatoes and cucumbers exceedingly early, as they may be seriously stunted by cold weather, even though there be no frost. Nevertheless, some gardeners set out a few plants very early, expecting to replace them if necessary.

Cultivation.—The word cultivation is a general term used in two or three different ways. As here applied, it refers to the maintenance of a thin layer of loose soil upon the surface of the garden throughout as much as possible of the growing season. This mulch is of great value in retaining moisture, in keeping the soil in good physical condition and in checking the growth of weeds. In small gardens the hand hoe and hand weeder will serve every purpose without undue labor. Even more universally used than the mechanical drill is the man-power wheel hoe, with its diversity of tools adapted for all sorts of soil stirring. Such implements are found useful, even though the rows be no more than forty feet long. In larger gardens, horse implements should be used as far as possible. In either case, it will be necessary to employ hand tools for maintaining the mulch between plants in the row. There is available a wider variety of tools and implements for cultivation than for any other type of garden work. These must be selected in accordance with the character of the soil, the crops, and the individual fancy of the gardener himself. There are a number of crops which spread over the ground comparatively early in the season and prevent cultivation from that time on. An increasing number of gardeners are securing the same results by means of a mulch of fresh, strawy manure, distributed between the rows. This conserves moisture and prevents weed growth as effectively as cultivation.

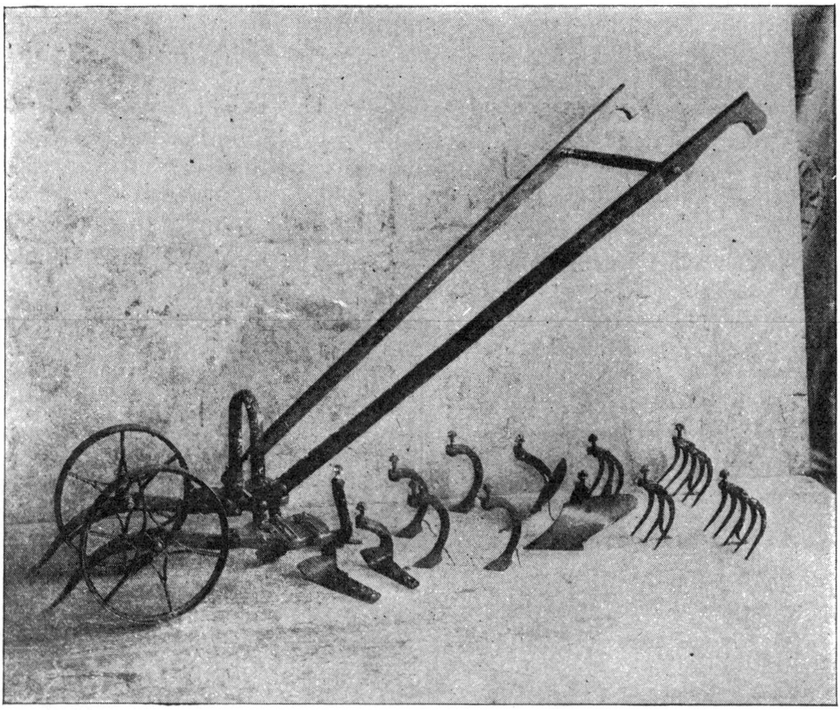

WHEEL CULTIVATOR AND ATTACHMENTS.[1]

Irrigation.—In spite of all these precautions, gardens often suffer from lack of water. It is not always possible to irrigate the rural garden,

[1] Courtesy of New York State College of Agriculture, Ithaca, N. Y. From Cornell Reading Courses, Vol. II.

but in some cases the plot may be so located that the water of a little stream may be so diverted as to flow between the rows when needed. An increasing proportion of country homes have water supply systems of their own. When this is the case, a line can be run to the garden for hose or overhead irrigation.

There is great danger that watering be done superficially, only the upper surface of the soil being moistened. This does more harm than good, as it dries out before it reaches the roots of the plants and at the same time it destroys the mulch which was fairly effective in conserving the moisture already present.

Pest Control.—No garden is free from the ravages of insect enemies and plant diseases. Each malady and each insect must be treated in its own way. Information as to methods must be sought in spray calendars or in special treatises upon such subjects. However, certain general principles must be borne in mind. As in the case of human ailments, an ounce of prevention is worth a pound of cure. The most important preventive measure is thorough cleaning up every fall. This removes from the garden the dormant forms of both insects and fungi, and so reduces the danger of infestation the next year. Crop rotation, or rather the refusal to grow the same crop on the same ground two years in succession, means much in the prevention of certain diseases. Plants which are making strong and vigorous growth are considerably less susceptible to attack than weaklings.

Cabbage, cauliflower, Brussels sprouts, kale and kohl-rabi are all subject to clubroot. This disease is caused by a slime mould which lurks in the soil and which attacks the plant through the roots. When the roots are thus affected, they are unable to secure either plant food or moisture, and the plants soon die. The plants wilt slightly at first and more seriously as the disease progresses. The roots become swollen, knotted and misshapen. There is no clear-cut method of control, but rotation, liming, thorough cleaning up each year and care to avoid the introduction of the disease by means of manure, tools and purchased plants is advised.

With beans, special precaution must be exercised against the rust or anthracnose. It is well to save one's own seed, choosing only pods which are naturally free from spot. Cultivation should be avoided when the plants are wet.

For insect enemies, plant diseases and their remedies, see chapters on same in Part VIII of this book.

Quality of Vegetables.—Quality in many vegetables depends to a large extent upon the stage of maturity. Peas and beans are more palatable, tender and digestible if gathered at a rather early stage of maturity than if allowed to get too large. As a rule, the protein content will be higher and the loss in actual nutritious value is more than counterbalanced by the good qualities above mentioned.

Few people realize how rapidly sweet corn deteriorates in value after it has been removed from the plant. It should go directly from garden to kettle. Garden beets, in like manner, lose considerable of their sweetness if allowed to stand long between time of pulling and cooking. Many of the garden vegetables suffer loss in a similar way.

Storage of Vegetables.—The character of storage that will give best results depends on the nature of the vegetable. Most vegetables, such as cabbage, root crops, potatoes and apples, keep best when stored under fairly dry conditions with some ventilation and a low temperature. The temperature cannot be too low so long as freezing is avoided. Low temperatures prevent the development of most fungous and bacterial activities which are directly responsible for various forms of decay.

There are a few of the vegetables, such as sweet potatoes and squash, that keep better at a temperature of about 50° F.

Literature.—The skilful gardener is always on the lookout for new ideas and new suggestions that will enable him to improve his garden from year to year. Many books have been published and a number of the experiment stations have issued bulletins dealing with the home lot. The following are a few references:

"Home Vegetable Gardening," by F. F. Rockwell.
"The Home Garden," by Eben E. Rexford.
"The Vegetable Garden," by Ida D. Bennett.
"Vegetable Gardening," by R. L. Watts.
"How to Make a Vegetable Garden," by Edith Loring Fullerton.
"Book of Vegetables," by Allen French.
"Manual of Gardening," by L. H. Bailey.
"Vegetables for Home and Exhibition." Beckett.
"Garden Making." Bailey.
"Principles of Vegetable Gardening." Bailey.
"Farm and Garden Rule Book."
"The Home Vegetable Garden." Farmers' Bulletin No. 255.
"The Home Garden in the South." Farmers' Bulletin 647.,
"Hotbeds and Cold-frames." Cornell Reading-Course Bulletin No. 30.
"Home-Garden Planning." Cornell Reading-Course Bulletin No. 34.
"Planting the Home Vegetable Garden." Cornell Reading-Course Bulletin No. 58.
"Summer Work in the Home Garden." Cornell Reading-Course Bulletin No. 92, and others.
"The Home Vegetable Garden." Illinois Circular No. 154.
"Farmer's Home Garden." West Virginia Bulletin No. 122.
"The Farmer's Vegetable Garden." Illinois Bulletin No. 105.

CHAPTER 30

VEGETABLE FORCING

By C. W. Waid
Extension Specialist, Michigan Agricultural College

Vegetable forcing is a term applied to the growing of vegetables in such a way that they mature or become suitable for use in a shorter time or at a different season than when grown under normal conditions. Cold-frames, hotbeds and greenhouses are used for this purpose.

Cold-Frames.—Cold-frames are wooden or concrete structures covered with glass or cloth. They are entirely dependent upon the sun's rays as the source of heat and serve as a protection against cold winds and too rapid radiation of the heat at night. The frames are usually built to run east and west with the south side about a foot above the surface of the ground and the north side a foot or so higher than the south side. They are about six feet in width and of any desired length. When glass is used as a cover the panes are fastened in sash. The standard size of the sash is three by six feet.

It is not possible to maintain a uniform temperature in cold-frames during very cold weather. Their use is, therefore, confined to relatively mild climates or to short periods in the colder climates. They are best adapted to the growing of cool-season crops, the starting of plants for late transplanting or the hardening off of plants started earlier in the greenhouse or hotbed. A cloth cover is sometimes used in the place of glass as a matter of economy.

Hotbeds.—Hotbeds are similar in construction to cold-frames. The chief difference is that in addition to the heat secured from the sun's rays other means are used to supply heat in the hotbeds. The common source of artificial heat is fermented horse manure. Hot-air flues and steam or hot-water pipes are also used for this purpose. When steam or hot water is used to heat a greenhouse or residence the same system can be used to advantage in heating the hotbeds. Hotbeds are more satisfactory than cold-frames for the growing of early crops or the starting of early plants in a cold climate, as the temperature can be made more uniform.

To prepare a manure hotbed, the dirt is removed from inside the frame to a depth of from one and one-half to two and one-half feet. Horse manure from grain-fed animals should be placed in a compact pile at least three weeks before it is to be put in the pit. As soon as the manure begins to ferment it should be forked over and thoroughly mixed. All lumps should be broken. A second forking over may be needed before it is ready for the pit, when it should be a steaming mass. Care must be taken not to let it overheat and burn, as this would reduce its value. When the pit is dug and the manure ready it should be placed in the pit a few inches at a time and evenly tamped. From one to two feet of solid manure is essential for best results; the greater amount being needed for the growing of heat-loving plants and for other plants as well in the colder sections of the country. Rich garden soil, preferably from a compost heap, should be placed over the manure to a depth of about six inches. The sash should be placed on the frame as soon as the manure and soil are put in to prevent the heat escaping too freely and to keep off the rain or snow. If the manure is well prepared it will raise the temperature of the soil so high at first that it will not be safe to sow seeds until several days later. The temperature of the soil should not be over 85° F. when the seeds are sown.

A Double Sash Steam-Heated Hotbed.

The Greenhouse.—Greenhouses are glass-covered structures, so built that the person who grows crops in them can work inside with ease. They are heated with hot-air flues, hot water or steam. When properly constructed it is possible to grow many kinds of crops to maturity in them at any season of the year. They are even more satisfactory than hotbeds for the starting of plants for early crops outside. It is possible for the gardener not only to give the plants better care in bad weather in the greenhouse, but he is not so much exposed, and thus can work more comfortably and to better advantage. The combination of a greenhouse, hotbeds and cold-frames is desirable when possible.

Growing Plants Under Glass.—The growing of plants under glass is very different from growing them in the open. To the inexperienced, it might seem easier to grow them inside than outside, because conditions are more nearly under the grower's control. This is not true, however, as the comparatively high temperature, excessive humidity and artificial conditions in general encourage the development of tender plants which are subject to attack by various insects and diseases. Vegetable forcing is perhaps the most exacting of all lines of intensive gardening. To be successful in this line of work when it is followed as a business, an individual must be able to apply himself and must have a knowledge of the needs of the crops to be grown. If he is in love with this particular line of work, his chances of success are much greater than when he forces himself into it because he thinks there is money in the business. This need not discourage the man who wishes to have a small greenhouse and a few sash to assist him in getting more money from his outside crops. Many successful greenhouse men have started in this very way.

A Greenhouse Suitable for Forcing Plants.

One of the most important things to keep in mind when starting in the forcing business even in a small way, is to try to supply as nearly as possible the needs of the crop to be grown as to temperature, ventilation, plant-food and water. If these demands are met there will be much less trouble from diseases than when they are not properly looked after.

The following is a list of cool plants which require a night temperature of from 45° to 55° F., and a day temperature of from 65° to 70° F. on clear days:

Lettuce	Peas	Onions
Radishes	Beets	Celery
Rhubarb	Asparagus	Parsley
Cauliflower	Carrots	

The warm plants demanding a night temperature of 55° F. or above and a day temperature of 75° F. or more on bright days are tomatoes, egg-plants, peppers, cucumbers, muskmelons and beans.

Too much importance cannot be placed on ventilation. In cold-frames and hotbeds poor ventilation is almost certain to induce the disease known as "damping off," while careful ventilation and watering will prevent it to a great extent. In a greenhouse such diseases as mildew and others which flourish in a moisture-laden atmosphere and high temperature will be much more liable to give trouble when the ventilation is insufficient than when it is given proper attention.

Watering is another important operation. As a rule, it is best to water only on bright days, and preferably during the forenoon to give time for the plants to dry off before night. The overhead system of watering is being used very commonly by progressive gardeners in the greenhouse and in hotbeds and cold-frames.

The most common source of plant-food in vegetable forcing is well-rotted stable manure. When this can be secured in sufficient quantity, little in the way of artificial fertilizers will be needed. In some cases the use of liquid manure or nitrate of soda in small quantities will produce good results. Wood-ashes, especially from the burning of hardwood and ashes secured from the burning of tobacco stems, can be used to good advantage.

It is not customary to make frequent changes of soil in the vegetable forcing business. Some soils have been in use for forty years and are still producing good crops. In some cases steam sterilization has been necessary to overcome certain soil diseases.

A brief treatment of this subject would not be complete without calling attention to the importance not only of good varieties but of good strains of vegetables for forcing. There is no line of gardening in which this matter is of greater importance. Much time and expense is incurred in the growing of plants under glass. It would certainly not be profitable to put so much expense upon varieties which even when well grown are inferior. Well-grown vegetables of good varieties and strains will demand the highest market prices. The forcing of vegetables is a profitable and pleasant line of work when properly done by the man who knows his business and delights in his work.

REFERENCES

"The Forcing Book." Bailey.

Illinois Expt. Station Bulletin 184. "Tests with Sodium Nitrate for Early Vegetables."

Farmers' Bulletin 460, U. S. Dept. of Agriculture. "Frames as a Factor in Truck Growing."

Canadian Dept. of Agriculture Bulletin 224. "Greenhouse Construction."

CHAPTER 31

Mushroom Culture

By H. M. Ware
Practical Mushroom Grower, Delaware

Over 5,000,000 pounds of the common mushroom (*Agaricus campestris*) are grown annually in the United States. Besides these, in 1914 we imported from Europe 9,188,177 pounds in cans, 30 per cent more than in 1910. Practically all of the $3,000,000 worth of mushrooms grown or imported by this country in 1914 were sold in a few of our larger cities. Hundreds of smaller cities and towns throughout the country offer undeveloped markets for this product, a fact which does not indicate that the supply will soon exceed the demand.

The uncertainty of mushroom growing as a business was eliminated when Dr. B. M. Duggar discovered the "Tissue Method" of manufacturing spawn in 1902. As a direct result of discarding the "Chance" spawn imported from England and France, the American industry has developed rapidly.

While it is true that the bigger the operation the lower will be the cost of production, nevertheless mushroom culture is adapted as a side line to many farms. When sold, mushrooms enter the same channels, wholesale or retail, as do other fancy products. Labor can be profitably employed in winter. The manure used in the houses is in ideal condition for application on the land. This point is better understood when it is considered that the composting of the manure is almost identical with the methods employed by market gardeners—that the fresh mushroom contains 90 per cent water and analysis has shown that, ton for ton, mushroom manure is more valuable than fresh stable manure, having lost little beside weight, water and weed-seeds.

It should be understood that much hard and some unpleasant work is unavoidable in mushroom culture. But that, with intelligent care in supplying the few essential details, success and a legitimate profit are assured.

The most common causes of failure are:

1. Poor spawn.
2. Heavy watering.
3. Unfavorable temperature.
4. Poor or improperly composted manure.

Houses.—The place in which mushrooms are to be grown must permit easy control of temperature, moisture and ventilation. While proper conditions may be afforded by caves, cellars or unused buildings, it will

A Fine Bed of Mushrooms Grown from Spawn of Pure-Culture Origin.[1]

[1] Courtesy of the U. S. Dept. of Agriculture, Bulletin 85, B. P. I.

generally be found advisable to build especially for the purpose. But no one should build a mushroom house without first inspecting the plant of a successful grower. Permanent walls can be made of hollow tile or other material that will not readily decay. Air space in the wall must be provided to maintain even temperature.

When grown in winter mushrooms require artificial heat. Hot-water heating, the system most economically and easily run, is in general use by all large growers. Five hundred square feet of pipe surface (1000 feet of $1\frac{1}{2}$-inch pipe) should be allowed to every 20,000 cubic feet of air space.

TURNING THE COMPOST.

Preparation of the Compost.—The best material is fresh horse manure, which contains plenty of the more resistant cereal straws. Care should be exercised to see that no disinfectant has been used. Build the pile with straight sides 3 or 4 feet deep throughout and 8 feet wide. This makes turning easy, and leaching of plant-food is prevented. In five days the pile should be turned; thereafter at weekly intervals, until rapid fermentation has stopped; usually in three or four weeks. Water the compost when turning and keep it moist. Heavy watering at first will do little harm, but when ready for the beds compost should be in such condition that when squeezed in the hand water will not readily drop from it. Some growers cover the piles with three inches of dirt before and after the first turning. Equally successful growers, however, use no dirt in the compost. Dirt seems only to shorten the time necessary for composting. When ready for the beds the manure has lost all objectionable odors, and the straw has changed from yellow to dark brown.

Filling the Beds.—The beds in common use are flat, 8 inches deep and 6 feet wide, built in tiers of shelves five or six beds to the tier. The boards used are generally chestnut, 1 inch by 8 inches by 12 feet. These are lapped loosely so they may be easily dumped.

The bottom beds should be filled first, so that the operator will have head room. They should then be firmed (*i. e.*, leveled by light pounding with back of fork); if not wet, the manure may be tramped. Firming lessens evaporation and prevents burning during the secondary heating. When filled the house is closed, and in a day or two the temperature rises, sometimes to 120° F., then slowly drops to normal. One ton of manure will fill approximately 65 square feet of bed 8 inches deep.

Spawning.—Only the best American brick spawn should be used.

A TYPICAL RANGE OF MUSHROOM HOUSES.

Of the several varieties, the White and Cream are most desirable; white is more salable, cream more prolific and hardy. Spawning should begin when the temperature of the beds has dropped to 70° F. The bricks should be broken into eight or ten pieces $1\frac{1}{2}$ to 2 inches square and placed evenly on

SIFTING THE CASING DIRT.

the beds. The pieces should then be inserted vertically one inch below the surface of the manure. After spawning, the beds should be firmed again.

Spawn should be kept in a cool, dry place. One brick costs from twelve to fifteen cents and will plant 8 square feet of bed.

Casing the Beds.—Two weeks after spawning, a piece of spawn should be dug up; if the mycelium appears as a mould running into the manure, the beds are ready to case. Casing consists in covering the beds with a layer of sifted loam 1 to 1½ inches deep. The loam causes the mushrooms to head, acts as a mulch and is the best medium for picking. The average farm wagon load of sifted loam will cover 250 square feet of bed.

Temperature.—Temperature is important because it regulates the competition of mushrooms with insects and with other fungi. It has been found that at 53° to 58° F. mushrooms grow slowly but strongly, while other growths are held in check. Even at freezing temperatures mushrooms lie dormant without apparent harm. Too much heat causes rapid

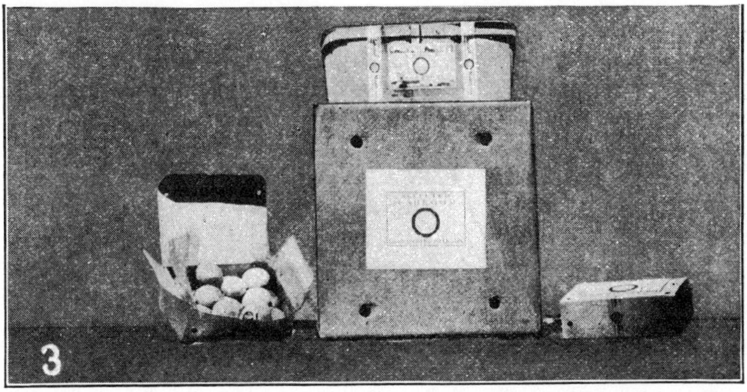

TYPES OF FANCY PACKAGES.

development, not only of mushrooms, but also of any other organisms present, so that the spawn soon "runs out." The temperature should be kept near 56° F. and sudden changes should be avoided.

Water.—Water should be applied to the beds only as a spray. The surface should never be allowed to dry out, nor should it be soaked. It is better to apply a little water every day than to water heavily at longer intervals. The air should be kept as moist as proper ventilation permits.

Ventilation.—Ventilation is of great importance, but must be accomplished without draughts. Draughts quickly dry out the beds and cause the mushrooms to crack and darken, especially after watering. Overhead ventilators give the most uniform ventilation with the least danger.

Picking and Marketing.—The first mushrooms appear six to eight weeks after spawning. When in full bearing they should be picked every day. Picking is an art. The yield and returns may be materially reduced by lack of judgment in this single operation. Experience only can teach one to pick properly. It should be remembered that mushrooms gain no weight after the veil begins to break and that an open mushroom is a third-class article on the market.

Mushrooms are a distinctly high-grade product. They deserve the most careful grading and care in the selection of a package. The standard grades in the New York market are Fancy, Choice, Buttons and Seconds. Too few growers use a label, their packages being known only by a number given in a commission house. The ventilated pound box will recommend itself for the retail trade. The four-pound splint basket is the standard wholesale package; these are tied in bundles of six for shipment by express.

After picking, delay and high temperatures are to be avoided. Even in cold weather some ventilation in the package is desirable.

The Following are the Most Troublesome Enemies.

DISEASES.

Enemy.	When Troublesome.	Injury.	Treatment.
Fogging off.	Warm weather.	Young mushrooms turn brown.	Lower temperature. Ventilation.
Black spot.	Improper watering and ventilation.	Discolored caps.	Proper conditions.
Mycogene disease.	Infected spawn or compost.	Abnormal growths, misshapen, unsalable, highly infectious.	Careful sanitation. Formaldehyde gas fumigation.

INSECTS.

Mites.	High temperature.	Troublesome; may affect spawn.	
Springtails.	Carelessness in cleaning house.	Similar to fogging off.	Fumigation with carbon bisulphide.
Larvæ of flies.	Poor manure. Warm weather.	Bore into mushrooms; unsalable.	Carbon bisulphide. Low temperature.

Mushroom Enemies.—By providing suitable growing conditions and exercising the utmost care in cleaning the house after a crop has been grown, little trouble from insects or fungous enemies will be experienced. Every speck of old compost must be brushed out. The bed boards and every part of the house should then be whitewashed and if possible fumigated.

Yield and Returns.—The profitable life of a mushroom bed averages three to four months. A yield of one-half pound per square foot will pay labor and expenses, but one pound per square foot should be produced under proper conditions.

The average wholesale price is 25 to 26 cents per pound. Mushrooms retail throughout the season from 35 to 75 cents per pound, depending alike on season and quality.

The cost of production depends mainly upon the yield and the cost of labor and materials in a given section. The large growers estimate the cost at 15 to 25 cents per pound.

REFERENCES

"Mushrooms: How to Grow Them." Falconer.
Ohio Expt. Station Circular 153. "Edible and Poisonous Mushrooms."
U. S. Dept. of Agriculture Bulletin 127. "Micogene Disease of Mushrooms."
U. S. Dept. of Agriculture, Bureau of Plant Industry, Bulletin 85. "Principles of Mushroom Growing."
Farmers' Bulletin 342, U. S. Dept. of Agriculture. "Mushroom Growing and Preserving Wild Ones."

CHAPTER 32

MEDICINAL AND AROMATIC PLANTS

BY W. W. STOCKBERGER
Physiologist in Drug and Poisonous Plant Investigations, U. S. Dept. of Agriculture

The market demand for the products of medicinal and aromatic plants when compared with the demand for staple products such as cereals, fruits or vegetables, is relatively very small, and is not sufficient to make them promising crops for general cultivation. Many such plants which can be grown and prepared for market with little difficulty, bring but a small return, and hence their cultivation offers little prospect of profit. A number of high-priced medicinal plants must be given care for two or more years before a crop can be harvested, and, since expensive equipment is usually required for their successful culture and preparation for market, the production of such crops offers little encouragement to inexperienced growers who are looking for quick returns and large profits from a small investment of time and money.

Requirements for Medicinal Plants.—Several medicinal and aromatic plants, for which the demand is fairly constant, have been profitably grown on a commercial basis, but the success of the growers has been due largely to the care which they have taken to produce a uniform product of high quality. However, the production of drugs of high quality requires skilled management, experience in special methods of plant culture, acquaintance with trade requirements and a knowledge of the influence of time of collection and manner of preparation on those constituents of the drug which determine its value. Small quantities of drugs produced without regard to these conditions are apt to be poor in quality and so unattractive to dealers and manufacturers that the product will not be salable at a price sufficient to make their production profitable.

The agricultural conditions generally prevailing in the United States and in Canada are far more favorable to the growing of medicinal and aromatic plants as a special industry for well-equipped cultivators than as a side crop for general farmers.

The growing of medicinal plants in the United States has hardly passed beyond the experimental stage, and although several of these plants promise satisfactory profits in suitable localities, any general attempt to grow them on a commercial scale would soon result in over-stocking the market. However, the demand for such plants as anise, belladonna, caraway, coriander, digitalis and sage is at present large enough to make them worthy of consideration.

Anise (*Pimpinella anisum*) is an annual plant grown for its aromatic seeds. It is cultivated on a small scale in Rhode Island, and is suited for localities similar in climate to that state. The best soil for anise is a light, moderately-rich and well-drained loam. The plant is very sensitive to unfavorable weather conditions, but in a good season the yield of seed should be from 400 to 600 pounds per acre. About 2000 acres should produce the average quantity of seed annually imported into this country. The price usually ranges from 6 to 8 cents a pound.

Belladonna (*Atropa belladonna*) is an important drug plant for which there is a steady demand. It has been cultivated in New Jersey, Pennsylvania and California, although not very successfully from a commercial point of view. It is apparently better adapted to the warmer states than to the colder regions where it is likely to winter-kill. Belladonna thrives best in deep, moist, well-drained loam containing lime. Sowing seed in the field usually gives very poor results, but sowing seed in the greenhouse and transplanting like tomatoes is usually successful. The cost of growing belladonna is high, owing to the large amount of necessary hand labor. Five hundred pounds of dry leaves per acre is considered a fair yield. At the end of the second year about 1000 pounds of dried root per acre may be harvested. The prices in the wholesale drug markets have been from 14 to 25 cents a pound for the leaves and from 9 to 18 cents a pound for the roots. Prices to growers have been proportionately less.

Caraway (*Carum carui*) is an annual, cultivated for its aromatic seeds, which are used medicinally and for flavoring. It grows and fruits well over a considerable portion of the United States, especially in the north and northwest, but its cultivation in this country has never assumed commercial proportions. Soil of a somewhat clayey nature and containing a fair proportion of humus and available plant-food is particularly suited to caraway, but the plant generally grows well in any good upland soil which will produce fair crops of corn or potatoes. The average yield of seed per acre is about 1000 pounds. At this rate about 2700 acres would be required to produce the quantity of seed annually imported. Anyone undertaking the cultivation of this plant might well consider growing dill and fennel also. Caraway seed is valued at about 6½ cents a pound.

Coriander (*Coriandrum sativum*) is also grown for its aromatic seeds and in its requirements and method of culture is very similar to caraway. The yield of seed is quite variable, but from 500 to 800 pounds per acre may be expected. If the average yield were 650 pounds per acre, 2000 acres would be required to produce the quantity of seed annually imported. The seed is valued at approximately 3 cents a pound.

Digitalis or Foxglove (*Digitalis purpurea*) is an important drug plant for which there is a constant demand. The leaves are used in medicine. Although widely grown in flower gardens as an ornamental, it has not yet been grown on a large scale in this country as a drug crop. This plant thrives best in ordinary well-drained garden soils of open texture. Sowing the seed in the field is usually unsuccessful. For good results they should be sown in seed-pans or flats in the greenhouse. When danger of frost is past the plants should be hardened off and transplanted to the field. Digitalis does not flower until the second year, when the leaves may be collected. Probably 600 pounds of dry leaves per acre may be obtained under favorable conditions. The wholesale price of leaves ranges from 8 to 40 cents a pound, averaging about 15 cents.

The Common Sage Plant (*Salvia officinalis*) is easily cultivated and will grow in almost any well-drained fertile soil. There is a good demand for American leaf sage, which sells at a considerably higher price than the imported article.

The dry herb or leaves of a number of aromatic plants form marketable products for which there is a small demand, but as a rule these plants are grown for the essential oils which they yield. The principal essential oils produced in the United States from cultivated plants are: peppermint, spearmint, tansy, wormwood and American wormseed. The price of imported sage is 3 to 5 cents a pound. American sage is usually a little higher.

Ginseng (*Panax quinquefolium*) is a fleshy-rooted herbaceous plant native to this country and formerly of frequent occurrence in shady, well-drained situations in hardwood forests from Maine to Minnesota and southward to the mountains of Georgia and the Carolinas. It has long been valued by the Chinese for medicinal use, though rarely credited with curative properties by natives of other countries. Under cultural conditions, ginseng should be shielded from direct sunlight by the shade of the trees or by lath sheds. The soil should be fairly light and well fertilized with woods earth, rotted leaves or fine raw bone meal, the latter applied at the rate of one pound to each square yard. Seed should be planted in the spring as early as the soil can be worked to advantage, placed 6 inches apart each way in the permanent beds, or 2 by 6 inches in seed-beds, and the seedlings transplanted to stand 6 to 8 inches apart when two years old. Only cracked or partially germinated seed should be used.

Ginseng needs little cultivation, but the beds should be at all times kept free from weeds and grass and the surface of the soil slightly stirred whenever it shows signs of caking. A winter mulch over the crowns is usually essential, but it should not be applied until freezing weather is imminent and should be removed in the spring before the first shoots come through the soil.

The roots do not reach marketable size until about the fifth or sixth year from seed. When dug, they should be carefully washed or shaken free of all adhering soil, but not scraped. Curing is best effected in a well-ventilated room heated to about 80° F. Nearly a month is required to properly cure the larger roots, and great care must be taken in order to prevent moulding or souring. Overheating must also be avoided. When well cured the roots should be stored in a dry, airy place until ready for sale. A market may be found with the wholesale drug dealers, some of whom make a specialty of buying ginseng root for export.

The price of cultivated ginseng root, as quoted in wholesale drug lists, has ranged during the past few years from $5 to $7.50 per pound.

A detailed account of ginseng culture is given in Farmers' Bulletin 551, entitled "The Cultivation of American Ginseng."

Peppermint (*Mentha piperita*) is frequently found growing wild throughout the eastern half of the United States, and can be grown under cultivation on any land that will produce good crops of corn. It is grown commercially with most success on the muck lands of reclaimed swamps in southern Michigan and northern Indiana. On good land the average yield of oil per acre is about 30 pounds, but as the yield is variable, approximately 15,000 acres of land are required to produce the annual market demand. It is valued at about $2.50 per pound.

Spearmint (*Mentha spicata*) is very much like peppermint in its requirements, but can be grown successfully on a wider range of soils. On ordinary soils the yield of oil varies from 10 to 20 pounds per acre, but on muck lands the yield is usually only a little less than that of peppermint. The annual market requirement for spearmint oil is about 50,000 pounds. The oil has an average value of about $3.30 a pound and the dry herb 3 to 4 cents a pound.

Tansy (*Tanacetum vulgare*) is a hardy plant which grows well on almost any good soil, but rich and rather heavy soils well supplied with moisture favor a heavy growth. The yield of oil varies, but about 20 pounds per acre is a fair average. The annual market requirement of this oil probably does not much exceed 3000 pounds. It is valued at about $2.60 a pound.

Wormwood (*Artemisia absinthium*) is a hardy plant which can be grown almost everywere, but commercially it is usually grown on fairly rich, moderately moist loams. It is cultivated on a small commercial scale chiefly in Michigan and Wisconsin. The annual production of oil is about 2000 pounds, which is apparently sufficient to satisfy market requirements. It is valued at about $2.40 per pound.

American Wormseed (*Chenopodium anthelminticum*) is a coarse weed which grows well in almost any soil. The yield of this oil varies, but about 30 pounds per acre is a fair average and the annual production is about 5000 pounds. It is gaining in importance largely through its use as a remedy for hook-worm. The price ranges from $1.40 to $5.50 a pound.

Additional Equipment.—In addition to the usual agricultural equipment the producer of essential oils must provide a suitable distilling apparatus, since such oils are usually derived from plants by steam distillation. The cost of setting up a still will depend upon what facilities are already at hand and the size and efficiency of the apparatus installed. It may easily range from a small sum to several thousand dollars.

Where successful production of medicinal plants has not been demonstrated it should be determined on small experimental plats before undertaking commercial plantings.

REFERENCES

Michigan Expt. Station Bulletin (Special) 72. "Some Gingseng Troubles."
U. S. Dept. of Agriculture Bulletin 26. "American Medicinal Flowers, Fruits and Seeds."
Farmers' Bulletins, U. S. Dept. of Agriculture:
531. "Larkspur or Poison Weed."
551. "Cultivation of American Ginseng."
613. "Goldenseal Under Cultivation."
694. "The Cultivation of Peppermint and Spearmint."
663. "Drug Plants Under Cultivation."

CHAPTER 33

PRINCIPLES OF FRUIT PRODUCTION, WITH SPECIAL REFERENCE TO THE HOME PLANTATION

BY M. G. KAINS

Professor of Horticulture, The Pennsylvania State College

The establishment of home orchards is as important as ever, especially in sections where fruit is not now grown but is shipped in. With the wealth of information available through government and experiment station publications, no one who owns land suitable for growing general farm crops need hesitate to plant fruit for home needs. Even for the cold sections hardy varieties are available.

The Main Factors to Consider.—Temperature decides as to the species, and sometimes the variety, that may be grown. That of a region and even of an orchard is determined mainly by latitude, altitude, physical character of the country and distance from large bodies of water. In the spring, lakes and rivers keep the air cool because they are cold. Thus, they hold back bud development and aid the plants in escaping late frosts. In the fall they continue warm and thus lengthen the season. Other sections even nearby, but beyond the reach of breezes from the water, are more likely to be frosted.

Moisture in the soil may be secured through rainfall or by irrigation. In the East enough rain generally falls to care for the fruit interests, provided proper tillage methods are practiced; in the West, irrigation has largely solved the water supply problem. Of more importance is the relative humidity of the air; for where the air is dry, crop growing is more difficult than where it is fairly moist. In the northern prairie states, where the winter air is both cold and dry, many fruits fail because the air sucks moisture out of twigs and branches while the ground is frozen. In the East, where the cold spells alternate with moist weather, the twigs have a chance to secure moisture either from the soil or from the air.

Soil.—In general, currants and European pears usually do best on heavy soils; peaches and strawberries on lighter ones. But there are countless successes on soils of other character. Because of this, it is evident that the distinctions drawn between soils adapted to certain varieties are perhaps too fine; and yet there are varietal preferences that should be considered for commercial orchards. For home and local market plantation these distinctions are of less importance than for big business orchards.

Subsoil is of even more importance than surface soil in fruit culture, especially of tree fruits. Many good business orchards are on thin soils that must be fed to keep the trees vigorous and productive. The secret is a deep, porous subsoil which insures good drainage and deep feeding; hence the ability of the trees to withstand seasonal vagaries. Since no business orchard should be planted without determining the nature of the subsoil, the prospective planter of a home orchard may well follow this practice.

GOOD NURSERY STOCK.[1]
(Note the extent of roots and form of tops.)

The Parasite factor is mainly controllable. Not that there are no difficult enemies to handle, but preventive or remedial measures are available and mostly effective where properly applied.

Site for the home farm orchard is as important in its degree as location is to the commercial fruit grower. Site pertains to the position of the orchard on the farm, as a gentle eastward or northern slope. Much of the success of the plantation may be in choosing a well-drained, elevated site protected from strong winds. Such a site allows the cold air as well as the ground water to drain away, thus preventing frost injury to buds and blossoms. It also favors holding fruit by the trees, whereas a site exposed to high winds would favor dropping.

Aspect formerly attracted far more attention than today. It was believed that southern and eastern slopes favor earliness—and they do—but the effect is less than commonly believed. Business fruit growers plant on all slopes and get good results from all.

Windbreaks may or may not be a benefit. No one should plant a

[1] Courtesy of Maloney Brothers and Wells Company, Dansville, N. Y.

windbreak without first studying the problem from all angles. Often the best windbreak is the outside row of fruit trees, especially if of a variety that grows large and holds its fruit tenaciously.

Nursery Stock is nowadays so low priced that no one should consider growing his own trees. Fruit trees need special care as to propagation, and also require too much time to grow to orchard planting size; so when the best standard varieties can be bought for thirty cents or less, why run the risk of failure in growing one's own? In buying stock, it is wise to insist upon getting straight, clean trees without Y-crotches, free from insects and diseases, and in plump, robust condition when received. Under no condition should fruit trees older than two years be considered. Peach trees should never be over one year. Trees older than these do not produce fruit sooner or make better orchard trees than young ones. Most commercial fruit growers prefer one-year trees of all kinds because these can be trained more easily than can older trees. The trees also make better progress because they have not lost so many roots.

Southern vs. Northern Grown Nursery Trees.—In the South "June budded" trees are popular. There they may be planted in the fall; but for northern fall planting they do not mature early enough to get a start before winter sets in. Therefore, in the North they should be bought only for spring setting. They are not inferior to northern trees when planted in spring.

Time to Plant.—Fall planting has decided advantages over spring planting. There is a far better chance to get the varieties ordered because nurserymen are not then sold out; if four or more weeks will elapse before winter sets in, the trees may be planted and thus the work done when time is not so precious as in the spring; nurserymen usually charge somewhat less for stock delivered in the autumn. Whether or not planting can be done in the fall, it is a good plan to have the trees delivered before winter so as to have them on hand for spring planting at just the proper time, thus avoiding possible delays of shipment in spring. Such trees may be "heeled-in" until spring.

To Heel-in Trees dig or plow a trench a foot or more deep, preferably running east and west. Make the north side vertical and the south with a long slant. Unpack the trees, prune the mangled and broken roots, and lay in the trench with their trunks on the slanting side. Bury both roots and tops with soil packed around the roots. Remove all litter that might favor mouse nests. In spring dig up and plant the trees as if just received.

Marking Out the Field.—This may be done by sighting, plowing or any other handy way that will get the rows straight. For convenience in handling it is a good plan in the home orchard to choose some unit measure that will suit all kinds of fruits. The rod is perhaps as good as any because peaches, sour cherries, plums, quinces, dwarf pears and apricots may be set that distance apart. Sweet cherries, standard pears and the smaller growing apple varieties require two rods, and the wide spreading apple varieties three rods.

Mixed Plantings are not considered wise by commercial orchardists. Each kind of fruit is kept in a block by itself. This favors uniform treatment. In home plantings, however, such a plan is not always feasible; so that by giving a little extra attention the general farmer may have all his fruit crops together in one area. Bush fruits and strawberries will not do well after the trees come into bearing, but up to that time they may be grown between the trees. Where the rod is the unit of measure, two rows of bush fruits may be placed between the tree rows five feet three inches from the trees, thus making them six feet apart and allowing for the planting of one row of strawberries or truck between them. The strawberries will give one good crop, perhaps two, before the bushes will need the space and the bushes will give two to perhaps four crops before they will have to be removed to get best results from the trees.

If desired one row of grapes may run between the trees, thus leaving eight feet three inches between it and the trees. But since grapes do well for ten or more years, they had better be placed at the side of the orchard. Besides strawberries, various vegetables may be planted between the tree rows for two to five or six years. It is a good plan to place the bush fruits in checks so cultivation may be given in two directions from the start.

The Operation of Planting offers no difficulty. The holes should be dug large enough to take in the roots without serious bending, though bending is not of much consequence. The largest roots should be turned toward the prevailing wind. When the holes are dug the top soil should be laid in one pile and the subsoil in another. Then when the tree is placed in the hole—never more than two inches deeper than it stood in the nursery row—the top soil should be worked among the roots and tramped down hard. Finally, the subsoil should be placed on top, tramped down and a few shovelfuls of soil scattered loosely on top to check evaporation of water from the ground.

First Pruning.—After the trees have been planted they should be pruned. All puny, inferior twigs should be removed, only three to five well-placed ones being left at least a hand's breadth apart on the trunk. If these are two hand-breadths apart, so much the better, because there will be less danger of splitting when loaded with fruit or ice. The frame limbs should be cut back a half or more. Usually, the leader should be cut out to make the tree open-headed.

The lowest limb should be fifteen inches to two feet from the ground to favor low heading with all its advantages of easy pruning, spraying, thinning and harvesting, to say nothing of lessened wind damage. Extension tillage tools will cultivate close to the trunks when the trees get large. Until then, ordinary harrows and cultivators will serve every purpose. During the first year, leaves should never be pulled from the trunk and branches. The tree needs them to ripen its wood. If removed the trees will develop longer limbs to get more leaves and these limbs will have to be cut off later to bring the tree within bounds. If there are twigs among the trunk leaves, they should be cut off the following spring.

How Fruit Buds are Borne.—Much of the success of fruit growing depends on intelligent pruning, and this on a knowledge of the way each plant produces its fruit buds. Apples and pears produce theirs mostly on short twigs in alternate years with leaf buds. These fruit spurs become gnarly as they grow old, but as long as they continue to bear they should be allowed to remain, unless the tree is producing too heavily. Then some may be cut out. Other trees that produce fruit more or less on spurs

Before and After Pruning.[1]

are cherry, plum, apricot, almond, currant and gooseberry. Some produce their buds on the sides of the shoots, not on spurs. Of these the peach is the leader, though almonds, Japanese plums, and apricots also do this more or less. All these trees develop fruit buds one year and blossom the following spring. These fruit buds may be distinguished from leaf buds during winter because they are round-topped and plump instead of pointed and thin.

There is another group, the plants of which develop blossom buds in the same season as they blossom and bear fruit. Quince and medlar each bear blossoms on the ends of short green shoots developed in early spring. Raspberries, blackberries, dewberries and oranges produce their blossoms more or less terminally on lateral summer shoots. Grape, mulberry, olive

[1] Courtesy of The Macmillan Company, N. Y. From "The Principles of Fruit Growing," by Bailey.

and persimmon produce strong shoots or canes from branch buds which have wintered over. On these the blossom buds are borne. The loquat bears its blossom buds at the tips of terminal shoots of the same season.

Pruning for Fruit.—In pruning for fruit, it is evident that the plants in these various groups must be pruned differently. Apples, pears and other plants which hold their bloom buds over winter may be encouraged to bear by summer pruning about the time the shoots have ceased to extend. This tends to develop blossom buds. Pruning of these plants during the dormant season, on the other hand, tends to produce wood at the expense of fruit production. (Consult bulletins of the Tennessee Experiment Station on "Summer Pruning.")

Plants in the second general group are usually pruned in spring, when the number of buds left will indicate approximately how many fruits or clusters of fruits will be produced—one for each quince bud, two or three clusters of grapes for each grape bud, and so on. Pruning of these plants, therefore, is equivalent to thinning, for it limits the number of fruits to be set and helps improve the size and quality of the specimens.

Pruning Older Trees.—In pruning trees great care should be taken to make the wounds close to the main trunk or limbs. If a limb to be cut off is large, the saw should first be used beneath it a foot or so away from the crotch. When the saw sticks, a second cut should be made above so the limb will drop off easily. Then the stub may be cut off close to the trunk without danger of splitting or tearing the tree and making an ugly, slow-healing wound. Beyond the removal of branches that cross each other young trees properly started and trained should need little or no pruning unless they break down.

Tillage.—Orchards in sod have in commercial practice practically given place to tilled orchards. Where success attends sod treatment, some other factor is usually evident enough upon study of the situation. The experiment station at Geneva, N. Y., has reported that a sod-mulched orchard under ten-year experiment yielded higher colored, earlier maturing fruit than a tilled orchard of the same variety and otherwise handled the same way, but that the tilled orchard yielded heavily and uniformly, gave fruit of better quality, larger size, longer keeping, less dead wood in the trees and better foliage and growth. Sod lowers the water supply and soil temperature, decreases certain plant-foods, reduces humus and air supply in soil, impairs work of soil bacteria, and forms substances that impair tree health. Sod, however, has special use where tillage is impossible either because of the steep slopes or stony land.

Tillage should start with the preparation of the land for planting and be done yearly while the plants remain profitable. The advantage of this is that the roots are encouraged to go deeply and thus withstand dry weather as well as escape the plow. Each year operations should be begun as early as the land can be worked and continue until the twigs have reached their full length about midsummer. Between mid and late summer, tillage should stop to give trees or shrubs a chance to ripen their growth to withstand the winter. Unless this is done, growth may continue too late in the fall, and the plants suffer during winter in consequence.

Fertilizing.—While it may be true that land which will grow any farm crops will grow fruit without manuring, yet most money is made from fruit crops fed to get higher quality, larger size, better color and the other points that make for higher prices. How much and what kind to apply will depend

PICKING APPLES IN THE ROGUE RIVER VALLEY, OREGON.[1]

upon the character of the soil, the kind of crop and so on. Many farmers and fruit growers put the question to the land itself by trying experiments with various combinations of fertilizers until they find out the one best suited to the desired end. In general, it must be remembered that nitrogenous plant-food tends to be lost by seepage and also to produce wood rather than fruit; hence, it must be handled with greater caution than either potash or phosphoric acid, neither of which is lost to any serious extent from the soil; nor does either jeopardize the ability of the plants to withstand winter injury.

[1] Courtesy of Portland Commercial Club, Portland, Oregon.

Thinning is steadily gaining popularity in the East, mainly because it tends to produce larger, finer specimens, to make the trees more hardy and to establish regular annual bearing. Even the Baldwin apple, perhaps the most notorious biennial cropper, has been made to produce profitable crops fifteen out of seventeen consecutive years.

Spraying has now become so general that no one thinks of planting fruit without counting upon it. The first point to remember is that it must be done with discrimination; for a plant disease cannot be combated with an insecticide nor vice versa. Second, spraying for plant diseases must be preventive; no remedy is known for diseases which have gained entrance to the plant tissues. Third, sprays for insects must be suited to the kind of insects. Those that bite off and swallow pieces of plant tissue must be poisoned internally, and those that merely suck the juice from the plant killed by some substance which chokes, burns or otherwise destroys them through their skins. Experiment station literature is rich in information on methods of control.

Harvesting and Marketing are rapidly becoming more businesslike. Growers are recognizing the advantages from grading their fruit and selling each grade for what it is. They are also learning that the laws which specify standard sizes for packages are steps in the right direction, so are adopting the new standards with profit to themselves and their communities.

The Value and Importance of the Home Fruit Garden to the general farmer lies mainly in the variety of pleasures as well as in the addition to the diet supplied. Such a plantation should contain all kinds of fruits so the table may be supplied from the time strawberries first ripen till the last winter apples are used the following year when strawberries come in again.

Two or three rows of strawberries one hundred feet long, one each of black, red and purple raspberries, one of dewberries, and one or two of blackberries or loganberries should supply an average sized family throughout the year with fresh and canned fruit, jelly, jam and preserves. Twenty-five plants each of gooseberries and currants should suffice. By choosing early and late maturing grape varieties, such a family should be able to eat the product of twenty or thirty vines, perhaps more. A dozen or a score of plum, peach and cherry trees, early and late, as many each of dwarf and standard pears, perhaps half a dozen quinces, and forty or fifty apples trees beginning with a few summer apples, continuing through fall varieties and ending with at least half or perhaps two-thirds of the trees of varieties that reach their best between Christmas and May Day will supply the needs of the average family.

Quality First for the Home.—In all cases the choice of varieties for the home should fall on fruits of best quality, either for dessert, for cooking or preserving. For local markets fewer varieties, preferably the best known kinds of the section, should be given preference. Never choose for business purposes varieties that have not been fully tested locally, no matter how famous they may be elsewhere. They may fail to come up to their standard established in some other sections.

REFERENCES

"Principles of Fruit Growing." Bailey.
"Popular Fruit Growing." Green.
"How to Make a Fruit Garden." Fletcher.
"Fruit Growing in Arid Regions." Paddock and Whipple.
"Beginners' Guide to Fruit Growing." Waugh.
"Propagation of Plants." Fuller.
"Fruit Harvesting and Storing." Austin.
"Nursery Book." Bailey.
Pennsylvania Expt. Station Bulletin 134. "Experimental Results of Young Orchards in Pennsylvania."
Canadian Dept. of Agriculture Bulletins:
211. "Fruits Recommended for Planting."
212. "An Orchard Survey."

CHAPTER 34

SMALL FRUITS

BY PROFESSOR L. C. CORBETT
In charge of Horticultural and Pomological Investigations, United States Department of Agriculture

The small fruit interests of the United States are made up of a diversity of fruits adapted to a wide range of territory and conditions. The cash value of these crops approximates $20,000,000 annually, two-thirds of which is derived from the strawberry, the most cosmopolitan of the small fruits. The second place is contested by the raspberry and the blackberry, both of which are important money crops, and the fourth crop of importance is the cranberry, which is restricted both by climate and by soil requirements. Each of the important small fruits is here given a brief but, it is hoped, clear and concise treatment.

THE STRAWBERRY

The garden strawberry is an American product. It adapts itself to a wider range of latitude and to greater extremes in environment than any other cultivated fruit. It is universally liked and is cosmopolitan in its adaptations.

Selection of Soil.—The soil best suited to the cultivation of the strawberry in the northeastern part of the United States is a sandy or gravelly loam. A warm, quick soil, although naturally poor, is to be preferred to a heavy, retentive soil well supplied with plant-food. The lacking plant-food can easily be supplied by the addition of fertilizers, while the physical characteristics of the soil can be modified only with great difficulty by cultivation, drainage and the addition of organic matter. Congenial soil and exposure are, therefore, important considerations.

Preparation of the Soil.—The land to be devoted to strawberries should, if possible, be planted in a cultivated crop, such as potatoes, beans or corn, at least one year previous to setting the plants, in order that the larvæ of such insects as wireworms, white grubs, cutworms, etc., may be as completely eliminated as possible.

Previous to setting the plants the soil should be deeply plowed in order that all organic matter of whatever nature on the surface may be completely turned under. Immediately following the plow the land should be thoroughly pulverized by the use of the harrow, and the surface should be reduced to a condition which would form an ideal seed-bed.

Fertilizers.—If the soil is not rich, for best results it should have a dressing of at least twenty cartloads of well-decomposed stable manure per acre, either plowed under or incorporated with the soil by surface culture after plowing. If stable manure is not available, plant-food should be supplied by a liberal use of fine-ground bone and chemical manures rich in nitrogen and potash. The use upon the plants at blooming time of highly nitrogenous manures, such as nitrate of soda, at the rate of about 100 pounds per acre often proves of great value. If it can be applied in solution it will give quicker results than if put on in the form of a salt.

A Spray of Good Strawberries.
Uniformity in size and form increases the market price.

Selecting and Preparing the Plants.—Plants with small crowns, *i. e.*, a moderate growth of leaves, and with an abundant development of fibrous roots, are most desirable. If the crown and the roots of the plant are in good condition, the success of the plantation is assured, provided the ground has been well prepared and the work of planting is done with care.

Perfect and Imperfect Flowered Plants.—Strawberries occur with imperfect (or pistillate) flowers as well as with perfect flowers (those containing both stamens and pistils). It is important to give careful attention to this point in planting a plantation, as a patch made up of pistillate sorts alone will be unproductive, while many such sorts when properly interspersed with perfect-flowered varieties have proved to be the largest fruited and most prolific sorts. A common practice is to set every fourth or fifth row with a perfect-flowered sort which blooms at the same period as the pistillate variety of which the plantation is chiefly composed.

When to Set the Plants.—The time to plant depends, in humid regions, more upon the rainfall than upon any other factor. If there are not timely rains at the planting season to give the plants an opportunity to establish themselves, the stand will be uneven, with the result that more work will be required to keep the land free from weeds and more trouble will be neces-

PLANTING A STRAWBERRY RUNNER.
On the right a plant correctly planted, showing roots spread out; on the left a plant put in in the wrong position with roots crowded together.

sary to fill the blank spaces with runners from the plants that survive. The plants that withstand the drought are checked and dwarfed. They seldom recover so as to make either satisfactory croppers or plant producers. It is most satisfactory and most economical, therefore, to choose that season which offers most advantages at planting time, other things being equal. It is impossible to specify the season for each locality or even for large areas, as local conditions of soil and climate necessitate different practices in localities only a short distance apart. In general there are only two seasons for planting—spring and autumn—but in some localities spring planting should be done in April or May by the use of the preceding season's plants, while in others it may be done in June from the crop of runners of the same season.

In irrigated regions planting can be done at whatever season the work will give best results in future crop production. In humid regions rainfall is a determining factor.

How to Set the Plants.—Success in transplanting strawberry plants depends, first, on the quality of the plant, and, second, upon the time and manner of doing the work. If the plants are good, the stand, other conditions being favorable, depends upon care in setting them. The success of this operation is measured by the degree of compactness of the soil about the roots of the plant. If the plant has many roots and these are thrust into a hole made by an ordinary dibble, it is more difficult to get the earth in contact with the roots than when the plant has fewer roots. The plant with the greatest number of feeding roots is, however, the most desirable if properly handled. Such plants should be set in a broad, flat hole where the roots can be spread out in natural form. By giving the crown of the plant a whirl between the thumb and finger to throw the roots out like the ribs of an umbrella and quickly putting it in place while the roots are still thrown out from the crown, the normal position of the root system can be closely approached.

Another very satisfactory method is to open a broad wedge-shaped hole by thrusting the blade of a bright spade into the soil and moving the handle forward. The roots of the plant are then spread in fan shape and placed in the hole back of the spade. The spade is then withdrawn and inserted about six inches further forward, and by a backward movement of the handle the earth is firmly pressed against the roots of the plant. Two persons—a man to operate the spade and a boy to place the plants—can set plants very rapidly in this manner. This practice is particularly well suited to localities with sparse rainfall, as it thoroughly compacts the earth about the roots of the plant and allows the roots to extend full length into the moist soil. Plants set in this way have their roots more deeply inserted in the soil than when the roots are spread out in umbrella fashion and as deeply as when set with a dibble. They also have the additional advantage of being spread out so as to have a larger percentage of their surface actually in contact with the soil than when set with a round dibble.

Depth to Set the Plants.—No plant which the gardener has to handle is more exacting in regard to depth of planting than the strawberry. As the plant is practically stemless, the base of the leaves and the roots being so close together, care is required to avoid setting the plant so deep that the terminal bud will be covered or so shallow that the upper portion of the roots will be exposed, either being a disadvantage which frequently results in the death of the plant.

Planting in Hills.—For the hill system of culture plants are set singly either 3 by 3 feet apart, or with the rows 4 feet apart and the plants 2 feet apart in the row, depending upon the character of the soil and the length of time the plantation is to be maintained. In Florida a common practice is to lay the land off in broad beds 8 to 12 feet wide, the rows of plants to run lengthwise of the beds, the rows 24 inches apart, with the plants 18 inches apart in the rows. Such beds afford sufficient drainage and hold the mulch better than narrow beds or raised rows, and the space between the plants admits light to all sides of the plant—an advantage in coloring the fruits which can not be secured by the matted row system early in the season in the climate of Florida.

A common practice is to set the plants in single rows 4 feet apart, with the plants 12 inches apart in the row. The runners which develop from these plants are then allowed to take possession of the area for 6 to 9 inches on either side of the original plants, thus making a matted row 12 to 18 inches wide; this leaves 30 inches between the rows, which allows ample space for cultivation and gathering the fruit. This space can be reduced from 30 inches to as little as 18 inches where land is valuable and it is necessary to secure maximum returns; on thin soil, however, the greater distance is most satisfactory.

Renewing Old Beds.—There is one advantage in the narrow cultivated space. After the second crop has been harvested the runners can be allowed to take possession of the cultivated middle, and when the young plants become thoroughly established the original rows can be broken up with a narrow turning plow or a sharp cultivator. In this way a patch can be very satisfactorily and cheaply renewed, and by a liberal use of suitable fertilizers the rotation can be kept up on the same soil for several years. Some planters prefer to set the plants for the matted row in a double row at planting time. The practice is to establish two rows 12 inches apart, 6 inches on each side of the center of the matted belt, setting the plants 2 feet apart in each row and alternating the plants in the row, so that the plants actually stand a little over a foot apart as shown in the accompanying diagram:

```
    *     *     *     *     *     *     *

 *     *     *     *     *     *     *     *
```

Cultivation.—Clean and shallow culture are the watchwords of successful cultivators. By conserving moisture, cultivation tends to counterbalance the evil effect of drought. A better stand of plants can be maintained during a dry period on well-tilled ground than upon ground that is poorly cultivated. The mechanical effect of grinding the soil upon itself during cultivation reduces it to smaller particles, thus exposing more surface to the action of soil moisture, and, as a result, increasing the available plant-food of the soil. The benefit from preserving a soil mulch, with its consequent economy in the use of soil moisture, is sufficiently important to justify thorough tillage.

Objects of Mulching.—Covering the surface of the soil with dead or decaying vegetable matter is the meaning of the term mulching as here used. Mulching serves different purposes, depending upon the locality in which the plants are grown. A mulch acts as a protection from cold, prevents freezing and thawing and the consequent lifting of the plants ("heaving out"); it retards growth in cold regions by shading the crowns and maintaining a low soil temperature longer than in soil not mulched; it acts as a conserver of moisture, discourages weed growth by smothering the young seedling, and finally protects the fruit from contact with the soil.

Materials for Mulch.—Whole or cut straw free from grains, strawy manure from the horse stable, and pine straw from the forest are among the more common mulching materials. In certain sections marsh hay, either from fresh or salt water marshes, is a common and very satisfactory mulching material.

When to Apply the Mulch.—At the North where the soil is likely to freeze and thaw several times in the course of the winter, it is the practice

AMERICAN QUART BOXES OF WELL-GRADED STRAWBERRIES.[1]
"Fancy" on the right, "No. 1" on the left.

to put on the mulch as soon as the ground is sufficiently frozen to allow driving upon it with a loaded cart or wagon. Where the freezing of the soil is only superficial or only temporary, if at all, the mulch serves the purpose of a protection from wind more than from frost, and in such sections the mulch is put on as soon as active growth ceases, usually early in December and is allowed to remain until after the crop is harvested.

Harvesting and Shipping.—The time of gathering the fruit, as well as the manner of handling, is governed by the use to which it is to be put. If intended for a local market, much riper fruits can be handled than when they are to be shipped long distances.

The most progressive growers of strawberries for local markets not only give particular attention to the ripeness of the fruit, but to assorting

[1] From Farmers' Bulletin 664, U. S. Dept. of Agriculture.

and grading as well, only large, perfect berries being placed in the first grade and all small or soiled fruits in the second.

Receptacles.—Whether it is to be shipped in crates or refrigerator carriers or to be carried to the local market, for best results the fruit should not be rehandled after it is picked. The pickers should be trained to do the necessary assorting and grading as they pick the fruit in the receptacles in which it is to be marketed.

The light splint-wood basket, holding one quart, is the most popular and most universally used. Many different forms of box or basket have been designed, and various materials other than wood have been used in their construction, but up to the present none has met with general adoption.

THE RASPBERRY

The name *raspberry*, as used in the United States, embraces four distinct species of plants, three of which are of American origin, thus placing to the credit of our native plants three important and widely cultivated culinary fruits. The two types of fruits represented by these species are known popularly as red raspberries and black raspberries or "blackcaps."

The red-raspberry group, as represented in cultivation, includes not only the native red raspberry but the European red raspberry, or bramble, and a type intermediate between the native red and black raspberry, which bears a purple fruit and is frequently spoken of as the "purple-cane" raspberry or as the "Schafer group." The red-raspberry group, besides having varieties which produce the characteristic red fruits, has another set of varieties which produce amber or yellow fruit. These horticultural varieties are recognized and are considered distinct sorts, but are not separated botanically into different species.

The black raspberry is distinct both in habit of growth and in the makeup of its fruit. It is recognized botanically as a species distinct from the three which enter into the red-raspberry group. The habits of this plant and the quality of its berries are such that it has gained an important place in certain sections of this country as a commercial fruit.

The fact that the varieties of the red-berry type have to be marketed from the bushes as soon as ripe confines their cultivation to the vicinity of large centers of consumption, where climatic and soil conditions favor their development. The black-raspberry industry, however, can be profitably and successfully carried on in regions more remote from the centers of consumption, because of the fact that a large proportion of the fruits are evaporated and are sold in a dry state there being ready sale for them when handled in this way.

Red Raspberries.—The red-raspberry group includes varieties which bear fruits of various shades of red, amber, yellow and purple, the last-named division being a hybrid between the red and the black types.

Selection and Preparation of Soil.—The soil upon which red raspberries thrive best is a sandy or clay loam of a glacial drift formation. They thrive well upon moderately rich, deep soils and yield largest returns under these conditions.

The preparation of the soil for red raspberries should be the same as for any small fruit, preferably one or two seasons' preparatory tillage in a "hoe crop," which will to a very large extent rid the land of weeds. Such crops as potatoes, beans, cowpeas and plants of this nature are good preparatory crops.

Planting.—The distance to plant will depend very largely upon the purpose for which the plantation is intended. If it is a commercial plan-

Land that will Produce Good Farm Crops will Produce Bush Fruits.[1]

tation upon soil which is not especially valuable, the plants should be 3 feet apart in the row, and the rows not less than 6 feet apart. This will allow of cultivation in both directions for two or three years, and will permit the use of horse-power implements, and consequently will lessen greatly the cost of tillage. On city lots or in a home fruit garden, where it becomes desirable to combine in the same plantation raspberries and other fruit-bearing plants, the distance can be somewhat lessened, but even under these conditions the plants should not be set closer than 2 feet apart in the row and the rows not less than 4 feet apart.

In home fruit gardens small holes can be opened with a spade, the plant roots spread in the ordinary fashion for planting larger plants, and the

[1] Courtesy of The Pennsylvania Farmer.

earth returned; but in all cases it should be the aim to firm the earth well over the roots of the plants as they are set.

Cultivation.—Clean cultivation is necessary with red raspberries, because, as above stated, they are themselves of a weedy nature, and, in order to hold them within bounds, implements which cut all the superfluous shoots and root sprouts from the cultivated area should be used. During the early life of the plantation it would be found most economical to keep the plants in check-rows so that cultivation by horse-power can be accomplished in two directions. Later, however, as the plantation grows older, it will be found advantageous, both in yield of fruit and for economy, to allow the plants to form a hedge or matted row, and to practice cultivation in one direction only. The space between the hedges should be plowed at least once each year, and whether this shall be done in the spring or in the autumn will depend upon the locality.

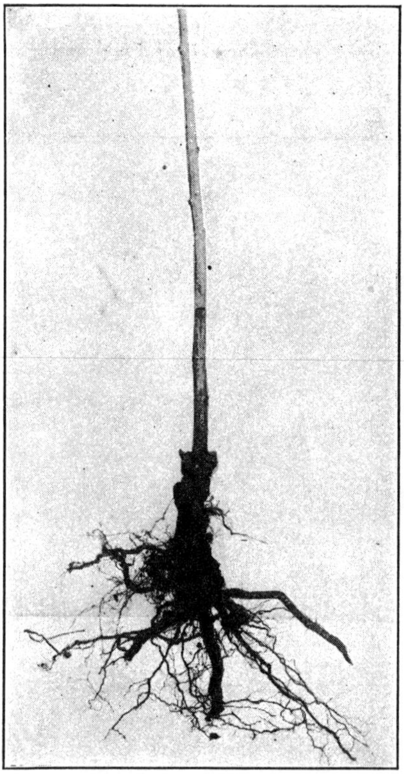

A Young Planting Cane of Raspberry Showing Fibrous Roots.

Fertilizers.—The liberal use of stable manure (20 tons per acre) will produce large yields of fruit, but the use of a complete fertilizer, containing nitrogen 4.5 per cent, phosphoric acid (available) 7.7 per cent, potash 13.3 per cent, at the rate of 500 pounds per acre gives a greater net profit at less outlay.

Pruning.—Red raspberries require attention to direct their growth and fruit production, at two seasons of the year. They should be pruned in the summer, during the growing season, to regulate the height of the canes and induce the formation of fruiting wood for the following season, and again during the winter or early spring for the purpose of eliminating the canes which bore last season. This will allow all the energy of the root of the plant to be directed to the production of fruit and the formation of the next season's bearing wood.

The summer pruning, which is not generally practiced with red raspberries, consists in topping the young shoots when they have attained a height of from 18 to 20 inches. This induces the development of side shoots and the production of additional sprouts from the root. Both these types of growth are desirable in order to insure as large a growth of wood as the plants can carry to advantage.

The winter pruning is a process of elimination. All canes which have served their purpose as fruit producers are removed, as are all dead or diseased canes, thus reducing the demands upon the roots of the plant and directing the energy to the wood intended for fruit production.

Harvesting the Fruit.—Because of the soft character of this fruit, it can be successfully harvested only by hand picking. Small receptacles holding not more than a pint, and preferably those made of wood, are best suited for handling this crop. Under favorable conditions, the yield of the better sorts of red raspberries, particularly of the native red and purple cane types, is very large, and where they can be placed upon the market quickly after being picked they are a very profitable crop.

Black Raspberries, or Blackcaps.—The black raspberry, or blackcap, because it lends itself to several methods of harvesting and marketing, is capable of a wider range of commercial cultivation than any of the types of the red raspberry, although it is not capable of withstanding so severe climatic conditions.

Propagation.—The black raspberry does not throw up root sprouts, and is propagated only from stolons or layers. In order to secure new plants the tips of the branches are bent over and slightly covered with earth during the month of August, after which they take root readily. The rooted tips are usually left attached to the parent stalk until the following spring, when the branch is cut 6 or 8 inches above the surface of the ground, the roots being lifted, tied in bunches and stored for use or carried to the place where they are to be replanted.

Character of the Soil.—Black raspberries grow best on a soil which is fertile and naturally well drained, rather than one which is moist. Strong loams of a clayey or gravelly nature are preferred to the lighter sandy soils.

Preparation of the Soil.—The same general preparation of the soil as outlined for the red raspberry is necessary for best results with the black raspberry. Preparatory treatment with cultivated crops in order to rid the land as thoroughly as possible of weeds is desirable.

Planting.—The distance at which black raspberries are usually set in commercial plantations is 3 feet apart in rows which are 8 feet apart. The same method of planting as described for red raspberries—that is, opening a furrow with the plow, placing the roots at the proper distances in the row and covering with a turning plow—is very convenient and satisfactory.

Cultivation.—Clean cultivation is equally as desirable for the black raspberry as for the red raspberry, because weeds between the rows interfere with the later operations in the berry field. While cultivation should not be carried on so late in the season as to interfere with the harvesting of the fruit, it should be sufficiently thorough and continued late enough to keep the ground free from weeds.

Winter Protection.—In some portions of the Northern states the raspberry can be successfully fruited only by giving it some form of protection during winter. One of the simplest methods of affording such protection is to bend the canes of the plant all in one direction along the line of the row and fasten them either by placing earth upon them or pegging them down. The roots are slightly loosened on one side of each plant and the canes are bent over the roots of its neighbor. After the tops have been properly placed a mound of earth is thrown over them. If after cold weather sets in the earth covering is deemed inadequate, additional protection may be provided by a layer of straw, strawy manure or corn fodder.

Fertilizers.—Stable manure in moderate quantities, supplemented by a fertilizer carrying 4 to 5 per cent of nitrogen, 10 to 12 per cent of phosphoric acid and from 6 to 8 per cent of potash, will prove beneficial. Such a fertilizer, if applied at the rate of from 300 to 500 pounds per acre, should so increase the yield as to make its use profitable.

Pruning.—Because of its manner of fruit bearing, the black raspberry requires care in annual pruning; in fact, pruning must be done at two seasons of the year in order to accomplish the best results. The young shoots as they appear from the roots in the spring should be tipped or disbudded when they reach the height of 18 inches. It is better to go over the plantation frequently, making three or four trips in all, in order to tip the canes when they are about the height mentioned, rather than to delay the operation until some of them have reached a height of 2 to 2½ feet. The early pinching or disbudding induces the development of more numerous lateral branches. Shoots which have been allowed to harden and to grow 2 or 3 feet in height will form few lateral branches. If tipped when 18 inches high, a cane should produce four, five, or six lateral branches. If allowed to attain a height of 3 feet and then cut back to 18 inches, it is probable that not more than two or three lateral branches will be formed; and, since these lateral branches form the fruit-bearing wood of the succeeding season, it is very desirable that the greatest possible number of branches be secured to insure a heavy crop of fruit. It is evident, therefore, that summer pruning predetermines the crop for the succeeding year more than does any other single cultural factor.

The second pruning, which is also important, consists in removing the canes which bore the last crop of fruit. This work can be done at any time after the crop has been harvested, but preferably during the spring following the crop. If the work is done in the spring the lateral branches borne by the canes which developed from the roots of the mother plant should at the same time be shortened to about 8 to 12 inches in length. From each bud of these short branches annual growth will be made which will terminate in a fruit cluster.

Harvesting.—Black raspberries to be marketed as fresh fruit for immediate consumption are always hand picked and placed in either pint or quart boxes similar to those used for strawberries. Those to be dried or evaporated, or to be marketed as dried raspberries, may be either hand picked or harvested with a mechanical contrivance called a "bat." This consists of a frame of light lumber a few inches deep backed up by strong cloth against which the ripe fruit strikes as it is jarred from the bushes by tapping them gently with a light stick or "bat," while the cloth-covered frame is held under the plants in such a position as to catch the fruits as they fall. Such fruits, after drying, are run through a fanning mill to separate leaves and stems, after which they are hand picked in much the same manner as beans, to remove all imperfect and green fruits, as well as those which still hold the receptacle.

THE BLACKBERRY

The blackberry in the United States is a native bramble of wide distribution over the eastern and northern part of the country. The fruit of the wild blackberry was an important factor in the supply of condiments provided by the early settlers. The esteem in which this fruit was held led to the cultivation of some of the wild plants producing berries of superior size or flavor, or those ripening in advance of the main crop, or such as lagged behind and thus extended the season for the fresh fruit. Such selections from the wild blackberries and their seedlings furnish the cultivated sorts of today. What may yet appear is suggested by some of the remarkable hybrids which have already been produced in this genus such as the Logan berry. The chief considerations in the selection of a location for a blackberry plantation are the facilities for harvesting and marketing the crop and the moisture condition of the soil. The fruit of the blackberry is highly perishable and will not endure rough handling in harvesting or long journeys over rough roads.

Few crops are more adversely affected by a lack of adequate moisture during the period of development and ripening than the blackberry, but an excess of moisture during the dormant period is equally as detrimental.

Soil.—The blackberry is not exacting as regards the general type of soil and will do fairly well on a clay, clay loam or sandy loam. The largest yields are on deep, rich soils which provide an extensive feeding area for the roots of the plants.

The preparation for blackberries should be such as to provide a deep, mellow root area and the best possible protection against rank growths of annual weeds. A hoe crop such as corn, beans or potatoes, if properly tended, leaves the area in the best possible condition for the small fruits.

While the roots of the blackberry are perennial, the canes or branches are practically biennial. The shoots spring up and grow one season from the fruiting canes of the following season, after which they die and should be removed to make room for the new growth of the following year. The fruit is borne only on wood of last season's growth in the standard high bush blackberries and dewberries, but the Himalaya and ever-bearing types have perennial canes and do not therefore lend themselves to this type of renewal.

Propagation.—The plantation of the standard blackberries can be increased in either or both of two ways, as follows: The plants, in addition to throwing up strong shoots or canes from the crown, throw up suckers or root sprouts, which may be allowed to develop and later be lifted as independent plants, or lateral roots of strong plants may be dug during the autumn or early spring and placed in sand much the same as are ordinary cuttings, except that blackberry root cuttings are cut to pieces 2 to 3 inches in length and should be entirely covered with sand or light soil to the depth of 2 to 3 inches. Nurserymen propagate their supply of plants largely by the root-cutting method. In one season root cuttings of this sort should produce strong plants for transplanting. The dewberry and certain blackberry hybrids take root at the tips, the same as do black raspberries, and new plants are secured by covering the tips of each plant with earth towards the end of the annual growth period.

Planting, Tillage and Fertilizers.—Blackberries are for the most part rank-growing plants and require liberal distances in and between the rows. A common planting plan is 4 feet in the row and 8 to 10 feet between the rows. In general, the best time for establishing a blackberry plantation is in the spring and, as growth normally starts early, the work of planting should be done as early as soil conditions will permit.

As the blackberry plants will not fully occupy the land the first season, it is customary to use some inter-crop, such as potatoes or beans, to contribute towards the cost of maintenance.

The tillage of the blackberry plantation should be such as to hold weeds and suckers in check and maintain maximum moisture and growth conditions, but cultivation should cease early enough to induce the plants to ripen their wood thoroughly before winter.

If the soil on which the blackberry plantation has been established appears to require fertilizer, experience dictates that the best results will in general be secured by the use of liberal applications of stable manure.

Pruning and Training.—The blackberry plant normally produces long, slender, non-branching shoots. These, where the soil is strong, grow long and produce less fruit than those which have been pruned. A common practice is to pinch the terminal bud of each shoot as soon as it reaches a height of $2\frac{1}{2}$ feet with moderate growing varieties, or 3 feet with robust growing sorts. This induces the formation of lateral branches which increases the number of buds from which fruit-bearing twigs will develop the following spring. The pruning causes the main stem of the shoot to thicken and stiffen and consequently make it better able to carry a large crop of fruit without a trellis. The lateral branches which are induced to develop on the pinched-back shoots should be shortened to 10 or 12 inches before growth starts in the spring.

Harvesting.—The fruit should be harvested as soon as well colored, and only firm, sound berries should be sent to market. A few over-ripe fruits in a box will shorten the marketing period of the whole box, as will rough handling in picking or transporting the fruit to market. Quart boxes are as large a receptacle as blackberries can be successfully marketed in, but the crates may run from 12 to 36 quarts capacity.

THE CURRANT

There are three general groups of currants cultivated to a greater or less degree in various parts of the United States. In general, however, the culture of the currant is confined to the northern half of the country,

Currants Should Find a Place in Every Home Garden.[1]

as none of the forms are able to withstand heat as well as they do cold. Of the three types represented by the common red, the Black and the Crandall, the Red is by far the most important from a commercial standpoint and is the form most generally cultivated. The other two are sparingly grown for special purposes. As currants are in little demand as fresh table fruits, but are almost universally used for the preparation of jellies, jams, preserves or for canning in mixture with sour cherries or red raspberries, they are restricted commercially. This should be borne in mind in planning a small fruit plantation. While the currant should be found in every home fruit plantation throughout the northern tier of states on

[1] Courtesy of The Pennsylvania Farmer.

account of its hardiness, and early and persistent fruit production, it would be an easy matter to carry the commercial production beyond profitable limits. Then, too, the currant is a fruit that is relatively expensive to pick, as the work must all be done by hand.

Soil Requirements.—The currant thrives best on a deep, moist, yet well-drained loam or sandy loam, but will thrive and produce on a great variety of soils, provided they are arable and neither too wet nor too dry. The soil should be well prepared by deep plowing and thorough fining for the reception of the young plants. In addition, it is well to give the land a year of preparatory treatment with crops which will tend to put it in good physical condition, and at the same time eliminate weeds, either through clean culture or by the use of a crop which is dense enough to smother the weed growth. Currants are usually set in rows 6 feet apart and the distance between the plants in the rows varies from 3 to 5 feet. If it is desirable to maintain cultivation in both directions throughout the greater portion of the life of the plantation, the plants should be allowed either 4 or 5 feet in the row. Strong one or two-year-old plants should be chosen and the planting can be done either in the autumn or spring, according to the prevailing practice of the locality. The usual care exercised in pruning the roots and tops of fruit trees at transplanting time should be carried out with the currant. The fruit-bearing habit of the plant should be carefully observed and the later pruning carried on in such a way as to provide as much bearing wood as the plant will carry and yet not overburden it or allow wood of too great age to accumulate in the bush to the detriment of high production or quality of the fruit. Wood more than three years of age should be removed. A little fruit is borne on the base of shoots of last season's growth, but the main crop is borne on wood two or three years of age.

WHITE CURRANTS.

Culture and Fertilization.—Clean culture so as to protect the plants from weed competition and for the purpose of conserving moisture should be the aim. Strong, vigorous plants are more profitable and are better able to resist the attacks of enemies and diseases. Stable manure, bone meal or other high grade fertilizers should be used to maintain the plants in a high state of growth and vigor.

Enemies and Diseases.—If the plants become infested with the currant worm, as the red sorts are almost certain to be, the plants should be thoroughly sprayed with a solution of Paris green, 5 ounces to 30 gallons of water, or dusted with white hellebore. If mildew is troublesome, Bordeaux mixture should be used. As a rule, however, currants are not as seriously affected by mildew as are the gooseberries.

Harvesting the Fruit.—Currants should be carefully picked so as to maintain the little grape-like clusters of fruit intact. Berries torn or stripped from the stems do not keep or ship as well as those carefully handled. The most popular receptacle for shipping currants is the quart strawberry box, but carefully picked currants will carry well in 4 or 10-pound climax baskets with scale board covers.

GOOSEBERRY

The gooseberry of Europe was early brought to this country by the colonists, but, like the grapes which they brought, it was not suited to the new conditions. An acceptable substitute was found in the wild gooseberry of the realm, and from these wild plants, or their seedlings, have developed the most valuable of the sorts adapted to eastern United States. The European sorts have proven better suited to the extreme northwest conditions in the United States and are there cultivated to a limited extent. In general, however, the basis of the commercial gooseberry industry is the American varieties.

The cultural range of the gooseberry coincides in general with that of the currant, but it is able to withstand a slightly higher temperature than the currant and its southern limit of cultivation extends somewhat farther than that of the currant.

Soil.—The gooseberry thrives well on a considerable diversity of soils, but rich, moist, well-drained loams or clays offer the most congenial conditions for the plant. Under a favorable environment the plants should continue in good condition long enough to produce seven to ten crops of fruit, after which the plants will be well spent.

Preparation of Land.—The area to be planted in gooseberries should receive at least one season of preparatory treatment if practicable, before the plants are set. This should consist of a crop which is well tilled and kept free of weeds, or one which, by reason of its density and rank growth, will smother the weeds.

Plants for Setting.—While the gooseberry can be propagated with a fair degree of success from cuttings as well as by layering and mounding, it will, in general, be found best either for the home fruit garden or for the commercial plantation to purchase strong one or two-year-old plants of the desired sort from a reliable nurseryman.

Planting.—As a rule the plants should be set in check rows so as to permit of cultivation in both directions. Satisfactory distances are 6 feet between the rows and 4 to 5 or 6 feet between the plants in the row. Planting can be facilitated by opening a dead furrow along the line of the row and by marking the field in the opposite direction so as to indicate the points in the row where the plants are to stand.

WELL-SET BRANCH OF GOOSEBERRIES.[1]

Cultivation.—Gooseberries form their root system near the surface of the ground. Cultivation should conform to the habits of the plants and be shallow enough not to be injurious to them. The main purpose of cultivation should be to conserve moisture, particularly early in the season while the fruit is forming and ripening.

Fertilizers.—Few tests have been made to determine the fertilizer requirements of the gooseberry. In general well-composted stable manure will prove to be a satisfactory fertilizer. On extensive plantations where fertilizers are evidently required it will be best to inaugurate a simple test to determine the combination and amount best suited to the needs of the particular plantation.

[1] Courtesy of The Pennsylvania Farmer.

Pruning.—The natural habit of the plant is to form a bush. Pruning should therefore be directed to checking the growth of rampant shoots at the proper time and to removing old branches which have served their purpose as bearing wood.

Enemies and Diseases.—The gooseberry suffers as severely from the currant worm as the currant itself and is only a slightly less desirable host plant. Paris green or hellebore should be applied the same as for currants.

The great drawback to the successful cultivation of the European gooseberry in eastern United States is, as has been pointed out, its susceptibility to mildew. This disease is so severe and so difficult to combat that resistant sorts are generally grown, although the mildew can be held in check by thorough applications of Bordeaux mixture or ammoniacal carbonate of copper.

Harvesting.—Gooseberries, because of their habit of growth, can be successfully harvested only by hand-picking. Those intended for pie making, which is one of the chief uses of the fruit, are picked before they have colored and ripened. They are, in other words, picked green, as it is the green fruit that is most prized for pie purposes. The usual receptacle for gooseberries is the one-quart splint box.

The ripe fruit is often seen in the American market. The preferences of the market should be determined in advance and the fruit harvested in the condition demanded, whether it be green or ripe.

THE CRANBERRY

The cranberry is one of the native fruits which has contributed an important product as well as a large share to the aggregate return from small fruits. Its restricted region of cultivation and the peculiar environment required by it place it outside the general list of garden small fruits, and in an exclusive class. The fact that it thrives only in swampy areas in high latitudes and elevations exclude it from this discussion. The general requirements of the crop are discussed in Farmers' Bulletin 176, of the United States Department of Agriculture.

REFERENCES

"The Strawberry in North America." Fletcher.
"Bush Fruits." Card.
"The Grape Culturist." Fuller.
Wisconsin Expt. Station Bulletin 248. "Strawberry Culture in Wisconsin."
Canadian Dept. of Agriculture Bulletins:
210. "Strawberry and Red Raspberry."
222. "Currants and Gooseberries."
Farmers' Bulletins, U. S. Dept. of Agriculture:
643. "Blackberry Culture."
664. "Strawberry Growing in the South."

CHAPTER 35

THE POME FRUITS

BY JOHN P. STEWART, PH.D.
Professor of Experimental Pomology, The Pennsylvania State College

These fruits, which include the apple, pear and quince as the principal members, constitute the most important group of fruits in temperate climates. In the United States, as indicated in the thirteenth census, their combined value during the year preceding this census was $91,659,335, or nearly two-thirds of the total value of all orchard fruits. The latter total was $140,867,000. Among the pome fruits, the apple is by far the most important. Its value in America in the above-named year was $83,231,492, or more than 90.8 per cent of the total for the group. The pear comes second in value with a total production of $7,910,600, or 8.63 per cent of the total for the group, while the quince showed a value of only $517,243, or but little more than one-half of one per cent of the group total.

THE APPLE

Origin.—All the true apples have descended from a wild form in Europe known as *Pyrus malus*. Most of the crab-apples have come from the wild *Pyrus baccata* of Siberia, which is commonly known as the Siberian crab. The Yellow and Red Siberian are probably as close to the original type as any varieties now grown. Most of the so-called crabs now in cultivation are hybrids, and are known botanically as *Pyrus prunifolia*. The Hyslop, Transcendent, Florence, Sweet Russet and Whitney are of this type. They are supposed to be hybrids between the true crabs and true apples. (See Budd and Hansen, *Horticultural Manual*, Vol. 1, pp. 161–62.)

The other source of crabs is the native American form, known as *Pyrus coronaria*, and especially the large western type which has been further distinguished by the name of *Pyrus ioensis*. The fruit of the latter often attains a diameter of two and a half inches or over, and keeps easily until the following summer. It is much like the quince in quality, however, and is suitable only for culinary uses. The principal varieties from this source are the Soulard, Kentucky Mammoth, Mercer and Howard. They are of chief value to the northwest section of the Mississippi Valley and northward. At present the number of apple varieties is very large. In America alone between 1804 and 1904 over 7200 distinct varietal names of apples were published, besides 383 named varieties of crabs. It is needless to say that the great majority are worthless.

Cultural Range.—In eastern America the apple is grown commercially from the plateaus of Georgia and Alabama on the south to Quebec and Nova Scotia on the north and east. On the Pacific Slope it succeeds well from the south-central portion of California to British Columbia. Between these regions it is grown more or less between parallels 33 and 46 degrees north latitude, except where the moisture is insufficient. With proper selection of varieties and care, good home orchards or moderate-sized commercial plantings can be grown successfully over practically all this region. The range of the crabs extends farther north.

Propagation.—Apples are propagated by root or whip grafting in

WELL LOCATED APPLE ORCHARD.[1]

winter on whole or piece roots, by crown grafting in the spring or by budding in late summer or early autumn. There is little or no difference between these methods so far as the growth of the resulting trees in the orchard is concerned.

The seeds to produce the roots used as stocks come largely from France, though some are also produced in Vermont. The former come from the so-called French crab, which is nothing but the wild native apple or *Pyrus malus* of France. The seedlings from them are produced chiefly in the soil of the Kansas River Valley.

In the central northwest these stocks are not sufficiently hardy, and

[1]Courtesy of The Macmillan Company, N. Y. From "How to Choose a Farm," by Hunt.

seedlings of the crab hybrids or of the pure *Pyrus baccata* are much preferred as the root stocks for those sections. Budding or crown-grafting is best when these stocks are used.

Dwarf apple trees are formed by grafting or budding on French Paradise stock, and semi-dwarfs by working on doucin stock. They are much used in Europe, but thus far have found little favor in America.

Location and Soil for the Orchard.—Many orchards are permanently handicapped by unsuitable locations, and many of their defects might easily have been avoided by proper foresight and care. The chief characteristics of the suitable location are good topography, proper soil, a convenient water supply and ready access to market or good shipping points if the orchard is to be commercial. A good topography is one that is moderately rounded or sloping and is enough higher than its immediate surroundings to give good drainage of cold air and water. Too much slope, however, is always objectionable, and a grade of two or three per cent is usually sufficient, especially if some sharper depression is near. The direction of the slope is of little or no importance, except possibly near the northern or southern limits of culture, in which cases the southern or northern slopes, respectively, are generally best.

The suitability of the soil seems to depend largely upon the character of the subsoil, as good orchards occur on all classes of top soils, from dense clays to light sands. A good subsoil is comparatively open and porous for about one to three feet below the surface, and then becomes compact enough to hold the moisture fairly well, but not so well as to give the trees "wet feet." For the first six or eight inches, a loamy soil with a moderate admixture of sand and gravel is usually very good. The so-called ironstone soils, or those derived from many of the red shales or sandstones, are often excellent. The presence of old and productive trees under similar conditions in the neighborhood is also a valuable indication.

Not all these conditions are needed, however. Many good home orchards have been made with some of the conditions less favorable, and their advantages are sufficient to warrant some risk in securing them.

Varieties.—The proper selection of varieties for the location involved is extremely important. Fortunately, much assistance can now be secured from the pomologists, horticulturists and horticultural societies of the various states, and also from the publications of the U. S. Department of Agriculture, such as Bulletin 151 of the Bureau of Plant Industry. Personal preferences and local experience should also be considered, whenever available in reliable form, and the following general advice should be useful.

For the home orchard or local market, a much wider range and greater number of varieties are desirable and generally available than for the commercial orchard. Among the early varieties, named in the order of ripening, the Yellow Transparent, Oldenburg and Wealthy are among the best, and they thrive practically across the continent. They are chiefly valuable for culinary use, and are all early bearers. For dessert use, the Early Harvest, Benoni, Maiden Blush, Gravenstein and Jefferis cover about the same season and are almost as widely adapted, at least for home orchards.

For the later varieties, more attention should be given to the section in which they are to be grown. In the general belt from New England to Ontario and Michigan, the McIntosh, Hubbardston, Northern Spy, Tompkins King, Baldwin, Rhode Island Greening and Roxbury are the leading sorts, although many others are also grown. This is known as the Baldwin belt. The varieties in it and those later are named approximately in order of maturity.

In the next area south, extending roughly from New Jersey and Virginia to Kansas and Oklahoma, the leaders are Grimes, Jonathan, Rome Beauty, Stayman Winesap, York Imperial, Ben Davis or Gano, Black-twig or Paragon, and Winesap. It is known as the Winesap belt. The first two or three varieties used in it are also frequently valuable in the Baldwin belt, and *vice versa*. The Red Astrachan, Primate, Summer Rambo, Fall Pippin, Smokehouse and Delicious also do well in many parts of both of these regions.

Still farther south, from North Carolina to Texas, the White Juneating, Red June, Horse, Kinnard, Buckingham, Terry, Buncombe and Shockley are the principal sorts. In the Colorado-Utah section, the leading varieties are much the same as those in the Winesap belt, with the Summer Pearmain, White Pearmain and Yellow Bellefleur in addition.

In the central northwest, or the general district including Wisconsin, Minnesota, the Dakotas and their immediate surroundings, only the hardiest varieties will succeed. For this district the first three early varieties named above are among the best. Others available are Tetofski, Borovinka, Charlamoff, Alexander, Hibernal, Gideon, Peerless, Okabena, Plumb Cider, Northwestern, Newell and Patten. This is rather a formidable list, both in names and quality, but in the latter respect the Wealthy, Peerless and Patten are best.

For the favorable mountain valleys of western Montana, Idaho, British Columbia, Washington, Oregon and Nevada, the following varieties are prominent in one or more sections: Gravenstein, Fall Pippin, Ortley, McIntosh, Grimes, Jonathan, Banana, Esopus, Wagener, Rome Beauty, Stayman Winesap, Delicious, Winesap and Yellow Newtown.

In California and northward along the coast, the more valuable sorts are the Red Astrachan, Red June, Gravenstein, Fall Pippin, Grimes, Jonathan, Esopus, Tompkins King, White and Blue Pearmains, Wagener, Yellow Bellefleur, Missouri Pippin, Gano, Yellow Newtown and Winesap.

These lists, supplemented with state and local inquiry to fit the immediate places concerned, should enable one to make satisfactory plantings almost anywhere in the apple-growing region of North America.

Purchase and Handling of Nursery Stock.—After deciding upon the varieties, the best way to get the trees is by direct order from responsible nurseries. It is immaterial where the nursery is located, provided the trees it furnishes are true to name, thoroughly healthy, entirely dormant, fully matured before digging, and free from all evidence of faulty storage or improper handling of any kind. The wood should not show any conspicuous blackening at the heart, and the roots should show entire freedom from woolly aphis, crown-gall, hairy root or borers.

One-year-old trees of good medium size are usually best, and in no case should they be older than two years from the bud or graft. One-year trees usually cost less, are more easily shipped and transplanted and their heads can be properly formed, which is not always the case with older trees.

It is well to order early, although the trees may be held at the nursery subject to shipment at planting time. Fall planting is often advisable where the winters are not too severe; otherwise, planting should be done in the spring as soon as the soil is fit. When received the trees should be examined and "heeled in" if practicable, with the dirt packed closely about the roots and the tops sloping toward the south or southwest to reduce the danger of sun-scald. Before planting, the roots should be shortened back to about six or eight inches and those broken or bruised should be removed with a smooth cut above the point of injury.

Laying Out the Orchard.—The orchard may be laid out either on the square or the hexagonal plan. The latter gives about 15½ per cent more trees to the acre at the same distance apart, or 15½ per cent more space for each tree at the same number per acre. The former, however, is rather better for inter-cropping, spraying, etc., and in general is rather more simple to care for.

A good planting distance is 40 by 40 feet for the permanent trees, with a semi-permanent or filler set in the center of the square. In the case of the smaller-growing varieties, the central trees may often remain indefinitely, without disturbing the general plan of the orchard. Where inter-cropping is desired, the permanents may be set at 32 by 48 or thereabouts, and then have the semi-permanents placed in the centers of the long sides, with very satisfactory results. The latter plan allows two more trees to the acre than the square at 40 by 40, or a total of 56 trees, including both fillers and permanents.

The number of trees allowed per acre for any distance in the square or rectangular arrangement may be readily found by determining the number of square feet in the square or rectangle formed by the nearest four trees, and then using this number to divide 43,560, the number of square feet in an acre. To find the number needed in the hexagonal arrangement, find the number allowed by the square plan at the specified distance and then increase this number by 15.47 per cent.

The square or rectangular arrangement can be laid out readily by plowing straight, deep furrows for the rows and then drawing a chain or other drag across them at the distances required for the trees. The hexagonal plan is best laid out by means of a couple of light wires or chains, with one end of each fastened to a single $2\frac{1}{2}$-inch ring and with a similar ring attached to the free end of each. These chains or wires must be exactly equal in length, and they should just reach over stakes set at the distance desired for the trees.

Planting the Trees.—The avoidance of all unnecessary root-exposure and thorough firming of the soil about the roots are the principal secrets of success in tree planting. The soil on the immediate surface, however, should be left rather loose. If the trees or soil are inclined to be dry, the roots may well be soaked in water for several hours before planting, but water is seldom or never needed in the holes themselves. Set the trees from one to three inches deeper than they stood in the nursery.

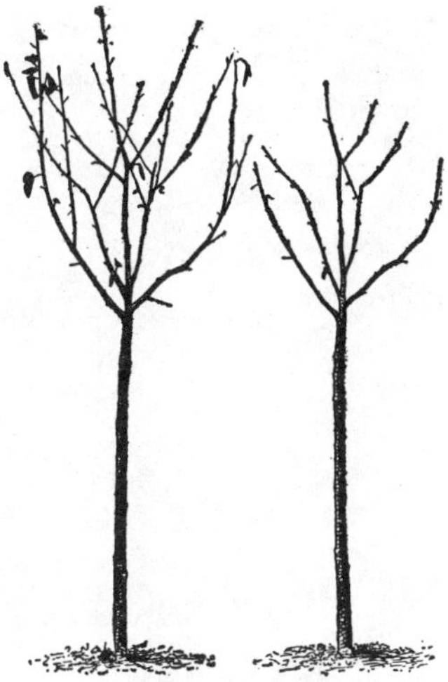

A Properly Pruned Young Apple Tree.[1]

Little or no fertilization is needed at planting as a rule. A good mulch of strawy stable manure, however, will often help greatly. It or any other mulch should be accompanied by proper protection against mice, and a screen of galvanized wire with two meshes to the inch and about eighteen inches square will probably prove most satisfactory for this in the long run.

Forming the Heads.—If one-year "whips" are used, the only pruning needed at planting time is to cut them off at the height of twenty to thirty inches. As soon as possible thereafter, four or five branches should be selected to form the general frame-work. The lowest of these limbs should be about 25 to 30 inches above the ground, as the original height of this limb is the permanent one.

The other three or four frame-work limbs should be selected above, at intervals of six or eight inches, if possible, and with an even distribution around the trunk so as to balance the top properly. This selection is probably best made in the early part of the season's growth, at which time the extra limbs should be rubbed off. In the open-center type of tree, which is preferable for most varieties and localities, the central leader should be eliminated at this time and should be kept from reforming later.

This is usually sufficient for the first season or two, unless some of the limbs get too long or begin greatly out-growing the others; in which case they should be headed back.

At the beginning of the next season some of the frame-work limbs will need heading back to keep the tree in balance and avoid too rangy a growth. Each of the primary limbs should develop two good branches during the season following their selection, and all the others should be rubbed off

[1]Courtesy of The Macmillan Company, N. Y. From "The Principles of Fruit Growing," by Bailey.

early. These branches in turn should produce not more than two branches each for the general frame-work, after which the tree-head may be considered formed.

Later Pruning.—The above work should be completed usually by the middle of the third season. After this the pruning should be reduced as much as possible until the trees come into bearing. A little thinning out in the dense places, removal of the crossing or plainly superfluous limbs, and an occasional heading back of the extra-vigorous branches will be

Apple Orchard Favored by Type of Soil.[1]

sufficient and all that should be given if rapid growth and early fruiting are desired.

The fruit spurs should always be saved and also the early blossoms, unless they become too numerous, in which case the fruit should be thinned. A little fertilizing of the right sort will avoid any possible injury from early fruiting and the early formation of the bearing habit is usually desirable.

In all pruning, make the cuts close to the parent branch and avoid trimming the limbs to poles. Also keep all blighting twigs off of the main limbs, so far as possible, to avoid the formation of the cankers in which the winter is passed by the blight organisms.

Soil Management.—Where tillage is advisable, the most practical

[1] Courtesy of The Macmillan Company, N. Y. From "How to Choose a Farm," by Hunt.

method of orchard development is the use of tilled intercrops followed by a winter cover. Potatoes, corn, vegetables or buckwheat are usually satisfactory for the intercrops, and rye or rye and vetch are good for the winter cover. When buckwheat is used, the rye and vetch combination can be sown at the same time, as it does not grow much until the buckwheat is taken off. In the other cases, the winter cover should be sown after the intercrop is removed, which should not be later than the 15th or 20th of September for best results.

Where the above plan is not desirable, the mulch system is generally best, especially for the home orchard. Strawy stable manure, at the rate of six or eight tons per acre annually, is probably the best mulch, unless the blight becomes too prevalent. Any other kind of vegetation is satisfactory, however, and it should be put on frequently and heavily enough to keep

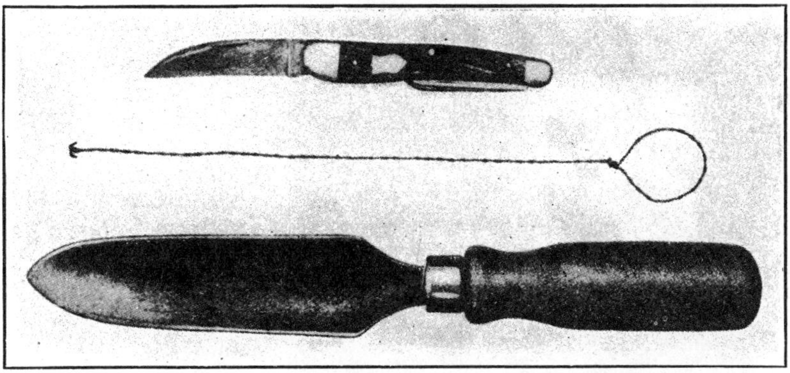

Tools for Use in Removing Roundheaded Apple-Tree Borer from Burrows.[1]

down most of the grass or other growth above the principal root-feeding area. Its chief function is moisture conservation, though it naturally also adds some plant-food as it decays. Any such mulch should extend out at least as far as the tips of the branches, and a clear space of at least six to twelve inches should be maintained immediately around the trees.

In many places all the mulch needed for the first eight or ten years can be grown between the trees by the use of alfalfa or other similar plants. When it begins to fail, manure or other outside resources should be drawn upon.

Fertilization.—The necessity for fertilization is largely a local problem. In general, young trees respond to it much less than those in bearing. In either case one of the safest and best applications that can be made is stable manure. On mature trees it can be applied at the rate of 6 or 8 tons annually per acre, while on young trees it may be reduced to as little as 50 or 100 pounds per tree.

[1]Courtesy of U. S. Dept. of Agriculture, Farm Bulletin 675.

When the manure is not available, similar benefits may often be secured with an application of about 500 pounds per acre of a commercial fertilizer carrying about 5 or 6 per cent of nitrogen, 8 per cent of phosphoric acid (P_2O_5), and 3 or 4 per cent of potash (K_2O). This is enough for bearing trees and it should be supplemented by a mulch or tillage to conserve the moisture. For younger trees, the application should be reduced approximately in proportion to the reduction in area covered.

A Power Sprayer Adapted to Large Trees.
A gasoline engine gives uniformly high pressure.

The area of application should be about the same as that described above for mulches, and the best time for the fertilizer is probably somewhat after the fruit has set. The manure may well be applied any time during late winter or early spring. In any orchard it is always best to leave a few typical trees unfertilized until the actual value of the fertilization is determined.

Protecting the Trees.—The chief enemies of young trees are mice, rabbits, borers and the San José scale. The screens described above are the surest protection against the first two, though poisoned syrup or shot-guns may also be useful.

For the borers, a protective covering, such as the lime-sulphur concentrate more or less mixed with sediment or lime, is often very helpful if renewed frequently enough. It can be readily applied either with a brush or a coarse spray nozzle and the trunk should be thoroughly coated to a height of eighteen inches or more. It should be kept in good shape from about the middle of June to the middle of September, after which the trees should be gone over carefully and any borers that may have entered in spite of the coating should be cut out or killed with a wire.

The scale is readily handled by thorough spraying during the dormant season with lime-sulphur at a density of 1.03 as indicated by the specific gravity hydrometer, or a dilution of about one part of the best commercial concentrates to nine parts of water.

Spraying During the Growing Season.—The materials needed are lime-sulphur, lead arsenate, and nicotine solution. The first is used at a density of about 1.008, which is obtained by diluting good home-made concentrates about 1 to 30 or the best commercial lime-sulphurs at about 1 to 38. The second material is used at the rate of 2 or 2½ pounds of the paste, or half as much of the powder, in 50 gallons of spray. The third is used at the rate of about an ounce of the 40 per cent nicotine preparations to 5 gallons of spray, which gives a diluted strength of about .05 of one per cent of nicotine. All these materials may be combined in a single application, or they may be applied separately as the case demands.

A GOOD CLUSTER OF APPLES, BUT WITH SOME SCAB SHOWING.

A little more thinning and spraying would have been useful here.

With these materials, the principal pests are controlled as follows:

1. If both scale and plant lice are present, defer the scale application mentioned above *until the first green begins to appear in the buds*, and add the nicotine solution above named for the aphids or plant lice, which are then just hatched.

2. Spray with the sulphur and arsenate *when the blossoms are just showing pink, or slightly before.* This is for scab, canker worms or the bud-moth. Also valuable against aphids and red-bugs if the nicotine solution is added.

3. *Repeat No. 2 immediately after the petals fall, to fill the calyx cups.*

This is the most important single spray. It is to control the apple worm, scab, curculio and the later species of red-bug if present.

4. *Repeat No. 3 in about two or three weeks.* This is for the same enemies as in No. 3 and is also useful against the apple maggot if present.

5, 6 and 7. In orchards infested by bitter-rot or apple-blotch, make

PICKING AND PACKING APPLES.[1]

three applications, preferably with Bordeaux mixture (3–3–50), at intervals of about three weeks, beginning eight or nine weeks after petals fall.

8. In the absence of sprays 5, 6 and 7, and where the second brood of apple worms, late scab or late summer caterpillars is bad, *repeat No. 2 about August 1st,* or somewhat earlier in the southern sections, depending upon the time of emergence of the codling moth. With the third and fourth

[1] Courtesy of Penn State Farmer, State College, Pa.

applications well made, this one is rarely needed, although much depends on the locality and season.

This schedule of sprays is all that is needed in the worst infested orchards, and it is seldom that more than those numbered 1, 3 and 4 need be given.

Thinning.—Whenever the crop on a tree is too large for normal maturity, it should be thinned. This should be done as soon as the June drop is largely over, or when the fruit has become about an inch in diameter. All defective fruit should be removed first and then the remainder thinned to a distance of at least four or five inches between fruits, unless they are on opposite sides of the limb and the limb as a whole is not well loaded. Grape shears or similar implements are sometimes used for this, but with a litte practice and by using the proper twist the work can probably be done faster without them and with as little damage to fruit spurs and fruit.

Fruit Picking and Storage.—The highest color and best eating quality in apples are generally secured by letting them ripen on the tree as far as possible. This can be done with the early apples and especially with those to be used at home. Too much ripening, however, interferes seriously with long keeping, and hence with the later varieties the best time for picking is when they are "hard ripe," *i. e.*, when they have reached their full size and redness, but have not yet begun to soften nor to show the yellow colors, except possibly in occasional specimens. In many cases two or more pickings are desirable to permit the immature fruits to develop.

Other ways of improving the keeping quality are to avoid bruises and broken skins and to transfer the fruit at once to cool conditions. Leaving the apples in piles in the sun is exceedingly hard on keeping quality. In storage the best temperatures range from about 30° to 35° F., though a range up to 40° or 45° F. usually does little or no damage. Well insulated cellars or storage rooms fitted with a good system of ventilation, which can be opened at night and closed in the daytime, are likely to be very serviceable except possibly in the southern third of the apple region. In that territory it may be necessary, to make use of commercial storage, at least temporarily, in order to insure satisfactory keeping of the fruit.

THE PEAR

Origin.—Practically all the present varieties of pears have come directly from the wild *Pyrus communis* of Europe and Asia. This fruit has been grown probably as long as the apple, but it was not until the great work of Von Mons of Belgium, in the early part of the nineteenth century, that any important dessert varieties were produced. The better varieties are now among the most delicious of fruits.

The other ancestor involved in a few of the commercial sorts is the Japanese or sand pear (*Pyrus sinensis*) of Asia. It is of no value in itself except for hardiness or ornamental use. Crosses between it and *communis*, however, have resulted in the hardy hybrids, of which the Kieffer, LeConte and Garber are most important. Although low in quality, they are usually very productive and are much used for canning. About 2300 names of pear varieties have appeared in American publications between 1804 and 1907. (See Bulletin 126, Bureau of Plant Industry.)

Propagation.—The pear is propagated in the same ways as the apple, but the stocks are different. In stocks, the pear has a very wide range of affinities. Those chiefly used are the seedlings of the small Snow pear (*Pyrus nivalis*) of Europe. This stock results in trees of the normal size.

Pear Tree in Blossom.

Note unusually spreading form of this tree. This is desirable, although difficult to secure in many varieties.

Dwarf pear trees are produced by budding on to quince stocks. The latter are secured from the Angers quince by mound layering. Some varieties, and particularly the Angouleme, are much benefited by this process. Such varieties are often planted with the stock below the surface to reduce injury from the round-headed borer, which attacks the quince, but rarely the pears. Some varieties do not unite well with the quince and they are "double-worked" by first using one that does. The Japanese quince is not satisfactory as a stock for any variety.

In the South, cuttings of the Kieffer pear are used to some extent, and in the more rigorous sections, the mountain ash (*Sorbus*) and even the shad bush (*Amelanchier*) have been used as stocks with fair success.

Cultural Range.—The pear resists cold about as well as the apple. Its great susceptibility to fire blight (*Bacillus amylovorus*), however, greatly restricts its profitable growth. In general, it does best in the Baldwin and Winesap belts in the East, and in the general territory from Colorado west and northward to the Pacific Coast.

Varieties.—The leading varieties in the eastern section, named approximately in order of ripening, are: Clapp Favorite, Bartlett, Seckel, Sheldon, Anjou, Angouleme, Kieffer, Lawrence and Winter Nelis. In the West, the same varieties are used, excepting the Kieffer, and with the addition of Flemish, Comice and Easter Beurre. In the South the three hybrids mentioned above are about all that have shown any profit. Very few varieties will succeed on the rich soils of the Mississippi Valley, but the LaMotte, Seckel, Dwarf Angouleme and Kieffer are most likely to succeed. It is always best to use more than one variety and to mix them somewhat in the planting, in order to insure satisfactory pollination. Further advice can

GOOD SPECIMENS OF WINTER NELIS.
The fruit, however, appears more nearly round here than it really is.

be secured from local and state sources, and from Farmers' Bulletin 208 of the U. S. Department of Agriculture.

Location, Soil and Culture.—Since blight is its worst enemy, the pear orchard should be located where the trees will not grow too rapidly. A fairly high and airy situation, with a well-drained and moderately fertile, clay or clay-loam soil, is therefore most desirable. For the same reason the amount of tillage and fertilization should be kept low or be eliminated entirely if the blighting becomes severe. The mulching method is often used with especial success on the pear, if not applied too heavily.

Trees, Planting and Pruning.—These are largely the same as described above for the apple. Pears, however, are planted closer. A distance of

20 by 20 feet is about right for the standard-sized trees, and 12 by 12 feet for the dwarfs. In forming the tops, it is customary to leave a central leader with most varieties of pears. Severe attacks of blight, however, are likely to be more serious in such trees than in those with three or more leaders, as in the open-centered tree. It is also important to keep all fruit spurs and sappy sprouts off the main branches, and to avoid any large amount of pruning at any one time. Special promptness is needed in pruning out and disinfecting blighted twigs whenever they appear.

Protection and Spraying.—The same general plans as stated for the apple will take care of the pear. The scab is especially bad on some varieties, *e. g.*, Flemish, but it can be readily controlled by lime-sulphur or Bordeaux mixture (4–4–50) applied just when the blossoms are showing pink. Most apple insects are less serious on the pear, but it has another important enemy in the pear psylla.

This insect attacks the buds and young leaves, sucking out the sap and blackening and often killing them. It is a very tiny insect, and when magnified looks like a minute cicada. It can be controlled by thorough spraying early in the spring, with nicotine solution and soap, at the rate of an ounce of the former and three or four ounces of the latter to five gallons of water. The rough bark should be scraped away before making this application. Lime-sulphur, at winter strength, just as the buds are swelling, is also effective in killing the eggs.

Picking the Fruit.—The pear is one of the few fruits that are improved by ripening off the tree. Both the grittiness and softening at the core are much reduced by this process. The "hard ripe" stage described for the apple is therefore especially applicable in picking the pear. After picking it should be stored at once in a cool place, free from drafts and preferably dark, to avoid wilting and bring out the full flavor of the fruit.

THE QUINCE

The quince has come down from the wild *Cydonia vulgaris* of Asia. It is still closer to the original type than any other orchard fruit. It is practically inedible raw, but has been used for at least 2000 years in marmalades and jellies. It is also used largely in preserves, canning and in flavoring other fruit products. It is especially adapted to home planting, as it is grown very little in a commercial way.

Cultural Range and Varieties.—The quince is less resistant to cold than the pear and is about equally susceptible to blight. Hence it is available in the less severe portions of the pear range.

In general, the leading variety is the Orange. The Champion is probably next, with the Rea, Missouri and Meech often useful. In the South the Chinese does best and in California the Pear is preferred.

Soil and Cultural Methods.—A heavy, retentive clay loam, with good drainage of both air and surplus water, is apparently best for the quince. Two-year-old trees are probably best, and they are set from twelve to fifteen feet apart. The quince is very shallow rooted; hence all deep tillage must be avoided, and winter covers are always desirable. The general method of culture and fertilization suggested for the pear are also advisable for the quince.

Pruning.—Quince blossoms and fruit are produced on the ends of twigs of the current season's growth. These twigs are developed largely from the terminal buds of spurs or branches, or from buds near the tips of the latter; hence too much cutting back may readily remove all the fruit-bearing wood of that season. The pruning of the quince, therefore, should be confined largely to the removal of dead or inferior wood, thinning out the dense places and heading back the extra vigorous shoots, to promote the development of fruit spurs and keep the trees in balance generally. Blighting twigs also need as prompt attention here as in the case of the pear.

Enemies.—The worst insect enemies of the quince are the quince curculio and the round-headed apple-tree borer. The latter can be controlled as described for the apple, although more attention is likely to be needed. The former is the chief cause of the "wormy" and knotty fruits. It is very difficult to control, but the best methods are: (1) thorough spraying with lead arsenate, at the rate of one ounce of the paste to a gallon of spray, when the first injury appears and again a week later; and (2) picking and destruction of all infested fruit about a month before the normal picking time.

The chief diseases are fire blight, leaf blight (*Entomosporium maculatum*) and rust (*Gymnosporangium germinale*). The first is controlled as in the pear, the second by spraying as for apple scab and the third by removal of all red cedars, or at least all diseased specimens, for a distance of at least one-half mile of the quince trees.

REFERENCES

"Productive Orcharding." Sears.
"Horticultural Manual," 2 vols. Budd & Hansen.
"The Pruning Book." Bailey.
"Systematic Pomology." Waugh.
Pennsylvania Agric. Expt. Station Bulletins 115, 121, 128 and 141.
New York (Geneva) Expt. Station Bulletin 406. "Dwarf Apples."
Canadian Dept. of Agriculture Bulletins:
176. "Bacterial Blight of Apple, Pear and Quince Trees."
194. "Apple Orcharding in Ontario."
Farmers' Bulletin, U. S. Dept. of Agriculture, 401. "Protection of Orchards from Spring Frosts by Means of Fires."

CHAPTER 36

STONE FRUITS

BY F. C. SEARS
Professor of Pomology, Massachusetts Agricultural College

Sites and Soils.—As a class the stone fruits do best on relatively high lands, principally on account of the effects of elevation on temperature. When peaches are injured by cold it is usually either by extremely low temperature in the winter or by frosts during the blossoming period. With cherries and plums the damage almost always comes at blossoming time. As all of these fruits bloom early, they are particularly liable to frost injury and it becomes necessary to guard against it.

Both these types of injury can be lessened and often largely prevented by placing the orchards on sites which are higher than the surrounding lands, thus allowing the cold air to drain away onto the lower levels. Occasionally the lower sites bring their crops through in better shape than the high ones. An instance of this kind was the winter of 1913–14 when extremely low temperatures were accompanied by very high winds. This combination did much more damage to orchards on high lands than to those on low lands. But on the average higher sites are much to be preferred.

As to the direction of the slope, two points are worth considering. A northerly slope retards the blossoming and so helps to escape spring injury. But, as just suggested, it may increase the danger from severe winter temperatures. Consequently, if one is in a section where the former type of injury is most likely to occur a northerly slope is to be preferred. But if the damage is generally done by low winter temperatures, a southerly slope is best.

For soils the stone fruits are not very exacting. Peaches prefer relatively light soils, but will do well in almost any soil up to a moderately heavy clay loam. Plums and cherries, especially the former, do best on medium to fairly heavy soils, heavy sandy loams to medium clay loams. Good soil drainage is absolutely essential.

Nursery Stock.—Medium grades of nursery stock of the stone fruits, from four to six feet tall, are to be preferred. This is especially important in peaches, for these are always set at one year old and where one wants to head them at all low and start a new top, the very heavy trees do not give as good results. Plums and cherries may be set at either one or two years from the bud. Where the soil is fertile and has been well prepared, one-year-old trees are to be preferred, particularly if one wants to head them low. But if the soil conditions are not good, then two-year trees are to be preferred, as the one-year trees will not usually form as good heads under poor soil conditions. Locally grown trees are always to be preferred if one can get good stock. They come fresher, the freight is less and it is easier to adjust differences with the nurseryman. On the other hand, there is probably nothing in the idea that either northern grown or southern grown stock is to be preferred. Southern stock will do just as well in the North, or northern grown in the South, if it can be landed at the orchard in good condition.

Varieties.—The variety question is always important and always difficult to decide. It can generally be decided best by referring to local authorities, but a few general considerations are worth keeping in mind.

A Typical Peach-Orchard Site, Allegheny Mountain District, Morgan County, W. Va.[1]

Good air drainage and proper exposure are important.

With peaches the important considerations are color of flesh, color of skin, quality, juiciness, whether they are clingstone or free, hardiness of fruit buds and season of ripening. There is a very strong prejudice (it is nothing more) in favor of yellow-fleshed peaches, especially for canning. It is best to respect this prejudice if possible, but many varieties which are leaders in all other respects have white flesh. It is often possible to educate consumers locally on this matter of color and convince them that in many cases the white varieties are to be preferred, but in the general market one is almost certain to be most successful with yellow sorts. Quality and juiciness are always desirable, though from the commercial standpoint the latter can be overdone, as very juicy peaches do not ship as well. Clingstones are never as popular, but some of the best commercial sorts among the early varieties are clings or semi-clings. Hardiness in the fruit

[1] From Farmers' Bulletin 431, U. S. Dept. of Agriculture.

bud is all-important in sections subject to low winter temperatures, and there is a very marked difference in this respect among different varieties. Greensboro, for example, will come through with a full crop when the fruit buds of Elberta will be largely killed and those of Crawford entirely so. As a class, the so-called Chinese Cling group, which includes such varieties as Greensboro, Carman, Belle of Georgia and Elberta, has much more hardy fruit buds than the Persian group, which includes such varieties as Early and Late Crawford and Old Mixon. As to season of ripening, it is well, of course, to have somewhat of a succession, particularly for local trade, but the very early and still more the late sorts are likely to be more profitable than mid-season varieties.

TYPICAL SWEET CHERRIES.

With plums one should consider the quality, the uses (whether for jelly, canning, preserving or eating in a fresh condition), the size and the color.

With cherries the all-important question is whether to grow the sweet varieties or the so-called sour cherries. As a class, the sour cherries are much more generally successful. In addition, there are the questions of size, quality and color.

With any of these fruits the number of varieties set must depend very largely on whether they are to go to local or distant markets. If the latter, then it is very important to restrict the number of varieties sufficiently to allow of shipping in car lots.

While realizing, as already suggested, that the variety question is very strongly local, the following lists may be helpful, including as they do those varieties which are most generally successful and popular:

PEACHES

Early Crawford	Champion	Belle of Georgia
Greensboro	Hieley	Elberta
Carman		

PLUMS

Burbank	Wild Goose	Bradshaw
Abundance	Reine Claude	Shropshire Damson
Lombard	Red June	

CHERRIES

(Sour)

Early Richmond	Montmorency	English Morello

(Sweet)

Black Tartarian	Windsor	Schmidt's Bigarreau
Yellow Spanish	Bing	Gov. Wood
Napoleon		

Planting.—Spring planting will be found most generally successful, particularly in those sections subject to variable winter climates. On the other hand, where soil conditions are ideal (soil well prepared and well drained both on the surface and beneath), planting in the autumn will give excellent results. A serious difficulty, of course, is getting the nursery stock in time to plant in the autumn and still have it well matured before it is dug by the nurseryman.

Peach trees are set all the way from 13 to 20 feet apart. A good average distance is 18 feet. The type of land and the variety will determine the best distance.

Plums can, on the average, be set closer than peaches, because they are more upright growing trees, but such sprawling growing varieties as Burbank will need fully as much room as any peach.

Sour cherries average about the same as plums, and sweet cherries somewhat larger.

In pruning the trees at setting there are two general methods used: In the case of one-year trees, they are simply cut back to the desired height, which varies with different growers from 6 to 30 inches, on the average perhaps 24 inches. With two-year trees, the head being already established, it is necessary to cut back the main branches rather severely. Generally from one-third to two-thirds of the last year's wood is removed.

Soil Management.—Stone fruits rarely succeed well in sod. Peaches practically never do, and cherries very seldom. Plums *can* be grown in sod, but are usually much better under cultivation. There is much less chance for discussion as to the relative merits of sod and cultivation in the stone fruits than with apples and pears. An additional reason for cultivating peaches is the fact that borers are much more troublesome where grass and weeds are left about the trees.

If the orchard is to be cultivated, the season's campaign would be to plow it as early in the spring as possible, and for this work nothing is more satisfactory than one of the "orchard gang" plows (usually a gang of three small plows). After plowing the soil is "fitted" in good shape with the disk and other harrows, and then is kept in good condition with some type of cultivator up to about July 1st. The essential points of such a cultivator are that it shall cover enough ground to do the work cheaply, that it shall leave the soil in good condition and that it shall work well under the trees without necessitating that the team get close to them. This latter point is particularly important with stone fruits, since they are almost universally headed very low. About July 1st the cover crop is sown in the orchard and the season's work on the soil is finished. The date of sowing this crop varies greatly with different growers and under different conditions. Where

BLOCK OF YOUNG PEACH TREES WITH STRAWBERRIES AS AN INTER-CROP.

trees are carrying a large crop of fruit and the soil and season are dry, cultivation may profitably be kept up considerably later in order to conserve the moisture, but it must always be borne in mind that moisture is saved in this way one season at the expense of the next season, because the longer the sowing of the cover crop is delayed the less growth it will make, and consequently the less humus it will add to the soil the following year. The chief functions of this cover crop are to prevent washing (and this is especially important in peach orchards, since they are usually on high and rolling lands), to check the growth of the trees in autumn and to add humus to the soil. If the cover crop is a legume, a large part of the required nitrogen may be secured. One of the best crops for this purpose is barley. Another is dwarf rape. Either may be combined with one of the clovers to advantage. Vetch is an ideal crop where the seed can be secured at a reasonable price. Some growers raise their own seed, sowing winter vetch with rye and cutting and threshing the combination the following season. One bushel of rye and a peck of winter vetch makes a good combination for this purpose. In this connection, it is very desirable to get all the humus possible into the soil before the orchard is set, since it is frequently difficult to get as much growth as desired from the cover crop and consequently the supply of humus in the soil soon runs low.

Peach Tree with Well-Formed Framework, Heavily Cut Back for Renewal Purposes.[1]

Fertilizers.—The fertilizer needs of stone fruits, as with other fruits, have not been worked out as fully as could be desired, yet it has been pretty well shown that reasonably liberal fertilizing is profitable. Practically all commercial peach growers fertilize their orchards and most of them very liberally. Plums and cherries are probably fertilized less freely on the average than peaches, largely perhaps because size with them is less important. There must be enough nitrogen added in some form so that, together with what can be gained through cover crops, the trees will be induced to make a good medium, well-ripened yearly growth. Peach trees ought to make from one to two feet on the leaders and plums about the same. Sweet cherries will stand perhaps a little more and sour cherries less. The foliage ought also to be kept in good vigorous condition. To accomplish this will require varying amounts of fertilizer and the orchard man must use his judgment as to what is required.

The following are formulas which are used by good growers, but even in different parts of the same orchard, and certainly in different

[1] Courtesy of Dept. of Experimental Pomology, Pennsylvania Experiment Station.

years, the applications may need to be varied. The formulas given are per acre:

1. 250 pounds high-grade sulphate of potash.
   400–600 pounds basic slag.
   Nitrate of soda as needed to produce proper growth—Usually 100–200 pounds per acre.
2. 100 pounds nitrate of soda.
   100 pounds dried blood.
   350 pounds slag.
   100–200 pounds high-grade sulphate of potash.
3. 25– 50 pounds dried blood.
   40– 80 pounds tankage.
   90–180 pounds bone meal.
   130–260 pounds basic slag.
   80–160 pounds high-grade sulphate of potash.

This is a more complicated formula than the others, but is used by a very successful grower.

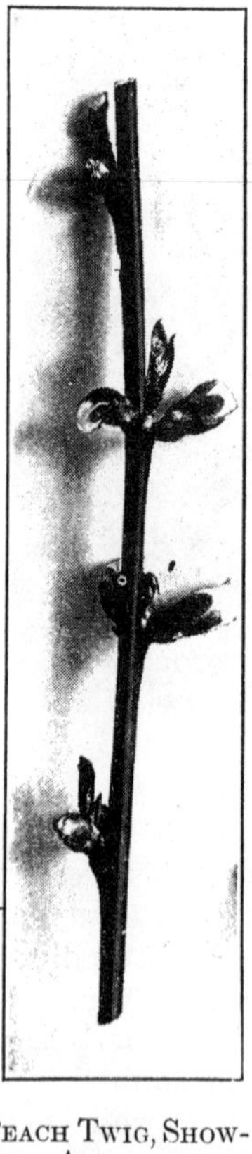

Peach Twig, Showing Arrangement of Leaf and Blossom Buds.

Pruning.—The most intelligent pruning of any kind of fruit tree requires that one should understand thoroughly the manner in which the fruit is borne by that tree. This is perhaps more emphatically true of the peach than of any other fruit, but is certainly a safe general principle. We will therefore consider this point first.

The peach bears only on last season's wood, the buds occurring normally in clusters of three on such shoots, the center one being a leaf bud and the two outside ones fruit buds. Shoots of medium size give the best results. If, for any reason, a peach tree makes a very rank growth it will be found that fewer fruit buds are produced on such wood and they are apt to be less hardy. In seasons when a large part of the fruit buds are killed by severe cold it almost always happens that the few buds which come through safely are on the smaller branches. The pruning of the peach, therefore, ought, first of all, to aim at keeping up a supply of new wood, and, except when one is trying to grow a new top on the tree, it should never be so severe as to give a very rank growth.

The following will be found a fairly satisfactory outline for the pruning of a bearing peach tree:

1. Do not allow the pruning of the tree as a whole to be severe enough to start a very strong wood growth.

2. Take out altogether any very high and very strong leaders. This is necessary because the fruiting wood tends to get very high if these leaders are allowed to remain. Less rank leaders may be headed back less severly or allowed to remain entirely.

3. Take out all dead or injured branches. It is sometimes a question whether one can afford time to take out all of the many small dead branches which are always to be found in the center of the tree, but as many of them as possible should be removed.

4. Thin the balance of the top as needed, taking out preferably no branches larger than one's thumb. The amount of this pruning is going to depend, of course, on how much has been taken out in other ways and on the type of tree. The amount of pruning should be varied somewhat according to the outlook for a crop that season. If the fruit buds are all killed it is a good opportunity to cut back rather severely and lower the tree down if necessary. If part of the buds are killed, it may be best to prune very lightly in order to save as much of the crop as possible. On the other hand, if there are plenty of live fruit buds the pruning may be fairly severe, as this helps to thin out the fruit.

A PROPERLY PRUNED PEACH TREE.[1]

Plums and cherries bear essentially alike, the fruit being produced on short lateral spurs and small twigs, and also to a considerable extent (especially with the sour cherries and the Japanese plums) on the last year's wood as with peaches. These spurs bear for several years, perhaps three to six, and then die away and need to be replaced by new wood. The pruning of such trees therefore should be moderate and should aim to keep the trees fairly open to encourage new growth. The following outline may serve as a guide for most trees of these two fruits:

1. They require relatively little pruning.

2. Cut back leaders if too high. This is especially important with cherries, since the picking of high trees is more expensive than with any other fruit.

3. Cut out dead, broken and diseased branches. This is particularly important with plums which are often badly attacked by the black knot.

[1] From Farmers' Bulletin 632, U. S. Dept. of Agriculture.

4. Take out crossing branches.

5. Thin the balance of the top slightly.

The following outline may be taken as reasonably accurate for pruning young trees of stone fruits—say trees two to four years old:

1. Examine critically the head of the tree. It should have three to six main branches and no sharp forks.

2. Shorten leaders that are running too high. Only very high leaders that throw the tree out of shape, or such as have made an exceptionally long growth the past season, need to be cut back.

3. Cut out bad (sharp) forks on all main branches.

4. Save all small shoots.

5. Take out only very large crossing branches..

6. Prune strong-growing trees less and weak-growing ones more.

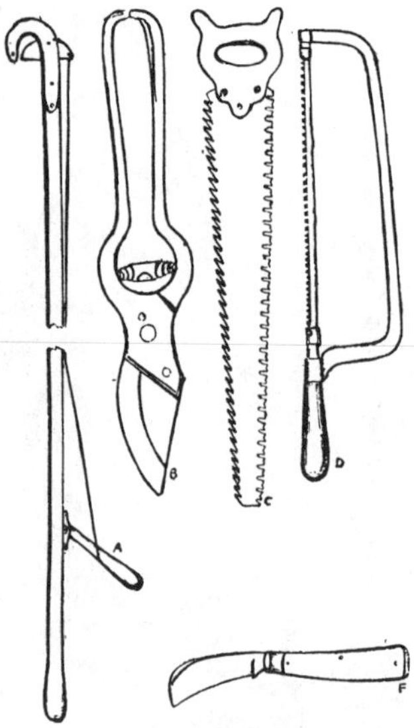

PRUNING TOOLS.

A—Waters' Tree-Pruner.
B—Pruning Shears.
C—Two-edged Pruning Saw.
D—California Pruning Saw.
E—Pruning Knife.

In pruning these fruits, especially the peach, a large pair of hand shears will be found most satisfactory. A ten-inch pair of the French wheel-spring shears will be found equal to almost any emergency, and much of the work can be done more rapidly with shears than with a saw. The operator will need a saw, however, for the heavier work and one of the following dimensions will be found very satisfactory:

Length	20	inches
Width at butt	$2\frac{3}{4}$	"
Width at point	$\frac{3}{4}$	"
Seven and one-half teeth per inch.		

Such saws may have to be made to order. Any hardware manufacturer will make them and they should always be of the best steel.

Probably the ideal time to prune these fruits is about a month or six weeks before they start into growth. But where one has much pruning to do, it is often necessary to greatly extend the time. It is largely a question of the economical use of farm labor. There are usually few expert pruners on the farm in comparison to the pruning to be done and it becomes necessary to keep these men at work over a relatively long period.

Diseases, Insects and Spraying.—Since the matter of diseases and insects has been treated fully in the general chapters on these subjects, it is necessary here only to give a very brief summary of the subject.

Among fungous diseases, the following are deserving of special consideration:

1. The brown rot which attacks all of the stone fruits and is to be dreaded far more than anything else. It attacks not only the fruit but the twigs as well, spreading to the latter from the former, and hence diseased fruits should be removed from the tree as soon as possible. It can be controlled largely by spraying.

2. Peach leaf curl, often serious but thorough spraying before the buds swell will practically eradicate it.

3. Black knot of plums and cherries. Often very serious but can be controlled by spraying and by cutting out and destroying the knots.

4. Peach scab. Often a troublesome disease, sometimes seriously so, but thorough spraying will usually control it, even in the worst seasons.

Among insects three are worth mentioning:

1. The plum curculio, which attacks both plums and peaches and is often a very serious menace, not only for its own attacks, but because it helps the spread of brown rot.

2. The peach borer, an ever-present pest where peaches are grown at all extensively. Digging out is the most commonly accepted method of combating.

3. The cherry aphis, often a serious pest and, like all aphids, difficult of control.

There are a number of other pests in both classes that are sometimes troublesome, occasionally very seriously so, but the above mentioned are the real standbys.

Thinning the Fruit.—A prerequisite to harvesting a satisfactory crop is thinning the fruit. Nothing is simpler to do and few things connected with fruit growing are more important. Cherries are not thinned, but peaches and plums ought always to be. The best time to do this is after the "June drop" has occurred, that is, after all the fruits which will fall "naturally" have fallen. The fruits will then be about the size of the first joint of one's thumb, and a safe rule, and one easily followed by those doing the thinning, is to thin so that no two fruits touch. In practice this works out so as to bring the fruits a good distance apart and the operator does not spend any time in wondering whether he ought to take off another fruit in order to bring them the required distance apart.

Thinning will help the crop wonderfully in several ways. Probably the most important is that it gets rid of all the small, defective fruits, leaving a crop which it is an inspiration to pick and a pleasure to sell. The work of sorting is reduced to a minimum because there are really very few poor fruits left. Moreover, one gets almost as much fruit in the aggregate, sometimes quite as much. The trees, too, will bear more regularly because they are relieved of the burden of maturing these extra fruits. And lastly it reduces greatly the loss from brown rot, because the rot can spread from one fruit to another where they are touching, and moreover an outbreak of it frequently starts where the moisture is held between the fruits at their points of contact. It requires a good deal of "hustle" to make a good "thinner," but boys who have that requisite will thin fully as well as, and more cheaply then, men.

Go over the trees systematically. Take off all defective fruits whether they touch or not. Don't be afraid of the cost. It will be paid back many times over in the better fruit and is really a small item. Peach trees that will bear four or five baskets can be thinned for not over three or four cents each. The writer has had this done in his own orchard.

Harvesting and Marketing.—To begin with, one must decide on the proper degree of ripeness. This is going to vary greatly with varieties and distance to market. Let the fruit get as nearly ripe as possible and still stand up well in transit, for stone fruits are never so good as when

PICKING PEACHES.[1]

allowed to ripen fully on the trees. Peaches ought to be picked for local markets as soon as they show signs of ripening on the shady side, that is when they begin to look edible. A little practice will soon teach one. Plums can be somewhat soft before picking, while cherries are picked just before they are fully ripe. Color and taste (of a few samples) should be the guide. All the above are for local markets. The more distant the market the greener the fruit must be when picked.

Have convenient receptacles into which to pick. For plums and peaches the ordinary round Delaware peach basket holding sixteen quarts is good. A strap with a hook at each end can be thrown over the shoulders and hooked into the rim of the basket so that it will hang just in front of the picker, leaving both hands free to pick. Cherries are often picked in the same way or may be picked directly into quart baskets if they are to be sold that way.

[1] Courtesy of Department of Experimental Pomology, Pennsylvania Experiment Station.

Do not allow the pickers to bruise the fruit in handling. This is a very important rule and one difficult to enforce. In handling the larger fruits like peaches and large plums, take the fruit in the hollow of the hand and grip it firmly with the entire hand. Never take it between the thumb and finger. With plums and cherries always have the stem attached. This means that the stem must be gripped by the finger and thumb.

Never pick these fruits when wet. This rule has very few exceptions. Fruit which is picked while wet looks badly and keeps worse. Brown rot is almost certain to develop in it.

So much for picking. Next for packing. Have a convenient packing room. If possible have the fruit brought in on one side, packed in the middle and delivered for marketing on the other side. There is then less confusion. Have a table for the packers and seats if they want them. They can work just as fast sitting down. See that the sorting is done rigidly. Nothing discourages customers like finding a few poor number two peaches in the middle of a basket of firsts. Be extremely careful that the best fruits do not gravitate to the top of the baskets. It is probably legitimate to turn the blush side up on the face, but this is as far as it is wise (not to mention honest) to go in facing.

Plums and peaches are sold for the wholesale market in the round Delaware basket of various sizes, and, for a more select trade, in the six-basket Georgia carrier or crate. The latter will not pay for cheaper grades on account of the greater cost of packing. To a limited extent these fruits are also sold in the Climax baskets. For strictly local trade both these fruits may be sold in the little baskets of the Delaware type with wire bails, holding two and five quarts.

Cherries are most commonly sold in strawberry baskets and crates, also in Climax baskets and for the large and finer sorts in boxes or cartons.

The desirability of roadside marketing where there is any great travel past the orchard should not be overlooked. The stone fruits lend themselves especially well to this type of traffic and one who has never tried it will be agreeably surprised at the amount of fruit which can be turned into cash in this way. Moreover, it offers an outlet for the over-ripe, soft grades which would not stand transit to market.

REFERENCES

"Plums and Plum Culture." Waugh.
Virginia Expt. Station Bulletin (Extension) 1. "Peaching Growing in Virginia."
New Jersey Expt. Station Bulletin 284. "Packing and Shipping Peaches."
Canadian Dept. of Agriculture Bulletins:
- 201. "Peach Growing, Peach Diseases."
- 226. "Plum Culture in Ontario."
- 230. "Cherry Growing."

Farmers' Bulletins, U. S. Dept. of Agriculture:
- 426. "Canning Peaches on the Farm."
- 440. "Spraying Peaches for Brown Rot, Scab and Curculio."
- 631, 632, 634. "Growing Peaches."

CHAPTER 37

CITRUS FRUITS AND THEIR CULTIVATION

BY HERBERT J. WEBBER, PH.D.
Dean of Graduate School of Tropical Agriculture, University of California

History.—The various species of the genus *Citrus* are natives of India and the Malay Archipelago. The date of the importation of citrus fruits to America is not known. They were apparently introduced into Brazil and the West Indies at a very early date, probably some time in the latter part of the sixteenth century. They were brought by the Spaniards to Florida at a comparatively early date and were apparently spread by the Indians over the state.

The commercial cultivation of oranges in Florida began in the early part of the nineteenth century; while in California the commercial planting cannot be considered to have started much prior to 1880. The first carload of oranges was shipped from California to St. Louis by William Wolfskill in 1877.

Citrus Species and Varieties.—The genus *Citrus* belongs to the family *Rutaceæ,* which is represented in the United States by the prickly ash *Xanthoxylum),* the hop-tree (*Ptelea*) and the like. The representatives of the family are mainly natives of tropical and sub-tropical countries. Following are the principal species and varieties cultivated in the United States.

The Sweet Orange (*Citrus sinensis*).—This is the species most generally cultivated throughout the world, and is the fruit commonly referred to as the orange. It has given rise to numerous cultivated varieties and exhibits a very wide range of variation in form, size, flavor, season of maturity and the like. Certain varieties have had marked influence in building up the industry in different sections.

This is particularly true of the Washington Navel in California. This variety originated in Brazil about 1820 near Bahia. It gradually became known for its good quality and seedlessness, and in 1870 twelve budded trees were imported into the United States by William Saunders of the United States Department of Agriculture. Other trees were propagated from these and sent to various of the orange-growing states for trial. The majority of these trials were apparently failures or attracted no notice. Two trees, however, were sent to Mrs. Luther C. Tibbet, at Riverside, Cal., in 1873, and were carefully cared for by her until they came into bearing. The stock of this variety in the world has been mainly taken from descendants of the Tibbet trees. It has been sent from California to Australia and South Africa, where it has become an important variety.

The next most generally grown orange in California is the Valencia, a late-maturing variety that can be held on the trees until July and August in interior sections of the state and until October or November in cool sections near the coast. This variety is also grown extensively in Florida as a late-maturing sort, but requires to be shipped considerably earlier than in California.

The orange plantings in California are made up largely of Washington Navels and Valencias with a few trees here and there of other varieties, such as Mediterranean Sweet, St. Michael, Blood, Joppa, Nugget, Ruby, etc.

EVER-BEARING ORANGE TREE.

In Florida a much larger number of varieties are grown, no two standing out as prominently as do the Navel and Valencia in California. The following are the leading sorts in their class in Florida, though other sorts are almost as extensively grown: Early varieties—Parson Brown, Boone, Early Oblong; mid-season varieties—Pineapple, Homosassa, Jaffa, Majorca, St. Michael, Ruby, Maltese; late varieties—Valencia, Bessie and Lue Gim Gong.

The Sour Orange (*Citrus aurantium*).—The sour orange is grown in the United States mainly as a root-stock on which other varieties are budded. A few varieties are cultivated to a limited extent for their fruits. Certain varieties are grown in some countries for manufacturing purposes.

The Lemon (*Citrus limonia*).—The lemon is grown extensively in California and to some extent in Florida. The commercial production in Florida has in recent years almost disappeared, primarily due to the damage caused by the disease known as scab. The principal varieties of the lemon are the Eureka, the Lisbon and the Villafranca. Of these, the Eureka, a nearly seedless variety that originated in California, is much the most extensively planted.

The Pomelo or Grapefruit (*Citrus decumana* or *Citrus grandis*).—

This fruit is grown extensively in Florida and the West Indies and to some extent in California. While the pomelo has been known for many years, it was first grown on an extensive commercial scale in Florida, first being introduced as a commercial fruit about 1885. The varieties most commonly grown are selected Florida seedlings, though one or two varieties, as the Pernambuco and the Royal, are importations respectively from Brazil and Cuba. Probably the most widely planted varieties in Florida are the Duncan, Josselyn, Walters, Pernambuco and Marsh. The Marsh, which is a nearly seedless variety is the most extensively planted of any variety in California.

The Lime (*Citrus limetta*).—The lime is grown throughout the citrus regions of the United States and the West Indies, but is produced commercially only in southern Florida and the West Indies. The demand for these fruits has rapidly increased in recent years and is assuming some importance. The principal varieties grown are the Mexican and the Tahiti.

The Mandarin Orange (*Citrus nobilis*).—This fruit, referred to frequently as the "kid glove orange" because of its loose, easily removable skin, is grown to a considerable extent in certain regions of the United States. It is in general rather more cold-resistant than the common orange, and this has led to its propagation to considerable extent in the Gulf states. The Satsuma or Unshiu, an early maturing sort of fair size, is grown rather extensively in northern Florida and southern Georgia, Alabama, Mississippi, Louisiana and Texas. The Dancy tangerine is grown to some extent in Florida and California and occasionally in some other states.

The Citron (*Citrus medica*).—The citron, the candied or preserved peel of which is a staple article of commerce, is not grown to any extent in America. A grove of about fifteen acres at Riverside, Cal., is the largest and only grove known to the writer in the United States. Another minor citrus fruit cultivated to some extent as an ornamental and for preserving is the kumquat (*Citrus japonica*).

Citrus Regions and their Production.—While the various citrus species are of tropical origin, the commercial development of citrus growing has taken place almost wholly in subtropical countries. The most important countries in the order of their production are the United States, Spain, Italy, Japan and Palestine. The normal citrus crop of the world is now equal to about 90,000,000 to 100,000,000 boxes of California capacity or from 230,000 to 250,000 carloads of California size.* The normal production of the United States is now about 78,000 carloads; Spain, about 68,148 carloads; Italy, 58,000 carloads; Japan, 10,896 carloads; and Palestine, probably about 9000 carloads. Small quantities of citrus fruits are, of course, produced in many other tropical and subtropical countries.

* "The World's Production and Commerce in Citrus Fruits and their By-Products," by F. O. Wällschlaeger. Bulletin No. 11, Citrus Protective League of California, Los Angeles, 1914.

According to the thirteenth United States census, there were in the United States in 1910, 11,486,768 bearing citrus trees and 5,400,402 of non-bearing age. The production in 1909 reached a grand total of 23,502,128 boxes valued at $22,711,448. This production was divided as follows: California, 17,318,497 boxes; Florida, 5,974,135 boxes; Louisiana, 153,319 boxes; Arizona, 32,247 boxes; Texas, 10,694 boxes; Mississippi, 3779 boxes; Alabama, 1201 boxes. A few boxes are also produced in Georgia and the Carolinas. The increase in yield and acreage since 1909 has been very great in California and Florida, so that the above data are very much below the present production.

Propagation of Citrus Varieties.—In the early days of the citrus industry, many seedling trees were grown in commercial groves. Now all groves are planted with stock budded with varieties of known excellence. It is important that the proper stocks be used. Orange and lemon varieties

Good Orange Seedlings.

are most extensively budded on sour orange stock, largely because of the resistance of this stock to foot-rot or gum disease. Wherever there is danger from this malady, the sour orange stock should surely be used. Sweet orange stock is also used widely, both in Florida and California. Trees on sweet stock probably in general grow rather more rapidly and rather larger than on sour stock, but the susceptibility of sweet stock to the gum diseases renders its use more limited. In dry, well-drained soils in Florida and in the dry interior regions of California it is a very satisfactory stock. Pomelo and Florida rough lemon stocks have some advocates, but have not been generally used. The Trifoliate orange is probably the best stock for the Satsuma and some oranges grown in the Gulf states, but has not given satisfaction in general. It has a very marked dwarfing effect on the Eureka lemon and some other varieties.

Orange Seedlings.—Sour orange seed for stock purposes is in general obtained from Florida; sweet orange seeds are usually taken from any sweet seedling grove. Seed must not be allowed to dry out. The seed is usually sown about one inch apart in a bed or may be drilled in rows. It is a good plan to cover the seed about one-half inch deep with clean river sand. It is desirable in most places to cover the seed-bed with a partial shade of some sort, as of cheesecloth or a lath shed. The seedlings are usually dug when they are about a foot high and transplanted to the nursery. Before transplanting the tops are cut back to about 7 or 8 inches above the crown.

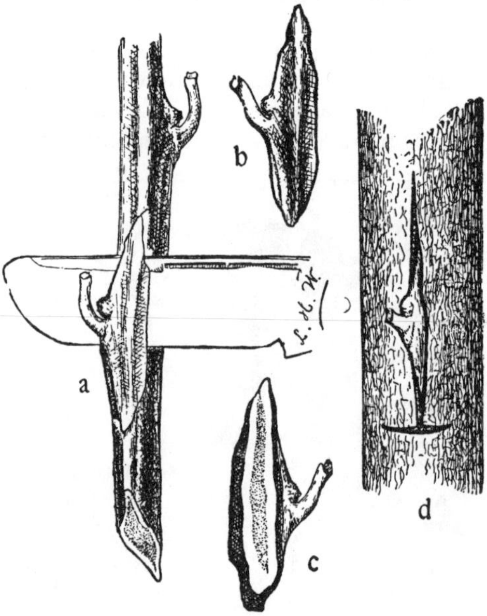

Shield Budding with Angular Wood.[1]

A—Cutting the bud. B—Bud cut ready to insert. C—Bud showing cut face. D—Bud inserted, bark on right side only being raised.

The Orange Nursery.—The nursery should be on a good porous soil that contains enough clay so that the trees can be balled if this method of transplanting is desired. The seedlings are set about 10 to 12 inches apart in rows $3\frac{1}{2}$ to 4 feet apart. The planting is usually done with a spade or dibble. Care must be taken not to let the roots get dry, and each tree should be set as nearly as possible at the same relative height it occupied in the seed-bed. The soil must be well firmed around the roots, and the plants should be watered as rapidly as planted. Small seedlings and those with imperfect roots should never be planted. Only the best and largest seedlings should be used. The nursery should be thoroughly cultivated, and the trees must be pruned occasionally to lead them to develop a single trunk for 6 or 10 inches above the ground. It usually requires from 16 to 18 months to grow trees to the right size for budding, an ideal size being a diameter of from three-eighths to one-half inch at about 3 to 5 inches above the soil.

Budding the Nursery Stock.—Trees should be budded from 4 to 8 inches above the ground. Budding is done mainly in the spring or in the fall. In the latter case, it is expected to keep the trees dormant until spring. Budding is almost universally done by the so-called eye-budding method,

[1] From Farmers' Bulletin 539, U. S. Dept. of Agriculture.

using a cut of an inverted T shape and pushing the buds up, being careful to have the leaf-scar of the bud downward. In citrus propagation, especially in the dry, arid sections of the Southwest, it is desirable to use strips of waxed cloth to wrap the buds, covering the buds entirely with the wrapping. The wrapping must remain on until the buds are thoroughly healed on, which will require about three weeks. The California method of forcing the buds is to cut the tops entirely off about an inch or so above the bud. In Florida the trees are cut half off above the bud and lopped over into the row, being allowed to remain until the sprout is a foot or so high.

Care of the Nursery Stock.—As the buds develop into sprouts, stakes must be set beside them, and the sprouts tied to the stake at frequent inter-

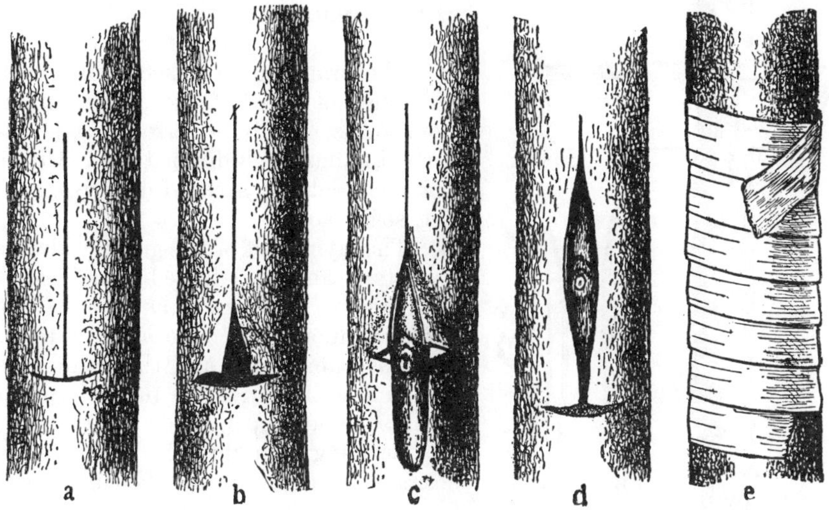

SHIELD OR EYE BUDDING.[1]

A—Incision on stock. B—Incision with lower ends of bark raised for inserting the bud. C—Bud partly inserted. D—Bud inserted ready to wrap. E—Bud wrapped with waxed cloth.

vals to insure straight trees. A single sprout is usually allowed to grow until it is about three feet high, and then it is headed back to about 30 inches or slightly lower. In forming the crown, three or four main branches are allowed to grow, and it is important for the strength of the tree that these should be on different sides of the young tree and 3 or 4 inches apart. Trees are set at one year or two years from the bud. In California and Arizona, owing to the dry conditions, nursery trees designed for shipment are usually balled. A trench about a foot wide and 14 inches deep is dug alongside of the row of trees, and the tap-root cut and the trees lifted with a ball of earth remaining around the roots. The ball and roots are then

[1] From Farmers' Bulletin 539, U. S. Dept. of Agriculture.

wrapped in burlap to hold the soil in place. When trees are to be removed for only a short distance, they may be planted with free roots, as is usually done in Florida. Great care must be exercised at every point to keep the roots moist, and they must be thoroughly watered after planting.

Planting the Orchard.—The site for a citrus orchard must be carefully selected to insure success. The warmest regions should be taken for the lemon and the slightly colder regions for the orange and pomelo. In California, the sloping sections next to the hills are usually considered the best and warmest, as they give good air drainage. In Florida lands in the southern part of the state or with water protection to the north are usually the warmest. The selection of a good site with reference to warmth is highly important.

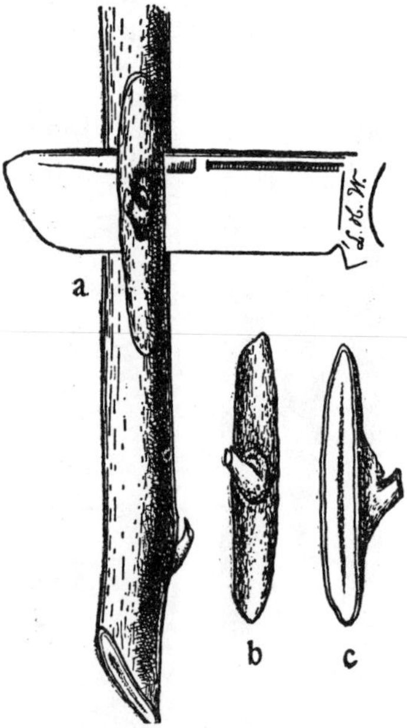

SHIELD OR EYE BUDS.[1]

A—Method of cutting bud from round twig. B—Bud cut ready to insert. C—Face of bud showing the cut surface.

The orange grows successfully on a wide range of soils, but a good, fine, sandy-loam soil is usually preferred. It is important to have well-drained land. Citrus fruits cannot be successfully grown in wet, soggy soil.

The laying out of the grove is always important and in irrigated countries requires very great care, as it is of the highest importance to be able to water the grove uniformly, and this cannot be done unless the irrigation furrows run at a uniform and proper slope.

Leveling or grading the land is rarely desirable, as this removes the surface soil from some places and makes it for some time unfit for tree growth.

The land for planting should be plowed deep and gotten in thoroughly good condition, finely pulverized and moist. Orange trees are usually planted from 20 to 25 feet apart, most commonly about 22 feet. Sometimes they are planted 20 feet apart one way and 22 or 24 feet apart the other way. Lemons and pomeloes are usually given rather more space than the orange.

There are four methods of arranging the trees known as the rectangular, triangular, quincunx and hexagonal. Of these the rectangular or square is the simplest and mostly commonly used. Planting in squares 20 by 20 feet gives 108 trees to the acre, and planting 22 by 22 feet, a very common distance, gives 90 trees to the acre. The land to be planted must be laid

[1] From Farmers' Bulletin 539, U. S. Dept. of Agriculture.

out accurately and the location of each tree staked. Before digging the holes a notched board with stakes or some other device should be used to insure the exact location and level of the tree in setting. The holes are dug of the size and depth necessary to accommodate the trees. The trees to be set should have their tops cut back severely and all but a few leaves removed. In arid regions, if the trees are not balled, it is not a bad practice to remove all of the leaves.

In planting the trees care should be exercised to plant them at nearly the same level as they were in the nursery. To insure this they must be set about 4 inches higher ordinarily to provide for settling. Many growers prefer to plant five or six inches higher than the level of the ground, having the trees on a slight ridge. This the writer believes to be a good practice. Balled trees are usually planted with the sacks surrounding the roots, these being opened at the bottoms and the strings cut. The sacks rot away in a short time. When trees with free roots are planted, the soil must be well sifted in around the roots and firmly pressed down. Thorough watering must immediately follow the planting. The watering of young trees is facilitated by forming small basins around the trees into which the water can be run.

It is important to protect the trunks of young trees from the sun in order to prevent sun-burning. This is usually accomplished by loosely wrapping several thicknesses of newspapers around them or by means of regular protectors that may be purchased for this purpose.

Cultivation.—Young orchards must be thoroughly cultivated around the trees. It is a common practice for two or three years to grow a strip about ten feet in width of alfalfa, beans or some other crop between the rows of trees, keeping a cultivated and irrigated strip immediately around the trees. As the grove comes into bearing, the normal cultivation of the whole area is taken up.

Many different systems of cultivation are followed in different places. In Florida the common practice is to grow a leguminous cover crop, such as beggarweed or cowpeas, in the grove during the summer, this being plowed or disked in, in the fall, followed by more or less frequent, shallow cultivations until the early summer when the cover crop is again sowed.

In California the most common practice is to grow a cover crop of some legume in the grove during the winter, from September to the first of March. The plants most commonly used for this purpose are the bitter clover (*Melilotus indica*) and the vetch (*Vicia sativa*). Of these the bitter clover is much the best. The purple vetch (*Vicia atropurpurea*), recently imported, is far superior to the ordinary vetch for this purpose, and will doubtless be much used when a sufficient supply of seed becomes available. The cover crop is plowed under to a depth of seven to ten inches during the early part of March before the trees begin to bloom and while the ground is still in condition and moist from the winter rains. Following this the land is harrowed and disked both ways. Very thorough working with the disk is believed to be preferable to plowing both ways across the grove, as is sometimes done. After this thorough disking the land is harrowed again and then left until it is necessary to furrow out for the first irrigation. In harrowing, either a knife harrow should be used, or if a spike-toothed harrow is used the teeth should be sloped backward in order not to pull up the cover crop.

The first irrigation is delayed if possible until after the blooming period, but the trees must not be allowed to suffer for water. After the irrigation, as soon as the soil has dried sufficiently, the land is harrowed and disked both ways and again harrowed. This should leave the surface soil thoroughly pulverized, and with a dry dust mulch. No other cultivation is necessary until after the second irrigation.

During the dry summer period an irrigation is necessary about every month. Following each of these irrigations, the land should be harrowed as soon as dry enough and about a week later cultivated both ways with some narrow, shoveled cultivator, running to a depth of 4 to 6 inches. These alternating periods of irrigation followed by cultivation are continued during the summer until the winter cover crop is sown in the fall.

In both Florida and California, the practice of mulching a portion or all of the land in the grove is gaining in favor.

Irrigation.—In the citrus regions of California and Arizona, irrigation is necessary and is one of the most expensive and difficult of all the various grove operations. Water in these sections is, however, the limiting factor of production, and an ample supply must be provided. Water is taken directly from flowing streams, is pumped from underground basins, or is taken from large, artificial storage reservoirs, filled mainly during the winter rains. Different locations and soils require different amounts of water. A porous, gravelly soil requires more water than a heavy clay or adobe soil, the latter being more retentive of moisture although more difficult to wet. Groves near the coast where there is more moisture in the air require less water than those in the drier interior regions. In general, enough water must be provided to be equal, when combined with the natural rainfall, to a depth of 35 to 45 inches. In a single irrigation it is ordinarily expected to apply enough water to cover the entire surface irrigated to a depth of about three inches. The supply of water usually provided for citrus orchards is one miner's inch to every four to eight acres.*

In the furrow method of irrigation the water is distributed over the grove by means of several furrows, usually four to six, between each row of trees. These furrows, which are made by a special furrowing tool or plow, should have a uniform fall, preferably not exceeding a grade of one-half of one to three per cent. The water should run through them slowly to give the best results. While these furrows are usually run straight, not infrequently they are curved in between the trees to water the middles. The

* The miner's inch most commonly used in California is the amount of water that will flow through a 1-inch square opening under a 4-inch pressure head. This equals 9 gallons per minute. The statute inch is 11¼ gallons per minute.

length of the furrows or of the "run" ordinarily ranges from 400 to 700 feet. While 600 and 700-foot runs are common, this is too long to give the best results.

In the basin system of irrigation, square or round basins, about eight to twelve inches deep, are formed around the trees, into which the water is run either by means of a single central furrow, from which it is turned into each basin successively, or by means of steel irrigation pipes fitted together like joints of a stove-pipe. In making the basins the soil should be left for a radius of about two or three feet around the base of the tree, so that the water will not come in contact with the trunk.

The water is brought into the grove usually either by open cement flumes or by buried cement pipes. These are run across the rows along the upper edge of the grove to be irrigated. With the open flume, gates are put in at intervals to discharge the water wherever a stream is desired. With the covered cement pipe flumes, a standpipe is placed at the end of each row of trees in which several gates are inserted according to the number of furrows or streams desired to be taken from it.

The length of time necessary to run the water is determined by the rapidity of penetration. The application should be continued until the water has penetrated to a depth of three or four feet.

Fertilization.—The great majority of soils on which citrus trees are grown require manuring to maintain the fertility, and yet no subject is so little understood as the fertilizer requirements. If the soil fertility is sufficient to provide for good growth in the beginning, then the addition of the materials removed by the crop, it would seem, should be sufficient to maintain the fertility. The following table shows the average percentage of nitrogen, phosphoric acid and potash in orange and lemon fruits and the pounds of these materials removed by a ton of fruit.

FERTILIZER ANALYSIS OF THE FRUIT OF ORANGES AND LEMONS.
(Computed from Bulletin No. 93, University of California Agricultural Experiment Station.)

	Nitrogen (N)		Phosphoric Acid (P_2O_5).		Potash (K_2O).	
	Per cent.	Pounds per Ton.	Per cent.	Pounds per Ton.	Per cent.	Pounds per Ton.
Oranges................	0.190	3.80	0.058	1.16	0.219	4.38
Lemons.................	0.151	3.02	0.058	1.16	0.253	5.06

Such a table as the above is suggestive only as a guide to fertilization, and the same may be said regarding soil analyses. The test of a fertilizer on the soil and the crop is the only safe guide.

In Florida a fertilizer containing about 3 to 4 per cent of nitrogen, 6 to 8 per cent of phosphoric acid and 8 to 12 per cent of potash is commonly used. In California the proportions commonly recommended are 4 per cent nitrogen, 8 per cent phosphoric acid and 4 per cent potash. In general, young trees are thought to require more nitrogen and a relatively smaller proportion of phosphoric acid and potash.

In Florida there is a tendency to avoid so far as possible the use of organic manures, such as stable manure, blood, cottonseed meal and the like, owing to the effect such materials apparently have in the production of the disease "die-back" or exanthema. Sulphate of ammonia, sulphate of potash and superphosphate are very largely used.

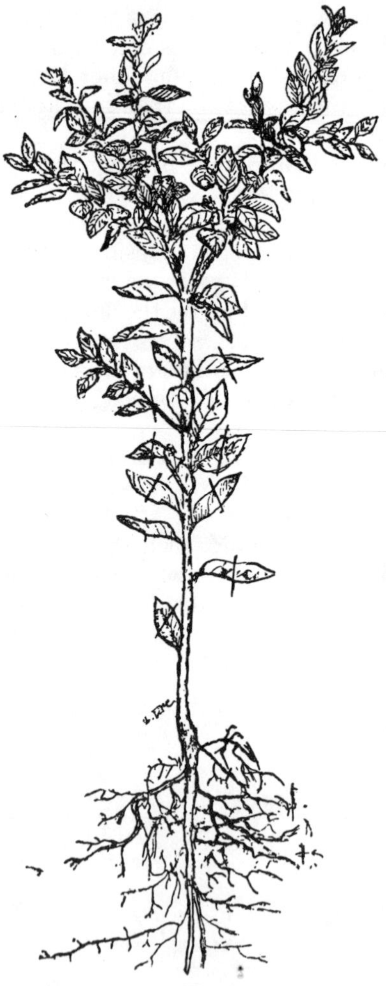

PRUNING AND ROOT TRIMMING OF CITRUS TREE AT TIME OF PLANTING.

In California, on the contrary, the tendency is to use organic sources to supply the various elements so far as possible. Experimental results indicate that organic matter and nitrogen are the most important elements to be added in the fertilization of citrus soils in California and minimize the importance of potash and phosphoric acid. Eight-year experiments show no gain over checks by the use of sulphate of potash with oranges and very slight gains with lemons. Similar experiments with superphosphate show but slight gains over check plats, while nitrogen plats give marked increase in growth and yield. California growers in general prefer stable manure to any other fertilizer and are also using large quantities of alfalfa hay and bean straw, both plowed under and as a mulch to supply nitrogen and organic matter.

The use of leguminous cover crops in citrus orchards to supply nitrogen and organic matter is recognized as good practice, both in California and Florida (see above under "Cultivation"), and a considerable amount of the necessary nitrogen can be produced in this way at very slight expense.

Pruning.—Ordinarily orange trees are pruned very little beyond the moval of dead brush and water-sprouts, but this results in the formation a very dense tree with the fruit distributed over the surface. The interior fruit is in general superior, and the removal of some of the interior limbs keeping the tree somewhat open is probably a desirable practice.

Lemon trees are generally pruned regularly. They should be cut back severely from the first and allowed to develop but slowly. The tendency of the lemon is to throw out long branches, which fruit at the end and are likely to bend over and break off or to be in the way. The principal purpose should be to cut back this rapid growth and develop a strong, stocky tree that will be open enough to bear considerable fruit on the interior branches. The lateral, crooked branches are much more fruitful than the upright, straight branches.

Trees are pruned at almost any season of the year, but the best time is in the spring after the danger of freezing is passed.

Frost Protection.—Many citrus-growing sections are occasionally visited by severe freezes that may cause a loss of the crop and even severely damage the trees. It has thus been found desirable, particularly with lemons, to provide some form of artificial protection.

In California this protection has been secured by the use of orchard stoves, burning crude oil, abundant quantities of which are available from nearby oil fields at reasonable prices. The principle of orchard heating, recognized as the most desirable, is to get the greatest amount of heat possible with the least soot and smoke. Direct, radiated heat is desired rather than a smudge.

In Florida oil heaters have been used to some extent, but there the burning of wood piled in the grove and other devices are also used.

Diseases.—The number of diseases affecting citrus trees is probably as great as those affecting any other similar group of plants. For many of these satisfactory treatments are known, but there are several maladies which are serious that are not as yet thoroughly understood. Only a few of the most important diseases can be mentioned.

The general group of gum diseases is important in most citrus-growing regions. Lemon gummosis, caused by the brown rot fungus (*Pythiacystis citrophthora*) causes considerable damage in California and is also present in Florida. This disease, which causes the exudation of gum and the decay of the bark on lemon trunks, is effectively controlled by cutting out the diseased parts and painting the injured surface with Bordeaux paste. Maldigomma or foot-rot, a closely related disease that occurs mainly in Florida, is controlled by use of the sour orange stock which is resistant to the malady, and may be cured usually by removing the dirt from around the crown roots, cutting out the diseased areas and painting them with Bordeaux paste. Another type of gummosis is the *scaly bark*, common both in California and Florida mainly on the limbs and trunks of orange trees. This disease is not understood at present, but is checked by cutting out and sterilizing diseased areas with Bordeaux paste.

Exanthema or die-back, a disease common in Florida and occurring to some extent in California, is apparently due to malnutrition, but is not understood. When caused by use of organic manure, such as blood, stable manure and the like, it is cured by stopping fertilization and cultivation for a period and mulching the tree. When caused by lack of drainage, tile drainage of the area frequently results in a cure.

Mottle leaf, a very common and injurious malady in California, is an obscure disease, the cause of which is not yet known. Very extensive investigations of this disease are now in progress.

Citrus canker, a very serious malady caused by a bacterium (*Pseudomonas citri*), has recently become epidemic in Florida and the Gulf states. It is now known to occur in Japan and the Philippine Islands and was apparently introduced into Florida from one of these sources. An extensive campaign is now being waged to eradicate this disease by burning all infected trees.

Verrucosis, or scab, and melanose are two important fungous diseases occurring in Florida that have not appeared as yet in California. Withertip, caused by the fungus *Colletotrichum glœosporioides*, is common both in California and Florida. It is controlled by pruning accompanied by spraying with Bordeaux mixture.

Many fruit rots caused by such fungi as the cottony mold (*Sclerotinia libertiniana*), brown rot (*Pythiacystis citrophthora*), blue mould (*Penicillium italicum*) and green mould (*Penicillium digitatum*), cause considerable loss in the packing-house and in shipment. These are controlled by careful handling, by the use of disinfectants in the wash water and the proper sterilization of the fruit boxes and packing-house machinery and the like. (For other diseases see Chapter 75.)

Insects.—Insect pests are very numerous in all citrus sections and require the systematic use of control methods to insure the financial success of the industry. By far the most serious pests are the scale insects of which there are numerous kinds.

In California the most common scale insects are the black scale (*Saissetia oleæ*), the citricola scale (*Coccus citricola*), the red scale (*Chrysomphalus aurantii*) and the purple scale (*Lepidosaphes beckii*). The control of these scale pests is aided to some extent by various parasites, but fumigation with hydrocyanic acid gas about once every two years, or more often if necessary, is almost universally practiced.

The cottony cushion scale (*Icerya purchasi*), which at one time was so serious as almost to threaten the life of the citrus industry in California, has been so thoroughly controlled by the introduction of the Australian ladybird beetle that it has ceased to be considered a serious pest.

In Florida the most common scale insects are the purple scale (*Lepidosaphes beckii*), the long scale (*Lepidosaphes gloverii*), the Florida red scale (*Chrysomphalus aonidum*), the chaff scale (*Parlatoria pergandii*), the soft brown or turtle-back beetle (*Coccus hesperidum*), the black scale (*Saissetia oleæ*) and the hemispherical scale (*Saissetia hemisphærica*). In the control of these insects in Florida, more reliance is had upon parasitic fungi and insect enemies than in California. While fumigation is used to some extent, when any treatment is used it is usually spraying with paraffin oil emulsion. Good's caustic potash whale oil soap, resin wash or kerosene emulsion.

The white fly (*Aleyrodes citri*), probably the most destructive insect pest in Florida, occurs in only one place in California and has not there become widely spread. It is controlled by fumigation or by spraying with paraffin oil emulsion, resin wash or kerosene emulsion.

The mealy bug, red spider, rust mite, thrips, aphis and numerous other insects cause damage both in California and in Florida and in other citrus sections, but are of minor importance.

PICKING AND PACKING ORANGES.

Picking, Packing and Marketing of Fruit.—The methods of picking, packing and marketing of citrus fruits are probably more highly developed than in any other fruit industry. The picking is in large measure done by carefully trained special picking gangs connected with the packing houses, rather than by the growers themselves. This insures the most careful work and handling and the employment of uniformly good methods.

The curing, grading and packing is also done by specially trained men working continuously under inspection to insure careful handling at every point. The special machinery devised for washing, drying, grading, sizing and boxing has reached a high degree of perfection and is almost universally used in the citrus sections of the United States. The watchword of all packing houses is *careful handling* to avoid bruising or puncturing the skin of the fruit and thus prevent decay.

The marketing methods have been developed with similar thoroughness, and a very large proportion of the fruit of California is marketed under the direction of a co-operative organization of the growers known as the California Fruit Growers' Exchange. This is probably the most successful co-operative organization of growers in the world.

REFERENCES

"Citrus Fruits." Cort.
"Citrus Fruits and Their Culture." Hume.
"California Fruits and How to Grow Them." Wickson.
"U. S. Dept. of Agriculture Bulletin 63. "Shipment of Oranges from Florida."
Farmers' Bulletins, U. S. Dept. of Agriculture:
538. "Sites, Soils and Varieties for Citrus Groves in the Gulf States."
539. "Propagation of Citrus Trees in the Gulf States."
542. "Culture, Fertilization and Frost Protection of Citrus Groves in the Gulf States."

CHAPTER 38

NUTS AND NUT CULTURE IN THE UNITED STATES

BY C. A. REED

Nut Culturist, United States Department of Agriculture

THE PRINCIPAL NUTS

The group of trees which bear edible nuts of commercial importance in this country includes a considerable number of species, some of which are important in both hemispheres. The most important of the world's nuts are the cocoanut, the peanut, the Persian (incorrectly called the English) walnut, the almond, the Brazil nut, the pecan, the hazelnut (filbert), the cashew, the pinon, the chestnut and the pistachio nut. Of these, with the exception of the Brazil nut (nigger-toe, Para nut, cream nut, castanea, etc.), which is strictly tropical in its requirements of culture, all are being grown to a greater or less extent, in continental or insular United States. The pili (pe-lee) of the Philippines and East Indies, characterized by its reddish-brown (artificial) color, its triangular form tapering to a point at each end, its very thick, hard shell and its single kernel, is now becoming fairly familiar in our principal nut markets. The pili nut is said to be very nutritious and pleasing to the taste when properly matured, but as it commonly appears in this country, it is inferior in quality to the majority of the better known nuts.

A choice nut occasionally seen in the American markets is the Paradise nut, a near relative of the Brazil nut, which also is indigenous to the lowlands of northern Brazil. Paradise nuts are somewhat longer than are Brazil nuts, but in the main are triangular in form. They are of a light buff color, irregularly grooved lengthwise, and have a close-fitting cork-like shell which encloses a single, delicately flavored kernel of fine texture.

Both the pili and the Paradise nuts are like the Brazil nut in that their tropical natures apparently preclude any likelihood of their ever becoming commercially important in any part of the United States proper.

The culture of the cocoanut, together with the drying and shipping of its dried flesh or copra, forms one of the leading industries throughout all tropics. The cocoanut produces the world's most important nut food supply. To some extent the cocoanut palm is grown in southern Florida, but thus far more largely as an ornamental and a curiosity than for commercial purposes. During the winter season cocoanuts are locally in lively demand as souvenirs among the tourists, who place postage and the addresses of northern friends on the smooth outer surfaces of the thick

Schley Pecan Tree.
In its seventh year and beginning to bear. Cairo, Ga.

husks and send the nuts through the mails. The expense of removing the husk from the nut has thus far made commercial cocoanut growing in this country in competition with the cheap labor of the tropics practically out of the question. Nevertheless, it is not unlikely that the devising of special machinery will soon overcome this problem, and that a more or less thriving industry will develop in the marshy borders of southern Florida. A few commercial cocoanut plantings recently set may be found off the Florida coast from Miami and near Cape Sable in Monroe County; but it appears altogether unlikely that the growing of cocoanuts will ever be of importance to American farmers outside of the southern parts of Florida, Texas and California.

The cashew nut likewise is of tropical nature. Trees of this species are rarely seen in the United States except in experimental plantings in Florida and in California. The nuts are borne singly at the apex of fleshy, pear-shaped fruits which form in clusters and which are known as cashew apples. The nuts are of much the shape of lima beans, but are both larger and thicker. In color they are between a purplish and an ashy-gray. They have a thin but stout, smooth-surfaced shell, within which is a secondary shell, also thin, and which encases the kidney-shaped kernel.

Between the two shells of this cashew nut there is a thin dark-brown fluid of an extremely caustic property similar to that of poison ivy and sumac, to which the species is closely related. Roasting entirely dispels this poison, and as the nuts are invariably prepared in this manner before being placed on the market, the consumer is in no danger of being poisoned. The kernels are among the most palatable of all nut products now found in our markets.

For the present, the cashew can hardly be said to be of commercial promise in any portion of this country.

The pistachio is much more hardy than is the cashew. To a considerable extent the two are now being grown in sections of southern California and west Texas, and single trees have been known to survive for a number of years in climates where zero temperatures are by no means uncommon. Thrifty trees are reported from Kansas and one tree several years of age near Stamford, Conn., was in a thrifty condition when seen by the writer in 1914. However, it is essentially a dry-land tree suited to the milder portions of the temperate zones. The nuts, which are encased in a thin leathery covering, form in loose clusters. They have thin but very stout, smooth shells which usually split open on one side of the suture while being roasted. To a considerable extent, the kernels, which are of greenish color and delicate flavor, are consumed with no preparation other than that of roasting and salting, but more largely they are ground and used in ice creams and other confections. The pistachio tree is a slow grower, requiring several more years to come into bearing than is the case with almond, Persian walnut or pecan trees. Propagation is by budding and grafting.

The Peanut.—The peanut is probably a native of tropical America.

It does well in light-colored, fertile, sandy loams in the warmer portions of the United States. Its principal centers of production in this country are in Virginia and the Carolinas, although it is common in the entire South, west to California. The peanut is common in the markets both in the shell or shelled and salted. Peanut butter and peanut oil are now among the most valuable of our common nut products. (The culture of peanuts is discussed in Chapter 20 of this book.)

The Pinon (*Pin-yon*).—The seeds of a number of pines of western and southwestern United States, variously known as pinons, Indian or Pine nuts and pignolia, form a very important article of food for the Indians

Franquette Walnut Orchard, near Santa Rosa, California.
This is the famous Vrooman Orchard.

and the Mexicans of the Southwest, who gather the nuts in enormous quantities. In this country the pines bearing edible nuts are not cultivated; the entire crop being obtained from the native trees in the mountains, which usually appear at altitudes of from 5000 to 7000 feet. The home product is largely consumed by the gatherers, and in the local markets of the West. The nuts are brownish in color, usually mottled with yellow, from an eighth to a quarter of an inch in length and have a thin but strong hard shell. The kernels are very fine in texture, rich in quality, of pleasant flavor and highly nutritious. The shelled seeds of the stone pine of southern Europe, greatly resembling puffed rice in form and color, form an important product in the nut markets of our Eastern cities.

The Persian Walnut.—For many centuries this nut, a native of Persia, has been under cultivation in southwestern Asia and in Europe, but with approximately a half century of serious cultivation in this country it has attained its greatest degree of perfection on our Pacific Coast. In the Old World, and until recently in the United States, propagation has been by seedage, but modern American orchards are comprised exclusively of budded or grafted trees. For its best development the species requires a deep, fertile, loamy soil, moist but well drained. However, it readily adapts itself with proportionate results to conditions less favorable. At the present time the chief centers of production in the United States are southern California near Los Angeles, the Sacramento Valley in northern California and the Willamette Valley of western Oregon. Varieties suitable for general culture in the southeastern quarter of this country have not yet appeared. In that area of the Eastern states lying between lower New England and the Potomac River on the Atlantic Coast and extending west to the Mississippi River, local varieties originating with trees reputed to be hardy and prolific bearers of desirable nuts are being given a fairly general trial. These are being propagated by budding on the black walnut stocks. To date, the chief eastern varieties are the Rush, Nebo, Barnes, Potomac, Holden, Hall, Lancaster and Boston. Thus far none of these have been given sufficient trial to determine their commercial value. For the present, planting should be limited to experimental numbers.

The most popular varieties of walnuts in southern California are the offspring of the Santa Barbara type, established during the late sixties by Mr. Joseph Sexton of Santa Barbara, with seed supposed to have come from Chile. Southern California walnuts are not sold under variety names, but under such trade appellations as "budded," "numbers one," "two," "three," etc.; the term "budded" applying to the large sizes which will not pass through inch squares of a wire mesh.

The leading varieties of northern California and Oregon are from French stock first introduced into this country by Mr. Felix Gillet of Nevada City, Cal., whose work closely followed that of Mr. Sexton, and these to a large extent are sold under their variety names. At present the more important are the Franquette and Mayette, direct introductions, and the Concord, San José and probably the Chase, seedlings of original introductions.

The Pecan.—The pecan is by far the most important nut indigenous to this country, and although at present its annual production is less than one-half that of the Persian walnut, the increased attention now being paid to the native bearing trees and enormous number of planted orchards in the south Atlantic and eastern Gulf states combine to make it fairly certain that this will soon become the leading nut grown in America. Its native range includes much of the lowlands of the Mississippi River and its tributaries from Davenport and Terra Haute on the north, south to near the Gulf and a large area extending southwest across Arkansas, Louisiana, Oklahoma and Texas to near the Rio Grande. Its requirements

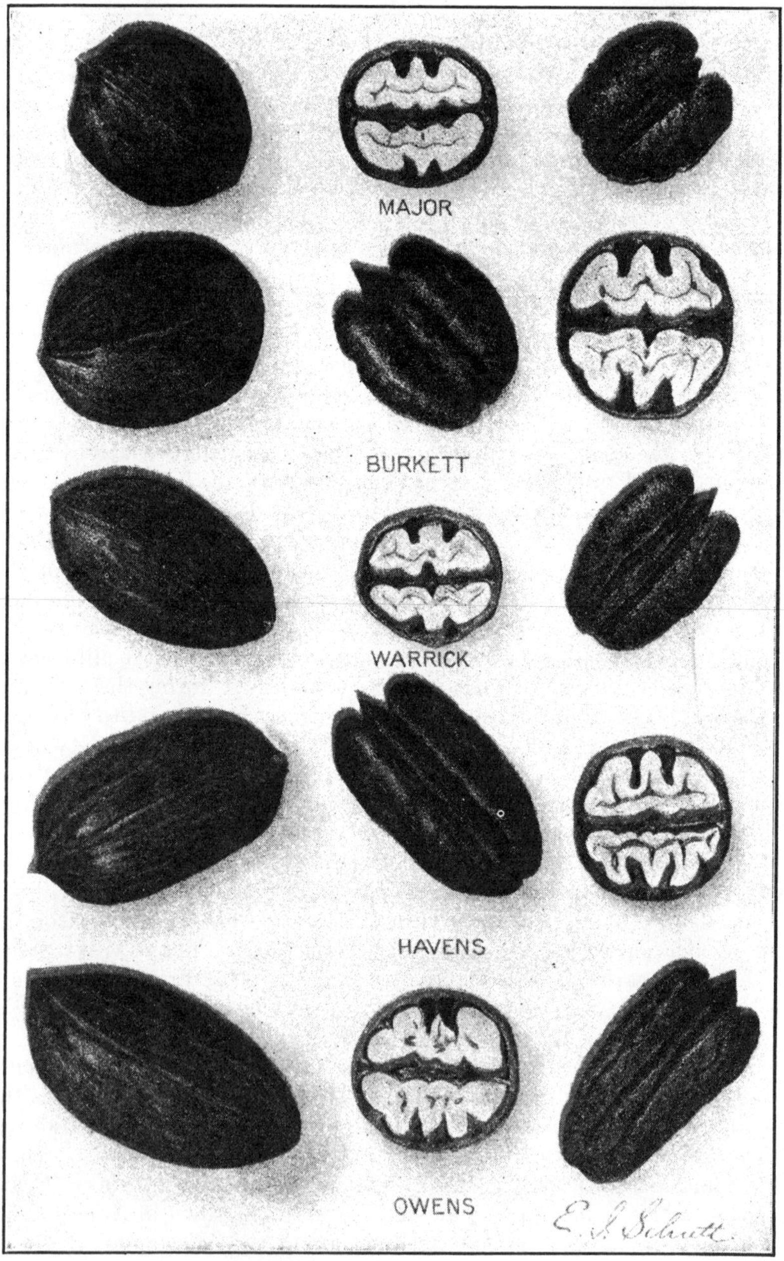

Major, Burkett, Warrick, Havens and Owens Pecans.[1]

[1] Year-Book, 1912 U. S. Dept. of Agriculture.

of soil and moisture are much the same as are those of the Persian walnut. However, it appears to be somewhat more exacting in its moisture requirement, for although being intolerant of improper drainage, it is less able to sustain itself proportionately in drier soils.

The pecan is propagated by budding and grafting on stocks of its own species. Under the most favorable conditions seedlings grown from nuts planted in midwinter may be budded when eighteen months old, and transplanted in orchard form by the end of the next season, or by the time the roots have been in the ground for three years.

Thus far a total of approximately one hundred varieties have been recognized in the South. The majority of these already have been eliminated. At present, the principal sorts of the south Atlantic and Gulf section, including Louisiana, are the Stuart, Schley, Curtis, Van Deman, Alley, Pabst, Moneymaker, Bradley, President, Russell, Delmas and Success. The leading varieties of central and southwest Texas are the San Saba, Halbert, Colorado, Sovereign and Kincaid.

The varieties of neither of these groups appear readily to adapt themselves to the climatic conditions of the other, nor do they seem capable of satisfactorily adjusting themselves to conditions in any of the inland states.

Varieties which have originated in southwestern Indiana and neighborhood, and which, therefore, now are thought to be well worthy of conservative planting in sections of fairly comparable conditions, are the Major, Niblack, Indiana, Posey, Busseron, Butterick, Greenriver and Warrick. However, for the present these should not be planted in latitudes greater than that of Vincennes, with the expectation of regular crops of nuts. If set in suitable soil the trees should thrive and live to a great age much farther north, but crops of nuts even from the forest trees are quite irregular beyond that point.

The Almond.—Culturally speaking, this nut is not of much interest to a great part of the United States. Its exactments for cultural success preclude its general planting over any large portion of the United States. It requires a fertile, moist, yet very well-drained soil and a dry atmosphere in a section quite free from late spring frosts. The commercial plantings of this country are in the Sacramento Valley of California, where the orchards are usually equipped with fire-pots as a protection against frost at blossoming time, and in northwestern Utah. A number of large young orchards just beginning to bear are on the highlands of Klickitat County in south central Washington near the Columbia River. The principal varieties are the Nonpareil, I. X. L., Ne Plus Ultra, Drakes and Languedoc.

The almond is propagated by budding on stocks preferably of its own species, although peach stocks answer nearly as well.

At least one variety of hardshell almond (the Ridenhower, of southern Illinois) is being propagated by eastern nurserymen for variety planting about the home grounds in sections adapted to the more hardy varieties of peaches. However, in no way does this almond compare with those in the market.

Nuts of Minor Importance.—In this class belong all of our native tree nuts, with the exception of the pecan. Our native hickories, the shagbark (*Hicoria ovata*), the shellbark (*Hicoria laciniosa*) and the pignut (*Hicoria glabra*), the butternut (*Juglans cinerea*), the American hazel (*Corvlus americana*), the beech (*Fagus grandifolia*) and certain foreign nuts, especially the Chinese chestnut (*Castanea mollissima*), and the Asiatic walnuts (*Juglans sieboldiana* and *Juglans mandshurica*), afford most inviting fields for the breeder and improver of nut trees.

The most of these species are capable of culture in the Eastern states from lower New England south to the middle Atlantic and west to the Mississippi. It is quite probable that this group also offers the most fruitful possibilities in nut culture in the states lying between this section and the Rocky Mountains.

In general, prospective growers of nut trees should obtain their stock from reliable nursery concerns, and in so far as obtainable, budded or grafted trees only should be planted. These are not now obtainable to any extent of the group just mentioned, although several varieties of hickory and black walnut are now being propagated by a few nursery concerns.

Nut trees should be ranked in the class with other kinds of fruit trees, and must be given the same degree of attention. Under the most favorable conditions commercial returns may be expected with almonds in from 6 to 8 years from the time of setting the trees; with Persian walnut trees in from 8 to 10 years; and with southern pecans in from 10 to 12 years. Almond trees may be set at from 28 to 30 feet apart, while walnuts and pecans should be set not nearer than 60 feet.

Each species of nut tree has its insect pests and fungous diseases, each of which is more or less serious. With the almond, the present most serious pest doubtless is the red spider; with the walnut, it is the walnut blight; and with the pecan, it is the rosette; although each species of tree has its other serious enemies.

REFERENCES

Georgia Expt. Station Bulletin 116. "Pecans."

Farmers' Bulletin 700, U. S. Dept. of Agriculture. "Pecan Culture, with Special Reference to Propagation and New Varieties."

CHAPTER 39

MISCELLANEOUS TROPICAL FRUITS*

The Pineapple.—As a tropical fruit the pineapple ranks second to the orange and banana. Originally a wild fruit, very small in size, it has by constant cultivation and improvement been developed into one of the choicest fruits in existence. Some varieties now produce very large fruit, weighing as much as twenty pounds.

Pineapples thrive best in Porto Rico, Cuba, Hawaii and the tropical islands, but can be grown easily in southern Florida and even further north, if not exposed to frost.

The pineapple resembles the cabbage in that it grows on a short, leafy stalk from one to three feet high.. The plant is very leafy, the leaves of most varieties being edged with spines.

Propagation.—Pineapples are propagated by means of ratoons, suckers, slips and very seldom, when only for experimental purposes, by the seed. A ratoon is an individual plant formed among the roots of the mother plant and appearing beside it from under the soil. A sucker is an individual plant coming from the side of the stem above the soil. A slip is the small plant that appears below the fruit on the fruit stalk. The small plants that grow on the apex of the fruit are known as the crown slips. There is no difference in the kind of plant produced by either the ratoon, the sucker or the slip. However, the sucker and the slip are to be preferred, because plants from ratoons will die easily if not handled properly. The main thing is to select a well-matured slip or sucker. Suckers have an advantage over the slips, inasmuch as they are several months older and, of course, they bear sooner. Whether suckers or slips are selected to be planted, they should be trimmed by cutting the base and stripping off the lower leaves. One inch and a half to two inches of stem should be left exposed. It is better to let them dry a little before planting. This is called curing.

Soil.—The pineapple will grow in a great variety of soils, but thrives best in light, deep, well-drained, sandy soils. Damp and heavy soils are unfavorable. The plant is a gross feeder and calls for a liberal supply of nitrogenous fertilizers. Experiments carried on in Porto Rico have demonstrated that the plant responds to commercial fertilizers. A small plant, although in poor soil, has attained astonishing proportions after the fertilizer has been applied. So, when enough plant-food is available and the roots may obtain all the air they need, the pineapple can be successfully grown on a wide range of soils.

*In preparing this chapter the author was assisted by Mr. F. G. de Quevedo, formerly of Porto Rico, now teacher of Spanish in Pennsylvania State College.

THE PINEAPPLE PLANT IN FRUIT.[1]

1—Main stalk. 2—Ratoon. 3—Sucker. 4—Head of fruit. 5—Slip. 6—Fruit. 7—Crown slip. 8—Crown.

[1]Courtesy of U. S. Dept. of Agriculture. From Porto Rico Bulletin No. 8.

From one to two thousand pounds per acre of blood and bone or cottonseed meal will improve the size and quality of the fruit, and maintain the fertility of the land. The following summary taken from Bulletin 104 of the Florida Experiment Station, will serve as the best guide for the fertilizing of pineapples.

(*a*) Fine-ground steamed bone and slag phosphate are best as sources of phosphoric acid; cottonseed meal, dried blood and castor pomace are best as sources of nitrogen; high-grade and low-grade sulphate of potash are best as sources of potash.

(*b*) Nitrate of soda, acid phosphate and kainit have not proven satisfactory. (While sulphate of ammonia was not used in the experiment, this material has in general practice been found unsuited to pineapple culture.)

(*c*) In case of shedded pineapples it has been found that it is profitable to use from 2250 to 3750 pounds per acre annually of a complete fertilizer.

(*d*) Analyses of a large number of fruits (Red Spanish) covering a period of four years show that the eating quality of the fruit is not affected by the kind of fertilizer used.

(*e*) The sugar content of the fruit (Red Spanish) is slightly increased by the heavier fertilizer applications.

(*f*) The large fruits contain a slightly higher percentage of sugar than the small ones.

(*g*) The analyses of a large number of pineapple plants show that they contain sufficient fertilizing materials, nitrogen, phosphoric acid, potash, lime and magnesia to make them of considerable value as a fertilizer.

(*h*) With an increase of nitrogenous fertilizers there was found an increase of nitrates in the soil.

(*i*) Nitrates are most abundant at the immediate surface. After a depth of one foot is passed the amount is very small.

(*j*) Where the surface of the ground is not protected, the nitrates are much less abundant than where there is a covering of plants and decaying leaves.

Preparation of Soil.—The essentials for the pineapple are a limited water supply, abundance of air for the roots and plenty of available plant-food. The selection and preparation of the soil should meet these requirements, as fully as possible. Sandy soils or sand, naturally most nearly meet the physical requirements. Such soil should be thoroughly plowed and freed from noxious weeds and grass before starting the plantation. If the soil is level and inclined to be wet after excessive rains, it should be made into rather wide beds on which the plants are set. The plants are set in rows 15 to 18 inches apart and as many as 20 rows to the bed. The advantages of this close setting lie in economy in the use of fertilizers, the support which the plants give to each other, and the thoroughness with which they shade the ground and prevent the growth of weeds and grass after they are fully established. With this system of planting, there should be ample room to pass between the beds for the purpose of carrying the fruits from the field when they are mature. There should also be roadways crossing the beds at intervals of a few hundred feet sufficiently wide to allow the passage of a wagon.

When planted on heavier soils the single-row or double-row systems of planting is preferred. This allows for horse cultivation by means of which

PINEAPPLES PLANTED IN AN ORANGE GROVE.[1]
This provides a revenue from the land while the trees are coming to the bearing age.

weeds and grass are subdued and the soil kept loose to facilitate thorough aeration.

Pineapple plants bear but one fruit, after which they die. The new crop is secured from the slips and suckers from the mother plant. Like most crops, pineapples will not succeed by continuous cultivation on the same land. A rotation of crops is therefore advised. On soils that are especially well adapted to the pineapple three consecutive crops can be grown before the soil is devoted to other crops.

Cultivation.—The cultivation should aim to maintain a loose condition of soil and prevent the growth of weeds and grasses. Hand cultivation will be necessary in case of level, sandy soils planted in beds as above

[1] Courtesy of U. S. Dept. of Agriculture. From Porto Rico Bulletin No. 8.

mentioned. The looser the sand the less stirring will be required and the greater the saving in labor. When planted in the single-row system the cultivation should also aim to support the plants from tipping over as much as possible. The fruits being borne at considerable height and being of considerable weight, cause the plants to tip. In this position the fruits are subject to sun-scald on one side which gives them a poor appearance when placed upon the market.

Varieties.—The leading varieties are the Cayenne, a conical, slightly yellow, aromatic, juicy fruit, weighing as much as ten pounds: Queen, an exceptionally aromatic fruit, very desirable and very extensively cultivated; it is a good keeper, ships well and weighs as much as eight pounds per fruit: Spanish, medium in size, juicy, good quality and early, fruits weigh as much as six pounds and are a favorite on many of the markets. There are many other varieties that are good for local consumption, but not all of them possess good shipping qualities.

Marketing.—The keeping qualities of pineapples depend largely upon the care with which they are handled. They are susceptible to injuries, especially bruises, and should be handled as carefully as strawberries or other perishable fruits. Stiff bushel baskets are recommended for collecting the fruit from the plants. Some of the varieties may be removed from the plants by giving the fruit a quick jerk across the knee. Others, like the cabezonas, must be cut off. In all events, the stem must never break into the fruit. Cutting with a long stem is advised.

The fruit is best shipped in crates. It should be graded to uniformity in size and appearance. Care should be exercised to so pack that the spines on the crowns will not puncture the fruit.

The Avocado is a tropical tree, adapted to climatic conditions in southern California and a considerable portion of Florida. Most of the varieties are injured by frost, but the more hardy ones will stand a few degrees below freezing. The tree is an evergreen with large, leathery leaves. It attains a height of from 25 to 60 feet, depending upon the variety and local conditions. The wood is brittle and easily broken by winds. The flowers and fruit are easily blown from the trees. For this reason the trees should be grown in sheltered locations or artificial windbreaks should be provided.

The fruit varies in size, shape and color. While it is usually pear-shaped, it not infrequently is round or oval. The color ranges through light-green, dark-green, brown, purple and red. The center of the fruit contains a single, large, round seed. The yellowish-buttery, fleshy portion between the seed and the skin is the edible part. It is rich in protein and oil, the percentage of the latter ranging from 10 to 30 per cent of the pulp.

Professor J. E. Higgins, in Bulletin No. 25 of the Hawaiian Experiment Station, describes the fruit as follows: "Its unique character reduces to a minimum its competition with other fruits, while its rich, nut-like flavor is almost universally enjoyed among those who have known it long enough to become familiar with its peculiar charm. It is a fruit and yet so unlike other fruits as to suggest a class of its own, and for this reason it has been called a 'salad fruit.' But this term seems too limiting, because it is used in so many other ways. There are many ways in which it might be served. The simplest treatment is to cut open the fruit longitudinally, remove the seeds and serve, affording everybody the opportunity to add salt, pepper, vinegar, olive oil, lime juice or other seasoning in any combination to suit the individual taste. Some prefer it as a dessert with sugar and cream, or with wine and lemon or orange juice. It may be served on the side with soup, and in this way is delicious. It is true that the taste for the avocado is an acquired one, yet there are few, if any, food products which so quickly overcome any prejudice and become so highly esteemed. The novice may pronounce the first fruit worthless, but the second is tolerable, the third good, the fourth better, the fifth a delight and after that the difficulty of learning to like them usually gives place to that of getting them often enough."

THE TAFT AVOCADO FRUIT.[1]

The avocado is adapted to a wide range of soils. It demands good drainage and plenty of organic matter. The trees do well in the southern coast district of California and in various sections of Florida. The geographic limits of successful avocado culture are at present undetermined.

The avocado responds to judicious applications of fertilizers. The texture and flavor as well as the yield of fruit are improved by fertilizers. Excessive amounts of nitrogen should be avoided.

The trees are propagated from the seeds. These must be fresh, as they soon lose their vitality when exposed to the air. It has been a common practice to produce bearing trees from the seedlings without grafting. The seed should be planted in the soil either in pots, in nursery rows or in

[1] Courtesy of University of California, College of Agriculture, Berkeley, Cal.

permanent position in the orchard. They should be planted so that the upper portion of the seed protrudes slightly above the surface of the soil. Best results are usually obtained by planting in pots and transferring the seedlings to the field when they are of the proper size.

When the trees are not to be grafted the seed should be selected with much care, only the largest seeds from trees that produce the best quality of fruit being used. Even this care will not insure a uniform good quality in the new orchard. Like all seedlings, they seldom come true to the parent stock. Best results are, therefore, obtained by propagating through some

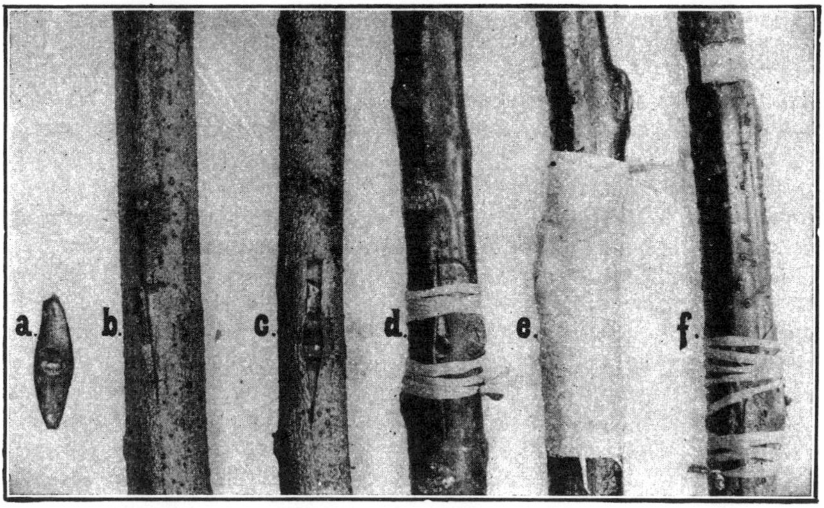

METHOD OF BUDDING THE AVOCADO.[1]

form of cuttings, selecting the scions from trees that are prolific and produce fruit of good quality.

Until recently it has been thought impossible to successfully bud the avocado. Careful study of the subject and numerous experiments have resulted in a successful method of budding, following what is known as the shield budding method. This is similar to that practiced in the budding of citrus trees. The success seems to depend chiefly upon the character of growth from which buds to be inserted are selected. Buds from what is called second flush in growth have been found superior to those in any other stage of development.

It is often desirable to re-work good-sized trees. This may be successfully accomplished by budding into new wood forced out for the purpose. The trees are usually cut back severely in the spring and the cut stubs

[1] Courtesy of U. S. Dept. of Agriculture. From Porto Rico Bulletin No. 8.

covered with wax or paint to prevent decay. Of the new shoots that start, only a few are allowed to grow and when these attain a size of three-quarters of an inch in diameter they are ready for budding in the manner above described.

Pruning should take place during the early growth of the trees to establish low heads and the proper form. After well grown, trees require very little pruning. The wood, being quite soft, will not stand abuse from pruning instruments. All cuts should be smoothly made and, on all larger branches, should be protected with a covering of paint or wax to prevent decay.

The seedling trees come into bearing between the fourth and eighth years, the average bearing age being about six years. The life of the tree in Florida and California is as yet not determined, although there are records of trees eighty or more years old in some parts of the American tropics. It will be safe to estimate the bearing life at not over twenty-five years.

Like the deciduous fruits, the avocado has a tendency to fruit in alternate years. This is generally due to setting more fruit than can be properly matured. Thinning is therefore advised. This will encourage larger size and better quality of the fruits that are allowed to remain and will not overtax the tree so as to prevent its bearing a crop the following year. Individual trees of the thin-skinned Mexican variety in southern California have produced as many as 5000 small fruits annually. Such fruits have little commercial value, but are of considerable value for their seeds, which are used for nursery purposes. Of course, these should all be grafted or budded before being set in orchards.

All fruits that are to be placed upon the market should be hand picked and handled with the greatest care. Orange clippers are advised for this purpose, about three-eighths of an inch of the stem being left on each fruit. The fruits, if to be shipped, should be carefully wrapped and packed in small packages, so that they will carry without injury. Fruits of fine quality in good condition on the large city markets in the United States sell for 30 to 75 cents each. The kind of fruits to ship will depend upon market demands and the shipping qualities of the different varieties.

The industry of growing avocados is comparatively new and a list of the most desirable varieties for different purposes is not available. At least twenty-five different varieties of California origin have come to notice. It is doubtful if the commercial variety of the future has yet appeared. At least, none have been found that may be considered good shippers.

As a fruit the avocado exceeds in food value all other species. A test of the food value of twenty-six varieties gave an average of 984 calories per pound of edible fruit. This is important, as it is more than twice the maximum noted for any other fruits. The fuel value is not far from twice that of average lean meat. Of course, they are much lower in protein.

The avocado is worthy of careful experimentation in those localities where climate will permit of its growth. It doubtless has great possibilities, although the demand for the fruit at present is limited.

The Mango.—It originated in India. There it has been cultivated for many centuries and the fruit is as important to the people of that country as is the apple to the people of North America.

The fruit of the mango is not well known outside of the regions in which it is grown. It is strictly a tropical fruit and under favorable condi-

FRUIT OF THE MANGO. SEED ON THE RIGHT.[1]

tions the tree attains a height of sixty feet or more and produces fruit for several decades. In the United States it is grown chiefly in the southern part of Florida. When in a dormant state the trees will withstand a temperature of seven or eight degrees below freezing, but if growing rapidly when freezing weather occurs, the trees are killed back to the ground.

It does best on fairly deep, rich, well-drained soils, but requires a liberal amount of moisture.

Mango trees are usually propagated from seeds. As with any other fruit, trees produced in this way are not true to the parent stock. More

[1]Courtesy of U. S. Dept. of Agriculture. From Annual Report of Porto Rico A. E. S., 1912.

recently, methods of grafting by inarching and patch grafting have come into use. The usual method of budding deciduous trees has not given satisfaction with the mango.

Inarching has long been in use in India. It has been adapted in Florida with many modifications.

The mango seeds are generally planted in pots four or five inches in diameter and eight to twelve inches deep. These may be made from cypress shingles or by using the internodes of rather large bamboos. The seeds are laid flatwise in the soil and covered to a depth of about an inch and a half. The pots must be watered at frequent intervals.

Best results are secured if seedlings are kept under partial shade. When they have attained a height of ten to twelve inches they are ready to be inarched. The pots are brought near the tree from which scions are to be secured. If the tree is so tall that its branches cannot be bent down to the ground, it will be necessary to provide a scaffolding to support the pots. The tree selected for this purpose should be one bearing the best quality of fruit. Branches for inarching should be in such condition that the bark will peel freely. A strip of bark about three inches long is removed from the side of the stock. A similar strip is also removed from the scion and the two are brought together so that the cut surfaces will fit closely and are securely held by wrapping. In about two weeks a cross-cut may be made in the stock two or three inches above the union and in the scion, just below the union. After two more weeks, these cuts may be deepened. At the end of six weeks all plants should be carefully inspected. If a good union has been effected the plant may now be severed from the tree and the top of stock removed. The plants should now be transferred to a plant house or the shade of a tree, where they should remain until one good flush of growth has matured. They are now ready to set in the field.

Budding the mango by the square patch method is also successful. For this purpose the stock or branch should be an inch or more in diameter. A patch of bark one and one-half inches long and three-quarters of an inch wide is removed with a sharp knife or chisel. Next the desired bud with an equal amount of bark attached is secured and fitted securely to the stock. The bud should be held in place with raffia or other wrapping material. The wrapping should not be too tight. A suitable form of grafting wax should be used to smear the cut edge and keep out water. The whole stem for several inches above and below the bud should be covered with waxed cloth, leaving only the bud open to view. Budding should take place when the sap is moving freely.

Mango trees should be planted about thirty feet apart each way. They should be properly cared for so as to form low-headed trees with strong branches from which the fruit can be easily gathered.

It is advisable to inter-till and during the early stages of growth inter-cropping may often take place.

The trees begin to bear from five to nine years of age.

For immediate use the fruit should be allowed to ripen on the tree. If it is to be stored or shipped long distances it should be gathered before it fully ripens. If hand picked, wrapped in paper and packed in small packages, it will keep for several weeks. The keeping period may be lengthened by cold storage.

The fruit is best prepared for eating by placing on ice, until thoroughly chilled. In this condition it may be readily peeled and sliced. The fruit is used chiefly in the fresh state, although in the tropics where grown it is frequently used for sauce or made into pies and has great possibilities for various forms of preserves.

A TOP-WORKED MANGO TREE IN FRUIT.[1]

The Banana.—The banana is strictly a tropical fruit. It is a large herb, with a perennial root stalk. The top grows rapidly and reaches a height of from ten to thirty feet, depending on variety. It requires from twelve to fifteen months from time of planting to the maturity of the fruit. Each plant bears one cluster of fruit, and upon its maturity the plant dies. Numerous shoots arise from the base of the original plant. Most of these are removed for use in establishing a new plantation, but some are left to take the place of the old plant.

Within the past thirty years the banana has become popular in the markets of the North and is quite extensively used. It excels in the ease

[1]Courtesy of U. S. Dept. of Agriculture. From Annual Report, Porto Rico Agricultural Experiment Station, 1913.

with which it is handled. On the plantation a cluster of from 100 to 200 fruits, equal in amount to a crate of other fruits, is severed from the plant with one stroke of the machete. The fruits are protected by a tough skin which readily separates from the rather dry meat.

The banana is cultivated in practically all tropical countries. Those countries leading in banana production are Jamaica, Costa Rica, Cuba and Honduras. The commercial supply for North America comes chiefly from the West Indies.

There are countless varieties of bananas, but very few of these are of commercial importance. Many of the most delicious ones are of local value only because of small size or poor shipping qualities. The varieties usually met with in the markets are the large yellow fruits, and, less frequently, the red ones.

The clusters of fruit are cut from the plant when quite green and hung up in a dark room to ripen. When shipped to distant markets, the fruit is sent directly from the plantations to the fruit steamers, and ripen while in transit. They generally reach their destination before they are sufficiently ripe to use.

The banana is grown in Florida and the southern portion of the Gulf states. It is found as far north as Charleston, S. C. In the extreme northern limits it is grown chiefly as an ornamental plant. In the southern half of Florida it has been grown commercially to a limited extent. For central Florida the Orinoco and Hart varieties are best. These are both early and hardy. In the southern part of the state the Dwarf Jamaica variety is successfully grown. If freezing weather occurs, the base of the plants to a height of two or three feet may be protected with earth or straw. If the tops are frozen they should be removed by cutting just below the frozen portion. A new growth will start almost immediately from the center of the stalk and will mature fruit before the close of the season.

With few exceptions the banana is seedless, and must therefore be propagated by planting suckers or sprouts. These are generally removed from the mother plant when several feet in height. By cutting the top of the sucker back to a foot in height it will keep for several weeks. These are set in the new plantation at intervals of ten to fifteen feet apart each way. The larger the variety, the greater should be the distance between plants.

Bananas require a fertile, well-drained soil, well supplied with humus. They develop best in a humid climate. Their extensive and tender foliage necessitates protection from strong winds. The soil between the plants should be cultivated to subdue weeds and grass and to conserve soil moisture until the plants are large enough to shade the ground.

The Fig.—While the original home of the fig tree is around the Persian Gulf, the tree will grow and thrive in any warm climate. It is very hardy and noted for its longevity, often remaining productive for a hundred years.

The fig can be cultivated in the warmer parts of the United States and will withstand considerable freezing. The young shoots are easily frosted, but, owing to its hardy constitution, a tree though severely frosted will send forth new shoots and will often bear fruit the following season. A heavy frost, however, while the sap is flowing freely, is apt to be fatal.

The warm interior valleys of Arizona and California, being dry, are much more favorable for fig culture than the Gulf states. The rainfall of the Gulf states, occurring at the time of fig ripening, often causes the fruit to burst and decay before maturity. The fresh fig is a delicious fruit, but on account of its perishable nature, has not been widely cultivated for commercial purposes.

The fig tree will grow and thrive in a variety of soils. It is a gross feeder and requires much moisture for its long spreading roots. Where frosts are liable to occur, rich, moist lands should be avoided, as this kind of soil promotes a late luxuriant growth, which is very easily killed by frost.

The fig is propagated by means of suckers and cuttings. Seldom is it propagated by seed, as seedlings have a tendency to revert to their wild state. Trees from seedlings require three years before beginning to bear, and several more years to come into full fruitage. Trees from cuttings may bear a few figs the first year and will be in full fruitage in two or three years. Cuttings six or eight inches long should be made from young, well-seasoned wood. These should be made in the spring before the sap begins to run, and hung inverted for a time until the ends are calloused over.

The trees mature more rapidly if the cuttings are planted in their permanent position. They should be planted in a deep hole, filled in with rich compost, and liberally watered. Cuttings, transplanted after growth has begun, are often retarded two or three years in growth. Barren trees may be successfully grafted.

Fig trees planted in orchards should be fifteen or twenty feet apart. This distance is sufficient for the Southern states, but in the Pacific Coast region, where the trees grow to a greater size, thirty to forty feet is a better distance. The low-branching varieties are best, as they are not so easily injured by winds.

Except to remove dead or decayed limbs, the fig tree requires very little pruning. In young orchards the cultivation must be shallow in order to avoid injury to the surface roots. Fertilizers scattered broadcast and worked into the earth near the trees are beneficial in the early stages of growth.

The Guava.—The guava, a native of tropical America, has spread to all tropical countries. In character of growth and fruit it most nearly resembles the quince of temperate regions. It is sometimes called the apple of the tropics.

The plant is a shrub, seldom attaining a height of more than twelve feet. The vegetative growth is easily killed by frost, but is renewed quickly from the roots. For this reason it can be successfully grown in sub-tropical localities.

The fruit varies greatly in size and color, ranging from an inch in diameter to the size of large apples. When ripe, it is white or yellow, with a sub-acid pulp of the same color as the skin. The color sometimes deepens into crimson. The fruit contains many small seeds. It is used chiefly for making jelly and preserves.

The guava is propagated from seeds and cuttings.

Recently this fruit has received considerable attention in Florida, where it finds a place in nearly every fruit garden. Where frosts occur, the tops may be protected during the winter by laying them down and covering them with straw and earth.

REFERENCES

"Text Book of Tropical Agriculture." Nicholls.

California Expt. Station Bulletins:
- 250. "The Loquat."
- 254. "The Avocado in California."

Florida Expt. Station Bulletins:
- 101. "Pineapple Culture, VI."
- 104. "Pineapple Culture, VII."

Hawaii Expt. Station Bulletins, O. E. S., U. S. Dept. of Agriculture:
- 28. "Effect of Manganese on Pineapple Plants and Fruits."
- 25. "The Avocado in Hawaii."
- 12. "The Mango in Hawaii."
- 20. "Shield Budding the Mango."
- 36. "The Pineapple in Hawaii."
- 29. "Management of Pineapple Soils."

Porto Rico Expt. Station Bulletin 11, O. E. S., U. S. Dept. of Agriculture. "Relation of Calcareous Soils to Pineapple Chlorasis."

O. E. S., U. S. Dept. of Agriculture Bulletin 8. "Pineapple Growing in Porto Rico."

CHAPTER 40

THE FARM WOODLOT

BY F. F. MOON

Professor of Forest Engineering, College of Forestry, Syracuse University, N. Y.

Need of Forestry.—To properly solve the land problem of any nation each acre should be put to its best permanent use. Field crops should be grown upon the tillable areas and the land which is too steep or stony for cultivation or too sterile for ordinary field crops should be made to produce repeated crops of timber. That is why the practice of forestry, which is "the raising of repeated crops of timber on soils unsuited to agriculture," is necessary to secure the proper use of *all* the land.

Forestry is not a part of agriculture. It is separate, but co-ordinate and interdependent. Agriculture has first call upon the land and selects the fertile and level areas for tillage. Forestry takes the remaining portion and raises the timber indispensable to our civilization. Both are concerned with crops, since the forester regards his timber-covered areas as fields to be sown (either by nature or artificially), tended and finally reaped, for forestry means *using* the products of the forest and does not mean locking up the woodlands for park purposes, as some people think.

The practice of forestry upon the non-agricultural soils is absolutely essential for three reasons:

(1) Timber is absolutely indispensable to our civilization.

(2) There are large areas of land which can never be used for agriculture.

(3) The indirect influence of the forest in moderating climatic extremes, in controlling run-off, etc., is necessary to the successful practice of agriculture and to the health and comfort of the people.

1. Next to food, shelter is most important. According to Fernow, over half our population live in wooden houses, and two-thirds use wood for fuel. The same authority estimates that 95 per cent of all the timber consumed in the United States is for necessities.

Our per capita consumption of wood is unusually high, and on the increase. (It is twice what it was fifty years ago.) We consume six times as much timber per capita as in Germany, and twenty times as much as in Great Britain.

2. Agriculture can never be practiced on a large part of this continent, and this land must not be allowed to lie idle. Of the 1,900,000,000 acres

of land in the United States, 550,000,000 acres are now covered with forests (65 per cent of the original forest area) and 415,000,000 acres are devoted to agriculture. Agricultural experts have estimated that within the next fifty years the forest area of this country will have been reduced to about 360,000,000 acres, and that the present area of forest land held in the form of farm woodlots (190,000,000 acres) will have diminished to 90,000,000 acres. So that with the exhaustion of virgin supplies of timber, the farm woodlot will be relatively much more important fifty years hence than it is at the present time.

3. The indirect influences exerted by forest cover are much greater than is generally supposed. Recent investigations have indicated that the rains in the interior of a continent are largely dependent on the presence of large bodies of timber situated in the track of prevailing winds.

In some parts of the Middle West the value of windbreaks in checking the force of hot southern winds may exceed their value as a source of timber, fuel and fencing. The influence of forest cover upon run-off—the drying up of springs, the increase in spring floods after extensive forest denudation—are well known. Water experts claim that the gradual lowering of the water in the soil is dependent to a large degree upon the absence of sufficient forest area.

Value of the Woodlot.—The value of a good woodlot to a progressive farmer is hard to measure in dollars and cents. It serves the following ends:

(1) It furnishes timber for home construction purposes, fuel, fence posts, etc.

(2) It should now, as in the past, furnish winter employment to horses and men. Domestic timber, telephone poles or railroad ties for the market, etc., can all be taken out during the winter months to the vast improvement of the bank account and woodlot.

(3) A good woodlot is like a bank account—it can be drawn on in time of need. After a fire, the barn may be largely rebuilt from home timbers, and in case money is badly needed, some logs or poles may be sold to tide matters over. A good farm woodlot is a fine nest egg.

(4) It vastly improves the appearance of the home place and makes it more salable.

Aside from sheltering the homestead and barns from wintry blasts, the woodlot covers the steep, rocky slopes or the marshy spots that would otherwise be most unsightly. Viewed from every standpoint—revenue, year-round farm management, appearance, real estate value and comfort—the woodlot is a splendid asset to an up-to-date farm.

Managing the Woodlot.—The average woodlot at present is suffering from the wrong point of view. It has been grazed and grazed again, burned, culled and culled again until in many cases the compact soil cannot suppor the growth of any desirable species or specimens. With the enormous stand of timber covering the agricultural land in colonial times, it is no wonder that the forest was attacked vigorously and even ruthlessly by the early settlers. It covered lands needed for tillage and it harbored enemies, beasts and redskins, of equal ferocity. With the end of the virgin timber supply less than five decades away, the farm woodlot is destined to play a still more important rôle in supplying the local markets with necessary timber. The reduction in the forest area and the increase in the value of forest products will make the woodlot more profitable each succeeding decade. Since a crop of timber cannot be grown over night, now is the time to start for the benefit of the next generation.

A Well-Protected Farm Homestead.

By protecting farm buildings with trees, comfort of the family is vastly increased and farm economy better maintained.

To make specific recommendations for the management of the farm woodlots in different parts of the country is impossible, for climate, soil conditions, species and markets are all different. General points only can be covered and if further details are necessary, bulletins from the Federal Forest Service at Washington or State Forest Office, or Manuals on Woodlot Management may be sent for.

At the outset the forest should be regarded as a crop of trees. It is sown by nature and is harvested only once every forty to sixty years, but if the crop idea is kept in mind the cultural methods to be pursued will be very easy to follow. The woodlot contains tree weeds, as well as desirable species, and the weeds as usual, should be exterminated. The laws of plant growth, as understood by the average farmer, apply to trees in the forests as well as to the plants in the field. There is only so much growing energy—light and heat, moisture and plant food—available for each acre of forest. This energy should be confined to a few valuable trees and not scattered among the several hundred additional weed trees that stand

FIELD AND WOODLOT.

Upon the fertile, level lands field crops should be raised, while the steep, rocky hillsides unsuited to agriculture should be made to yield corps of timber.

upon each acre. It should be the aim to raise a crop of valuable timber and not forest weeds.

Improvement Cuttings.—Under ordinary circumstances no improvement cuttings are attempted until the material to be cut is large enough to pay the cost of removal. Cuttings to improve the composition are sometimes made in very young stands where intensive management is possible. Such cuttings, or *cleanings* as they are called, are ordinarily beyond the pale of woodlot management, as the average farmer cannot afford to make the investment ($1.50 to $3 per acre in young sprouts) which such cleanings would cost. Therefore, it is better to postpone the cutting until the undesirable specimens reach cordwood size (say twenty-five to thirty years), when a *thinning* may be made.

The general idea in such a thinning would be to remove competing trees which take light, food and moisture from the straight, thrifty trees of more desirable species. Every farmer knows which trees are valuable

A WOODLOT AFTER THINNING.

By removing dead and diseased trees and those of less desirable species, the remainder of the stand will greatly increase its growth rate.

in his neighborhood and which individuals are not thrifty. In the Eastern states, for example, such trees as ash, basswood, tulip-poplar, red oak, etc., are generally favored over the slower-growing and less desirable beech, maple, black oak, horn bean, etc. Rapid-growing conifers, like pine and spruce, are to be preferred to slower-growing and less valuable species like hemlock and white cedar. As a rule, conifers should be encouraged upon poorer soils, since they make less demand upon the site for plant food and moisture.

A method of thinning a woodlot which foresters term the French method, can be used in many stands to advantage. The idea is to select from 200 to 250 trees per acre, depending on the species, soil, etc., to form your final crop, and to remove all weed trees or defective specimens which are in any way interfering with the growth of these selected trees. By cutting away the trees crowding and competing with them, all of the growing energy will be forced into the straight, thrifty stems which remain, with the result that the succeeding years' growth rings will be laid on the trees of greatest value. In this way railroad ties may be secured at thirty-five years, whereas if left untouched, they would not reach sufficient size until forty-five or fifty years.

For the final result, the technical quality of the species (including local demand), the growth rate and the condition of the individual tree determine whether or not it should be removed. Briefly summarized, the points to be kept in mind in making a thinning are as follows:

1. Leave straight, fast-growing, thrifty trees of most valuable species.
2. Avoid making holes in the canopy that will not be filled within five years by the natural growth of the crowns. (Excessive exposure of the soil to sunlight causes drying out of the soil, a rapid growth of weeds and diminished volume growth.)
3. In case of doubt, leave a tree, as it may be taken out at the time of the next thinning.

Reproduction Cuttings.—The previously described cuttings are designed primarily to hasten growth and to improve the composition of the stand. The reproducing of the stand is not intended, although a heavy improvement cutting in a woodlot old enough to produce seed may result in a fine stand of young seedlings the next spring. This is by accident rather than by design.

In certain of the Middle Western states where grazing is permitted in the woodlot as a matter of course, where fires and bad cuttings have exhausted and compacted the soils, reproduction cuttings are out of the question. Only weed trees or old and decrepit specimens of desirable varieties are still standing. The best, the only way, in cases like this, is to cut clean and replant with species suited to the region.

Where the soil is in good shape and good seed trees are found, a light cutting to prepare the soil, followed two or three years later by another thinning to give more light to the seedlings on the ground, will provide sufficient stand of reproduction. The thinnings, to be successful, require considerable care in removing the defective trees and specimens whose seed is not wanted. Great care should be exercised to prevent excessive light coming in at first, as weeds may then choke out desirable seedlings. After the seedlings have gotten started the trees overhead are gradually removed, the cuttings being located where light is needed for proper development of the young growth. When the leaves of the seedlings turn a yellowish-green, more light is needed and a few nearby trees should be cut.

In regions where the sprout hardwoods are found (chestnut, oaks maples, etc.), reproduction may be secured by clear cutting, allowing the woodlot to spring up from stumps. The best time for sprout reproduction is under thirty years of age, but ordinarily good sprouting species will retain this quality until fifty or sixty years of age. This type of management, *coppicing* as it is called, should not be practiced too many times in succession, as the soil becomes exhausted and the vitality of the stand lowered.

Pruning.—In certain parts of the East farmers have attempted to secure a higher quality of lumber by artificially pruning coniferous stands.

Good Work in Piling Brush.

Advocates of this plan claim that the clear lumber thus produced will bring a sufficiently larger yield to pay for the cost of this intensive process. On the other hand, men who have sawed second growth white pine, which was artificially pruned, claim that loose knots are produced by too rapid drying of the stub. If pruning is desirable to improve the looks of a piece of woodland—to open up a vista beneath the crowns—it may be done, but let the cost be charged against landscape improvement and not added to the cost of the forest crop.

Planting.—Where it is desired to cover an unsightly area or abandoned pasture with trees, planting may be resorted to, as the proper species are immediately started at the correct distance. The question is often raised, "Why is not nature's method followed and seed scattered broadcast on the soil?" The answer is this: It has been found after repeated experiments that broadcast seeding is not only extremely expensive on account of the high price of seed, but the results obtained are decidedly uncertain, owing to the activity of squirrels or field mice and the frequent drying out of the seed. Placing young seedlings in the ground six feet apart is more certain and cheaper in the long run. Planting six feet apart each way, an acre containing 1210 trees can be planted at a cost of $7 to $10, depending on price of labor and whether seedlings or transplants are used.

If the woodlot has been very much run down as a result of injudicious cuttings, excessive grazing or repeated fires, it may be desirable to plant under the openings with fast-growing, shade-bearing species. In this case it is desirable to first make as heavy a thinning as circumstances will permit, and then, after the timber has been removed, plant the open spaces immediately with the chosen species before grass and weeds take possession of the soil. Underplanting a run-down woodlot of broad-leaf trees with four-year transplants of spruce or pine is a splendid way of injecting new blood. The trees will cost about one cent each in the ground, and from three to four hundred per acre is generally sufficient.

Financial Results.—The best measure of the success of any farm activity is the financial yield obtained, and it is safe to say that the difficulty in marketing the forest crop and the long waits between receipts are largely responsible for the slight attention paid the woodlot. Forest management must be financially profitable before it will be accepted by the farmer.

At the present time forest products are not sold as easily as grain, potatoes or fruit, and this fact often causes discouragement. While the average farmer will scan the market reports very closely to find out the prevailing price for his field crops, the same man is apt to sell the standing timber on his woodlot to the first mill owner who offers him real money. If the selling of forest products can be simplified and the farmer can be assured a reasonable return from his non-agricultural acres, it is certain that the practice of forestry by the individual owner will advance rapidly. These small holdings are destined to play a more important part in supplying local timber markets in the coming generation, but it is necessary that proper marketing facilities be provided in order that the owners receive a fair return.

When an offer is made for "all the timber on the woodlot," great care should be exercised before it is accepted. Such a sale usually results in parting with the cream of the trees at a meagre price and leaving the land in the possession of forest weeds, for the local mill man generally "skins the lot." It is far better to designate by axe marks the mature trees and those that should be removed for the good of the remainder, and thus sell a known amount at a fixed price per thousand board feet on the stump.

A sale contract covering methods of cutting, payments, fire protection, provision against waste and excessive damage should be drawn.

The New York State College of Forestry at Syracuse has provided a selling service for the private timberland owners of the state. Two years ago a study of the wood-using industries was made in connection with the Federal Forest Service, and at that time considerable data concerning stumpage prices, costs of manufacture and value of manufactured products, etc., were secured. This information is on record in a card catalog, and when a farmer or small land owner writes for information concerning the management of his woodlot and a possible sale, he is put in touch with the nearest manufacturer, and the dimension and grades in demand and average selling price given him. By this means the College of Forestry is acting as a clearing house for information and is endeavoring to secure a fair price for the man who raised the forest crop. In the extension of this scheme of co-operative marketing of forest products lies the future profitable management of non-arable lands by the farmers of this country.

As previously stated, forestry must be financially profitable, else it will not be practiced by the business men of this country. The farmer, however, is in the best possible position of all owners of forest land to practice forestry, for he has the land, he has an annual income from his arable land, and finally, he has the winter season to work in his woodlot.

European experience proves beyond a doubt that forestry does pay good dividends—from $2.50 to $7 per acre per year *net revenue*—while from the woodlots of this country, a revenue of $109,000,000 in 1899 and $195,000,000 in 1909 was obtained.

Forest plantations will yield from four to five per cent compound interest upon the value of the land, plus the cost of planting. Thus it can be proven conclusively that the practice of forestry is a paying proposition at present stumpage values, while the reduction in the timbered area will cause an increase in these values and much higher yields will be obtained.

Summary.—The farm woodlot should be treated as a *producing portion of the farm,* and the following points should be borne in mind:

1. Tend your woodlot during the slack periods. It will pay handsomely.
2. Cut your firewood and fence posts where cuttings are needed, and not where it is easiest to cut.
3. Do not permit fires to run rampant through the woods. It kills the little trees and checks the growth of the big ones.
4. Do not permit extensive grazing in the woodlot. If more pasture is needed, clear-cut the best land and sow to grass. You can't raise good grass and good timber on the same piece of land.
5. Use the same energy and business sense in selling a crop of trees as you would in selling a crop of apples. Know how much you

have and about how much it is worth. If you can't get your price, hold on, as your woodlot keeps on growing in bulk and value while you sleep.

REFERENCES

"The Farm Woodlot." Cheyney and Wentling.
"Principles of Handling Woodlots." Graves.
"Principles of American Forestry." Green.
Canadian Dept. of Agriculture Bulletin 209. "Farm Forestry."
U. S. Dept. of Agriculture, Forest Service, Bulletins:
42. "The Woodlot."
52. "Forest Planting."
U. S. Dept. of Agriculture, Forest Service, Circulars:
97. "The Timber Supply of the U. S."
117. "Preservation Treatment of Fence Posts."
138. "Suggestions to Woodlot Owners in Ohio Valley Region."
U. S. Dept. of Agriculture, Year-Book 1914. "The National Forests and the Farmer."

CHAPTER 41

BEAUTIFYING HOME GROUNDS

BY A. W. COWELL

In charge of Landscape Gardening, The Pennsylvania State College

How ridiculous would be the man who proceeded to build his house by first buying up a lot of lumber, bricks, pipes and paint, and then going ahead to put them together without first having a very definite working plan! Too often that is the way the home surroundings are arranged and ornamented—and don't they appear so? Whether of houses or homes, which is a broader term and includes the house and all its immediate surroundings, it is essential to good results to have a definite working plan and stick to it. If you cannot plan it yourself, you will save time and money by obtaining expert advice.

The Survey.—To make such a plan for the grounds, first measure up the boundaries of the area and note all the features contained therein, including buildings, standing trees with their approximate spread, steep banks, rocks, swampy places and other natural features, besides roads and walks. Next, indicate the fine views and the views of undesirable character that should be eliminated. This accomplished, you are ready to plan changes and alterations and record your desires and ideals. Using an ordinary foot ruler, adopt an eighth or a sixteenth of an inch to represent a foot of your actual measurements and thus accurately draw on paper the survey you have made. Draw the new scheme on the same scale. It is likely that practical and ornamental considerations will be thought of together in this way. This study of the place as a whole should aim at a systematic arrangement, an effective appearance, and provide for convenience and comfort. Beautification should start back in the practical first arrangement of buildings, roads, paths, windbreaks and screens, and not be confined to the little patch of ornamented front lawn.

Planning for Convenience.—Speaking of the farmhouse, one located upon the north side of an east and west public road will most nearly approach the ideal in matters of arrangement of parts. The house should stand not less than 150 nor more than 400 feet from the road, somewhere near the center of the farm lands; for all operations begin and end at the house, and it should, therefore, be most conveniently centered. It should face the south. Behind it at a distance of about 150 feet, or less, if fire hazard is minimized, may stand the barns and other service buildings arranged perhaps most conveniently for work around a hollow square or barn court. A windbreak upon the west and north of this group of buildings, while sheltering them, will likewise protect the house and home garden and orchard from prevailing winds. To reach this court, the entrance drive would pass the house, preferably upon the west side, but not nearer than fifty feet—a little spur being provided from it for the house visitors. For convenience, arrange the buildings with the chicken house nearest the kitchen, and for comfort place the hog pen or other more obnoxious necessity farthest to the northeast.

A Convenient and Attractive Farmstead.[1]

Provide a vegetable garden, hedged in if possible, very near the house and let it be tastefully laid out and contain the small fruits such as currants, raspberries, strawberries, grapes, asparagus and rhubarb; possibly dwarf fruit trees as well as kitchen vegetables and hardy flowers for cutting. Such a garden need not, in fact, it *should* not, be relegated to the back of the place, but may lie toward the front road and form the east side of the remaining area of the house lawn. All of this makes for convenience of operation of the farm plant and affords opportunity to ornament it with the greatest ease and effectiveness. But it is only an ideal, and most places are

[1] Courtesy of Doubleday, Page & Co., Garden City, N. Y. From "Farm Management," by Card.

very unlike it. Others may profit by such a picture and it will give them something to work toward along the line of home ornamentation.

Formal Ornamentation.—Before planning and planting for ornamentation, have a landscape ideal. If the place is in the city surrounded by straight streets, shade trees in avenue rows, massive architecture and other conventions of one kind or another, the formality of straight walks, terraced lawns, clipped trees and bushes, and even architectural gardens and statuary is quite in keeping. The object is to provide a setting appropriate to the building and in harmony with its environment. Formal landscape treatment requires expert knowledge beyond the scope of this chapter. Simple "old-fashioned" flower gardens with box-bordered paths, and rose gardens with grass walks laid out in simple geometrical fashion can be successfully designed by the amateur, but they should be set away by themselves and in close relation to the house or other buildings, or else isolated and secluded from any general view.

Informal Ornamentation.—For farm homes in the open country it is much more effective and harmonious to arrange the home grounds with naturally sloping lawns, convenient curving paths and trees and shrubs grown in their natural form in groups and masses. The simplicity of nature's masses of foliage as seen in copses and fence rows, of her trees standing in splendid dignity alone or in groups of soft outline; her wood edges that are irregular in outline and of material of different heights rising from the ground line to high trees of the background; her colors, so subdued and so gracefully blended together—these should be our ideals. A close observation of natural landscape in general, and little bits here and there in particular, may properly instruct us in the proper arrangement of the simple home grounds as regards the planting. Very few homes depend for their charm upon their natural surroundings. More often is all natural beauty destroyed when man takes possession and adapts the land to his economic necessity. But hints for the changes and for the embellishment as well should be taken from the place itself and its environment if it is to be in harmony with its site and become what we call "charming."

In the plan, considerations of convenience rule, but beauty may be served also. The paths, which are not in themselves things of beauty, however well constructed they may be, should if possible be kept out of the center of the picture, and should not divide the open lawn more than necessary. They should pass from house to road toward the side of greatest travel, which satisfies the consideration of convenience while also creating a graceful curve in course of the path and leaving unbroken the central area. Do not interrupt any path by a flower bed, flag pole or fountain, except in pleasure gardens, and do not cause its course to become circuitous and tiresome in order merely to introduce curves. Where the distance is less than fifty feet, introduce no deviation from a perfectly straight course. Walks should not be lined by ribbons of flower beds, but

AN EXAMPLE OF GOOD INFORMAL ORNAMENTATION.[1]

A small country place in which trees, vines, flowers and shrubbery are used to good effect.

[1] Architect, D. Knickerbacker, F. A. I. A.

a few good specimens or a group of bushes or a tree may properly stand in the bend of a path.

Lawn Planting.—The lawn also should serve the considerations of practicability with beauty. It should therefore be rather open and unbroken. It should be somewhat enclosed by a frame of shrubbery, but it must not, without defeating both considerations, be planted all over with trees and bushes standing alone. This is a "spotty," not effective use of material and is hard to maintain. Arrange the bushes—they may be wild ones taken from the woodside, flowering kinds from the nurseryman, or both—planted in groups together, in bordering beds at sides of

A Desirable Method of Planting Daffodils, Showing the Bulbs Before Covering.

the lawn area. Such a bed should be dug over, no grass should be maintained between bushes, and its outline against the lawn planned in long, flowing curves like that of the native woodland. Set the tall-growing species generally toward the center or rear of the bed, allow the bushes to grow together in a natural way, cut out the dead wood, but do not trim them into rounded formal shapes. There should be a bed made against the base of the house and other buildings. Plant this with shrubs of a moderate height of growth and of good bushy habit. More homes look bare and uninteresting, almost inhospitable, because of the lack of this planting which lends a warming influence to the building, than from any other reason. Against unattractive objects or views noted in the preliminary survey should of course be arranged a heavy plantation. It may take on a little different character and contain many trees, especially the smaller growing kinds, as well as evergreens and closely planted shrubs. Do not forget the softening influence of clinging vines in helping to harmonize houses and landscape and to afford privacy to porches and service buildings. Shade trees do not clothe the earth and in this dissertation are left to the last for the reason that shrubbery and vines and grass are all-important in home ornamentation; shade trees are not so often forgotten or so badly used by the amateur planter. Arrange them in groups, not rows, of different species, and for lawn specimens, endeavor to preserve the lower limbs. Street and roadside trees are of a different ideal.

Hyacinth Bed.[1]

Use of Flowers.—The use of flowers and flower beds in home ornamentation is not to be discouraged, although it harbors much danger in chances of introducing colors and material difficult to place and to harmonize with most natural landscape. If the advice be confined to that type of flowers called "old-fashioned" hardy plants, the matter is simplified. They add charm to most shrubberies and lawns when planted along in front of the shrub beds, arranged in and out among the shrubs. The other class of flowers known as "bedding plants," which includes geraniums, cannas, coleus, salvia and so forth, is more difficult to blend, more foreign to simple places and more predominant in its color note. Such bedding can be best used directly against the house, but never in beds, stars, crescents and bologna sausage shapes, in the middle of the lawn,

[1] Courtesy of The Countryside Magazine, New York City.

and seldom in front of shrubbery, as effectively or so practically as hardy perennials of the other class.

These are all principles and ideals to observe in drawing a plan for home ornamentation. As to detail, each place is a problem unto itself, to be solved with due regard to two services—convenience of use and landscape charm. Nature is a good instructor in principles. From her examples in field and wood we learn of the "open center" of lawn with borders of massed foliage, of the beauty in flowing, rounded outline, both of foliage and of ground. We cannot copy nature, but we can and should derive much inspiration and many ideas in the uses of trees, shrubs, vines, flowers and grasses, and how to combine them into good groups and masses. A few uses and combinations follow. They are merely catalogued. Perhaps they will suggest details in the comprehensive plan.

SUGGESTED MATERIALS

Street Trees for roadside or driveway should consist of one species upon one road, but different species upon different roads.

Maples.—Sugar, Red, Norway, distances, 45–35–40 feet.
Oaks.—Red, Pin, Scarlet, Mossy Cup, distances, 45–30–35 feet.
Elm.—American, the ideal American tree, distances, 45 to 60 feet.
Linden.—American Basswood, European or Crimean, distances, 45–35–32 feet.
Plane.—European (or Oriental), distance, 35 feet.
Ash.—American white, distance, 35 feet.
Gingko.—Chinese Maidenhair Tree (narrow streets only), distance, 25 feet.

Trees for lawn planting, besides those mentioned for street use:

Oaks.—White, English, Golden, Pyramidal.
Maple.—Weir's Cut-leaf, Purple, Norway, Cork-barked, Tartarian.
Elm.—Cork-barked, Scotch, Japanese.
Linden.—Silver, Weeping Silver, Broad-leaved.
Mountain Ash.
Empress Tree (*Paulonia*).
Larch.—European and Japanese.
Bald Cypress.—An excellent.
Magnolias.—Chinese species.
Buckeye.
Japanese Maples.
Pine.—White, Swiss, Dwarf Mountain, Austrian.
Fir.—Douglass, Colorado Silver.
Spruce.—Englemann, Colorado Blue, Eastern, Norway.

For screen planting, to obscure objectionable views:

Poplars.—Lombardy, Bolles Silver, White.
Willows.—White, Laurel-leaved.
Mulberry.—White.
Maple.—Weir's Cut-leaf, Water or Box Elder.
Birch.—White, Red.
Ailanthus.
Spruce.—Norway.
Pine.—Austrian, Scotch, White.
Arborvitæ.—Western.

Shrubs for screen:

Sumacs, Privet, Nine-bark, Elder, Alders, Dogwood, Witch Hazel, Red Bud, Shad Bush, Bush Honeysuckle.

Shrubs suitable for the base of the house:

Japanese Barberry, Thunbergs, Waterer's and Van Houttes' Spirea, Red-Twigged Dogwood (*C. alba*), Dwarf Deutzia, Hydrangea, Kerria, Lespedeza, St. John's Wort, Regel's Privet, Japanese Rose (*Rugosa*), Snowberry, Stephenandra, Mahonia, Rhododendron, azaleas, Eulalias (ornamental Grasses), and hardy perennials.

Shrubs suitable for general border plantings:

Blooming in early spring: Amelanchier, azaleas, daphne, calycanthus, forsythia, cercis, cornus mas, cydonia, lindera, lonicera fragrantissima, almond.

Blooming in late summer: Althea, baccharis, aralia spinosa, caryopteris, cephelanthus, clethra, hydrangea, hypericum, lespedeza, Sambucus canadensis, spirea Bumalda, tamarix, vitex agnus castus, rosa rugosa.

For winter berries: Rosa rugosa, berberis, corylus, crateagus, euonymus, ilex, cephelanthus, callicarpa, physocarpos, symphoricarpos, viburnum opulus, ligustrum, rhodotypos.

For winter bark color:—Cornus alba, stolonifera, lutea, Euonymus alata, kerria, eleagnus, tamarix.

Suitable hedges:

Japan barberry, privet-california, common and for untrimmed hedge, Regellianum.

Flowering hedge.—Spirea van houttei, Althea, rosa rugosa, cydonia, deutzia gracilis, lilacs.

Protective hedge.—Barberry, rhamnus, cratægus, gleditsia.

Evergreen hedge.—American arborvitæ, hemlock, white pine (for a broad hedge), Norway spruce.

Suitable for windbreaks:

Evergreens.—Norway spruce, Douglas spruce, Scotch pine, Austrian pine, arborvitæ.

Deciduous.—Poplars, willows, box elder, larch, birch, ailanthus, mulberry, osage orange, and other tree species set out about ten feet apart to form a belt at least twenty feet wide.

REFERENCES

"The Practical Flower Garden." Helen Ely.
"A Woman's Handy Garden." Helen Ely.
"Gardening for Beginners." Cook.
"Landscape Gardening." Waugh.
"Landscape Gardening." Maynard.

CHAPTER 42

WINDOW GARDENING

By A. W. Cowell
In charge of Landscape Gardening, The Pennsylvania State College

The prime requisites in raising plants in the house are proper soil, good drainage, equable temperature, the correct amount of sunlight and regular care in watering and re-potting. Contrary to superstition, no better geraniums can be grown in a tomato can than in a piece of fine pottery. So you may choose your own receptacle so long as it fits the plant it is to house—being neither too large nor too small.

Drainage.—Good drainage is brought about by having an opening in the bottom of the receptacle—at least half an inch in diameter, and for very large jars or tubs, three or more openings. Over these lay pieces of broken pottery to prevent the dirt from falling through. Good drainage allows any excess of moisture to escape and provides for free circulation of air through the soil. This prevents it from becoming soggy and sour.

Soil and Exposure.—Good soil is often difficult to secure. Many planters take chances and use what is handiest. This is a mistake. Even the blackest woods earth is not always most suitable to use. Soil which is clayish and bakes is not good; neither is light, sandy soil. A combination of the three types, however, is satisfactory, and a soil recommended by a practical florist is one made up as follows:

Skim off the sod thinly from a bit of pasture land and take the loam directly under the sod for the ground matter of your soil; mix together 32 quarts of this loam with 4 quarts of black woods earth and 4 quarts of sharp sand. For the plant-food, mix together 8 quarts of decomposed manure, 1 quart of air-slaked lime and 1 quart of ground bone (bone meal). Now mix and mix and *mix* these two piles together, sift through a sieve of a quarter-inch mesh, and you have a soil suitable for the most "persnickity" of plant tastes.

As to light, for flowering plants generally, a south or east window is best. Some foliage plants and ferns like the sunless windows or interior of a room. Their numbers are few, however, and this is unfortunate.

Method of Potting.—To pot up the plants, cover the drainage material in bottom of the flower pot with an inch or two of the soil prepared as above described. Then place the plant roots flatwise into the soil, holding the stem erect while soil is sprinkled in until the pot is nearly full, and press down firmly but not too hard. Now sprinkle a light covering of soil (*not* firmed in) over top of the pot to prevent excessive evaporation and drench with water.

A plant which has made a vigorous growth may need more root room. It fills up the receptacle and becomes "pot bound," as the florist says. It should be "shifted up" to a pot the next size or two sizes larger. There is failure in pots larger than necessary. Reverse the plant with the palm of the left hand against the top of the pot and the stem passing through the fingers and with a slight tap the pot may be removed. The ball of roots should be put into the new quarters, setting the old surface about level with the top of the new pot. Chink in new soil around the ball of roots and then water the plant plenteously. In potting up plants from the summer garden—geraniums, snapdragon, ten weeks' stocks, petunias, scarlet sage—set them in a shaded corner for a few days and *syringe* the tops daily before placing in the sunny window.

Removing the Plant from Old Pot.[1]

Nothing is more unsightly than a lot of "leggy" old plants or puny weak ones. Grow few plants and have kinds which will thrive. Make cuttings and keep the plants vigorous and shapely. Cut back the old plants, remembering that flowers are on new wood, and that it is "easier" for an old plant to grow a lot of new shoots than to carry leaves on the tips of long, lanky branches. So cut the old plants back vigorously once in a while.

In selecting plants at the florist's for home window gardening, do not be interested in those of his hottest house; choose plants from a night temperature of about 50 degrees. Plants like equable temperatures as well as

[1] Courtesy of The Countryside Magazine, N. Y.

regularity of other conditions. Do not allow the room temperature to get above 70 degrees in daytime nor below 50 degrees at night.

Watering.—The watering of plants is largely a matter of judgment. It is offered as good advice that a plant should be watered when it needs it, and contrariwise *not* when it does not need it. Water copiously once in two days rather than a little each day, unless the earth has become dried out. This can be determined by tapping the flower pot with the finger nail; a clear, ringing sound will indicate dryness; a dull sound shows a damp condition and water *not* required. Watering at the roots is not sufficient, strange to say. Plants respond also to a wetting of the leaves. This can

A WELL-PROPORTIONED FERN.[1]

be accomplished by turning them half over in a tub and syringing the tops. Do not allow the sun to play upon wet leaves; it may injure them severely.

Feeding Plants.—Pot-grown plants respond to "feeding up"—the application once in a while of liquid manure—which is merely stable manure and water allowed to stand a few days and strained. Apply the liquid once a month for two successive waterings. Bone meal worked in at the top of a pot is slow in its action, but beneficial. There are prepared plant-foods which are valuable and convenient, but more expensive than these two.

[1]Courtesy of The Countryside Magazine, N. Y.

Ferns and Foliage Plants.—Plants should fit the purpose for which they are intended. If a green and growing plant for house decoration during the winter months, one that can be moved from place to place, is wanted, the aspidistra, dracæna, cocos and other palms, asparagus plumosus, rubber plant, auraucaria (Norfolk Island Pine), and with restrictions, the Boston, Scott's and crested ferns should be chosen. Maidenhair ferns do not generally succeed with house culture, but of them all *Adiantum gracillimum, cuneatum,* and *Capillus-Veneris* are best. For the table, small ferns in a fern dish are as good as anything except the pots of spring bulbs as they are brought in from the cellar. The fern dish should have a porous earthen dish in which to grow the plants, regardless of the ornamental character of the dish in which it rests. Ferns, purchased as "table ferns," are but baby big ferns, and are good to use in a fern dish. As they become larger, they should be transplanted to larger pots or to a fern box and placed in a sunless window.

A Large Boston Fern.[1]

Flowering Plants.— For the sunny window flowering plants may be used. A shelf on castors is the best stand, as it may be turned around occasionally. A box the length of the window and from six to eight inches deep may be used. Set the plant jars up an inch above the bottom of the tray in order that they may not be too wet. For plants there is a good variety: Abutilon, flowering begonias; fuchias, swainsomia, billbergia, Quenista, geraniums (especially "Christmas Pink''), cuphea, lobelia, oxalis (also for hanging basket), cyclamen (in shaded spot), Chinese, starry, and "Baby'' primroses, stevia, Marguerites, candytuft, alyssum, ageratum, heliotrope, bouvardia, balsam ("touch-me-not''), cactus, and plants mentioned later which may be brought from the outdoor summer gardens. Among bulbs, amaryllis, calla and the so-called "Dutch bulbs'' are probably the most satisfactory of all flowering plants for the house. A dozen Paper white narcissus may be grown in an eight-inch deep glass dessert-dish half full of sand, above which the bulbs rest, held firmly in place by

[1] Courtesy of House and Garden, Published by Robert J. McBride & Co., N. Y.

pebbles sprinkled in among them and covered with water. Freesias, hyacinths, tulips, daffodils, single narcissus, crocus, even the Easter Lily in any of the many named varieties listed in seedsmen's catalogues, may also be grown successfully by the amateur. These Dutch bulbs should be purchased in September, and excepting freesias, Paper white narcissus and Roman hyacinths, which may be started at once, planted and put away for about six weeks to *form roots* before any top growth is allowed. Set them in a cool place—buried in coal ashes in a corner of the cellar or out-of-doors in a box buried in cinders for one inch above the pots and protected from freezing too hard by a layer of straw, leaves and boards. Keep them moist and cool. They may be brought into flower a pot at a time and furnish pleasure from Thanksgiving to April—a gamut of color and delightful fragrance.

Bulbs Grown in Water-tight Receptacle.[1]

Plant Lice.—The most prevalent insect pest attacking house plants is the plant louse, a little green insect feeding upon the under side of leaves and tender shoots. Another form is black. Both forms are combated by tobacco concoctions obtainable ready-made at the seed store; also, the plants may be fumigated with burning tobacco, dusting the leaves with tobacco dust, and by spraying the leaves with soap in solution. The insect must be *wet* with the solution, so care must be exercised in spraying to reach the under side of the leaves. Another common pest is the brown scale which attaches itself firmly to branch or leaf and resists water and fumigation. It can be removed by brushing the leaves and by kerosene emulsion, which, however, may injure a tender plant.

REFERENCE.

"Manual of Gardeinng." Bailey.

[1]Courtesy of The Countryside Magazine, N. Y.

BOOK IV

LIVESTOCK FARMING

(Animal Husbandry)

CHAPTER 43

Advantages and Disadvantages of Keeping Livestock

Without the aid of domestic animals as beasts of burden, man would have a sorry existence. The horse, ass and camel have been of great service in past ages in aiding man to conquer new regions, and by their aid he has been enabled to very materially increase his productive power.

Animals have also been a great aid to man as a source of food and clothing. Those countries that depend upon animals and animal products the most are, as a rule, the most productive and highly civilized. In North America animal products, such as meat, milk, butter, cheese, lard, eggs, etc., constitute fully one-half of the value of the products of human consumption.

A large part of the vegetation on the earth is unsuited for human consumption. Of this, such by-products as straw and stover are converted into milk, butter, cheese, meat and animal fats. It is estimated that 80 per cent of the corn produced in the United States is consumed by livestock in the county where produced. This conversion of crude farm products adds greatly to the quality of man's diet.

The essential characteristics of domesticated animals are: (1) their ability to convert food into energy and animal products for human use, (2) the readiness with which they become subject to the will of man, and (3) their prolificacy or ability to breed abundantly.

Value and Importance of Livestock.—The United States and Canada with 28,000,000 horses, 63,000,000 cattle, 51,000,000 sheep and more than 62,000,000 swine, is pre-eminently a livestock country. South America leads in the production of sheep with 115,000,000 and ranks third in cattle with 48,000,000. It falls to India to lead in cattle production, which, including the water buffalo, numbers 125,000,000 head. The United States, however, far outranks all other countries in its numbers of horses, mules and swine. It is second in production of cattle and sheep.

During the past half century, the livestock in the United States has increased about three times in numbers and about six times in value. While numbers have not quite kept pace with increase in population, the value per capita has steadily increased. This increase in value has been due chiefly to two factors: (1) the improvement in livestock, and (2) the increased value per unit of weight of animals and animal products. In 1850 the average fleece of a sheep weighed 2.4 pounds; in 1900 it had increased to 6.9 pounds. During the fifty years sheep nearly doubled in number, while the yield of wool increased five times. This increase was due chiefly to breeding rather than feeding. If statistics were available we would doubtless find that the increase per cow in milk, and particularly in butter-fat, would not be less striking.

Thirty-five years ago, the usual work-team in the corn belt consisted of two 1000-pound horses. Today, the prevailing team is three 1500-pound horses. This increase in the size of the team has been an important factor in increasing the man unit of production on the farm, and has undoubtedly been one of the factors instrumental in the increase in land values in that region. The following table gives the numbers, value per head and total value of the principal classes of livestock in the United States for 1880 and 1915, as reported by the Bureau of Statistics of the United States Department of Agriculture:

NUMBERS AND VALUE OF LIVESTOCK ON FARMS IN THE UNITED STATES IN 1915 AS COMPARED WITH 1880.

Class of Animals.	1880.			1915.		
	Number.	Value per Head.	Farm Value.	Number.	Value per Head.	Farm Value.
Horses	11,202,000	$54.75	$613,297,000	21,195,000	$103.33	$2,190,102,000
Mules	1,730,000	61.26	105,948,000	4,479,000	112.36	503,271,000
Cows	12,027,000	23.37	279,889,000	21,262,000	55.33	1,176,338,000
Other cattle	21,231,000	16.10	341,761,000	37,067,000	33.38	1,237,376,000
Sheep	40,766,000	2.21	90,231,000	49,956,000	4.50	224,687,000
Swine	34,034,000	4.28	145,782,000	64,618,000	9.87	637,479,000
Total			$1,576,908,000			$5,969,253,000

From the above table it will be noted that the total value of livestock in the United States increased from a little more than $1,500,000,000 in 1880 to nearly $6,000,000,000 in 1915. During that period, horses and mules doubled in number and quadrupled in value. The increase in numbers of cows and other cattle did not quite double, while the value per head of the former considerably more than doubled and the latter slightly more than doubled. The increase in numbers of sheep and swine was slightly less marked, but in both of these classes the value per head slightly more than doubled.

ADVANTAGES OF LIVESTOCK

Animals Furnish Food, Labor and Clothing.—Even when not profitable to rear anmals for market, the cost of living on farms may be greatly reduced by the judicious production of livestock and livestock products for the home food supply. The difference between the purchase price of animals and animal products and the price which the producer receives has materially increased during recent years. The value of these products to the farmer for his own consumption is equal, whether bought or produced on the farm. Furthermore, animals and animal products may be produced on a small scale on most farms on what otherwise would be wasted.

The acres of land cultivated by each horse depends on the size of the horse, character of farming, the type of soil and the topography of the land. In England, two horses are generally required for 80 acres of light, sandy soil or 60 acres of heavy, clay soil. In the United States, there is about one horse or mule of working age to each 30 acres of improved land. Formerly, many oxen were kept for work, but these have been largely replaced by the horse and mule because of their more rapid movements and consequent greater efficiency. The draft of the ox is larger in proportion to his weight, but his slowness has caused his displacement with the increase in the value of human labor.

With the introduction of cotton and silk, the value of animal products as sources of clothing decreased relatively. The value of leather, wool and

UTILIZING WOODLAND FOR PASTURE.[1]

hair is very large, however, and plays an important part in the clothing of the human race.

Animals Make Use of Land Otherwise Unproductive.—According to the last census, only about one-half of the farm area in the United States was improved land, and only about two-thirds of the improved land was in farm crops, including meadows. The other one-third, together with considerable of the unimproved portion, is utilized as pasture for animals. On most farms there are areas more or less extensive which may be steep, stony, partly wooded, undrained or otherwise unprofitable for cultivated crops, that may be utilized for grazing purposes.

Animals Utilize Crops that would be Wholly or Partly Wasted.—The straw of the cereals, the stover of corn, have little value on most farms except as roughage and bedding for livestock. Low grades of hay, damaged

[1] Courtesy of E. K. Hibshmann, Pennsylvania State College.

by rains or delay in harvesting, often are unsalable, but may be utilized for feed for stock. In the same way corn and small grains are sometimes damaged by exposure to the weather or early frosts, and may have considerable feeding value, but no value on the market.

Animals Transform Coarse, Bulky Products into Concentrated Form.—Animals convert coarse, bulky, raw materials into a more concentrated and valuable finished product, and one that may be marketed with less cost and to much better advantage. It requires about 10 pounds of dry matter to produce 1 pound of beef or 30 pounds of dry matter to produce 1 pound of butter. The farmer in transforming such coarse products to a more refined one not only reaps the profit in the process of manufacture, but the pound of butter may be sent to a market a thousand miles away, when the material from which it was made could not be profitably sent to a market ten miles distant. One cent a pound for transporting butter

LIVESTOCK AND THE SILO INCREASE THE PROFITS ON HIGH-PRICED LAND.[1]

would be but a small percentage of its value, but one cent a pound for transporting hay would be prohibitive.

Animals Return Fertility to the Soil.—In the manufacture of these finer products on the farm, animals leave much of the fertilizing material to be returned to the soil. This has been thoroughly discussed in the chapter on farm manures, and need not be emphasized at this point. In considering livestock farming from this standpoint, it is only necessary to determine whether it has been successful in maintaining soil fertility. A study of the crop-producing capacity of the soil in different regions shows conclusively that crop yields are largest where large numbers of livestock are maintained.

Livestock Facilitate Good Crop Rotations.—A good crop rotation should include inter-tilled crops, small grains and grasses and clovers.

[1] Courtesy of Webb Publishing Company, St. Paul, Minn. From "Field Management and Crop Rotations," by Parker.

Livestock make possible the production and profitable utilization of grasses and clovers. When these are fed to livestock and the manure is returned to the land, the fertility of the soil is increased. Good sods, plenty of manure and animals to utilize by-products extend the range of crops that may be grown on the farm and thus provide for better crop rotations.

Capital More Fully Used.—The wheat farmer in the Northwest is very busy from spring until fall, but is generally idle from September to March. When livestock is kept, labor of men and teams is more fully employed and equipment more fully utilized.

Livestock Call for Higher Skill.—Animal husbandry, including keeping of dairy cattle, poultry, etc., may be made to require higher skill than ordinary extensive production of crops. It calls for the same requirements so far as the care of the soil and the production of crops are concerned, and there is added to this the skill of the breeder and the feeder. The products of skilled workmen command a higher price than do those of the unskilled workmen. In this country those communities that have given most attention to livestock are in general the most prosperous. There are, of course, some exceptions to this.

More Land may be Farmed with the Same Labor.—This is true only in the extensive grazing of livestock, as exemplified in the ranches of the West, notably in the breeding and rearing of cattle and sheep. When these are brought to the farm of the feeder, they really reverse the process and call for increased labor and skill on the unit of area.

DISADVANTAGES OF LIVESTOCK

Animals Require Larger Capital—This is especially true when kept in connection with the production of hay and grain. On a 160-acre farm 40 head of cattle worth $1500, 40 sheep worth $300 and 20 hogs worth $300 may be kept, and the farm made to raise all the necessary food for them. This would increase the capital of the farm by $2100. It would also call for additional capital in buildings, and this would all be an increase over what would be required if the same land were used only for cash crops. On a farm that supplies all the feed for livestock, $10 per acre invested in livestock may be considered as moderate. If only the coarse feed is grown it may carry stock to the value of $25 to $30 per acre. This is exemplified in many dairy farms close to market, and sometimes on farms where stock are fattened for market.

Capital of Perishable Nature.—Animal diseases, such as tuberculosis or foot and mouth disease in cattle, cholera in hogs, and internal parasites in sheep, may quickly wipe out the animals on any particular farm. This entails a loss not only of the product for a single year, but also of all the capital that may have been invested in feeds and labor to bring the stock to that stage of maturity at which it was destroyed by disease.

Formerly, it was not uncommon in the corn belt to find farmers keeping 100 or more head of hogs in a single herd, but it is now deemed best to keep them in herds of small units, not more than 20 or 30, as protection against cholera. More recently, of course, methods of control have been developed, which, if properly administered, hold the disease in check.

Products Cannot be Indefinitely Held.—The holding of livestock for a considerable time after reaching the proper stage of fattening for the market entails considerable loss. It may sometimes result in actual decrease in quality with little or no increase in weight, and a loss of both food and labor for maintenance. In this respect livestock for meat is sharply contrasted with wheat and some other cereals that may be held almost indefinitely with very little deterioration. It is true that the development of better markets, systems of cold storage and methods of preserving meat have lessened somewhat this difficulty.

Crop Failures may Cause Loss on Livestock.—A low production for the staple crops used largely for livestock food results in a marked advance in price. This frequently causes a loss to the farmer on his livestock. This is especially true in case of swine that depend so largely on concentrates for their production. A decrease of one-quarter in the yield of a staple crop for the whole country often causes an increase in price so marked that if the whole crop were sold it would bring more than a normal crop or an extra large one. Since, however, so large a percentage of the crop is fed, this does not mean much to the farmer unless there is a corresponding increase in price of meat animals. A number of instances may be cited whan a marked advance in price of corn without a corresponding advance in hogs has induced farmers to sell their hogs before fully ready for market, thus causing the hog market to decline in the face of advancing prices on corn. This condition once under way will often continue for a full year before normal prices again prevail.

The advantages and disadvantages of keeping livestock have been presented without prejudice, and it must be apparent that the advantages seem to outweigh the disadvantages, especially from the standpoint of permanent systems of agriculture. It is, of course, recognized that with increasing population there should be a tendency for people to depend more and more upon the direct products of the soil in the form of cereals, vegetables and fruits rather than to depend so largely upon animal products; and doubtless the increase in land values and high prices of animal products will gradually tend in this direction.

CHAPTER 44

Breeding, Care and Management of Farm Animals

By W. H. Tomhave

Professor of Animal Husbandry, The Pennsylvania State College

BREEDING OF LIVESTOCK

History of Animal Breeding.—The first systematic work in animal breeding was done among the Arabians. This is indicated by the character of the Arabian horses that were developed during the sixteenth and seventeenth centuries. Following the Arabians, the French did the next constructive breeding of animals, which was at that time encouraged by the French Government in the developing of their breeds of horses. The most important animal breeding from the point of view of the American farmer of today was done by the people of the British Isles during the last half of the eighteenth century, and throughout the entire nineteenth century. Robert Bakewell is known as the foremost early breeder of livestock, having begun his work about 1764 and continued it until the time of his death. He was followed by noted men such as Collings Brothers, Booth and Bates, all of whom were early breeders of Shorthorn cattle. Amos Cruickshank was probably the most noted breeder of recent years, and was recognized as the peer among the Shorthorn breeders of Scotland during the nineteenth century. Great interest was then shown in developing the various classes of livestock and this has resulted in giving us our present breeds of pure-bred livestock.

The foundation work in animal breeding in America was done largely during the last half of the nineteenth century. The foundation animals used by most of the noted breeders were imported into the United States and Canada from Europe. Large importations of well-bred animals were made into the United States from 1880 up to 1900. Since that time only limited importations have been made into this country, as most of the noted animals in America at the present time are the product of American breeders. While a great deal of work has been done in both Europe and America, less than two per cent of all the farm animals in the United States and Canada at the present time are of pure breeding. This seems to indicate that there is a fertile field for livestock breeding for the American farmer.

Lines of Breeding.—There are three distinct lines of breeding that can be followed by the American farmer. These may be enumerated as follows: breeding of pure-breds, grading and cross-breeding. The breeding of pure-bred animals is by far the most important system of breeding, and the one that should be followed to a greater extent by farmers in the United States and Canada. The greatest improvement can be made in a herd of livestock by this system of breeding. The use of both pure-bred sire and dam enables the farmer to follow a more rigid system of selection and cull out undesirable individuals, which is not always possible in grading and cross-breeding. There is one weakness, however, that every breeder of pure-bred animals is apt to encounter, and that is a certain degree of hesitation about eliminating an animal from his herd that may be pure-bred and yet not up to the standard which he has set for building up his herd.

Two Pure-bred Bulls. Polled Angus on the Left, Shorthorn on the Right.[1]

Sires of this character should head the herd of all well-regulated stock farms.

Grading is another means of making a marked improvement on the average farm herd. By grading is meant the mating of a common or relatively inferior animal with one that is more highly improved, usually a pure-bred. This pure-bred may be either the sire or dam, but it is usually the sire, as the sire can be used upon a number of females in the herd and thus exercise greater influence in making the improvement. If the pure-bred dam and a grade sire are used, very little improvement is made; besides, such improvement is restricted to one mating. If a pure-bred sire is used for five generations, it will mean that at the end of that time the herd is practically pure-bred, but can never be registered. Rigid selection and the use of a pure-bred sire should always be continued.

By cross-breeding is meant the mating of two pure-bred animals of different breeds. Nothing is to be gained by such method of breeding, as it destroys the pure lines that may have been established and also has a tendency to cause a greater variation. Cross-breeding is sometimes profitably carried on in producing market animals, but it should never be carried beyond the first generation. Cross-bred animals should never be

[1] Courtesy of Dept. of Animal Husbandry, Pennsylvania State College.

retained as breeders in the herd, as this has a tendency to cause sterility in the breeding animals, besides retarding progress in building up the herd.

Selection of a Breed.—The selection of the breed of animals must be determined by the farmer or livestock grower, as there is no such thing as the "best breed." All breeds of livestock have been developed for a definite purpose and among all breeds are found desirable and undesirable individuals. In deciding upon a breed, the farmer should secure all data available about the breeds in which he is interested and adopt the one that

PURE-BRED SHORTHORN BULL.[1]

will best suit his conditions. It is highly important that he select good individuals of the breed adopted and that he continue with that breed indefinitely. To change breeds at the end of one or two years is not conducive to improvement, and means a loss of time. It is important to select representative animals that possess pronounced characteristics of the breed, and if possible to secure animals with a known ancestry. In the selecting of a brood sow as an illustration, such sow should come from a prolific

[1] Courtesy of U. S. Dept. of Agriculture.

strain. The same thing is true in the selection of a herd boar or any other animal that is to be used for breeding. It is a wise precaution to visit the herd from which the animals are to be selected, in order to study the prepotency of the sire that is at its head. It is also very important to avoid the introduction of barrenness or sterility in the herd. The sires selected should be strong, vigorous and in thrifty condition. Since the sire will be used on a number of animals, it is important that he be given the greatest consideration, both as to individuality and pedigree.

Pedigree.—The mere fact that an animal is pure-bred and has a pedigree is not an indication of its being a desirable animal. The pedigree is not a guaranty of excellence in the animal. There are many poor pure-bred individuals as well as desirable individuals. A combination of good individuality, together with a pedigree tracing back to known ancestry, will usually result in the securing of desirable animals.

Gestation Period.—The farmer or livestock breeder must keep a record of the breeding dates of his animals. This should be done so that he may know at what time they are to produce their young. The gestation period varies with the various classes of animals. For cows, it is about 9 months, or approximately 280 days; for mares, 11 months, or approximately 340 days; for ewes, 5 months, or about 150 days; for sows, 4 months, or about 112 days. The gestation period for all animals as stated is only approximate, and has been known to vary a number of days from this period. It is well, however, to watch the animals closely at the end of the number of days given for each class of livestock.

CARE OF LIVESTOCK

The breeding, feeding and management of livestock must be combined for the greatest success. Each class of livestock must be given special care and attention, and a system worked out to meet the needs of the farm. The feeding of the young animals, for instance, should not begin at the time of birth, as is so often the case, but should be properly carried on during the gestation period. The young life begins at the time of breeding and for the greatest development must be properly nourished throughout the gestation period. A well-bred animal does not guarantee the production of a desirable individual unless the animal is properly fed, so that the growing fœtus may be properly nourished. Young growing animals must have an abundance of food that is rich in protein and mineral matter for the development of muscle and bone rather than fattening material. This, combined with proper exercise and plenty of fresh air and sunlight, will result in a properly developed individual.

Preparation of Feeds—In feeding livestock, it is necessary to economize on the use of grains; yet at the same time, it is not a wise plan to carry this economy to an extreme. The method of preparing the feed for stock will vary with the different classes of livestock and the different kinds of feeds used. Cooking feed for hogs was at one time considered a desirable practice, but hog-feeding experiments conducted in Canada and the United States for the purpose of comparing the merits of cooked and uncooked grain all show an actual loss from cooking. There was a saving of labor and larger gains for uncooked feed.

The grinding of grain for farm animals will depend upon the kind and price of grain and the animals to which it is to be fed. Small grains, such as wheat, barley and rye, should always be crushed or ground before they are fed. The kernels of these grains are hard and some of it, if fed whole, will pass through the system of an animal without being masticated or digested. There is a saving of about six per cent in feeding value of corn when fed ground or cracked instead of whole. Generally speaking, when corn is worth more than 75 cents per bushel, it will pay to grind it or have it cracked for all classes of farm animals, except when fed to cattle where hogs follow in the feed lot.

Feeding Condimental Stock Feeds.—The feeding of proprietary stock foods or condition powders should be avoided. These preparations usually cost from ten to thirty cents per pound and contain nothing that cannot be secured by using standard feeds. They are usually made up of ground screenings, weed seeds, bark of trees, a little oil meal, and such materials as charcoal, copperas, epsom salts, etc. The feeding of such "foods" will do more harm than good. When animals are out of condition, the addition of a little oil meal to the regular feed will usually give fully as good results. Salt, usually found in these preparations, should always be supplied to farm animals in liberal amounts.

Care of the Breeding Herd.—The breeding herd must be properly cared for if the best results are to be secured. It is not necessary to keep the animals fat, but they should be kept in a thrifty condition, so that they can supply the nutrients necessary to properly develop their young during the gestation period. Breeding animals should have exercise, plenty of nutritious feed and good water. They should be fed largely on farm-grown feeds where the right kind can be produced cheaply.

Care of Work Animals.—The term work animals applies usually to horses and mules. These animals are the principal beasts of burden in the United States and Canada. The best results can be secured only through proper feeding and care. Work horses and mules should receive the largest portion of grain ration during the morning and noon meals, and be allowed the bulk of their roughage at the evening meal. The reason for this is that the horse and mule do not possess large stomachs, and thus cannot carry a large amount of bulky feed without seriously interfering with their ability to work. The amount of grain and roughage to supply depends upon the work that is being done. For a horse doing heavy work, about 1¼ to 1½ pounds of grain to 100 pounds liveweight daily should be allowed, and approximately the same amount of roughage. This amount should be reduced to about one-half the regular allowance when the horses stand idle over Sunday or any other day. Over 90 per cent of all cases of azoturia in horses taking place on Monday morning result directly from carelessness in over-feeding. Work horses should not be watered when overheated, but a horse accustomed to drinking water from which the chill has been removed will usually suffer no injury if allowed to rest a short time before watering. The usual and common practice is to allow the horse all the water he cares to drink before feeding in preference to heavy watering after feeding.

Assist Animals at Time of Giving Birth to Their Young.—There is probably no time when breeding animals require assistance and watching as much as at the time of giving birth to their young. It is well to watch the animals at this time and provide them with comfortable quarters and the proper feed. It is a good practice to allow only a limited ration at this time. The system will be in a much better condition to give birth to the young than where full allowance of feed is supplied. If the animal has difficulty in giving birth to its young, assistance should be given, which in case of horses and cattle, can best be secured by calling in a competent veterinarian.

MANAGEMENT OF LIVESTOCK

The management of livestock increases in importance with the rise in the value of livestock and the increase in the cost of feeds, labor and, building materials. The three most important factors to be kept in mind in the economical production of livestock is to keep down the cost of shelter, labor and feed. The buildings or housing facilities for all classes of farm animals should be adequate, yet not expensive. If they can be made convenient and comfortable, that is all that is necessary. Too many farmers insist on making their buildings too warm. This is seen in many cases where large basement barns are built that become extremely hot during the winter. Such barns favor the development of livestock diseases, rather than keeping the animals in a healthy condition. Farm animals will thrive much better and be healthier if they are put in open sheds that offer protection from cold winds, rain and sleet. This is especially true in case of cattle and sheep. Hogs and horses can also be kept in open sheds the same as cattle and sheep if they are given plenty of bedding and are kept dry. The sleeping quarters for all farm animals should be kept well bedded.

Open Sheds.—A number of experiments have been conducted to compare open sheds and warm barns for cattle and sheep. In nearly every case it has been found that beef cattle fed in open sheds made greater daily gains, consumed less feed per pound of gain, and were in healthier and thriftier condition than those kept in warm barns. The housing of cattle and sheep in open sheds is a saving to the farmer, as it does not require as much capital to construct a shed as it does to construct the usual expensive barn. It is also a saving of labor, as the cattle are not tied like they are in the barn. Open sheds should be built to face the south so the interior will not be exposed to the severe north winds. They should be built high enough so that the manure can be taken out by driving into the shed with the wagon or manure spreader. Feed carriers should also be provided in order to save carrying a large amount of feed.

Arrangement of Labor.—The amount of labor necessary to care for the livestock should be reduced to a minimum. This can be done by arranging convenient quarters in which to feed the livestock. The farmer's and livestock producer's business should be so arranged that the bulk of the labor connected with the livestock comes during the winter. If this is done it means that the labor employed upon the farm can be distributed more equally throughout the entire year. It can be used to work the fields during the summer and care for the livestock during the winter. Very

Open Sheds for Steer Feeding.[1]

Shelter of this character is less expensive than warm barns, and wherever the climate is not too severe steers make better gains for feed consumed than when sheltered in warm barns.

little labor is required during the summer if plenty of pasture of the proper kind is provided. Such distribution of labor also makes it possible to secure more competent help than where it can be employed only during a portion of the year.

The Kind of Farm Animals.—The class of farm animals to keep will depend entirely upon the location and equipment of the farm. On farms where a large amount of pasture and rough feed is produced, beef cattle and sheep are best adapted. This is also true of farms where there is no

[1] Courtesy of Dept. of Animal Husbandry, Pennsylvania State College.

adequate means of transportation. With good transportation facilities or near cities where there is a good demand for dairy products, dairying may be advisable. In many sections of the United States and Canada where cream only is sold from the dairy, hogs make an admirable addition to the dairy. Hogs, on the other hand, are well adapted to most all types of farming, and provide a source of quick returns from the feeds fed. The number of farm animals to keep upon a farm depends entirely upon the size of the farm and the feeds that can be grown. It is a good practice to produce as much as possible of the feeds necessary to maintain or fatten the livestock produced on the farm. This does not mean that feeds should not be purchased. The purchase of nitrogenous supplements to feeds grown on the farm is not as universally practiced as it should be.

Regularity in Feeding and Watering.—The best results from farm animals cannot be secured unless the feeding and watering is done with system and regularity. Plenty of clean water should always be supplied. The more water consumed by an animal, the more of the feeds supplied will it consume, thus producing heavier gains or larger amounts of milk. The cost of the feeds supplied is a factor of importance. The cost of the feed bill should be kept as low as possible. This can be done only by the use of farm-grown feeds. In many cases a large amount of roughage or grain is grown that does not have a ready sale, possibly on account of being slightly damaged by weathering or improper curing. Such feeds can best be used upon the farm. Not only does it provide a desirable place to dispose of them, but the fertility which would be lost if the feeds are sold from the farm is thus saved. Such practice makes the land more fertile and more productive than where such crops as hay, stover and corn are sold from the farm.

Observing Individuals.—Every owner of livestock should study the individuals in the herd and see that they are in good condition of health. It frequently happens that animals are not doing well, and upon investigation it is found to be due to internal or external parasites. Usually an unthrifty animal is infested with internal parasites, which, if noticed in the early stages, can often be destroyed. External parasites, such as lice, are a source of annoyance and should be destroyed. In the case of sheep, it is an excellent practice to dip all of the flock in a coal-tar dip at least once a year. This is usually done following shearing in the spring. It is also well to provide new pasture for young lambs at weaning time, as at that time they are more subject to stomach worms than at any other time. This is due to the fact that they become more easily the prey of worms on account of the change from nursing the dam to depending entirely upon food supplied for their maintenance. Hogs should frequently be sprayed or dipped with a coal-tar dip so as to destroy lice that are often found on their bodies. Hogs are also often unthrifty as the result of stomach worms.

Keep up Records.—It is highly desirable for a farmer or livestock breeder who is breeding pure-bred animals to keep his records up to date.

It frequently happens that desirable pure-bred animals are grown on the farm, but their registration is not completed. Such practice is well enough where only market animals are being produced. There may come a time, however, when the breeder will desire to sell animals as breeders. Buyers of pure-bred cattle require the registration to be complete in order that they may sell any offspring produced from such animals for breeding purposes. Registration involves only a small amount of time and expense, but is a practice that is well worth while.

Preparation and Shipping of Livestock.—All livestock, whether breeding animals or market animals, should be in the very best of condition when shipped. If pure-bred stock is shipped by express, it should be properly crated. If shipped by freight, it should be properly tied and bedded. If the animals arrive in good condition, the purchaser will gain a good impression of them upon first inspection. If they arrive in poor condition due to careless preparation, the buyer as a rule will not be satisfied and probably will not make another purchase. In selling pure-bred livestock by mail, it is always a wise plan not to praise too highly the animals that are offered for sale. It is much better to have the purchaser find the animals that are shipped him better than he expected. Such practice usually makes more sales and is a good means of advertising. If a customer is not satisfied with the animals shipped, the breeder should always make it a point to satisfy his customer either by refunding the purchase price and the expense of shipping or by shipping another animal.

Cattle, hogs or sheep when shipped to market should be started in as near normal condition as possible. Some farmers salt heavily before shipping in order to get the proper "fill" on the market. Cattle salted just before they are shipped will arrive on the market in poor condition. They will be feverish, will drink very little water, will not eat much hay and will also be apt to scour. Cattle in such a condition usually sell at a discount. The car in which the livestock is to be loaded should be well bedded and in the case of cattle, the racks should be filled with hay so they can eat while en route. Always ship the livestock so as to reach the market early in the week, as there is usually more active buying at that time than later in the week.

REFERENCES

"Manual of Farm Animals." Harper.
"Types and Breeds of Farm Animals." Plumb.
"Beginnings in Animal Husbandry." Plumb.
"Productive Feeding of Farm Animals." Woll.
"Animal Breeding." Shaw.
"Feeding and Management of Farm Animals." Shaw.
Farmers' Bulletin 350, U. S. Department of Agriculture. "Dehorning of Cattle."

CHAPTER 45

FEEDS AND FEEDING

BY DR. H. S. GRINDLEY AND SLEETER BULL
Professor and Associate of Animal Nutrition, University of Illinois

Introduction.—A knowledge of the scientific principles of stock feeding is important to the stockman. This knowledge is not absolutely essential, as many have achieved success in feeding as a result of years of experience. However, "experience is a dear teacher" and if one combines a study of the scientific principles of feeding with the experience gained in the barn and feed lot, he will learn the art of successful feeding more quickly, more thoroughly and with less expense than if he depends upon experience as his only teacher.

Chemical Composition of Feeding-stuffs.—All feeding-stuffs are composed of a great number of different compounds which are grouped into five classes, viz., water, mineral matter, crude protein, carbohydrates and fats. These classes of compounds are usually spoken of as "nutrients," because they are used for the nutrition of the animal.

Water is found in large amounts in such feeds as green pasture, silage, beets and milk, while such feeds as hay, bran, corn, middlings, etc., contain from 10 to 20 per cent water. A knowledge of the water content of feeds is important for two reasons: (1) feeds high in water content are lower in feeding value, pound for pound, than feeds low in water; (2) feeds containing more than 18 or 20 per cent water usually ferment and spoil when stored in bulk.

Mineral Matter, or ash as it is sometimes called, is that part of the feed which remains as ash when the feed is burned. In the animal body, mineral matter is used principally for the repair and growth of bone. It is also used in the growth and repair of the muscles and vital organs. It is found in the blood and other body fluids. A certain amount of it is absolutely essential to proper growth and development, or even for life itself.

Most of the roughages, especially the legumes, as clover, alfalfa and soy beans, are quite high in mineral matter. Also such feeds as tankage, middlings, cottonseed meal, linseed meal and bran are high in mineral matter. The cereal grains, especially corn, are low in mineral matter. Consequently, in feeding horses, cattle and sheep, little account need be taken of the mineral matter of the ration, except to provide salt, as these animals are nearly always given feeding-stuffs, some of which are high in mineral matter. However, in case of hogs, the ration may be deficient in mineral matter, especially if considerable corn is used in the ration. The hogs should have access to a mineral mixture consisting of charcoal, air-slaked lime, salt, wood-ashes and rock phosphate or "floats."

Crude Protein includes all the compounds of the feed which contain the element nitrogen. Familiar forms of protein are egg albumen (the white of the egg) and casein (the curd of milk). Protein is found in all living matter and is absolutely essential to life. It is found in every plant cell, but in larger amounts in the seeds. It also occurs in every animal cell and makes up a large part of the solid matter of the blood, muscles and organs of the body. Thus the crude protein of the ration is absolutely essential to the animal for the repair and growth of the muscles, bones, organs, etc. It is also essential for a pregnant animal for the formation of the fœtus and, later, for milk production. If there is any surplus of protein in the ration above the requirements of the animal for the purposes just mentioned, the surplus may be used to produce energy or to liberate heat. If there is still a surplus, it may be used for the production of body fat. However, protein is not an economical source of energy or body fat, as it usually is the most expensive nutrient and the one which it is most often necessary to buy. Hence, no more protein should be fed than needed by the animal for repair, growth and milk production. Tankage, cottonseed meal, linseed meal, gluten feed, distillers' grains, brewers' grains, bran, middlings and soy beans are high in protein. The legume hays are also relatively high in protein. Corn, timothy hay, the straws, fodder, stover and silage are low in crude protein.

Carbohydrates are the chief constituents of all plants. However, they are not found to any large extent in animals. Familiar forms of the carbohydrates are starch, sugar and vegetable fiber, such as hemp, paper and cotton. As the carbohydrates contain such a variety of compounds which differ considerably in nutritive value, they are often divided into two sub-classes: "nitrogen-free extract" and "crude fiber."

Nitrogen-free extract includes those carbohydrates which are high in feeding value, as starch and sugar. In the animal body these substances are used as a source of energy to do work or for heat to keep the body warm. If there is any surplus, it may be used for the production of energy and the formation of body fat. As carbohydrates are considerably cheaper than protein, it is more economical to use them for these purposes than to use protein. Feeds high in nitrogen-free extract are corn, wheat, barley, rye, rice and oats. The flour by-products, the oil meals, the straws and hays contain medium amounts: while the pastures, silage and packing house by-products are low.

Crude fiber includes the tough, woody, fibrous portion of the plant. Owing to the fact that it is not very digestible, the nutritive value of crude fiber is less than that of the other nutrients. In the animal the digested crude fiber is used as a source of energy and as a source of body fat. Feeds high in crude fiber are the hays, straws, fodders, stovers and roughages in general. The cereal grains, the oil meals and most mill feeds are low in crude fiber.

The Fats, sometimes called "ether extract," include all the fats and oils found in the feed. Practically all plants contain some fats, although usually in only small amounts. In animals, fats occur much more abundantly, occurring in nearly every organ and tissue. Fat animals often contain 40 or 50 per cent of fat. The fat of the ration is used in the animal as a source of energy and as a source of body fat. It is about two and one-quarter times as valuable as protein and carbohydrates for these purposes. Tankage and the oil meals contain the largest amounts of fat of the ordinary feeding-stuffs.

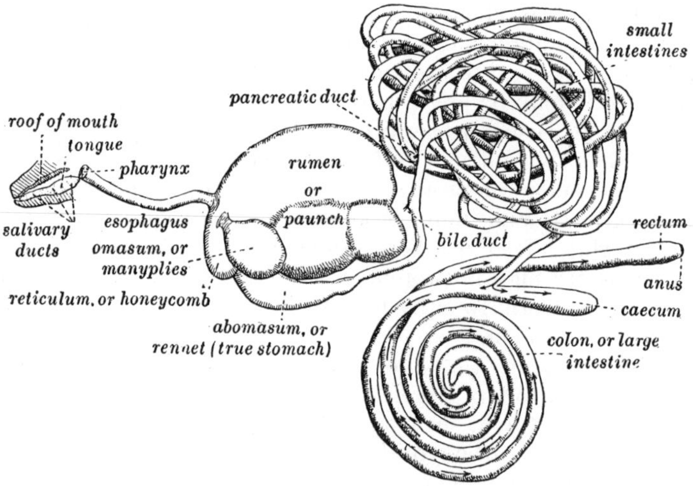

THE DIGESTIVE TRACT OF A COW.[1]

Digestion of the Nutrients.—Before the nutrients can be of any use to the animal they must be digested and taken up by the blood. Digestion is the process of separating the useful constituents of the feed from those that are useless, and changing the useful constituents to such form that they may be taken up by the blood. These processes take place in the mouth, stomach and intestines. Inasmuch as only the digestible nutrients of a feed are of value to an animal, the amount of digestible components of the feed are of special interest to the stockman. Table I shows the percentages of the digestible nutrients in the ordinary feeding-stuffs. (See Appendix to this book.)

The Nutritive Ratio.—Proteins are used primarily for growth and repair, while carbohydrates and fats are used for energy and fat production. The nutritive ratio expresses the value of a feed or ration as a flesh pro-

[1] Courtesy of Iowa State College.

ducer or as an energy and fat producer, *i. e.*, it is the ratio of digestible crude protein to digestible carbohydrates and fat in the feed or ration. Inasmuch as fat is two and one-quarter times as valuable as carbohydrates, the amount of digestible fat is multiplied by two and one-quarter and added to the amount of digestible carbohydrates. The sum is then divided by the amount of digestible protein. The first term of the ratio is always "1," while the second term is obtained by the following formula:

$$\frac{\text{digestible carbohydrates} + 2\frac{1}{4} \times \text{digestible fat}}{\text{digestible protein}} = \text{second term of the ratio.}$$

The nutritive ratio is written as "1 : 6" or "1 : 14," or whatever it may be. It is read as "one to six," or "one to fourteen," Thus one finds the nutritive ratio of corn as follows: from Table I it is found that 100 pounds of corn contain 7.8 pounds of digestible protein, 66.8 pounds of digestible carbohydrates and 4.3 pounds of digestible fat. Then, substituting in the above formula:

$$\frac{66.8 + 2\frac{1}{4} \times 4.3}{7.8} = 9.8$$

Therefore, the nutritive ratio of corn is 1 : 9.8. The nutritive ratio of a ration containing two or more feeds may be calculated in a similar manner.

Ordinarily, a nutritive ratio of 1 : 6 or less is called a narrow ratio; *i. e.*, the feeding-stuff or ration contains a relatively large amount of protein and a relatively small amount of carbohydrates and fat. A ratio of 1 : 7 to 1 : 9 is called a medium ratio; *i. e.*, there is present a medium amount of protein and a medium amount of carbohydrates and fat. A ratio of 1 : 10 or greater is called a wide ratio; *i. e.*, the proportion of protein to carbohydrates and fats is relatively small.

The Energy Values of Feeding-stuffs.—One of the functions of the ration of an animal is to act as a source of energy to do work or form heat. Also the formation of body fat may be looked upon as a storage of energy, because it may be used as a source of energy to do work or for heat at any time when the ration is insufficient for these purposes. Hence, in addition to a knowledge of the digestible nutrients in feeds, the scientific stock-feeder should have a knowledge of the energy values of feeds, *i. e.*, the value of different feeding-stuffs for doing work, storage of fat, milk production, etc. Energy values of feeding-stuffs are expressed in "therms." A therm is the amount of energy in the form of heat necessary to raise the temperature of 1000 pounds of water 4° F. The energy values of some of the common feeding-stuffs are given in Table II.

Feeding-stuffs.—In general, feeding-stuffs may be divided into two classes, concentrates and roughages, according to the amounts of digestible nutrients and their energy values.

Concentrates are feeding-stuffs which contain a relatively large amount of digestible nutrients and energy in a small bulk. They usually are highly nutritious in nature. Concentrates usually have an energy value of 60 or more therms per 100 pounds. Concentrates may be subdivided into *nitrogenous and non-nitrogenous concentrates.*

A nitrogenous concentrate is one which is relatively rich in protein. It usually contains 11 per cent or more of digestible protein. Common examples are tankage, cottonseed meal, linseed meal, gluten feed, dried distillers' grains, dried brewers' grains, soy beans, bran, middlings and shorts. As a rule, but few nitrogenous concentrates are produced on the

THE RESPIRATION CALORIMETER IN USE FOR AN EXPERIMENT.[1]

farm and therefore they must be purchased. Nitrogenous concentrates are almost essential in the rations of all growing animals in order to furnish protein and mineral matter so essential to the proper development of muscle and bone. Likewise the milk cow requires nitrogenous concentrates in order to provide the large amounts of protein and mineral matter which she excretes in her milk. Fattening animals and work horses often need small amounts of nitrogenous concentrates, especially if they are still growing.

A non-nitrogenous concentrate is low or only medium in protein content, but is usually rich in carbohydrates. It generally contains less than

[1] Year-Book, U. S. Dept. of Agriculture, 1910.

11 per cent of digestible protein. Examples are corn, barley, oats, wheat, rye, molasses and dried-beet pulp. Ordinarily the farmer raises all the non-nitrogenous concentrates necessary, and usually it will not pay him to buy such feeds on the market. All classes of fattening animals require large amounts of non-nitrogenous concentrates in order to furnish the carbohydrates and fats which, as has already been stated, are the cheapest sources of body fat. Also work horses must have large amounts of non-nitrogenous concentrates in order to furnish energy for doing their work. Milch cows need medium amounts, while they should be used more sparingly in the rations of growing and breeding animals.

Roughages are feeding-stuffs which contain a relatively small amount of digestible nutrients, or net energy in a large bulk. They usually contain less than 40 therms of energy per 100 pounds. Roughages contain a large amount of crude fiber which lowers their feeding value considerably. Roughages, like concentrates, may be sub-divided into *nitrogenous and non-nitrogenous.*

Nitrogenous roughages usually contain 6 per cent or more of digestible protein. Examples are clover, alfalfa, cowpea, soy-bean hay and alfalfa meal. In general, all the legume hays fall under this sub-class. Nitrogenous roughages should be grown on nearly every farm, not only for their feeding value but also for their fertilizing value in the crop rotations. It will rarely pay to buy nitrogenous roughages on the market, as they can be produced more cheaply at home. The nitrogenous roughages are valuable in the rations of nearly all classes of animals except hogs, and even they make valuable use of some of them at times. Fattening animals, with the exception of hogs, should have nitrogenous roughage. Unless they do, it will be necessary to feed large amounts of nitrogenous concentrates, and even then the results will be only fair, unless corn silage is used. Growing animals should have nitrogenous roughage, as it furnishes much of the protein and mineral matter so essential to their proper development. Even pigs may make use of some alfalfa or clover hay. It is almost impossible to produce milk economically unless nitrogenous roughages are used. Breeding cattle and sheep need little or no other feed than nitrogenous roughages in winter. Brood sows and boars will eat considerable of the leaves. Horses may have nitrogenous roughages if they are clean, well cured and free from dust. Often, however, they are too dusty for horses.

Non-nitrogenous roughages usually contain less then 6 per cent of digestible protein. Examples are timothy hay, corn silage, corn stover, oat straw, wheat straw, barley straw and rye straw. Silage is the best form in which to get all the feeding value of the corn crop. It may be used to advantage in the rations of practically all classes of animals except hogs, if it is properly supplemented with other feeds. The other non-nitrogenous roughages have little value except in the rations of mature breeding animals, stockers and work horses.

The Requirements of Farm Animals.—Knowing the digestible nutrients and the energy in the different feeding-stuffs and the amounts of nutrients and energy required by farm animals, one can formulate approximately a proper ration for different farm animals under different conditions.

The Balanced Ration.—A balanced ration is a ration which contains all the nutrients in such proportions, forms and amounts as will nourish properly and without excess of any nutrient, a given animal for one day. Extended study of the amount of each nutrient required by the different farm animals for the various purposes for which they are kept has led to the formation of so-called "feeding standards." Theoretically, feeding standards may be looked upon as formulas which tell at a glance the amount of each nutrient necessary to produce a given result. In practice, however, feeding standards cannot be regarded as such, but only as a guide to be used in connection with one's practical knowledge of the amounts, proportions and combination of feeds which are used in stock feeding. Although there are a number of valuable feeding standards, the limits of this chapter will permit a discussion of only a few.

The Wolff-Lehmann Standards show the requirements of farm animals under different conditions, expressed in pounds of total dry substance, digestible crude protein, digestible carbohydrates and digestible fat per 1000 pounds live weight. The nutritive ratio required by the animal also is given.

The calculation of a ration according to any feeding standard consists essentially of three steps: (1) Having given the requirements for an animal of a given weight, usually 1000 pounds, the requirements of the animal under consideration are determined. (2) A "trial ration" is assumed, using the amounts and proportions of concentrates and roughages which, in the opinion of the feeder, are necessary. (3) The trial ration is modified by adding or deducting concentrates or roughages of such composition as to furnish approximately the required amounts of nutrients.

Thus, for example, one calculates a ration according to the Wolff-Lehmann standard for a 1200-pound horse at light work as follows: According to the standard (see Appendix, Table III) the requirements of a 1000-pound horse at light work are as follows: dry substance, 20 pounds; digestible protein, 1.5 pounds; digestible carbohydrates, 9.5 pounds; and digestible fat, 0.4 pounds. The first step is to calculate the requirements of a 1200-pound horse, which are found to be as follows: dry substance, 24 pounds; digestible protein, 1.8 pounds; digestible carbohydrates, 11.4 pounds; and digestible fat, 0.5 pound. The second step is to assume a trial ration which will meet approximately the requirements as determined in the first step. From the amount of dry substance required and from practical experience, one judges that a ration consisting of 12 pounds of oats and 14 pounds of timothy hay will about answer the purpose. Calculating the dry substance and digestible nutrients of this ration from Table I, the following results are obtained:

	Dry Substance, pounds.	Digestible Protein, pounds.	Digestible Carbohydrates, pounds.	Digestible Fat, pounds.
Oats, 12 pounds.........	10.8	1.1	5.9	0.5
Hay, 14 pounds.........	12.2	0.4	5.9	0.2
Total ration........	23.0	1.5	11.8	0.7

Comparing the nutrients of the trial ration with the requirements of the standard, it is seen that the trial ration is a little below the standard in dry substance and protein, and a little above it in carbohydrates and fat. Thus the third step is to modify the trial ration so that its nutrients conform to the standard. Consequently, a feed which is high in protein and low in carbohydrates should be substituted for part of the ation. Inasmuch as it is not desirable to lessen the bulk of the ration as the dry substance is already a little low, one may substitute two pounds of linseed meal for two pounds of the oats of the ration. The ration then contains the following nutrients:

	Dry Substance, pounds.	Digestible Protein, pounds.	Digestible Carbohydrates, pounds.	Digestible Fat, pounds.
Oats, 10 pounds.........	9.0	0.9	4.9	0.4
Oil meal, 2 pounds......	1.8	0.6	0.6	0.1
Timothy hay, 14 pounds.	12.2	0.4	5.9	0.2
Total ration........	23.0	1.9	11.4	0.7

The nutritive ratio is:

$$\frac{11.4 + 2.25 \times 0.7}{1.9} \text{ or } 1:6.8$$

This ration, except that it is a trifle low in dry substance, comes very close to satisfying the standard. Of course, in many cases, especially until one has had considerable practice in the calculation of rations, the trial ration may have to be modified several times before the ration conforms with the standard. However, by applying his practical knowledge, the stock feeder should not have much difficulty in calculating balanced rations.

In view of modern investigations, certain modifications must be made to the Wolff-Lehmann standards to adapt them to American conditions. In practically every instance the amount of dry substance prescribed is 10 to 20 per cent too high. The protein prescribed is from 10 to 40 per cent too high, the greatest difference being in the case of fattening and working animals, and, consequently, the nutritive ratio is too narrow. Very little attention should be given to the fat content of the ration, it being considered satisfactory if the requirements for protein and carbohydrates are fulfilled.

The Armsby Standards express the requirements of farm animals in pounds of digestible protein and in therms of energy. Instead of giving separate standards for all the different classes of farm animals, Armsby gives standards for maintenance and growth. Inasmuch as any excess of feed above maintenance may be used for fattening or milk production, he gives the amount of nutrients above the maintenance requirements necessary to produce a pound of gain or a pound of milk. Thus, the standards for fattening and for milk production vary with the amount of gain or with the amount of milk produced. To determine the standard for a fattening animal, one adds 3.5 therms per each pound of daily gain to the energy requirement for maintenance, as all the energy above the maintenance requirement may be used for the production of flesh and fat. Armsby recommends that a 1000-pound ruminant should receive 20 to 30 pounds, or an average of 25 pounds of dry matter per day. A horse should have somewhat less. The amounts of digestible protein and of energy in the common feeding-stuffs as presented by Armsby are given in Table II. His feeding standards are given in Table IV. For example, if one desires to calculate a ration for a 1000-pound steer gaining two pounds per day, the first step is to determine the requirements. From Table IV it is seen that the requirements of a 1000-pound steer gaining two pounds per day are 1.8 pounds of digestible protein and 13.0 therms of energy. As the second step, we will assume a trial ration consisting of 10 pounds of corn and 8 pounds of clover hay. Referring to Table II, it is found that the digestible protein and energy in this ration are as follows:

	Dry Substance, pounds.	Digestible Protein, pounds.	Energy therms.
Corn, 10 pounds	8.91	0.68	8.88
Clover hay, 8 pounds	6.78	0.43	2.78
Total ration	15.69	1.11	11.66

Comparing the trial ration with the standard, we find that it is low in both protein and energy. As the third step, we will add 2 pounds of cottonseed meal, as it is high in both protein and energy. The ration then contains the following nutrients:

	Dry Substance, pounds.	Digestible Protein, pounds.	Energy, therms.
Corn, 10 pounds	8.91	0.68	8.88
Clover hay, 8 pounds	6.78	0.43	2.78
Cottonseed meal, 2 pounds	1.84	0.70	1.68
Total ration	17.53	1.81	13.34

This ration, although a trifle low in dry substance, fulfils the requirements of the Armsby standard.

In calculating a ration for a dairy cow according to the Armsby standard, one adds to the requirements for maintenance, 0.05 pounds of digestible protein and 0.3 therm of net energy for each pound of milk produced. For example, one wishes to calculate a ration for a 900-pound cow giving 22 pounds of milk. According to Table IV the requirements are as follows:

	Digestible Protein, pounds.	Energy, therms.
For maintenance of 900-pound cow	0.45	5.7
Additional for 22 pounds milk	1.10	6.6
Total requirement	1.55	12.3

The ration is then calculated in the manner previously described.

The Haecker Standard for Dairy Cows holds that the requirements of the dairy cow vary not only according to her weight and the quantity of milk yield, but also according to the quality of the milk. According to Haecker, a 1000-pound cow requires for maintenance 0.7 pound of digestible crude protein, 7.0 pounds of digestible carbohydrates, and 0.1 pound of digestible fat. For each pound of 4 per cent milk the Haecker standard requires the addition of 0.054 pound of digestible crude protein, 0.24 pound of digestible carbohydrates, and 0.021 pound of digestible fat in addition to the maintenance requirement. If the milk contains less than 4 per cent of fat, smaller amounts of nutrients are prescribed. The amounts of digestible nutrients to produce one pound of milk containing various percentages of butter fat are given in Table V.

For example, to calculate a ration according to the Haecker standard for a 900-pound cow giving 20 pounds of milk daily containing 5 per cent of butter fat, the process is as follows: (1) determine the maintenance requirement for a 900-pound cow; (2) add to the maintenance requirement the requirement to produce 20 pounds of 5 per cent milk; and (3) calculate a ration to conform with this standard. Thus a cow weighing 900 pounds requires 0.63 pound of digestible protein, 6.30 pounds of digestible carbohydrates and 0.09 pound of digestible fat for maintenance. According to Haecker, to produce one pound of 5 per cent milk requires the consumption of 0.060 pound of digestible crude protein; 0.28 pound of digestible carbohydrates, and 0.024 pound of digestible fat, in addition to the maintenance requirements. Thus the total requirement to produce 20 pounds of 5 per cent milk is calculated as follows:

	Digestible Protein, pounds.	Digestible Carbohydrates, pounds.	Digestible Fat, pounds.
For maintenance	0.63	6.30	0.09
To produce 20 pounds of 5 per cent milk	1.22	5.60	0.50
Total	1.85	11.90	0.59

The ration is then calculated in the same manner as described under the discussion of the Wolff-Lehmann standards.

REFERENCES

"Principles of Stock Feeding." Smith.
"Feeds and Feeding." Henry.
"First Principles of Feeding Farm Animals." Burkett.
"Principles of Animal Nutrition." Armsby.
"Feeding of Animals." Jordan.
"Productive Feeding of Farm Animals." Woll.
"Profitable Stock Feeding." Smith.
California Expt. Station Bulletin 256. "Value of Barley for Cows Fed Alfalfa."
Illinois Expt. Station Bulletin 172. "Study of Digestion of Rations for Steers."
Minnesota Expt. Station Bulletin 140. "Investigations in Milk Production."
Missouri Research Bulletin 18. "Maintenance Requirements of Cattle."
Nebraska Expt. Station Bulletin 151. "Corn Silage and Alfalfa for Beef Production."
New Hampshire Expt. Station Bulletin 175. "Analysis of Feeding-stuffs."
South Dakota Expt. Station Bulletin 160. "Silage and Grains for Steers."
Texas Expt. Station Bulletin 170. "Texas Feeding-stuffs, Their Composition and Utilization."
Wisconsin Expt. Station Circular 37. "The Feeding Unit System for Determining the Economy of Production by Dairy Cows."
Wisconsin Research Bulletin 26. "Studies in Dairy Production."
Wyoming Expt. Station Bulletin 106. "Cottonseed Cake for Beef Cattle."
Pennsylvania Expt. Station Bulletin 111. "Feeding."
Farmers' Bulletins, U. S. Dept. of Agriculture:
346. "Computation of Rations for Farm Animals."
655. "Cottonseed Meal for Feeding Beef Cattle."

CHAPTER 46

HORSES AND MULES

BY E. H. HUGHES
Assistant Professor in Animal Husbandry, College of Agriculture, University of Missouri

The horse even today plays a very important part in moving merchandise and performing other labor. The work on our farms is largely accomplished by the horse, and in spite of the motor truck the horse is

MORGAN STALLION, "GENERAL GATES."[1]

considered indispensable in a large amount of business in the city. Modern methods of transportation move enormous quantities of freight, yet the demand for the work horse does not diminish.

Development of Type.—The usefulness of a horse depends upon his power of locomotion and the characteristics which adapt him to the different

[1] Courtesy of U. S. Dept. of Agriculture.

kinds of service determine his type. Whether he moves with power, speed, extreme action and style or to carry weight, will determine whether he is a draft, a race, a show or a saddle horse.

Our modern breeds of light horses have probably been developed with the Arabian horse as foundation stock. The Arabians developed a light horse with endurance and courage for desert travel, and intelligence and friendliness for companionship on the long journey.

The low-lying, luxuriantly vegetated Flanders led to the development

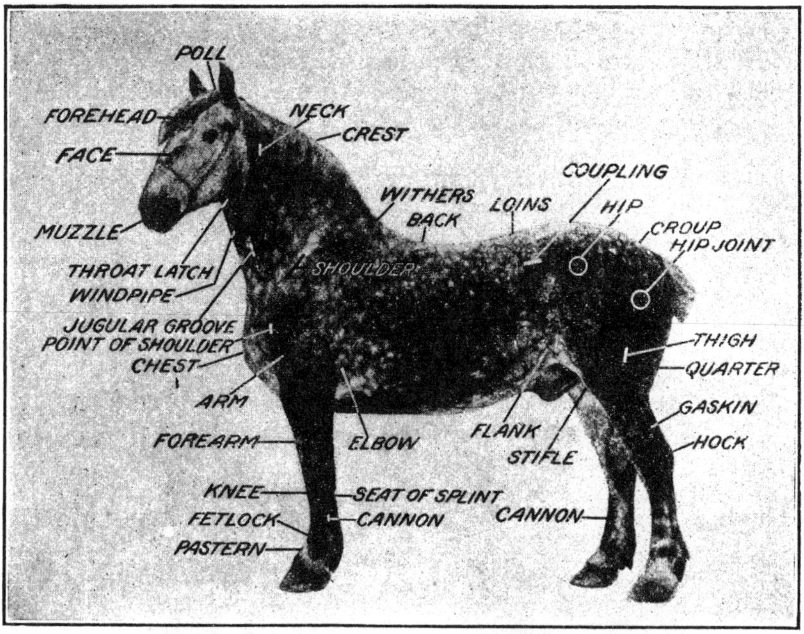

A HIGH-GRADE WORK HORSE OF FINE QUALITY AND GOOD CONFORMATION, ILLUSTRATING THE "POINTS" OF A DRAFT HORSE.[1]

of the patient, sluggish Flemish horse with plenty of power to accomplish the heavy work required of him. The Flemish blood is the most important basis of the draft types.

The Light Horse.—It is essential that the horse of this class show refinement in all his parts. His conformation, action, style, finish and endurance should be such that he can meet the requirements for a distinct purpose.

Action is essential in this class. The coach horse should show high action; the roadster must be able to haul a light vehicle at a rapid trot; and the saddle horse is required to give the rider satisfaction.

[1] Courtesy of U. S. Dept. of Agriculture. From Farmers' Bulletin 451.

Weight is not so important. The carriage horse must necessarily be heavier than the roadster, because he is required to draw a heavier vehicle and the saddle horse must be able to perform the gaits of his class. Size will depend upon the use to which he is put.

Draft Type.– Horses of this type are used in hauling heavy loads at a comparatively slow gait, and should possess strength and endurance. A

PERCHERON STALLION.[1]

draft horse should be massive, relatively close to the ground and weigh at least 1600 pounds. He should have a heavy body; a short, strong back; a strong constitution; a sloping shoulder and a long, level croup. He should also have plenty of bone of good quality and large, sound feet. His legs should set properly under him and his pasterns should be long and sloping.

The important gait of the draft horse is the walk. The stride should be long and straight. A combination of weight, muscle and good feet and

[1] Courtesy of The Field, New York City.

Light Horses.

Breeds.	Native Country.	Origin.	Height, hands.	Weight, pounds.	Color.	Other Characteristics.
Arabian.	Arabia	Native horses.	14–15.2	850–1000	Bay, brown, chestnut, occasionally black or gray.	Good action; intelligent.
Thoroughbred.	England.	Arabian mixed by English people.	14.2–16.2	900–1100	Bay, brown, chestnut, black or gray.	Running horse; great speed; intelligent, sometimes difficult to manage.
Standard bred horse.	America.	English horse.	14–16.2	900–1200	Bay, brown, chestnut or black; few grays and roans.	Fastest of harness horses; remarkable endurance; intelligent.
Morgan (branch of standard bred)	America.	Thoroughbred (Justin Morgan foundation horse).	14.3–16	950–1150	Bay, chestnut, brown or black.	Not extreme action or speed; noted for endurance; intelligent.
Saddle.	U. S.: Virginia, Kentucky, Missouri.	Thoroughbred.	15–16	900–1200	Chestnut, black, bay, brown, gray and roan.	Action, style, manners, five gaits, walk, trot, canter, rack, running-walk, foxtrot or slow pace.
Hackney.	England.	Arabian; thoroughbred native horse.	13.2–16	750–1200	Chestnut, bay, brown, black, and roan, white markings common.	Extreme high action.

German coach, French coach and Cleveland bay horses have not greatly influenced the horse business in this country.

Draft Horses.

Breed.	Native Country.	Origin.	Height, hands.	Weight, pounds.	Color.	Other Characteristics.
Percheron.	France.	Native horses, Flemish, Arabian.	15.3–17	1600–2200	Gray, black, bay, brown, roan, chestnut.	Good action; intelligent.
Belgian.	Belgium.	Flemish.	15.3–17	1600–2400	Roan, chestnut brown, black, gray.	Compact, deep and wide; heavily muscled.
English Shire.	England.	Native horses, Flemish.	16–17.3	1700–2400	Bay, black brown, gray, chestnut, roan, white on legs and face.	Largest of draft breeds; heavy feather on legs.
Clydesdale.	Scotland.	Native horses, Flemish.	16–17	1600–2100	Bay, black, brown, chestnut, roan, gray, white on face and legs.	Very good action; intelligent; feather on legs.
Suffolk-punch.	England.	Native horses.	16–16.2	1600–2000	Chestnut.	Smallest of the draft breeds.

Mules.

Class.	Height, hands.	Weight, pounds.	Color.	Other Characteristics.
Draft.	16–17.2	1200–1600	Bay, brown, gray or dun. Most desirable color is black with a tan nose and flank.	Large, heavy boned, heavy set mules.
Sugar.	16–17	1150–1300		Tall, with considerable quality and finish.
Cotton.	13.2–15.2	750–1100		Small and compact, with quality.
Mining.	12–16	600–1350		"Pit" mules, small; "miners," large and rugged.
Farm.	15.2–16	1000–1350		Plain and thin, with good constitutions.

legs should be an indication of a horse's ability to haul a load at a fair rate of speed.

The Mule is a hybrid, having for parents a mare and a jack. Because of this fact it shows many parental characteristics which are common to both ancestors. It has longer ears than the horse, a Roman nose, heavy lips, clean legs, small, narrow feet of good quality, and a scanty growth of

ENGLISH SHIRE STALLION.[1]

hair on the tail and a scanty mane. The sexual organs of both mare and horse mules are undeveloped, consequently they do not breed.

The mule is generally smaller than the draft horse, being from 14 to 17 hands high, and weighing from 600 to 1600 pounds. Mares of good quality weighing about 1350 pounds when bred to a heavy-boned jack with long ears produce mules which have good size, quality and action. As a rule, mare mules bring better prices on the market than do horse mules.

America has done more towards the economical development of the

[1] Courtesy of The Field, New York City.

mule than any other country, and more than one-half the mules in the world are in the United States.

Due to his hardiness and his ability to take care of himself, the mule is adapted to most climates and to kinds of work for which it would not be practical to use a horse. In most contagious and infectious diseases, however, the mule has no more resistance than a horse.

Market Requirements.—The market requires that a horse shall fill some definite purpose. There is a demand for good horses that fill a definite

CLASSIFICATION OF MARKET CLASSES OF HORSES.*

CLASS.	SUB-CLASS.	HEIGHT, HANDS.	WEIGHT, POUNDS.	OTHER CHARACTERISTICS.
Draft.	Light draft.	15.3–16.2	1600–1750	Heavy, rugged, compactly built, denoting strength and endurance.
	Heavy draft.	16–17.2	1750–2200	
	Loggers.	16.1–17.2		
Chunks.	Eastern Export.	15–16	1300–1550	The same type as draft, except that he is more compact and lighter in weight.
	Farm.	15–15.3	1200–1400	Low down, blocky horses not as heavy as the Eastern chunk. Possess quality finer and not so heavy as the other sub-classes.
	Southern.	15–15.3	800–1250	
Wagon.	Expressers.	15.3–16.2	1350–1500	Upstanding, deep-bodied, closely coupled, with good bone quality, energy and spirit.
	Delivery.	15–16	1100–1400	Conformation similar to express; not so large.
	Artillery.	15.1–16	1050–1200	Sound, well bred, with quality; prompt action in walk, trot or gallop. Free from vicious habits, without blemish, and broken to harness and saddle.
	Fire horses.	15–17.2	1200–1700	More rangy in conformation than expressers; ability to take long runs.
Carriage.	Coach.	15.1–16.1	1100–1250	Smoothly turned, full-made horses with high action combined with beauty of form.
	Cobs.	14.1–15.1	900–1150	Small horses of stocky build with plenty of quality.
	Park.	15–15.3	1000–1150	Excellent quality; high action.
	Cab.	15.2–16.1	1050–1200	Similar to coach horses; calk in finish; good feet and legs and endurance.
Road.	Runabout.	14.3–15.2	900–1050	Not so stockily built as cob, having more speed.
	Roadster.	15–16	900–1150	Conformation more angular than runabouts, denoting speed, stamina and endurance.
Saddle horses.	Five-gaited.	15–16	900–1200	Conformation denoting style, action, with strong back; possesses five distinct gaits under the saddle.
	Three-gaited.	14.3–16	900–1200	Size depending on weight to be carried with ability to walk, trot and canter.
	Hunters.	15.2–16.1	1100–1250	Large, strong; must be jumpers; stand long country rides.
	Cavalry.	15–15.3	950–1100	Sound, well bred; have quality; broken to saddle; easy gaits.
	Polo Ponies.	14–14.2	850–1000	Smallest saddle class; used for playing polo.

purpose, but misfits sell at a low figure. The horse should be sound, at least serviceably sound, with a conformation adapted to the work required of him. He should be in good condition in order to look well and be ready for hard work. Condition is also an indication of the health and feeding quality of the horse. The market requires that a horse be broken and of good disposition. Horses between five and eight years old sell the best. Solid colors are preferred because they can be matched more easily, and many firms use their teams of two; four or six horses and equipment as a part of their advertising.

* Illinois Experiment Station Bulletin No. 122.

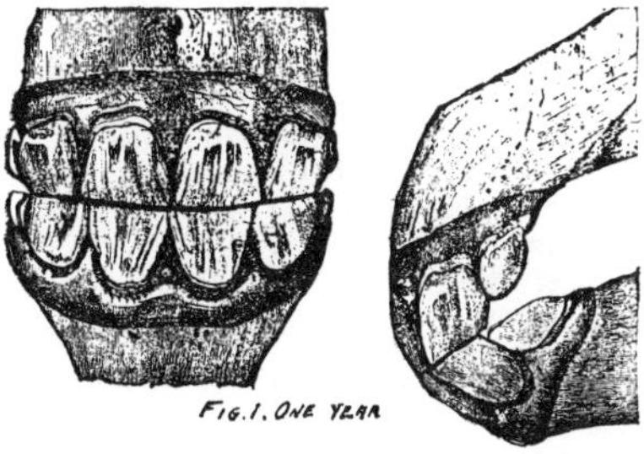

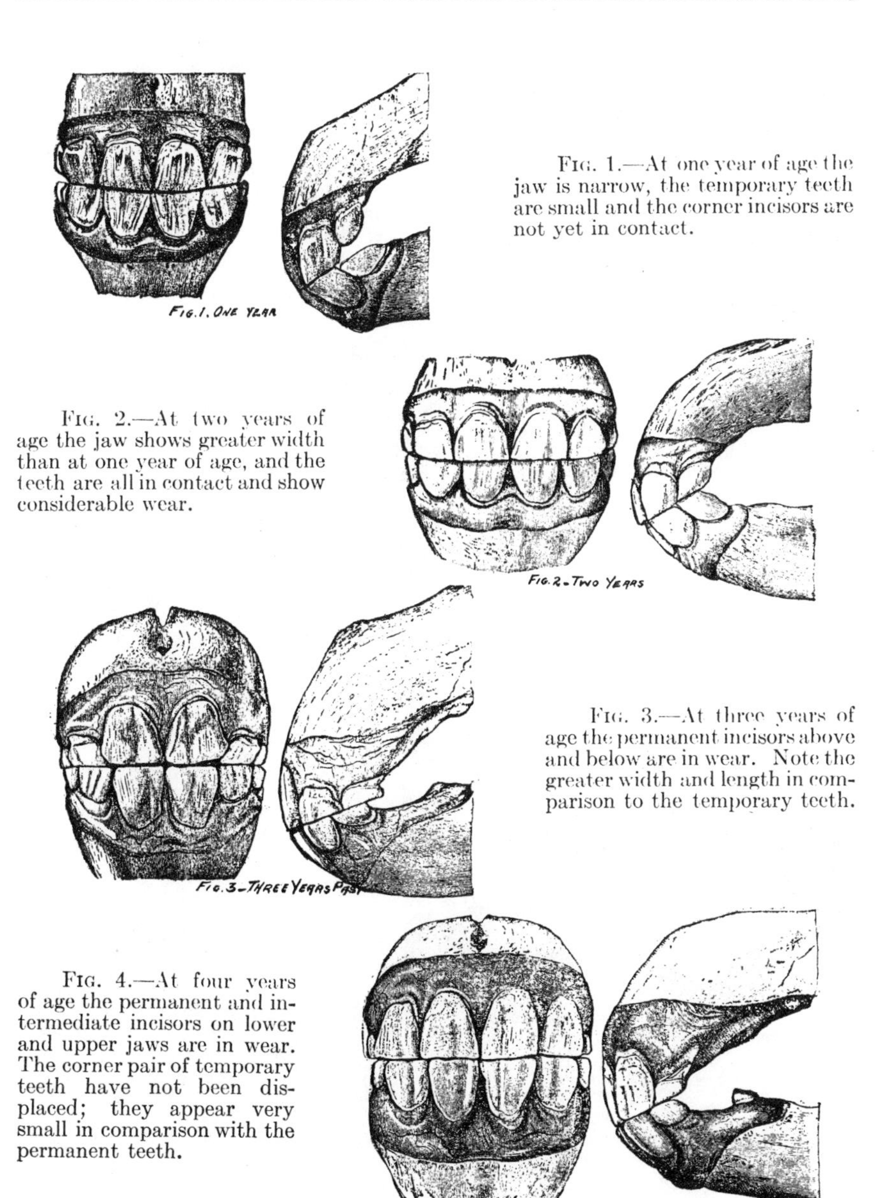

FIG. 1.—At one year of age the jaw is narrow, the temporary teeth are small and the corner incisors are not yet in contact.

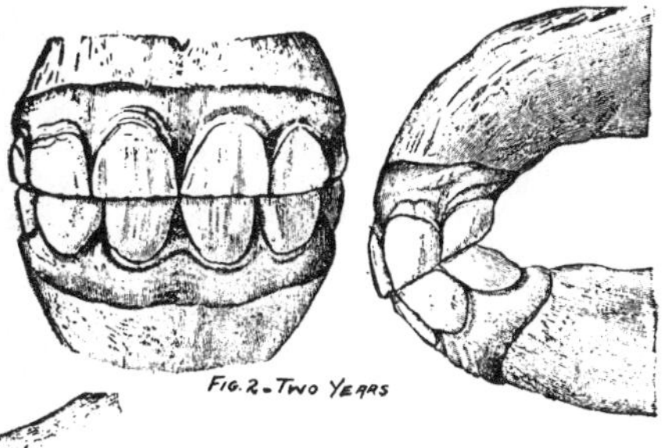

FIG. 2.—At two years of age the jaw shows greater width than at one year of age, and the teeth are all in contact and show considerable wear.

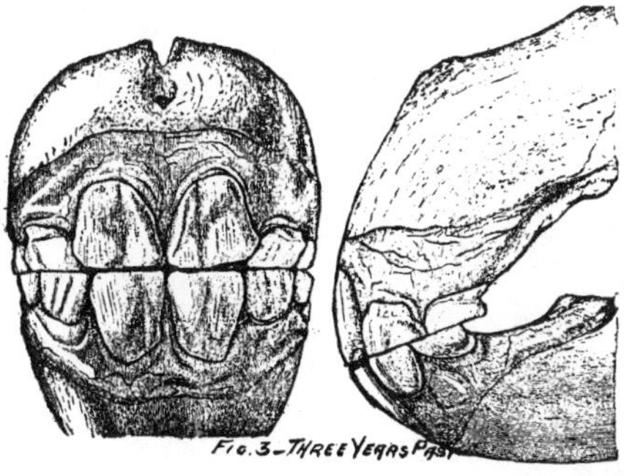

FIG. 3.—At three years of age the permanent incisors above and below are in wear. Note the greater width and length in comparison to the temporary teeth.

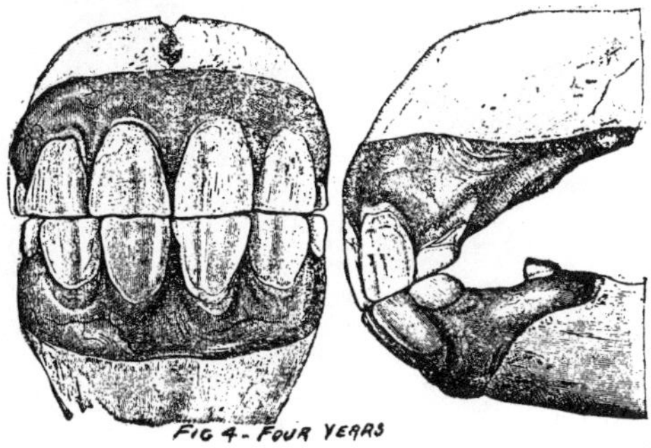

FIG. 4.—At four years of age the permanent and intermediate incisors on lower and upper jaws are in wear. The corner pair of temporary teeth have not been displaced; they appear very small in comparison with the permanent teeth.

NOTE.—Photographs showing teeth at various ages, by courtesy of Prof. S. T. Simpson, Agricultural Extension Service, Missouri Experiment Station.

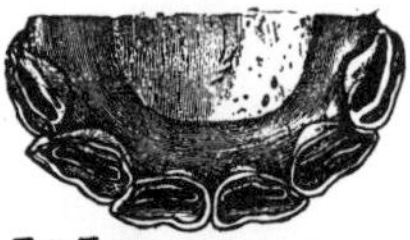

FIG. 5 FIVE YEARS LOWER INCISORS

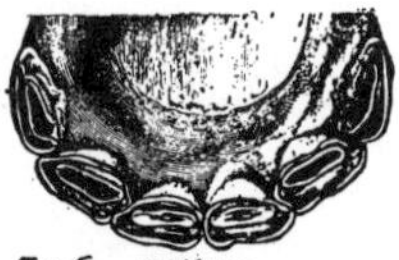

FIG. 6- SIX YEARS LOWER INCISORS

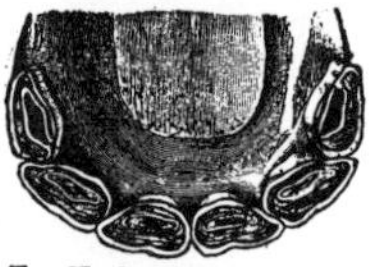

FIG. 7- SEVEN YEARS LOWER INCISORS

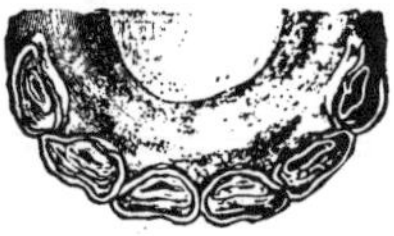

FIG. 8- EIGHT YEARS LOWER INCISORS

FIG. 5.—Complete set of lower permanent incisors showing deep cups at five years.

FIG. 6.—The cups in the lower central incisors have nearly disappeared and the tables are smooth at six years.

FIG. 7.—The cups in the lower intermediate pair of incisors have disappeared at seven years.

FIG. 8.—The cups in the lower corner pair of incisors have disappeared and the tables are all worn smooth at eight years.

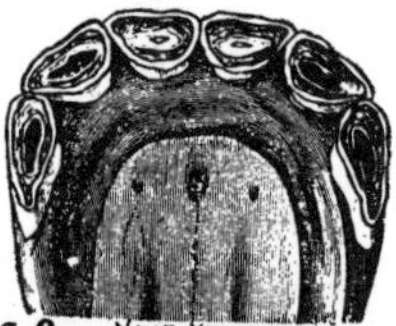

FIG. 9 NINE YEARS UPPER INCISORS

FIG. 10 TEN YEARS UPPER INCISORS

FIG. 11 ELEVEN YEARS UPPER INCISORS

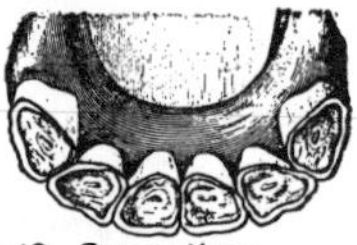

FIG. 12 FIFTEEN YEARS LOWER INCISORS

FIG. 9.—The cups in the central incisors above have practically disappeared at nine years.

FIG. 10.—The cups in the intermediate incisors above have disappeared at ten years.

FIG. 11.—At eleven years the tables on the upper jaw are nearly smooth.

FIG. 12.—Note the smooth tables and the length of the teeth showing considerable wear at fifteen years.

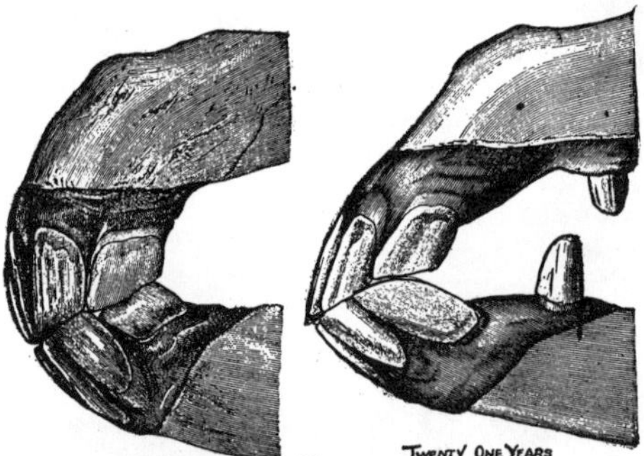

FIVE YEARS — FIG. 13 — TWENTY ONE YEARS

FIG. 13.—Showing a comparison of the angles of the jaw at five and twenty-one years. Note the acute angle of the teeth at twenty-one.

FIG. 14. ORDER OF APPEARANCE OF PERMANENT INCISORS

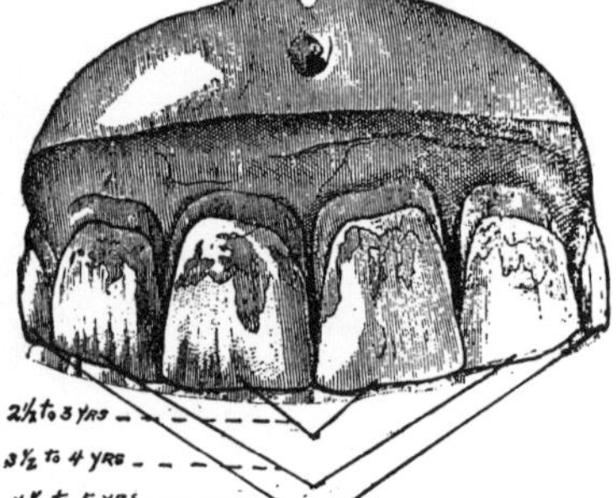

FIG. 14.—Showing order of appearance of the permanent incisors. The central pair at $2\frac{1}{2}$ to 3 years. The intermediate pair at $3\frac{1}{2}$ to 4 years. The corner or outside pair at $4\frac{1}{2}$ to 5 years.

NOTE.—Photographs showing teeth at various ages, by courtesy of Prof. S. T. Simpson, Agricultural Extension Service, Missouri Experiment Station.

The Age of the Horse.—The teeth form the most accurate basis for estimating the age of a horse. The first teeth which appear are called colt, milk or temporary teeth. As the horse grows older these are replaced by broader, thicker, darker-colored permanent teeth.

The central incisors in the upper and lower jaws usually appear within two weeks after foaling. The intermediate incisors, one on each side of the central incisors, make their appearance between the second and fourth week, and the corner or outside incisors are in at six months of age (Fig. 1).

The central pair of permanent incisors will displace the temporary incisors and be in use at three years of age. (Fig. 3). The permanent intermediate ones will be in use at four (Fig. 4) and the corner pair of permanent incisors will displace the temporary corner or outside incisors at five years of age.

The five-year-old horse has a full mouth of permanent teeth (Fig. 5). These have large cups that wear smooth as the age of the horse advances. The cups or tables of the central incisors below have worn smooth at six years of age, the intermediate incisors below are smooth at seven and the corner pair at eight. (See Figs. 6, 7 and 8.)

The cups of the central pair of incisors on the upper jaw have disappeared at nine, the intermediates above are smooth at ten and the cups in the corner incisors are worn smooth when the horse reaches his twelfth year. (See Figs. 9, 10, 11 and 12.)

There is no accurate method of estimating the age of a horse after he is twelve years old; however, the angle of the teeth becomes more acute as he becomes older (Fig. 13).

HORSE FEEDINGS

Proper management in feeding and caring for the horse is an essential for his best health and development. The digestive system of a horse is not large, therefore a comparatively small amount of roughage and a relatively larger amount of grain is required. Sudden changes in feed should be avoided, as the digestive system requires time to readjust itself to the new conditions.

Grinding or soaking of feed is not economical except in the case of colts or horses doing exceptionally hard work; however, many horsemen favor rolling the oats they feed. Salt should be accessible at all times.

Feeds for the Horse.—It is economical under most conditions to use the feeds at hand. The most common feeds for horses are oats with timothy hay, or a mixture of timothy and clover. In many sections corn is substituted for a part or all of the oats in the ration and prairie hay or alfalfa is substituted for the timothy.

A combination of oats and timothy hay forms an excellent ration for work horses. The nutrients are in about the proper proportions and the bulk seems to fit the needs of the horse. Both are usually free from dust. For the light horse that is required to make long, hard drives no satisfactory substitute for oats and timothy hay has been found.

Grain.—Corn and barley are used extensively in some sections for a part or all of the grain ration. Because of its hardness the barley should be ground or cracked.

Bran and oil meal are often used to supplement corn or barley, and for growing animals or brood mares corn, oats and bran form an excellent ration. The bone and muscle-building elements in bran and its laxative effect are considered indispensable by many horsemen.

Roughages.—A mixture of timothy and clover is considered an excellent roughage for horses. Either clover or alfalfa hay is good, except for horses doing heavy or rapid work. Oat hay, when cut a little green, forms an excellent roughage and sheaf oats are often fed with good results. Millet hay is considered unsafe to feed by most horsemen.

Corn stover and oat straw are used with success when properly balanced with a grain ration.

Good clean silage that is not too acid is an excellent feed for brood mares, idle horses and growing colts, though it should be fed in limited quantities and with some dry roughage. It has been fed with good results, but great care must be exercised in feeding. No mouldy or musty silage should be fed.

Watering.—Horses, under natural conditions, drink frequently. The most common practice among horsemen is to water the horses before feeding, although many practice watering before and after feeding. Horses that are heated should be compelled to drink very slowly. The value of good running water in the horse pasture cannot be overestimated.

The Work Horse.—A horse at work should receive ten to eighteen pounds of grain daily, depending upon the kind of work performed and the size of the horse. On days when idle the grain ration should be reduced and the roughage increased. The addition of a small amount of bran is recommended.

The Foal.—The foal should be taught to eat grain and hay as early as possible. Oats and bran with some clover or alfalfa hay of good quality are the best feeds because they contain the muscle and bone-forming elements required for growth.

While the mare and colt are in the pasture some grain can be fed very satisfactorily in a small creep. After weaning, at about five or six months of age, feed for growth rather than condition.

The Orphan Foal.—Milk from a fresh cow, one whose milk is low in butter-fat, is well adapted to raising an orphan foal. To a dessert-spoonful of granulated sugar should be added enough warm water to dissolve it. To this three tablespoonsful of lime water and enough fresh milk to make a pint should be added. A small amount, one-half pint, should be given each hour. In a short time the amount should be increased and feed should be given every two hours, more being given gradually and the time between feeding lengthened.

The Brood Mare, used for breeding purposes only, does well without grain when on good pasture. In winter, if she is in foal, she should be given feeds high in protein and mineral matter for the best development of the fœtus. She should receive plenty of exercise at all times.

The Stallion.—Good whole oats and bran with plenty of clean timothy hay is a very good ration for the stallion. The addition of corn or barley to the ration lends variety and increases its palatability. Exercise is at all times absolutely essential for the best results with any stallion. When standing for service he should be required to walk six to ten miles per day.

STANDARD RATIONS

Foals:

	Parts.		Parts.		Parts.
Ground oats	6	Oats	4	Oats	4
Ground corn	2	Corn	4	Bran	4
Bran	2	Bran	2	Corn	2

Whole oats, Shelled corn, Bran .... equal parts.

With either of the above rations, feed clover, alfalfa, or timothy and clover mixed.

Work Horses:

Oats	5	Oats.		Corn	9
Corn	5	Hay.		Oatmeal	1
Hay.				Hay.	

Timothy and clover mixed or just timothy is recommended as roughage.

Brood Mare:

Corn	4	Corn	7	Corn	8
Oats	4	Bran	2	Linseed oil meal	1
Bran	2	Linseed oil meal	1		

Clover or alfalfa of good quality, or timothy and clover mixed are good roughages to feed with the above grain rations.

Grooming.—For the best health of the horse he should be groomed before he is harnessed and at night after the harness has been removed. A good currycomb, a stiff brush and a soft woolen cloth are the only utensils ordinarily needed. The currycomb is used to loosen the dirt and sweat in the hair and skin over the body and is followed by the brush. The woolen cloth is then rubbed very firmly over the entire body to take up the fine dust and to put the coat in good condition.

REFERENCES

"Productive Horse Husbandry." Gay.
"The Horse Book." Johnston.
"The Horse." Roberts.
"Breaking and Training Horses." Harper.
"Management and Breeding of Horses." Harper.
Farmers' Bulletins, U. S. Dept. of Agriculture:
451. "Draft Horses and Care of Horses."
667. "Breaking and Training Colts."

CHAPTER 47

Beef Cattle

By W. A. Cochel
Professor of Animal Husbandry, Kansas Agricultural College

Beef production is associated with the best type of farming in every country. A careful survey of any community shows that the cattlemen are leaders in public matters, are financially responsible, farm the best land and are considered among the best citizens. Counties and communi-

Pure-bred Hereford Bull.[1]
A hardy, early maturing, beef breed of good quality.

ties noted for their production of beef are also noted for their large yields of agricultural crops and their great productive wealth. There never has been a permanent and profitable system of farming established on an extensive scale in any country where beef cattle have been eliminated from

[1] Courtesy of The Field, New York City.

the farms. Beef cattle make the greatest and most profitable use of roughage and grass, are comparatively free from disease, require less shelter and attention than other farm animals, enable the farmer to distribute his work uniformly throughout the year and are easily marketed.

Sources of Profit.—The cattleman has four sources of profit: (1) from growing crops; (2) from feeding crops; (3) from using by-products which otherwise have no market value, such as straw, stover, damaged hay and grain; and (4) from increasing soil fertility and the yield of crops. It frequently happens that the greatest profit comes from the use of farm by-products and the increase of soil fertility. The successful cattleman of the future must be as good a farmer as the man who produces grain and hay for the market, and also have the ability and judgment to select and feed animals that can convert grain and hay into meat profitably.

There are four distinct methods of handling beef cattle, dependent upon the amount of capital available and the kind of crops adapted to the farm, as follows: (1) breeding pure-bred cattle, (2) producing stockers and feeders, (3) grazing cattle, and (4) fattening cattle.

Breeding Pure-Bred Cattle.—This is the highest type of beef production and requires the investment of a large amount of money for a series of years. The breeder must not only understand and practice the best methods of breeding, feeding and developing livestock, but must also follow the best methods of farming. He should keep the buildings and grounds neat and attractive to impress customers with the fact that breeding pure-bred livestock is profitable and attractive.

Excellent pasture should be available for summer grazing and the best methods of feeding must be practiced during the winter to develop the inherited type and form to the maximum. More breeders fail because of poor feeding than of any other one factor. In addition to the ability to select the approved type of the breed and to feed successfully, the breeder of pure-bred cattle must be a business man and a salesman so that he can successfully dispose of what he produces. It is usually better for the beginner to start with grade or market cattle and, if he succeeds, to purchase a few pure-bred animals and go into the business gradually, than to invest all his capital in a specialty with which he is unacquainted.

Producing Stockers and Feeders.—The production of stockers and feeders should be confined to those parts of the country where the larger part of the land cannot be plowed profitably, and grass is the principal crop. This class of cattle is kept on grass during the summer season and fed on roughage, with little or no grain, during the remainder of the year. Lying east of the Rocky Mountains is a large area which is peculiarly adapted to the production of grass and roughage, such as Kaffir and sorghums on the uplands, and alfalfa on the bottom land, and which logically should be the great stocker and feeder producing section of the United States. Where both legumes and silage crops are produced, little or no commercial feeds are required. If it is impossible to grow legumes, protein should be supplied in the form of linseed meal, cottonseed cake or some other protein concentrate.

It is essential that cattle of the best beef type be used in producing stockers or feeders, because the chief profit comes from producing animals of superior merit for which there is always a keen demand. It is very important that the herd of cattle used for this purpose be uniform in type, color, size, breeding and quality and that the animals have large feeding capacity, because buyers prefer to buy feeders or stockers as nearly alike as possible.

Grazing Cattle.—The business of grazing cattle is generally followed in those sections where the area of land in cultivation is very small compared with that which must be left in grass. The cattle are seldom produced in the grazing sections, but are usually shipped in by the train-load about the first of May, and are pastured on grass until they are fat enough to be marketed as grass-fat cattle during the late summer and early fall.

The cattle used to convert grass into fat are usually older, coarser and plainer than cattle selected to convert corn into the same product. Not so much attention is paid to quality and breeding as in pure-bred cattle, stockers or feeders, because the profit comes from the increase in value secured by fattening rather than in the final price per hundredweight. Very thin steers, three years old or older, make much larger gains than younger or fatter cattle. However, it frequently happens that when fleshier cattle are used, they may be shipped from grass earlier in the season, thus avoiding extreme heat, flies, water shortage or a heavy run of cattle on the market, which will more than overbalance the larger gains made by thinner cattle.

Fattening Cattle.—This has proven profitable in sections where corn is the leading crop and the area devoted to permanent pasture is relatively small. The kind of cattle selected for the feed lot depends upon the season of the year, the feeds available, the probable demand for the cattle when fat and the experience of the feeder. Young cattle make cheaper gains than older cattle, but they require a longer feeding period to become fat, because they use a large part of their feed for growth.

Calves that are to be fattened should show quality and breeding. They should have short legs and blocky, broad, deep bodies, otherwise they will grow rather than fatten. It will require from eight to nine months from the time calves are weaned to make them prime even when on full feed. An excellent ration is ten pounds of silage, five pounds of alfalfa hay, one pound of linseed meal or cottonseed cake per head daily, and all the corn they can eat. Older cattle consume more roughage in proportion to the grain and are fed where corn is relatively scarce.

To fatten cattle successfully and to secure satisfactory gains, the ration should be improved as the animals become fat. The customary farm practice is to start the cattle on roughage, such as silage, hay and fodder, with about six pounds of corn per thousand pounds liveweight daily, and to increase the amount of corn as they become fatter. This makes the period when they are really on full feed very short.

Fitting Show Animals.—The production of show animals is in reality a form of advertisement, and is restricted largely to the breeders of pure-bred cattle. Every art known to the feeder is utilized to develop such animals. The ration is quite similar to that fed to fattening animals during the last part of the feeding period, and is improved by grinding the grain, cutting the hay and adding a greater variety of feeds. Sometimes barley is boiled and fed at the rate of one gallon per day and sugar or molasses is mixed with the grain to increase the palatability. In fact, everything possible is done to keep up the animal's appetite.

THE SELECTION OF CATTLE FOR THE FEED LOT

The selection of cattle for the feed lot is probably the most vital question before the cattle feeders today. Upon this one problem depends the ultimate financial success of those who make a business of converting grain and roughage into beef. There are three factors which should always be given consideration: (1) the purpose for which the cattle are to be used, (2) the ability of the individuals to consume feed over and above that required for maintenance, and (3) the probable demand for beef when the cattle are returned from the feed lots.

Methods of Feeding.—Cattle feeders may be divided into different groups according to their methods of feeding: (1) those who produce market-topping animals, (2) those who handle shortfed cattle, and (3) those who produce the great bulk of beef which usually finds its way to market after a period of grazing or roughing followed by a finishing period of either short or long duration.

Characteristics of Good Feeders.—It makes little difference which method is followed. The essential characteristics of a good feeding steer remain constant. He must have good constitution and capacity associated with as much quality and type as it is possible to secure. A wide, strong, short head; short, thick neck; and deep, wide chest indicate constitution, and a deep, roomy barrel indicates capacity. These characteristics may be found in steers of plain as well as of excellent breeding, which accounts for the fact that individual dairy and scrub steers frequently make as rapid gains in the feed lot as beef-bred steers. The type, quality, form and finish as indicated by the deep covering of muscle, even distribution of fat, high percentage of the higher priced cuts of meats, high dressing percentage, smoothness and symmetry of carcass, and quality and texture of meat, are always associated with beef blood.

The success of a feeder buyer depends largely upon his ability to see in thin cattle the possibility of improvement which results from the deposit of fat. As a general rule, there is little change in the skeleton proper. A feeder with a low back will finish into a fat steer with a low back. A feeder with a high tail, head or prominent hook-bones will finish into a fat steer

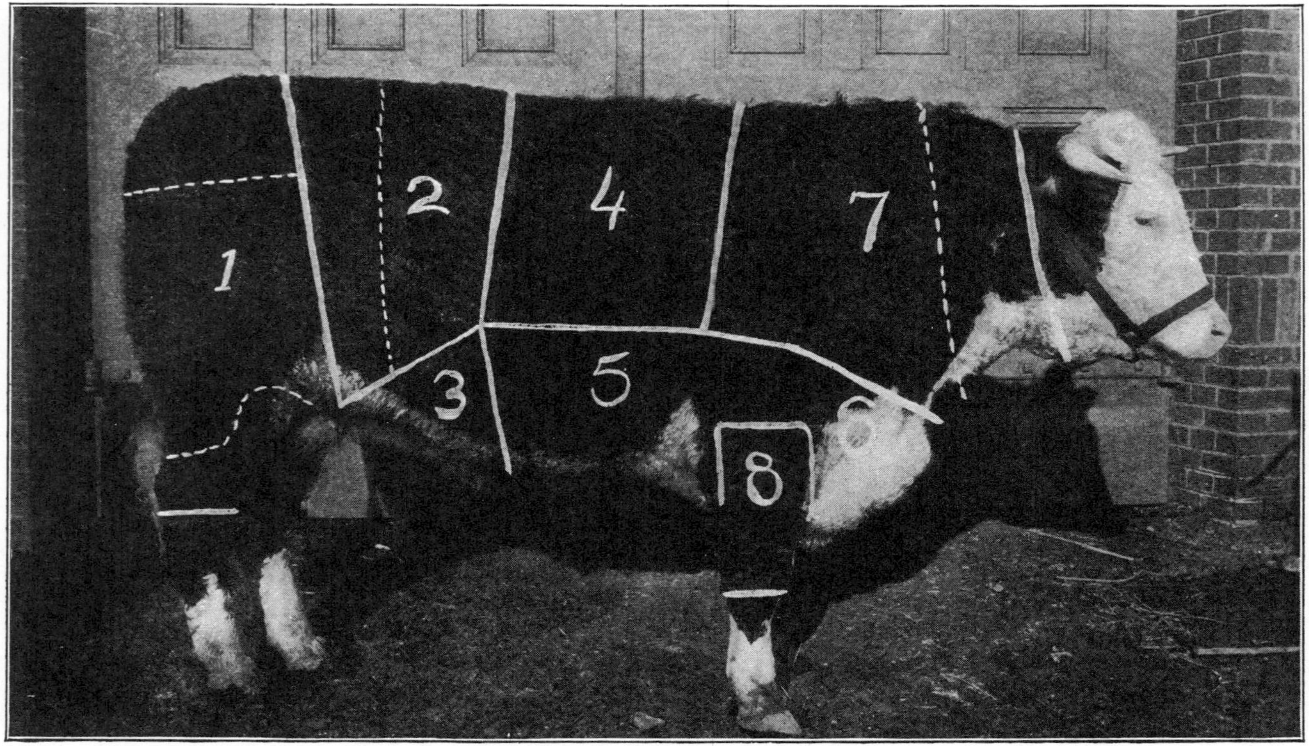

THE PRINCIPAL CUTS OF BEEF.

1—The round; upper portion, rump roast; middle portion, round steak; lower portion, for stews and soup. 2—Loin; fore portion, porter-house or tea-bone steak; rear portion, sirloin steak. 3—Flank; used mostly for stewing, considerable of it is waste. 4—Prime ribs; roasts. 5 and 6—Plate; used for stews and corned beef. 7—Chuck; rear portion used for steaks, roasts and stews; fore portion used for hamburg, minced meat, etc. 8—Shank; soup bone.

with these same deficiencies. A feeder with a long, narrow head, long legs, or shallow body will not alter his type in the feed lot. The greatest improvement comes in those regions of the body where the natural covering of muscle is thickest, in the shoulder, crops, back, loin and round. The body will increase more in width than in length and will decrease in apparent paunchiness due to the greater proportional increase in the width of the upper half of the body than in the lower half. The quality of meat will be improved by the deposit of fat within the bundles of muscle fiber, and the tenderness of meat will be improved because of the distention of all cells with fat, and the proportion of edible to non-edible parts of the animal will increase during the fattening period. These are potent reasons for the immense industry represented by the cattle feeders.

Kind of Feed Related to Class of Cattle.—In addition to these factors which are inherent in the steer, the successful feeder buyer must give attention to the kind of feeds at his disposal. If he intends to use a large amount of grass or roughage in proportion to grain, he should select thin steers carrying some age. Older and thinner cattle will make better use of roughage than those which are younger and fleshier. If the feeder has a large acreage of corn and comparatively little pasture and roughage, he should select either heavy, fleshy feeders which he can return to market within a comparatively short time, or fancy calves of the best possible type and breeding which will develop into prime yearlings. If heavy fleshy feeders are selected, their quality and type should determine their market value, as compared with that of the plainer sort. The probable demand for the various grades of beef at the close of the feeding period is also a determining factor. The feeder should limit his selection to those cattle which will make the greatest improvement in value per hundred pounds while in the feed lot.

Calves and Yearlings.—Quality and type are essential in the selection of calves for feeding purposes. They should be bred for early maturity, otherwise they will grow rather than fatten and the cost of production will exceed their market value. The majority of yearlings are marketed from sixty to ninety days before they are fat, which indicates that it is essential to secure calves of the type that will fatten. The feeder should realize that he is entering into a proposition that requires eight to twelve months to complete and that he must feed the best of feeds in a concentrated form to secure satisfactory gains and finish.

Time to Market.—The time to market fat cattle is when further gains will not result in an increase in the value per hundredweight. For this reason plain, rough steers which will not produce attractive carcasses should be sold before they are thoroughly fattened. When fancy cattle of quality and type are fed, it is a general rule that they are more profitable the fatter they become, because there is usually a demand for fancy finished beef.

The season of the year also controls to some extent the quality of cattle that should go into the feed lot. Where grain-fed steers are to be marketed from the middle of July to the first of December, a better grade of cattle and a higher finish are demanded than at any other season of the year. In the late summer and early fall the markets are usually well supplied with beef that has been produced cheaply on grass with which the half-fat grain-fed cattle cannot compete profitably. After the Christmas holidays all the cattle come from dry lots and have been fattened on expensive feedstuffs so that the plain, rough cattle can be marketed to better advantage than during the grazing season, because the competition of grass-fed cattle is eliminated.

The reasons for feeding beef cattle are that they reduce farm crops into a more concentrated market product and they are a means of permanently maintaining the soil fertility. All feeding operations should be conducted with these facts in mind. The selection of feeding cattle which will serve the purpose and at the same time produce an immediate profit is the mark of the successful cattle feeder.

THE DEFICIENCY IN THE MEAT SUPPLY

Statistics need not be presented to substantiate the assertion that there is a decided deficiency in the supply of meat. The shortage is the result of a long-continued series of years during which the final value of the finished animal was less than the market value of the crops necessary for its production. During the early development of the country there were a considerable number of meat animals bred and fed in the Atlantic states. When the territory west of the Alleghenies and east of the Mississippi River was settled, the breeding industry moved to this section because cattle were the only means of marketing the grass, grain and forage. When transportation facilities were provided for the shipment of grain and other farm products, the breeding industry moved on to Missouri, Iowa, Kansas and Nebraska, where more favorable conditions existed. Here it dominated the agricultural practice until the free range in the West was made available through the suppression of lawlessness. The trend of the cattle-breeding industry has been westward toward the less expensive grazing lands, until there is now no cheap land available. With the decline of breeding operations, finishing or fattening for market became a well-established practice in those sections where the breeding of livestock was unprofitable. The result of this condition is that the demand for animals suitable for the feed lot has finally become so great that the West is no longer able to furnish an adequate supply of feeders, with subsequent high prices.

In recent years the papers and magazines have kept up an almost continual agitation against the high price of meat. The high price has been attributed to the avarice of the farmer, the packer, the stockyards or the retail dealer, rather than to the laws of supply and demand. It has discouraged many from entering into a legitimate business venture for fear that unfavorable public opinion might at any time crystallize into the form of laws of such restrictive nature as to obliterate profits.

Reliable data in regard to methods of meat production are insufficient to enable us to recommend practices which can be substantiated by records of unquestioned reliability. There is, however, so great an abundance of information as to methods of fattening that it is possible for one familiar with the publications and the general farm practices to recommend rations which are certain to produce rapid and economical gains in the feed lot with acceptable dressing percentages.

Tenant Farming Unfavorable to Beef Production.—The rapid growth of tenant farming has eliminated the production of meat from thousands of acres of land which should never have been plowed, and will probably continue to exert a depressing influence upon the business until the value of farm lands is based upon production rather than upon speculation. Under the present system of renting, it is almost impossible to handle beef cattle profitably on a tenant farm. The cattle business requires a number of years to develop and a system of farming that will produce the feeds necessary to maintain a herd of cattle during the winter. A further reason is that the chief profit in cattle farming is the increase in the fertility of the soil and the yield of crops which comes from using the manure on the land. Where land is rented annually there is no incentive to build it up and increase crop production when a different renter may farm it the next year. A system of longer leases must result which will give the tenant an incentive to increase rather than exhaust the fertility of the soil.

Breeding Cattle Requires Capital.—If means of financing breeding operations were provided, the supply of breeding animals on both farms and ranges would be increased tremendously. It is possible for a farmer who has produced a crop of corn or has pasture, to go to almost any bank and secure funds with which to purchase steers to consume these products. Money is loaned for ninety to one hundred and eighty days with the privilege of renewal. It is impossible, however, for him to borrow the same money with breeding females as security, because three to five years must elapse before the increase will be marketable. This is probably the greatest problem to be solved if breeding operations are to be materially increased in the near future.

Breeding herds should be established in the South, the East and in the cut-over districts near the Great Lakes on the land that is adapted to the production of pasture grasses. More attention should be given to pastures to increase their carrying capacity by fertilizing them with manure or fertilizers, by thickening the stand of grass by natural or artificial means and by using silage during unfavorable periods. While grass is the most important crop produced in the United States, more land being devoted to its production than to all others except trees, there is not an important investigational project on the subject reported which the meat producer can use in a practical manner. Throughout the great grazing areas of the country something of definite permanent value must be done to re-establish pastures or the supply of feeding stock will diminish rather than increase in the next few years.

The tremendous waste of the farm by-products of the cereal crops, corn, oats and wheat, which takes place annually throughout the entire country is sufficient to maintain thousands of animals in good breeding condition. This material has not, as yet, been successfully used on a large scale, but recent investigational work indicates that the use of a succulent feed during the winter makes these dry, coarse feeds palatable to a large extent. Refinement in the methods of feeding will in the future enable us to utilize other waste products which are now considered almost worthless.

In the sub-humid sections, the use of the silo to preserve drought-resisting crops, such as Kaffir, milo, feterita and sorghums, and the introduction of new crops, such as Sudan grass, will make it possible to more than double the livestock production of that area. In all parts of the United States at least 300 pounds increase in weight can be secured on the average two-year-old steer by furnishing him an abundance of grass in the summer and an abundance of roughage in the winter. A limited amount of high protein feed should be used to make up the deficiency of the ordinary roughages usually produced where legumes cannot be successfully grown.

It is probable that the loss of livestock from infectious and contagious diseases will be greatly reduced by the practice of sanitary measures, that a more careful study of breeding will result in the production of animals of greater efficiency, that a better knowledge of feeding will result in decreasing the cost of production, but the most potent remedy for the present deficiency in the meat supply is now being administered in the form of market values which leave a reasonable profit to the man who has courage to invest his capital in breeding cattle and the feeds necessary to maintain them. The farmer, as a business man, increases his operations along those lines which promise to return the greatest profit.

REFERENCES

"Beef Production." Mumford.
Indiana Expt. Station Circular 29. "Livestock Judging for Beginners."
Farmers' Bulletins, U. S. Dept. of Agriculture:
588. "Economical Cattle Feeding in the Corn Belt."
580. "Beef Production in the South."
612. "Breeds of Beef Cattle."
Pennsylvania Expt. Station Bulletin 133. "Steer Feeding Experiments."

CHAPTER 48

SWINE

BY JOHN M. EVVARD
Chief in Swine Production, Animal Husbandry Section, Iowa Experiment Station

The hog is one of the most valuable and profitable domestic animals the farm can produce.

In the selection of the herd these factors need to be considered:

1. **Personal Preference** is a most important consideration.

CHESTER WHITE BOAR.[1]

2. **The Feeds Available.**—In the corn belt lard type hogs are best because of their adaptation, whereas in Canada a bacon type will utilize the northern grown feeds to better commercial advantage.

3. **Location and Climate.**—The hog that is best for a certain county in

[1] Courtesy of The Field, New York City.

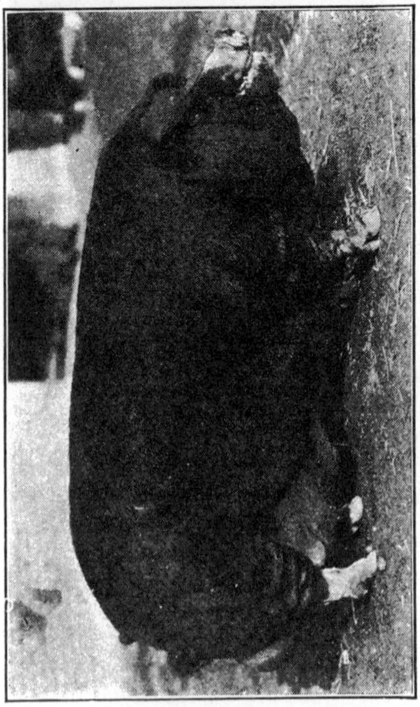

Poland-China Boar.[1] Lard Type Hog.

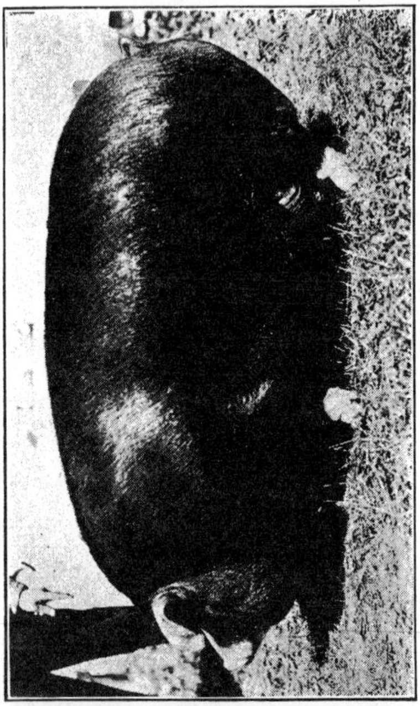

Poland-China Sow.[1] Lard Type Hog.

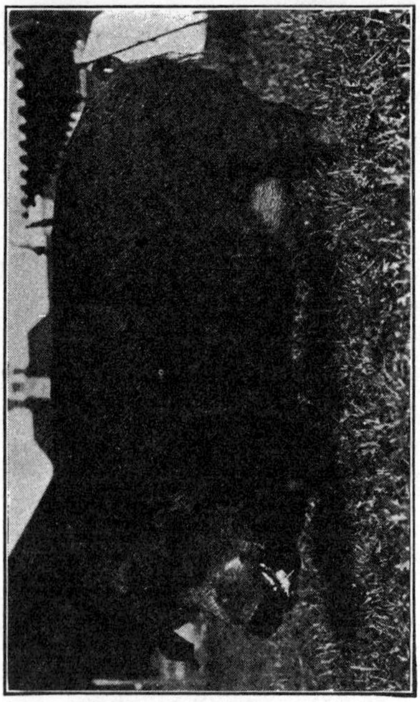

Duroc-Jersey Boar.[1] Lard Type Hog.

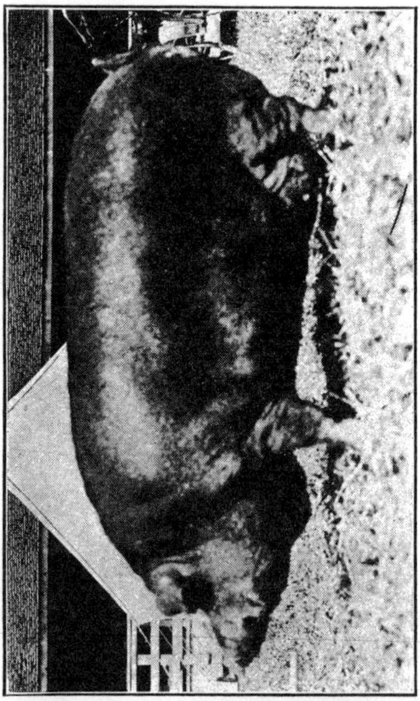

Duroc-Jersey Sow.[1]

[1] Courtesy of The Field, New York City.

Iowa may be ill-adapted to a county in Maine because various community conditions, such as customs, pasture range and cattle raising have their unmistakable effects. The climate in the South, because of the hot, long hours of piercing sunshine, puts the white hog at some disadvantage, whereas in the northern country he gets along exceptionally well.

4. **Distribution.**—A large number of swine of one type in a certain district usually indicates that they are well adapted. When in doubt, that breed which is well distributed in the community should be adopted. To raise Poland Chinas in a county where practically none but Tam-

Chester White Sows.[1]
Lard Type Hogs.

worths were raised, may result in disappointment, this being especially true if one depends upon local buyers for the sale of hogs.

5. **Markets.**—A nearby market which demands the bacon type, discriminating against the lard type, pound for pound, would have much influence in determining the kind of swine to raise in that particular section.

Breeds of Swine.—The two principal types of hogs are the lard and the bacon. Lard hogs are noted for their great depth, breadth, general compactness, smoothness, short legs, large hams, heavy jowls, relatively heavy shoulders, mellow finish (due to heavy fat layers) and docile

[1] Courtesy of The Field, New York City.

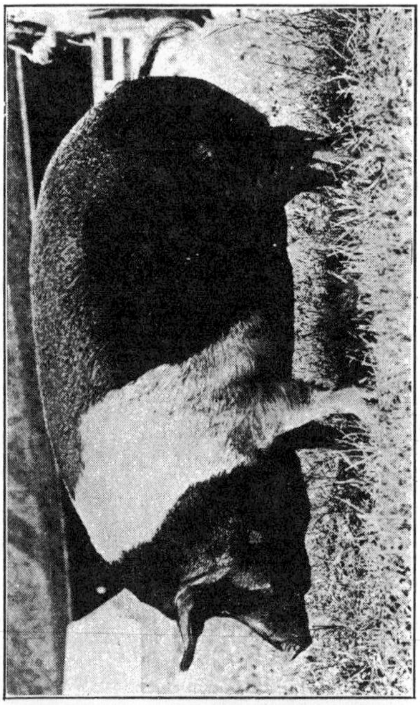

Hampshire Sow.[1]

Hampshire Boar.[1]

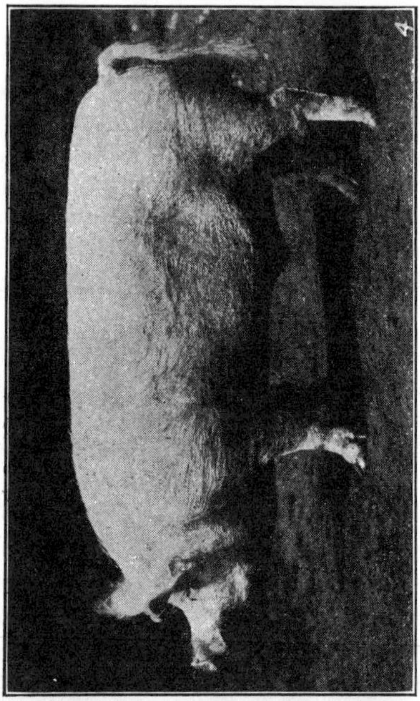

Yorkshire Boar.

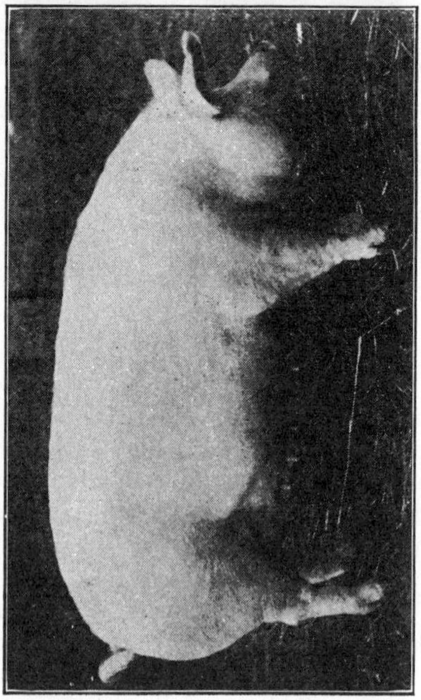

Yorkshire Sow.[1] Bacon Type Hog.

[1] Courtesy of The Field, New York City.

temperament. Bacon hogs stand in marked contrast in that the typical representatives have greater relative length, medium depth and breadth, similar smoothness but more trimness, long legs; small, trim, tapering hams; very neat, tidy jowls; very light and trim shoulders; exceptionally firm finish (with slight external fat layers) and active temperament.

The general or dual purpose breeds are a combinaton of the bacon and lard types, emphasis being placed upon the development of suitable market hams, bacon, ribs and loin, as well as the tendency to produce marketable animals suitable for lard.

The most typical lard type representatives are the Poland Chinas, black with white markings or spotted black, white and sandy; the Duroc Jerseys, entirely of a cherry red; and the Chester Whites, wholly white. These three breeds are especially popular in the corn belt, and deservedly so. Other lard type breeds are the Mulefoots, black, sometimes with white markings; the Victorias, white; the Cheshires, white; the Suffolks, white; the small Yorkshires, white; the Essex, black; and the Sapphires, blue (sometimes white markings).

The typical bacon type representatives are the large Yorkshires, white; and the Tamworths, red; both being especially prominent in Canada and the northern United States.

The dual purpose representatives are the Berkshires, black with white markings; the Hampshires, black with white belt; and the middle white or middle Yorkshires, white but little known in this country.

Of the breeds mentioned, seven are white, six all black or black with either white or sandy markings, two red and one blue. The most widely distributed pigs in the Canadian country are white, whereas in the corn belt and southern districts they are either black or red. This probably represents climatic adaptation.

Grading Up the Herd.—An ordinary farm herd composed of native individuals may be advantageously graded up by using successive purebred sires of the same breed. The first-cross animals are especially vigorous for market as well as for breeding purposes; they gain very rapidly and economically, and likewise make very good mothers.

In a grading-up program, assuming that a Duroc Jersey is used for the first cross and the offspring of the first cross are again bred to Duroc Jersey sires, it is surprising how quickly the offspring approach the Duroc types. In a few years, providing gilts only are kept each year for breeding, a typical Duroc Jersey herd, resembling closely the typical pure breed, will be a reality. The same grading-up process may be followed with any pure breed. This is an excellent practice and one to be followed with profit in the production of a uniform, dependable market type.

The crossing of breeds already crossed is to be discouraged, largely because of the heterogeneous individuals which result, these being of various types, sizes, colors and so on; this dissimilarity of offspring being

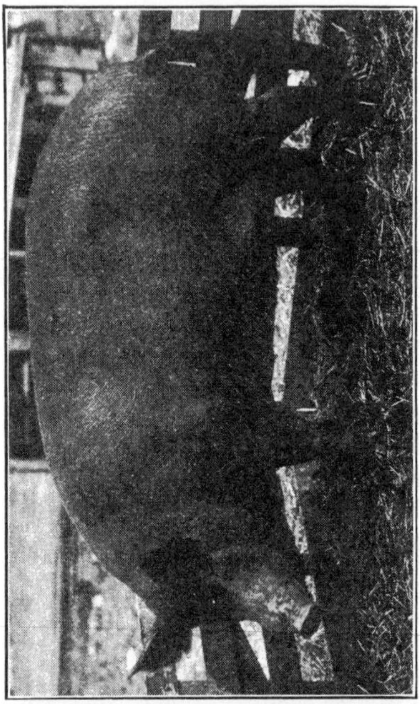

Tamworth Sow.[1] Bacon Type Hog.

Tamworth Boar.[1] Bacon Type Hog.

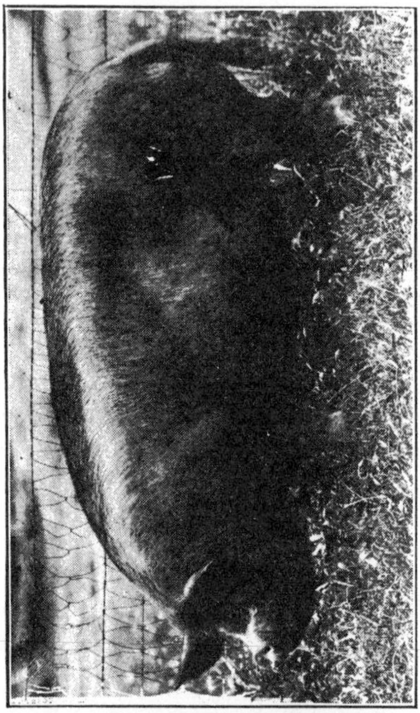

Berkshire Sow.[1] Dual Purpose Hog.

Berkshire Boar.[1]

[1] Courtesy of The Field, New York City.

all the more marked if the original pure-breds used are very different and less true if they are very similar.

Age of Breeding Stock.—Mature sows as compared to gilts enjoy some very marked and practical advantages, as follows:

1. A larger number of pigs at farrowing time.
2. Heavier, stronger, bigger-boned new-born pigs.
3. More pigs usually saved to each sow up to weaning time, hence more reach the market.
4. They are tried mothers, the undesirable brood sows being naturally eliminated.
5. Less high-priced protein feeds are needed to supplement the cheaper carbohydrates.
6. More rough feeds may be used, such as alfalfa hay and pasture.
7. Matured and tried-out sires can be used to advantage; this ofttimes not being feasible with young gilts unless a breeding crate is used.
8. Immunized, cholera-proof sows may be continuously kept, and the expense and bother of the annual immunization of young sows thus eliminated.
9. Two litters a year are raised with less difficulty. All young gilts cannot raise two litters successfully the first year.
10. Less loss in condition during the suckling period.
11. More dependable as breeders.

The disadvantages of sows older than gilts are not to be overlooked, and are as follows:

1. Require more feed.
2. More house room necessary.
3. If the "one litter a year" practice is followed these sows must be carried through a six months' unproductive period, which is relatively expensive.
4. Greater capital investment imperative.
5. Greater risk involved because of the greater capital invested.
6. Swine money not turned so often because the sows are kept longer and not sold annually as are the gilts.
7. Docked more on marketing, ofttimes twenty-five cents per hundred, than "trim-bellied" gilts.
8. Gains while "fattening off for market" more expensive.
9. Usually need an older, mature boar, because the younger ones are not so handily used; hence, the breeding more difficultly managed.
10. Apt to become overfat, clumsy and awkward, and hence overlie the new-born pigs. This tendency to overfatness must be carefully guarded against.
11. More difficult, generally speaking, to manage.

In profitable practice, a happy combination of both gilts and old sows may be kept to good advantage. The breeder of pure-bred swine obviously may well keep more old sows proportionately than does the market man.

In the selection of individual sows for the herd it is well that they show:

1. Trueness to the particular type and breed wished, having a desirable ancestry and being preferably from a prolific, tested family.
2. A well-formed udder with active teats and no blind ones.
3. Feminine characteristics of refinement.
4. Roominess and capacity.
5. A kindly disposition.
6. Good breeding record as manifested in their offspring.
7. Absence of overfatness and flabbiness.
8. A good, healthy constitution free from vermin and diseases.

The boar should possess most of these general requirements, emphasis being placed upon his masculinity as indicated in the well-developed crest, shield, tusks and general ruggedness. A mature, tried boar is more acceptable than an immature, untried one.

Housing.—In the housing of swine emphasis should be placed upon warmth, dryness, abundance of light and direct sunlight, shade, ventilation, sanitation, safety, comfort, convenience, size, durability, low first cost, minimum cost of maintenance and pleasing appearance of the structure to be used. This applies to the large centralized community or the small movable individual type.

The selection of a correct site for the location of the hog house is very important. Emphasis should be placed upon the economy of labor and time in management, drainage, exposure, slope, windbreaks, nearness to pasture and shade, elevation, prevention of odors reaching home dwelling and risk from disease infection. To place any hog house in an undesirable, ill-adapted place is to invite loss, dissatisfaction and possible failure.

The large community house as compared with the small movable one has some advantages in that the time and labor required for some operations is less; durability is usually greater; lighting from direct and diffuse sunlight better arranged; ventilation made more simple and systematic; general equipment usually less and more compactly arranged; close attention to the herd easily and practically possible; herdsman experiences minimum of exposure; feed storage, water supply and general rooms may be conveniently arranged; sanitation in some respects may be more encouraged; vermin more largely eliminated; site selection is simplified because only one site is needed; the heating problem is comparatively easy; common feeding floor and water wallow may be more handily arranged; danger of loss less than with large number of houses in common yards; provides headquarters for the swine farm; fire and other risk may be minimized through masonry construction; a number of swine under one cover become better acquainted; makes possible adjustable pens; facilitates collection of liquid manure; and advertising value may be greater.

On the other hand, the community house is a disadvantage in that location is not easily changed; isolation is practically impossible; sanitation may be sometimes discouraged; construction is more complex; it is not so practical for beginners; the first cost is somewhat high; more fencing is required to provide similar range conditions; it is likely to be used solely for a farrowing house and thus decrease serviceability; and fire and other hazardous risks may be greater if it is built of wood and is in close proximity to other buildings.

A combination of the two types of houses, *i. e.*, the large centralized or community one supplemented with the small movable one, deserves favor in practice. Each type has its own peculiar advantages and disadvantages, whereas the two together counterbalance each other so as to make a very complete efficient practical combination system.

Feeds for Swine.—Feeding swine has to do with the balancing of the grain ration to make it most efficient. The shortcomings of corn as feed for swine have their counterpart in other grains used less extensively in pork production, namely, barley, wheat, rye, sorghum seed, Kaffir corn and milo maize.

The predominating deficiencies of corn as a grain for growing swine are:

1. **Low in Protein.**—A young growing pig should have a pound of protein with every three to four pounds of carbohydrates. Corn has only one pound of protein to about eight and one-half pounds of carbohydrates.

2. **The Quality of Protein is Only Fair.**—Corn products alone, partly because of the protein content being of low quality, are inefficient in dry lot feeding, even though an abundance of protein be supplied as in the form of gluten meal.

3. **Lacking in Mineral Elements.**—Corn is particularly low in calcium, which comprises 40 per cent of the dry ash of bone. The young pregnant gilt would have to eat something like thirty pounds of corn a day in order to get enough calcium to supply the growing fœtus. Common salt and calcium and potassium phosphate have been found beneficial when added to a corn diet. The deficiency of minerals in corn has been largely responsible for the widespread general use of condimental material, such as wood-ashes, charcoal, bone phosphate, rock phosphate, cinders, slaked coal and others being used in practical hog feeding.

4. **Presents an Acid Ash.**—When corn is high in protein this acidity is especially marked. To make corn more productive this acidity should be counteracted. This is made possible by the judicious use of efficient and proper supplements.

The most acceptable, practical supplements to corn and the other similar starchy grains may be briefly enumerated as follows: skimmed milk, buttermilk, tankage, blood meal, linseed oil meal, cottonseed meal, gluten meal, wheat middlings, Canada field peas, soy beans, alfalfa and clover hay.

The production of pastures is an economical proposition and is to be encouraged on every American as well as Canadian swine farm in order to obviate the necessity of purchasing high-priced protein concentrates. The most profitable supplemental pastures in the corn belt in the order of merit are: alfalfa, rape, red clover, blue grass and sweet clover of the first year's growth. In the South cowpeas, soy beans, Spanish peanuts and Lespedeza clover may be added, inasmuch as these plants give excellent success in warm climates and on soils that are sandy and relatively unproductive as compared to the corn belt soils of Iowa and Illinois.

Swine feeding and dairying, along with judicious use of green forages in a good corn, barley or other similar grain country, is a most excellent swine-farming proposition. If alfalfa can be raised, so much the better.

Preparation of Feeds.—Hard, tough, fibrous-shelled seeds such as Kaffir corn, sorghum, milo maize and millet will be much more efficient if fed in the ground condition. If grinding is impossible, soaking is the next best possible procedure.

Wheat, rye and barley likewise give better results when ground, and can also be soaked as an alternative. Some experiments show as much as 20 per cent increase in the efficiency of wheat through the grinding as compared to feeding the grain dry and whole.

The general herd, young pigs and sows on a maintenance ration do better on ear corn than any other form. However, fattening sows and heavy fat hogs in the final stages of fattening make more economical gains on the soaked shelled grain. If any preparation should be used other than dry ear, it should be shelled soaked corn rather than the ground grain dry or soaked.

Corn-and-cob meal has little to commend it to any class of swine except possibly the brood sows on maintenance, and even with these the ear corn is the most profitable.

Hays, such as ground alfalfa, may be ground ofttimes in order to facilitate their mixture with the grain rations and to encourage their consumption.

Wetting and cooking of feeds is not ordinarily profitable, although to produce rapid gains these procedures are sometimes permissible, this being especially true in the production and finishing of show stock.

Hand vs. Self-Feeding.—Better results will be secured by the self-feed method than by hand-feeding twice daily. However, feeding three times a day is the most efficient, considering rapidity of gains and economy in feed required for 100 pounds of gain. Under our ordinary high-priced labor conditions, thrice a day is not enough better to excel self-feeding.

The "Free-Choice" scheme of feeding consists of allowing acceptable feeds before swine in such a manner that they can balance their own rations. In 1914 a group of pigs fed at the Iowa station according to this scheme, receiving shelled corn, linseed oil meal, oats and meat meal (or tankage), limestone, charcoal and salt in separate feeds, weighed 316 pounds when 8 months 7 days old. They did as well as if they had been fed according to accepted feeding standards. Tests now in progress (1915) at the Iowa station indicate that pigs can feed themselves better than a trained animal husbandman can feed them if he follows the customary feeding standards.

When pigs are allowed starchy corn and high-protein meat meal (tankage) in separate feeders, this feed being kept before them from weaning time until they reach a weight of 300 pounds, they will eat of these two dry lot fed feeds approximately as follows:

	Approximate Age, days.	Approximate Weight, pounds.	Pounds 60 Per Cent Protein Meat Meal (Tankage) Eaten with Every 100 Pounds Corn.	Pounds Protein Eaten with Every 100 Pounds Starches.
Weanling	60	35	20	3.38
Shote	120	100	15	3.84
Fattening hog	180	210	8	4.79
Fat marketable hog	240	300	1	6.50

The self-feeding method is excellent in dry lot feeding, on pastures and where skim milk or buttermilk is used.

Feed for the Brood Sows.—The brood sow should have good feed in order to produce strong, healthy pigs. Skim milk, tankage, alfalfa pasture, etc., combined with corn or other grains, often increase the litters as much as one pig over corn alone.

To demonstrate the advantage of supplementing the starchy grain feeds such as corn with the proper protein supplement, the resulting average weight and strength of the pigs secured is given for a few typical rations:

Ration Fed.*	Average Weight per Pig, pounds.	Strong Pigs, per cent.	Cost of New-born Pigs, cents.
Corn only	1.74	68	41
Corn plus 4 per cent tankage	2.01	92	18
Corn plus 16 per cent tankage	2.23	93	22
Corn plus alfalfa in rack	2.12	98	31

These gilts were charged at the rate of 50 cents a bushel for shelled corn, $2.50 for meat meal and $15 a ton for the alfalfa hay; yet in spite of the comparative high cost of the supplements, the pigs produced were not only cheaper but much better pigs were secured, the animals being much larger and stronger than where corn only was fed.

* Animal Husbandry Section—Iowa Experiment Station Results.

The unborn pig must be well fed indirectly through its host the brood sow if a strong, vigorous, active pig carrying big bone and strong muscle at the time of farrow is desired.

In general, it is well to emphasize in brood sow management:

A. Acceptable balanced feeds in sufficient quantity.
B. Maximum growth without excessive fattening.
C. Abundant exercise, especially the latter two months of gestation.
D. Riddance of lice and worms.
E. Moderate laxativeness, because constipation is a menace.
F. Gentleness in handling the herd.

The suckling sow and litter should be fed better than any other class of swine. The demand for growing feeds by both the sow and pigs at this time is great. Such feeds as corn, barley, skim milk, buttermilk, tankage, wheat middlings, alfalfa, rape, clover, blue grass and other pastures, and similar feeds equally as good are to be particularly commended.

Feeding the Pigs.—From time of weaning to maturity pigs should have plenty of suitable forage. Nothing is better than pastures of clover, alfalfa, rape, etc. Here they may be given a full or limited ration, depending on circumstances.

If the usually better early fall markets are the goal, full-feeding will be in order; whereas if the later but somewhat lower markets are most acceptable, the grain ration can be limited. Ordinarily, it is not a paying proposition to feed less than three pounds of grain with every 100 pounds of pig daily. A lesser amount, especially if the pasture be poor, will cause the pigs to become stunted.

The fall pigs which are raised in winter dry lot must be fed a relatively high-priced ration; in other words, one high in protein. The fall pigs need warm shelter, and it is best to feed them inside the shelter. They must be protected from the cold winds, snows, hails, and general wintry conditions, while the spring pig should be protected from the hot sun and the flies. The fall pig lives at a time when attacks from worms are at low ebb and are not readily passed from one host to another.

The "hogging-down" of corn deserves much emphasis. It is practiced profitably in all of the corn belt states. It may be likened to dry lot feeding if the field is clean and free from weeds, and supplemental protein feeds should be supplied accordingly. It is well to have an alfalfa, rape or similar pasture field adjoining in order to supply this protein at the lowest cost.

Rape, winter rye or winter wheat in the northern corn belt may be sown in the corn at the last cultivation or shortly thereafter with excellent success; in the more southern districts cowpeas and soy beans may be included with profit.

Successful swine rations for general American conditions, and suitable for the various sizes, ages and classes of hogs are suggested in a ready reference table presented herewith:

The Swine to be Fed.	Pounds of Tankage* (60 per cent Protein) to be Fed along with every 100 Pounds of Corn to Swine of Various Classes in		
	Dry Lot.	Low-Protein Pasture.†	High-Protein Pasture.‡
I. *Growing and Fattening for Market.*			
1. Suckling pigs (a creep) 5–40 pounds..	25	25	20–12
2. Weanling pigs, 30–100 pounds.......	25–18	23–16	12–5
3. Shoats, 100–175 pounds............	18–10	16–9	5–2
4. Hogs, 175–250 pounds..............	10–4	9–4	2–2
5. Fat Hogs, 250–350 pounds..........	4–1	4–1	0
II. *Fattening Sows for Market.*			
1. Yearlings (gilts) after weaning.			
A. In poor condition, run-down.....	11–8	11–8	5–0
B. In good condition, thrifty........	9–5	8–5	0
2. Two years or older.			
A. In poor condition, run-down.....	6–4	6–4	4–0
B. In good condition, thrifty........	2–0	2–0	0
III. *Stags, Fattening.*			
A. Young.........................	9–4	9–4	0
B. Old...........................	5–0	5–0	0
IV. *Carrying Sows, Breeding.*			
1. Breeding swine, flushing.			
A. Gilts..........................	14	14	10
B. Yearlings and older.............	11	11	8
2. During pregnancy.			
A. Gilts..........................	14–10	10–7	0–5
B. Yearlings and older.............	10–6	6–4	0–4
V. *Suckling Sows.*			
A. With large litters...............	25–18	25–18	10
B. With small litters...............	20–8	20–8	3–5

* If corn is not available, it may be substituted pound for pound in these proportions with barley, wheat, rye, sorghum seed, Kaffir corn, milo maize, or feterita, or a combination of any of these. If 60 per cent protein tankage is not available, linseed oil meal or soy bean meal may be substituted, 2 to 2½ times as much being used. For example, the suggested dry lot ration for growing and fattening shoats is "corn 100, tankage 18 to 10;" now substitute oil meal 2 times as much and we have corn 100, linseed oil meal 36 to 20. To substitute wheat middlings, allow 17 times as much, skim or buttermilk 20 times, and blood meal 60 per cent as much, or almost two-fifths less. Blood meal runs about 85 per cent protein and but little is required, but blood meal is not so good a supplement as tankage, everything considered.

† *Low-Protein Pastures.*—Dry, hard, fibrous blue grass; sorghum; feterita; millet; Sudan grass; milo maize; timothy when over four inches high; rye or wheat over eight inches; or oats and barley over five inches, or beginning a couple of weeks before beginning to joint; and sweet clover of second year's growth after two feet high.

‡ *High-Protein Pastures.*—Alfalfa; rape, Dwarf Essex; medium red, mammoth, alsike, and white and other clovers; young, tender, sweet clover, first year's growth; quite early, tender, new coming timothy, rye or wheat; short, "shooting," tender, green, succulent blue grass, cowpeas; and soy beans.

REFERENCES

"Productive Swine Husbandry." Day.
"Swine in America." Coburn.
"Swine." Dietrich.
"Forty Years' Experience as a Practical Hog Man." Lovejoy.
"The Hog Book." Dawson.
Alabama Expt. Station Bulletin 185. "Dipping Vat for Hogs and Dips;" "Hog Worms, Lice and Mange;" "Hog Lot, Houses and Water Supply."
Kentucky Expt. Station Circular 4. "Mal-Nutrition of Hogs."
Nebraska Expt. Station Bulletin 147. "Pork Production."
Ohio Expt. Station Bulletin 268. "Fattening Swine with Substitutes for Corn."
South Dakota Expt. Station Bulletin 157. "Rape Pasture for Pigs in Cornfield."
Wyoming Expt. Station Bulletin 107. "Swine Feeding."
Canadian Dept. of Agriculture Bulletin 225. "Swine."
Farmers' Bulletins, U. S. Dept. of Agriculture:
411. "Feeding Hogs in the South."
438. "Hog Houses."

CHAPTER 49

SHEEP AND GOATS

By T. C. Stone
Instructor in Animal Husbandry, Ohio State University

Early Importance of Sheep.—There is evidence that sheep were under domestication in Europe in prehistoric times. The primitive man used the skin for clothing and the meat and milk for food. As man has advanced in civilization, sheep farming has become an important branch of agricul-

A Typical Cotswold Ewe.[1]

ture. Sheep and their wool were very early acknowledged to be the foundation of the national prosperity and the wealth of Great Britain and other European countries. The more recent introduction of silk manufactures and the establishment of the cotton trade have lessened the demand for woolen goods; still, the sheep and its fleece are of great importance.

[1] Courtesy of The Field, New York City.

The Sheep of Spain.—The Spanish Merino, the only type of sheep in Spain, are noted for: (1) the production of a very fine wool, (2) hardiness and ability to travel, and (3) the disposition to stay close together when feeding, resting and traveling. These characteristics have had an important influence on their later history.

The Sheep of England.—In England were developed several types of sheep, and each type or breed was adapted to a certain locality. These breeds were quite unlike in fleece. The wool found favor on the market because of its variety in length and quality, which made it adaptable to

A Typical Lincoln Ewe.[1]

different uses. The Royal Agricultural Society of England in its show catalogue recognizes twenty-five breeds. These were all developed on the British Isles. Some were developed in the lowlands, some in the hills and others in the midlands. They were developed principally for meat; fresh meat in England, with its great population, being of greater consequence than wool. The various breeds were divided into four classes, namely, the long-wool breeds, the middle-wool breeds, the highlanders or mountain breeds and the upland breeds.

Breeds of Sheep.—Two distinct types of sheep have been produced, namely, the mutton and wool types. The former are valued chiefly on

[1] Courtesy of The Field, New York City.

account of their ability to make mutton economically, although the wool-producing ability of the mutton sheep constitutes no small part of their value to the farmer. The wool type, however, is raised mainly for the wool it produces.

In conformation, the mutton sheep are compact, with a short head and neck, a broad, level back, a full leg of mutton, a deep body and short legs. The wool ranges in length from $2\frac{1}{2}$ inches in the middle-wools to 10 inches in the long-wools. The fleece does not cover the body so compactly as does the fleece of the fine-wool sheep. The medium-wool breeds greatly

A TYPICAL SHROPSHIRE.[1]

excel the long-wools in this respect. The fleece of the medium-wool breeds is much less fine in quality and has much less yolk or oil in it than does the fleece of the Merino sheep.

LONG-WOOL BREEDS

Leicester.—Very large sheep, wool 6 inches long at 12 months, being bright and lustrous; face and legs white; no wool on head. Weight of mature rams ranges from 225 to 250 pounds; ewes from 175 to 200 pounds.

Cotswold.—Wool 8 inches long at 12 months; pronounced tuft of wool on forehead; face and legs white. Rams weigh from 250 to 275 pounds; ewes from 200 to 225 pounds.

[1] Courtesy of The Field, New York City.

Lincoln.—No breed furnishes so long a fleece as the Lincoln. It ranges from 8 to 12 inches; tuft of wool on forehead. Rams weigh about 385 pounds; ewes about 275 pounds.

MEDIUM-WOOL BREEDS

Southdown.—They are smallest of the middle-wools, very low-set and compact, with steel-gray or mouse-brown markings on face and legs. Fleece is $2\frac{1}{2}$ inches long at 12 months. Rams weigh from 185 to 200 pounds;

A Typical Cheviot.[1]

ewes from 125 to 140 pounds. Criticised for lack of wool production and insufficient size. Much improvement has been due to this breed.

Shropshire.—They are stylish sheep with pronounced extension of wool over face and legs; color marking is a deep, soft brown. Wool 3 inches long at 12 months. Rams weigh about 225 pounds; ewes from 140 to 160 pounds. Rank high as a dual purpose breed.

Oxfords.—They resemble the Shropshire, but are larger and do not have as great wool extension over face and legs. Lighter brown is the color marking, and usually are more upstanding. Wool is 4 inches long at 12

[1] Courtesy of U. S. Dept. of Agriculture.

months. Heavier than Southdowns and Shropshires, equal to Hampshires. Rams weigh from 275 to 300 pounds; ewes about 175 to 200 pounds. They give size and weight when crossed on short-wools and quality and better mutton when crossed on the long-wooled breeds.

Hampshires.—They have darker color markings than the Oxfords, and a very pronounced Roman nose. Wool is $2\frac{1}{2}$ inches long at 12 months. Very early maturing sheep.

Dorset Horn.—Have white color markings; very little wool on face

A Typical Merino.[1]

and legs and it does not extend well over lower parts of the body. Both ewes and rams have horns. Wool at 12 months is 3 inches long. Weight of rams from 250 to 275 pounds; ewes 175 to 185 pounds. A mutton breed of merit; valued as early lamb raisers.

Cheviot.—They are very alert, stylish sheep with white markings. Face and legs are free from wool. Wool is 4 inches long. Rams weigh from 200 to 225 pounds; ewes from 125 to 140 pounds. It is a very hardy breed and individuals graze independently of each other.

Fine-Wool or Merino Sheep.—This type is the result of efforts to

[1] Courtesy of U. S. Dept. of Agriculture.

produce a fleece of finest quality. In developing this type some breeders did not overlook the mutton qualities, while others did. The Spanish Merino was the foundation of the three classes of Merinos as they exist today. The three classes are A, B and C. This classification is based on differences in conformation, character of fleece, and number and disposition of wrinkles or folds on the sheep.

The Merino blood must predominate on our western ranges because of the gregarious nature of this breed. They have great constitution and vigor and are much less susceptible to parasitic trouble than the breeds of the mutton type. They can be kept in smaller quarters and the ewes do not need as much care at lambing time as ewes of the mutton breeds. They are lacking greatly in mutton qualities, and there is a strong demand for

A Typical Flock of Sheep in Pasture.[1]

the dual purpose animal. The Merino will not be supplanted, but as the demand for mutton becomes stronger, they will no doubt be supplemented very largely by the mutton breeds.

Establishing a Flock.—Sheep may be kept profitably on either high or low-priced land. On the high-priced lands of England sheep are found in great numbers and they would certainly not be kept if they were not profitable. Sheep do best on slightly rolling land where dry footing prevails. They get more sustenance and at the same time do the land more good than any other class of livestock. The manure from sheep contains more fertilizing value per ton than any other kind of farm manure with the exception of poultry.

[1] Courtesy of The Macmillan Company, N. Y. From "Crops and Methods for Soil Improvement," by Agee.

Very little capital is needed to start a flock of sheep. They need not be housed in expensive buildings. Nature has fitted them to endure cold weather. A small flock requires very little labor, especially during the busy summer. These advantages, along with the fact that sheep destroy weeds, thereby helping to beautify the farm, make the sheep a valuable asset to the American farmer. These advantages are not mentioned with the view of urging the farmer to give up other classes of farm animals, but to remind him of the advantage of supplementing his stock with a small flock of, say, forty ewes or even less.

Essentials to Success.—One should choose the breed best adapted to local conditions, especially the climate and market. There is no best breed for all conditions. It is best that a man gain his experience with grade stuff. One may purchase either Merino or mutton breeds and then grade them up by using a pure-bred ram. The latter is of great importance. A ram having a good pedigree and good individuality should be selected. He should be purchased from a reliable breeder and the stockman should not hesitate to pay a good price for a desirable ram. The ram should possess good breed type and be masculine. An effeminate ram should have no place in a flock. Masculinity is indicated by a short, broad head, large, broad nostrils, ruggedness in appearance and a lack of too great refinement throughout. Rams should have a good conformation, and those which have been very highly fitted should be avoided, as they often prove non-breeders. There are only a few instances where it would be permissible to use a ram lamb to head the flock. This is done more often in the case of the Hampshire breed than others. Older rams usually make the best breeders. A ram of the middle-wool breeds is sufficiently developed and fit for service at the age of 1½ years.

Only ewes that are sound in their mouths and udders, and that possess feminine characteristics and good general conformation should be purchased. It must be remembered that the ewes are half the flock.

One should not make the mistake, after establishing a flock, of allowing the sheep to care for themselves. Suitable but inexpensive shelter and plenty of forage should be provided and plenty of salt and water should be kept before them. It is necessary to be on the lookout for internal parasites, especially in lambs, during the summer months.

The Breeding Season.—The breeding season of the year in this country commences in September or just as soon as the cool nights begin. The heat periods of the ewes last from one to two days and normally appear at intervals of 16 days. The Dorset Horn and Tunis will breed at any time.

Period of Gestation.—The usual period is 146 days. Ewes, however, are very irregular about bringing forth their young. Shepherds in the old country figure on 140 days. The period of gestation is often longer for Rambouillets than for other breeds.

Care of Ram During Breeding Season.—Not more than 40 ewes should

A Good Flock of Sheep.[1]

[1] Courtesy of U. S. Dept. of Agriculture.

be allowed to one ram. The last born lambs are often weaker than those born earlier in the season. This indicates that it is not advisable to breed the ram to too many ewes. In a large flock, the ram should be put with the ewes for an hour at the end of each day. In a small flock, he may be allowed to run with the ewes all the time. Where hand coupling is not practiced, one should paint the brisket between the ram's fore-leg with paint. Red lead and linseed oil make a desirable paint for this purpose. This mark will indicate that the ewes have been bred. After 16 days the ram may be painted another color. By this means the breeder may know whether the ewes are returning. The ram should be fed liberally during the breeding season, but not too well. A mixture of equal parts of oats, bran and oil-cake, say one pint, both mornings and evenings, will prove a good ration.

Winter Care of Ewes.—A lamb gets its start on the right or wrong way before it is born. The pregnant ewes should be sufficiently fed, but not overfed during winter. They should be given plenty of exercise; the more they get, the healthier the lamb crop will be. The feeding of too much grain just previous to lambing time should be avoided. Bran, oats, oil meal and clover make an ideal ration for the breeding ewe. Silage and roots are good succulent feeds. but must be fed in small quantities and must be of good quality.

Care of Young Lambs.—Lambs should be weaned when 3½ to 4 months old, and put on fresh pasture. The secret of successful and profitable lamb raising is to keep them growing and in good condition from birth to maturity. Lambs should be given grain as soon as they can thoroughly digest it. In order that they may eat at will, it is necessary to build creeps for them. The feeds given and the amount will depend largely on the purpose for which they are being prepared. Those being fitted for the market should be fed liberally with grain until they are of market age. Their ration may consist almost wholly of corn. A good grain ration for lambs just beginning to eat is ground corn, one part; crushed oats, one part; linseed oil meal, one part; and wheat bran, two parts.

All lambs should be docked and all males intended for the open market should be castrated. Lambs that are not castrated often sell for at least $1.50 per 100 pounds less than castrated lambs. This does not take into consideration the loss of flesh due to activity of ram lambs. The lamb that is not docked gets filthy around the dock and presents a poor appearance on the market. They may be docked and castrated when about two weeks old. It is much more convenient to do both at the same time, and no evil results will follow if the operations are performed in the right way.

Marketing the Lambs.—It is usually best to market the lambs at weaning time. This will occur about July 1st. There is great demand for lambs weighing from 65 to 70 pounds. They furnish a superior product for the consumer and make very economical gains for the producer. There are other reasons for marketing lambs at this time. First, lambs gain very little during hot summer months; second, there is risk of losing them through the internal parasites; third, one avoids heaviest run of western lambs; and lastly, one gets the use of his money earlier.

Shearing the Flock.—Time of shearing depends on the weather, the season and the locality and equipment. It is advisable to shear as soon as warm weather begins in the spring. Late shearing is unadvisable, as the sheep will lose in weight if compelled to carry heavy fleeces. They

AN ANGORA BUCK.[1]

are also liable to lose some of their wool during the later months. Well-fed ewes with comfortable sheds may be sheared fairly early. They will not suffer if the days should become a little cool. Wethers fed under the same conditions may often be sheared as early as March. They will gain faster when fleeces are removed. Care should be exercised to see that they do not overeat at this time.

Both hand shearing and machine shearing are practiced.

Dipping the Flock.—All sheep should be dipped for three reasons. First, to promote healthy condition of the skin; second, as a remedy for scabies in sheep; third, to kill the lice and ticks.

[1] From Farmers' Bulletin 573, U. S. Dept. of Agriculture.

The time for dipping depends upon the time of shearing. It is best to dip five or six days after shearing. The ticks and lice leave the shorn ewes and go to a more sheltered place on the bodies of the young lambs. If one delays dipping for any length of time after shearing, the lambs will suffer a great deal with these pests. A second dipping should take place during the fall.

Any of the recommended coal tar dips may be used. In using these, one should see that they have the approval of the Department of Agriculture and should follow the directions carefully.

A flock thus handled will afford the owner much pleasure and profit for capital and labor invested. The earnings from sheep will compare very favorably with those of any of our domestic animals.

GOATS

Goats are very valuable as a renovator of brush lands. They are not naturally grazing animals, but rather browsers. In some states, the cost of clearing large tracts of land has been greatly reduced by pasturing with flocks of goats.

Besides this, many goats, especially the representatives of the breeds of milch goats, are noted as milk producers. They have held a recognized place as such for a great many years among the poorer people of the world. In some countries varieties of goats are bred especially for their milk-producing qualities.

In this country, the Angora goat and the common goats give milk, but milking families have not been produced.

The Angora goat yields a fleece which is valued highly on the market. It is commercially known as mohair. It is coarser than fine wool, but longer and stronger.

When sold on the market, goats bring a lower price than sheep. The mutton from goats is not considered nearly as good as mutton from sheep.

Angora and common goats are found in almost every state in this country . They seem to do well under a wide range of climatic conditions. A dry climate, however, seems most favorable for them.

REFERENCES

"Sheep Farming." Craig and Marshall.
"Sheep Farming in America." Wing.
"Productive Sheep Husbandry." Coffey.
"Sheep Farming." Kleinheinz.
"Sheep Feeding and Farm Management." Doane.
"The Winter Lamb." Miller and Wing.
"Angora Goat Raising and Milch Goats." Thompson.
Nebraska Expt. Station Bulletin 153. "Fattening Lambs."
U. S. Dept. of Agriculture, Bureau of Animal Industry, Bulletin 68. "Information Concerning the Milch Goat."
Farmers' Bulletins, U. S. Dept. of Agriculture:
573. "The Angora Goat."
576. "Breeds of Sheep for the Farm."
652. "The Sheep Killing Dog."

CHAPTER 50

THE FARM FLOCK (POULTRY)

BY M. C. KILPATRICK
Instructor in Poultry Husbandry, Ohio State University

Improved methods of production and the establishment of large specialized poultry farms have greatly increased the supply of poultry and eggs during recent years. The demand for these products, however, has been increasing even more rapidly than the supply. This increasing demand is due both to the rapid increase of the consuming population and to a growing preference for these products as food. The increase in the demand for eggs is especially marked, due largely to the increased price of meats and the fact that modern transportation facilities, storage warehouses and improved methods of handling eggs have resulted in a better distribution of the supply throughout the year and a higher standard of quality upon the large city markets.

Importance of the Farm Flock.—The farm flocks of the country furnish 90 and possibly 95 per cent of the total supply of poultry and eggs. It is natural that the general farms should be the principal source of supply, because poultry husbandry is essentially a livestock industry, and for this reason, best adapted to development under farm conditions. The farm provides those conditions which are essential to profitable poultry production, viz., ample range and pasture at low cost, cheaper feeds, the opportunity to make use of waste materials and convert them into marketable products, low labor cost, and of greatest importance, natural conditions which tend to increase rather than to decrease the health and vigor of the flock.

Unfortunately, the average farm flock falls far short of its productive possibilities. This is due largely to the fact that fowls are kept on the farm primarily for the purpose of supplying the home table with fresh meat and eggs and have not been regarded as an important source of income. This has resulted in flocks of small size and poor quality, inadequate equipment and a general indifference toward poultry on the farm. The increasing demand for poultry and eggs, and the general increase in the farm price of these products have resulted in making the farm flock of good size and quality, and properly equipped and handled, an important source of income. In addition it performs its primary function in supplying poultry and eggs for the home table.

The Size of the Farm Flock.—The size of the farm flock is an important factor in determining whether poultry is to be a profitable farm enterprise or not. The optimum size of the flock for a particular farm depends upon a number of conditions. These conditions are so variable that it is impossible to set a definite standard which will be applicable to all farms. It is evident, however, that the flock should number at least 100 fowls, and, except under very favorable circumstances, should seldom exceed 500 fowls. As many fowls should be kept as possible without allowing the poultry work to come in direct competition with more important farm enterprises. For the average farm, this will mean a flock of 300 to 500 fowls.

Sources of Income.—The principal sources of income from the farm flock are poultry and eggs for market. The production of eggs for market is the more important because of the relatively greater demand for them and the greater convenience with which they may be produced and mar-

A Typical Farm Flock.

keted. It is impossible to separate the two and, under some conditions, the production of market poultry may become the more important. Other possible sources of income are the sale of eggs for hatching, fowls for breeding purposes, day-old chicks, and the production and sale of pullets for egg production. The relative importance of each of these sources of income and the extent to which they may be combined will be determined by the personality of the poultryman and the organization of the farm business.

Advantages of Pure-Bred Poultry.—A second factor of greater importance in determining the value of the farm flock is the quality of the fowls. Pure-bred poultry is superior to mongrel, cross-bred or grade fowls because of greater reliability in breeding, more attractive appearance, ability to feed more efficiently, greater uniformity in the size, shape and color of the eggs, and greater uniformity in the appearance and condition of the dressed fowls. The first cost of pure-bred fowls is greater than of inferior stock, but no greater investment is needed. The best practice in starting

WHITE PLYMOUTH ROCKS.[1]

Winners of First and Second Prize Exhibition Pens, Madison Square Garden, N. Y., December, 1911.

BUFF ORPINGTONS.[1]

First Prize Exhibition Pen, Madison Square Garden, N. Y., December 31, 1915–January 5, 1916.

[1] Courtesy of Owen Farms, Vineyard Haven, Mass., Maurice F. Delano, Proprietor.

a flock of pure-bred fowls is to purchase a pen consisting of a male and four to ten females. These should be housed apart from the main flock and all of the good eggs laid during the breeding season should be incubated. Pure-bred fowls of good quality may be purchased in the late summer or early fall for $3 to $5 each for males and $2 to $4 each for females. Yearlings or two-year-old stock should be bought. After the pure-bred flock has been established, the many advantages of the pure-bred fowls are obtained without additional cost.

Grading Up a Farm Flock.—While pure-bred poultry are always to be preferred, it is possible to improve the quality of the average farm flock by the use of a pure-bred male. If a pure-bred male of the desired variety is mated with ten or twelve of the best hens on the farm, the offspring will carry one-half the blood of their sire. If the male is a strong, prepotent individual, a large percentage of the offspring will resemble him in many of his characteristics. Ten or a dozen of the best pullets resulting from the original mating should be selected and mated to their sire for the second season. The offspring from this mating will carry 75 per cent of the blood of the pure-bred male. For the third season, ten or a dozen of the best of these pullets should be mated to another pure-bred male of the same variety and of similar breeding. It is advisable to obtain the second male from the same breeder as the first one. If the fowls used have been carefully selected, the offspring from this third mating will be practically as uniform in size, shape and color as pure-bred fowls.

The Choice of a Variety.—The choice of a variety for the farm depends upon the purpose for which poultry is kept and the type of product most in demand in the best available market. The efficiency of the various varieties depends more upon the breeding and handling of the fowls than upon breed or variety differences.

The most popular fowl for the production of white eggs is the Single Comb White Leghorn. It is not a good market fowl, however, because of its small size, nervous temperament, and greater loss in dressing. The cockerels make good broilers at weights of 1¼ to 1½ pounds, but do not make good roasters or capons.

The Plymouth Rocks, Rhode Island Reds and Wyandottes are the most satisfactory breeds for the production of both eggs and meat. The solid-colored varieties of the Plymouth Rock and Wyandotte, particularly the white and buff, are preferable on account of the absence of dark-colored pin feathers. The Columbian varieties are rapidly increasing in popularity. The most popular farm fowl in the past has been the Barred Plymouth Rock. It is slowly being replaced by some of the newer varieties. The three breeds mentioned are good layers, hardy, easily handled; the chicks grow rapidly, making them well adapted to the production of broilers. They make superior roasters and capons. Where the market prefers brown eggs or will not pay a premium for white eggs, one of the many varieties of these three breeds should be chosen.

WHITE WYANDOTTES.[1]
First Prize Pen, Chicago Show, December, 1912.

SINGLE COMB RHODE ISLAND REDS.[1]
First Prize Young Pen at Boston Show, January, 1915.

[1] Courtesy of Owen Farms, Vineyard Haven, Mass., Maurice F. Delano, Proprietor.

Selection of the Breeding Stock.—It is seldom necessary and never desirable to use all of the fowls on the farm for breeding. Special matings are necessary each season in order to make any definite improvement in the quality of the flock. It is seldom necessary to use more than 20 per cent of the entire flock for breeding. The fowls used for this purpose should be the choicest on the farm. They should be strong, healthy and vigorous, above the average in size for the variety, good layers and fully matured. Hens are always preferable to pullets, because the eggs from hens are larger, hatch better and produce larger and more vigorous chicks. Strong, vigorous, early-hatched cockerels may be used, but yearling or two-year-old cocks of proven breeding ability are to be preferred. Care should be taken to avoid using for breeding purposes any fowl which has had any sickness at any time, no matter how well it may appear to have recovered.

Housing the Breeding Stock.—It is not necessary to house the breeding flock separately during the entire year. The fowls to be used for breeding should be separated from the main flock three or four weeks before it is necessary to save eggs for hatching. They should be housed in portable colony houses during the breeding season, and may be returned to the main flock as soon as the last eggs needed for hatching are gathered. The colony houses may then be used for the growing chicks or for some other purpose.

INCUBATION

Selection of Eggs for Hatching.—Eggs for hatching should weigh not less than two nor more than two and one-half ounces each. They should be of a medium type, neither very long and pointed nor very short and rounded. The shells should be clean, smooth and strong, free from ridges, cracks, transparent spots or lime deposits. The eggs selected should be as uniform in color as possible. Dead chalk-white or uniform brown eggs are to be preferred. Careful selection of the eggs to be incubated will aid greatly in improving the general quality of the eggs produced by the flock.

Care of Eggs for Hatching.—Eggs for hatching should be gathered frequently, two or three times daily, and immediately removed to a clean, dry place where the temperature is less than 68° F. A temperature of 50° to 60° F. is best. Eggs for hatching should not be held longer than two weeks, as there is a rapid loss of vitality after that time. They should not be washed. Eggs hatch better if they are turned once daily from the time they are laid until set.

Natural or Artificial Incubation.—Whether hens or incubators should be used depends upon local conditions. If chicks are wanted before April 1st, or if non-setting varieties are kept, or if more than 150 chicks are to be reared each season, incubators should be used. There is no apparent difference between the vigor and vitality of hen-hatched and incubator-hatched chicks.

Hatching with Hens.—Hens of medium weight, from five to seven pounds, and of quiet disposition should be selected. They should be kept where they will be comfortable, easily controlled and free from annoyance by other fowls. A small brood coop is advisable for each hen during warm weather. These coops may be placed in a cool, shady location and the nest made upon the ground, a bottomless box about five inches high being used to confine the nesting material. During cool weather, a comfortable room should be provided. The nests used should be approximately 14 inches square. They should be constructed so that each hen may be confined to her own nest. In this way a number of hens may be set in the same room, all being released for food and water at the same time. It is necessary to see that each hen returns to her nest as soon as through feeding. Several hens should be set at the same time. This will save labor and allow the chicks hatched by two or three hens to be given to one for brooding. Hens should be removed from their regular nests to the nests in which they are to be set after dark. If handled quietly and given a few decoy eggs they may usually be moved without difficulty. The hen should be allowed to become accustomed to her new surroundings before setting her. This usually requires two to three days.

Setting hens must be kept free from lice and mites. The nest box and the walls of the coop or room should be painted or sprayed with a good lice killer a few days before the hens are set. The hen should be well dusted with a good insect powder two or three days before the eggs are placed under her and again two or three days before the chicks hatch.

The feed for setting hens should consist of hard grains. No wet or dry mashes should be given. A constant supply of fresh water, grit and shell should be provided.

One hen should not be given more than twelve eggs during cold weather or more than fifteen during warm weather.

Should any eggs become broken in the nest, the nesting material should be renewed and all badly soiled eggs washed in water at a temperature of 90° F.

Hatching with Incubators.—There should be no difficulty in hatching chicks with incubators if a good machine and good eggs are used. Different types of incubators require different care. Each manufacturer has compiled a set of directions for the operation of his incubator under average conditions. These directions should be carefully followed and an exact record kept of the operation of the machine throughout the hatch. If results are not satisfactory, variations should be made in the operation of the incubator during the following hatch as the judgment of the operator indicates. Poor hatches are more often due to poor eggs than to any failure on the part of the incubator.

BROODING

Importance of the Brooder.—The greater part of the mortality among young chicks occurs during the first four to six weeks. The losses during this period are very great, careful observers placing the total mortality as high as 40 to 50 per cent of all chicks hatched. The greater part of this loss is due directly or indirectly to poor brooding. In order to reduce the mortality among chicks to a minimum, good brooders must be used.

Qualifications of a Good Brooder.—A good brooder for farm use should be capable of maintaining a temperature of 90° to 100° F. under the hover and a temperature of 70° to 85° F. outside of the hover. The chicks should be allowed to choose the temperature in which they are most comfortable, and should not be compelled to submit to any given temperature.

A Brooder Heated by Oil Lamp.[1]

The brooder must be well ventilated, providing an abundant supply of pure, fresh air without drafts striking the chicks. Fresh air is as essential for growing chicks as good food and water. A two-compartment brooder is advisable, as it permits of feeding the young chicks in fairly cool, fresh air and they are not required to pass directly from the warm hover into the outside atmosphere.

The brooder for farm use should be portable. Chickens should not be reared on the same ground year after year. The most satisfactory results will be obtained by rearing them in the orchard, in the cornfield after the last cultivation, or on the hay and grain fields after the crops are harvested, moving the brooders from place to place frequently. If handled in this manner, the chicks will make use of a large amount of waste material and will be more healthy and vigorous and make more rapid growth than if confined to small yards.

The brooder should be usable for some purpose during the entire year. Any brooder which can be used only for brooding chickens is unsatisfactory for farm use. It should be capable of housing the chicks

[1] Courtesy of Prairie State Incubator Company, Homer City, Pa.

from the time they are hatched until fully matured, and should be readily convertible into a breeding house or fattening pen.

The brooding device which best meets these requirements is a portable colony house 6 by 8 feet to 8 by 15 feet in size, equipped with portable hovers, gasoline brooder heater or a coal-burning brooder stove.

Management of the Brooder.—During the first two weeks a temperature above 90° and below 100° F. should be maintained two inches above the floor in the warmest part of the brooder, that is, beneath the hover. After the second week the temperature should be gradually reduced, the exact temperature to be maintained being determined from the actions of the chicks. If the temperature is right, the chicks when at rest will be spread out around the outer edge of the hover. Any evidence of crowding is an indication of a lack of heat. If the temperature under the hover is kept a degree or two higher than the chicks actually need, there will be very little crowding.

The brooder must be kept absolutely clean at all times. The floor should be covered to a depth of several inches with clean, dry litter, such as short-cut clover, alfalfa, straw or chaff. The litter should be removed whenever it becomes damp, dusty or soiled.

Ration for Chicks.—A good ration for chicks consists of a grain mixture of 30 pounds finely cracked corn, 20 pounds cracked wheat and 10 pounds pin-head or cracked hulled oats. With this should be fed a mash consisting of 30 pounds wheat bran, 30 pounds wheat middlings, 30 pounds corn meal, 20 pounds fine beef scrap or granulated milk and 10 pounds of bone meal. This ration should be supplemented by a liberal supply of succulent food such as alfalfa, clover, sprouted oats or beets. Fine grit, finely crushed oyster shell, charcoal and clean fresh water should be before the chicks at all times. If skim milk is available, the chicks should have all they will consume.

The grain should be scattered in the litter on the floor of the brooder in order to induce the chicks to exercise. Grain should be fed early in in the morning, at noon and later in the afternoon. As much should be fed as the chicks will clean up from one feeding time to the next. If any considerable amount remains in the litter, a feed should be omitted and the amount reduced. No definite information can be given as to the exact amount to feed, as the needs of the chicks vary from day to day. The poultryman must study the appetite and actions of the flock in order to feed intelligently.

The mash should be fed dry. Shallow pans may be used for feeding the mash while the chicks are small. Small feeding hoppers should be used as soon as the chicks are large enough to feed from them. Chicks should never be without the dry mash.

This method of feeding should be continued until the chicks are large enough to do without artificial heating or are weaned from the hen, with the exception that the cracked wheat should be gradually replaced by whole wheat, and the finely cracked corn by the coarse cracked corn, when the chicks are six to eight weeks old. After the chicks have free range, the grain mixture may be changed to equal parts of cracked corn and whole wheat. The same dry mash should be continued until the chicks are mature. The grain may also be fed in hoppers after this time.

The Care of Growing Chicks.—The age at which chicks may be deprived of artificial heat will depend upon weather conditions and the condition of the chicks. This should not be done until all danger of sudden changes in temperature is past and the chicks are well feathered out. During the brooding period the brooders may be kept close to the farmstead and small, portable runs provided for the chicks. As the chicks increase in size, the brooder should be moved farther away and the size of the yards increased. As soon as the chicks no longer require artificial heat they should be given free range. They must have plenty of shade, abundant pasture, be kept free from lice and mites and protected from their natural enemies. The brooder should be proof against rats, weasels, etc., and should be closed every night. The chicks should be confined to the house in the morning until the grass is well dried off. This practice should be followed at least until they are half grown. The cockerels should be separated from the pullets as soon as the sex can be determined. It is advisable to caponize all males except a few of the most promising to be reserved for breeding purposes. The pullets will be hindered in their development if the cockerels are allowed to remain with them. The cockerels, if not caponized, should be put together in a separate field or on another part of the farm.

The Care of the Pullets.—The pullets should be transferred from the colony house on the range to their permanent winter quarters as soon after the first of September as possible. This will give them an opportunity to become accustomed to their new surroundings before cold weather sets in. Careful attention must be given the pullets at this time. There is usually a tendency for them to crowd on the roosts at night or to roost above the open doors and windows. This should be prevented, as it may result in colds which will hinder egg production. The bulk of the eggs received from October 1st to March 1st are produced by the pullets.

Feeding Mature Fowls.—The principal object in feeding should be to use the poultry on the farm for the purpose of converting grains, mill by-products and waste materials not suitable for human consumption in their raw state into concentrated, easily handled, nutritious food products. For this reason the farmer should make use of grains grown on his own farm and of mill products which are easily obtained at comparatively low prices, supplementing them with the necessary protein concentrates.

Suitable rations may be made from a great many combinations of grains and mill feeds. There is no one combination which is superior to all others under all conditions. For this reason it is possible for the farmer to adjust any suggested ration to meet his own conditions without seriously impairing its efficiency.

The ration should contain in proper proportions the various food elements required by the fowl. It should be easily digested and assimilated, palatable, economical, suitable for its intended purpose, easily obtained, easily handled and conveniently fed. It should be a two-part ration consisting of a grain mixture of scratch feed and a mash. It is not possible to obtain a maximum of production with either grain or mash alone. They should be fed in combination with grain constituting approximately two-thirds of the ration.

The following ration and method of feeding is particularly adapted to farm conditions. The ration as given is based on feeds at normal prices and may be varied with a variation in the price of any feed. The grain mixture consists of 200 pounds corn, preferably cracked, 200 pounds wheat, and 100 pounds heavy oats. If buckwheat is available, 100 pounds may be added during cold weather. The mash consists of 200 pounds corn meal, 100 pounds wheat bran, 100 pounds wheat middlings and 100 pounds of beef scrap containing not less than 55 per cent protein.

The grain should be fed by hand, being scattered in clean litter six to twelve inches deep. The grain should be fed at least twice daily, preferably early in the morning and late in the afternoon. If it is necessary to keep the fowls confined to the house, it is advisable to give additional light feeds in the middle of the forenoon and in the middle of the afternoon in order to keep the fowls busy.

The amount to be fed will vary with the variety, the weather conditions, the egg production and various other factors. It should be determined by the actions and appetites of the fowls. They should be well fed. Endeavor should be made to regulate the feeding so that they will consume approximately twice as much grain as mash. Fowls of medium size when in full lay will consume from 2 to 2½ ounces of grain daily.

The mash should be fed dry. Self-feeding hoppers should be used. For Leghorns and similar varieties and for pullets of the dual purpose varieties, such as Plymouth Rocks, Wyandottes, etc., the hopper should be open during the entire day. For yearlings and older hens of the dual purpose varieties, the hopper should be opened at noon and closed when the evening feed is given.

This ration should be supplemented by a constant supply of clean, fresh water, grit and oyster shell. Sour skim milk should be fed as a drink if it is available, allowing the fowls to consume all they will. Succulent feed of some sort is necessary. During the late fall and winter, mangels, sprouted oats, unsalable cabbage, beets, apples, potatoes, steamed clover or alfalfa, or any other succulent food available may be used. The yards should furnish all the green feed required during the spring and summer.

The Care of Market Eggs.—The quality of market eggs is determined by their size, shape, appearance and freshness or interior quality. All of these factors may be controlled by the poultryman to a considerable degree through breeding and the care with which the eggs are handled. Improvement of the quality of the eggs produced is fully as important from a financial point of view as increased production. If the following suggestions are observed, there should be no difficulty in producing eggs of a quality that will meet the requirements of the best grades in any market.

Breed only from hens which lay eggs of the desired size, shape and color.

Provide for at least one clean, convenient, well-ventilated nest for every four or five hens in the flock.

Renew the nesting material whenever it becomes damp, dusty or soiled. Planer shavings make excellent material for nests, but soft hay and clean straw may be used.

Shipping Cases for Eggs.[1]

Gather eggs at least twice daily and more often if convenient. This is particularly important during cold weather to avoid freezing, and during warm weather to avoid the development of the embryo and to retard evaporation.

From the time eggs are gathered until marketed, keep them in a clean, cool, dry place. Fertile eggs will begin to develop at any temperature over 68° F.

Do not put eggs into a box, basket, carton or case until all the animal heat has escaped. When gathered, place them on a wire tray similar to an incubator tray for ten to twelve hours and then grade and pack them in standard cartons or cases.

Market eggs at least once weekly and more often if possible. Nothing is ever gained by holding eggs for a rise in price. The egg is a perishable food product and should be marketed as soon as possible in order to avoid deterioration and loss.

Market eggs in standard egg packages. The standard thirty-dozen egg case is preferred. If production is not great enough to enable a case or two of graded eggs to be shipped weekly, use the smaller, returnable cases which may be secured from any dealer in poultry supplies.

[1] Courtesy of Missouri State Poultry Experiment Station, Mt. Grove, Mo.

When eggs are being transported from the farm to the market or shipping point, they should be protected from the rays of the sun.

Do not wash eggs. The washing of eggs greatly impairs their keeping qualities and spoils their appearance. Market eggs should never be allowed to become wet. Moisture dissolves the protective bloom or covering of the shell, opens the pores and allows bacteria and moulds to enter. Avoid the necessity for washing by providing sufficient nests and keeping the house and yards clean.

Remove all males from the flock as soon as the hatching season is over and keep them away from the hens during the warm weather. The male has no influence on the number of eggs produced. His only function and use on the farm is to fertilize the eggs to be used for hatching. Fertile eggs spoil very quickly during warm weather. Approximately 18 per cent of all eggs produced upon farms become unfit for food before reaching the consumer. At least half of this loss could be avoided if only infertile eggs were produced.

REFERENCES

"Productive Poultry Husbandry." Lewis.
"Turkeys." Reliable Poultry Journal Co.
"Principles and Practices of Poultry Culture." Robinson.
"How to Keep Hens for Profit." Valentine.
"The Beginner in Poultry." Valentine.
"Farm Poultry." Watson.
"Races of Domestic Poultry." Brown.
"Poultry Production." Lippincott.
"Poultry Breeding." Purvis.
"Our Domestic Birds." Robinson.
North Carolina Expt. Station Bulletin 233. "Common Diseases of Poultry."
Ohio Expt. Station Bulletin 284. "Rations for Roosters and Capons."
Purdue Expt. Station Bulletin 182. "Poultry Investigations."
West Virginia Expt. Station Bulletin 102.
Canadian Dept. of Agriculture Bulletins:
- 189. "Farm Poultry."
- 193. "Tuberculosis in Fowls."
- 217. "Farm Poultry."

Farmers' Bulletins, U. S. Dept. of Agriculture:
- 309. "Incubation of Eggs;" "Causes of Young Chicks' Death;" "Snow for Chicks."
- 317. "Water Pans and Catching Hook for Poultry."
- 357. "Methods of Poultry Management at Maine Station."
- 452. "Capons and Caponizing."
- 528. "Hints to Poultry Raisers."
- 530. "Important Poultry Diseases."
- 585. "Natural and Artificial Incubation of Hen's Eggs."
- 624. "Natural and Artificial Brooding of Chicks."
- 656. "The Community Egg Circle."
- 682. "A Simple Trap Nest for Poultry."

CHAPTER 51

BEES

Many farmers are unaware of the great service rendered them by the honey bee; especially in horticulture and vegetable raising is he a necessary asset. Estimates from reliable data show that bees in the United States produce $25,000,000 worth of honey and beeswax annually. Their value as agents in the pollinization of fruits and vegetables is many times their worth as producers. Many small fruits are entirely dependent upon insect visitors for fertilization. Cucumbers, squash, melons and tomatoes are also dependent upon the bees for the production of fruit. Pear trees especially need the bees for cross-pollinization.

Aside from the service rendered as pollinators, bees, if properly handled, make a most profitable side line in the business of farming. While they need intelligent care, and care at the proper time, yet much of this can be given at odd hours and at times when the regular farm work is not pressing. Even the time of swarming can be anticipated and to some extent regulated.

Bee keeping furnishes a most pleasant recreation and one that pays its own way as well as produces a profit. There is so much of marvel in the economy of the honey bee that the most casual observer becomes an enthusiast.

One disadvantage may be mentioned, however. Many orchard and garden diseases are easily spread by means of spores carried by insects. The bee plays no small part in the distribution of plant contagion. Pear-tree blight, the brown rot of plums and the wilt of cucumbers and melons are diseases spread through the agency of bees and other insects. The danger of infection may be reduced to the minimum by exterminating all diseased plants and trees; thus giving the bees no opportunity to carry contagion.

Breeds of Bees.—The German bee is the most common in the United States. Although not very attractive in color, being black, they winter well and make whiter honey combs than any other race. At times they are inclined to be cross and frequently use their stings. They are not easily handled by the novice.

The Cyprian bees are handsome, being yellow in color, but have not come into wide popularity on account of their extreme sensitiveness. When once aroused, they will not even be subdued by smoke.

The Carolina bee is one of the most gentle of all bees. It is gray in color and very prolific. The chief objection to this bee is its ever-ready tendency to swarm.

The Caucasian bee has only recently been introduced into this country and has not yet established wide popularity. It is prolific and so gentle that some report it to be without sting. This, however, is not the case.

The Italian bee is the most satisfactory and profitable. It is more gentle than either the German or Cyprian, and quite prolific. It is handsome in color, having yellow bands, and is an energetic worker in gathering honey. It is also most active in defence of its home against marauders. In order to winter well, the Italian bee must be well protected.

Personnel and Activity of Colony.—A bee colony consists ordinarily of one queen bee, who is the mother of the colony, and a multitude of females (sexually undeveloped), who carry on the work of the hive. The

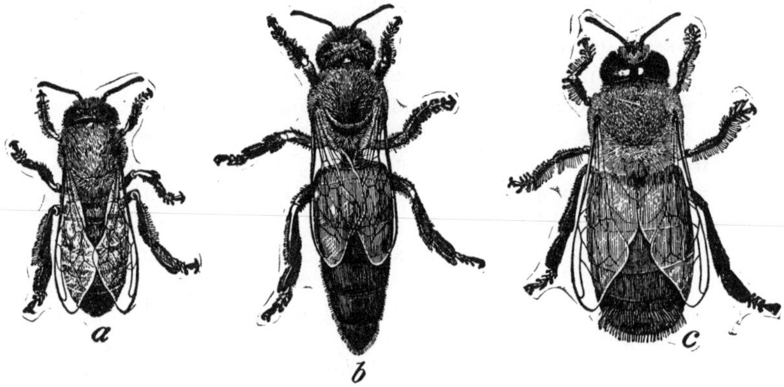

THE HONEY BEE.[1]
A—Worker. B—Queen. C—Drone. Twice natural size.

queen bee lays all the eggs. The female workers lay no eggs at all. It is their duty to gather honey, feed the young, keep the hive clean; in fact, perform all the labors of the hive.

During some parts of the year, hundreds of males, commonly called drones, live in the colony. These perform no labor. Their mission is to mate with the young queens. Their number should be restricted by the keeper.

The bee hive permits of no idlers after the young queens are mated. The drones are then destroyed by the workers. Even the queen bee is killed or superseded by a younger queen as soon as she lays no more eggs. In fact, any individual in the colony who ceases to be useful is immediately put to death or thrown out to perish.

The length of life of any bee depends much upon the time of year and amount of labor performed. In summer, which is the working season, a worker bee will live about 45 days. During the winter months, while

[1]Courtesy of U. S. Dept. of Agriculture. Farmers' Bulletin 447.

dormant, time of life will extend from 6 to 8 months. It is, therefore, necessary to maintain a strong, prolific queen in order to repopulate the colony.

Size and Location of Apiary.—Authorities agree that for the most intensive bee culture, 100 colonies are all that can be managed with profit.

The beginner will do well to start with a colony or two and gradually build up as he becomes more familiar with the work. A year or two will prove his success or failure. While the necessary initial capital is small,

General View of an Apiary.[1]

still a plunge into the bee business without previous experience and a thorough knowledge of bee habits is very apt to end in disaster.

The ideal location for an apiary is in an orchard or near fields where bloom is plenty; although colonies have been successfully maintained in city back yards and even on housetops.

Although bees travel a distance of two miles in search of nectar, it is best to provide for it nearer home. The time wasted in transit is negative, as the bee flies very rapidly; but if far from home, sudden rain or wind storms bewilder the bees and cause loss of life. In rainy or cold weather, bees do not travel far from the hive. Should the nectar be far afield, continued unfavorable weather necessarily decreases their activity.

The hives should be placed a few feet apart so that in working with

[1]Courtesy of U. S. Dept. of Agriculture. Farmers' Bulletin 447.

one, the adjacent hive is not disturbed. They should be far enough away from roads or walks so as not to annoy passersby.

In the North, hives should be placed on a sunny slope, facing away from the prevailing winds. Some shade is desirable, but the hives should be so placed as to catch the morning sun. This encourages bee activity early in the day, thus gathering the best of the nectar.

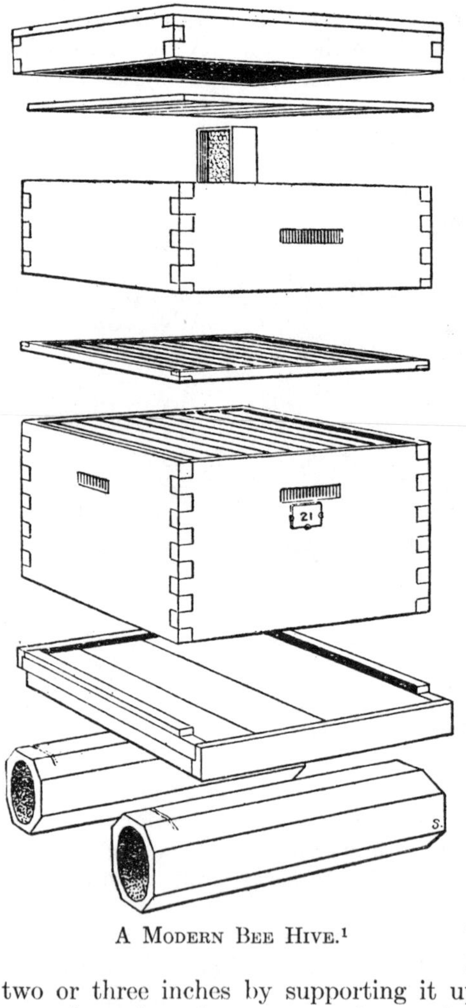

A Modern Bee Hive.[1]

The colony must be located in a dry place and kept free from weeds, each hive being raised a few inches from the ground by means of a stand. These stands may be of wood, stone or concrete, and serve to keep the hive dry.

Shade and Ventilation.—A reasonable amount of shade is beneficial, although dense shading of the apiary is disastrous. It promotes dampness and encourages disease. If a natural shade is not possible, a temporary shade of boards or canvas should be used during the heated portions of the day. Newly swarmed hives should be kept well shaded and cool. Temperature influences the swarming habit; a colony subjected to the burning rays of the sun will swarm much sooner than one well shaded.

Roomy, well-ventilated hives are necessary for comfort and health. During warm weather, ventilation is improved by raising the front of the hive two or three inches by supporting it upon small blocks of wood. Care must be taken, however, to lower the hive in case of a change in temperature. Most authorities do not approve of opening the upper part of the hive. It is apt to cause a draft through the hive, and also encourage robber bees. A wide entrance at the bottom is much preferred for ventilation purposes.

[1]Courtesy of U. S. Dept. of Agriculture. Farmers' Bulletin 503.

Stocking the Apiary.—Bees may be secured more easily at swarming time and the colonies are apt to be stronger at that time. Usually the purchaser provides a hive into which the apiarist puts the new swarm. This may be moved at night and, if taken a distance of a mile or more, there is no danger of the bees returning. A good strong colony purchased at this time will yield a second swarm if the season is favorable.

Introducing a New Queen.—The prosperity of the colony depends much upon the strength of the queen. Bees from a strong queen winter better than those from a weak one, and are more prolific in spring. If the queen becomes weakened, it is best not to wait until the workers destroy her, but to make away with her and introduce a new one at once. Queens may be purchased from any dealer in bee-queens. They are sent through the mail in a small cage, accompanied by a few workers.

Many methods of introducing a new queen are used, but if the queen is a valuable one, it is best to use a perfectly safe method. Remove the old queen in the evening. In the morning lay the cage containing the new queen and attendants, wire side down, on the frames under the quilt. Close the hive and leave it alone. In a short time the bees will have eaten their way into the cage and released the queen. The wait over night is necessary on account of the excited condition of the bees when their queen is removed. This excitement might cause them to destroy the new queen. Queens introduced in this manner are generally at work in two or three days laying eggs.

Some introduce by first blowing tobacco smoke down the hive to drive the bees down, then release the queen and allow her to run down between the combs, blowing a little smoke after her. This not only obscures all strange odors about the queen, but stupefies the bees.

Introducing a queen makes the opportunity to change breeds of bees, as the new queen is usually mated when purchased. Queens are sold under one of three labels: tested queens that are mated with a drone whose race is known; untested queens mated with an unknown drone; and breeding queens, those that have shown superiority for breeding purposes before leaving their home. The bees in the colony have no influence on the progeny of the new queen already mated. By the time the new brood hatches out, the old ones begin to die, and soon the race is changed.

Uniting and Transferring Colonies.—It often becomes advisable to unite two weak colonies, making one strong one. Some fundamental facts about bees must be understood in order to make this a success. Every colony has a distinct odor and resents bees from other colonies. It is necessary, therefore, to obscure this odor by using smoke. Smoke also stupefies the bees and renders them more docile. Both colonies should be smoked, but care should be taken not to use too much smoke, or the bees will be completely overcome. One queen should be destroyed; the one saved should be caged for a day or so to prevent the bees killing her. At swarming time when the bees are full of honey, it is a simple matter to unite colonies. If the two colonies are not near each other, one should be moved nearer the other, a few feet each day, that the bees may not notice the changed location. When side by side the change can be made without difficulty.

Transferring a colony from a box hive to one with movable frames often becomes necessary. This should be done during the honey season and while the larger number of bees are in the field. The two hives should be adjacent. The new hive should contain combs or sheets of foundations. Turn the box hive upside down and fit over it a small empty box, inverted. Then drum on the hive until most of the bees desert their combs and go into the empty box above. These may be carried to the new hive and put at the entrance. Care must be taken to secure the queen, as the bees will not remain without her. If there is brood in the old hive, turn it right side up again and after twenty-one days this will be hatched out. These bees may then be gathered in the same manner and, by smoking both colonies, reunited in the new hive.

General Methods of Handling.—Certain general rules will apply at all times in handling bees.

Hives should never be jarred or disturbed more than necessary. Rapid movements should be avoided. Bees have a peculiar eye structure which enables them to see movements more readily than objects. Quick movements irritate them, causing them to sting. Stings are not only painful, but the odor of the poison irritates the other bees, thus making them difficult to manage. The novice should wear a veil over a broad hat, and use a good smoker. A few puffs are sufficient to subdue the bees. Gloves generally prove a nuisance, but rubber bands on the arms prevent the bees crawling up inside the sleeves. Black clothing is particularly objectionable to bees. Do not handle bees at night or on cold, wet days, unless absolutely necessary. The middle of the day, particularly during the honey season, is the best time to manipulate bees. Always stand to the side or back of the hive, never in front of the entrance. In handling frames, care should be taken not to let the bees drop off on to the ground.

Swarming.—Swarming is the exit of the original queen with part of her workers to seek a new home. In this manner, new colonies are formed. An abundant supply of honey and a crowded condition of the hive are the immediate causes of swarming. Swarming may occur in May, but is more apt to occur during July and August, or when the honey flow is at its best.

The only outward indication preceding swarming is a partial cessation of field work and the loafing of many bees about the entrance, as if waiting for some signal. Suddenly the bees all rush forth, accompanied by the old queen, and after circling about for a time, cluster on a nearby limb. This is the critical time for the bee keeper. If he has made no previous preparation to house his departing swarm, he may lose them altogether.

A wise keeper will have clean hives in readiness. These should be kept in a shady place, so as to be cool as possible for the incoming swarm. Newly swarmed colonies will not remain in overheated hives. For this reason the hive should be kept well shaded and well ventilated for several days after the swarm goes into it. Some recommend giving a frame of brood to the newcomers, as bees are less apt to desert this.

Bees rarely fail to cluster after swarming. If they light on a limb that can be spared, it may be sawed off and the bees carried to the new hive. If this is not practical, the bees may be shaken off into a basket or box and taken to the hive. A box with a long handle is useful for swarms on high limbs. It is not necessary to secure all the bees. If the queen is hived the rest will follow. If she is not hived, however, the bees will leave the hive and join the cluster again. Bees are usually peaceful at swarming time, having filled themselves with honey before starting out. A little smoke blown into the cluster usually subdues them.

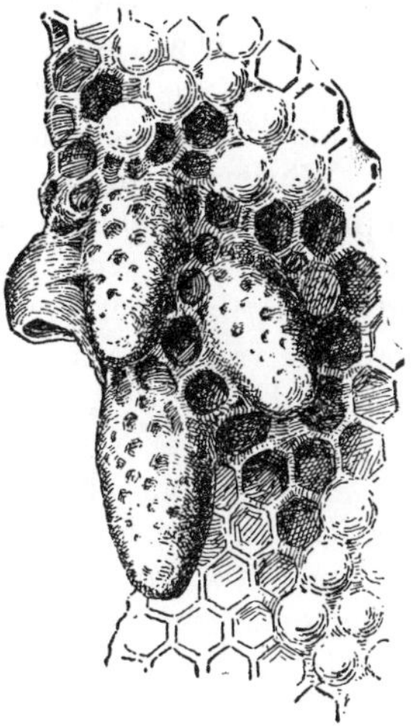

QUEEN CELLS.[1]

Great care must be used in handling the bees that none be crushed. The odor from a crushed bee excites the living bees and makes them difficult to handle.

Soon after hiving the bees resume their normal duties. The queen begins to lay eggs and the workers store honey in anticipation of the new brood. Extra frames should now be placed for the storage of honey. If there were incomplete supers on the parent hive, these should be lifted over on the new hive.

The departing swarm leaves behind several queen cells which will hatch in a few days. All but one of these will be destroyed by the workers. Two or three days after the remaining queen bee has been fertilized she begins to lay eggs and the colony resumes its normal routine.

How to Prevent Swarming.—A reasonable amount of swarming is desirable, as in this manner new colonies are started. However, much swarming weakens the colonies. Weak colonies do not store an abundance of honey or winter well. Neither do they resist moths and disease. An overcrowded colony is the most common cause of swarming. As a preventive, plenty of room should be kept in the hive by removing the honey supply often and furnishing extra supers. The hives should be kept well shaded and ventilated.

One queen to a colony is the rule. Too many queens cause swarming. If the queen cells are carefully watched and cut out, the number can be regulated. The queen cells are readily recognized by the keeper, as they

[1]Courtesy of U. S. Dept. of Agriculture. Farmers' Bulletin 447.

are larger than any other cells. They are rough on the outside and hang vertically on the comb, having much the shape of a peanut. The supply of queens should not be entirely cut off, however, as a vigorous colony needs requeening at least once in two years.

Artificial swarming is sometimes resorted to by dividing an overcrowded colony and furnishing a new queen to the queenless portion. This process is expedient, however, only after indications of swarming are observed. Otherwise, the bees may swarm naturally later on.

Clipping the queen's wings to prevent her flying is sometimes resorted to to prevent swarming. In this event, she will be found near the hive when the swarm issues and can be recaptured and put into a new hive. The parent colony should be removed and the new hive put in its place. The swarming bees will then enter the new hive. The bees afield at the time of swarming will also return to the new hive, thus strengthening the new colony and relieving the congestion of the parent colony. The same shifting of hives should take place in event of a natural swarming.

Wintering of Bees.—Queens showing lack of vitality as winter approaches should be replaced, in order that the colony may begin the inactive period with young and vigorous bees. Cellar wintering is not advisable unless under the direction of an experienced bee keeper. The dangers from moths, sweat and other bee troubles make the practice doubtful. A dry, well-ventilated cellar with an even temperature is imperative.

Throughout the South, where the winters are mild, no packing is needed for outside wintering. The entrance should be closed enough, however, to keep out cold drafts and prevent the entrance of mice and other enemies. Enough space must be left for the passage of the bees.

In the North the hives must be well packed to retain the heat generated by the bees. Heavy building paper tacked around the hive, leaving the entrance open, makes a good winter protection for bees. A piece of burlap, tacked over the front of the hive and hanging over the entrance, makes a good shield from snow and wind. This may be lifted on fair days to permit the passage of the bees. Dark wrapping paper should be avoided, as it absorbs the rays of the sun. This creates a rise in temperature within the hive, resulting in too much bee activity. Dampness is more fatal than cold to bees. It is advisable to place burlap or other absorbent material on top of the frames to absorb the dampness which otherwise might condense and dampen the cluster of bees.

Bee Feeding.—A colony of bees should enter the winter with from 25 to 40 pounds of honey stored for food. The quantity depends upon the length and severity of the winter.

Fall and spring feeding is often resorted to in order to continue activity in the colony late in the season and stimulate it early in the spring. Honey from unknown sources should not be fed, on account of introducing disease. Syrup made from granulated sugar makes a satisfactory food.

A small pan filled with shavings or excelsior saturated with the syrup may be placed on top of the frames.

Hives.—There are many good hives on the market, but the one most widely used is the Langstroth hive. Unless one is skilled in making hives, it is best to purchase them ready-made. All hives in the apiary should be of the same style and size, so that the frames are interchangeable.

Foundation Combs.—Foundation combs should be furnished either as starters or as entire sheets. The finished product will then be beautifully uniform. If the bees are left to furnish their own wax, much time is consumed and the resulting comb is irregular. Full sheets of foundation produce the finest quality of comb. When one super is half full or more, it should be raised and an empty one put under it. Care must be taken not to furnish too many sections at once or some will be left unfinished.

Handling and Marketing.—In handling the honey combs, care must be taken to keep the frames in a perpendicular position. If placed on their sides, the combs will be broken. The same caution applies in packing for market or in handling foundation or brood frames.

Honey should not be stored in a cool, damp cellar, but kept in a warm, dry room. Honey taints easily and care must be taken to use as little smoke as possible in the hives in handling the bees.

The home market is the best for the small honey producer. The product deteriorates rapidly in shipping, and much care is needed to pack, so as to ship without loss. Unless handled in large quantities the added expense of packing will offset the higher price at a distant market.

Wax from extracted honey and that scraped from frames can be melted and made into beeswax. Beeswax not only has a market value as wax, but if sent to a foundation factory, new foundations can be made from it at a cost much less than the purchasing of new foundations.

Diseases of Bees.—Moth is not a disease, but is a common enemy of the bee. The presence of moth denotes a weak colony, for a strong colony will destroy moth webs and keep them out. Once in, not much can be done save to so strengthen the colony, that it rids itself of the moth.

Foul brood is the most common bee disease. It is a germ disease. much to be dreaded, as it spreads rapidly from one apiary to the other, the first trace is noticeable in the grubs. They turn yellow and stretch out in their cells instead of being white and curled up. Later a stench arises from the hive. Drastic measures must be taken at once to keep the disease from spreading. The bees should be removed to a clean hive without comb and kept for thirty-six hours with the hive closed. At the end of that time they may be put into a new hive with clean comb and a fertile queen. Sugar syrup must be furnished them for a time. The infected hive and all its parts must be burned.

So serious has this disease become that many states have passed laws governing its control, and provide inspectors to see that the laws are enforced. It is to a bee keeper's advantage to co-operate in every way possible with these inspectors in controlling this disease.

REFERENCES

"Bee Keeping." Phillips.
"How to Keep Bees for Profit." Lyon.
"Bee Book." Biggle.
Canadian Dept. of Agriculture Bulletins:
213. "Bee Diseases."
233. "Natural Swarming of Bees; How to Prevent."
U. S. Dept. of Agriculture, Bureau of Entomology, Bulletin 14. "Diseases of Bees."
Farmers' Bulletins, U. S. Dept. of Agriculture:
442. "Treatment of Bee Diseases."
447. "Bees."
503. "Comb Honey."
652. "Honey and Its Uses in the Home."
695. "Outdoor Wintering of Bees."

BOOK V

DAIRY FARMING

(Dairy Husbandry)

CHAPTER 52

THE DAIRY HERD; ITS SELECTION AND IMPROVEMENT

BY F. S. PUTNEY

Assistant Professor of Dairy Husbandry, The Pennsylvania State College

The dairy cow of today has been so long domesticated that it is impossible to identify her exact origin. Several possible origins have been written about, but one thing we are sure of is that the original cow gave milk only for her young for a few months. The modern dairy herd is the result of selection and improvement by man.

Scrubs, Grades, Crosses and Pure-Breds.—A dairy herd which is the result of accident and which has never been improved is called a common or scrub herd. Such a herd usually has the blood of several breeds, but has been bred without thought. Occasionally a scrub dairy cow is profitable, but it is rare indeed to find a scrub herd that is profitable. A large percentage of the dairy cattle in the country today are high-grades. A grade animal carries over 50 per cent of the blood of some particular breed. The pure-bred sire is now believed to be an essential of a good dairy herd, hence the result is that most of the cows are now high-grade, carrying over 75 per cent of the blood of one breed. A cross-bred animal has the blood of two pure-bred animals of different breeds in its veins. Such breeding is good to produce vitality, but is not good for milk production; especially is this true in the crossing of such distinct breeds as the Holstein-Friesian and the Jersey. Comparatively few pure-bred dairy herds exist. However, the number is sufficient to permit of every one owning a pure-bred sire, and the number of pure-bred animals is on the increase. A pure-bred animal does not have the blood of any other breed since the founding of that breed.

Value of Pedigrees.—A pedigree is a list of the names and registry numbers of the ancestry of an animal. A dairy farmer who keeps pure-bred animals should exercise care in keeping his animals registered in the herd-book of the breed association. This is profitable because pure-bred animals sell better than grade animals, as the offspring are more uniform, especially in type and color. The latter fact adds a great deal to the selling price. Further, the pure-bred dairy animals have been developed to higher milk production than any other class of farm animals and naturally the dairyman is willing to pay for their production ability. The more high producing animals in the ancestry of an animal, the better is that pedigree.

Breed Differences.—Within dairy cattle are several definite strains of a special type. These definite strains are called breeds. Some breeds have been developed for the large amount of milk they give, other breeds for the large percentage of fat which the milk contains. The size of the different breeds also varies a great deal. These breeds are quite largely the result of conditions that exist in different countries. Great as is the difference in the quantity and the quality of the milk and size of the breeds, the individual variations within a breed are nearly as great.

The following table, from Bulletin No. 114 of the Pennsylvania Experiment Station, shows the difference in percentage of fat of breeds:

	Per cent.
All Jerseys or Guernseys of high-grade	5.0
Mixed herd with some Jersey or Guernsey animals	4.5
Common mixed herd	4.0
Mixed herd with some Holstein animals	3.5
All Holsteins	3.0

A Standard of Production Necessary.—In order to select and improve animals for the dairy herd, it is necessary to have a standard of production. The standard is, of course, the lowest limit for profitable production. Since production of milk varies with the age of the animal, it is necessary to have a standard for the first few lactation periods. A heifer with first calf usually gives about 70 per cent of her future production as a mature cow. A cow makes her maximum production at about seven years of age. The standard of production varies with each community, but in a very general way, where up-to-date dairying is followed, a cow should produce between 6000 and 8000 pounds of milk and 250 to 300 pounds of fat to stay in the herd.

Individual Selection.—If it is necessary to have a standard of production for each cow, it is equally necessary to have some way of selecting animals that should come up to this standard. This will be discussed in the paragraphs on Records which follow. In order to improve a herd properly, one must keep more than records of production. The individuals must be selected for size, vigor and trueness to type. This selection must begin with the calf. Only calves of the right type and vigor should be raised. Size in the animal is important, but vigor is even more essential. Vigor and lung capacity are essential to enable the cow to resist all the diseases to which the dairy cow is heir. It is desirable to raise farm animals that have shown prolificness, as this quality is reproduced to a marked degree in dairy animals. Having decided to raise a particular breed, it is necessary to know the characteristics of that breed better than any other.

In starting a new herd, the females should be selected for uniformity of type, and should be typical of the breed they represent. In selecting a bull, some breeders prefer one that is strong, where the females in the herd are weak. If possible, this is a good practice. The bull should always be from as long a line of high producing animals as is possible to secure. In starting a herd, do not allow passing fads to have undue weight. To illustrate, the Jersey cattle have been greatly hurt by the solid color fad that went over the country. The breed was not solid colored at its foundation, and whatever fad comes into a herd after it has been founded reduces the number of animals to select from for production and hence weakens the herd instead of strengthening it. The Guernsey fad of light-colored noses and the white color of the Holstein-Friesian and Ayrshires are illustrations

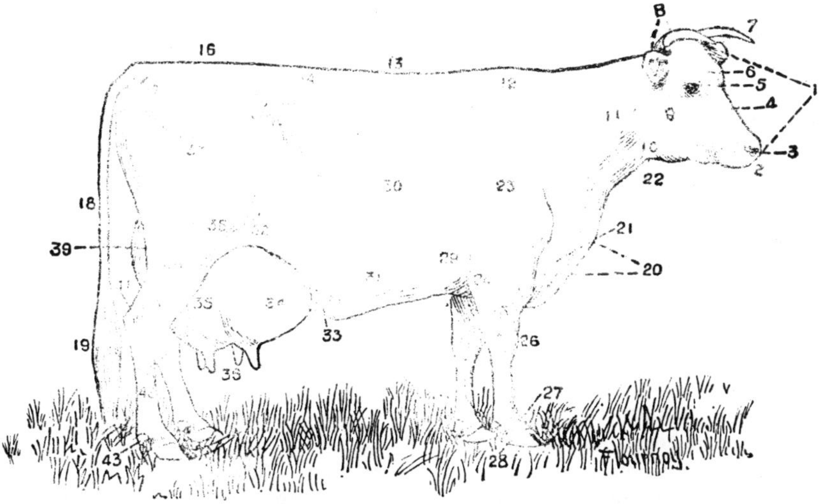

A Typical Cow, Marked to Show Points in Judging.[1]

1—Head. 2—Muzzle. 3—Nostril. 4—Face. 5—Eye. 6—Forehead. 7—Horn. 8—Ear. 9—Cheek. 10—Throat. 11—Neck. 12—Withers. 13—Back. 14—Loins. 15—Hip Bone. 16—Pelvic Arch. 17—Rump. 18—Tail. 19—Switch. 20—Chest. 21—Brisket. 22—Dewlap. 23—Shoulder. 24—Elbow. 25—Forearm. 26—Knee. 27—Ankle. 28—Hoof. 29—Heart Girth. 30—Side or Barrel. 31—Belly. 32—Flank. 33—Milk Vein. 34—Fore Udder. 35—Hind Udder. 36—Teats. 37—Upper Thigh. 38—Stifle. 39—Twist. 40—Leg or Gaskin. 41—Hock. 42—Shank. 43—Dew Claw.

of this fad. In order to select animals wisely, one should be a good judge of the breed in which he is interested.

Records.—While a breeder can select cows by the eye for many good and desirable points, the only real test of a dairy cow is the record of her milk and butter-fat yield. This should be kept for every year that a cow stays in the herd. If the farmer has the time, he should keep other records, such as list of offspring, feed records and the like. The greatest improvement is possible only when complete records have been kept.

In order to ascertain the production of a cow, a pair of scales, a Babcock testing outfit and milk sheets are necessary. The most popular scale today is the Chatillon Improved Spring Balance, which can be hung

[1] Courtesy of U. S. Dept. of Agriculture. B. A. I. 15th Report.

in some handy place in the barn or milk room. The two hands on the dial enable one to read the amount of milk directly. The milk sheet can be made for the month, week or any convenient length of period. The monthly record is the most popular. It is desirable to have a space for tabulating ten-day periods for the reason that grain is usually fed in accordance with the yield of milk. The amount fed should be adjusted at least as often as every ten days. Some adjust it every week, but when added for ten days the amount can be read directly without division.

Records show that about one-third of the cows in the United States are "boarders," or cows that do not even pay for their feed. When it is remembered that so many cows are unprofitable, and that if records are not kept, the daughters of these same unprofitable cows will be retained in the herd, and in turn more than likely become unprofitable, the value of records in dairy herd improvement is readily understood. Records show that one-third of the dairy cows in the country should be killed. The net profits of the herds remaining would then be greater than is now the case.

Cow-Testing Association Records.—Since it takes time to keep records, groups of farmers find it economical to organize and employ a man to keep records for them. This man is called a supervisor, and his services enable a group of both small and large farmers to practice selection based on production. Since the supervisor must visit each farm at least one day in a month, only about twenty-five farmers can co-operate in the hiring of one man. A supervisor can be had for from $500 to $600 a year with board and room. If these twenty-five farms keep 500 cows, the expense of keeping records by the supervisor method is less than though the owners paid themselves for the time that they would take to keep the same records. The supervisor weighs all feed given to the cows during the day on which he visits the farm. From this data he figures the cost of the feed by the month. In the same way he weighs the milk from each cow and tests it for butter-fat. This enables him to calculate the production for a month. He figures for the owner the value of the product from each cow for the month at the price that the owner is receiving.

Each cow-testing association is bound together by by-laws, contracts and some sort of articles of confederation. In some cases the association buys feed in carload lots so as to reduce the cost to the members. Such an association must be gathered from a community covering a small area. Some cow-testing associations stretch over considerable territory.

Bull Associations.—One of the outgrowths of the cow-testing association is the bull association. These associations are often formed from a group of men within the cow-testing association. It is necessary that the members keep the same breed of stock. These men own a bull, or several bulls, together. The bull is kept in the community as long as he is a good producer. A good producing bull is one that is a sure getter, and whose heifer calves prove to be better producers as cows than their dams. Such a bull should be given a herd as long as he will breed. Through the bull associations, it is often possible to bring into a community a bull of better breeding than any single member of the community could finance alone. It is not the cost of the bull that determines its value, but rather the producing ability of his daughters.

The bull association, to be of value, needs records, and the cow-testing association assures the records. They work well together.

Advanced Registry Records.—Any daughter or son of a registered dam and sire can be registered in the herd-books of that breed association. Unfortunately, many registered animals are no better producers than scrubs. In order to improve the animals within a breed, the different breed associations have started Advanced Registry Requirements. These

A Good Dairy Herd.[1]
Exercise is beneficial to the health of the cows.

requirements are based upon performance, and hence only worthy animals find their names on its lists. The different breeders have different names for the books in which such animals are listed, but all serve the same purpose. Representatives from the different experiment stations vouch for the production of the animals after personal visits. Such records have done much toward developing the modern, wonderful milking cow. Pure-bred sires should have some near relatives whose names appear in the advanced registry of the particular breed. All breeders of pure-bred stock should be encouraged to make advanced registry tests so as to improve the sale of bull calves from their herds.

The Bull is Half the Herd.—It is one of the sayings of breeders that "the bull is half the herd." Where in-breeding is practiced, he is even

[1] Courtesy of the Department of Dairy Husbandry, Pennsylvania State College.

AN OPEN STABLE FOR HEIFERS.[1]

[1] From Maryland Agricultural Experiment Station Bulletin 177.

more than half. Even if the cows are scrubs, there is no place in the herd for a grade or scrub bull. Only a pure-bred bull should head a herd of cows. The bull should possess quality and type and come from a long line of good producing females. In order to be sure that a bull can improve a good herd of cows, only tested bulls should be used. In order to test a bull he should be bred when young to a few of the good cows in the herd and the resulting heifer calves watched. If they are better than their dams, a good herd sire is indicated.

Professor W. J. Fraser, of the University of Illinois, calculates that in a herd of thirty-five cows it costs $3 per heifer more to have them sired by a pure-bred bull than by a scrub. This, then, is the total cost of providing each heifer calf with one good parent. If this same heifer calf produces only three pounds of milk more a day than her dam, this, at 80 cents a hundred pounds, means that in six years of milking, for 300 days a year, she would bring the owner $43 more than her dam. On this basis the rate of interest on the investment is better than anything else on the farm. Professor Fraser believes his figures to be conservative.

The University of Missouri has a Jersey herd that has had the fortune of having some excellent bulls at its head and the misfortune of having had some sires of very poor quality. To illustrate: ten daughters from Lorne of Meridale, one of their bulls, would have produced in six years $900 more than their dams, while ten daughters of Missouri Rooter in the same time would have produced $980 less than their dams. This shows that two farmers of equal ability living on farms side by side, and of the same size, would differ $2000 in wealth at the end of six years with only ten daughters from such different character bulls. The necessity of records is seen when it is remembered that the "bull is half the herd."

Buying Cows or Raising Calves.—One cannot build up a dairy herd and continue to improve it by buying cows. The only way to improve a herd is by raising calves that are better than their dams. Near large cities it is a common practice to buy cows to replenish the herd. In this country, far from large cities, an excess of calves is raised. If all the cows in this far-removed section had good records this method could continue. The farmer who gets his herd free from tuberculosis and contagious abortion can hope to keep it so, providing he raises his own calves. It can never be done if he buys cows.

The new-born calf must be well fed and made to grow. The feeding of the calf undoubtedly has some effect on the later usefulness of the cow. A stunted calf will never be as good a cow as though she had never been stunted.

For purposes of record it is necessary that every calf be marked before being taken from its mother. This often seems unnecessary, but when the young heifer spends her first summer on pasture, the owner is liable to forget the particular animal unless he visits the pasture frequently, or unless the heifer has some very distinguishing mark. If the habit of putting a tag in the ear, or some other good system of marking is established, trouble is avoided.

The calf should be fed so as not to develop scours or disorders of any kind. The best way is to feed the mother's milk for a few days, and see that all milk is warmed to blood heat. It should be fed only from clean pails. For the first few days it is well to feed three times a day; after that, twice a day is sufficient. The calf should be fed liberally, but more danger comes from over-feeding than under-feeding. Modern milk substitutes grow good dairy calves.

Developing the Young Animal.—Dairy cows are developed successfully in several different ways. The essential point is that the bone must be nearly grown at the time of dropping the first calf. Some feeders simply give large amounts of roughage to heifer calves during the winter after weaning from milk. In this way the frame grows, but little fat is put on. Other feeders give some grain, up to four pounds per animal per day, and this assures the heifer being in good flesh. When pastures are excellent, the first method is all right, but when pastures are only good or fair, better results are obtained by feeding some grain to heifers. A well-developed growing heifer gives more milk than one poorly developed, since she requires less feed for growth.

Open Stables for Heifers.—Heifers over one year old are today kept in open sheds facing the south. It is believed that this open-air method develops a stronger constitution and more hardiness, two qualities of great value in warding off disease later. This method of housing is much cheaper than housing in expensive closed quarters, and the results are at least equally good.

REFERENCE

Nebraska Expt. Station Bulletin 149. "Raising the Dairy Calf."

CHAPTER 53

Dairy Herd Management

By C. W. Larson
Professor of Dairy Husbandry, The Pennsylvania State College

The dairy cow is more sensitive to her treatment than any other of our productive animals. By care and breeding she has been developed into an animal of habit, and upon the care she receives depends largely the profits of the herd. The feed is an important item in the cost of milk production, but the systems practiced also materially affect the profits of the herd. It is no longer profitable to keep a cow all the year for the small amount of milk that she produces during the summer months while on pasture. The cheapest method of keeping a herd is not always the most profitable. This chapter deals only with the heifers from breeding time.

Age to Breed.—The age at which a heifer should be bred depends largely upon her size, but in general, an animal that has grown well can be bred to have her first calf when two years of age. During the last three months of the gestation period a heifer grows very little, so that it is not advisable to breed a small heifer too young, and some prefer to wait until the heifer is two and one-half years old before she has a calf. A heifer bred too young will not attain a large size, which is desirable in a dairy cow.

Gestation Period.—The gestation period of a cow is from 280 to 285 days. It is a good practice to keep a record of service, so that the cow can be properly taken care of before calving time.

Regularity.—A regular routine of work should be planned for the herd so that the cows will receive the same treatment each day. Any disturbance or irregularity affects both the amount and quality of milk. The cows should be milked at the same time each day. The milker should start at the same end of the row and be as regular in the treatment of the cows as possible. There are a number of points to keep in mind in planning the routine of the cow stable. Grain may be fed before milking, but hay should not be, because of the length of time it takes to eat it and because of the dust it will raise. Silage also should not be fed immediately before milking, because of the effect in the flavor of the milk. The stable should be cleaned before milking, if possible, and if the cows are kept in the stable, the grooming should also be done before milking. The cows need not be watered until after the morning feed is given. Hay should be fed late in the evening.

Care of Cow at Calving Time.—A cow should be carefully watched and fed during calving time. She should be provided with a clean, well-bedded stall. For several days previous to calving she should be fed a bulky ration and one that is laxative. She should not be given heavy grains. A mash of ground oats and bran is good. For two or three days before calving time she should be given slightly warmed water. Do not give cold water. After two or three days the cow can gradually be put on the regular grain and roughage feed, but the feeding should not be too heavy to start with.

Rest for Dairy Cows.—It is desirable to give a cow at least six weeks of rest each year. Most cows dry off before this time, but occasionally persistent milkers give a considerable supply up to the time of their next calving. When this is allowed, it is at the sacrifice of the milk in the next

A Good Cow Stable.[1]

Convenience in arrangement, ease of cleaning, plenty of light and good ventilation are essential to the health of cows and the production of clean, pure milk.

lactation period, and also sometimes at the sacrifice of the calf. It is sometimes difficult to dry off a cow, but usually by cutting down the grain and giving straw or timothy hay she can be reduced to a sufficient amount to be safe to stop milking her. It is sometimes desirable to milk once a day for a while and then stop altogether. It is not safe to stop milking her if she is giving too much, although cows producing as much as six to eight quarts have been dried off without injuring them.

Care of Cows when Dry.—A cow should be well taken care of when dry, for she is then preparing for her next milking period, besides growing the calf. Nearly all of the development of the calf takes place during the last few weeks. She should be given succulent and laxative feeds and should be well fed.

[1] Courtesy of the Department of Dairy Husbandry, Pennsylvania State College.

Exercise.—Many dairymen believe that a cow receives all the exercise she needs in producing milk, but on many farms it is desirable to turn the cows out for a part of the day. It not only gives a better opportunity to clean out the barn, but also gives the cows an opportunity to rub themselves, and their feet and legs keep in better condition. Too much exercise, of course, requires energy at the expense of milk production. Cows that are required to walk long distances do not do as well as those that are more confined. Cows should not be turned out during bad weather and exposed to rains and cold winds.

Grooming.—Cows kept in the stable all, or nearly all, of the time should be carefully groomed at least once a day. It is believed by many that grooming has an effect upon the milk flow. Cows seem to do better for having been groomed.

Milking.—A good milker has a fairly rapid, uniform stroke which he continues throughout the milking period. The whole hand should grip the teat and the pressure should come from the whole hand. The practice of using the thumb and first finger is not recommended. The milking of diagonal teats is thought to give best results. The Hegelund method of the manipulation and milking has been found to stimulate milk production. A cow milked by this process gives more milk. The steps are described as follows:

"First Manipulation: The right quarters of the udder are pressed against each other (if the udder is very large, only one quarter at a time is taken) with the left hand on the hind quarter and the right hand in front on the fore quarter, the thumbs being placed on the outside of the udder and the forefingers in the division between the two halves of the udder. The hands are now pressed toward each other and at the same time lifted toward the body of the cow. This pressing and lifting is repeated three times, the milk collected in the milk cistern is then milked out, and the manipulation repeated until no more milk is obtained in this way, when the left quarters are treated in the same way.

"Second Manipulation: The glands are pressed together from the side. The fore quarters are milked each by itself by placing one hand, with fingers spread, on the outside of the quarter and the other hand in the division between the right and left fore quarters; the hands are pressed against each other and the teat then milked. When no more milk is obtained by this manipulation, the hind quarters are milked by placing a hand on the outside of each quarter, likewise with fingers spread and turned upward, but with the thumb just in front of the hind quarter. The hands are lifted and grasped into the gland from behind and from the side, after which they are lowered to draw the milk. The manipulation is repeated until no more milk is obtained.

"Third Manipulation: The fore teats are grasped with partly closed hands and lifted with a push toward the body of the cow, both at the same time, by which method the glands are pressed between the hands and the body; the milk is drawn after each three pushes. When the fore teats are emptied, the hind teats are milked in the same manner."

Difficult Milking.—Occasionally cows are difficult to milk because of defective teats. Sometimes the openings are too small, in which case an instrument known as the bistoury may be used, but there is danger of greatly injuring the teat, and it should be used only by those experienced in its use. Only with especially good animals does it pay to spend much time with such cows. Sore teats, caused sometimes by teats becoming wet and exposed to the cold, can best be treated by rubbing them with vaseline or some antiseptic grease. A cow that has developed the kicking habit is a great annoyance. Sore teats and abuse, however, are often the cause. Most cows, by gentle treatment and care of the teats, will cause little trouble. Some, however, are naturally vicious, but these are few in number. A strap tied around the body of the cow just in front of the udder, and drawn fairly tight, will prevent most cows from kicking. A clamp made of wood with two straps, long enough to reach around the leg of the cow, will prevent her from bending her leg, making it impossible for her to kick.

Abuse.—A dairy cow should always be handled gently, for any disturbances affect her. Loud noises or any unusual disturbances should be avoided. A cow should never be struck or mistreated, nor should she be talked to in a loud voice.

Water and Salt.—A cow requires considerable salt, and this should be given regularly. One practice is to mix it with the grain, but the maximum requirements should not be given in this way, for a cow may be required to eat more than she wants of it. A little in the grain is all right, but a small amount should be given regularly, perhaps once a week, so that she can get all she wants. A cow will eat about one-half pound of salt a week.

An abundance of good water should be provided for dairy animals. A cow producing large quantities of milk will consume as much as one hundred pounds, or more, per day. Heavy milkers should be watered twice a day. The water should not be too cold, but at the same time it is well not to have it too warm. A uniform temperature is desirable.

Stabling.—With most large dairy herds it is customary to have regular stanchions in which the cows can all be tied up in rows. This seems the best system where high-class milk is being produced. For the small herd, the practice of allowing the cows to run in an open shed is being followed. This method of housing, however, makes it possible to keep the cows in a more healthy condition and to produce milk more economically. An experiment has been conducted by The Pennsylvania State College of housing cattle in an open shed as compared with a closed stable, and is summarized as follows:

"1. From the data presented it appears that cows kept under an open shed have keener appetites and consume more roughage than those kept in stables.

"2. There was sufficient protein consumed when either Van Norman's or Eckles' Standard was used to account for the yield of milk in addition to maintenance.

"3. Figured on Eckles' Standard, there was a slight excess of energy consumed above maintenance and milk production the first two years, and a small deficiency the last year. When computed on Van Norman's Standard, there was a deficiency in energy consumed for maintenance and milk production, except for one group the second year.

"4. The milk yield of the outside group decreased more rapidly each winter than that of the inside group.

"5. Sudden drops in atmospheric temperature caused corresponding decreases in milk yield for both groups, the outside group having a slightly greater decrease.

"6. More bedding was required outside, but less labor was necessary to keep the animals clean.

"7. Both groups finished each winter's trial in good health with the exception of one that reacted to the tuberculin test in April, 1914. She had shown no reaction in two previous tests. The hair of the animals kept outside was longer and coarser the first two winters. The third winter this was noticeable in only one ainmal."

Flies.—In the management of a milking herd, the problem of flies is a difficult one. Not only are they annoying to the cow and the milker, but they also carry disease. They should be reduced to as small a number as possible. It is believed that they do not travel a great distance, so that a farmer may have them fairly well under his control. Manure should not be allowed to accumulate, and if it does, it should be treated with some spray or disinfectant that will kill the flies. There are a number of sprays on the market that can be used for killing flies in the barn. Some have found traps to be practical.

Marking the Cow.—For the purpose of identification, dairy animals should be marked. Some have a system of clipping the ears with certain notches to represent the various figures and thus of keeping records. This, however, is not very satisfactory. An ordinary hog ring with a metal or composition tag fastened to it makes a satisfactory marker. Occasionally these are torn out, but if they are properly put in and the tag is small and round, they will stay a long time. The tattoo is also being used successfully when good tattoo material is used.

Dehorning.—In the general milking herd all cows should be dehorned. There is more or less pain connected with the operation, but it does not, in the estimation of the writer, compare with the pain due to the cows being gored day after day. It prevents the possibility also of one animal that may be "boss" depriving others of their rightful share of food and water. The dehorning, however, should not be done until the animal has reached the age of two years, for if it is done before this, growth takes place and scurs will be formed. The dehorning should be done in cold weather and when there are no flies. The horns should be cut or clipped as quickly as possible.

CARE OF THE BULL

A young bull should not be used too much for breeding purposes. He should be kept growing and should be well cared for, but not overfed. A good, thrifty young bull may breed six or seven cows before he is one

LEADING A BULL.[1]

year old with no injury to him. Even during the second year he should not be used too much. Often a young bull is injured by overuse. A cow should be served only once during a period of heat. A bull should never be allowed to run with the herd, but should be kept in a separate inclosure. He should be given exercise and be kept out in the open as much as possible. Where two bulls are needed in a herd, it is a good practice to dehorn them and then turn them together, or even train them to drive. A yard in which a bull is kept should be strongly fenced, for they are powerful, and once they break through a pen, it is very difficult to get anything that will hold them. They should be sheltered from the winds and rain, but can stand the cold. Bulls sometimes become vicious, due to treatment, although

[1]Courtesy of Orange-Judd Company, N. Y. From "The Young Farmer," by Hunt.

some bulls are naturally cross. In any case, great care must be take with them. They should never be trusted. They should always have a ring in their noses and be led by a stock from the ring. Bulls seem to know when a man is afraid and are more apt to attack such a one than one who is more courageous. A bull that becomes vicious is often subdued by being thrown with a rope. He then learns that he is under the control of man. The amount of service that a bull may have depends upon his age and condition. During the second year, a good, thrifty bull can be used once a week. A mature bull may serve one hundred to two hundred cows a year if the periods are distributed well throughout the year. In general, however, because of the variation in the intervals in which cows come in heat, a bull should be provided for each forty to fifty cows.

REFERENCES

"Dairy Cattle and Milk Production." Eckles.
Iowa Expt. Station Circular 16. "Care, Feed and Management of the Dairy Herd."
Minnesota Expt. Station Bulletin 130. "Feeding Dairy Cows."

CHAPTER 54

DAIRY BREEDS OF CATTLE

BY GEORGE C. HUMPHREY
Professor of Animal Husbandry, University of Wisconsin

Dairy Breeds Essential.—Choosing a dairy breed of cattle is fundamental to successful dairying. The modern improved breeds of dairy cattle are the result of high ideals, carefully laid plans and systematic effort on the part of many generations of dairymen who realized there were great possibilities in the development of breeds of cattle especially adapted for large and economical production of milk and butter-fat. Cattle which are true representatives of the recognized dairy breeds are very distinct from ordinary native cattle and cattle of the improved beef breeds, both in conformation and production of milk. They also tend to reproduce themselves from generation to generation with such marked degree of uniformity that one familiar with their history and characteristics would reject any other kind if he were engaged primarily in dairying. Natural laws that govern the reproduction of plant and animal life and preserve forms of like character from generation to generation and the experience of a vast number of dairymen teach the value of preserving and utilizing the distinct dairy breeds of cattle for dairy purposes.

Dairy Type Common to All Dairy Breeds.—The development of dairy breeds has established a distinct dairy type that is naturally correlated with extensive milk production. Dairy type refers to the conformation and peculiarities of the body that are characteristic of animals capable of producing large and economical yields of milk and includes the following:

1. Medium to large size of body for the breed.
2. Large feed capacity, as indicated by a roomy and capacious abdominal cavity, a large mouth and sufficient strength of body to consume and utilize a large quantity of feed.
3. Dairy temperament or a disposition to convert the larger portion of feed consumed into milk rather than body flesh. It is indicated by the absence of surplus flesh and a comparatively lean and refined appearance of the entire body.
4. An udder that is large, carried well up to the body, evenly and normally developed in all quarters and of good quality.
5. A strong, healthy flow of blood to all parts of the body, giving vigor, alertness and constitution. These characteristics are indicated by prominent facial, udder and mammary veins, abundant secretions in the ears, skin of the body and at the end of the tail and a coat of fine straight hair.

Ignorance of breeds and breeding and of proper feeding and management cause a great many cows to fall below the standard embodied in the foregoing qualifications for dairy type. This fact, however, is no argument against the merit of improved breeds and should not cause one to question the value of well-established dairy breeds.

Recognized Dairy Breeds of America.—Ayrshire, Brown Swiss, Guernsey, Holstein-Friesian and Jersey breeds of cattle are recognized and have been exhibited at the National Dairy Shows of America as specific dairy breeds. Dairy cattle of the Dutch Belted, French Canadian and

A Typical Ayrshire Cow. "Auchenbrain Hattie."
Medium in size, usually red and white, horns upturned and pointed.

Kerry breeds are bred and maintained in America in comparatively small numbers. The unimportance of these breeds in well-developed dairy districts, however, does not warrant more than mention and a very brief discussion of them in the limited space of this article.

AYRSHIRE CATTLE

Origin and Development.—The County of Ayr in southwestern Scotland is the native home of the Ayrshire breed. The land in this section is rolling and more or less rough, the climate moist and the winters extremely cold, except for being somewhat tempered by the Irish Sea. The hills produce rolling pastures in most parts, while the better lands grow grain crops and grass in abundance. The conditions, on the whole, demand a hard yrustling breed of dairy cattle, and Ayrshires have been developed to suit the needs of their native country. Early history records the use of several different breeds of cattle which undoubtedly have contributed to the establishment of the Ayrshire breed. Teeswater, Shorthorn, Dutch, Lincoln, Hereford, Devon and West Highland breeds are mentioned by various authors as having been used. Whatever the true origin may have been, the breed has been bred pure for many years, and its character fixed after the manner of other pure breeds of livestock developed by the Scotch people. The production of a breed of cattle suited to the condition of environment of that country, and especially adapted for the production of large yields of milk, was the standard which guided the breeders in fixing the characteristics of this breed. The breed has found favor in other countries and to a greater or less extent in all dairy sections of America, especially in the New England states and the provinces of Canada.

Characteristics of Ayrshire Cattle.—Ayrshire cattle are medium in size. Cows should weigh on the average 1000 pounds and bulls 1500 pounds. The color is a combination of white, red, brown and black. White predominating with red or brown markings is the more popular color. There are black and white Ayrshires in Scotland whose purity of blood is not questioned. A neat head with horns of medium length, inclining upward, a body with straight top line, well-developed chest, arched ribs, deep flank, and comparatively smooth hind quarters and an udder that is symmetrical and well balanced in form and well carried up to the body characterize the typical Ayrshire cow. The size of teats in many cows is subject to the criticism of being too small and one will do well to bear this in mind in making selections. The milk production of mature cows has in a few instances, under official tests, exceeded 20,000 pounds of milk per annum. An Ayrshire cow should be expected to yield 6000 to 8000 pounds of milk under ordinary conditions. The milk tests in the neighborhood of four per cent butter-fat. The highest official yearly production for an Ayrshire cow to date was made by Auchenbrain Brown Kate 4th, 27943, owned by Percival Roberts, Jr., Narberth, Pa. Her yearly production amounted to 23,022 pounds of milk testing 3.99 per cent and 917.6 pounds of butter fat.

BROWN SWISS CATTLE

Origin and Development.—The Brown Swiss breed of cattle has its origin in Switzerland and the cattle by virtue of their native home are strong, rugged and hardy. In this country they have been developed with reference to their dairy qualities to the extent that they have become recognized as one of the distinct dairy breeds. Up to 1907 they were bred and largely advertised as a dual purpose breed. In the meantime, however, the American breeders have given careful attention to selecting types and developing strains which excel more particularly in yield and economy of milk production. In the eastern and middle sections of the United States the breed is gaining favor and promises to have a place sooner or later of equal rank with older and better recognized breeds of dairy cattle.

Characteristics of Brown Swiss Cattle.—The breed is noted for its large size and ruggedness. Due to comparatively large bones and robust appearance, it is sometimes regarded as too coarse for economy of production. Cows will vary from 1200 to 1400 pounds in live weight at

A Brown Swiss Cow.[1]

maturity, and bulls quite frequently exceed 2000 pounds in weight. There is a tendency toward refinement and less size where dairy type is sought and selected to take the place of the former dual purpose type. Breeders aim, however, to maintain good size and large capacity for milk production in their efforts to develop herds of this breed.

A dark-brown or mouse color with a line of gray along the back, a mealy ring about the muzzle, a light fringe of hair on the inner side of the ear and more or less light hair on the under side of the body, constitutes the characteristic color of the cattle of this breed. Quite frequently the lighter gray color covers the entire body.

[1] Courtesy of The Field, New York.

A GUERNSEY BULL.[1]

A TYPICAL GUERNSEY COW.

[1] Courtesy of The Field, New York.

The head and neck are comparatively heavy; the males, and quite frequently the cows, carrying more or less dewlap. Well-developed udders, proportionate in size to the size of body, are sought in the selection of mature cows. A register of production for animals of superior merit has been formed by the American Brown Swiss Cattle Breeders' Association and there is a loyal effort on the part of breeders to make records that will compare favorably with records of other breeds. The breed has demonstrated its ability to make very profitable productions of milk and butter-fat. The milk tests on the average about 4.0 per cent. The highest official yearly record for a Brown Swiss cow at the present time is 19,460.6 pounds of milk, testing 4.1 per cent and 798.16 pounds of butter-fat. This record was made by the cow College Brauvura 2d, 2577, owned by the Michigan Agricultural College, East Lansing, Mich.

GUERNSEY CATTLE

Origin and Development.—Guernsey cattle take their name from Guernsey Island, located in the English Channel not far from France. This island and two smaller ones, Alderney and Sarnia, belong to the Channel Islands group, and is where the Guernsey breed originated and has been developed. These islands, of which Guernsey is the largest, contain only 2600 acres. On Guernsey the land is more or less hilly and rough, and the farms are small and devoted exclusively to horticulture and dairying. Many of the crops, such as grapes, melons and flowers, are grown in greenhouses. The cattle are owned in small herds and, in order to make the best use of the available pastures, are tethered or staked out when allowed to graze.

The people devote their attention to the one breed of cattle and exclude all other cattle from the island, except those which may be imported for immediate slaughter. The breed undoubtedly has its origin in stock of early French varieties known as Brittany and Normandy cattle. The production of a rich quality of high-colored milk and butter has always been the principal object in breeding and developing this breed, and naturally this has resulted in excellence of performance on the part of well-grown cattle of the breed.

Guernsey cattle were introduced into America early in the nineteenth century, but not until 1893, when the dairy qualities of Guernsey cattle were brought to the attention of the general public by records made at the World's Columbian Exhibition at Chicago, did Guernsey interests develop to the extent they deserved. The American Guernsey Cattle Club was organized in 1877, and of late years many importations of Guernsey cattle have been made and much enthusiasm has been aroused on the part of dairymen in exploiting and developing the interests of the breed. The breed ranks at the present time as one of the most popular.

Characteristics of Guernsey Cattle.—The standard weight for Guernsey cows is 1050 pounds, and for bulls 1500 pounds. Standards which

A Holstein-Friesian Bull.[1]
The largest of dairy breeds—color, black and white.

A Typical Holstein Cow.[1]

[1] Courtesy of The Field, New York.

demand excellence in conformation and characteristics pertaining to dairy type are fulfilled by many cattle of the breed. Development for usefulness rather than for beauty of form has resulted in a lack of refinement and neatness of outline in a good many of the cattle. The comparatively few Guernsey cattle in the country encouraged breeders to retain all pure-bred animals and this accounts for much of the lack of uniformity that exists. The American Guernsey Cattle Club was first to establish an advanced registry for official annual productions of milk and butter-fat and this again has been a standard toward which breeders have worked to a greater extent in many instances than they have for excellence of form. Marked improvement, however, in uniformity and excellence of dairy form has been noted in the show herds exhibited during the past few years.

In color the Guernsey is a shade of fawn, varying from dark-red to light-yellow with white markings. The color of the muzzle in most instances, which is regarded as most desirable, is buff or flesh color. A dark muzzle is permissible but undesirable on the part of critical judges. More emphasis is laid upon rich yellow secretion in the skin, especially in the ear and at the end of the tail, together with a yellowish appearance of the horns and hoofs than is laid upon the color markings. The rich orange secretions of the body are believed to indicate a rich yellow color of the milk, which is regarded as a most important Guernsey characteristic. Guernsey milk is not only yellow but of good quality, testing in the neighborhood of five per cent. The yield of milk under ordinary conditions should be 6000 to 7000 pounds per annum. Under official tests, many Guernseys have far exceeded this amount. In three instances Guernsey cows have held the world's championship record in butter-fat production. The highest official yearly record of milk and butter-fat production held by a Guernsey cow was made by Murne Cowan, 19597, owned by O. C. Barber, Akron, Ohio, her production amounting to 24,008 pounds of milk, testing 4.57 per cent and 1098.18 pounds of butter-fat.

HOLSTEIN-FRIESIAN

Origin and Development.—Holstein-Friesian cattle, commonly called Holsteins in America, have their origin in Friesland, a province of Holland bordering on the North Sea, where low, fertile dyke lands have been favorable for the development of a large breed of cattle capable of making large productions of milk. History records that for a thousand or more years these cattle had been bred and utilized for dairy purposes. Since 1885 they have been extensively introduced into most of the dairy sections of America and because of their large size and the large quantity of milk which it is characteristic of them to produce, the breed ranks as one of the most popular.

Characteristics of Holstein-Friesian Cattle.—The type and size of the cattle of this breed varies considerably and the terms "beef," "beef and milk," "milk and beef," and "milk forms'' are used to describe the different types. The milk and beef form is the most generally accepted type and should be the aim of men engaged in the breeding of these cattle. Extreme milk form is usually the result of improper growth on the part of young animals or selection of breeding stock which produces too much refinement. The following quotation characterizes true Holstein type and owners and breeders of Holstein-Friesian cattle base their claim for the superiority of this breed on the following points:

1. "That the Holstein-Friesian is a large, strong, vigorous cow, full of energy and abounding in vitality.
2. "That her physical organization and digestive capacity is such that she is able to turn to the best advantage the roughage of the farm, converting the same into merchantable products.
3. "That she produces large quantities of most excellent milk fit for any and all uses, and fit especially for shipping purposes.
4. "That heredity is so firmly established through her long lineage that she is able to perpetuate herself through strong, healthy calves.
5. "And that when, for any reason, her usefulness in the dairy is at an end, she fattens readily and makes excellent beef."

Cows of this breed should weigh 1200 to 1400 pounds. Mature bulls ordinarily weigh 1900 to 2000 pounds or over.

Black and white is the characteristic color in America. More or less white should extend below the knee and at least some black should be present where white predominates. The two colors should be distinct from one another. In Holland red and white is characteristic of many cattle of this breed and occasionally in America there are cattle born of this color. Such cattle, however, are not eligible to register in the herd books of the American Holstein-Friesian Association.

The breed excels in quantity of milk rather than quality, the fat in the milk under ordinary conditions being 3 to 3.5 per cent. A higher test is unreasonable to expect where the large flow of milk characteristic of this breed is maintained. In some instances, the fat falls below 3 per cent, which is regarded as too low, even in cheese districts where this breed is very popular. A low percentage of fat should be avoided by the careful selection of sires whose dams yield milk of a higher percentage of fat. Naturally this breed with its large size and natural tendency to produce milk of low percentage of fat has always excelled all other breeds in milk production. Cows of this breed have in four instances won the championship record for both milk and butter-fat production, and hold the world's record at the present writing, with a production amounting to 28,403.7 pounds of milk testing 4.14 per cent and 1176.47 pounds of butter-fat, made by the cow, Finderne Pride Johanna Rue, 121083, owned by Somerset Holstein Breeders' Company, Somerville, N. J.

JERSEY CATTLE

Origin and Development.—Jersey cattle were originally developed on the Island of Jersey, the largest of the Channel Islands group, where a delightful climate, a rich soil and a people united in their effort to excel in the production of a single breed of dairy cattle combined to make conditions most favorable for perfecting and preserving the breed. In 1793 enactments began restricting the importation and maintenance of cattle other than Jerseys, which finally resulted in its being a crime to keep cattle of other kinds on the island for a longer period than twenty-four hours when they had to be slaughtered for beef.

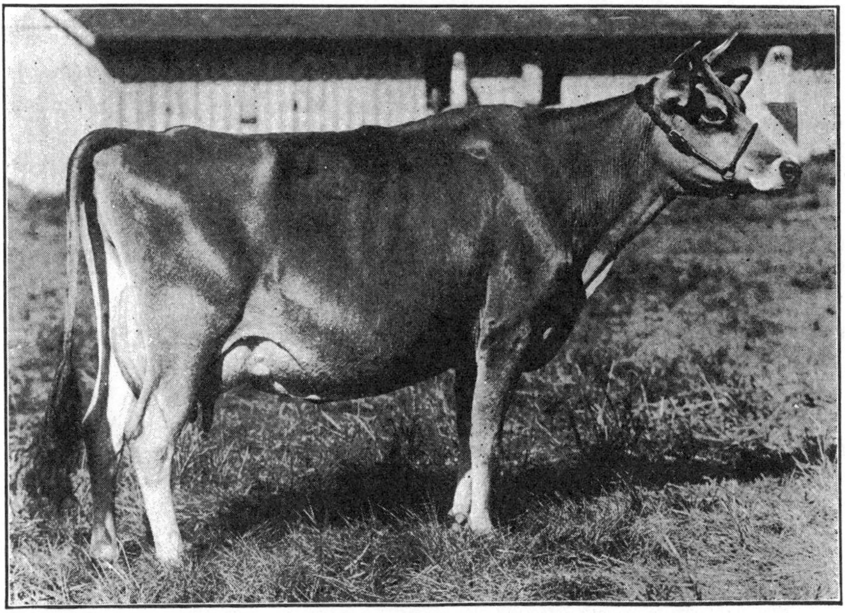

A Jersey Cow.[1]

Jerseys, as nearly as history reveals, share with Guernseys the blood of the old Brittany and Normandy cattle of France, in which they undoubtedly have their principal origin.

The Jersey breed early attracted the attention of England's aristocracy, who introduced them into England to beautify parks and furnish the rich milk that it was characteristic of them to produce. Beauty of form has been as much a part of the standard of excellence that guided the breeders in the development of their cattle as has production of milk, and has resulted in cattle of marked refinement and beauty.

[1] Courtesy of The Field, New York.

Remarkable herds were produced in England, From the herd of Philip Dancey of that country, the bull Rioter, 746E, was brought to America and to him the St. Lambert family of Jersey cattle, so prominent in this country, all trace.

The American people have imported many Jersey cattle both from the Isle of Jersey and from England, and have always regarded cows of the breed most excellent butter producers. Practical dairymen whose choice of breeds has been the Jersey, have favored the larger-sized cows and persistently worked to secure large productions of milk and butter. As a result, many of the American-bred Jerseys are larger and more robust and productive than the so-called island type. The greater size and production of the American type of Jerseys has enabled the breed to hold a popular place in dairy states and districts with other dairy breeds.

Parallel with the development of the American-bred type of Jersey, many people have taken great pride and pleasure in maintaining and preserving the refined and smaller sized island type. Jersey cattle have been quite universally distributed over the world and under proper care and supervision give excellent satisfaction.

Characteristics of Jersey Cattle.—Jersey cattle conform to a dairy type that is usually extreme. They are regarded as most economical producers of butter because of the marked dairy capacity they possess in proportion to their size. The size varies according to the strain or family and for cows ranges from 700 to 1000 pounds. Bulls will range from 1100 to 1500 pounds. The American-bred families, more especially the St. Lambert's, are larger than the imported stock from the Isle of Jersey or from England.

The breed matures early and as a result many mistakes have been made in carelessly and intentionally having young heifers produce their first calves at too young an age. This practice, together with scant feed rations, not only reduces the size but the constitution and usefulness of any breed and, for a breed that is naturally small, results in severe criticisms that are unfair when they apply to a breed rather than to individuals. Jersey cattle that are properly reared and well cared for tend to be long lived and very satisfactory dairy cattle. They have ranked high in economy and production tests at many shows and expositions and the production of cows admitted to the Jersey register of merit verify the fact that cows of this breed have highly developed powers for dairy production. The milk is of rich quality, testing ordinarily around 5 per cent. It is reasonable to expect a production of 300 pounds of butter-fat annually as an average per cow in herds that are well selected and managed. Jacoba Irene, 146443, an American-bred cow, owned by A. O. Auten, Jerseyville, Ill., in three consecutive years produced 42,900 pounds of milk and 2366.1 pounds of butter-fat. The present highest yearly record of butter-fat production made by a Jersey cow is 999.14 pounds, the amount of milk being 17,557.8 pounds testing 5.69 per cent, a record made by Sophie 19th of Hood Farm, 189748, owned by C. I. Hood, Lowell, Mass.

OTHER DAIRY BREEDS

The Dutch Belted, French Canadian and Kerry breeds of cattle heretofore mentioned rank as dairy breeds, but representatives of them are comparatively few and in many sections unknown.

Dutch Belted cattle are so-called from their peculiar marking which is black with a white band about the middle of the body. This characteristic color is uniformly found in all pure-bred herds of the breed, and is the result of scientific breeding experiments in Holland where the breed had its origin, and was known as Lakenfeld cattle from the word "Laken," meaning blanket or sheet about the body.

The usefulness of the breed was not a primary object in its development and for that reason it does not enjoy a popularity common to more prominent breeds.

Marked general improvement in type and production and an increase in the number of cattle is the ambition of those who are promoting the breed in America.

French Canadian cattle are a local and popular class of dairy cattle in the somewhat rough country and severe winter climate of the province of Quebec, Canada. Here the breed has been developed from early French stock and bred for over two hundred years. The characteristics of the cattle resemble very much the Jersey breed and lead to the belief that they have the same origin in blood. Their hardiness and adaptability to withstand Canadian winters and make economical yields of rich milk are commendable.

A production of 5000 pounds of milk testing 4 per cent or more is regarded as a fair average annual production for cows of this breed.

Kerry cattle originated in the Kerry mountains of Ireland under most adverse conditions of soil, climate and people. They have been called "the poor man's cow." They are very small as a natural result of their poor environment, bulls weighing 800 to 1000 pounds and cows 400 to 700 pounds. There are two types of the breed resulting from a cross which resulted in the type called the Derter-Kerry, which is smaller and more beefy than the original true Kerry. The economic value of Kerry cattle is best appreciated in its native home, where its adaptability, hardiness and ability to rustle and thrive recommend it. The novelty of the breed has led to a very limited distribution of the breed, a few herds having been introduced into Canada and the United States.

DAIRY BREED ORGANIZATION IN AMERICA

The welfare and preservation of breed interests are secured by responsible national breed associations that are recognized and approved by the United States Department of Agriculture, Washington, D. C., and the Canadian Department of Agriculture, Ottawa, Ont. All the dairy breeds of cattle except the Kerry have such organizations which are supported by a membership composed of the cattle breeders whose interest prompts them to become members, and by all who register pure-bred cattle of the respective breeds. Each association registers only cattle that are eligible by virtue of their purity of breeding, proper identification and being formally and regularly presented for registration on forms of application furnished by the association and certified to by the breeder or owner. Certificates of registry are furnished the breeders or owners and all transfers of ownership of registered animals where the identity of subsequent offspring is to be preserved must be formally reported. Upon being reported it is recorded and a certificate of transfer issued to the owner.

The associations all publish herd books containing a complete list of all registered animals and in most instances also publish literature that is useful and helpful in promoting its cattle interests. One who is particularly interested in a given breed will do well to avail himself of such literature, which is usually furnished gratis to those who apply for it. The location of the office and the secretaries of the respective associations can be readily determined by writing the national departments of agriculture, heretofore mentioned, if not by acquaintance with breeders of registered stock.

In addition to a registry of the names of pure-bred animals, the five more important breed associations maintain an advanced registry or register of merit for cattle which have excelled in production and made official records of milk and butter-fat equivalent to or surpassing definite standards fixed for periods varying from seven days to one year.

Following is a tabulated statement of the requirements for respective breeds, ages and periods of production:

Age.	Ayrshire.		Brown Swiss.		Guernsey.	Holstein.	Jersey.		
	Year Record.		Year Record.		Year Record.	7-Day Record.	7-Day Record.		Year Record.
	Pounds Milk.	Pounds Butter Fat.	Pounds Milk.	Pounds Butter Fat.	Pounds Butter Fat.	Pounds Butter Fat.	Pounds Butter Fat.	Pounds Butter Fat.	Pounds Butter Fat
2 years.....	6000	214.3	6000*	222.0*	250.5	7.2	12.0	14.0	250.5
3 years.....	6500	236.[illegible]	6430	238.5	287.0	8.8	12.0	14.0	287.0
4 years.....	7500	279.0	7288	271.3	323.5	10.4	12.0	14.0	323.5
5 years.....	8500	322.0	8146	304.2	360.0	12.0	12.0	14.0	360.0
6 years.....	....	.....	9000	337.0	.....	....	....	....	.....
Requirements increase each day by pound	1.37 and 2.74	0.06 and 0.12	2.35	0.09	0.1	0.00439	....	....	.1

* Two and one-half years.

Great advancement in the appreciation and breeding of pure-bred cattle has been and is being brought about by volunteer state and community organizations. The closer contact which these associations have with the masses engaged in dairying make their opportunity greater than that of national associations for giving encouragement to men to use the very best breeding animals, especially sires, that it is possible to secure. In fact, such organizations cannot be encouraged too much, for in the work of local breeders and community effort lies the success of maintaining high standards of excellence and the preservation of all breeds.

REFERENCES

"Types and Breeds of Farm Animals." Plumb.
"Study of Breeds." Shaw.

CHAPTER 55

CLEAN MILK PRODUCTION

BY C. W. LARSON
Professor of Dairy Husbandry, The Pennsylvania State College

More than half of the milk produced in the United States is used for direct consumption. Pure, clean milk is an excellent food and is cheap. It contains all the essential elements for a complete and balanced ration for man. On account of its being used by infants, children and invalids who are least able to resist the effects of unclean food, and because milk is so easily contaminated, it is essential that great care be taken in its production and handling.

CLASSES OF MILK

Sanitary Milk is no definite class of milk. It is simply a term used to designate good, clean milk produced with extra care. It is usually sold at a price somewhat above prevailing milk prices.

Guaranteed Milk is milk that the producer guarantees to be produced under certain conditions and usually with some standard of fat and bacterial content.

Standardized Milk is milk which has been altered in its amount of butter-fat by skimming or the adding of cream.

Certified Milk is milk that has been produced under certain conditions prescribed by a commission, usually consisting of a veterinarian, a physician, a chemist and a bacteriologist. The prescribed conditions include scrupulously clean methods, healthy cows, healthy milkers and carefully sterilized utensils. Such milk should not contain over 10,000 bacteria per cubic centimeter. It is usually sold at a considerably higher price than ordinary milk.

Inspected Milk is produced from healthy cows that have been inspected. The inspection involves an examination by a city or state inspector of premises and methods.

Pasteurized Milk is milk that has been heated to a sufficiently high temperature to kill the harmful bacteria, or germs, and then immediately cooled. The temperature to which it is heated varies with the length of time it is held. For market milk, it is customary to heat to 140° to 145° F. for twenty minutes.

Modified Milk is high-class milk, such as certified or sanitary milk, altered in composition to suit certain uses. Such milk is used for infants and invalids.

EQUIPMENT AND METHODS

Clean, Healthy Cows.—The first essential in the production of clean, healthy milk is to have cows that are clean and free from disease. The milk from emaciated animals, or those suffering from any disease, should not be sold. The milk from cows having inflamed udders or sore teats should not be put into the general supply. The cows should be comfortable in order to produce normal milk. Any unusual condition or disturbance will cause them to produce abnormal milk. They should be kept in dry, clean, properly-bedded stalls. The food should be free from mustiness and no decomposing silage or wet foods should be given after they become mouldy. The drinking water should be clean and fresh.

Most of the dirt that finds its way into the milk pails falls from the bodies of the cows. It is essential, therefore, that the cows be kept clean. One of the most important factors in keeping cows clean is to have the platforms on which they stand the proper length, so that the manure will drop into the gutter. Adjustable stanchions are also desirable, so that the proper alignment can be made on the platform. Cows kept in the stable should be groomed at least once a day, but this should be sufficiently long before milking time to permit the dust to settle. Wiping the udder and the flanks with a clean, damp cloth requires only a short time and will do much to remove dirt that would otherwise fall into the pail. It is practical, where clean milk is being produced, to clip the udder and flanks occasionally. This prevents the dirt from sticking, and makes it possible to keep the cows cleaner.

Stables.—Expensive barns are not essential to the production of clean milk. The health of the cows and the methods of the milker are of greater importance and have more effect upon the finished product. Good construction and convenient arrangement of the stable may lessen the work, keep the cows more comfortable and have a beneficial effect upon the milkers.

The barn should be located on well-drained land and be free from contaminating surroundings. Horses, chickens, stagnant water and manure piles, when near the stable, may pollute the air. Odors are easily absorbed by milk. The stable floor should be of concrete or some other material that does not absorb the liquid manure, and at the same time should be sufficiently smooth to be easily cleaned. The walls should be smooth and free from ledges to avoid collecting dirt. The occasional use of whitewash on the walls and ceiling is recommended.

The barn should not be overcrowded and at the same time should not have an excessive amount of space in cold climates. From 500 to 1000 cubic feet per cow is satisfactory. Too much light cannot be provided. Sunlight destroys bacteria and also makes a healthy atmosphere for the cows. The more light the better, and it is well that it be evenly distributed and that the windows be located, if possible, so that the light can shine into the gutter.

The dairy barn should be well ventilated. Experiments at the Pennsylvania Experiment Station have shown that cows will do well even in an open shed, providing they are kept dry and out of the wind. Since, therefore, it is not necessary to have the dairy barn warm, the problem of ventilation is greatly lessened. It is not difficult to get fresh air into the barn, but it is difficult to get sufficient fresh air without cooling the atmosphere. The air in the barn should be changed, even if it does become cold. Cows must have fresh air in order to produce their maximum of milk and keep healthy. Have many and small intakes and few and large outlets. The capacity of the intakes and the outlets should be equal and provide about one square foot in cross section for each four or five cows.

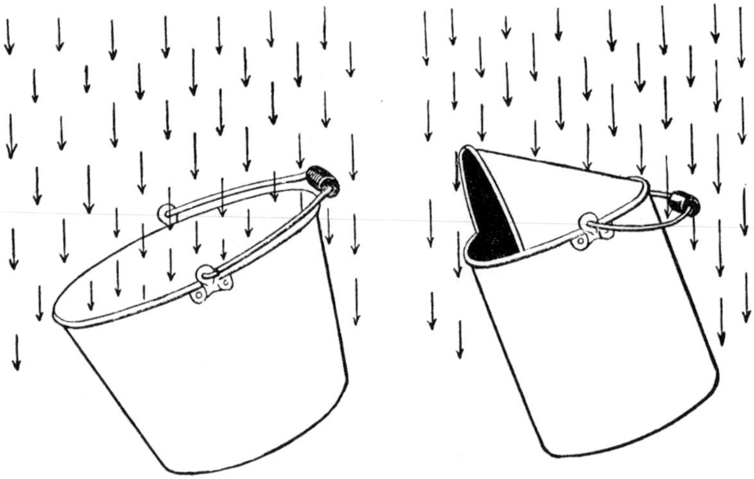

MILK PAILS OF BEST DESIGN.[1]

Milkers.—A clean and careful milker can produce clean milk in a poor barn, but an unclean milker cannot produce clean milk in any barn. The milker must be clean and healthy and, above all things, should milk with dry hands. The practice of wetting the hands with milk is deplorable. It is unnecessary. The milker should always wash his hands before starting to milk. The air, during the milking, should be kept free from dust and odors. Manure should not be removed from the barn, nor should any dusty feed be given during the milking time. Silage or other feeds that have an odor should be fed at least three hours before milking, so that the odor will not be taken up by the milk.

Small-top Milk Pails.—Most of the dirt that gets into the milk drops from the cows during milking time. If, therefore, the opening at the top of the pail is closed to one-sixth the size of an ordinary pail, only

[1] Courtesy of U. S. Dept. of Agriculture.

one-sixth as much dirt gains access to the milk. With a little practice, the small-top milk pail can be used as easily as the large-top pail.

Clean Tinware.—All the cans and pails that are used for milk should be of metal, and all of the joints and corners should be completely filled with solder. Wooden pails should not be used. To wash the tinware, it should first be rinsed with lukewarm water, then thoroughly scrubbed with brush, hot water and washing powder, and finally, either steamed or rinsed with boiling hot water. It should not be wiped with a cloth, but should be allowed to drain and dry. The heat of the steam or boiling water will soon dry the tinware.

Strainers.—Where milk is produced in a clean way it is not necessary to have a strainer. It is usually not desirable to have a strainer on the milk pail, for the dirt collected will have all the injurious effect washed from it into the pail during the milking. A strainer may be used on the can or milk cooler. For this a cloth strainer, made especially for that purpose and used only once, is satisfactory. The cheesecloth strainer that is used from day to day contaminates the milk instead of purifying it. A metal strainer is satisfactory.

Handling the Milk.—As soon as the milk has been drawn it should be removed from the stable so that it will not absorb odors. A convenient milk-room should be provided. This room should be clean and free from dust and odors. The milk should be cooled at once. Under the best of conditions, some bacteria or germs get into the milk, and the problem, therefore, is to prevent these bacteria from increasing in number. At a temperature of 70° F. one bacterium may increase to two in twenty minutes, but at 50° F. or lower it requires a much longer time. One bacterium at 50° F. may increase in twelve hours to six or seven, while at 70° F. it may increase to six or seven hundred. Since, therefore, there are several hundred bacteria in every cubic centimeter of good milk, some realization may be had of how many thousands of bacteria will be developed in ten or twelve hours at 70° F. The following table, prepared by Stocking, shows the importance of cooling milk at once to a low temperature. The milk that was used in this experiment contained a low percentage of bacteria when produced.

Effect of Different Temperatures Upon the Development of Bacteria in Milk.

Temperature Maintained for 12 Hours.	Bacteria per c.c. at end of 12 Hours.
40° F.	4,000
47° F.	9,000
50° F.	18,000
54.5° F.	38,000
60° F.	453,000
70° F.	8,800,000
80° F.	55,300,000

Coolers.—There are various styles of apparatus on the market for cooling milk. These are called coolers. They are arranged so that the water passes on the inside of the tubes and the milk passes over them. By having a supply of cold water passing through the tubes, the milk can be cooled to within two or three degrees of the temperature of the water. Unless the cooler is placed in a room free from dust, the milk may become contaminated. Coolers with a hood or covering are preferred. Those having few joints so that they may easily be cleaned are also preferable. When it is not necessary to cool the milk immediately for shipment, or otherwise, it may be cooled by placing the can in a tank of cold water. Unless the water supply is plentiful and the water cold, it is desirable to have ice.

Suggestions for Improvement.—A list of suggestions and instructions of good methods and practices placed in a conspicuous place in the barn does much to improve the quality of the milk. A list of twenty-one suggestions, composed by Webster, gives the essential points to be followed in the production of clean milk. These suggestions are as follows:

"I. Cows.

"1. Have the herd examined frequently by a skilled veterinarian. Remove all animals suspected of not being in good health. Never add an animal to the herd unless it is known to be free from disease.

"2. Never allow a cow to be abused, excited by loud talking or other disturbances. Do not unduly expose her to cold and storm.

"3. Clean the under part of the body of the cow daily. Hair in the region of the udder should be kept short. Wipe the udder and surrounding parts with a clean, damp cloth before milking.

"4. Do not allow any strong-flavored foods such as cabbage, turnips, garlic, etc., to be eaten except directly after milking.

"5. Salt should always be accessible.

"6. Radical changes of food should be made gradually.

"7. Have plenty of pure, fresh water in abundance, easy of access and not too cold.

"II. Stables.

"8. Dairy animals should be kept in a stable where no other animals are housed, and preferably one without a cellar or storage loft. Stables should be light—four feet of glass per cow—and dry, with at least 500 cubic feet of air for each animal. The stable should have air inlets and outlets so arranged as to give good ventilation without drafts over the cows. It should have as few flies as possible.

"9. Floors, walls and ceilings should be tight and the walls and ceiling should be kept free from dust and cobwebs and whitewashed twice a year. There should be as few dust-catching ledges and projections as possible.

"10. Allow no musty or dirty litter or strong-smelling material in the stable. Store the manure under cover at least forty feet from the stable and in a dark place. Use land-plaster in the gutter and on the floor.

"III. Milk House.

"11. The can should not remain in the stable while being filled. Remove the milk from each cow at once from the stable to a clean room. Strain immediately through absorbent cotton or cotton flannel; cool to 50° F. as soon as possible. Store at 50° F. or lower.

"12. Utensils should be of metal with all joints smoothly soldered. If possible, they should be made of stamped metal. Never allow the utensils to become rough or rusty inside. Use them for nothing but milk.

"13. To clean the utensils, use pure water. First rinse them with warm water, then wash them inside and out in hot water in which a cleaning material has been dissolved. Rinse again and sterilize in boiling water or steam. Then keep them inverted in pure air, and in the sun as much as possible, until ready to use.

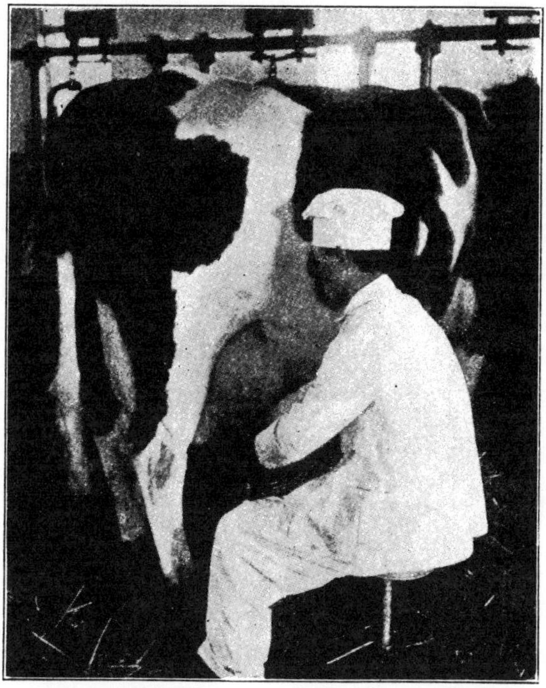

A Clean Milker in a Clean Stable at Milking Time.[1]

Note the clean suit, sanitary milking stool, small-top pail, cow with clean flanks and udder, and sanitary stable construction. Under these conditions clean milk can be easily produced.

"IV. Milking and Handling Milk.

"14. A milker should wash his hands immediately before milking and should milk with dry hands. He should wear a clean outer garment, which should be kept in a clean place when not in use. Tobacco should not be used while milking.

"15. In milking be quiet, quick, clean and thorough. Commence milking the same hour morning and evening. Milk the cows in the same order.

"16. If any part of the milk is bloody, stringy or not natural in appearance, or if, by accident, dirt gets into the pail, the whole should be rejected.

[1] From Farmers' Bulletin 602, U. S. Dept. of Agriculture.

"17. Weigh and record the milk given by each cow.

"18. Never mix warm milk with that which has been cooled. Do not allow milk to freeze.

"19. Avoid using any dry, dusty feed just previous to milking.

"20. Persons suffering from any disease, or who have been exposed to any contagious disease, must remain away from the milk.

"21. The shorter the time between the production of the milk and its delivery, and between its delivery and its use, the better will be the quality."

REFERENCES

"Dairy Chemistry." Snyder.
"The Milk Question." Rosenau.
"Bacteria and Country Life." Lipman.
"Modern Methods of Treating Milk and Milk Products." Van Slyke.
"Practical Dairy Bacteriology." Conn.
Kentucky Expt. Station Circular 6. "Inexpensive Appliances and Utensils for Dairy."
Farmers' Bulletins, U. S. Dept. of Agriculture:
348. "Bacteria in Milk."
366. "Effect of Machine Milking and Milk Supply of Cities."
413. "Care of Milk and Its Use in the Home."
602. "Production of Clean Milk."
608. "Removal of Garlic Flavor from Milk."

CHAPTER 56

DAIRY BUTTER-MAKING

By Ernest L. Anthony

Assistant Professor of Dairy Husbandry, The Pennsylvania State College

Farm dairying has attracted public attention to an uncommon degree in the last few years. This is due largely to the modern development in the dairy field as well as to the adaptability of dairy farming or certain phases of it to average farm practices.

Adaptation.—Dairy farming is especially adapted to farms located near markets, because of the regular demand for fresh dairy products. Dairy products are, as a whole, perishable and must be marketed soon after being produced. For this reason easy and frequent access to markets is very desirable. Dairying is also adaptable as a side line in general farming, fruit raising and poultry farming. It provides for the utilization of waste products on the farm as feed for cows and aids in the continuous and economical employment of labor.

It is also particularly adapted to the person starting in to farm on a small scale, as it is possible with a comparatively small capital to start a dairy business which enables the dairyman to live while his business grows.

The Need of Dairy Farming.—According to late authorities, the people of the United States consume over seven-tenths of a pint of milk per capita daily. To this should be added the enormous consumption of butter, cheese, ice cream, condensed milk and other minor dairy products. This gives an idea of the possibilities which are before the American dairyman today.

Types of Dairy Farming.—Several types of dairy farming are pursued in the United States; they are: (1) the production of milk for wholesale and retail trade; (2) the production of cream for creameries and ice cream factories; (3) the manufacture of cheese on the farm; (4) the manufacture of butter upon the farm, or farm butter-making.

Market Milk.—The production of market milk is one of the leading types of dairy farming. It requires easy access to reliable markets, and is most successful when conducted on a fairly extensive scale. It requires less labor than most other types of dairy farming. Clean milk production is discussed in the preceding chapter.

Farm Cheese-making.—This type is especially adapted to dairy farms not located close to dairy markets. Cheese is less perishable than the other dairy products and this enables the farmer to engage in dairying and market his products at his convenience. Farm cheese-making is most extensive in the eastern part of the United States, especially in the rougher sections. Cheddar and brick cheese are largely made. Much soft cheese, such as schmier kase, cottage and Dutch hand is also produced in many localities. Successful farm cheese-making requires some special cheese apparatus, as well as a fair understanding of the principles which govern cheese manufacture.

Farm Butter-making.—On the general farm more attention is paid to the making of farm butter than to any other phase of farm dairying. This is true because of the large market for the product and the adaptability of farm butter-making to average farm conditions.

A Good Type of a Dairy House.[1]

Control of Products.—The production of good butter of uniform quality starts with the cow. Milk from unhealthy cows can never be made into first-class products. Neither can cows that are kept in unclean, unsanitary places produce clean milk.

One making butter on the farm can have complete control of his milk from the time that it is drawn from the cow until it is made into butter. This is not true of the creamery man or manufacturer, who has to secure his product from outside sources over which he has no supervision. This advantage means much to the farm butter-maker if he realizes it and makes the most of it.

Cleanliness Necessary.—The cows should always be brushed off and kept clean at milking time. Care should be taken that all utensils be kept clean and in good condition, so that the cream, whether skimmed or separated, shall be good, sweet and not absorb any undesirable taints or odors. Much butter which would otherwise be good is damaged in flavor because

[1] Hygienic Laboratory, Washington, D. C.

care is not taken to keep dirt and impurities out of it. Milk not separated by a cream separator should be at once cooled by some suitable method and held as cold as possible until the cream has risen. Cream should be cooled as soon as it is separated.

Percentage of Fat in Cream.—If a separator is used the percentage

A Good Type of Cream Separator.[1]

of fat in the cream may be regulated. When it is impossible to test the cream for its percentage of fat, the separator so regulated that about 12 to 14 per cent of the total milk is separated and comes out as cream, will give approximately the proper richness to the cream. The best results

[1] Courtesy of the Sharples Separator Company, West Chester, Pa.

will be obtained when the cream has about 28 to 30 per cent of fat. Cream with too high a percentage of fat has a tendency to adhere to the sides of the churn, which causes difficult churning and increases the danger of loss of fat in the buttermilk.

Thin Cream Undesirable.—When cream is too thin or has too small a percentage of fat in it, as in the case of hand-skimmed cream containing from 12 to 20 per cent of fat, good, uniform churning is hard to secure. Such cream loses too much fat in the buttermilk and also requires longer churning.

Methods of Ripening Cream.—Poor quality in farm butter is most frequently due to a lack of proper ripening of the cream previous to churning. On the farm it is often necessary to store the cream from two or or more days' milkings in order to secure a sufficient amount for a churning. The common method now in use on most farms is simply to collect in a cream can or jar successive creamings, until enough has been secured for a churning. Meanwhile the cream is held in the cellar, milkhouse, back porch or springhouse. The temperature at which it is held varies with the weather, season of year and other conditions. Under these conditions the cream usually ripens or develops acid until at the end of three or four days it becomes sour and is then stirred and churned. If it is kept too cold for ripening during this holding period, it is warmed for several hours and allowed to sour before churning.

This is a bad practice and is responsible for many of the taints and off flavors found in farm butter. The reason for this is that the temperature of the cream is usually about 55° F., which is a little too low to secure a good growth of the lactic or acid-forming bacteria which produce the proper flavors in the cream. Some claim that this average cellar temperature favors the proper conditions for the growth of the bacteria that produce objectionable flavors and taints in cream. These undesirable bacteria produce no acid, will not grow well in the acid medium and seem to grow best at a temperature of 50° to 60° F.

The Pennsylvania Experiment Station, Bulletin 135, has conducted some experimental work to determine the best way to ripen cream on the farm. This work indicates that there are three other methods, any one of which will give better results than the storing of cream at cellar temperatures. They are: (1) holding or storing the cream at a very low temperature (below 45° F.) until enough is secured for a churning, and then warming it up to 70° to 80° F. and ripening; (2) ripening the first collection of cream at once and adding each skimming to it, from day to day, until a churning is secured; (3) adding a portion of buttermilk to the first cream gathered and then adding each skimming until enough is secured for a churning.

The first method is a good one for butter-makers who have ice for keeping the cream cold. Immediately after separating each day's cream, it should be cooled to 45° F. or below, and held at this low temperature until enough is secured for a churning. It is then warmed up to 75° F. and held at that temperature until the proper amount of acid is developed in it. At this temperature about twelve hours is required to develop the proper percentage of acid.

The second method is to ripen the cream of the first separation that is to form the new churning at about 75° F. until 0.3 per cent of acid is developed. It is then cooled to the temperature of the springhouse or cellar, and each subsequent creaming, after it has been cooled, is added to this lot until enough is secured for a churning. Under average conditions this will give enough acid development in the whole churning for best results. The ripening of the first separation of cream develops a large number of lactic acid bacteria and produces some acid, which serves to hold in check the undesirable types of bacteria.

The third method is to add a portion of buttermilk of good quality to the first separation, and then add each succeeding creaming and hold the whole amount at cellar or springhouse temperature until a sufficient quantity is secured for a churning. If the ripening has not sufficiently developed by that time the temperature can be raised to 75° F. and the cream allowed to ripen until the proper amount of acid has developed.

The object in the last two methods is essentially the same, namely, to hold in check the undesirable bacteria by having developed or introduced into the cream a preponderance of the desirable bacteria and a small amount of acid. The last two methods are simple, handy and require no special apparatus. Care must be taken, however, in the last method to make sure that the buttermilk comes from butter of a good flavor and quality. The using of buttermilk of medium or poor quality is very likely to produce butter of much the same kind as that from which the buttermilk was secured.

Amount of Acid to Develop, or Degree of Ripening.—Large amounts of farm cream are ripened or soured too much before churning. Because of this, an old and tainted or stale flavor is developed. Cream ripened until it is sharply sour usually contains from 0.6 to 0.8 per cent of acid, which is too much. The best flavors and keeping quality are secured when it is ripened so as to contain about 0.4 to 0.5 per cent of acid. Where no acid test is used, this amount of acid may be approximated. The cream should taste only very mildly sour. Cream naturally ripened at 70° to 75° F. will develop about this amount of acid if held ten hours.

The Use of Starters.—Starters are not much used on the farm and when used are generally of the natural kind, that is, made up of buttermilk or good sour milk. They are very desirable, if care is taken to use only good buttermilk or sour milk, and in most cases will improve the quality of the butter produced. They are especially desirable when cream is hard to churn because of improper ripening, and where it is difficult to secure proper ripening. The amount to use varies with the condition of the cream, but in most cases from 10 to 20 per cent is a suitable quantity.

Natural Starter.—The natural starter made from sour milk is perhaps the best for farm conditions. To make it, set several samples of good, clean skim or whole milk in small jars until the milk becomes sour. The holding temperature should be about 70° F. When the samples have become sour they should be examined. They should have formed a good, smooth curd, free from gas bubbles. The flavor and taste should be clean and sharply sour. The sample showing the best flavor and condition of the curd should be selected for the starter. It may be built up in larger quantities by adding the sample to about ten times its volume of clean, sweet skim milk and allowing the mixture to stand at about 70° F. until it has coagulated. The coagulated milk is then the starter to use in the cream-ripening process. It contains a preponderance of the desirable lactic bacteria which are necessary for that process.

The Amount of Starter to Use.—The amount of starter to add to cream varies from 8 to 50 per cent. If the starter is a good one, the more added the better, but if too much be added it will dilute the cream too greatly and make it hard to churn. About 10 per cent is a common amount to use.

Churning Temperatures.—The temperature at which cream is churned is very important. Properly ripened cream should be cooled down to the temperature at which it is to be churned and held at that temperature at least two hours to allow the fat to become cool and firm enough to churn.

The churning temperature varies widely. It is affected by the season of the year, kind of feed given the cows, condition of the cream and temperature of the churning room.

Variations in Churning Temperature.—In the spring and summer, when the cows are fresh and the feeds succulent and soft, the butter-fat is naturally softer than later in the season. Under average conditions temperatures ranging from 52° to 56° F. will give best results for these seasons. This temperature should be increased to about 56° to 60° F. in the winter. Much cream is now churned on the farms at above 60° F. Experiments seem to indicate that the lower temperatures are to be preferred, as butter is much firmer when coming from the churn, does not so easily incorporate buttermilk, and will stand more working, thus producing a better body and a more uniform quality. Because of the cream being poorly ripened or abnormal in some way, it is often necessary to use higher temperatures than are here given. When difficulty in churning is experienced, the cream should never be raised in temperature by adding hot water to the churn, but should be poured from the churn into a can and gradually raised a few degrees in temperature by setting the can in a pan of warm water.

Care of the Churn.—The proper care of cream in the ripening process, although very essential, does not insure good butter. Good cream can easily be spoiled in churning. Unless the churn is kept in good condition it is impossible to make good butter with it. The churn should always be well scalded out and well cooled down before using. There are two reasons for this: first, the hot water will scald out and kill all moulds that may be growing in the wood and will close the pores of the wood so that the cream or butter will not adhere to it; second, the churn should be cooled so that the temperature of the cream will not be raised while churning and yield soft, greasy butter.

Length of Time to Churn.—The length of time best for churning varies with the condition of the cream, but ranges from 15 to 30 minutes.

Farm Butter-making Apparatus.[1]

If the cream churns in less than 15 minutes, the butter is very likely to be too soft to work well and will have a poor body when finished. Cream that requires much longer than 30 minutes may be improperly ripened or abnormal in some way. Taking the cream from the churn and raising the temperature in the manner suggested above will in most cases overcome the trouble.

The churning should stop when the butter begins to collect in the buttermilk in granules from the size of a pea to that of a grain of corn. Granules of this size do not contain so much buttermilk as do larger ones. The butter is easier to wash, salt and work.

Washing Butter.—It is a common practice on the farm to wash butter through several wash waters. This is unnecessary if the churning has been stopped at the right time. If the granules are about the size of peas

[1] Courtesy of Pennsylvania Agricultural Experiment Station.

or grains of corn, one washing will remove all the buttermilk. Too much washing has a tendency to remove the finer flavors and give the butter a flat taste. The amount of wash water should be about equal to the volume of cream churned.

Temperature of Wash Water.—The temperature of the wash water may vary considerably, but it should not be much above or below the churning temperature. Very cold wash water is to be avoided. Cold water absorbs the flavors of the butter readily, causes brittleness of body and lowers the quality.

When a low churning temperature is used, the washing temperature may be the same, and should never be more than 4 to 6 degrees less. Where a higher temperature is used for churning, the washing temperature may differ as much as 4 to 10 degrees from that of the churning. The wash water should be pure and clean and free from odors or taints, as these will be readily absorbed by the butter.

Preparation of Working-Board.—After the wash water is drawn from the butter—unless a combined churn and worker is used—the butter should be taken out in the loose, granular form and placed on the working-board or table. This table should be clean and thoroughly wet with cold water. Butter will stick to a dry, warm or dirty board.

Salting.—Fine dairy salt of the best quality should be used. The quantity varies with the taste of the maker and the markets on which the butter is sold. Under average conditions where the butter is worked on a hand-worker, three-quarters of an ounce of salt to each pound of butter-fat is a desirable amount to use. Butter made in a combined churn requires heavier salting, and as much as one and one-quarter ounces of salt per pound of butter-fat may be required. This larger amount is necessary because of the wash water which is held in the churn.

The salt should be evenly distributed over the granules of butter on the working-board, and the working may begin at once. It is a common practice to let the butter stand with the salt on it for a while before working. This is unnecessary if the butter is in a good granular condition, firm in body and the salt fine and of a good grade.

Working of Butter.—The working should begin by first using the sharp edge of the worker to cut and flatten the butter out into a thin sheet. This sheet should then be folded to the center of the working-board, and the process repeated.

The working of butter accomplishes three important things: It evenly incorporates the salt, removes the excess water and makes the body compact. The working should be continued until the excess water no longer appears and the salt is worked evenly through the mass. The texture of the body may be ascertained by breaking off a piece of the butter. The break should show a brittle, grainy appearance, similar to that of broken steel.

When the butter has been sufficiently worked it should be printed into some desirable shape. The common rectangular one-pound mould is the best, as it makes a neat, attractive print and is easy to handle.

Wrapping of Butter.—After the butter is printed it should be wrapped in a good grade of parchment butter paper. This is very essential. Much butter is wrapped in cloth or oiled paper. This is a very bad practice, as the cloth holds moulds, which readily grow and produce taints and odors. The oiled paper, if kept for any length of time in a warm place, becomes very rancid and imparts undesirable flavors.

Value of Standard Product.—It is always advisable to have the name of the producer or his farm name on the wrapper of the butter, if it is sold on the market. If the butter is of good quality, this will tend to increase the sales and be an incentive to the highest effort for maintaining uniformity in quality. The attractiveness and neatness of the package always helps to sell the butter, often at much above the average market price.

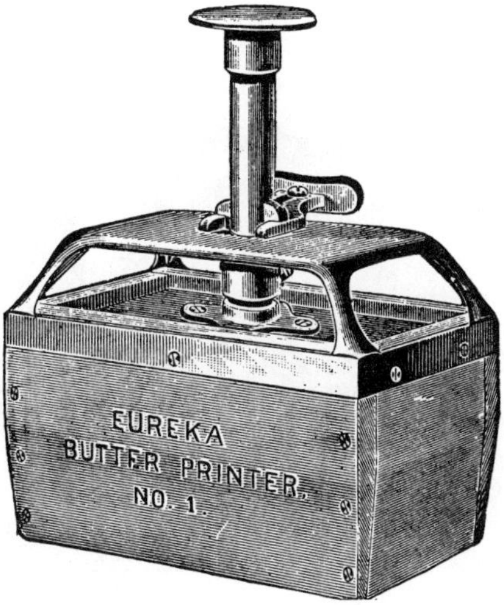

BUTTER PRINTER.

Care of the Farm Churn. After the butter is taken from the churn, the latter should be rinsed out with warm water and the rinsing followed by a thorough washing with very hot water. The rinsing out with warm water will remove any buttermilk which may remain in the pores of the wood. The hot water will remove any fat which may be left in the churn.

It is never well to use soap powders on the interior of the churn, but the occasional use of a small amount of dairy washing powder or lime-water is beneficial.

To keep the churn sweet and free from odors and taints a small handful of lime placed in some water in the churn or in the last rinsing of the churn is very effective. It is essential in good butter-making to see that all apparatus used is absolutely clean and free from undesirable odors and taints, as these are quickly absorbed by the butter.

Dairy Apparatus.—In the selection of dairy apparatus there are several things which must be taken into consideration. They are: Simplicity of construction, ease of cleaning, durability and first cost.

Care of Other Dairy Apparatus.—All other dairy apparatus should at all times be kept scrupulously clean and free from rust. Pails, buckets, crocks, etc., after being used should be rinsed out and washed well with a brush and a dairy washing powder. After they are carefully cleaned they should be rinsed out and then either scalded with very hot water or steamed if steam is available.

The cream separator should be taken apart and well cleaned after each milking and left apart until its next use. If it is left unclean, or is not well aired, bad taints and odors will develop in the cream, causing a poor quality of finished product.

All dairy apparatus should be placed in the sun after it is washed, as the sun will quickly dry it. Sunlight also acts as a powerful disinfecting agent. However, care should be taken to see that the apparatus is so placed that there is no danger of dust and dirt blowing in on it.

BUTTER READY FOR MARKET.[1]

Churns.—The farm churn should be of ample size for the largest churning made during the year. The common barrel churn is the most practical for farm use, as it is simple, easy to clean and very durable as well as economical in the first cost. On farms where large amounts of butter are made a small combined churn, as illustrated, is very desirable.

On farms where more than three cows are kept a cream separator, of a size depending upon the number of cows kept, is advisable. It is best to select a make of separator that is sold in the community, so that the purchaser can always quickly secure necessary repairs. Cream separators have been so well perfected that there is practically no difference in the skimming efficiency of the several machines. They all skim sufficiently

[1] Courtesy of Hinde & Dauch Paper Co., Sandusky, Ohio.

clean, but one should look to simplicity of construction and durability of wearing parts.

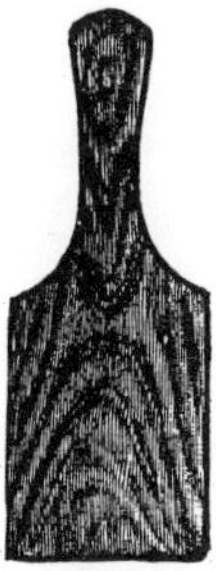

WOODEN LADLE.

Buckets and Tinware.—All buckets should be made of heavy stamped metal, heavily tinned and with all joints and corners smoothly soldered so as to leave no place for dirt or impurities to collect. Buckets like those shown in the preceding chapter are desirable for milking purposes, as they admit the smallest amount of dust and dirt and are still simple in construction.

Wooden Apparatus.—Wood is best suited for the construction of certain dairy apparatus such as butter ladles, butter moulds, workers, etc., because, by proper treatment, butter will not adhere to wood as it will to other materials.

REFERENCES

"Principles and Practice of Butter Making." McKay and Larson.
"The Business of Dairying." Lane.
"Milk and Its Products." Wing.
"Dairy Farming." Michels.
"First Lessons in Dairying." Van Norman.
"Science and Practice in Cheese Making." Van Slyke and Publow.
"Farm Dairying." Laura Rose.
Pennsylvania Expt. Station Bulletin 135. "A Study of Manufacture of Butter," "Methods of Making Farm Butter."
Purdue Expt. Station Circular 51. "Producing Cream for Good Butter."
Farmers' Bulletins, U. S. Dept. of Agriculture:
349. "Dairy Industry in South."
541. "Farm Butter Making."

BOOK VI

FARM BUILDINGS AND EQUIPMENT

CHAPTER 57

FARM BUILDINGS, FENCES AND GATES

Farm buildings should be located and constructed with a view of meeting the needs of the farm and farmer's family. They should harmonize with the natural surroundings and have sufficient room for the housing of the farm animals, equipment and the storage of forage, grain and such other crops as may be grown. The number, character and size will be determined by the size of the farm and the type of farming. They should be as fully adapted to the type of farming as possible.

(For further details relative to the location of the buildings and their relation to each other, see Chapter 68.)

The Farm Residence.—With some farmers the housing of the live-stock is considered of more importance than the housing of the farmer and his family. Where capital is very limited and the farmer is accustomed to an exceedingly simple life, this may prove advantageous for a short time, in order to get a start. At the present time and in most localities, the housing of the farmer and his family properly receives first consideration. The farm residence should be the most important building of the farm. It should occupy a conspicuous place in the farmstead and bear a convenient relationship to the other buildings of the farm. There is more latitude relative to the direction the farm house should face than there is in case of the city house. This feature should be carefully considered in the construction of the house, the position of verandas and the location of the living rooms. Size of windows and the entrance of sunlight should also be considered in this connection.

The foundation and the roof of the house are two important features. These should be constructed with reference to durability and strength as well as appearance. The height of the house or the height of the rooms may be increased with little additional cost, since this will increase the cost of neither foundation nor roof. There is little excuse, however, for tall houses in the country. Land is cheap and comparatively low structures harmonize better with country surroundings.

It pays to paint a farm residence thoroughly immediately after its construction, and to re-paint whenever paint is needed. Paint lengthens the life of a house and makes it warmer. Light colors are generally preferred for country dwellings. The smoke and dirt which make bright colors impracticable and expensive in cities are not present in the country. Such colors harmonize with the green foliage that should surround a country residence. On new lumber, the first or priming coat should be mixed very thinly and applied promptly after the house is constructed. At the time

AN ATTRACTIVE FARM HOUSE.[1]

A house of moderate size and cost in which comfort and convenience have been carefully considered. The roof and walls are shingled. Architects, Mrs. Myron Hunt and Elmer Grey.

[1] From "Distinctive Homes of Moderate Cost," by H. H. Saylor.

of priming, the boards should be reasonably dry in order that the paint may enter the wood and fill any cracks that are present. It should be worked well into the wood with the brush and allowed to become thoroughly dry

PLANS OF FARM HOUSE.

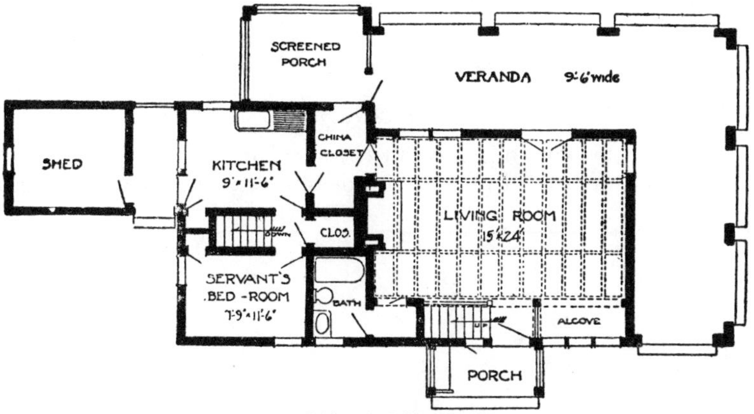

FIRST FLOOR PLAN

In warm weather the dining table is set in the screened porch, convenient to the kitchen. During the winter one end of the living-room takes the place of a dining-room.

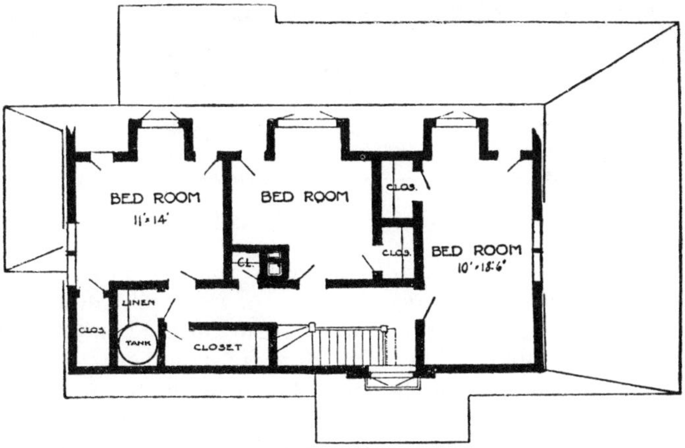

SECOND FLOOR PLAN

There are three good bedrooms on the second floor, and the end ones have cross ventilation through the gable windows.

before the second coat is applied. The second coat should be somewhat thicker, smoother and of the proper color. A third coat will generally be required, but the application should be deferred from three to six months.

This allows time for the second coat to become hard and any small cracks that may open in the meantime by shrinking of the boards will be filled with paint.

Whether the farmer does his own painting or hires it done, it is generally advantageous for him to purchase his own paint, and to be careful to select durable materials.

The interior arrangement of the house and its ventilation, equipment and appointment are discussed in Chapter 79.

BARNS

The principal barn of the farm is second in importance only to the house. In case of noted livestock breeders or some large stock farms, the

A GOOD TYPE OF BARN.[1]

barn becomes the most important structure on the farm. The prime requisites for a good barn are convenience, especially in arrangement, comfort for the animals, ample storage room for feed, proper light and ventilation, and durable but not expensive construction.

Whether all livestock on the farm should be housed in one structure or in several structures must be determined by the kind and number of stock reared. It is generally advisable to house the cows in a separate structure. The noise and odor of swine is detrimental to both the yield and quality of milk. Swine should not be kept in the main barn. If horses and cows are stabled in the same structure, they should have separate compartments. It will frequently be convenient to house the cows in the basement and the horses on the floor above them. This is the usual

[1] Courtesy of Wallace's Farmer, Des Moines, Iowa.

arrangement in case of bank barns. Where all stock is on the same floor, cows should be in an extension to the main structure. This should be only one story in height with no storage above.

Bank Barns.—The chief advantage in the bank barn is in the ease with which materials are stored by driving the loaded wagons onto the upper floor. This obviates the necessity of hoisting materials to the height necessary in the other forms of barns. The ideal location for the bank barn is on a southern slope, thus facing the barn toward the south with exercise yards also to the south. When so situated the more elevated land to the

INTERIOR OF COW STABLE.[1]

north brings the north wall of the stable below the surface, thus protecting the stable from cold north winds. The chief objection to the basement barn lies in its lack of light and thorough ventilation. This, however, may be largely overcome by not setting the basement too low in the earth and by providing plenty of windows, especially in the east and west walls.

Dairy Barns.—Great improvement has been made in the housing of cows, and much attention is now given to the health of the animals and the production of clean milk, low in its content of bacteria. Best dairymen demand that the cow quarters shall be separated entirely from those of all

[1] Courtesy of The Macmillan Company, N. Y. From "Crops and Soil Management," by Agee.

other stock. The structure should be narrow, housing not more than two rows of cows. The walls, floor and ceiling should be smooth and easily cleaned. For this reason concrete floors that can be frequently washed are preferred. Such floors do not absorb liquids, and if properly cleaned, avoid the objectionable odors so common in stables with wooden or earth floors. Milk is the most widely used uncooked food, and those producing market milk need conditions approaching the ideal for cleanliness in order to secure a high-grade product. Furthermore, the modern dairy cow is bred and fed for efficiency in milk production. This often taxes her health and shortens her life. It calls for the best sanitary surroundings to overcome this drawback.

Storage Capacity.—The storage portion of the barn should connect with one end of the cow barn and should have posts of ample height to store a year's supply of roughage and concentrates for the dairy herd. It should be moderately narrow and have sufficient length to meet the storage requirements. The hay chutes and feed bins should be conveniently placed and connected with the cow stable by suitable carriers, conveyed on overhead tracks.

Economical and Practical Manure Shed.

Silos.—Silos will generally be needed and may be connected with the cow stable through a portion of the storage barn. This prevents the silage odor from permeating the stable and contaminating the milk. It is usually considered best to have the storage structure extend east and west. This permits the cow stable to extend north and south, thus admitting sunshine from both the east and west, enabling it to sweep across all the floor surface during the day. When there is one extension it should connect near the center of the storage barn. When there are two they should connect one at each end of the storage structure, thus leaving an open and protected court between the two cow stables.

Floor Space and Arrangement.—The width of the cow stable should be 36 feet and of sufficient length to accommodate the desired number of cows. The two rows of cows face each other with a spacious feed alley between. Manure alleys of requisite width are located between the gutters and the outside walls. The width and depth of manure gutters, the form of feed troughs and the kind of stanchions, together with many other details, may be obtained from bulletins on this subject.

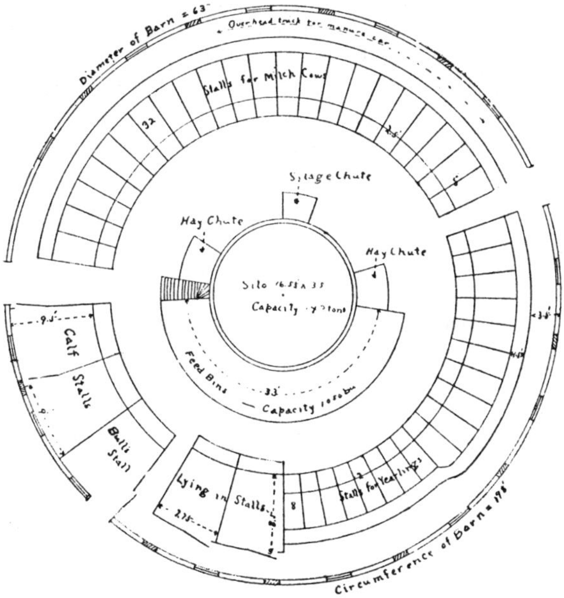

PLAN FOR A CIRCULAR BARN. FLOOR PLAN.[1]

Stable Floors.—Floors that absorb urine and are difficult to clean should be avoided in cow stables. Of all floor materials within reach of the average dairymen, concrete holds first place. It is durable, non-absorbent and can be disinfected without injury. Its chief objection is hardness and smoothness; the former may be partially overcome by the liberal use of bedding. Precautions should be taken when making the floor to leave its surface slightly roughened without interfering with the ease of cleaning. Concrete conducts cold more freely than other floor materials. For this reason it should be underlaid with eight inches or more of rather coarsely broken fragments of rock. The conductivity may be

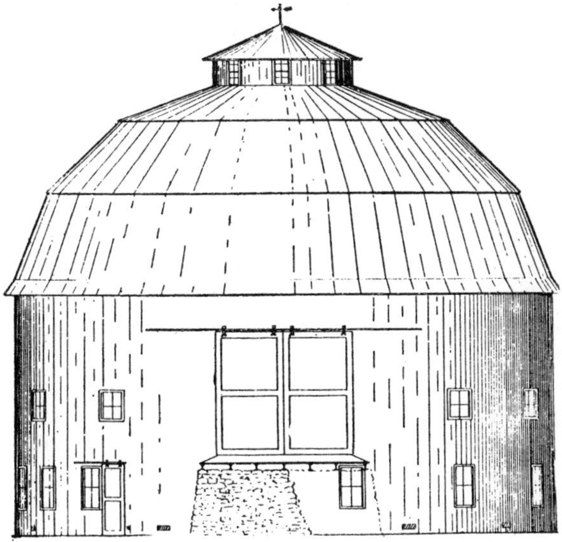

ELEVATION PLAN.[1]

[1] Courtesy of The Pennsylvania Farmer, Philadelphia, Pa.

still further reduced by introducing a thin layer of asphalt or other non-conducting material an inch beneath the surface of that portion of the floor on which the cows lie.

A four-inch thickness of concrete is sufficient. The usual proportion of materials are 1 part of cement, 2½ parts of sand and 5 parts of crushed stone by measure. Screened gravel may be substituted for the stone, or good bank gravel may be used unscreened. Screening is to be preferred, unless the proportion of fine material and gravel is about 1 to 2. A bag of cement is equal to 1 cubic foot. The concrete should be laid in sections, similar to the manner of constructing walks. This provides for seams at reasonable intervals and allows for shrinkage without cracking the cement.

Lighting.—Plenty of light is essential in all portions of a stable where animals are kept or work is performed. Its absence is not only inconvenient, but allows the unobserved accumulation of dust and bacteria. Not only should there be good light, but direct sunshine should also be admitted as much as possible on account of its sanitary effect. The size and location of the windows should permit an abundance of both light and sunshine and provide as great a distribution of the latter as possible. North and south windows are not as effective in this respect as those on the east and west. Windows in cow stables should be screened against flies.

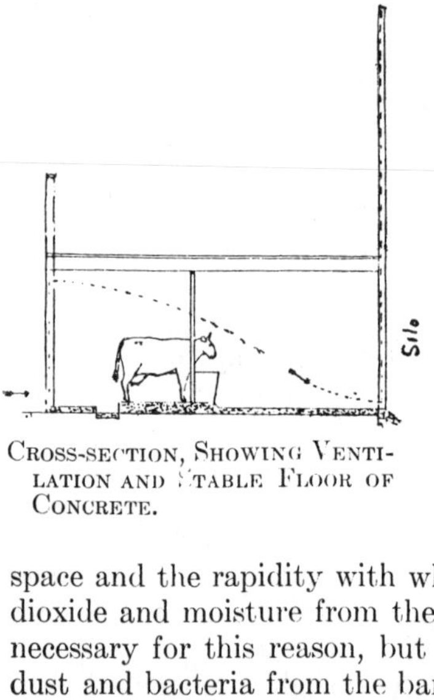

Cross-Section, Showing Ventilation and Stable Floor of Concrete.

Ventilation.—Fresh air is as essential to the health of cows as it is to man. It is necessary to have much better ventilation in cow stables than in dwellings, because of the number of animals within a given space and the rapidity with which the air becomes charged with carbon dioxide and moisture from the lungs of the cows. Not only is ventilation necessary for this reason, but it also sets up currents of air that convey dust and bacteria from the barn.

The King system of ventilation is the one generally used in barns. It is described in Chapter 60.

Professor King, in his book on Ventilation says, "A cow requires six full pails of pure air each minute of the day and consumes twice the weight of air that she does of food and water combined." This gives a basis for calculating the volume of air required daily by each cow, and is used in determining the number and size of ventilating flues necessary.

Conveniences.—The tendency of the times is toward the saving of labor. This should be seriously considered in connection with the arrangement of the stable and the conveniences that should be therein. Canvas extensions to both hay chutes and ventilators are convenient. The former prevents the distribution of dust from hay while feeding. These extensions for both hay chutes and ventilators may be folded and hung against the wall or ceiling so as not to interfere with the stable work.

Closets for harness should be provided. They will prove economical in keeping the harness clean and preserving it. In some instances, a small room in which to hang, clean and repair harness is advantageous.

It will pay to have water delivered by pipes directly to the barn. If it has considerable pressure, a hose can be used in washing the walls and

ENSILAGE CUTTER AND FILLER.[1]

floor of the cow stable. This will necessitate a drainage pipe leading from the stable floor to a suitable outlet.

Silos.—Silos have come into quite general use as a means of storing roughage for cows, steers and sheep. The product of an acre of land can be stored in less space when made into silage than when cured in any other way. Hay stored in the mow will take up about three times the space and cornfodder about five times the space of the same quantity of food material placed in the silo

[1] Courtesy of The International Harvester Company, Chicago.

Corn can be made into silage at less cost than when cured as fodder. There is not only a saving of time, but there is less waste of the crop and it goes to the feed trough in a succulent and more digestible condition than when dry. Crops may be put into the silo under weather conditions that will not make possible the harvesting for putting in the shock or mow. The silo enables the farmer to keep more stock on a given area of land, and is a step in the direction of greater intensity.

There are many forms of silos, but the essential of a good silo is a strong, durable, tight wall that will permit of thorough settling of the stored material. Silos of the circular form are preferred. The greater the depth, the more compactly the material settles, the better it keeps and the larger the quantity that may be stored in a unit of capacity. The monolithic concrete silo is coming into extensive use. It is fireproof, and when properly constructed should last many years. Its first cost is a little greater than a good wooden silo, but it should prove cheaper in the long run. Concrete blocks and tiles are also used for silos and have proven both satisfactory and durable.

The size of the silo will depend on the number of stock to be fed out of it and the length of the feeding period. In northern latitudes this period is seldom less than 200 days. It is usual to feed cows 30 to 40 pounds of silage daily. On the above-mentioned basis, 3 to 4 tons per animal will be required. These figures give a rough basis for calculating the amount of silage required and the capacity of the silo to construct. It is estimated that there should be fed from the surface of the silage about two inches daily in order to prevent the material spoiling. A feeding period of 200 days would, therefore, call for a silo 400 inches in depth, or about 35 feet deep. Silos are often constructed to a greater depth. The following table gives the height and inside diameter of silos in feet, together with the capacity of silage in tons. This will be helpful in connection with determining the size to build.

Height of Silo, feet.	Inside Diameter of Silo.						
	10 Feet.	12 Feet.	14 Feet.	15 Feet.	16 Feet.	18 Feet	20 Feet.
	Tons.	Tons.	Tons.	Tons.	Tons.	Tons.	Ton .
20	26	38	51	59	67	85	105
21	28	40	55	63	72	91	112
22	30	43	59	67	77	97	120
23	32	46	62	72	82	103	128
24	34	49	66	76	87	110	135
25	36	52	70	81	90	116	143
26	38	55	74	85	97	123	152
27	40	58	78	90	103	130	160
28	42	61	83	95	108	137	169
29	45	64	88	100	114	144	178
30	47	68	93	105	119	151	187
31	49	70	96	110	125	158	195
32	51	73	101	115	131	166	205

It should be borne in mind that the deeper the silo the more compact the silage becomes and the greater the weight per cubic foot. In silos of ordinary depth the weight ranges from 30 to 50 pounds per cubic foot, depending on the position in the silo. On an average, a cow requires one cubic foot of silage daily.

Details concerning the construction of different forms of silos may be secured from bulletins issued by a number of state experiment stations and also by the manufacturers of cement.

OUT-BUILDINGS

The out-buildings of the farmstead, consisting of sheds, cribs, milk house, pig houses, poultry houses and other minor buildings, should be

A Good Implement Shed.[1]

grouped with reference to accessibility and appearance. It is worth while in this connection to consider the possibility of fire and fire protection.

The Implement House.—The first essentials of a good implement house are a good, dry floor and a roof and walls that will keep out rain and snow. It should have sufficient strength to withstand winds, ample size for the storage of all machinery without taking much of it apart and freedom from interior posts or obstructions. Such a building need not be expensive. In fact, it should not be expensive if it is to prove a profitable investment. If a comfortable workshop is provided in one end of it where odd jobs of repairing can be done and where a stove can be installed so much the better. Such a provision encourages the proper repair and care of the tools and makes this work possible in weather unsuited to outside work.

[1] Courtesy of Wallace's Farmer, Des Moines, Iowa.

The building should have several wide, rolling doors, and in most instances should be provided with eave-troughs to conduct the water away from its foundation.

Corn Cribs.—The essentials of a good corn crib are a good foundation and a good roof, together with ample capacity and convenience for filling and emptying it To this might be added protection of grain from the ravages of vermin, especially rats and mice. Where much corn is grown, the double crib is preferred. The usual width of each crib is eight feet and the length is made to conform to the amount of corn raised. The advantage of the double crib is that one or more loads may be driven under shelter and unloaded in stormy weather or at leisure. The driveway, after husking time, may be utilized for storing farm wagons or farm implements.

Since corn dumps and elevators have come into quite general use, corn cribs are constructed much taller than formerly. This is economical, since the capacity is materially increased without enlarging either the foundation or the roof, which are the most costly parts of the structure.

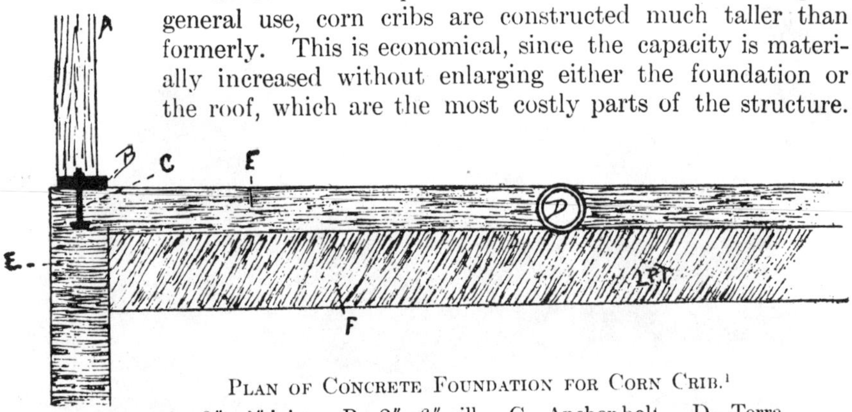

Plan of Concrete Foundation for Corn Crib.[1]
A—2″ x 6″ joist. B—2″ x 6″ sill. C—Anchor bolt. D—Terra cotta ventilator. E—Concrete. F—Broken stone.

Extending the posts and walls from four to eight feet adds very little to the cost in proportion to the increased capacity.

Concrete floors are coming into general use for corn cribs. These are so constructed as to afford no harbors for rats and mice. It is necessary to provide against dampness in such floors by thorough drainage about the walls or by building them up on a considerable thickness of coarsely broken stone. It is also advisable to provide floor ventilation by the use of hollow terra cotta tiles laid in the concrete. The accompanying sketch shows the construction of such a floor. It will be noted that bolts ¾ inch in diameter are set in the concrete to a depth of 4 inches, a 3-inch washer being on the inserted end. The thread end should project above the concrete sufficient to pass through a 2-inch sill and allow a good washer and tap to be attached. The sill fastened in this way holds the crib secure to its foundation.

[1] Courtesy of Wallace's Farmer, Des Moines, Iowa.

Hog Houses.—The profitable production of swine demands dry, sanitary, comfortable housing. Warmth is also essential, especially at the time of farrowing. Early pig production is impossible without warm shelter. The hog house should be conveniently located, but should take an inconspicuous position in the group of farm buildings. Whether the house is stationary or movable, it should be well ventilated and admit plenty of sunlight. The movable type of hog house is coming into quite general use, and has several advantages over the stationary one. In case of disease the houses may be disinfected and moved to new lots, thus escaping the infected ones. They are also very convenient where pasture is depended upon and is changed from year to year. To be serviceable, such houses should be suited to all seasons of the year. During the summer they should be open and afford shade. During the winter or the farrowing season they should be closed and still admit direct sunlight. The accompanying illustrations show two views of the Iowa gable roof hog house. This house meets the requirements named.

A bill of material and estimate of cost of this type of individual house is as follows:

INTERIOR OF DOUBLE CORN CRIB.[1]

BILL OF MATERIAL AND ESTIMATE OF COST.[2]

THE IOWA GABLE ROOF HOUSE.

1 piece 4″ x 4″ x 16′ for runner, fir, $21\frac{1}{3}$ board feet, at \$55 per M	\$1.17
4 pieces 2″ x 12″ x 12′ for floor, No. 1 white or yellow pine, 96 board feet, at \$30 per M	2.88
1 piece 2″ x 4″ x 8′ for floor stiffeners, No. 1 white or yellow pine, $5\frac{1}{3}$ board feet, at \$28 per M	.15
3 pieces 2″ x 4″ x 8′ for rafters, No. 1 white or yellow pine	
1 piece 2″ x 4″ x 8′ for girt, No. 1 white or yellow pine	
1 piece 2″ x 4″ x 10′ for ridge, No. 1 white or yellow pine	
2 pieces 2″ x 4″ x 10′ for plates, No. 1 white or yellow pine	

[1] Courtesy of The Pennsylvania Farmer, Philadelphia, Pa.
[2] Courtesy of Iowa Agricultural Experiment Station.

2 pieces 2″ x 4″ x 8′ for studs,* No. 1 white or yellow pine	
2 pieces 2″ x 4″ x 10′ for studs,* No. 1 white or yellow pine	
2 pieces 2″ x 4″ x 8′ for fender, No. 1 white or yellow pine	
1 piece 2″ x 4″ x 10′ for fender, No. 1 white or yellow pine, $82\frac{2}{3}$ board feet, at $28 per M	$2.32
1 piece 1″ x 4″ x 12′ for brace, No. 1 white or yellow pine, 4 board feet, at $30 per M	.12
5 pieces 1″ x 10″ x 16′ shiplap for ends and sides, No. 1 white or yellow pine*	
1 piece 1″ x 8″ x 8′ No. 1 white or yellow pine	
3 pieces 1″ x 10″ x 10′ No. 1 white or yellow pine, 97 board feet, at $30 per M	2.91
11 pieces 1″ x 10″ x 8′ shiplap for roof, white or yellow pine, $72\frac{2}{3}$ board feet, at $30 per M	2.21
3 pieces 1″ x 4″ x 16′ for bottoms, 16 board feet, at $30 per M	.48
12 eye-bolts at 5 cents	.60
8 U-bolts at 8 cents	.64
5 pairs 12-inch strap hinges at 22 cents	1.10
1 pair 8-inch strap hinges at 18 cents	.18
1 door pull	.10
1 wire for holding door open	.10
12.5 pounds nails at 4 cents	.50
0.6 gallon to paint double coat 150 square feet, at $2 gallon	1.20
Cost of material	$16.66
Labor, 15 hours at 25 cents	3.75
Total cost	$20.41

Further details of this and other forms of movable hog houses may be found in Bulletin 152, Agricultural Experiment Station, Ames, Iowa.

Poultry Houses.—The poultry house should be well lighted and ventilated. The walls should have only one thickness of boards. Double walls afford a harboring place for lice. In cold climates, the boards may be covered on the outside with prepared roofing. This will make a fairly warm house. Chickens can stand much cold if protected from drafts. The interior walls should be smooth and occasionally whitewashed. Good perches should be supported from the rafters and in such a way as to prevent harboring places for lice. A concrete floor is durable, sanitary and easily cleaned. Ventilation may be provided by substituting a muslin-covered frame for one or more of the windows. These may be hinged at the top so as to be swung up out of the way in warm weather. Perches should be at least twelve inches apart and on the same level, otherwise, there will be crowding on the higher perches. A good dropping board should be beneath the perches, and the droppings should be frequently removed with a hoe or scraper. The perches should be in the warmest and lightest part of the house. The nests should be removable and should rest on supports in the darkest portion of the house. If the dropping board is not too low, some of the nests may be beneath it.

Milk Houses.—No matter what type of dairying the farmer follows, if he has many cows, a milk or dairy house becomes a necessity. Milk is easily contaminated by dust and by absorbing odors. It should, therefore, be kept in a pure, clean place. The milk house should not open

* If the sides of the house are built higher than specified to allow of large doorway for tall swine, make due additions in lumber.

directly into the cow stable. The size and equipment of the house will depend on the amount of milk and the manner of disposing of it. When the milk is made into butter or cheese, the size of the house should be sufficient for the proper installation of the separator, churn, butter worker and for the storage of utensils and butter. If steam or gasoline power is used, it should be located outside and a shaft or steam pipe extend into the dairy house. Steam has the advantage of affording heat for warming water and for sterilizing utensils.

The walls of the building should be constructed with reference to keeping as uniform a temperature as possible. These may be of concrete. The floors should always be of concrete.

Two Views of Iowa Gable Roof Hog House.[1]

Ice Houses.—Ice is essential to the proper handling of milk during the summer months. Every dairy farm should have an ice house. In good-sized dairies a thousand pounds of ice per cow yearly is required to cool the milk. In smaller dairies the waste would be greater and proportionately more per cow would be required.

So far as possible the ice house should be located in the shade. It should have double walls and be sufficiently large to store the required

[1] Courtesy of Agricultural Experiment Station.

amount of ice and allow a space of twelve inches between the walls and ice, which should be filled with sawdust or other non-conducting material. Fifty cubic feet should be allowed for each ton of stored ice. The doors should close tightly to exclude air. Windows are unnecessary. A ventilator should be provided at the roof to allow the escape of vapors.

Wooden structures, because of the continual dampness of the wood, are short lived. For this reason ice houses of concrete blocks or hollow tile are preferable. They keep the ice well and are much more durable than wood.

Roofing.—Wooden shingles have long been the chief roofing material. They embody lightness, ease of construction, good appearance and, when made of the right kind of wood and properly treated or painted, are reasonably durable. It is folly to put thirty-year shingles on with five-year nails. The new process nails rust out more quickly than the type made in former years. It is, therefore, recommended that good galvanized wire nails be always used for shingles of any material that is reasonably durable.

A Concrete Block Ice House.[1]

Slate and tile roofing are much heavier than wood shingles, but when of good quality are more durable and generally of better appearance. They have the advantage of affording fire protection from sparks and cinders falling on the roof. Any kind of shingles demands a roof of ample pitch to make them durable. If the roof is too flat, more water is absorbed, snow is held, and consequently decay occurs more rapidly.

There is now on the market prepared roofing of many types, much of which is cheaper and more easily placed in position than slate, tile or shingles. The type of building and its permanence should in large measure determine the kind of shingle. Heavy, expensive roofing is out of place on a cheap, temporary building.

[1] Courtesy of The Pennsylvania Farmer, Philadelphia.

Use of Concrete.—Concrete is durable, easily cleaned, simple of construction and finds many good uses on the farm. It makes excellent foundation for all kinds of buildings, is well suited for silos, outside cellars, water troughs, walks, feeding floors and stable floors. The essential in concrete constructions consists in the use of clean sand and gravel, mixed in the proper proportions with a good quality of cement. The greater the strength required and the more impervious the structure is to be, the larger should be the proportion of cement. For building foundations and walks, the 1 : 2½ : 5 mixture is used. Where more strength is required the 1 : 2 : 4 mixture is preferred. Strength is further increased by iron or steel reinforcement. All overhead work—water tanks, silos, bridges, etc.—calls for reinforcement, the extent of which will be determined by the strain to which the structure is to be subjected. The reinforcing material should be placed where it will be most effective. Concrete is most durable if allowed to dry slowly. It should never freeze until thoroughly dry.

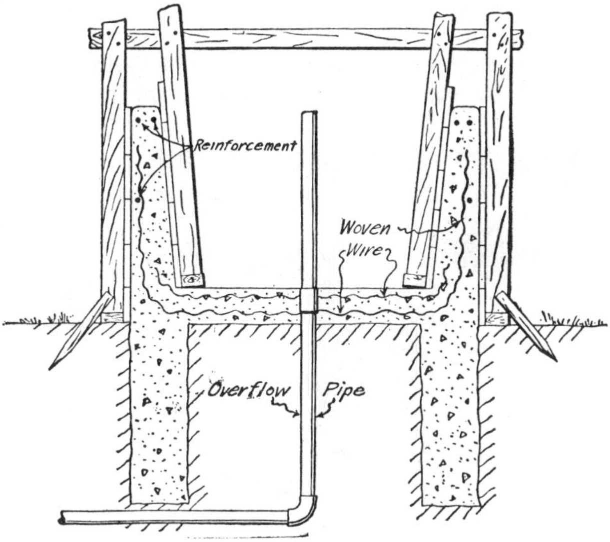

How to Construct a Concrete Water Tank.[1]

Watering troughs should have thick walls and the sides and ends should be sloped on the inside to lessen the danger of bursting by freezing water. It is safest to provide a means of draining the water off during cold periods. The accompanying sketch shows the foundation, drainage pipe, forms and reinforcement necessary in the construction of a concrete water tank.

Both wooden and metal forms are used. The latter are preferable in the construction of silos and round water tanks. Metal forms, when used repeatedly, are cheaper than wooden ones. They leave a smoother concrete surface than wooden forms. The latter should be soaped or greased on the surface next to the concrete to prevent the material sticking to the forms. Wooden forms should also be sprinkled with water before being filled with concrete, lest they absorb water from the mixture too rapidly.

[1] Courtesy of The Pennsylvania Farmer, Philadelphia.

The concrete materials should be thoroughly mixed and enough water used so that the mixture will flow slowly. The smaller the forms into which it is placed, the more liquid it should be. Where much work is to be done, mechanical mixers facilitate the work and do it more thoroughly than can be done by hand. In the absence of a mechanical mixer, a strong, tight board platform, about 8 by 10 feet in dimension, is convenient on which to do the mixing. A square-pointed shovel, a rake and two or more hoes may be advantageously used in mixing the material. If running water is not available, water in barrels or a tank should be convenient to the mixing board. The cement usually comes in bags of 100 pounds each, equal to one cubic foot. Bottomless boxes for measuring sand and gravel are most convenient. They should be constructed of a size suitable for a bag or two-bag mixture of the proportions desired.

One desiring to build should first estimate the cubic space to be occupied by concrete. This known, the amounts of sand, gravel and cement can be easily estimated. For a 1 : 2 : 4 formula, the cement required will equal .058 times the cubic feet in the structure. For the 1 : 2½ : 5 formula, it will be .048 times the cubic feet in the structure. The amounts of sand and gravel will be relatively as much more than the cement as the formula specifies.

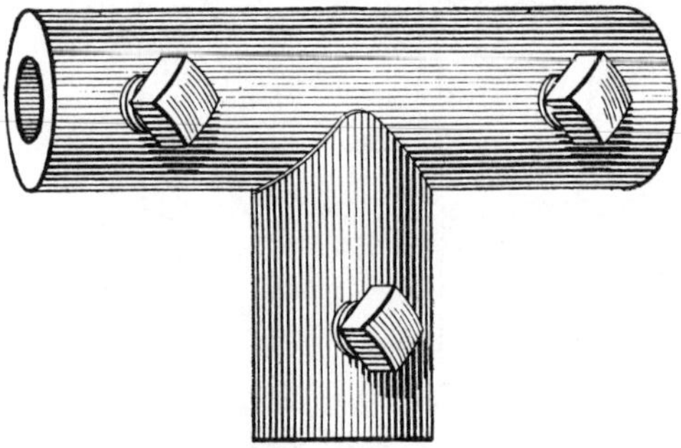

A "T" Connection for Heavy Wire Lightning Rods.[1]

Plans and specifications for structures of different kinds may be obtained from any cement manufacturing company, as well as from bulletins of many of the state experiment stations and from the United States Department of Agriculture.

Lightning Rods.—The larger buildings of the farm group should be protected with lightning rods. The building most likely to be struck by lightning is the barn. Observations show that many barns with entire contents have been burned as the result of lightning. The greatest danger occurs for one or two months immediately after filling the mows with hay. This is due to the accumulation of moisture from the newly-made hay. This moisture fills the peak of the loft, often escaping through the cupola, and increases the conductivity of the air, and in case of a passing thunderstorm attracts the lightning.

Investigations during recent years by insurance companies show that properly installed lightning rods are quite effective as protection against

[1] From Farmers' Bulletin 367, U. S. Dept. of Agriculture.

lightning. Eight years' investigations in Iowa show $4000 worth of damage done to rodded buildings as compared with $340,000 damage to buildings having no rods. In Canada and Michigan investigations show similar results. Professor Day, of the Ontario Agricultural College, states that out of $1000 worth of damage by lightning to unrodded buildings, $999 would be saved if the buildings were properly rodded.

Effective lightning rods for a barn may be installed without much cost. The expensive copper rodding and elaborate system of points and insulators formerly used by lightning rod companies are not necessary. The essentials of a rodding system are metal rods of any good conducting material, sufficiently large to carry a heavy charge of lightning. These should have good contact with moist earth at all times. It is, therefore, well to have the lower ends buried to a depth of three feet or more. On the ends should be a coil at least a foot in diameter. The rods should extend one up each side of the building and over the roof, connecting with a horizontal rod extending along the entire length of the ridge. There should be perpendicular extensions to the horizontal ridge wire at intervals of 15 to 20 feet. These need not be more than 18 inches in length and should be sharpened at the upper end. A terminal point should extend above each cupola, ventilator and chimney on the structure.

No. 3 and No. 4 double galvanized iron telegraph wires make good lightning conductors. The wire may be fastened directly to the building by staples or by means of small wooden blocks and screw eyes. Blocks 1½ inches thick, 2½ inches wide and 4 inches long may be nailed to the side of the buildings and roof at intervals of ten feet or less. The wire can be passed through the eyes screwed into these blocks. The vertical wires and terminals may be connected with the horizontal ridge wire by means of galvanized T's.

The quality and type of rodding system should conform to the nature and character of the building. An attractive system of rodding adds much to the appearance of the building.

Fences and Gates.—The need for farm fences is probably less than formerly. The chief purposes are for the confinement of stock and poultry and for ornamentation. The extensive use of farm machinery and the adoption of systematic crop rotation have reduced the number of fields on the average farm. The increase in the price of land has reduced the acreage used as pasture. As a rule, highway fences, except where pastures border the road, may be omitted. Nothing mars the appearance of a farm more than an untidy fence grown up with weeds. The farmer is benefited and the appearance of the farm improved if unsightly fences are removed and the fields cropped to the border of the road.

The type of fence selected depends much on the service to be rendered. A hog-tight fence is cheapest and most effective when constructed of well-galvanized woven wire. The posts should not be too far apart and the bottom wire should be fastened close to the ground at intervals sufficiently frequent to prevent hogs from springing it and crawling beneath. Woven wire 36 inches high is sufficient to turn the hogs. If the fenced field is to be used for cattle or horses, two barbed wires may be placed above the woven wire. With a little additional expense, a fence 48 or 52 inches high may be secured which will turn all kinds of stock. A single strand of barbed wire, three inches above the top of the woven wire will prevent horses reaching over and stretching the fence.

The top wire of a 48 or 52-inch fence should be of No. 9 wire. Wires below this may be of No. 10 or No. 11 material. Perpendicular wires are sometimes even smaller. The lighter wires are less durable and more easily stretched and broken; consequently, it is economy to pay more fcr the fence and secure a heavier wire. This is especially true if the fence is to be permanent. For temporary fences to be moved from time to time, the lighter wire is more easily handled and stretched.

Stone fences, plank fences and hedge fences, once thought desirable, are now seldom advisable and will not be discussed.

Wooden posts will probably continue to be extensively used, but are being replaced to some extent by metal posts and reinforced concrete posts. Metal posts should be set in concrete. Both metal and concrete are somewhat more expensive then wooden posts and have not been used sufficiently long to determine extent of their durability. Much greater durability is claimed for them than for wooden posts. The chief advantage of the wooden posts is in the ease with which the wire may be fastened to them.

Red cedar posts are to be preferred, chiefly because of their straightness and long durability. Next to red cedar comes the black or yellow locust, catalpa and white oak. Many other kinds of wood may be used. The kind to select depends chiefly on the cost, together with the feasibility and cost of treating the posts to increase their durability. For permanent fences, the best posts are usually the cheapest. Posts of short duration must be replaced frequently, and this adds much to the upkeep cost of the fence.

It generally pays to treat the bottom ends of posts with creosote. The material for this purpose will cost from four to eight cents a post, depending on size. The outfit for treating consists of a metal tank sufficiently large to hold a number of posts, under which a fire may be built and the creosote heated to about 220° F. The well-seasoned posts should remain in the solution two or three hours, after which they are put into cold creosote for an hour or two. Only the lower three feet of the posts need be treated. Posts decay most rapidly at or just beneath the surface of the soil. Such treatment is claimed to add ten to fifteen years to the usefulness of ordinary soft wood posts.

Every farmer should have a wood lot that will supply posts for the farm. Trees cut for posts should be cut the last of July or during August. Trees felled at this time need not be cut into posts at once. In fact, it is an advantage to let them lie until the leaves draw the water from the sap, thus leaving the starch to preserve the wood. At a convenient season the trees may be cut into posts and the posts set on end to further cure. Posts cut in this way last much longer than when the trees are cut in the winter or spring.

The interval between posts in fence construction depends on the size of the posts, the depth to which they can be conveniently set, the weight or strength of the wire and the strain to which it will be subjected.

A Good Type of Farm Fence.[1]

It will often prove economical to alternate small posts with large ones. With exceptionally good strong posts, the intervals may be as much as from 25 to 30 feet. The usual distance, however, will be from 15 to 20 feet.

Woven wire should be stapled to the posts so that the wire will move freely beneath the staple. With barbed wire the staples may be driven tightly so as to prevent the wire from slipping. The length of the staples used and the number per post depend on the hardness of the post and the number of wires. With woven wire it will usually be sufficient to staple alternate wires at each post, although the top and bottom wire should be stapled at every post. When so stapled, the staples should

[1] Courtesy of The American Steel and Wire Co.

alternate on the intermediate wires. For example, the second wire from the top should be stapled to the first, third and fifth post, while the third wire should be stapled to the second, fourth and sixth post, etc.

Woven wire calls for the strongest and best braced end and corner posts. This permits stretching the wire tightly, thus increasing its efficiency. These posts should be set to a depth of four feet in the ground, have cross pieces on the bottom to prevent them pulling up and be securely braced and anchored as shown on preceding page.

It pays to provide substantial, durable gates of light material that may be easily opened and closed. The style of gate should conform to the fence. There are on the market comparatively cheap, tubular, framed woven-wire gates that are light, neat and durable. They may be easily attached to wooden posts. If wooden gates are preferred, 1 x 4-inch material, well braced, is generally better than heavier material. The weight and strength of material, however, will depend on the strain to which the gate is likely to subjected.

REFERENCES

"Successful Houses and How to Build Them." White.
"Farm Structures." Ekblaw.
"The Care of a House." Clark.
South Dakota Expt. Station Bulletin 154. "Pit Silo."
Canadian Dept. of Agriculture Bulletins:
207. "Ice Cold Storage on the Farm: How to Provide."
220. "Lightning Rods: How to Install on Farm Buildings."
Farmers' Bulletins, U. S. Dept. of Agriculture:
367. "Lightning and Lightning Conductors."
387. "Preservative Treatment of Farm Timber."
403. "Construction of Concrete Fence Posts."
405. "Cement Silos."
438. "Hog Houses."
457. "Reinforced Brick Silos."
461. "The Use of Concrete on the Farm."
469. "The Plaster Silo."
474. "Use of Paint on the Farm."
475. "Ice Houses."
574. "Poultry House Construction."
589. "Home-Made Silos."
623. "Ice Houses and Their Use on the Dairy Farm."

CHAPTER 58

FARM MACHINERY AND IMPLEMENTS

During the past century the invention and introduction of farm machinery and implements has almost revolutionized methods of farming. The great change from the simplest of tools to the almost perfect farm machines has had a marked effect upon the life of the farmer. It has shortened his hours of labor, increased his efficiency and brought to him better wages. It has reduced the necessity of brute strength and increased the demand

A GOOD TYPE OF WALKING PLOW.[1]

for a better developed intellect. Mechanical ability is now an essential in farming.

Advantages of Farm Machinery.—Farm machinery has decreased the percentage of people living upon farms in North America. In 1800, 97 per cent of the people lived on farms. In 1850 this proportion had decreased to 90 per cent. In 1900 it was 36 per cent and is now about 33 per cent. At the present time one-third of our population produces the bulk of food supplies and the raw materials for clothing. Consequently the remaining two-thirds are free to engage in constructive work for the advancement of the race.

This decrease in the proportion of people on farms has been accompanied by a great increase in production per capita. In 1800 in the United

[1] Courtesy of Doubleday, Page & Co., Garden City, N. Y. From "Soils," by Fletcher.

States 5.5 bushels of wheat were produced per capita. In 1850 it had fallen to 4.4. About this time improved harvesting and threshing machinery was developed and the production per capita increased rapidly. In 1880 it was 9.16 bushels per capita, and in 1915 it was 10 bushels per capita.

Although the wage of farm labor has doubled or trebled, the cost of production has decreased. The amount of labor required to produce a bushel of wheat by hand implements was a little over three hours. Improved machinery has reduced it to less than ten minutes.

Machinery has also improved the quality of farm products. Shortening the time of operations enables the farmer to plant his crops at the proper time, thus insuring full maturity. Shortening the harvesting period enables him to gather the crop when fully matured and with the minimum loss.

ONE TYPE OF SULKY PLOW.[1]

Tillage Machinery. — The plow takes first rank in tillage implements. It is estimated that more power is required to plow the fields of North America than is used in all the factories. While the plow is a very old implement, the steel plow, the sulky plow and the disk plow are implements of recent development. These are modified in form and construction to adapt them to different kinds of soil and the power available for doing the work. The mold-board plow is most universally used. It should be highly polished and kept reasonably sharp in order to perform its work with the minimum power. Rolling coulters, standing coulters and jointers are attached to more completely cover trash, prevent clogging or reduce the draft.

Disk plows are adapted to a dry soil and to land heavily covered with vegetation. They have been recently modified so that one disk follows the other in such a way that it increases the depth of plowing to 12 or 14 inches and mixes the subsoil with the surface soil.

[1] Courtesy of The Janesville Machine Company, Janesville, Wis.

AN ADJUSTABLE SMOOTHING HARROW.[1]

Mold-board plows are made in sizes ranging from 6 inches to 18 inches. The 12 and 14-inch sizes usually prevail. Where larger plows are needed gang plows are substituted. A gang plow of two 12-inch bottoms will turn 25 to 26 inches of soil at one passage of the plow and generally requires four good horses. It is essential to have the center of draft fall directly back of the center of the team, otherwise there will be a side draft that will increase

SPRING-TOOTHED HARROW.[1]

[1]Courtesy of The International Harvester Company, Chicago, Ill.

the draft of the plow. This necessitates adjusting the team, and if five horses are used better results will be secured by placing two in the lead and three in the rear, rather than five abreast.

Next in importance to the plow comes the harrow. The leading forms of harrows are the smoothing harrow, the spring-toothed harrow and the disk harrow. There are a number of forms and many makes of each. The steel-frame smoothing harrow, made in moderate sized sections, with levers to adjust the angle of the teeth, is most efficient. The teeth should be sharp

DOUBLE DISK HARROW.[1]

in order to do effective work. They should be held in place by clamps that do not easily loosen. When one side of the teeth is badly worn, they may be turned half way around and a new surface brought into use.

The spring-toothed harrow is made with both wooden and steel frames. The better forms also have either adjustable runners or wheels to regulate the depth of harrowing and to hold the teeth out of the ground in passing from one field to another. Without these adjustments, the harrow may be turned upside-down when taken from shed to fields or from one field to

[1] Courtesy of The International Harvester Company, Chicago, Ill.

another. This form of harrow is adapted to stony land, for the destruction of weeds, for a thorough loosening of the soil and for covering broadcasted seeds rather deeply.

Disk harrows are made in two forms: the full disk and the cutaway disk. The former is most extensively used, while the latter is best adapted to stony land and for light work. Double disks frequently combine both forms. They provide for the use of large teams and increased rapidity of work without increasing man labor. Disks of the several forms are used, especially for pulverizing the soil. They should generally be followed with a smoothing harrow. Disks are generally best adapted for preparing the

A CORRUGATED ROLLER.[1]

seed-bed on fall plowing or early spring plowing. They are also extensively used in preparing corn land for the seeding of spring oats without plowing. The disks of these harrows should be kept sharp to do effective work. This is especially true when there is trash on the surface of the soil. The depth of disking is adjusted by the angle at which the disks are set. Levers are provided for setting at different angles. A disk truck reduces the weight on the horses' necks, and is generally advised.

On most farms a combination of the three forms of harrows above mentioned is advantageous.

Under this heading should also be mentioned the roller and the drag. The chief purpose of the former is to compact the soil and crush clods.

[1] Courtesy of The Dunham Company, Berea, Ohio. From pamphlet "Soil Sense."

Seldom should the soil be rolled, except when very dry. Under these conditions it brings the moisture nearer the surface and helps to germinate newly planted seed. The roller is most frequently used in preparing the soil for seeding winter wheat. Rollers of large diameter compact the surface soil without much pulverizing effect. Those of smaller diameter have more pulverizing effect.

The drag or planker is a cheap implement, usually home-made. It is generally constructed of four 8 or 10-inch planks. These are fastened together with two or three cross pieces, to which the planks are securely nailed or bolted in such a way that one plank overlaps the next about one inch. The width may vary from eight to twelve feet. Such a drag requires two or three horses, depending on length. For light work it may be loaded with stones or bags of earth. For heavier work the operator may ride upon it. The drag pulverizes the surface soil, fills up depressions and levels the surface. It is most effective when the surface soil is rather dry.

A HOME-MADE PLANKER.[1]

Cultivators. — There are numerous forms of cultivators requiring from one to four horses, depending on size. These are used for many of the truck crops, for orchards and for general farm intertilled crops such as corn, cotton, cane, potatoes, etc. Cultivators are made both for riding and walking. The number and form of the shovels are determined by the crop to be cultivated and the character of the soil. The size and prevalence of weeds and grass are also determining factors. The large single and double shovels formerly used have largely given place to smaller shovels, disks and sweeps. The small shovels and sweeps are designed for shallow tillage, and are extensively used for both corn and cotton. Such cultivators do little damage to the roots of the crop, make an effective soil mulch, and, if used in the nick of time, destroy all small weeds.

The disk cultivator is better suited for larger weeds and for throwing the earth either to or from the plants.

Numerous forms of hand cultivators are available for garden work. There are also several forms of one-horse cultivators extensively used on truck farms.

The weeder consists of numerous flexible teeth and is designed to break the soil crust and destroy very small weeds when the plants to be

[1] Courtesy of Orange-Judd Company, N. Y. From "Soils and Crops," by Hunt and Burkett.

tilled are small. A variety of tillage implements is advantageous, and the selection should meet the needs of the owner.

Seeding Machines.—Until within the last century much of the sowing and planting of seeds was done by hand. Recently the broadcast seeder has taken the place of broadcasting by hand, and the drill and planter have supplanted hand planting of seeds either in hills or rows. The end-gate seeder, used extensively for seeding oats, and the knapsack seeders, used for grasses and clovers, are an improvement over hand seeding, but are subject to much the same defects as hand seeding. The speed of the distributor, the weight of the seed and the condition of the

A Much Used Form of Corn Cultivator.[1]

wind all affect the distance seed will be thrown. Great care is, therefore, necessary in the spacing of the passages back and forth across the field in order to avoid uneven seeding.

Broadcast seeders with long hoppers carried on two wheels give much better results than the sorts above mentioned. They are provided either with the agitator feed or the force feed. The latter is the more satisfactory. The former has a revolving agitator that passes over each opening from which seed issues and prevents stoppage. The rate of seeding is controlled by adjusting the size of the openings in the bottoms of the hoppers. The seed either falls on a vibrating board or passes through

[1] Courtesy of The International Harvester Company, Chicago, Ill.

fan-shaped spouts that distribute it evenly over the ground. The wheelbarrow seeder used for grasses and clovers has the same arrangement, but is usually without the vibrating board or spouts.

Seeders of the same form, provided with a force feed, are most satisfactory. The force feed can be set to seed at any desired rate and makes uniformity reasonably certain.

Broadcast seeders are sometimes attached to disk harrows. The seed may be sown either in front of or behind the disks. In one case it will be rather deeply covered; in the other it will lie on top of the ground and the disk must be followed with a harrow to cover the seed.

Grain drills came into use to some extent in England soon after 1731, at which time Jethro Tull advocated a system of seeding and tillage called "Horse Hoeing Husbandry." In the United States drills worthy of mention were not perfected until after 1840. Drills are more expensive than seeders, are heavier of draft and seed more slowly. As they have become perfected they have displaced broadcast seeders to a large extent. The chief advantage lies in a uniform depth of planting that may be controlled to suit the kind of seed and the condition of the soil. This insures more perfect germination and requires less seed than when broadcasted. Nearly all wheat is now drilled, and the best farmers also drill oats, rye and barley. Even alfalfa and the clovers are now being drilled with good results.

A Wheelbarrow Seeder in Operation.[1]
An even distribution of grass seed is secured by its use.

There are now several forms of furrow openers for drills. The hoe drill was the first to be developed. It has good penetration and works well on *clean* land, but clogs badly in trash. The shoe drill was next to be developed, but has not been so extensively used as the hoe. Disk furrow openers are of more recent use and both single and double disks are used. They are especially good in trashy ground. Press wheels are sometimes provided to follow the disks and compact the soil over the seed. Covering chains are also used, their sole purpose being to insure covering all of the seed. The several forms of furrow openers are provided

[1] Courtesy of Lowery's Summer School Report.

with a tube through which the grain passes, and these are connected with the seed box by flexible tubes either of rubber or of steel ribbon. Spaces between furrow openers vary from 6 to 9 inches, 7 inches being the most common distance.

Drills are provided with both fertilizer and grass-seed attachments if desired.

The drill compels the farmer to put his land in good condition before seeding and this is another of its advantages. For oats, the drill has very little advantage over broadcasting in wet seasons. On an average, however, drilling oats has increased the yield about three bushels per acre. It will save from one-half to one bushel of seed to each acre.

Grass and clover generally do better with drilled grain than with that broadcasted. The drill should be run north and south so the sun

THE USUAL TYPE OF GRAIN DRILL WITH SINGLE DISK FURROW OPENERS.[1]

can get into the grass. With winter wheat, north and south drill rows generally hold snow better and heave less than rows running east and west. All seed used in drills should be thoroughly cleaned to avoid clogging and insure even distribution. Care should be exercised to adjust the furrow openers so that the seed will be deposited at the most desirable depth. The smaller the seed, the shallower it should be covered. Seed may be covered more deeply in a dry, loose soil than in a wet, compact one.

Corn Planters.—These are strictly an American invention and have been developed within the last sixty years. They have reached the highest stage of development of any of the seeding machinery. The corn crop is so important and is grown on land of such high value that the importance of accuracy in planting is greater than with the small grains. The

[1] Courtesy of The International Harvester Company, Chicago, Ill.

tillage demanded by this crop makes it essential that the rows be straight, and in case it is check-rowed, that the hills be reasonably compact.

The dropping device should be carefully adjusted and the plates selected to drop the desired number of kernels. It pays to grade the seed for uniformity in size. No device can do perfect work with seed corn, the kernels of which vary greatly in size. There are two forms of plates: the round-holed plate and the edge-selection plate. Whichever form is used, the adjustments should be such that the kernels of corn will not be broken.

A Good Corn Planter.[1]

There are four forms of furrow openers for corn planters, viz., the curved runner, the stub runner, the single disk and the double disk. Each has its advantages, depending on character and condition of soil and presence or freedom from trash. Whatever form is used, the seed should be deposited at a uniform depth and properly covered.

There are several forms of planter wheels. Their purpose is threefold: (1) to support the frame of the machine, (2) to cover the corn, and (3) to compress the earth about it. A solid wheel is made both flat and concave on its surface. The concave surface is superior, because it more completely closes the furrow and leaves the track slightly higher in the center than at the sides. The open wheel is also used. This leaves a

[1] Courtesy of Emerson-Brantingham Implement Company, Rockford, Ill. From pamphlet "A Book About Emerson Planters."

narrow ridge of loose earth directly over the corn. This prevents crusting of the soil directly over the seed in case rains follow planting.

Check-rowers are attached to corn planters for the purpose of having the corn plants in rows in both directions. This provides for cross cultivation and is desirable on weedy soil. There are two forms of check-rowers, one in which the wire enters the device on one side of the planter and is left on the ground on the opposite side, where it is gathered up by the planter upon its return. In the other form the wire remains on the side of the planter next to the planted portion of the field. In the first form, the knots on the wire are twice as far apart as the hills of corn, each knot dropping two hills as it passes through the mechanism. In the second form the distance between knots on the wire is the same as the distance between hills.

The best planters are so constructed that the distance between furrow openers and wheels can be adjusted. The adjustment generally ranges from 3 to 4 feet in width. On good soil, corn is generally planted with rows $3\frac{1}{2}$ feet apart.

The seed boxes should have tight covers with good latches. The boxes should be hinged so that they can be inverted to change the plates without removing the corn. This also provides for the quick removal of corn when one wishes to change from one variety of seed to another.

HARVESTING MACHINERY

In no phase of farm activity has there been a greater saving of labor than through the introduction of improved harvesting machinery. In less than three-quarters of a century this phase of farm work has passed from the use of the cradle by which two men by long hours of back-breaking work could cut and bind an acre and a quarter of grain in a day, to the eight-foot self-binders, by which one man and three horses can cut and bind fifteen acres in a day. Not only is much more accomplished, but the work is better done.

Mowing Machines.—The side-cut mowing machine, in spite of its side draft, has not been displaced by the direct cutting machine. The two-horse mowing machine with a six-foot cutting bar is generally preferred. While there are a number of makes of mowing machines, selection should be made to fit the character of work to be done. The machine should be no heavier than is required for the work it is to do. The important parts of the mowing machine are the cutting device, consisting of the cutting bar, guards and sickle, and the transmission gearing which transmits the power of the team from the wheels to the cutting device. Ample adjustment should be provided for regulating the height of cutting and also for quickly elevating the bar to avoid obstructions in the field.

It is important to keep all bearings tight and thoroughly oiled. This increases the length of life of the machine and promotes efficiency. The sickle knives should be kept sharp and should be held firmly against

CORN HARVESTER WITH BUNDLE ELEVATOR.[1]

[1] Courtesy of The International Harvester Company, Chicago, Ill.

the ledger plates. Damaged plates or badly worn and broken knives should be promptly replaced by new ones.

The Pittman bearings are the ones most likely to become loose. This will give rise to pounding, which will wear the bearings rapidly. The bearings of the Pittman at both the sickle head end and the Pittman crank end should, therefore, be of easy adjustment.

Self-Rake Reaper.—This machine soon followed the improvement and development of the modern mower. It was extensively used for a short period, but was soon displaced by the self-binder. The self-rake reaper is still a desirable machine for harvesting such crops as flax, buckwheat and clover for seed. These crops, when harvested, cling together

A Mowing Machine with Pea Vine Attachment.[1]

and there is little advantage in having them bound into bundles. This machine, therefore, does the work of harvesting these crops at less initial cost of machine and a further saving in twine. Since the mowing machine and the modern self-binder are both required on most farms, the self-rake reaper is now generally dispensed with, unless the acreage of the above-mentioned crops is large.

Self-Binder.—This machine has been developed since 1875, and is now almost universally used in harvesting small grains. There are a number of different makes, but the most satisfactory ones are built principally of steel, combining strength with lightness of weight and durability. The essential parts consist of the cutting device, the elevators and the

[1] Courtesy of F. Blocki Manufacturing Company, Sheboygan, Wis.

binding apparatus. To these may be added the reel with its several adjustments and the bundle carrier. There are numerous details which will not be described here. The precautions advised relative to the working parts of the mowing machine apply with equal force to the self-binder. Various parts of the binding apparatus must work in harmony and be so timed that each part will do its work at exactly the right moment. In order to operate the self-binder satisfactorily, one should understand the working of the various parts and be capable of adjusting them.

The canvas elevators should be neither too tight nor too loose to insure good work. They should be loosened when the machine stands in the field over night. If rain threatens, it is wise to remove them or cover the machine to keep them dry. Their usefulness will be greatly lengthened by removing them from the machine, rolling them so mice cannot enter the folds and storing in a dry place at the close of the harvesting season.

The best way to keep the self-binder in first-class condition is to oil all wearing parts as soon as the harvest is over and store the machine under shelter at once. If work is not rushing at this time, repairs should be made while the farmer knows how the machine has been running and what parts need repairs. If these precautions are not taken, three or four times as much labor will be required to remove the rust and get the machine to operating smoothly the following season.

One should always have on hand a small supply of knife blades and rivets, extra links for the chains that are likely to break and a few extra small bolts and taps. It is essential to have with the machine suitable wrenches, pliers, a cold chisel, screwdriver and hammer. The frequent oiling of all bearings is necessary.

Corn Harvesters.—The modern corn harvester is the outgrowth of the self-binder. It combines the same principles in both cutting and binding apparatus. The apparatus for conveying the stalks to the binder is very different from that of the self-binder. The various parts of the machine are much stronger than those of the self-binder, in order to handle heavy green corn without straining or breaking the machine. It is designed to cut one row of corn at a time and is now extensively used in cutting corn for the silo as well as cutting more mature corn for shocking in the field.

This machine costs equally as much as the self-binder, and is an economical investment where there are twenty acres or more of corn to be harvested.

Threshing Machines.—The modern threshing machine has reached a high stage of development and does all the work of separating the grain from the straw, cleans the grain of chaff and foreign material, delivers the grain to bag or wagon and the straw to stack or mow without its being touched by the hands of man after it is forked from the wagon to the self-feeder and band cutter.

Since the average farmer does not own a threshing outfit, it is not necessary for him to understand the details of it. Threshermen would not be satisfied with the brief description that space will permit in this chapter. They can secure ample information from the threshermen's books published by threshing machine manufacturing companies.

The clover huller is a modified threshing machine and is generally owned and operated for a community by the owners of a general thresher or corn-sheller outfit.

Small threshing machines are manufactured for individual farmers, and may prove economical for farmers in the eastern section of the

AN UP-TO-DATE THRESHING MACHINE.[1]

United States, where it is the custom to store the sheaf grain in large barns and thresh it in the winter time. The essential points in operating the thresher are the speed of the cylinder, which should be uniform, the setting of the concaves, and the number of teeth in it so as to remove all grain from the heads, the speed of the fan, and the selection and adjustment of the sieves, so as to clean the grain without blowing any into the straw. Rapid and satisfactory work necessitates ample power. The power may consist of steam, gasoline or electric motors, and should be adapted to as many other uses as possible.

[1]Courtesy of The International Harvester Company, Chicago, Ill.

Corn Shellers.—In the corn belt, large corn shellers are used for shelling nearly all corn that goes to market. They are owned and operated for community work the same as threshers.

Many small hand and power corn shellers are used on farms for shelling corn for feeding purposes. There are two general forms, viz., the spring sheller and the cylinder sheller. All hand shellers are of the first-named type, but some of the power shellers are of the second type. The latter are cheaper and of simpler construction, and seldom get out of order. They break the cobs badly and small pieces of cobs are more numerous in the corn than when spring shellers are used. For this reason,

Four-hole Mounted Belt Corn Sheller with Right Angle Belt Attachment.[1]

the spring sheller is considered superior. The unbroken cobs are much better fuel.

The larger shellers of both types are provided with a cleaning device which separates chaff, husks and cobs from the shelled corn, and elevators which elevate both shelled corn and cobs.

In order to do good work, corn should be reasonably dry when shelled. It is impossible for the sheller to do satisfactory work when corn is so damp that the kernels are removed with difficulty. Furthermore, such shelled corn will heat or spoil when placed in storage. Corn

[1] Courtesy of Sandwich Manufacturing Company, Sandwich, Ill.

shells most easily when the temperature is below freezing, especially if inclined to be damp.

Silage Cutters.—A silo may now be found on nearly every dairy farm; consequently, silage cutters are in much demand and have been greatly improved in recent years. The essential parts of the silage cutter are the feeding table, provided with an endless apron which feeds the corn into the cutting apparatus, the cutter head and the elevator. There are two types of cutter heads: one with radial knives fastened directly to the flywheel; the other with spiral knives fastened to a shaft. The modern elevator consists of a tight metal tube, through which a blast of air is driven by a fan. This blows the cut corn to the top of the silo, frequently having an elevation of 40 or more feet. It is a good plan to have a movable cylinder, either of metal or canvas to descend in the silo nearly to the surface of the filled portion. A man in the silo can move this to any point, thus keeping the surface level and avoiding a separation of the lighter and heavier portions. This not only saves labor, but provides for uniform settling of the silage.

The cutter knives should be kept sharp and be carefully adjusted so as to have a close shearing effect. If they are too loose, the material will be broken instead of cut, thus requiring more power. If the knives press against the ledger plate with too much force, there is undue friction and wearing of the knives.

The cut corn leaves the silage cutter coated with juice, and acids frequently are developed, thus causing rapid erosion and rusting of all metal parts. It is, therefore, advised to run a few forkfuls of hay or straw through the cutter to remove this material, thus leaving it in a dry condition.

Manure Spreader.—A manure spreader should find a place on every farm where there are 100 loads of manure to spread annually. It not only reduces the work of spreading the manure, but spreads it more evenly and with more rapidity than can be done by hand. Careful experiments show that light applications of manure for general farm crops bring better returns per unit of manure than heavier applications. Manure spreaders make the manure cover more land, thus increasing the returns.

The essentials of a good manure spreader are strength, ample capacity and an apron that will not clog or stick, together with a beater that will spread the manure evenly. The machine should be capable of adjustment so that any desired amount may be applied. The gearing should be covered so as to protect it from the manure. Spreaders are of heavy draft, and may be provided with shafts so that three horses may be used.

It saves time to have the spreader so placed that the manure carrier may be dumped directly into it. When filled, it may be hauled to the field, the manure spread and the spreader returned for refilling. Good farmers find it economy to provide a cement floor, slightly hollowed in the center, on which the spreader stands. This saves the liquid which may drain from the spreader, and the overflow of manure that sometimes occurs. If this is covered with a roof the spreader is protected and leaching is prevented. If such a shed is sufficiently large, it may serve as a storage place when there are no fields on which manure may be spread.

Milking Machines.—These have been rapidly improved within the last few years, but have not come into very general use. For economical use, they require power and tubing for suction in addition to the apparatus proper. They should, therefore, be most economical in large dairies where

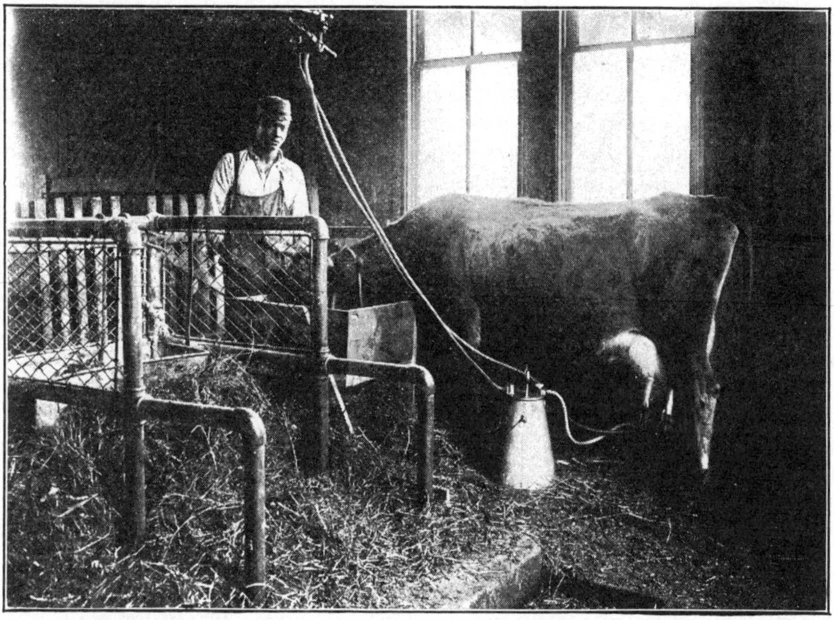

MILKING MACHINE IN OPERATION.[1]

the power can be utilized for other purposes as well. The chief advantages of the milking machine are the saving of time in milking and cleaner milk. Cleanliness of milk demands that the apparatus be kept sterilized and clean. The machine should be washed with soda and hot water and all metal parts boiled for half an hour. The rubber parts will not permit of boiling. It is recommended that they be hung in a tank of water containing about 7 per cent of salt and 0.75 per cent of chloride of lime.

The labor saved in milking by the use of the machine may be offset by the extra work in operating and caring for the apparatus. In large dairies, where stablemen are required to do no other work, this is not a

[1] Courtesy of The College of Agriculture and Kentucky Agricultural Experiment Station, Department of Animal Husbandry, Lexington, Ky.

serious objection, since the average man can feed and care for more cows than he can milk by hand during the milking period.

Spraying Machines.—On all truck and fruit farms spraying machines are a necessity. The size and kind of outfit will depend on the size of business and character of plants to be sprayed. Wherever there are more than eight or ten acres of orchard, a power sprayer mounted on wheels is recommended. Those which develop power from the wheels are cheapest, but are not so satisfactory for spraying large trees. A high-grade gasoline engine and a good tank for compressed air provide a uniform pressure under all conditions. Good work demands a pressure of from

A Power Sprayer Routing Orchard Pests.

90 to 125 pounds. Good nozzles that will give a fine spray without clogging are essential. There should be an agitator in the receptacle that holds the spraying material. The hose attachments should be ample in length to reach all parts of the trees.

Horses attached to the sprayer should be protected by suitable covering.

For small orchards or for small fruit, the barrel sprayer with hand pump, mounted on a sled, will serve the purpose. Knapsack sprayers may meet the needs for garden purposes, and are also useful in connection with larger outfits. They are suited to spraying the base of trees for mice, rabbits and borers. They are also good to spray young plants and for shrubs and bushes around the home.

Tractors.—The rapid development of small tractors adapted to a wide range of uses on the moderate sized to small farm is certain to displace considerable of the horse power within the next decade. The advantages of tractors lie in the saving of time and in the fact that they are of little or no expense when not in use. With present prices of horse feed and fuel for tractors, whether it be coal, crude oil or gasoline, the tractor furnishes power at less cost than the horse.

The motor truck is recommended for farmers having much marketing to do, especially if the distance from market is great and roads are suitable for such a vehicle.

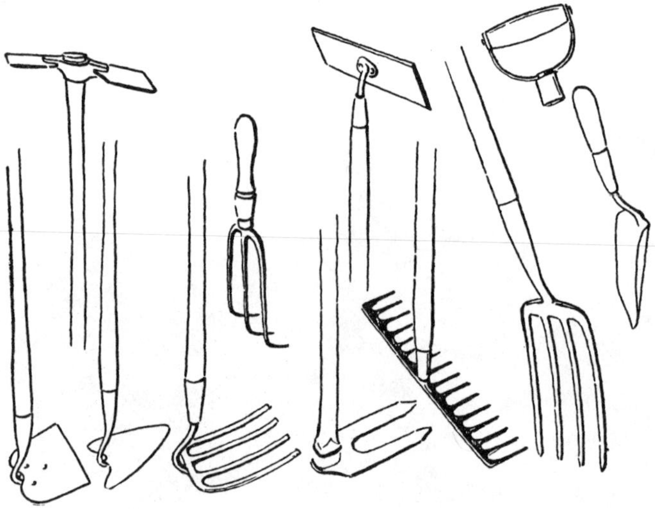

A Collection of Useful Hand Implements.[1]

For a fuller discussion of farm motors and tractors, see the following chapter.

Farm Vehicles.—Farm wagons should be selected to suit the character of work to be done, and be adapted to the character of roads in the vicinity. Wide tires are recommended for farm use and for dirt roads. Under most conditions they are lighter of draft and injure roads and fields less than do the regulation narrow-tired wagons. It pays to buy the best makes of wagons, to provide shelter for them and to keep both running gear and boxes well painted.

A low-wheeled running gear on which may be placed the regulation wagon box or hay rack finds favor on most farms. It saves much lifting.

[1] Courtesy of The Macmillan Company, N. Y. From "Soils," by Lyon and Fippen.

A light runabout, suitable for one horse, is useful on nearly every farm. A carriage or surrey should be provided for the pleasure of the family.

The automobile is now displacing the carriage or surrey to a considerable extent. It serves for both business and pleasure and is a great saver of the farmer's time where considerable distance and frequent trips are involved. The automobile costs little or no more than a good driving team and carriage, and should be less expensive to maintain.

INTERIOR OF A WORKSHOP WITH A $25.00 OUTFIT OF TOOLS.[1]

Hand Implements.—The number and variety of hand implements found on a farm will be determined by the type of farming. They will be most extensively needed on truck and fruit farms. Several forms of hoes, suited to the different kinds of work, are necessary. The hand rake, spades and shovels should be of a type best suited to the work to be done. It pays to keep hand implements sharp and well polished. One can not only do more work with a sharp, well-polished hoe than one can with a dull, rusty one, but pleasure is added to the work.

There should be an ample outfit of barn implements suited to the kind of feed to be handled and the cleaning of the barn. These should

[1] From Farmers' Bulletin 347, U. S. Dept. of Agriculture.

include suitable brooms and brushes for sweeping dry floors, shovels of the size and form suited to the kind of floor and also the gutters. Good currycombs and brushes, always in their place when not in use, insure better care of the stock.

Tools.—The most used forms of carpenter's tools should be found on every farm. There should be a small shop in which to keep them and where they may frequently be used. The ax, hatchet and two or more kinds of hammers, the cross-cut and the rip saw, a brace and suitable outlay of bits, and one or more good planes will frequently be needed. There should also be a suitable collection of files, punches, pliers and wrenches. Both flat and three-cornered files will be found useful. The bastard and second-cut are the grades of files most needed for general work. Cold chisels and a few wood chisels will also be useful. There are many other small tools that can be added to the outfit as needed. The extent of the outfit will be determined by the extent and character of the farm machinery, the mechanical ability of the farmer and the accessibility to local repair shops.

HOME-MADE BARREL CART FOR HAULING LIQUID FEED.[1]

Handy Conveniences.—There are innumerable conveniences, many of which are home-made, that find much use on the farm. Among these may be mentioned the various forms of eveners and double-trees, suitable to three horses or more, and made to suit the character of machinery on which used.

A pump with hose attachment, fastened to a board, may be placed across the wagon bed and is very handy in filling barrels from a stream or shallow well. A derrick of suitable height is useful in the home butchering of hogs, sheep, calves or beef animals. A hoisting apparatus suitable for putting hay into the mow or stack should find a place on nearly every farm.

The wagon jack will make the work of greasing wagons and other vehicles easy.

A hand cart and a wheelbarrow are frequently needed. Suitable

[1] Courtesy of The Pennsylvania Farmer.

carriers operated on tracks in the barns are superior to the wheelbarrow for conveying feed to mangers and manure to the spreader or manure pit, but are more expensive.

Standard measures for carrying and measuring grain are always useful. These may be in the form of good splint baskets or as metal measures with handles.

Machinery for the House.—The weekly wash for the average farm family, when done in the old-fashioned way, is a laborious task. It can be greatly lightened by the use of the washing machine, wringer and mangle that are operated by mechanical power. A laundry, with modern equipment, is of more urgent need in the country than in the city. Power for such a laundry may be used for other purposes, such as pumping water for a pressure system, operating the cream separator, churn and possibly a suction cleaner. There are too many farmers who are able to supply such an equipment who are content to permit their wives to do this work in the old-fashioned way. It is safe to predict that if these duties were to fall to the lot of the farmer himself, he would find a way to do the work more easily and quickly.

HOME-MADE DUMP CART TO MAKE STABLE WORK EASIER.[1]

There are on the market many labor-saving household implements, including power churns, cream separators, sewing machines, meat cutters, vacuum cleaners, etc. Wherever electricity is available, electric irons and other electrical devices help to lighten the work.

If water must be pumped or drawn from the well by the housewife, no reason exists why a pipe could not be extended and a pump placed in the kitchen or a pump house connected with the kitchen.

[1] Courtesy of The Pennsylvania Farmer.

Buying Farm Machinery.—The farmers of the United States spend more than $100,000,000 annually for the purchase of farm machinery. The average life of such machinery is about ten years. Its durability could doubtless be much lengthened if it had better care.

It generally pays to buy the best makes of machines, even though the initial cost is greater than that for cheaper ones. Whether or not it pays to buy a machine depends on the amount of work for which it can be used. If the amount of work is small, it is frequently cheaper to hire a machine than to own one. In some localities the more expensive machines are owned jointly by two or more farmers.

It requires good judgment to know when to replace an old machine with a new one. Frequently machines apparently worn out may be made to work as good as new by replacing badly worn parts. On the other hand, some machines go rapidly out of date because of important improvements. A new machine may, therefore, be purchased to advantage and the old one discarded even though not worn out. There is a tendency on the part of too many farmers to get along with the old machine at a sacrifice of much time spent in continual repairing.

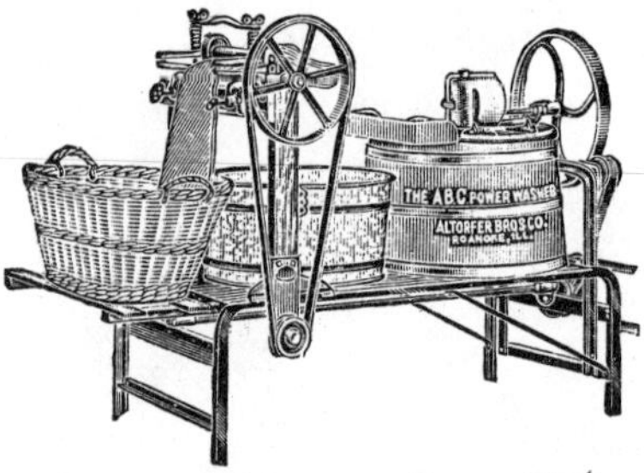

A WASHING MACHINE SAVES MUCH HARD WORK FOR THE HOUSEWIFE.[1]

Care of Machinery.—Every farmer should have a shed large enough to house all his farm implements. This may be a cheap structure, the two essentials being a dry floor and a good roof. There should be sufficient room to store the implements without taking them all apart. It is well to arrange them in the shed when time is not pressing, so that those first needed in the spring are most accessible.

The woodwork of all machinery should be painted whenever it shows need of it. This should be done in leisure time. All machinery should be examined and nuts and bolts tightened. The metal parts, such as the surface of plow bottoms, cultivator shovels, the disks of disk harrows, drills and cultivators should be greased, either with kerosene and tallow or cheap axle grease, as soon as their work is done. This prevents rusting and is easily removed when the machine is again needed for use. Although paint is sometimes used for this purpose, it is not advised, as it is too difficult to remove.

Condition of Machinery.—Every farmer realizes the importance of having all machinery and implements in good working order. This pertains to the adjustment of all complex machinery and applies also to the adjustment of clevises on plows, so that they will run at the proper depth. A machine out of adjustment not only does its work poorly, but

[1] Courtesy of Altorfer Bros., Roanoke, Ill.

generally requires more power to operate it. Some one has well said, "Constant vigilance and oil is the price of smooth-running, efficient farm tools, and to spare either is dangerous as well as expensive." Saws that will saw, knives that will cut, hammers that will stay on their handles, are much to be preferred.

Utilizing Machinery.—A full equipment of farm machinery costs so much that interest and depreciation are a burden for the small farmer. This may be overcome by joint ownership of the more costly machines. Large farms can own a complete outfit and utilize it quite fully. The smaller the farm the greater the machinery cost per acre. On small

WHERE DO YOU PREFER TO KEEP YOUR IMPLEMENTS? UNDER THE SKY?

farms the use for certain machinery may be so small as to make ownership unprofitable.

The greater the skill and higher the wage of workmen, the greater the necessity of using the best and most efficient machinery.

For the general farmer tools that are adjustable and can be used for several purposes are advantageous. A combined spike and spring-toothed harrow that may be changed from one to the other by the use of two levers often saves an extra trip to the house or prevents one being used where the other would have served better. The same principle applies to cultivators where gangs or shovels can be changed for disks or sweeps.

Cost of Farm Machinery.—The principal items in the cost of farm machinery are depreciation, interest on the capital invested, cost of repairs,

[1] Courtesy of Wallace's Farmer.

oil and labor in caring for machinery, together with the proper housing of it. When these costs are figured on the acre basis the rate varies inversely in proportion to the acres covered. Low cost, therefore, is associated with the fullest possible utilization of the machines. It is significant that the high-priced machines are usually those used for the shortest period.

The method of computing the cost of farm machinery is well illustrated in the accompanying table taken from the *Tribune Farmer:*

TABLE SHOWING METHOD OF FINDING THE COST OF USING FARM MACHINERY:

IMPLEMENT.	Date of Purchase.	Purchase Price.	Estimated Life.	Actual Life.	Approximate Average Value for Life.	ANNUAL COSTS.				1910, Hours Used.	Cost per Hour.
						Five Per Cent Interest.	Depreciation.	1910, Repairs.	Total Cost.		
Two two-horse walking plows	1902–04	$24.00	16.4	..	$13.00	.65	$1.46	$0.95	$3.06	344	$0.0089
Spring-tooth harrow	1902	14.40	11.4	12	7.50	.38	1.26	.33	1.97	242	0.0073
Spike-tooth harrow	1903	12.00	20.0	..	6.50	.33	.60	.25	1.18	78	0.0151
Roller	1903	12.00	15.0	..	6.50	.33	.80	3.25	4.38	82	0.0534
Weeder	1906	9.00	17.5	..	5.00	.25	.52	.32	1.09	25	0.0436
One-horse plow	1905	7.50	16.4	..	4.00	.20	.46	.18	.84	10	0.0840
Two riding cultivators {	1903	31.00	13.7	..	38.00	1.90	5.33	4.30	11.53	154	0.0749
	1906	42.00	....	..	.....	....	....	....	.....	...	......
Grain binder	1902	125.00	12.5	..	64.00	3.20	10.00	2.98	16.18	28	0.5773
Grain drill	1904	70.00	14.8	..	37.00	1.85	4.73	3.50	10.08	59	0.1710
Fanning mill	1904	25.00	21.8	..	13.00	.65	1.15	.65	2.45	35	0.0700
Mower	1903	38.00	12.8	..	20.00	1.00	3.00	1.37	5.37	16	0.3356
Hay rake	1903	18.00	12.8	..	9.50	.48	1.40	.50	2.38	16	0.1488
Orchard sprayer	1908	80.00	10.0	..	42.00	2.10	8.00	5.25	15.35	44	0.3490
Gasoline engine	1908	200.00	13.5	..	105.00	5.25	14.83	5.10	25.18	128	0.0200
Harnesses	1902–03	83.50	16.2	..	43.00	2.15	5.15	5.60	12.90	3192	0.0037
Wagons, boxes, racks	1903	110.25	20.5	..	58.00	2.90	5.38	4.75	13.03	250	0.0521
Hay slings, fork track	1909	50.00	20.0	..	26.00	1.30	2.50	.50	4.30	40	0.1075
Miscellaneous minor equipment	1902	386.00	10.0	10	197.00	9.85	38.60	1.38	49.83	78*	0.6400*
Total cost	....	$1,337.65	12.7	..	$695.00	34.77	$105.17	41.16	$181.10	....	......

Numerous records of the cost of farm machinery show that the annual cost per farm is about one-quarter of the actual value of the machinery for the year involved.

Farm surveys in Wisconsin indicate that too many farmers economize on their farm equipment to such an extent that efficiency is sacrificed and profits are below what they would be with a more modern and efficient equipment.

Duty of Farm Machinery pertains to the amount of work each machine will do daily or for the season. Manufacturing concerns standardize different operations in their shops as much as possible. This enables them to estimate very closely the amount of work that can be turned out in a given time, and makes it possible for them to state to customers when a stated task can be completed. It is just as essential for the farmer to

* Miscellaneous minor equipment charges are distributed on the basis of the total productive area of the farm, 78 acres. In this group all machinery and small tools not specifically mentioned are included.

standardize his various machines in order to know what machinery will be required for his various operations.

There are many factors influencing the duty of a given machine, such as the speed of the team, the weather conditions and the condition of the ground. On an average, the daily duty of a machine in acres is equal to the width in feet times 1.4. In other words, a 12-inch plow will average 1.4 acres per day. A 6-foot mower will cut 8.4 acres per day. The size of fields will also influence the duty, since small fields require more turning and loss of time.

Careful investigations in Minnesota and Ohio show that in the former state the acre cost of corn machinery is $1.07, while in the latter it is only 49 cents. The lower cost in Ohio is due chiefly to the relatively larger acreage of corn per farm and the fuller utilization of machinery.

REFERENCES

"Farm Machinery and Farm Motors." Davidson and Chase.
Kentucky Expt. Station Bulletin 186. "Mechanical Milker."
New York Expt. Station Bulletin 353.
Ohio Expt. Station Bulletin 227. Circular 98.
U. S. Dept. of Agriculture, Bureau of Plant Industry, Bulletins: 44, 212.
Farmers' Bulletin 347, U. S. Dept. of Agriculture. "Repair of Farm Equipment."

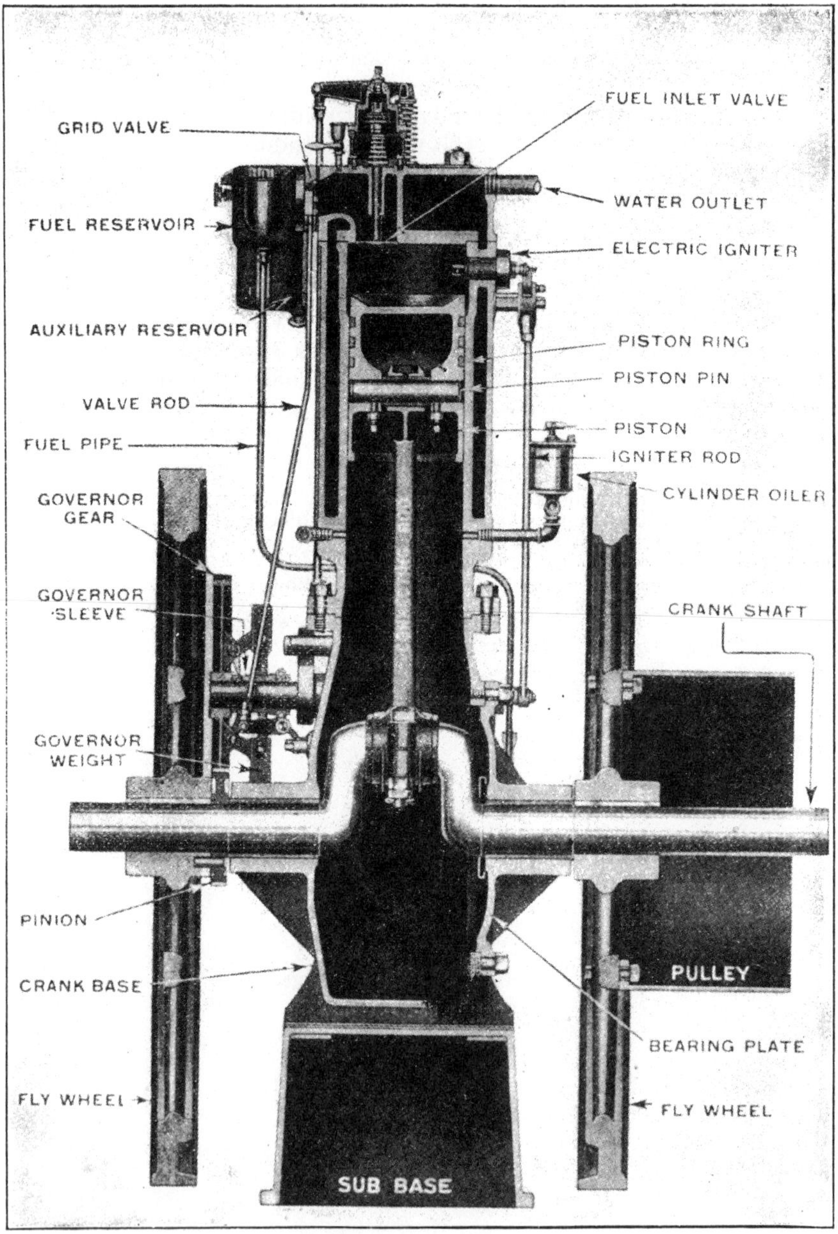

Sectional View of a Four-Cycle Vertical Gas Engine.[1]

[1]Courtesy of Fairbanks, Morse & Co., Chicago, Ill.

CHAPTER 59

ENGINES, MOTORS AND TRACTORS FOR THE FARM

By R. U. BLASINGAME
Professor of Agricultural Engineering, Alabama Polytechnic Institute

THE REAL POWER FOR THE FARM

The real call of the farm is for power, some means by which the skill of a single man can direct a force that will do as much work as a score or more men could do unaided. From plowing to the feed trough, it takes $4\frac{1}{2}$ hours work to raise one bushel of corn by hand. The use of improved machinery and the multiplicity of power has reduced this figure to 41 minutes.

Various forms of power, such as the treadmill, the sweepmill and the windmill, have all failed in many respects. Windmills are objectionable because they are not portable, they are not steady in power and are often wrecked by the wind. The sweep power is hard to move, cumbersome and requires the operators to be exposed to many storms.

The steam engine, but for the close attention it requires, might be the real power needed for farm purposes. Electricity, when correctly installed, is safe, efficient and convenient, but for farm purposes where all jobs are not under one roof as in factories, the lack of portability makes it inconvenient.

The gasoline engine is the only power at the present time that embodies all the requirements for farm purposes. The operator of such power needs no greater mechanical training than should be necessary to properly operate a grain binder. If power is needed in the laundry room, a small engine might easily be transported to run a washing machine. If it is needed in the furthest corner of the wood lot, it can be conveyed to that place without a second or third trip for water and coal, as would be required for a steam engine. In the coldest, driest and calmest weather the gas engine produces power without delay. It can be obtained in units of from one-half horse power to any size that might be required for any farm job.

In parts of the West where the gas engine is best known, it is plowing, harrowing and seeding in one operation by the square mile instead of by the acre, and is doing the work better quicker and cheaper than it could be done by horse or steam power.

Gas Engine Principles.—There are two distinct types of gas engines on the market at the present time which are used for agricultural purposes; the four-stroke cycle and the two-stroke cycle engine.

The four-stroke cycle or four-cycle engine requires four strokes in order to get one working stroke. These strokes are as follows: The intake stroke, in which the charge of air and gas is mixed in the right proportions to give an explosive mixture. The second stroke compresses the charge of air and gas which was previously drawn into the cylinder. The third stroke is the working one in which the compressed charge of air and gas is exploded and the energy hurled against the piston head. The fourth stroke is the exhaust, or elimination of all the old gases which were burned. Therefore, the four-cycle engine requires two revolutions of the fly wheel to complete the four strokes necessary for obtaining power from this type of engine. The four-cycle engine requires two openings which are provided with valves held tightly in place by springs. These valves are operated by mechanical means, although in some engines the intake valve is operated by suction.

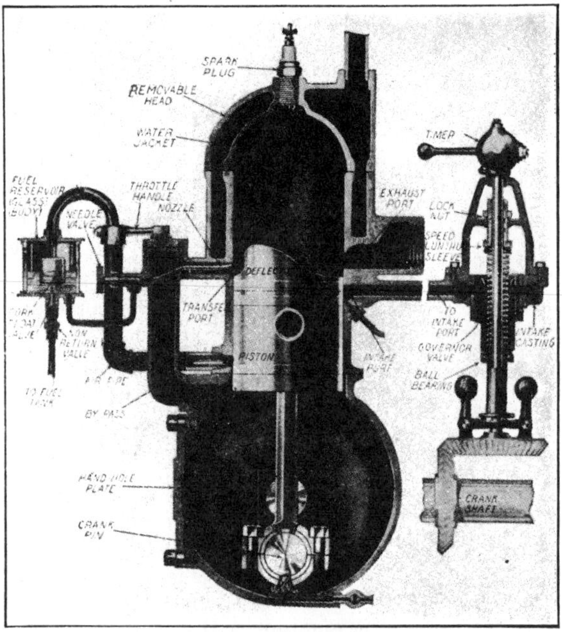

Sectional View of a Two-Cycle Engine.[1]

The two-stroke cycle or two-cycle engine requires two strokes of the piston in securing one working stroke. Therefore, this engine theoretically receives twice the power per square inch hurled against the piston that the four-cycle engine does. The crank case of such an engine must necessarily be airtight, because the charge of air, or sometimes a mixture of air and gas, is brought into this part on the up-stroke of the piston and on the downward stroke the burned gas passes out of the exhaust port while the new gas from the crank case enters the combustion chamber. It is, therefore, entirely necessary that the crank shaft which runs through the crank case fit airtight in its bearings. This is a condition which is difficult to maintain, especially in an old engine. This type of engine does not operate with valves at the intake and exhaust, but operates with ports or openings which are opened and closed by the piston passing over them.

About 90 per cent of all the gas engines used for agricultural purposes

[1] Courtesy of Ellis Engine Company, Detroit, Mich.

at present are of the four-cycle type; also all but a few of the automobile engines are of this type. By experience, users and manufacturers have found the four-cycle engine the most successful.

Vertical and Horizontal Engines.—Either four-cycle or two-cycle engines may be vertical or horizontal in appearance. The horizontal engine, especially of the four-cycle type, is much easier to repair than the vertical one. However, the vertical engine requires less space for its installation, but may not lubricate as well as the horizontal engine with the oil flowing from the top of the cylinder.

Ignition.—There are three types of ignition used in gas engine operation: high tension, low tension and compression ignition.

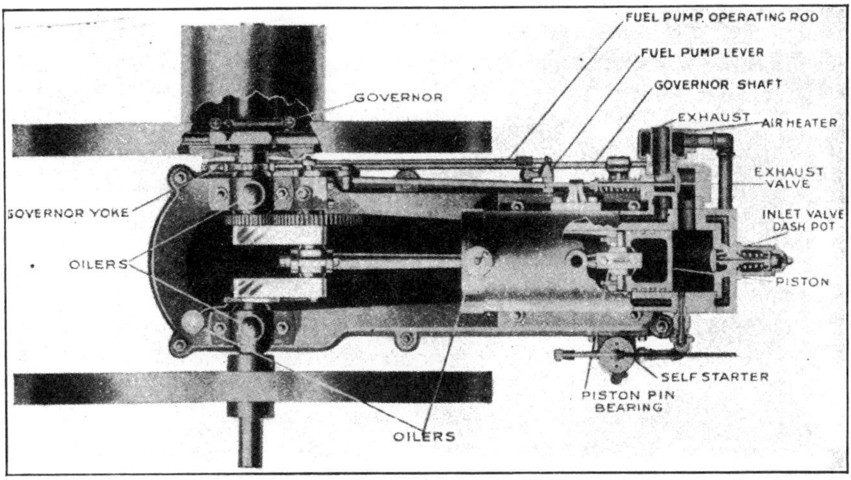

SECTIONAL VIEW OF A FOUR-CYCLE HORIZONTAL GAS ENGINE.[1]

The high tension system requires a current of electricity with a voltage sufficiently high to cause a spark to jump from one point to another of a spark plug. This system is used, as a general rule, on high-speed motors.

The low tension system requires a low voltage for ignition of compressed air and gas mixed together in the compression chamber. The spark is produced by the separation of two points in the cylinder which have been brought together and caused to separate.

The source of current for these two types of electric ignition may be from dry or wet batteries or from magnetos. A very successful means of ignition is the battery to start the engine and the magneto to furnish the source of current after it is in operation. In no case should any one purchase a modern engine without a magneto. It is not heir to the many

[1] Courtesy of Fairbanks, Morse & Co., Chicago, Ill.

diseases which render battery ignition worthless. The most modern engines do not require batteries even for starting the engines.

Compression ignition is not so common at present in gas engine operation. It may be found upon several recent crude-oil engines, some of which are being used very successfully and cheaply for agricultural purposes. The principle of this ignition depends upon the separation of the heavy and light gases as the fuel is vaporized and drawn into the cylinders with the charge of air. In the compression stroke the lighter gases are ignited by the heat generated by the compression caused by the advancing piston. The light gases in turn ignite the heavier ones. This type of engine not only burns a very cheap grade of fuel, but may be operated with gasoline, kerosene or most any mixture of the fuels used in internal combustion engines.

Cooling Systems.—When a mixture of gas and air is exploded in a gas engine the temperature rises to about 3000° F., which would melt the cylinder of such an engine if a part of the heat was not conducted away in some manner. Some manufacturers use water, some oil and others air for cooling gas engines. Also a mixture of several liquids is sometimes used in extremely cold weather to prevent freezing and the consequent bursting of the water jacket. Oil, when used for this purpose, takes the place of an anti-freezing mixture.

Some engines are cooled by water poured around the cylinder in a hopper and the heat conducted from the engine by means of evaporation. Other engines require a circulating pump which causes some liquid to be circulated through the water jacket and thence over a screen where it is partially cooled and used again. There are other types of liquid-cooled engines which depend entirely upon the liquid circulating after the engine is warm enough to cause convection currents.

The air-cooled engines for agricultural purposes have not proven altogether satisfactory on account of the small radiating surface; also the poor material which enters into the make-up in order that it may sell at a cheap price.

Lubrication.—Graphite is the true lubricant. It is not affected by heat or cold. The reason it is not used more than it is, is because of the inconvenience it offers in passing through small openings which are ordinarily used for oils. A mixture of powdered graphite and oil might be occasionally placed in gas engine cylinders to aid in lubrication, but this could not be depended upon entirely because the operator may forget when it is time to replace the lubricant.

All bearings may be lubricated with a cheap grade of animal or vegetable oil, but the cylinders of a gas engine must not be lubricated with any except the best grade of gas engine cylinder oil. The temperature in the cylinder of a gas engine is extremely high; therefore, a vegetable or animal oil would burn and be worthless for lubricating. More gas engines are sacrificed to the god of friction each year than from any other legitimate cause. It should be remembered by all who operate gas engines that oil is cheaper than iron.

The gravity system is the most common means of lubrication. It consists of a glass cup placed above the highest point to be lubricated. The splash system is very often used and consists of a crank case filled with oil to the point that the crank touches the oil at each revolution. The force feed type of lubrication is very successful; however, it adds a few more working parts to an engine, which complicates and may cause an added trouble. There are other systems of lubrication which will not be mentioned because of the infrequency of their use.

Gas Engine Parts.—The base of a gas engine supports the cylinder and all other parts of the engine structure. It should be in proportion

THREE H.P. GAS ENGINE OPERATING BINDER.[1]

to the rest of the engine. The cylinder serves the purpose of a container and a receiver. It should be smooth and free from irregularities or dark spots. The cylinder contains the piston and receives the charge and its walls receive the force of every explosion. The piston transmits the power to the connecting rod which is similar to the pitman of a mowing machine. The crank shaft receives the sliding motion from the connecting rod and changes it into rotary motion.

Governors.—There are two distinct types of governors used in gas engine operation at the present time. The hit-miss governor causes the exhaust valve to be held open mechanically when the engine begins to run above speed. So long as the exhaust valve is held open fresh air is drawn in and blown out; therefore, no power is obtained. As soon as

[1] Courtesy of Fairbanks, Morse & Co., Chicago, Ill.

the engine begins to operate below the rated speed, the exhaust valve closes and a charge of air and gas is drawn into the cylinder through the carburetor. This type of governor, of course, gives an uneven speed, but it is all right for ordinary agricultural purposes. It would not do for furnishing electric lights direct from the dynamo, because the lights would flicker with every variation in speed. This type of engine would do for charging batteries from which lights may be taken.

The throttle governor regulates the amount of air and gas mixture which enters the combustion chamber. This is done automatically in the stationary engines. This type of governor may be relied upon to give a more even speed than the preceding one, and especially is this true if extra heavy flywheels are used.

Gas Engine Troubles.—Gas engine troubles are almost unlimited. They are generally from two causes: the things we forget and the things we don't know. Troubles most frequently occur in the ignition system or from lack of proper lubrication. The first is easily remedied, but the latter usually means a new part. If dry batteries are used they may become wet and deteriorate, or a connection may be loose in the wiring. A drop of oil or water may be over the point of the spark plug. Points of the spark plug may be too far apart or too close together. There may be a loss of compression due to leaking valves or piston rings which do not fit tightly against the walls of the cylinder. Leaking may take place also around the spark plug or igniter. The mixture of air and gas may not be proper, in which case, either the gasoline supply is not regular or the air is not properly supplied. In cold weather the fuel often refuses to vaporize. Such a condition may be remedied by pouring hot water in the water jacket in order to warm the cylinder enough for good vaporization.

TRANSMISSION OF POWER

The best farm motor on the market is of no value on the farm unless the power which it develops is transmitted to some other machine doing useful work. Power is transmitted by shafting, belts and gear wheels. While there are other methods of transmitting power, they are only modifications of these three.

Shafting.—The shafting should transmit to the pulleys which it carries whatever energy it receives minus the amount consumed by friction at its own bearings. Shafting should be of the very best material in order to reduce the friction in the bearings by reducing the size. It should be absolutely straight, because much power is required to spring even a two-inch line shaft into line during each of two hundred or four hundred revolutions per minute. A shaft should be driven from the center if possible and between two bearings, and transmit its power to a series of pulleys on either side of the main drive. If possible, heavy shafts should have their bearings or hangers rest upon posts which are directly connected with the ground, because there is always more or less "give" in the average floor, especially if heavy storage should be above. Line shafting hangers should not be over 8 feet apart and if the shaft is light, not more than 6 feet apart. The horse power of a good shaft may be figured in the following manner:

Multiply the cube of its diameter by the number of revolutions per minute and divide the result by 82 for steel and 110 for iron. In other words, "The amount of power that can be transmitted by two shafts of similar quality varies directly with the speed and with the cubes of their diameters.'

The twisting strain on a shaft is greatest near the main drive; there-

ENGINE OPERATING PUMP JACK.[1]

fore, the nearer the main drive is to the hanger, the more nearly will its strain be counteracted. A disregard of any of the above principles is calculated not only to waste power, but gives an unsteady energy to the machine driven and affects both the efficiency and life of the machine being driven by it.

Speed of Shafting.—If only one machine is to be driven by a shaft the problem of shaft speed is very simple. With the operation of a cream separator at a speed of 60 revolutions per minute and a wood saw at a speed of 400 to 600 revolutions per minute as well as other varied speeds, the problem is more difficult. It is at this point that many very large, expensive pulleys and a number of very small pulleys upon which belts

[1] Courtesy of The Christensen Engineering Company, Milwaukee, Wis.

do not work very successfully are used. It is best to average all the speeds of machines and operate a line shaft at a medium speed.

The Size of Pulleys.—From the following formulas and conditions one may figure the speed or diameter of any given pulley.

With the speed of the driver, the speed of the driven and the diameter of the driver given, the diameter of the driven may be found.

EXAMPLE NO. 1.

$$\frac{\text{Diameter of the driver} \times \text{speed of the driver}}{\text{Speed of the driven}} = \text{Diameter of driven.}$$

EXAMPLE NO. 2.

Given the

$$\frac{\text{Speed of the driven} \times \text{diameter of the driven}}{\text{Speed of the driver}} = \text{Diameter of driver.}$$

EXAMPLE NO. 3.

Given the

$$\frac{\text{Diameter of the driven} \times \text{speed of the driven}}{\text{Diameter of the driver}} = \text{Speed of the driver.}$$

EXAMPLE NO. 4.

Given the

$$\frac{\text{Diameter of the driver} \times \text{speed of the driver}}{\text{Diameter of the driven}} = \text{Speed of the driven.}$$

Kind of Pulleys.—Pulleys on the market at the present time are manufactured from cast iron, steel, wood and paper. Of these, iron is the most commonly used. It is more compact than wood and is cheaper than steel, although wood can stand much higher speed than the average iron pulley of similar size and design. Wooden pulleys have the advantage of holding to a belt better than steel or iron, especially if a belt begins to slip upon the iron pulley, thus wearing its face very smooth. For light work the split pulley, or the pulley which can be divided into two parts, is the most convenient upon the market, especially if machines are changed from time to time for different purposes.

Straight and Crown Faces.—Iron pulleys are usually made crowning or slightly oval across the face. Where belts do not require shifting, this form holds belts to place in good shape. If the load is not heavy the crown pulley does not weaken the belt to a great extent, but with heavy loads the main strain comes upon the center of the belt and this causes a stretching and often develops splits.

Covering Steel Pulleys.—If steel pulleys are used and their surface becomes slick to the point where belts slip badly, they may be covered with a leather face. This can be accomplished in the following manner:

Clean the surface of the pulley with gasoline and apply a coat of varnish upon which a layer of soft paper is placed. Upon this paper a second coat of varnish is applied. A piece of leather belting is cut to fit the diameter of the wheel and while the varnish is still moist the section of belting is laced as tightly as possible upon the surface. The size of the pulley has now been materially changed; therefore, the effect upon other machines must be corrected.

Pulley Fasteners.—Pulleys may be fastened to line shafting either by a key fitting into a key seat both in the pulley and the shafting or by means of a set screw. The set screw arrangement is convenient and is often used where light work is to be done. The set screw may be a source of danger, especially in machines run at a high speed and where they are exposed and likely to catch the clothes of an operator. Also if the set screw once slips and grooves the shafting, it becomes necessary to shift the pulley to a new place.

BELTS AND BELTING

About 90 per cent of all the power transmission in the United States is accomplished by means of belts.

Advantages of Belts.—In the first place, belts are noiseless. Energy may be transmitted by them at a much greater distance than by direct gears. There is less risk of accident than by any other means of transmission. They are simple and convenient and are applicable to a great many conditions. In case of breakage they can easily be repaired, and in case machines are moved this means of transmission is the most convenient. For these reasons belting is especially adapted to farm uses.

Disadvantages.—Belts are expensive because they wear very easily. They are not always economical of power and unless carefully adjusted and of ample size they are likely to slip.

Essentials of a Belt.—If a belt has strength, durability, the absence of stretch and pulley grip, it has four very valuable qualities. Other qualities, such as flexibility and resistance to moisture, should also be considered.

Leather Belting.—The oak-tanned leather is the best material for belting. It has strength and durability, but has a disadvantage in that it comes to the manufacturer in short lengths and if especial care is not taken in cementing the ends together, it goes to pieces very early. It has been found by experience that as high as 25 per cent more power and greater wear may be obtained from a leather belt by running it with the grain or hair side next to the pulley. That is to say, there is a rough and smooth side to leather belts. The smooth side should be run next to the pulley because this side would crack more readily if placed outward, especially in passing over smooth, small pulleys.

Rubber Belts.—Rubber belting is manufactured by placing several layers of cotton duck and rubber alternately together and vulcanizing the mass into one. The strength of this kind of belt depends entirely upon the quality of the fabric which goes into its make-up. This belting has the advantage of being waterproof and may be made endless and in any length. Endless belts are not always best in a power house where every machine and pulley is stationary, because the length may change slightly with use. For outdoor work where machines may be moved, it gives excellent service.

Oil of any kind is detrimental to almost every kind of belt, and care should be exercised to keep rubber belts free from it. Rubber belting is resistant to steam and is, therefore, used to a great extent in creameries.

Belt Slipping.—All manner of belt dressings should be avoided because they often contain some material which shortens the life and hardens the surface of a belt. The hardening of a belt finally causes it to crack. Any sticky material put upon a belt will cause a loss in power due to an excess adherence to the pulley. If a large pulley drives a small one, it is best to pull with the lower side which is kept horizontal and allows the upper side to sag. This brings a greater surface of the belt in contact with the pulley.

To twist a belt, as in pulleys to run in opposite directions, often prevents slipping by a greater exposure of the belt to the pulley.

WATER MOTORS

Overshot Wheels.—The overshot wheel receives its power from the weight of water carried by buckets which are fastened to the circumference of the wheel. The water enters the buckets at the top of the wheel and is discharged near the bottom. A wheel of this character is made by placing between two wooden disks a number of buckets or V-shaped troughs. The wheel may be supported upon a wood or steel shaft supported on concrete piers. Motors of this type can be built to operate under falls as low as four feet and may be expected to supply anywhere from 3 to 40 horse power, depending on the head of the fall and the water available.

Undershot Wheels.—The undershot wheel is propelled by water passing beneath it in a horizontal direction, which strikes veins carried by the wheel. Such wheels are often used for irrigation purposes where the fall is too slight for other types of wheels. Most of the undershot wheels have straight, flat projections for veins, but the most efficient wheels are built with curved projections. This form of water motor operates satisfactorily where the water current is rather swift and in places where the volume of water is kept constant. They will not operate in streams that are ever flooded.

Breast Wheels.—Under conditions where little fall may be procured, a breast wheel may be employed to develop power from running water. This type of wheel receives the water near the level of its axis, but in most features it is similar in its action to the overshot wheel. The veins may be straight or slightly curved backward near the circumference.

The wheels mentioned above are very awkward and cumbersome for the amount of power that they are capable of developing. In other words, they are not what is known as efficient; however, they are cheap in construction and often may utilize water where other types of more efficient wheels cannot be employed.

Impulse Water Motors.—Impulse water motors are provided with buckets around the circumference of the wheel against which a small stream of water under high pressure operates. The Pelton wheel is one of the most efficient of the water motors, but requires for successful operation a head of water considerably higher than is required by most of the other water wheels. This type of wheel may be secured in sizes under one horse power and up to several hundred horse power.

Pelton Water Wheel.[1]

Turbine Wheels.—The turbine is a water motor which is built up of a number of stationary and moveable curved pipes. It consists of the following parts:

A guiding element which consists of stationary blades the function of which is to deliver the

Turbine Water Wheel.[2]

[1] Courtesy of Pelton Water Wheel Company, New York.
[2] Courtesy of J. and W. Jolly Company, Holyoke, Mass.

Three-plow Tractor in Operation.[1]

[1] Courtesy of Advance-Rumely Company, Inc., La Porte, Ind.

water to the rotary part under the proper direction and with the proper speed.

A revolving portion which consists of veins or buckets which are placed in a certain position around the axis of the motor.

The last two mentioned are the most efficient and up-to-date water motors on the market. Power obtained in this method is dependable, inexpensive, safe and sanitary.

The Hydraulic Ram.—This device, although very wasteful of water, is one of the most economical motors for pumping water. It serves both as a motor and a pump. It is not only used for furnishing water for the farm house, barn and dairy, but it is used in many cases for irrigation purposes. Only about one-tenth of the water passing through a ram

HACKNEY AUTO-PLOW.[1]

is finally delivered to the water tank. There is a ram on the market at present which will operate on impure water which may be secured in large quantities and made to pump a pure supply of water. This is commonly known as the double-acting ram.

THE FARM TRACTOR

Farm tractors have been placed upon the market in the past in such large units that they were practical only on extremely large level farms in the Middle West. This type of tractor is being driven from the field by smaller and more compact tractors which are finding a place also on the small farm of 160 acres or less.

The Size of Tractors.—A tractor of less than five tractive and ten-

[1] Courtesy of Hackney Manufacturing Company, St. Paul, Minn.

belt horse power has no place under average farm conditions on the small farm. This size should operate one fourteen-inch or two ten-inch plows. It should operate a small threshing machine and also the small silage cutter for silos not taller than thirty feet. This size tractor may operate a line shaft from which power can be secured for pumping, grinding feed, separating cream, churning, for electric lights and for many other farm operations at one time.

In hilly land where irregular fields are sure to be prevalent and rocky ledges are very likely to occur, the tractor has little place. As plowing is the biggest job in farm operation, the tractor should in this case have its greatest usefulness and should replace about one-third of the horses ordinarily employed upon the farm. It generally takes about one-third less horse power to cultivate, harvest and haul to market the crop of any farm than it takes to plow and prepare the seed-bed in a thorough fashion. Under ordinary small farm operations, the writer believes that an 8–16-horse power tractor is the most economical size.

CREEPING GRIP TRACTOR.[1]

Tractor Efficiency.—The tractor has been used for agricultural purposes long enough for this fact to become well established; where a tractor of repute is employed, more depends upon the intelligence of the tractioner than upon the ability of the machine to do good work. This does not mean that one has to have a college training in engineering or to be a master mechanic, but one should know the principles upon which a gas engine operates and the intelligent remedy of all diseases to which this mechanism is heir.

Type of Tractor.—It has long been proven that a multi-cylinder engine is the most successful on the road for speed and power and it is becoming recognized by the best tractor manufacturers that more than one cylinder is more dependable and gives more constant power than the one-cylinder type of motor. More cylinders mean more working parts,

[1] Courtesy of The Bullock Tractor Company, Chicago, Ill.

but it also means that a steady pull may be secured, where with one cylinder the power is secured in large quantities at fewer intervals, which is not calculated to give the best efficiency.

The multi-cylinder engine costs more at first, but the efficient service which it will render will more than compensate for its greater initial cost.

REFERENCES

"Power and the Plow." Ellis and Rumely.
"Agricultural Engineering." Davidson.
"Heat Engines." Allen and Bursley.
"Farm Gas Engines." Hirshfield and Ulbricht.
"Power." Lucke.
,'Farm Motors." Potter.

CHAPTER 60

FARM SANITATION

By R. U. Blasingame
Professor of Agricultural Engineering, Alabama Polytechnic Institute

Farm sanitation ordinarily includes five distinct branches, namely: lighting, heating, ventilation, water supply and sewage disposal. Following is a brief consideration of each of the above mentioned:

LIGHTING

There are several sources of light for isolated farm homes at the present time. They are as follows:

1. **Kerosene Lamps.**—These are cheap in initial cost. The fuel may be obtained at any cross-roads store. They are quite safe. There are a few disadvantages to such a source of light, namely, the odor they emit, the soot which they produce and the fact that they burn more oxygen than other forms of lighting. Lastly, the light is not a white light.

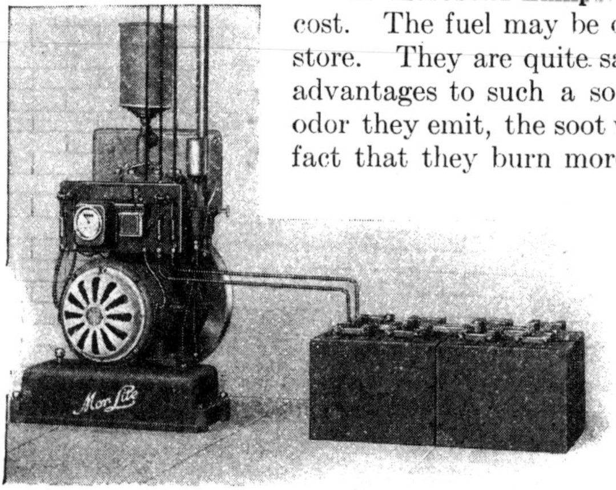

Mor-Lite Electric Plant.[1]

2. **Gasoline Lamps.**—These may be divided into two groups, the cold process and the hot process. The former system requires a lighter grade of gasoline for the production of light and is more expensive to operate. The cold process lamps are much safer than the hot process lamps which may be operated with heavier, cheaper gasoline. While cheaper, the latter are more dangerous than the former.

3. **Acetylene Gas.**—This gas is produced by water and calcium carbide being brought together. The safest system of acetylene lighting may be had by feeding calcium carbide in small quantities to a large quantity of water. The heat produced is conducted away too fast for any danger of explosion. While this system is reasonably safe, there have been many explosions which have cost both life and property. This gas may cause

[1]Court sy of Fairbanks, Morse & Co., Chicago, Ill.

death if inhaled. It has a characteristic odor which any one can easily detect if it is escaping from the system. The light produced from this system is white and considered excellent.

4. **Electrical Lighting.**—The lighting of isolated homes by a private electrical system is generally thought to be an expensive luxury. However, during the past twenty years the cost of living has increased about 20 per cent and the cost of farm labor has increased about 35 per cent, but for the

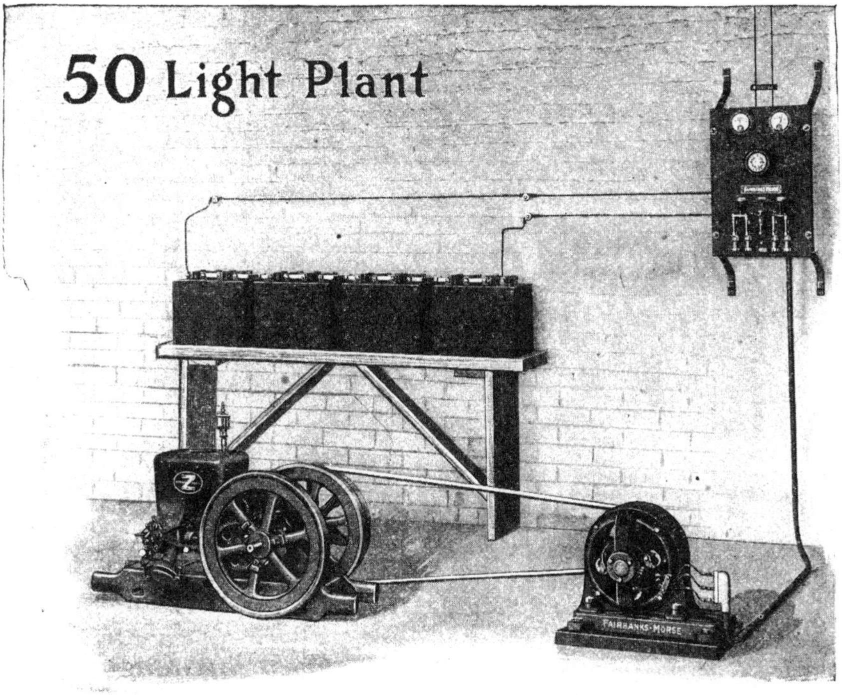

ELECTRIC LIGHTING PLANT FOR FARM HOUSE.[1]

same period the cost of lighting by electricity has decreased about 85 per cent. This method of lighting, if correctly installed, is the safest, most sanitary, most convenient and most efficient of all modern lighting systems. There are manufacturing companies who are building very successful private electrical lighting systems for farm homes. These operate on different voltages, namely: 30 volts, 60 volts and 110 volts. If the system is to furnish power for home conveniences such as operating churns, sewing machines, etc., the writer would recommend the 110-volt system. A storage battery will supply about two volts of electrical energy; therefore the 110-volt system would require about 56 cells, whereas, the 30 and 60-

[1] Courtesy of Fairbanks, Morse & Co., Chicago, Ill.

volt systems would operate at a less cost for such equipment. In most cases these systems receive their power from small gasoline engines; however, it is becoming popular in mountainous regions to use small streams to furnish motive power. Where water is used, the storage battery is not necessary, because water forces through the wheel at a steady rate which will in turn produce a steady light. This is not true of a small gasoline engine, although some companies are making very sensitive engine governors and heavy flywheels which are calculated to run very smoothly.

Heating.—There are three distinct heating systems from one central plant, namely: hot air, hot water and steam. These systems are used mostly in extremely cold countries.

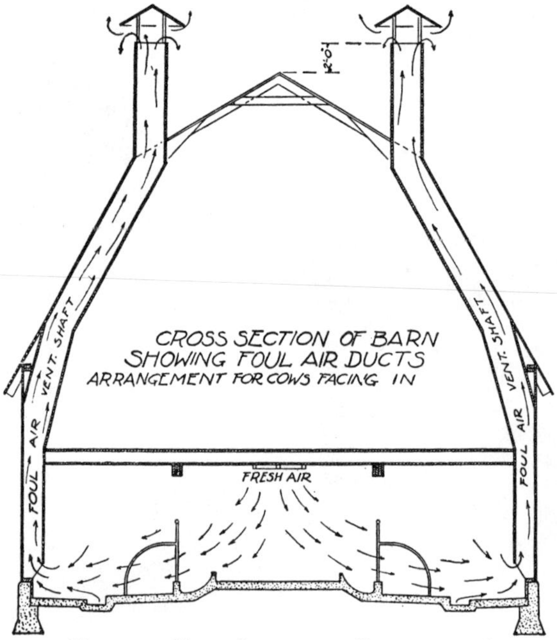

Modified King System of Ventilation.[1]

1. The hot-air system, if properly installed, gives the best ventilation, and in most cases is the cheapest of the three. In cold, windy weather this system is rather hard to control on account of the leeward side of the house receiving the greater part of the heat.

2. The hot-water heating system is the most expensive to install on account of two systems of piping, one for feed, the other for return. It has been found that the Honeywell generator or the Mercury-Seal system causes the hot water to flow more rapidly than without, thus increasing the efficiency of the system.

3. Steam heat is entirely satisfactory. It gives quicker heat, but does not retain its heat as long as the hot-water system.

Ventilation.—There are two influences which cause ventilation, namely: (1) the force of the wind, which causes more or less suction from any opening in a building; (2) the difference in outside and inside temperatures, the warm air inside rising and escaping through any opening, thus causing ventilation. The "King system" is generally used in farm buildings at the present time. It consists in admitting fresh air near the

[1] Courtesy of Louden Machinery Company, Fairfield, Ia.

ceiling and conducting the foul air from the interior through an opening sometimes located at the highest point of the building.

Dampers should be placed at the intake and the outlet in order that this system may be thoroughly controlled. For horses and cows the area of cross section of outlet flues should not be less than 30 square inches for each animal when the flue is 30 feet high, and 36 square inches for each when only 20 feet high. The cross section of the intakes should aggregate

A Pneumatic Water Tank.[1]

approximately the same as the outlets. Ventilating flues should be airtight and with as few bends as possible.

There is a system of using double sash windows for dairy barns, in which the top sash is hinged at the bottom so as to permit the entrance of air when the top of the sash is drawn into the barn a few inches. The air entering is deflected upward, thus avoiding a draft of cold air upon the cattle in the barn. This is one of the absolute essentials of a good ventilating system. Deflectors should be placed at the sides of the windows, which will also prevent air from blowing directly upon the stock.

Water Supply.—Water can be supplied to a home under pressure from an elevated tank, also from a pneumatic tank into which water is pumped

[1] Courtesy of Fairbanks, Morse & Company, Chicago.

against a cushion of air. An elevation may be procured by placing the water tank upon a silo, upon a tower or upon a hill. In extremely cold climates water in an elevated tank is likely to freeze, and in hot climates it becomes warm and is not palatable. Where it is not too expensive, a reservoir placed on the side of a hill and well protected supplies water under pressure at an even temperature the year around. Such an elevation is permanent and the pipes are placed beneath the ground so they do not freeze. It is considered, after first cost, the most satisfactory

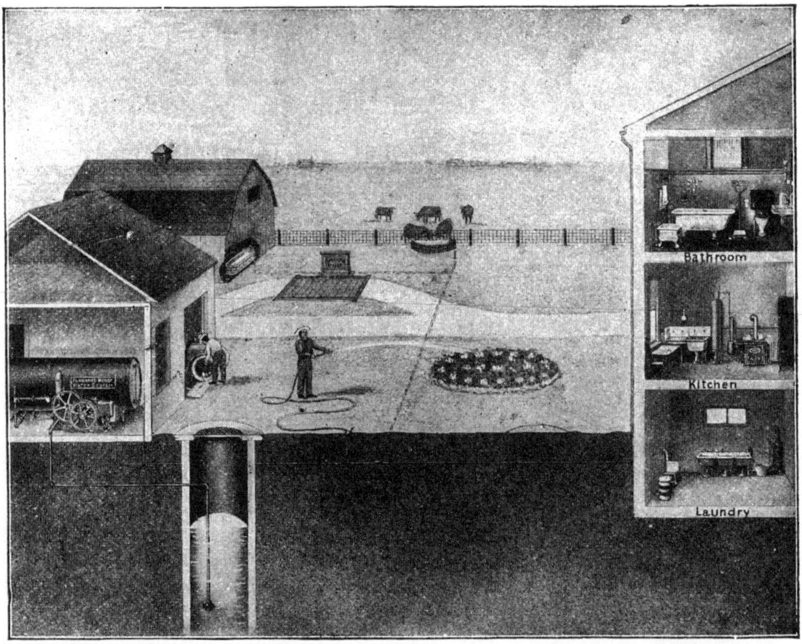

FAIRBANKS-MORSE WATER SYSTEM FOR FARMS AND SUBURBAN HOMES.[1]

system of water supply. In recent years the pneumatic tank which may be buried in the ground or placed in the cellar is considered an excellent method for supplying water under pressure to the farmstead.

In installing a system of this kind, one should be sure he is dealing with a responsible company. It is very necessary that the pump supplying the water to this tank should be provided with a small air pump as well. This will supply air as well as water, thus insuring the air cushion at all times. Such a system should be operated under about 50 pounds pressure.

Sewage Disposal.—In some states there are laws which prohibit the discharge of sewage from even a single house into a stream of any size,

[1]Courtesy of Fairbanks, Morse & Company, Chicago.

even though the person discharging the sewage may own the land through which the stream flows. Such a law should not require legal machinery for its enforcement, but should appeal to the sense of justice and intelligence of all good citizens.

Vital statistics show that the death rate from typhoid fever in New York State since 1900 has decreased in the cities, while it has remained about constant in rural districts. This reduction in the death rate in the cities may be accredited in large measure to the improved methods of sewage disposal and close attention to pure water supply intended for human consumption.

It is, therefore, desirable to purify sewage before its discharge into any place where it may contaminate food or water intended for human consumption.

The art of sewage treatment when purification is carried on in septic tanks consists in two distinct forms of decomposition.

The first form of decomposition takes place in the absence of oxygen or air, and is called anærobic, or without air. Under ordinary circumstances it is accompanied with disagreeable odors. The second decomposition process takes place in the presence of air and is called ærobic, or with air. It is accomplished without disagreeable odors.

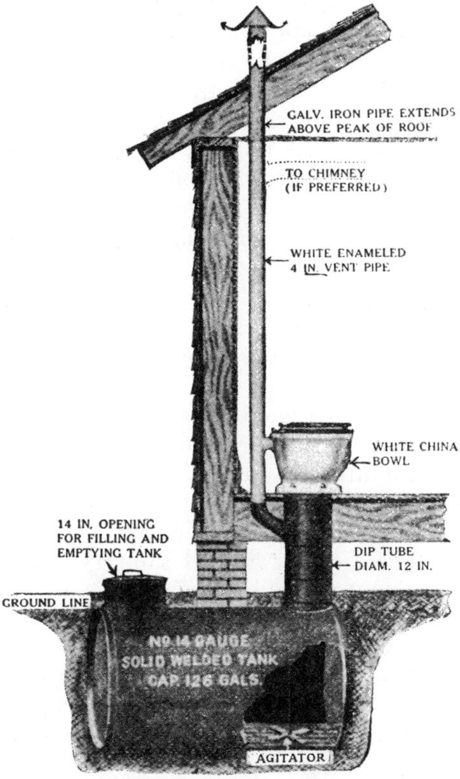

THE KAUSTINE CLOSET.[1]
A germless water closet.

The first treatment consists in allowing the fresh sewage to enter a water-tight septic tank, and remain for twenty-four or forty-eight hours. During this period, in the absence of air, the organic matter of the sewage is broken down into small particles. The purpose of this treatment is to get the sewage in such a condition that it can be purified No purification is accomplished during this process. The secondary treatment consists in exposing the effluent from the septic tank to the atmosphere, where

[1] Courtesy of The Kaustine Company, Inc., Buffalo.

the mass of small particles may be oxidized after the water has been strained from it. This process is accomplished generally in two ways. First, the effluent from the septic tank is flushed upon filter beds which are made by excavating in the ground about two feet deep and filling with sand after placing four-inch drain tile on the bottom. The drain tile should have an outlet from whence the filtered liquid may escape. The air and sunshine decompose the organic matter which is left upon the filter bed. The second method of final disposition of sewage consists in flushing the sewage from the septic tank into a series of drain tile which are placed under ground and have a slope of about 1 inch in 100 feet. In sandy soil about 150 feet of pipe should be allowed for each person living in the home. In clay soil about 400 feet of pipe should be provided for each person. It is necessary to ventilate these lines of pipe at intervals in order that the material left in the pipes after the liquid has escaped into the soil may be oxidized by the air. The size of the tank should be determined by the size of the family, allowing twenty-five gallons of water per day for each person.

By writing the Department of Agriculture at Washington, D. C., one may receive farmers' bulletins which describe and illustrate different systems of sewage disposal. It is often thought and sometimes stated in literature that after sewage has remained in a septic tank for twenty-four hours it may be dumped into a stream without fear of pollution. This is absolutely wrong, for the sewage may contain disease germs which are not affected in the least by the decomposition in the septic tank.

There is a patented sanitary closet which is manufactured by the Kaustine Company, Buffalo, N. Y., which is giving good satisfaction. The principle upon which this method of sewage purification operates is as follows:

The excrement enters a steel tank containing a very strong chemical which is mixed with water. This chemical destroys all bacteria and odor and also disintegrates all solid matter to the point that it may be drained or pumped from the tank and disposed of without fear of contamination. This tank will hold the sewage produced by a family of five during a period of six to eight months. The contents of the tank rates high in fertilizing value.

REFERENCES

"Electricity for the Farm." Anderson.
"Rural Hygiene." Ogden.
Canadian Dept. of Agriculture Bulletin 78. "Ventilation of Farm Buildings."
U. S. Dept. of Agriculture Bulletin 57. "Water Supply and Sewage Disposal for Country Homes."
U. S. Dept. of Agriculture, Year-Book 1914. "Clean Water on the Farm and How to Get It."
Farmers' Bulletin 463, U. S. Dept. of Agriculture. "Sanitary Privy."

CHAPTER 61

FARM DRAINAGE AND IRRIGATION

Water is the first essential to plant growth, and yet either too much or too little prevents a normal growth of most farm crops. The removal of water from the soil is known as drainage, while the adding of water is called irrigation.

LAND DRAINAGE

The need for drainage and the advantages of it are discussed in Chapter 7. Only the engineering features of it will be discussed here.

Co-operation.—Wherever large tracts of farm land are to be drained, co-operation among the land owners is necessary for the establishment of an economic drainage system. The laws of most states provide for an equitable appraisement of benefits derived by the land owners in a drainage district and make possible the establishment of the district when the majority of land owners ask for it.

The first step in the formation of a district is an accurate survey of the natural water course and an estimate of the size and length of the system of open ditches necessary for the proper drainage of the land. The ditching is generally done by a contractor making a specialty of this kind of work. His services are secured through the ditch commissioners, three or more in number, who are elected by the land owners of the district. Bids are usually let in order to secure competition and get the work done at an equitable price.

The dredged ditches, when completed, usually provide each land owner with an outlet. All subsequent drainage is done by the individual owners, each for his own farm. The individual farm drainage consists chiefly or wholly of tile drains that empty into the open ditches.

The old plow-and-scraper method of making ditches is applicable only when the soil is fairly dry. It will not be described here. Except for very small jobs, it is more expensive than excavating with one of the several forms of large ditching machines.

Of the several types of ditching machines, the floating dredge is the most common and the most successful in level land and for large jobs. It begins at the upper end of the drainage course and works down stream so that the excavation is always well filled with water and easily floats the dredge. This style of dredge is adapted to a large channel, varying from 12 to 60 feet in width. The earth is excavated by large scoops on immense steel arms, operated by steam power. The earth is deposited on either side of the channel and at a distance of 6 to 12 feet from the edge of it. In the absence of stones, roots or other obstructions, ditches may be excavated at a cost of from 7 to 13 cents per cubic yard. The contract is frequently made on the basis of material removed.

It is essential that such water courses be made as straight and as deep as conditions will permit. The straight course makes the shortest possible ditch and provides for the maximum fall. Good fall and straightness both accelerate the flow of water and make possible adequate drainage with a smaller ditch than would be possible with a longer and more circuitous route.

The ditch embankments, after weathering for a year, may be gradually leveled down and worked back into the adjacent fields by the use of plows and scrapers. The banks of the ditch need not be as sloping, as formerly thought, although the slope will depend on the character of soil. In heavy, tenacious soils, a slope of $\frac{1}{2}$ to 1 is sufficient, that is 6 inches horizontal to 1 foot vertical. The fall of the ditch may range from 6 inches to 3 feet or more per mile. With 3 feet of fall per mile, the velocity of the water will keep the ditch fairly free from sediment, provided it is not allowed to become filled with growing grass, weeds or willows. If these grow in the ditch during the dry portion of the year, they should be cut and removed annually. Where the fall is too great, the banks of the ditch are apt to erode and cave in. The caved earth will be carried and deposited in lower portions of the stream course and cause trouble. The banks of the ditch should be kept covered with grass to prevent erosion.

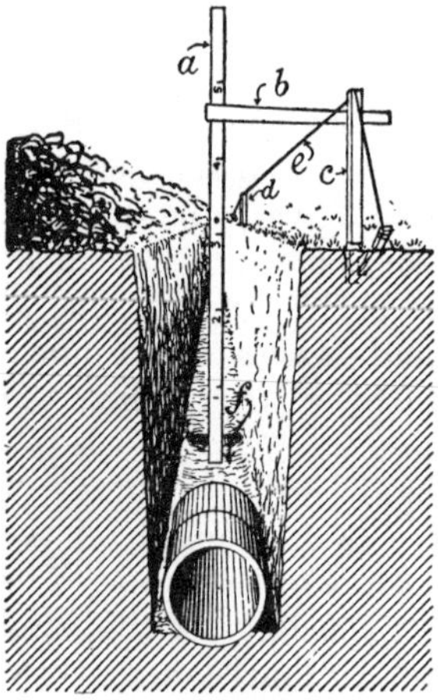

GRADING THE DITCH AND LAYING TILE.[1]

A—Depth gauge. B—Cross piece. C and D—Stakes driven in ground to give proper slope to grading line. E F—Hollow tile drain.

Tile Drains.—The first step in tile drainage is an accurate survey of the land to be drained. This will determine the fall and the best position for the main drains. It should also include an estimate of the water shed, that is, the amount of water to be carried away, whether falling on the land to be drained or flowing on to it from adjacent higher lands. The lines of drainage should be as straight as conditions will permit. The mains should be in the lowest portions of the field. Laterals may extend from them into more elevated portions. In case of very level land, this makes provision for the greatest possible fall in the drainage lines.

Running the Levels.—This work may be done by the farmer. In large systems or on very level land, the employment of an engineer is advised. A farm drainage level that is sufficiently accurate may be pur-

[1] Courtesy of U. S. Dept. of Agriculture, Farmers' Bulletin 187.

chased for about $15. For very small jobs a home-made water level will serve the purpose. This consists of a section of gas pipe about three feet long, with a glass tube attached to each end by means of corks or rubber tubing. The glass tubes should be at right angles to the pipe. When filled with a colored solution and held approximately level, the operator sights across the top of the colored solution as it appears in the two glass tubes.

Establishing the Grades.—The drainage lines are laid out by driving stakes at intervals of 50 to 100 feet, about 18 inches to one side of the center of the ditch. These stakes are driven into the ground until the tops

A Low-priced Tile Ditcher.

are only two or three inches above the ground level. By use of the level, the elevation of each is ascertained. The next step is to calculate the total fall of the line and determine whether the grade is to be uniform or whether it must be changed for a portion of the course. This will depend on the variation in the slope of the surface of the ground. If the slope varies much, two or more grades may be necessary in order that the drainage pipe may be placed at the desired depth beneath the surface of the ground. A single grade may result in the tile being too deep over a portion of the course, thus necessitating expensive excavating, or it may be too shallow to provide effective drainage. These difficulties are avoided by suitable changes in the grade.

Grade stakes projecting about 18 inches above the surface of the ground are set one beside each of the stakes designating the level. These are driven so that the tops are a uniform distance above the bottom of the ditch as it is to be excavated. This may be $4\frac{1}{2}$ feet or any convenient height. A cord or wire is next stretched tightly over the top of the grade stakes. By means of a gauge, the ditcher can control the depth of the ditch. Care should be exercised not to get it too deep, or to make the bottom wider than necessary.

The sketch on page 766 shows the method of gauging the depth, the character of excavation and the position of the tile.

Small Ditching Machines.—These may be used to facilitate the work

THE DITCHER IN OPERATION.

Can be operated by one man and six horses. It will excavate 100 rods of dirt to a depth of 3 feet daily.

of excavation. They do it more rapidly than can be done by hand and at less cost. They are adapted only to fairly long courses. It will generally be necessary to grade the bottom of the ditch by hand.

Size of Tile.—In any system the major portion of the tiles will be three inches in diameter. All lines not exceeding 500 feet in length and having no branches entering may be of this size. When such lines exceed 500 feet the lower portion should be 4-inch tile. The capacity of pipes is in proportion to the square of their respective diameters, plus something for the relatively lesser amount of friction in the large diameters. In practice, one 4-inch line will accommodate two 3-inch lines. One 8-inch line will accommodate five 4-inch lines, etc.

The removal of one-quarter inch of rainfall in 24 hours will generally provide adequate drainage. On this basis the area in acres drained by given sizes of tile and grades are as follows:

Diameter of Drain.	Grade 1 Inch to 100 Feet.	Grade 3 Inches to 100 Feet.
5	19.1	25.1
6	29.9	39.6
7	44.1	58.9
8	61.4	80.9
9	82.2	108.4
10	106.2	140.6
12	167.7	221.1
16	341.4	449.9

To double the fall for steeper grades than those given in the above table will increase the carrying capacity of the tile one-quarter to one-third.

IRRIGATION

Water, wisely used, has converted many desert acres into fruitful fields and orchards. This has made possible thriving settlements in many parts of the arid West, and encouraged the development of industries other than agriculture, especially the mining of useful metals.

Water Rights.—In regions of limited water supply, laws for the control of water become essential. These laws should be understood and obeyed by all users of water. It is a principle that rather definite shares in the water supply of a region shall be apportioned to specific areas of land. When the water supply is insufficient for all available land, priority of appropriation receives first consideration. A new settler is prohibited by law from sharing in the water supply at the expense of early settlers. In many irrigation districts, the extravagant use of water has prevailed. A more economical use on the part of the older settlers would produce equally as good crops. In fact, the extravagant use of water is more often injurious than otherwise.

Co-operation.—This is a necessary feature in most irrigation districts, because the water supply must serve the entire community, and in order to do so most advantageously, co-operative action is called for in its use and conservation. Co-operation means that the farmers on an irrigation ditch must take turns in using the water. The larger the volume of water the shorter the time each may use it and the greater number of farmers can be supplied. The apportionment of the water should correspond to the acreage of crops to be irrigated by each farmer. This rotation of the allotment of water to the farmers on a ditch is advantageous from two standpoints. First, it gives each farmer sufficient water to cover his land in a very short time, thus economizing on the time spent in irrigating. Second, it overcomes the loss of water by seepage and evaporation which takes place when he has a constant small stream.

Sources of Water.—The chief sources of irrigation water are perennial streams, springs and wells. The first named is by far the most important. The first consideration in the development of an irrigation supply from a stream is the volume of water carried at all times during the year; and second, whether or not the water can be brought to the land to be irrigated at a reasonable expense. This will depend principally upon the length of ditch to be constructed and the character of land that must be traversed by it. In some cases, pipe lines may take the place of ditches without great additional expense and with much less waste of water.

The larger the ditch and the more porous the soil through which it passes, the smaller should be the fall. If, however, the grade is too small, the ditch must be larger in order to carry the supply of water. In ordinary soils, a grade of one foot in 600 feet may be given. In clay soils, it may be increased to two feet in 600 feet. A slow movement of water in the ditch prevents scouring and encourages the settlement of fine sediment. This ultimately forms an impervious lining and prevents seepage.

Springs offer an excellent irrigation water supply, and although the volume is much less than that from perennial streams, it is subject to less fluctuation in volume and is consequently more dependable.

Wells form a considerable source of irrigation water supply in many of the irrigation districts. They are virtually artificial springs secured by boring deep wells provided with iron casings. In some instances, as in case of wells that do not flow, and in elevating water from lakes and streams to land lying above the water level, pumping is resorted to.

Dams and Reservoirs.—Perennial streams are subject to great fluctuation, due to periodic rains and melting snow. Their direct diversion for irrigation purposes, therefore, fails to utilize much of the water during high stages. This has led to methods of storing the water to be used as needed, thus increasing the area irrigated. While dams are necessary for diverting water from streams into canals, much larger and more expensive ones are required in the building of reservoirs. It is important to select the dam site with a view of securing the largest possible water storage capacity with the minimum expenditure for construction. Such sites are most usually found in the upper courses of a stream where it passes through a narrows or canyon. Rocky, impervious abutments to which to connect the dam are essential. On large projects the reinforced masonry or concrete dam that will be permanent is advised. The deeper the water in a storage reservoir the less will be the relative loss by evaporation.

Methods of Transmission.—The census of 1910 gave an aggregate of over 125,000 miles of irrigated ditches in the United States. At that time, less than four per cent of this mileage was *lined* or otherwise made impervious to water. A limited amount of irrigation water is conveyed through pipe lines of different types, of which wood, terra-cotta and cement predominate. It is important to construct the irrigation ditch of the proper size to convey the maximum amount of water that will be available or the maximum that can be used by those who irrigate. In this connection it is advised to secure the services of an engineer. It should be understood that the amount of water conveyed depends on the cross section of the canal and the rate of movement of the water. In a small ditch capable of carrying 50 miner's inches, a fall of 2 inches to the rod will give a velocity of 2 feet per second. In a ditch carrying 20 times as much water, a fall of $\frac{1}{4}$ inch to a rod will give an equal velocity. Except in hard clay or a mixture of gravel and clay, a velocity greater than 3 feet per second is likely to cause serious erosion. A velocity of 2 to $2\frac{1}{2}$ feet is the maximum that should be permitted for ordinary sandy loams or loams. Where the fall of the land is such as to cause a greater velocity of the water, checks in the canals should be provided. These may be wooden dams or obstructions of cobblestones, causing a drop in the water.

In lined canals erosion is overcome and the velocity of the water may be much greater. Where there is ample fall, such a canal may be much smaller than an ordinary earth canal. The transmission of water through pipes has a still greater advantage in this respect and may be conducted down very steep grades.

Losses in Transmission.—Much water diverted from streams for irrigation is lost from the ditches by seepage and evaporation, and is still further wasted by over-irrigation and by allowing the water to penetrate the soil beyond the reach of crops. Water lost in these ways often causes serious damage to the lower lying land in the irrigation district. Numerous water measurements and experiments have led to a conservative estimate that not more than 35 per cent of the water diverted from streams is effective in plant production.

The efficiency of irrigation water can be greatly increased by the substitution of pipe lines for open ditches and by greater care in the distribution of water in the fields.

Head Gates.—Head gates are necessary at the point of diversion from a stream into the main irrigation canal, and also at points along the main canal at the juncture of laterals. Such gates are usually constructed of plank with a gate that slides up and down to control the volume of water. A simple form is shown in the accompanying illustration.

Preparing Land for Irrigation.—The preparation of the land consists in clearing it of the native vegetation, which in the arid region is usually sage-brush, rabbit-bush, cacti and native grasses. Plowing frequently precedes the clearing operation. This makes easy the gathering and burning of the vegetation. The plowing and clearing should be followed by a thorough harrowing, grading and smoothing of the surface. The supply ditch should be above the highest portion of the land to be irrigated. After the field is cleaned and leveled, farm ditches should be

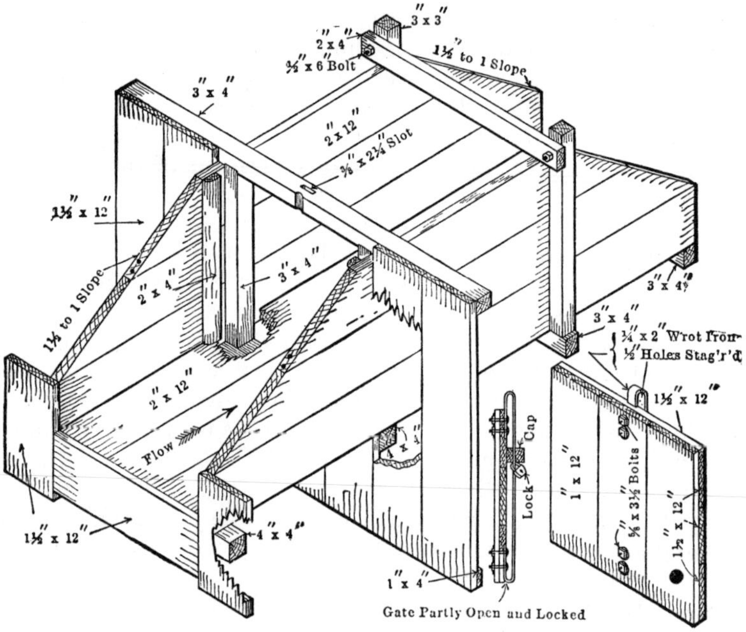

Delivery Gate to Farm Lateral.[1]

conducted over the higher portions of it. From these ditches the water may be conducted to all portions of the land. As far as possible these ditches should extend along the borders of the fields in order to avoid

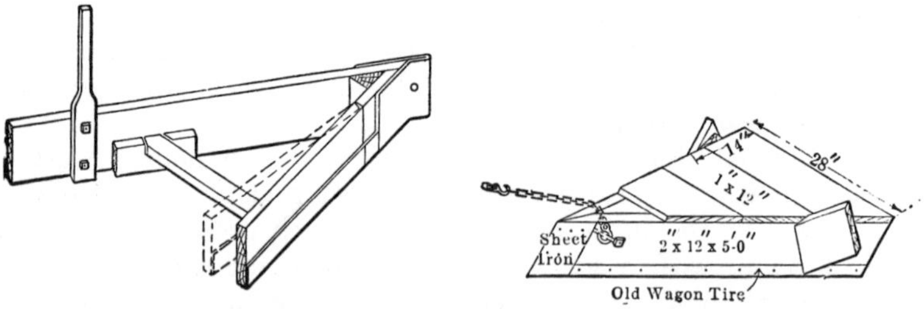

The V-Crowder is Excellent for Making the Farm Ditches.[1]

obstructions to cultivation. When necessary to cross fields with open ditches, they should be so placed as to avoid as far as possible irregularity in shape of fields.

[1] Courtesy of The McGraw-Hill Book Company, N. Y. From "Use of Water in Irrigation," by Fortier.

Farm Ditches.—The size of the farm ditches will be determined by the acreage of land irrigated by each, the fall in the ditches and the amount of water that must be cared for in a unit of time. On uneven land it is necessary to bridge over the depressions with levees or flumes. The levee is usually the cheaper, but should be allowed to settle. It will be subject to wash-outs during the first few years.

Wooden flumes are more satisfactory, but wood soon decays when used for this purpose. Metal or concrete pipes cost most, but are durable and generally cheapest in the end. The method of constructing the farm ditches depends on their size. Most of the work on them may be done with the plow and the V-crowder. The crowder makes a ditch with a triangular bottom. This bottom becomes rounded by usage. It is important that the ditch be made in the proper place at the outset. The older the ditch, the more impervious its banks and bottom become and the more satisfaction it gives. Leaky ditches may be greatly improved by puddling the earth of the sides and bottom. This may be done by drawing off the water and driving a flock of sheep the length of the ditch while it is muddy. Dragging the bottom with a brush harrow may be resorted to for the same purpose.

CANVAS DAM TO CHECK WATER.[1]

On well-established ditches the chief items of maintenance are the removal of silt, weeds and aquatic plants that may grow in them.

Distributaries.—These consist of small wooden, metal or rubber tubes, imbedded in the bank of the ditch so that the water will pass through the embankment and be uniformly distributed on the adjacent land. These need not be permanent, but may be imbedded temporarily, and moved from field to field as needed. Square boxes, made of lath cut in half, are cheap, light and serve the purpose as well as more expensive metal tubes. Being square and rough, they stay in the embankment better than the smoother metal or rubber tubes.

Small syphons of rubber hose are also used. These obviate the necessity of disturbing the ditch bank. The chief objection to these is the starting of the flow of water.

Distributing the Water.—The method of distribution will depend on the slope of the land, the character of the soil and the kind of crop. Level land is easily irrigated by flooding the whole surface. This method

[1] Courtesy of The Macmillan Company, N. Y. From "Principles of Irrigation Practice," by Widtsoe.

is applicable to the irrigation of alfalfa, grass and small grains. The surface, however, should be divided into areas that may be covered in a comparatively short time with the water available. When one area has received sufficient water, the flow is then directed to the next one, and so on until the irrigation is completed. If the field to be irrigated is large, it necessitates a network of ditches or parallel ditches at intervals of 300 to 400 feet, extending across the field. The distance to which the water may travel over the surface of the ground depends on the character of soil and the ease of penetration. The more porous the soil, the shorter the intervals should be. If the intervals are too long, the soil

ORCHARD IRRIGATION BY FURROW METHOD.[1]

nearest the ditch becomes over-irrigated before the water reaches the further portions.

With this method of irrigation the water is generally made to flow over the embankment by use of a temporary dam. The most convenient form consists of a strong piece of canvas four or five feet square with one edge securely nailed to a tough but light piece of wood that will reach from bank to bank of the ditch. When this is laid in the ditch with the canvas upstream and a few shovels of dirt thrown on its edges, it completely dams the water. It is easily moved from place to place as needed.

All crops planted in rows, such as vegetables, sugar beets, potatoes and fruit, are generally irrigated by the furrow method. Where the rows are close together, the furrows alternate with the rows, being midway

[1] Courtesy of The McGraw-Hill Book Company, N.Y. From "Use of Water in Irrigation," by Fortier.

between them. If they are further apart, as in orchards, two or more furrows for each row of plants are desirable. The length of furrows will depend on the character of soil. If very porous, they should not be more than 300 feet long. In heavy soils, the length may be as much as 600 feet. In this type of irrigation the rows extend at right angles to the ditches, and the water is most conveniently taken from the ditch by distributors previously described. It is usually desirable to turn the water into as many as 50 furrows at one time.

The Check System.—It consists of dividing the field into a number of small compartments, surrounded by low levees. The water is turned in these to the desired depth. This gives a rather complete control of

Celery Under Irrigation, Skinner System.[1]

the amount of water applied to each unit of ground. The size of the checks depends on the slope of the land, small checks being necessary where the slope is severe. This method is adapted to orchard irrigation.

Where water is conveyed through pipes and there is sufficient waterhead for pipe pressure, spraying irrigation may be resorted to. The Skinner system is probably the most successful of the several spray methods. It consists of a series of pipes at intervals of about forty feet, extending across the field to be irrigated. These are connected with a water main which is closed by a valve when not in use. The lines of pipe are supported at a height of about seven feet on posts, in such a way that the pipes may be turned. The pipes are fitted with small nozzles at intervals of about three feet. These should be in straight lines. The water issuing from them under high pressure is thrown a considerable

[1] Courtesy of The Pennsylvania Farmer.

distance in a fine spray. By turning the pipe, the water is directed to either side of the pipe line at the desired angle.

With the pipes parallel and the supporting posts in line at right angles to them, cultivation may take place in either direction beneath the pipes. While this system is rather expensive to install, it is well adapted to small areas intensively farmed, to truck crops and small fruits. Such systems are common along the Atlantic Seaboard and in some parts of the South.

Duty of Water.—This pertains to the area of land that may be irrigated with a unit of water, such as a "second foot" or a "miner's inch." The wasteful methods of irrigating and lack of knowledge on the part of the farmer result in a low duty. Under favorable conditions the duty should be about 200 acres for each "second foot." It would seem wise that the duty of water should be fixed within reasonable limits by some competent authority for a particular state or irrigation district. Local conditions, such as rainfall, length of growing season and the intensity of agriculture, should be taken into consideration in fixing the duty of water.

When to Irrigate.—How often to irrigate and how much water to apply will depend on local conditions, such as character of soil, kind of crop and weather conditions. Economy in water as well as the labor of irrigating, should make the intervals as long as feasible. Water should be applied until the soil is wet to the full depth to which the roots of the crop in question penetrate. The deeper the soil is wet, the longer may be the interval between irrigations. Lighter and more frequent irrigations penetrate the soil to less depth, increase the labor and result in greater loss of water by direct evaporation. Water should be applied when the crops need it and irrigation cease when the need is fully met. Enough water is better than too much.

Where there is a bountiful winter supply of water and a scant supply during the summer, winter irrigation is recommended. It stores the soil with water and lessens the need during the summer.

Water should be applied to crops abundantly when they are growing most rapidly. Irrigation may be withheld as they approach maturity.

Irrigation Waters.—Irrigation water sometimes becomes so heavily charged with salts that it proves harmful to tender plants. This condition arises either from concentration through evaporation in shallow reservoirs or from passing through alkali soil. Along stream courses, the reckless use of water gives rise to much seepage which returns to the stream lower down. This frequently becomes so plentiful that it forms a supply for another irrigation district further down the stream course. Such water is frequently unsuited for irrigation purposes.

Alkali Troubles.—The rise of alkali is generally caused by over-irrigation. An excess of water causes the ground water table to rise until the gravitational water can reach the surface by capillary attraction. This causes excessive evaporation at the surface of the soil and results in the accumulation of alkali salts. In time, the concentration will prevent the growth of crops. This can usually be avoided by greater care in irrigating. Where conditions are such that it cannot be avoided in this way, under-drainage should be installed. The alkali may now be washed out of the soil through the underdrains, by flooding the surface with fresh water. The use of alkali waters also stocks the soil with alkali salts. The use of such water should be avoided as far as possible, or the difficulty overcome by drainage and flooding as above mentioned.

REFERENCES

"Practical Farm Drainage." Elliott.
"Principles of Irrigation Practice." Widtsoe.
"Irrigation and Drainage." King.
"Irrigation Institutions." Mead.
"Practical Irrigation." Bowie.
"Irrigation." Newell.
"American Irrigation Farming." Olin.
Utah Expt. Station Bulletins:
115. "The Movement of Water in Irrigation."
118. "Method of Increasing Crop Producing Power of Water."
U. S. Dept. of Agriculture, O. E. S. Bulletins:
177. "Evaporation Losses in Irrigation and Water Requirements of Crops."
248. "Evaporation from Irrigation Soils."
Farmers' Bulletins, U. S. Dept. of Agriculture:
373. "Irrigation of Alfalfa."
371. "Drainage of Irrigated Lands."
392. "Irrigation of Sugar Beets."
394. "Use of Windmills in Irrigation."
399. "Irrigation of Grain."
404. "Irrigation of Orchards."
524. "Drainage on the Farm."
673. "Irrigation Practice in Rice Growing."
698. "Trenching Machinery Used for the Construction of Trenches for the Drains."

BOOK VII

FARM MANAGEMENT

CHAPTER 62

FARMING COMPARED WITH OTHER OCCUPATIONS

FARM MANAGEMENT

Columella, noted writer of the first century, who wrote extensively on agriculture, stated that there are three essentials to success in agriculture, namely, love for the occupation, knowledge and capital. These three are just as essential now as they were in the time of this noted writer.

Farm management is the application of the science and practice of soil management, crop production and the raising of farm animals, to the business of farming. It includes the planning of crop rotations; feeding systems; the employment, distribution and direction of labor; the equipment of the farm; the keeping of records and accounts of each farm enterprise; and the application of modern business methods. Farm management pertains to the farm as a unit, while rural economics, to which it is closely related, pertains to the community. How to produce any particular crop is a question of crop production, but whether to produce alfalfa or potatoes is farm management. Whether to sell a crop for cash or to feed it to livestock on the farm is farm management. The effect of tenantry on a community is rural economics, but the form of a lease and the equitable distribution of the proceeds between the tenant and the landlord is a problem in farm management. Farm management is concerned with the profitable relationship of all the factors of the individual farm. If any particular farm enterprise is over developed or insufficiently developed, it should be readjusted so as to bear the proper relationship to the other enterprises, and thus increase the profits of the farm as a whole.

Farm management calls for good judgment, practical experience and a knowledge of the principles underlying farming. A change in the management of a farm involves many questions, such as the determination of which crops are most profitable; should certain crops be fed or sold; when will it pay to purchase a machine to perform certain farm work; or when will it pay better to do it by hand; can soil fertility be maintained by the use of green manures and fertilizers as economically as by keeping livestock; will a three-course or four-course rotation be most economical; how many horses and men will be required to handle a certain acreage of crops; what is the most economical roughage on which to raise young animals; and when will a soiling system most economically replace pasturing of animals? These are some of the many questions in farm management.

Rather extensive investigations have been under way for the past few years with the view of ascertaining some of the fundamental facts in farm management. These investigations have given rather exact information relative to the cost of producing various crops, and of feeding and maintaining the different classes of livestock. It also has shown what it costs to produce one pound of butter, one dozen eggs or a hundred-weight of meat in different classes of animals. Likewise, it has shown the cost per hour of horse labor in many farm districts. This data is all available to the student of farm management and to the man who is about to embark in practical farming. It gives to him the means of calculating the probable cost and returns in a particular line of farming on a given scale.

Farming is so different from most other occupations that its advantages and disadvantages as compared with others call for careful consideration.

The Farmer as a Naturalist.—In no other occupation are men brought so intimately in contact with nature and nature's laws as in the occupation of farming. A knowledge of the laws of plant and animal growth add greatly to the interest in crop and animal production, even though the farmer may not be able to apply this knowledge. The fund of knowledge acquired by many generations of farmers is now being organized on a scientific basis, and new knowledge acquired through investigations is being added.

The Farmer as a Mechanic.—The rapid development of agricultural machinery and its extensive use in farming has increased rather than diminished the necessity of the mechanical ability of the farmer. This mechanical skill is a necessity, whether the farmer work by himself or whether he directs the labor of others. The farm laborer without mechanical skill may do more damage in an hour's time, if charged with operating a complicated machine, than his month's wages would cover. Machines often do work much better than when done by hand, and do it at a reduced cost, but their efficiency calls for proper adjustment and operation.

The Farmer as a Laborer.—While the introduction of machinery has reduced the necessity somewhat for brute strength and endurance, it has not and never will obviate the necessity of the farmer being able to perform manual labor with skill and dexterity. Farm labor is so diversified and calls for such a variety of motions on the part of the laborer that the highest skill in it is attained only through a number of years of practice. This is most readily acquired during early life, and men who have never been accustomed to farm work seldom find it advantageous to make the change after middle life. The same rule applies to men who have always lived on farms and who contemplate changing to some other form of business as conducted in cities.

The Farmer as a Business Man.—A little more than half a century ago farms were largely self-supporting. Farming then was not capitalized. It was diversified and most farmers produced nearly everything required for their livelihood. The wool produced on the farm was carded, spun and made into cloth on the farm, and the cloth in turn was made into garments, either by the housewife or the neighborhood tailor, who went from house to house to ply his trade. In the same manner the hides of the farm animals were tanned and made into gloves, boots and shoes for members of the family by the shoemaker who went from farm to farm for this purpose.

Conditions, however, have now greatly changed. Land values have risen, farming is capitalized and much more specialized than formerly. The standard of living has also risen, so that farmers sell and buy much more extensively than formerly.

The business ability required of the farmer is not so much that of the merchant or trader as it is that of the organizer who can organize the enterprises of the farm in such a way as to make production profitable. It is true that there is a constantly increasing need for the buying and selling ability of farmers also.

Personal Traits of the Farmer.—Success is most easily acquired in the line of work one loves best; and the first problem is to get into that character of work as soon as possible. Men cannot always advantageously estimate their own abilities, but so far as possible, one should engage in the occupation which he likes and for which he is best fitted by nature, experience and training. It is important for the young man to reach his decision as early as possible and then acquire the knowledge and experience as rapidly as possible. While men are sometimes quite successful with no particular qualifications except strength and industry, this is no argument that they would not have succeeded even better with knowledge and the application of science in their occupation. A good executive may have fair success without doing manual work, but in farming the highest success is usually attained by those who combine executive ability with labor. Scientific knowledge, experience, business ability, manual and mechanical skill, and hard work make a combination that is successful.

Farm Experience.—Experience is one of the essentials to success in farming. City industries are specialized. Farming calls for experience with the weather, diseases of plants and animals, insect pests, the feeding of plants and animals, breeding, machinery, business affairs and numerous other things that it takes time to learn. For one not reared on a farm, experience is best acquired by seeking employment under the direction of a successful farmer. If engaged in farming in a new locality, one should follow the practice of the most successful neighbors. When he can duplicate the results of his best neighbors, it will then be time enough to undertake changes that will make for improvement.

The Farm Hand.—Most of the farm hands in the United States are young men, a large percentage of whom later expect to become farmers. The farm hand receiving $25 a month with board, room and washing, will generally be able to save more than he would if employed in the city at $50 per month. Not only are his living expenses greatly reduced, but the temptation and opportunity to spend for pleasure is also reduced.

The clothing requirements of the farm hand are less expensive than those of his city brother.

It is true that the farm does not call for as many grades of men as do most other occupations, and there is, therefore, less opportunity for advancement. Only here and there are there opportunities for high-priced men as wage-earners. The farm hand who proves his worth will generally receive his reward by soon being able to rent a farm advantageously, and this will form a stepping stone to farm ownership.

Farm Ownership.—Farm ownership embraces two forms of investment: (1) a speculative and (2) a purely business venture.

A Good Rural Scene Showing an Attractive Farming Country.[1]

With the rapid increase in land values and the rapid improvement of farm equipment, the capital required in farming is much greater than formerly. A large part of the capital generally consists of land, and this, as a rule, is one of the safest investments. It is not subject to the severe losses that occur in commerce and manufacture. There may be periods of depreciation in land values, but these are not likely to continue long. As a rule the reverse will be true, and land values will increase and may comprise a large part of the farmer's increase in wealth. During the decade

[1] Courtesy of The Macmillan Company, N. Y. From "Crops and Methods for Soil Improvement," by Agee.

ending 1910, farm ownership in most parts of the United States was exceedingly profitable from the investment standpoint. During that period, land values in many of the states of the Union increased more than 100 per cent.

Ordinarily, the farmer invests in land in order to go into business with his capital. In the purchase of a farm both the speculative and business aspects should be carefully considered. Where the rise in land value is likely to be large and continue for some time, one may be justified in buying rather extensively and going in debt to the upper limit. On the other hand, where land is not increasing in value, it is much better to have ample capital for stock and equipment. Plenty of working capital is essential to good and profitable farming.

That part of the investment consisting of buildings, livestock and equipment is subject to destruction and depreciation just as rapidly as is the investment in buildings and equipment in any line of manufacture.

One should not expect as large a percentage when he invests money as when he goes into business with his money. Money in the form of mortgages on land is safe, calls for little attention, and, therefore, should demand only a moderate rate of interest. Such rate of interest ranges from 5 to 7 per cent at the present time in different sections of the country.

The Occupation of the Farmer.—In comparing farming with other occupations one should consider the capital invested, the safety of the investment, the probable returns, the hours of labor and the standard of living. It is somewhat difficult to compare these directly with the same items in other occupations.

While the hours of labor on farms are frequently longer than in other occupations, it should be borne in mind that there are generally periods when the hours are either quite short or there is almost nothing to do. This, together with the fact that the home is so closely associated with the work, compensates for the more regular and exacting hours in most other occupations. In large cities much time is often consumed in going to and from one's place of business.

Independence of Farming Occupation.—The farmer enjoys independence to a greater degree than men in most other occupations. He is his own boss, and is sure of self-employment. In nearly all occupations men who are employed for wages have fixed hours of labor and are in danger of being out of employment. These items often mean more than a man can fully realize until he has begun to work for others.

It Furnishes Employment for Children.—Modern conditions have made necessary the enactment of laws to regulate child labor. The steady employment of children in mills, factories and shops at the age when they should be in school tends to lower the standard of citizenship. It is advantageous, however, for boys and girls to have some light tasks to perform at regular intervals, but it is difficult to provide desirable work for children in cities. The farm offers an opportunity for just such employment, and is decidedly educational in its value, and may at the same time be made productive. Farm boys have the companionship of their fathers, and are often made to feel that they are taking a part in the work that he is doing.

The introduction of manual training into the schools of the cities is but a reflection of the necessity for training the city boy to use his hands in productive work. Habits of industry formed early in life are essential to a successful life.

Healthfulness of the Country.—Recent investigations by several life insurance companies show that longevity and diminished liability to disease are decidedly favored in the country. Typhoid, malaria, influenza, dysentery, apoplexy, peritonitis, paralysis and heart disease are found to be slightly more common in the country than in cities. A number of these diseases, however, are recognized as diseases of old age. The marked tendency during recent years for the young people of the farm to go towards the cities has resulted in a much greater percentage of old people on the farms than in cities. In cities venereal diseases, cancer, meningitis, enteritis, bronchitis, pneumonia, appendicitis, Bright's disease, tuberculosis, alcoholism and violence are much more common than in the country.

Sanitary conveniences and regulations are much better in the cities than in the country, but there is no reason why they should not be equally as good in the country, and the tendency is now marked in that direction.

The Farm a Home Enterprise.—In no other business is the home so closely related to the business as in farming. Success on the farm depends to a considerable extent on the home. Much of the farm work is done in the home, such, for example, as making butter, caring for the milk and preparing many of the articles for market. The care of poultry, the garden and the fruit often falls largely into the hands of the housewife and the children. For this reason, farming is a family business and unmarried farmers are at a decided disadvantage. Statistics show a very small percentage of farms operated by unmarried people. Such are usually the result of death of either the husband or wife, and frequently result in a change of occupation.

The Farm as a Home.—The advantages of the farm as a home are recognized to such an extent that we find close to cities and villages many who live on small farms, but who are engaged in some other occupation. The advantages lie largely in the reduced cost of living. Land, being less expensive, may be had in sufficient quantity to produce a large part of the products necessary for family use. These consist of fruits, vegetables, poultry, eggs, milk and butter, and in many cases a considerable part of the meat. Then, too, there are the advantages previously mentioned of affording employment for members of the family that would otherwise have no work to do. The work in this case becomes profitable, both from the educational and financial standpoints.

The farmer frequently fails to give the farm full credit for the prerequisites he receives from it. The farm is entitled to credit for the house rent saved and the provisions furnished to the farmer.

What the Farm Supplies.—The house which shelters the farmer and his family is considered a part of the farm investment, and in so far as the work of the farm is carried on in the house this is justifiable. On the other hand, it furnishes him a home which in any other occupation would have to be paid for either in the form of rent or as a separate investment. The average house rent value of the farmhouse is probably worth about $150 a year.

The farm supplies much besides money. It gives food, shelter and in most instances fuel. The United States Department of Agriculture has recently made investigations of between 400 and 500 farm families in different parts of the United States with the view of ascertaining the cost of maintaining persons on farms. It was found that the cost of main-

A WELL-PLANNED AND NEAT FARMSTEAD.

taining a grown person averaged $176 a year. Of this sum, only about twenty-two per cent was paid out in cash, the remainder being furnished by the farm.

Cost of Living on Farms.—The reduced cost of living on farms is due not only to the reduced cost of what is used, but, also, too frequently, to a reduced living. We often find that the farmer disposes of all the best of his products, while the family uses that which cannot be sold. This reduced standard of living is also manifest in the absence of such conveniences as the bathroom, running water, sewage disposal and modern systems of heating and lighting. It is true that some of these items are cheaper in cities than they are in the country. The individual lighting system for a country house may make such lighting more expensive than in the city, where the electricity or gas may be obtained from a company or municipality manufacturing it on a large scale. Likewise, running water may be cheaper in the city home than in the country. This reduced standard of living in the country is more often a necessity than otherwise, since most farmers who are financially able are installing such improvements. It is true that habit or custom has much to do with the providing of modern conveniences, and the introduction of such conveniences into any community generally spreads rapidly, especially among the younger people, as soon as means are available with which to supply them.

Uncertainties in Farming.—No other occupation is so largely dependent upon weather conditions, insect pests and plant and animal diseases as farming; consequently, it is not so fully controlled as is most other lines of business. In a shoe factory the output of each machine may be estimated within a fraction of one per cent. Knowing the cost of materials, the number of machines in operation and the capacity of each, the output of a shoe factory of any size may be closely calculated. Heat and cold, drought and rain, are likely to have little or no effect. With the farmer the season is so important that he may have an unusually large crop or he may have less than half a crop.

Then, too, his plan of operations may be frequently changed as a result of the change in weather. The men and teams that were to plow or cultivate in a certain field today may have to be directed to some other line of work, because rains prevent working in the field. It is, therefore, necessary to have plans that will include more than one line of work for any particular period. Fortunately, this uncertainty in the farmer's cash income is largely offset by his assurance in a livelihood, and a certainty that if some years are failures, others are just as certain to be doubly successful.

Preparation for Farming.—The opportunities on the farm hinge largely on the preparation of the farmer for the occupation. It is now recognized that farming calls for a greater diversity of knowledge and a greater variety of skill on the part of the workman than any other occupation. The unit of production of American farmers has increased rapidly as a result of better methods and the utilization of better farm equipment, and the tendency is to require a relatively smaller number of farmers in proportion to the entire population of the country. It seems probable that the time will soon come when one farmer will produce sufficient to feed four or five other families. This, of course, will mean that only fifteen or twenty per cent will be farmers, unless exports of farm products continue to be important.

One who is to farm should prepare for the business. The man who seeks an agricultural education as a preparation for farming is preparing not only for today, but for many years to come. The necessity for such training will increase rapidly from decade to decade. The establishment of state agricultural colleges and experiment stations for the education and enlightenment of the man on the farm is not only an opportunity that should not be neglected by the farmer, but, too, is indicative of the importance which agriculture plays in the welfare of the nation.

Back to the Farm.—The back to the farm movement naturally originated in cities as the result of increased cost of living and the apparent better opportunities for the man on the farm. Much has been written on this subject and a word of caution should be sounded, lest those that are illy prepared for farm life should engage in it, only to be greatly disappointed. They should thoroughly understand that the farm is no place for the inefficient. Those who are not capable of successfully directing their own efforts will do better to serve under the direction of others who are competent. In farming, one generally must direct his own operations. One without experience is necessarily handicapped when thrown into competition with those born and reared on farms. It is, therefore, suggested that those without experience who are anxious to acquire farms with a view of farming them successfully should first get experience by actually working on farms under the direction of the best and most practical farmers.

Back to the Village Movement.—There is no doubt but that cities have become over-populated, resulting in great congestion and sacrifice on the part of many of its residents. There are many industries crowded into great cities that could be operated to better advantage in villages. The demands of a great city call for so much in the way of food products that much of it must necessarily be shipped long distances, while the products manufactured in the city must in turn be sent back to the people in the country, thus entailing much expense in transportation. The cheaper living in the country would do much to reduce the cost of the manufactured article and should at the same time afford the farmer equally as good prices for his products. It is true that it might reduce somewhat the demands on the farmer, since the provision of small parcels of land for families in towns and village would enable each family to produce part of its own food supply. Whatever might be the disadvantage on the part of the farmer, it would certainly be an advantage to the country as a whole.

The Farm Manager.—The purchase of farms by men in other forms of business has given rise to a demand for men trained as farm managers. In many cases such farms are used chiefly as country homes, and cannot be expected to pay interest on over-capitalization. Farming, to be profitable, must be placed on a business basis. It is not possible for the ordinary cow to make a profit on the feed consumed, the labor required, and at the same time pay 6 per cent interest on a barn that costs $1000 per cow. The young man who engages as a farm manager is cautioned to accept such a position only after careful consideration and with a clear understanding as to what he is expected to accomplish. He should know at the outset whether or not he is merely to operate a country estate for the pleasure of its owner and his friends, or whether he is expected to run the farm on a business basis with the expectation of making it pay. The man trained as a farm manager or superintendent should be given much latitude and allowed to plan the operations of the farm. The general policies of operation may be outlined by the owner, but if it is to be successful, he cannot expect to direct the management in any detail without being thoroughly conversant with the business of farming.

The Farmer's Labor Income.—Farm surveys in many sections of the country have recently given us data showing the farmer's income. It has never been possible to figure from statistical returns as printed by the United States census exactly what the average farmer of the United States makes. The last census, however, was somewhat more complete than former ones, and when supplemented with a few facts established by the office of farm management of the United States Department of Agriculture, it has been possible to estimate closely the average labor income of farmers of the United States. According to the last census there were in round numbers 6,361,000 farms in the United States. The average size of the farm was 138 acres, 75 acres of which were improved and 50 acres of which were devoted to crops. The average farm investment was $6444, of which amount $994 was in buildings and $199 in implements and machinery. The average gross farm income per farm was $981 and the total expenses $340, leaving a net farm income of $640. If interest on the investment at the rate of 5 per cent is deducted from the net farm income, it leaves a labor income of only $318. This differs very little from the average hired man's wages.

There are many farmers having large farms that are living on the interest of their investment. On the other hand, there are many who are making very good labor incomes.

The following table shows the average labor income of a considerable number of farmers, taken just as they come in different sections of the country for the years indicated in the table:

LABOR INCOME OF FARMERS IN DIFFERENT PARTS OF THE UNITED STATES.

Operated by Owners.

Number of Farms.	County.	State.	Average Size, Acres.	Average Value.	Labor Income.	Years.
6,361,502	United States[1]		138	$6,444	$318	1909
193	Chemung	New York	...	4,642	190	1911
615	Tompkins	New York	108	5,712	423	1907
670	Jefferson	New York	143	9,006	609	1910
178	State	New York	132	11,137	981	1906–7
578	Livingston	New York	148	12,037	666	1909
410	Chester	Pennsylvania	88	11,815	743	1912
123	Clinton and Tipton	Indiana	105	17,535	310	1911
77	Green and Guthrie	Iowa	176	23,193	291	1911
73	Menard and Cass	Illinois	253	51,091	622	1911

[1] Calculated from census figures.

Operated by Tenants.

Number of Farms.	County.	State.	Average Size, Acres.	Average Value.	Labor Income.	Years.
127.......	Chester...........	Pennsylvania	104	....	$778	1912
83.......	Clinton and Tipton.	Indiana.....	128	Cap. $1,728	755	1911
93.......	Green and Guthrie.	Iowa........	187	Cap. 2,667	716	1911
71.......	Menard and Cass..	Illinois......	202	Cap. 2,867	1,139	1911

In studying the above tabulation it should be understood that the year in which the survey was made may or may not have been normal, and the labor income for any particular year is influenced greatly by the character of the season. Many facts are brought out by these surveys which will not be fully discussed at this point. In general, however, the larger the farm or the larger the investment, the larger the labor income. Increase in working capital, either in the form of livestock or equipment, also makes for larger incomes. There is considerable difference in labor incomes from farms as the result of location. This is due partly to soil and climatic conditions and partly to transportation and markets. How much is due to methods of farming is difficult to ascertain, since there are so many factors that influence the final result.

Profits in Farming.—These vary from time to time as the result of cycles of over and under-production. From 1870 to 1895 farm profits were small because of over-production during that period. As a result, prices of all farm products were exceedingly low. This helped to make the cost of living in cities low, and since there was a great development along the lines of manufacturing and merchandising, city opportunities were apparently better than on the farm. Many farm people consequently moved to the cities. This was especially marked in case of the younger people. From 1895 there has been a gradual change. Prices of farm products have increased greatly. Farm profits consequently have been larger. The farm has become more attractive, and there has been a tendency for city people to move back to the land.

It should be borne in mind that farming is a conservative business. With good management, it yields good returns, but fortunes can hardly be expected from farms of average size. Farmers may hope to make a comfortable living, be able to travel some, educate their children as well as city folk and save up a competence sufficient for old age. It is doubtful if more than this is of any particular advantage.

REFERENCES

New York Expt. Station Bulletin 295. "An Agricultural Survey of Tompkins County, N. Y."

Farmers' Bulletin 432, U. S. Dept. of Agriculture. "How a City Family Managed a Farm."

CHAPTER 63

Factors that Determine Best Type of Farming

Type of farming pertains to the enterprises that make up the farm business. The farm may be devoted to a single crop, such as wheat, or it may consist of a number of crops in combination with the production of dairy products or meat. In the one case it would be a special type; in the other, a general type.

It is important that the many factors that influence the type of farming harmonize with all the farm enterprises. When a combination of crops and animals can be selected, all of which are adapted to the climate, soil and topography of the farm, and when first-class transportation facilities or good markets for all of the products are at hand, farming is comparatively easy, and success is assured if the farmer does his part.

The Man.—By nature, training and experience, men are better adapted to some lines of production than others. It is a significant fact, however, that men soon learn to like most any type of farming that is financially successful, and to dislike that type that is unsuccessful. While success may depend to a considerable extent upon the man, there are many types of farming that are impractical in certain localities because some of the factors necessary for success are lacking.

Climate.—Sunshine and rain are important factors in plant growth. They are beyond the control of man, and this makes it necessary for him to work in co-operation with climatic conditions as far as possible.

Cotton cannot be grown successfully beyond the limits of the cotton belt, chiefly because climatic conditions are unfavorable farther north. Corn requires plenty of rainfall and warm weather during its growing season. It does poorly in the arid regions, even though it bé irrigated. It is also an unsatisfactory crop where the summer season is short and the summer nights are cool. Potatoes, Canada peas and other crops will not succeed under the warm weather conditions of the Southern states. It is, therefore, important to have the crop adapted to climatic conditions.

The Soil.—Soil adaptation has been thoroughly discussed in the first part of this book. Soils may be somewhat modified, and there is more range in soil adaptation than there is in climatic adaptation. Where markets are especially good, one may be justified in growing a crop to which the soil is not as well adapted as it may be elsewhere, but the crop should always be adapted to the soil as far as other conditions will justify.

Topography.—Comparatively level land may be devoted to any kind of crops to which the soil and climate are adapted. It permits the use of any kind of machinery. A rugged topography, on the other hand, may make impossible the use of a self-binder, and, consequently, makes impossible the successful growing of small grains in competition with their production on the level lands adapted to machinery. The use of machinery has reduced the cost of production of these crops to such an extent, and transportation facilities have so reduced the differences in price in different localities, that the whole country is brought into competition in the production of most of the cereals.

Erosion on steep land is often so severe as to make it necessary to grow only those crops that tend to hold the soil in place. Cultivated crops, such as corn, encourage the loss of soil by erosion, and should not be grown on steep hillsides.

Location.—In general, the nearer one is located to good markets, the

TYPICAL CORN LAND.[1]
A 320-acre cornfield, Illinois.

more intensive the type of farming may be. Fruits and truck crops close to good markets may be produced profitably even though the soil and the climatic conditions are not ideal. The farmer who can secure high prices can afford to spend more in modifying the soil in order to produce crops to which it is naturally not best adapted. He may irrigate his crops if there is a shortage of rain, providing prices are uniformly good.

Far away from good markets the type of farming should become more general and more extensive. It will generally tend more toward the production of livestock and livestock products that may be marketed with the minimum expense.

Neighbors.—The character of neighbors is generally considered chiefly from the social standpoint, but neighbors also have an economic value. The type of farming that prevails in any neighborhood will generally be the one to adopt. There will be advantages in several ways.

[1] Courtesy of E. L. Worthen.

(1) Neighbors have an advertising value. The neighborhood noted for the production of fine apples or peaches attracts buyers for these products. Prices in such neighborhoods are generally higher than in regions that have no reputation for the same product.

(2) The labor in such a neighborhood is trained in the operations that pertain to the industry and work more rapidly and with greater care than untrained labor in that kind of work. New crops often fail because labor adapted to their production is lacking. Cotton picking becomes an art and calls for cheap labor. The introduction of Sea Island cotton in the island of Porto Rico a few years ago failed solely because the available

THE UTILIZATION OF LAND TOO STEEP TO PLOW.[1]

labor was not trained in cotton picking, and the crop could be harvested only at large expense.

(3) The prevailing breeds of livestock in the neighborhood facilitate securing the services of the best sires.

(4) The experience of the neighbors is of value to all who are engaged in the prevailing type of farming.

Occasionally, changed economic conditions may justify a change in the type of farming. The growth of a town, creating a new local market, will make possible successful truck and fruit farming on a scale sufficient to meet the local demand, where previously it would not have been successful. Then, too, new enterprises may occasionally be discovered that will prove more profitable than the old ones. The introduction of such enter-

[1] Courtesy of Department of Animal Husbandry, Pennsylvania State College.

prises often entails considerable expense in improvements or in the equipment of the farm, as well as necessitating the training of labor to meet the demands of the new enterprise. Changes in the type of farming must, therefore, be rather gradual.

Markets.—The distance from market or shipping point often determines the character of product to be grown. Soil and climate may be admirably adapted to the production of potatoes, but if the farm is some distance from the market or shipping point the cost of marketing may be too great to make production of potatoes profitable. The character of road between the farm and market is also a factor. The cost of hauling on poor, hilly roads may be two or three times as great as on level macadam

INTENSIVE FARMING ON A LARGE SCALE.[1]

roads. The better the road, the further one may be from market and yet be successful. The prices of wheat and hay near large Eastern markets are better than in the West. Local conditions may be favorable to the production of both crops, but the difference in prices of the two commodities in the two localities will be greater in case of hay. This often makes it advisable for the eastern farmer to produce hay rather than wheat.

Market milk which must be delivered daily cannot be hauled long distances to market with profit. Many farmers spend too much time in the delivery of small quantities of milk. Often the time spent is worth as much as the receipts from the milk sold. A part of this difficulty may be overcome by co-operation and hauling milk in full loads, thus reducing the cost of hauling.

Transportation.—It is a fundamental principle that products cheaply

[1] Courtesy of Wallace's Farmer, Sioux City, Iowa.

and easily transported may be successfully grown some distance from markets, while bulky and perishable products should be grown near points of consumption. The value of a product is also a factor in this connection. The greater the value per unit of weight, the less important will be the relative cost of transportation in connection with placing it on the market. The greater the distance between points of production and market, the greater the difference in price; and the more bulky the product, the greater the relative difference. To illustrate: hay in the Boston market may be worth $20 per ton, whereas 1000 miles away it would be worth less than $10 a ton. At the same time and for the same places the difference in concentrated products, such as butter or cheese, would probably be no more than 8 to 10 per cent in price.

One bushel of corn will produce ten pounds of pork, but the ten pounds of pork can be shipped to market at considerably less cost than the bushel of corn. Consequently, the best place to raise hogs is in the corn belt where corn is cheapest. Iowa excels all other states in pork production.

Sheep may be reared for their wool in regions remote from market, because wool is a high-priced product and may be hauled and shipped long distances, and the cost of marketing is comparatively small in comparison with its value.

Horses are generally more valuable, weight for weight, than cattle or sheep, but they are more difficult to ship, because they are more subject to injury in transportation.

It is very difficult to get eggs to market in prime condition, and the egg producer near the markets has a decided advantage. In country districts eggs are generally produced on so small a scale that they are marketed by the farmer to advantage only in connection with his going to the market place for other purposes.

Fresh fruits, such as strawberries, raspberries and cherries, must be marketed promptly, and are, therefore, best produced near points of consumption. On the other hand, apples may be shipped long distances.

Supply and Demand.—The supply and demand of all farm products are subject to greater fluctuation than any other line of production. Favorable seasons over large areas often cause such abundant crops that the market is oversupplied and prices may be so reduced that the large crop nets the farmer less than a small one would. One needs only to study the crop statistics to be impressed with the violent fluctuations in both yields and prices, even of our staple crops. Corn in Illinois has varied from 21 bushels to 40 bushels per acre; wheat in Kansas from 10 to 18 bushels; cotton in Texas from 125 to 225 pounds per acre. The fluctuation in truck crops and fruits is even more striking, while that in special crops, such as broom corn, is greater stilll Prices of broom corn may range all the way from $25 to $200 per ton, due chiefly to supply and demand.

Crops are frequently subject to cycles of over and under production. When prices are high, farmers are apt to think they are going to continue to be high; consequently, they increase their acreage of those products which command the best price. This generally results in over production and the consequent decrease in price. Farmers then think that low prices will continue, and restrict their production to such an extent that prices again rise. This process repeats itself periodically and nearly all products are consequently subject to cycles of over and under production. These cycles vary greatly in their length of duration. They are longest in case of crops that require much time in coming into full fruitage, such as apples and other tree fruits.

In apple production the cycles range from twenty to twenty-five years in length. In annual crops they are subject to very short cycles of two or three years in length. With livestock, the longer cycles are with horses that require four or five years to come into maturity, while the short ones are with swine that mature in a year or less.

It is worth while for farmers to study crop statistics and crop prices with a view of meeting the demands as nearly as possible. Such forethought on the part of the farmer will lessen the variation in both supply and price, and make more certain the probable price.

Animals.—Livestock production is affected by climate and the grazing period. Beef cattle have never been successfully raised in the Northern states, because of the short grazing period and the long winter feeding period. With higher prices for meat it is possible that beef cattle may be successfully raised in that region.

Long, cold winters require better housing facilities for livestock and more labor in the feeding of them. Both add to the cost of production.

Horses are not adapted to hard work in tropical regions. Mules and oxen are better adapted to such regions.

Labor.—The type of farming is dependent to a considerable extent upon the character and supply of labor. Improved livestock and the extensive use of agricultural machinery call for skilled labor. Dairying calls for men trained in milking. Attempts to establish dairying in the South have often failed because of the unreliable nature of available help and its lack of adaptation for the dairy business. Sugar beets and most vegetables call for much hand labor, and cannot be extensively grown unless labor is comparatively cheap and abundant. The weeding of beets and the picking of berries, if done economically, must be done by cheap labor.

Competing Types.—The crops which pay best are the ones that should be given prominence. Corn may be profitably grown near cities, but it cannot compete with potatoes, sweet corn and other crops requiring similar soil and climatic conditions, and for which there is a good market. Root crops make excellent stock feed, but cannot compete with corn which produces equally as good winter feed in the form of silage and which can be grown more cheaply.

Natural Enemies.—Plant diseases, insect enemies and weeds may seriously affect the type of farming. In some instances an entire change in system has been necessary because of these factors. The peach yellows in Delaware practically wiped out the peach industry in that state at one time. Cattle raising and dairying in the South have been held back because of the prevalence of Texas fever. In parts of the cotton belt, the cotton-boll weevil has caused the introduction of new crops and crop rotations. The weevil in this case has doubtless been a blessing in disguise. The corn-root worm has forced rotations in portions of the corn belt. An abundance of noxious weeds may prevent the profitable production of such crops as beets.

Land Value.—The rise in land values necessitates gradually increasing the intensity in the type of farming and the replacing of crops of low value with those of higher value. There must be a rather definite relationship between land values and values of crops produced. Where land values are low it may pay better to own more land and farm more extensively than to apply intensive methods. It should be remembered that the product per man is more important than the product per acre so long as land is fairly abundant. In time and with increase in population the product per acre will become relatively more important, and may be of more benefit to the community as a whole.

Capital.—The type of farming may, to a certain extent, be adjusted to the available capital. With small capital it may be advisable to depend chiefly upon crops, rather than engage in both crop and livestock production. The crops alone will require less capital and generally bring quicker returns.

Types of farming that require a long time for returns demand more capital than those bringing quick returns. Orcharding or the raising of horses are examples of slow returns and expenditure of much capital. The production of cash crops, dairy products and swine are examples of quick returns, although considerable initial capital will be required in land and stock. The permanent investment should be large, but the working capital may be very small. Men who engage in orcharding must either have some other source of income on which to live until the orchard comes in bearing, or devote the orchard to crop production until it begins to bear fruit.

Changing Type of Farming.—The wrong type of farming may continue for some time because of the difficulty in making a change. A wrong type of farming usually results in bad financial condition of the farmer. A radical change in type will generally call for increased expenditures. This may be in the form of new buildings, additional equipment or the purchase of livestock. This makes the problem difficult for the poor man, even though he may fully realize the benefits to be derived from such a change. The opening of the country in the Central West and the development of cereal production there necessitated a readjustment of Eastern agriculture, but it took much time to complete the change, and even yet some farmers cling to the old enterprises when they would do much better by making a change.

Successful Types of Farming.—Agricultural surveys conducted in all parts of the United States reveal the fact that there are many successful types of farming, but in the field of general agriculture, diversity is usually the keynote of success. It has many advantages. It provides for the most uniform distribution of labor; it facilitates crop rotations; crop rotations help to maintain soil productivity, reduce plant diseases and overcome the depredations of insects.

It is very important to have the type of farming fit all local conditions as far as possible, and farmers are urged to make a careful study of the subject.

REFERENCES

"How to Choose a Farm." Hunt.
New Mexico Expt. Station Bulletin 59. "Forty Years of New Mexico Climate."
North Dakota Expt. Station Bulletin 52. "The Length of Growing Season in North Dakota."
U. S. Dept. of Agriculture, Bureau of Soils. "Field Operations."
U. S. Dept. of Agriculture, Weather Bureau. "Weather Reports."
U. S. Dept. of Agriculture, Year-Book, 1908, pages 289-300. "The So-Called Change of Climate in the Semi-Arid West."
U. S. Dept. of Agriculture, Year-Book 1906, pages 181-189. "The Use of Soil Surveys."
U. S. Dept. of Agriculture, Forest Service Circular 159. "The Future Use of Land in the U. S."
U. S. Dept. of Agriculture, Biological Survey, Bulletin 10. "Life Zones and Crop Zones."
U. S. Dept. of Agriculture, Biological Survey, Bulletin 10. "Distribution of Cereals in North America."
Colorado Expt. Station Bulletin 127. "Climate of Colorado."
Farmers' Bulletins, U. S. Dept. of Agriculture:
454. "A Successful New York Farm."
472. "Systems of Farming in Central New Jersey."

CHAPTER 64

COST OF PRODUCTION

Profits in farming depend on many factors, among which the cost of production and the selling price are the most important. The chief difficulty at the present time lies in the fact that few farmers actually know what it costs to produce a pound, bushel or ton of the various farm products. The cost of production is more fully under the farmer's control than is the selling price.

The method of procedure in ascertaining the cost of producing different crop and animal products on the farm will be outlined in another chapter.

Cost Depends on Yields.—The cost of production per acre varies less than the cost per unit of product, chiefly because of the wide variation in yields. It costs a definite amount for the rent of land, taxes, labor of men and teams, use of equipment and purchase of seed and fertilizers for an acre of land, whether it produces a full crop or a half crop. It is, therefore, important that yields be made as large as possible by every economic method. This will necessitate the use of the best strains of seed, the application of the optimum amount of manures and fertilizers and the right amount of labor per unit of land. The cost of producing 60 bushels of corn per acre is very little greater than that of 30 bushels per acre. Dividing the cost by 60 and 30 respectively, the cost per bushel is found to be nearly twice as great in case of the smaller yield.

The largest possible yield may not be the most profitable one. Yields are affected not only by soil fertility, but by the amount of labor and fertilizer applied to the land. In general, the more valuable the crop, the more intensive may be the cultivation. Intensive methods applied to the production of hay and cereals may actually reduce profits. A recent farm survey of 135 farms ranging from 80 to 120 acres each, gave an average labor income of $421 per farm. On 14 of these farms crop yields were 16.3 per cent above the average, but the labor income was $1 below the average. In this case the cost of production was so much greater than the average that the labor income fell somewhat below the standard of the community. On 26 other farms where the crop yields were 12 per cent below the average, the labor income averaged $835 per farm, or 98 per cent above the average. In this case the cost of production was so reduced as to increase the profit. Evidently, this survey was in a district where some were employing methods too intensive for the type of farming. This is probably unusual. Generally, crop yields may be increased 20 to 25 per cent with a corresponding increase in profits. Farm surveys of over 1300 farms in New York State showed the crop yields on 13 farms making labor incomes of over $2000 each, were 27 per cent better than the average. Increased yields are not difficult to secure, but increased profits are a different matter. It requires very careful study to know just what degree of intensity will bring the best results. It is granted that weather conditions, insect enemies and plant diseases are factors that may materially modify what would be otherwise normal yields.

Product per Animal.—The difference in the cost of rearing and maintaining animals is less than the difference in the value of their products. It costs nearly as much to feed and care for an unproductive cow as it does for one that is highly productive. A man spends as much time in driving a little horse or a small team as he does in driving a large horse or a team consisting of several horses. The one will do much more work than the other, and hence increases the efficiency of the unit of man labor. The small horse usually occupies a stall in the stable equally as large as the larger one, and requires the same amount of labor in his care, although he may not consume as much feed. The cost of producing a quart of milk or a pound of butter is, therefore, determined to a considerable extent by the natural productivity of the animal. Productive cows are generally profitable, but there are many in nearly every herd that do not pay for their feed and labor. The problem becomes one of eliminating the unproductive cows.

The following table, taken from "Farm Management," by Warren, shows the relation of size of cows to value of product above food cost:

Weight of Cows, pounds.	Average Weight, pounds.	Number of Cows.	Butter Fat, pounds.	Value of Product.	Value of Feed.	Value of Product for $1 in Feed.	Value of Product Above Cost of Feed.
900 and under	847	87	366.2	$114.52	$60.32	$1.90	$54.20
901–1000....	952	82	417.8	131.22	69.86	1.88	61.36
1001–1100....	1071	53	447.8	142.56	76.28	1.87	66.28
1101–1200....	1175	60	477.7	155.02	82.81	1.87	72.21
1201–1300....	1276	31	506.2	163.52	91.51	1.79	72.01
1301–1400....	1379	26	525.8	171.79	92.15	1.86	79.64
Over 1400....	1556	16	566.6	184.61	96.60	1.91	88.01

Data furnished by F. W. Woll, for cows whose records are reported in Wisconsin Bulletin No. 226.

The cost of producing dairy products varies greatly with location, and is dependent upon cost of feed, price of labor and cost of housing, as well as upon the character and efficiency of the animals. The cost of producing dairy products under different conditions has been worked out at a number of experiment stations, as well as by farm surveys for farm conditions. In general, feed constitutes just about 50 per cent of the cost of producing milk or butter. Labor constitutes about 25 per cent, and the other items, such as interest, depreciation and housing make up the other 25 per cent.

Labor of Men and Teams.—In the production of most crops, the largest items of cost are the labor of men and teams. The more intensive the character of farming, the larger becomes the relative importance of these items. In the grazing of animals, the crop is harvested by the animals with very little expense on the part of the farmer. In the production of some of the more extensive crops, such as oats and wheat, on land of high value, land rental becomes almost as large as that of labor. Since labor is so important, it becomes essential to utilize labor as fully as possible. This means that there should be diversity of enterprises in order to afford continuous employment. It means that each enterprise should be sufficiently large so that time will not be wasted in numerous changes from one piece of work to another, and so that fields will be sufficiently large to avoid waste of land and waste of time in the turning of teams and implements in the process of tillage.

An Efficient Team.[1]

The efficiency of man labor in field operations, as previously indicated, is greatly increased by driving large teams and using large machines. A man plowing with a two-horse plow does work valued at about $4 per day. With a four-horse team and a gang plow he may accomplish work valued at just twice as much.

Equipment.—The completeness of the farm equipment affects the cost of production in two ways. Machines of various kinds increase the work that may be accomplished by the unit of labor, and in many cases do the work more uniformly than it can be done by hand. The manure spreader spreads more perfectly than can the man spreading with a fork. Corn planters drop with much accuracy and seeding machinery seeds more uniformly than seeding can be done by hand.

The cost of machinery and the extent of its utilization determine its effect upon cost of production. Much of the farm machinery is used for a very short period of time. The self-binder costing $125 is frequently used only two or three days in the year. The interest on the investment and the depreciation, therefore, make the cost of harvesting even with a binder rather expensive. The same will be true of many other machines on the farm. The farmer can better afford to cut fifteen acres of corn by hand than he can to own a corn harvester for cutting so small an area. Where the acreage is large, there can be much saving on the cost of the

[1] Courtesy of Virginia-Carolina Chemical Company, Richmond, Va. From V.-C. Crop Books.

operation. A plow costing $15 if used to plow only 10 acres a year, would cost 24 cents per acre, but if the same plow is used to plow 50 acres, the cost is reduced to 5 cents per acre. It is necessary, therefore, to estimate carefully whether or not, in the purchase of farm machinery, it will be used enough to make it a paying investment.

Land Values.—The value of land has a rather definite relation to cost production. Since the interest on its value or the rental is definite for any particular locality, the higher the price of land, the greater is the interest or rental as an item in the cost production. Increased land values generally call for a gradual increase in the intensity of the type of farming. The more intensive the type of farming, the less important land becomes as a factor in the cost of production. In the production of such crops as mangels and potatoes, land rental usually constitutes only 10 or 15 per cent of the cost production. In the raising of hay it may be 50 per cent or even more of the cost production. For such crops as the cereals, it will usually range from 20 to 35 per cent of the cost.

Taxes, Insurance and Depreciation.—These items, while of minor importance, have a direct effect upon the cost of production, and vary with the nature of the product in question. Taxes and insurance will be fairly uniform, but depreciation is quite variable. In case of buildings and equipment, it is usual to allow about 10 per cent annually for this item. For some farm implements 5 per cent is sufficient, while for others that wear out quickly or go out of date because of rapid change in improvement, 10 per cent is not sufficient.

The depreciation of work animals and cows is also variable, and is largest in case of the more valuable animals. In figuring depreciation on cows, it is necessary to first take into account the difference in the value for milk and beef purposes. A cow as a productive animal may be worth $200, while she would not bring more than $40 when sold for beef. The average milking life of a cow is about seven years. The difference between the milk and beef value, which in this case would be $160, divided by 7, the number of years in milk, would give the major portion of the annual depreciation. To this should be added the mortality in cows, which averages about 1.2 per cent annually. Likewise, the interest on the valuable cow is much greater than on an ordinary one. This all increases the cost of production.

Intensity.—In every locality and for each type of farming there is an optimum degree of intensity that will bring maximum profit or minimum cost of production per unit of product. In general, the higher the price of land and the higher the value of product, the greater may be the intensity. High-priced labor has just the reverse effect. Cheap labor encourages intensity. Each producer must carefully consider the amount of labor and the value of fertilizer that can be applied to each acre of land in order to bring the largest profit.

Size of Business.—Careful investigations show that it costs less per

AN EXAMPLE OF CHEAP LABOR.[1]

[1] Courtesy of Virginia-Carolina Chemical Company, Richmond, Va. From V.-C. Fertilizer Crop Books.

acre to produce crops on large farms than it does on small ones. The same is true in the cost of producing livestock and livestock products. The larger farms are enabled to make a fuller use of the labor of men and horses, as well as equipment. It costs nearly as much to equ p an 80-acre farm as it does a 160-acre farm, so that the equipment cost per acre is little more than half as much for the larger size as it is for the smaller. The buildings on an 80-acre farm will cost nearly as much as on a 160-acre farm, so that the cost of housing per animal is relatively less in case of the larger farm. Larger farms make for the use of large teams and large implements, thus reducing the cost for labor. They also offer better opportunities for more continuous employment and greater diversity than do small farms. They reduce the waste of time that is encountered in the cultivation of small fields.

An agricultural survey of 586 farms in Tompkins County, N Y., showed that the labor cost per acre of producing crops ranged from $3.33 for farms averaging 261 acres to $19.90 for farms averaging 21 acres in size. In other words, the area farmed with $100 worth of labor ranged from only 5 acres in the case of small farms to as much as 30 acres in case of the largest farms. It is true that the receipts per acre on the smaller farms were somewhat larger than on the larger farms, but the difference was nowhere in proportion to the difference in labor cost. On the larger farms there was a profit, while on the smaller farms there was a decided loss.

On 578 farms in Livingston County, N. Y., $100 worth of labor farmed only 4.4 acres in case of farms of 30 acres or less, and 21.8 acres in case of farms of over 200 acres in size. The acres per horse ranged from 15 in case of small farms to as much as 49 in case of the large farms. Size of business or size of farms, therefore, is rather important in keeping down the cost of production.

Character of Feed.—In the production of animals and animal products the character of feed that can be most economically used should be carefully considered. While balanced rations are desirable, the value of the different classes of feed must be considered, and those selected that will give good results at the minimum cost. When concentrates are high and roughages low, the amount of roughage may be increased and a small sacrifice made on the production side with enough saving in the cost of feed to make the profits greater than they would be by using large amounts of concentrates at high cost in order to secure high production. When concentrates are cheap and roughages relatively high, the reverse may be advisable.

In general, young stock should be fed largely on cheap feeds, much of which will be roughage. Feeds that are not marketable at good prices can generally be used for some kind of livestock production.

Class of Labor.—The skill of the labor employed should correspond with the character of work to be done. Skilled labor at high cost would very materially increase the cost of production of most intensive crops where much hand labor is required. Women, children and other cheap forms of labor may be utilized more economically in much of such work.

In the production of livestock and livestock products, including poultry, there is considerable work that may be performed by the women and children on farms, such as the care of the milk, the making of butter, the feeding, care and marketing of poultry and poultry products. This all tends to reduce the cost of production.

Utilization of Time and Products.—On all farms a flock of 50 to 100 chickens, a few pigs and a few larger animals may be kept on what would otherwise be absolutely wasted. The time required in caring for these will not necessarily interfere with other lines of production. Such items when attended to on over 6,000,000 farms, become of much importance, and materially reduces the cost of production of a vast amount of food products. More than 90 per cent of the eggs are produced in this way on general farms. Much of the dairy products come from small farms that keep only a few cows. Those products are consequently produced with cheap feed and low-priced labor. This necessarily makes the margin of profit for the man who engages in either poultry or dairying as a specialty, very narrow.

Amount of Waste.—Perhaps no business permits as large waste as is incurred in farming. There is waste in labor of horses and men, waste in products through delay in harvesting and marketing or through improper storage. The farmer, as well as other classes, must live, and these losses must be made up by a little better price on that which finally reaches the market in good condition. Waste of whatever nature, therefore, tends to increase the cost of production.

Fertility of Land.—Since land values play a minor part in the cost of production of most crops, fertile soil, even though the land may cost more than where the soil is poor, generally makes for reduced cost of production. Land that will produce one-fourth more than average land should command a price that is much more than one-fourth above the average price of land. The most successful farmers are usually on highly productive land, regardless of its money value.

Weather Conditions.—The cost of producing both crop and animal products is affected by weather conditions. Long periods without rain so reduce the yield that the cost per unit of product is necessarily greatly increased. In the same way, periods of wet weather may prevent proper tillage and reduce yields to such an extent that the cost is greatly increased. Unusually long, severe, stormy winter periods increase the amount of feed required for livestock, and thus increase the cost of their products. In like manner, droughts which affect pastures necessitate supplementing with feed at some increase in cost of production. Seasons are uncertain and play an important part in both cost of production and the probable price. Fortunately, reduced production and increased cost of production are generally offset to considerable degree by increased price of products.

Weeds, Insects and Diseases.—These all reduce the yields in proportion to their prevalence and, consequently, increase the cost of production. They not only increase the cost in this way, but they call for extra expense in production. Weeds necessitate additional tillage; insects call for the use of spray materials, as do also diseases. Diseases increase not only the cost of production in crops, but may in the same manner affect the cost of producing animals and animal products, since they likewise are subject to disease which may cause loss and call for additional expense in treatment.

Efficiency.—A lack of efficiency increases the cost of production. Efficiency is made up of little things, and includes the wasted moments and hours, as well as the waste of products. Efficiency calls for the proper care of implements and animals on the farm, the preparation of land and the planting of crops at the most favorable time. Destruction of weeds is much less expensive when tillage is given in the nick of time. Efficiency calls for the utilization of time, of men and teams to the fullest possible advantage. The loss entailed by hunting for some missing part of a machine or mending a long-neglected break is often serious and adds materially to the cost of production.

Reducing Expenses.—Expenses in the cost of production may be reduced by an actual reduction in the expenditures for labor of horses and men and machinery, or the business may be increased in order to more fully utilize the present force and equipment. On many farms horses work on an average only three hours per day, whereas by proper organization a number of work horses might be dispensed with and the hours per horse increased. This reduces the actual expenses. In the same way the expense for labor may be reduced. The following tabulation taken from the *Tribune Farmer* shows the distribution of horse labor for a year on a New York farm:

DISTRIBUTION OF HORSE LABOR ON A NEW YORK FARM.

Month.	No. 1.	No. 2.	No. 3.	No. 4.	No. 5, Extra.	Total, Four Horses	Average Hours a Day Each Horse Worked.	Average Rate an Hour at $10 a Month Cost.
January	8½	12½	..	12¾	..	33¾	.3	$1.180
February	3½	7½	..	4	..	15	.2	2.666
March	23½	25½	..	..	..	49	.5	0.816
April	78	86½	..	3	..	167½	1.5	0.240
May	198	183	110	83	9½	574	5.3	0.070
June	223½	222½	175	149	45½	770	7.7	0.052
July	220	201	122¾	75¾	20	619½	5.7	0.064
August	125	132½	56½	27½	36	341½	3.2	0.117
September	112½	120	17	28	23	277½	2.8	0.144
October	125	120	51½	52	10	348½	3.2	0.115
November	95	92½	33	36½	..	257	2.4	0.156
December	22	27½	10	9	..	68½	.7	0.586
Year total	1,234½	1,231	575¾	480½	144	3,521¾	33.5	......
Averages	3.9	3.9	1.8	1.5	..	880½	2.8	$0.136

NOTE.—A cost of $10 a month or $120 a year is assumed.

"This table shows the actual amount of work done by each of four horses during each month in the year on a western New York farm of 104 acres. The extra horse is not counted in the averages, as it was a colt just being broken into work. On this farm there were, in one year, 25 acres of mixed orchard plantings and about 72 acres of crops—wheat, hay, beans and oats. While two of the horses worked more than 1200 hours in the year, the average total work of all four horses was 880½ hours each, or 2.8 hours a day each for the 313 working days of the year. At an average cost of $10 a month a horse, this amounted to 13.6 cents an hour for horse labor. The variation in cost was from $2.66 an hour in February, when the four horses worked only fifteen hours during the whole month, to a little over 5 cents an hour in June, the busiest month. The personal use of horses on Sunday for driving is not counted; neither is the time of the horses, though the cost of maintenance on Sundays is included in the $10-a-month charge. It should be possible to utilize this horse labor to better advantage in the winter, thus relieving the congestion later and possibly lowering the rate. This is a good illustration of the uneven distribution of horse labor on a farm."

Reducing the cost of production calls for careful information relative to the items that make up the cost. The introduction of records and accounts on farms in such a way as to account for the time and expenditures incurred for each enterprise will doubtless do much to stimulate better methods and more attention in the production of farm products.

REFERENCES

"Manual of Practical Farming." McLennan.
Connecticut Expt. Station Bulletin 73.
Illinois Expt. Station Circular 177. "Relation Between Yields and Prices."
Minnesota Expt. Station Bulletins:
97. "Cost of Producing Farm Products."
124. "Cost of Producing Dairy Products."
117. "Cost of Producing Farm Products."
Missouri Expt. Station Bulletin 125. "Cost of Production on Missouri Farms."
Texas Expt. Station Bulletin 26. "Cost of Cotton Production and Profit per Acre."
U. S. Dept. of Agriculture, Bureau of Statistics, Bulletin 48. "Cost of Producing Farm Products."
U. S. Dept. of Agriculture, Bureau of Statistics, Bulletin 73. "Cost of Producing Minnesota Farm Products."
Farmers' Bulletin 364, U. S. Dept. of Agriculture. "A Profitable Cotton Farm."
U. S. Dept. of Agriculture, B. P. I., Bulletin 49. "Cost of Raising a Dairy Cow."

CHAPTER 65

Intensive and Extensive Farming

Intensive farming is frequently applied to the production of truck and fruit crops on small farms. This, however, is not a true definition of intensive farming, since any crop may be produced by intensive methods, and on either a small or large scale. It is true that hay and cereals are generally produced on a rather large scale and with extensive methods. This type of farming is looked upon as extensive farming.

Intensive farming may mean any or all of the following methods: (1) the application of more labor to a unit of area in preparation of the soil and the cultivation and handling of the crop; (2) the use of more capital in the form of machinery and fertilizers on a given area of land, thus enabling the same labor to produce larger yields; (3) the application of better methods for the improvement and maintenance of soil fertility.

Extensive farming calls for the smaller amount of labor and capital per unit of area, although considerable capital is frequently invested in machinery for extensive farming.

Intensity Depends on Available Land.—Farming, to a greater extent than any other occupation, necessitates ample surface area. There must be room for the development of plants. They require sunshine and rain. The roots must have room to develop and sufficient soil in which to forage for plant food.

A manufacturing or mercantile establishment needs very little area. If land values become high, the business may be enlarged by increasing the height of the building. This is exemplified in the big factories near cities, and in the big mercantile and office buildings, frequently as much as twenty stories in height.

By intensive methods it is possible to make one acre produce 100 bushels of wheat, but under existing conditions it is much more profitable to use two acres or possibly four acres in the production of 100 bushels of wheat. In every locality there will be for every crop and for every agricultural product an optimum of intensity that will bring best returns. This will be determined by many factors.

So long as there is waste land that may be brought under cultivation by reclamation or irrigation, there will be little occasion to severely magnify intensity in the process of production. When lands are no longer available, then intensity must gradually and continually increase in order to meet the demands of a growing population for food.

Economizing Land.—In the development of any country the more easily tilled and most productive lands are the first to come under cultivation, providing markets and transportation facilities are available. The least productive are the last to come into use; in fact, cannot be economically used until economic conditions are such as to enable their cultivation with profit. The variation in the location and character of land necessarily determines the character of crops and the intensity of tillage that prevail. Two of the most extensive types of farming consist of the utilization of land for grazing and forestry. The former has been popular because it requires small capital and gives quick returns. The latter requires so much time for returns that but little has been done in reforestation.

There is little doubt, however, as to the large possibilities and good

ECONOMIZING LAND.

An example of intercropping. Pennsylvania State College student gardens.

profits that may accrue from reforesting lands that are adapted to tree growth and that are of little value for other purposes. They offer inducements for long-time investments by people who have capital for investment purposes.

The grazing and reforestation of land is a most expensive manner of economizing land area. Other methods of economizing pertain to increasing the product per acre, and this involves an increased intensity in agricultural methods. The necessity for this economy comes gradually with the increase in population.

Many of the Old World countries practice intensive methods, and produce two or three times as much yield per acre as is secured in North America. It is interesting to note, however, that the returns per man are much lower in the Old World countries. What we really desire to know is how to increase the productivity of our land per acre without reducing the productivity per farmer. Productivity is increased by the substitution of more productive crops and by the more intensive cultivation of each crop. Increase in productivity of crops has been gradually brought about by plant selection and breeding. The more usual method, however, is to substitute a more productive crop for a less productive one. Such an example would be the substitution of corn for oats, or potatoes for corn.

Economizing Labor.—The prosperity of agriculture and the standard of rural life depends more on the character of labor than upon the character of land, although fertile land makes easier a high standard of living. History shows that nations have declined in the midst of fertile lands and favorable surroundings. This has been due to the human factor, and especially to the fact that labor has been allowed to run to waste. Other nations have grown rich and powerful in spite of sterile soil and poor surroundings. Such people were intelligent, industrious and painstaking. The natural conditions of New England were far inferior to those of other portions of North America more recently settled, and yet the early people of New England prospered in the absence of transportation facilities and inventions that are so abundant today.

Economizing labor means a larger product per man. A large product per acre is desirable only when it means a large product per man. An abundance of cheap labor sometimes facilitates securing a large product per acre, but this means large numbers of families to be supported on very low wages. It gives rise to widespread poverty, a condition which true political economy aims to avoid.

Labor is economized by the introduction of scientific methods and the utilization of the most modern labor-saving implements. These include all the best agricultural machines and the use of large teams and mechanical power.

Increasing, Stationary and Diminishing Returns.—The niggardly application of labor and capital to a piece of land in the cultivation of any crop is little better than wasted. It generally produces very little in proportion to itself. With a more generous application a much larger crop yield is secured. One day's labor with man and team on ten acres of land would give no crop at all. Five days' labor on the same area might produce a very poor crop. Ten days of labor would certainly produce more than twice as much as five days of labor, and twenty days of labor might produce a good crop and one more than twice as large as that produced with ten days of labor. Up to this point we have what is known as increasing returns. The addition of another ten days of labor might result in an increase just sufficient to pay for the increased labor. This would give us what is known as stationary returns. To go beyond this the returns for additional labor would not be equal to the added cost of labor, and would give us diminishing returns. The point of stationary returns should not be passed, and will vary with locality and character of crop. It will be determined largely by land values, cost of labor and value of product.

Starting with land that is reasonably productive and with a degree of intensity that is moderate, an increase in the labor and capital applied to land will increase the yield of crops, but not in the same proportion as the labor and capital applied. For every crop there will be an optimum of labor and capital that can be applied for best results. If this is exceeded the value of the returns at once begins to diminish as compared with the value of labor and capital which is applied.

Danger of Under-Production for Growing Population.—An increase in population means that there must either be a corresponding increase in agricultural producers or the art of agriculture must improve so that each worker will increase his output. An increase in agricultural workers ultimately means a marked reduction in size of farms, and in order to maintain the labor income, this demands increased intensity. A great danger lies in the reduced labor income and the lower standard of living for the agricultural worker. The avenues of escape from this situation are: (1) in improving the art of production by discoveries in the science of agriculture; (2) by reduced population by migration; (3) by acquiring of new land either peacefully or by war; (4) a reduced standard of living. The most logical of these seems to be the first, namely, increasing production by a better knowledge of the art of agriculture.

Profits per Acre vs. Profits per Man.—The gross receipts per acre from crops give very little indication of the possible profits per man. For example, it often is more profitable to produce hay that brings a gross return of not more than $20 per acre than it is to grow cherries, lettuce or some other high-priced crop that produces $200 or $300 per acre. The larger area and more extensive methods on the one hand bring a larger return per hour of labor than does the small acreage and large yield of the intensive crops that require a great deal of labor. While the profit per acre may be largest on the intensive crop, the greater number of acres that can be farmed in the more extensive crops will more than offset the difference and makes the latter the more profitable.

Extensive investigations relative to the returns per hour of labor on different classes of crops in the State of New York show that very few crops return as much per hour of labor as does hay. Of course, it would be impossible to depend solely on the production of hay, because it would afford employment for too small a proportion of the year. Under prevailing conditions in New York, farmers are justified, however, in growing as large an acreage of hay as the farm force can take care of during the haying season. The accompanying table is taken from a set of cost accounts on a 90-acre farm in New York.

From the table it is evident that timothy hay giving a return of $15.67 per acre gave a larger profit per hour of labor than did the orchard which gave a return of $101.75 per acre. Studies of this character enable the farmer to know what crops may be developed most extensively to his advantage.

The law of supply and demand will, of course, prevent any very extensive changes in the class of products grown by farmers. There will be a limit to the amount of hay for which there is a market, and if the supply becomes too great, market prices will fall and naturally hold in check the amount produced. This will be true of any product.

COMPARISON OF RECEIPTS AND PROFIT PER ACRE WITH PROFIT PER HOUR OF LABOR.

Crop.	Receipts per Acre.	Profit per Acre.	Profit per Hour of Labor.
Orchard	$101.75	$38.28	$0.23
Oats	26.42	6.84	.33
Timothy hay	15.67	6.37	.63

Intensive and Extensive Enterprises.—It frequently happens that the crop that pays the largest profit per hour of labor also gives the largest profit per acre. Such crops are doubly desirable. So long as such crops prove to be staple products, the acreage may be considerably increased by many farmers. It is well, however, to avoid a large dependence upon speculative enterprises that are subject to violent fluctuations in price. The majority of farmers should continue to produce the staple crops. There is a more uniform market for hay than for strawberries, cotton is needed as well as apples, and the more extensive crops generally pay as good wages as the intensive ones. In fact, there is no crop or group of crops that offers for a long-continued period of time over a wide territory any striking advantages over other crops. Such a condition may prevail for a short time, but the advantages offered in this way induce many to engage in that particular enterprise and this naturally causes a decline in profits. Where one crop pays much better than others, it is generally a question of adaptation. It may be soil, climate, markets or a combination of these. Intensive enterprises do not appear to require much less capital than extensive ones. They may require much less land, but this is generally offset in larger expenditures for fertilizers, labor, seed and equipment. The advantage of one over the other will often depend largely on the trend of land values. Where land values are increasing, the extensive enterprises that demand more land combine profits from products and profits due to increase in land values.

Relation of Intensity to Land Values.—High land values necessitate large gross returns and fair profits in order to pay normal rates of interest on rent of land. This means intensive farming and usually the production of crops of high value, such as vegetables, tobacco or fruits. In general, the higher the value of land the greater the degree of intensity in farming.

Relation of Intensity to Labor.—A low price for labor encourages intensity, although so far as machine labor is concerned, high-priced labor may be used advantageously in intensive enterprises. Intensive farming will generally not only employ more labor, but can advantageously use a greater variety of labor. Skilled and unskilled men, women and children frequently can all be utilized to advantage in intensive farming. The more extensive farming generally calls for skilled labor and less variety. There is little opportunity for the employment of women and children in the production of cereals. Most of the work on the larger farms is done with complicated machinery and large teams that call for skilled and experienced workmen.

Rape Seeded in Standing Corn at Last Cultivation, Protects the Soil and Increases the Returns.

Relation of Intensity to Type of Farming.—Intensity depends largely on type of farming and character of enterprises included in it. The growing of forest trees, pasturing of animals and the production of cereals are necessarily extensive, requiring large areas of land, and because of the low returns per acre, require much land to bring a satisfactory labor income. The production of cultivated crops, such as corn, potatoes and roots, involves more labor and better class of land, and requires less area. These crops are consequently cultivated in a more intensive manner and often this type of farming may fall in the class of intensive farming. The next step towards intensity would be the production of the usual run of truck crops, small fruits and tree fruits. This will almost always be strictly intensive farming. The degree of intensity will be determined by the value of the land and the price of the products. The most intensive form will be represented in greenhouse production where the returns per acre of land may be several thousand dollars annually.

The Most Profitable Yield.—Increasing the yield of any crop involves additional capital in the form of irrigation, drainage, manures or fertilizers. Increased cost per acre may not only increase the net income per acre, but may also reduce the cost per unit of product. There will always be a limit to the amount of labor and capital that can be applied most economically. The following quotation and tabulation from the *Tribune Farmer* illustrates the point in question:

"The following figures show how two fields on the same farm in New York State were operated; the first under a very intensive system of management, the other under a comparatively extensive system of management. The figures in the first column are an average made up from careful records for ten years. The figures in the second column are made up from careful estimates, as no records are available:

Items of Expense.	Cost per Acre.			
	Intensive Method.		Extensive Method.	
Plowing		$3.00		$3.00
Preparation of land, very thorough		2.00		1.50
Planting		1.00		1.00
Cultivating	14 times	6.30	5 times	2.50
Spraying	8 times	6.30	Once	1.00
Harvesting	282 bushels	7.00	125 bushels	4.50
Marketing		3.00		1.50
Total labor cost		$28.60		$15.00
Seed	22 bushels	$9.32	15 bushels	$6.36
Fertilizers	1500 pounds	25.63	300 pounds	5.13
Spraying material	8 times	6.00	Once	.75
Use of equipment	$28 per acre	1.68	$14 per acre	.84
Interest on land	$150 at 6 per cent	9.00	$100 at 6 per cent	6.00
Total cash cost		$51.63		$20.08
Total cost		$80.23		$35.08
Total cost per bushel		$0.284		$0.281
Total receipts, 282 bushels		119.56	125 bushels	53.00
Total receipts per bushel		.424		.424
Net profit		39.33		17.92
Net profit per bushel		.14		.143

"In this case, where the markets and supply of capital and labor apparently warranted it, the increasing of the intensity of the methods of production more than doubled the net profits an acre. The cost and the profit a bushel, however, remained practically the same, the increased income an acre being wholly due to increased yields, which in turn were due to better and more intensive methods."

Given $2000 to raise potatoes, which of the above degrees of intensity would give the best returns on the capital in hand? On the intensive plan $2000 would finance as many acres as the cost of production per acre, $80.23, is contained in $2000, or 25 acres. At the net profit received, $39.33, per acre, this would give a profit of $983.

On the extensive plan $2000 would finance as many acres as the cost per acre of production, $35.08, is contained in $2000, or 57 acres. At the net profit per acre, $17.92, this would give a profit of $1022.

Where plenty of land is available, a little greater profit on the capital in hand could be secured from the extensive method. In the absence of available land, one would be fully justified in increasing the intensity of farming to the limit given in the above table. To go beyond this point in the production of potatoes under the conditions that prevailed would likely have resulted in a reduction in profit.

Crop Yields on Successful Farms.—Agricultural surveys show that there are some farms that produce crop yields much above the average, but fail to make good labor incomes. This may be due to any one of several causes. It frequently is due to poor management and too great a cost in production. It is sometimes due to feeding the crops to unprofitable cows. A man may produce large yields at low cost, which, if sold for cash, would bring good returns, but when fed to unproductive cows, may cause a loss instead of a gain. It requires a careful consideration of all the factors of production in order to know how much above the average yields it is safe to go and still reap the largest profits. It is well not to attempt to increase the yields more than 25 per cent over the average of the locality, providing one's land is of no more than average fertility and other conditions are no better than the average. This applies to general farm crops. In case of more intensive crops, the upper limit is likely to be higher rather than lower. Yields may apply to animals as well as crops. As previously stated, crop yields in several counties in New York where the labor income was much above the average ranged from 18 to 27 per cent better than average crop yields. There were a number of instances when large crop yields were accompanied by low labor income.

Intensity in Dairying.—The degree of intensity in dairy farming depends chiefly upon the relative prices of feed, labor and dairy products. Near large cities with high-priced feed, costly land and excellent markets, intensity may be quite marked. Under such conditions it may pay to resort to the growing of soiling crops, roots and silage, and provide no pasture whatever. Under such conditions it will be most profitable to produce market milk. Butter, which is easily transported long distances at low cost, can be more economically produced in sections remote from markets where land and feed are much less expensive. One represents intensive dairying, the other extensive dairying. In one case it is essential to use very productive cows and to feed heavily and carefully. In the other, the poorer grade of cows may be used and a poorer quality and smaller quantity of feed may give fair results. This fact seems to be recognized by dairymen, and milk production is carried on extensively near the large cities, whereas the production of butter is being most extensively developed in regions rather far from cities, such as northern Iowa, Minnesota and to some extent the Dakotas.

Formerly, much butter was produced in the summer time because pastures afforded the cheapest form of feed for cows. Cows were generally allowed to go dry at the approach of winter, and maintained as cheaply as possible on dry feeds during that period. This resulted in marked difference between summer and winter prices for butter. More recently farmers have found it advantageous to have cows freshen in the fall or early winter, and feed more heavily during the winter period. While this involves the storage of more feed, which is somewhat more expensive than pastures, it affords employment during the winter season, and avoids so much conflict with the production of crops during the growing season. This distributes the labor and enables the farmer to grow more crops to feed more cows and consequently increases the total of production. It also lessens the fluctuation in the price of dairy products.

RYE AND WINTER VETCH MAKE AN EXCELLENT EARLY SOILING COMBINATION FOR COWS.

Receipts per Cow and Profits.—The difference in the value of cows should be relatively greater than their relative milk or butter-fat production. It costs considerable to maintain a cow even when she is producing nothing. The cost of maintenance is increased slightly with the increase in production. Production involves work in the transformation of crude products into milk and butter-fat. Cows, like types of soil, differ in productivity, and like the soil, the yield may be increased by better care and feeding. The same principle applies to both relative to the degree of production that will be most profitable. The more valuable the cow and higher the price of her products, the greater the intensity in care and feeding may be.

On the most profitable farms in Tompkins County, N. Y., the receipts per cow were 48 per cent better than the average of the region.

In Jefferson County, N. Y., the receipts per cow from the best farms were 56 per cent more than the average. In these same counties the receipts per acre on best farms exceeded that of the average by a much smaller margin. This indicates that the degree of intensity may be carried further with cows than it can be with crops.

Relation of Cows to Size of Farm.—The number of dairy cows that may be maintained on a farm will depend on the productivity of the land, the feeding system best adapted and the proportion of the total feed that is to come from the farm. It will also depend on the extent to which cash crops may be grown.

Knowing the average yield of the crops to be grown for the dairy and having a feeding system for the year, it is possible to calculate just how many cows may be maintained on the farm, and just what acreage of the several crops will be required. Such a problem is worked out and presented herewith.

How many cows can be kept on a 110-acre farm when the yields of crops and the feeding system are as follows:

Yields:

Pasture, one acre for each cow and bull.
Soiling corn, 7 tons per acre.
Silage corn, 10 tons per acre.
Corn for grain, 50 bushels per acre = 1.4 tons.
Oats for grain, 50 bushels per acre = 0.8 ton.
Wheat for grain, $26\frac{2}{3}$ bushels per acre = 0.8 ton.
Hay, 3 tons per acre.

Feeding System:

Five horses, one for delivery and four for work.
365 days at 15 pounds hay each daily.
365 days at 10 pounds grain each daily.
Grain, one-half oats and one-half corn.

COWS AND ONE BULL, DAILY REQUIREMENTS OF EACH.

Date.	Days.	Pasture, acres.	Soiling Corn, pounds.	Silage, pounds.	Hay, pounds.	Grain, pounds.
May 10 to July 31	83	1	..	..	..	3
August 1 to September 15	46	1	25	..	5	3
September 16 to November 15	61	1	..	20	5	5
November 16 to May 9	175	..	..	35	10	8
	365					

Grain for cows and bull, 3 parts corn, 2 parts oats, 5 parts purchased concentrates.
Wheat sold. Acreage equal to acreage of oats.
Oats and wheat straw for bedding.

Acres of Each Crop.—From the food requirements calculate the amount of each crop required for the year for each cow and for the five horses. This is expressed decimally in tons and entered in the first and second columns of the following table. The figures are obtained by multiplying the daily requirement by the days in the feeding period, and dividing by 2000, the pounds in a ton. To illustrate: 20 pounds silage × 61 days, + 35 pounds silage × 175 days, ÷ 2000 = 3.672 tons of silage per cow, thus:

$$\frac{(20 \times 61) + (34 \times 175)}{2000} = 3.672.$$

It is next necessary to ascertain the acres of each crop as given in the third and fourth columns of the table. To illustrate, the silage requirement per cow, 3.672 tons, divided by the yield of silage per acre gives the acres of silage corn required per cow (3.672 ÷ 10 = .367 acres).

Let x equal the number of cows plus one (bull) and the tabulation is as follows:

Crop.	Tons of Feed Required for		Acres of Crops Required for	
	1 Cow or Bull	5 Horses.	1 Cow or Bull.	5 Horses.
Pasture	.....	.....	1.000x +	.....
Soiling corn	.575	.....	.082x +	.....
Silage corn	3.672	.....	.367x +	.....
Corn for grain	.314	4.562	.224x +	3.259
Oats	.209	4.562	.261x +	5.704
Wheat	.....	.....	.261x +	5.704
Hay	1.142	13.688	.381x +	4.562
Area of farm, 110 acres, equals	....	....	2.576x +	19.229

The equation is solved for the value of x, the number of cows plus one, as follows:

2.576x = 110 — 19.229 = 90.771 ÷ 2.576 = 35.24 or 34.24 cows and 1 bull.

With the crop yield assumed and feeding system given the 110 acres will maintain 34 cows, 1 bull and 5 horses, and will provide for an acreage of wheat, as a cash crop, equal to the acreage of oats.

Substituting the value of x (35.24) in the third column of the above table and adding the acres of the several crops for horses given in the fourth column, the acres of pasture and crops are as follows:

Crop.	Acres.	
Pasture	35.240	
Soiling corn	2.889	
Silage corn	12.933	26.975 (Soiling corn, Silage corn, Corn for grain)
Corn for grain	11.153	
Oats	14.900	
Wheat	14.900	
Hay	17.988	
Total	110.003	
Pasture	35	
Cultivated crops	75	

It is necessary to next outline a rotation that will best provide for the crops needed. By inspection, we find the requirements to be as follows: corn 27 acres, oats and wheat each 15 acres, hay 18 acres. This could be provided for in several ways, but since the acreage of corn is nearly twice as large as that for oats and wheat, it will likely be best to grow corn two years in succession. Five fields of 15 acres each are, therefore, provided and the rotations become as follows:

Field.	1915.	1916.	1917.	1918.	1919.
1	12 Corn 3 Hay	Corn	Oats	Wheat	Hay
2	Corn	Oats	Wheat	Hay	12 Corn 3 Hay
3	Oats	Wheat	Hay	12 Corn 3 Hay	Corn
4	Wheat	Hay	12 Corn 3 Hay	Corn	Oats
5	Hay	12 Corn 3 Hay	Corn	Oats	Wheat

If young stock are to be included, let x equal the number of cows plus half as many young, one-half calves and one-half yearlings. If the number of young stock are known, their requirements may be included with the horses. The bull and other stock may also be included with horses.

The Soiling System.—This provides for the production of crops to be cut green and fed to cows in stables. It calls for a succession and variety of crops that will afford continuous, succulent feed in the best stage of maturity, throughout the greater part of the summer. Such systems are quite extensive in European countries, especially for the production of market milk on high-priced land near cities. It enables the keeping of the maximum number of cows on a limited area. In this country the system is but little used. Farmers here are more concerned with the profits of the year's work than they are in entertaining the largest possible number of cows. There doubtless always will be a tendency to increase the soiling system for milk production, with increased price of milk and advance in land values near centers of large milk production.

Proper Balance of Intensity.—Intensity depends on many factors. These should be properly balanced. The high-priced cow responds to more and better feed, to a greater extent than the low-priced or poor cow. There will always be a limiting factor, and this factor is the one that should first be intensified. If, for example, land is poorly drained, no amount of fertilizer will insure maximum crops. Drainage is the first deficiency to overcome. The various factors must be kept in proportion. It requires more capital, feed and better care in the rearing of pure-bred stock than it does for scrub stock. This relationship should run throughout the farm business. The best horses and most modern implements call for the most skilled labor. It would be more serious to employ inefficient labor for the operation of such equipment than it would in case of the one-mule farm in the cotton belt.

Intensity Related to Citizenship.—In spite of the arguments frequently presented in the agricultural press for smaller farms and increased intensity, most of the farming of America continues on the extensive basis. There is no country in the world in which farmers are more prosperous or of a higher order of intelligence. This type of farming encourages the use of large areas of land, much horse power, the best farm equipment, and develops the highest form of rural civilization. It provides for an income above physical needs and affords means for the procuring of broader culture. It is in marked contrast to the very intensive systems that prevail in many of the European countries.

REFERENCES

"The Small Country Place." Maynard.
Colorado Expt. Station Bulletins:
89. "Wheat Raising on the Plains."
117. "The Colorado Potato Industry."
North Carolina Expt. Station Bulletins:
84, 112. "Trucking in the South."
New York Expt. Station Bulletin 226. "An Apple Orchard Survey of Wayne Co., N. Y."
New York Expt. Station Bulletin 229. "An Apple Orchard Survey of Orleans Co., N. Y."
Oregon Expt. Station Bulletins:
99. "An Orchard Survey of Wasco County."
101. "An Orchard Survey of Jackson County."
Texas Expt. Station Bulletin 94. "Horticultural Survey of Gulf Coast."
U. S. Dept. of Agriculture:
Year-Book 1908, pages 351 to 366. "Types of Farming in the U. S."
Year-Book 1904, pages 161 to 191. "Growing Crops Under Glass," "Fruit Growing," "General Farming."
Farmers' Bulletin 519, U. S. Dept. of Agriculture. "Intensive Farming in the Cotton Belt."

CHAPTER 66

SIZE AND DIVERSITY OF FARM RELATED TO EFFICIENCY

In every locality and for every type of farming there is a degree of diversity and a size of business that will prove most advantageous. It should be the aim of farmers to attain this so far as capital and ability will permit. The size of the farm does not necessarily pertain to area, since a large business in certain types of farming may be carried on on a comparatively limited area. The most striking example of this is found in the

ONIONS AS A SPECIALTY.[1]

production of vegetables and flowers under glass. An acre of this kind of farming often employs a number of men and brings a large gross return.

Diversified farming is a somewhat indefinite term, and may mean the producing of a few well-chosen products or may include a large number of products.

DIVERSIFIED FARMING

Diversified farming is often the production of a little of everything and not much of anything. It is at once evident that such farming will

[1] Courtesy of Wallace's Farmer.

not pay. Successful diversification consists in the production of enough of a few products to make it worth while.

Special farming is generally looked upon as the production of one commodity. The subject is somewhat confused, because of such terms as dairying, fruit farming and vegetable farming, that are often thought of as special farming. Dairying, if confined to the production of wholesale milk, would be a special line of farming, but it frequently includes the retailing of milk, and the sale of pure-bred stock, in which case the farmer has three lines of production. In fruit farming or truck farming, there will usually be a variety of fruit and truck crops, so that it will not be specialized in the true sense of the word. The same may be true of grain farming where a variety of grains are grown in a systematic rotation and in such a way that there is an equitable distribution of labor and a variety of cash crops.

General View of Specialized Wheat Farming in Canada.

Advantages of Special Farming.—Special farming, or the production of one product, has an advantage in that there is usually enough of it to make it quite worth while. With a given area, it will provide for fields sufficiently large to economize on the labor of men and teams and make possible the use of the best machinery adapted to the crop in question. With small capital one can provide for enough of one crop, whereas with a number of crops it might be impossible to have enough to secure a desirable equipment. One becomes more skilled when dealing with only one line of production.

Special farming is subject to many disadvantages. It does not provide for crop rotations. In case of failure from whatever cause, there is nothing to fall back on. In diversified farming, if one crop fails there are others that will bring fair returns. A single crop does not provide for distribution of labor. There are times and places for this type of farming, but it generally consists in the production of some special crop for which there is a limited demand. Wheat, as grown in the Northwest and in Canada, is strictly special farming. In many places cotton is also grown as a specialty. These are staple crops, and have proven successful for a number of years, but we know that better results may be obtained by diversity and the introduction of crop rotations. We find now and then farmers near large cities who make a specialty of producing hay. This necessitates prices that are far above the average and calls for land that is not too high priced.

Advantages of Diversified Farming.—The majority of farmers in North America are engaged in diversified farming. A large percentage of them combine the production of crops and animals or animal products.

General View of a Good Diversified Farm.[1]

There are many advantages in such farming. Cotton farms excepted, probably 95 per cent of the remaining farms of the country would be classified as general farms that derive their income from a combination of crops and animal products. The advantages of diversity so far as crops are concerned are discussed in the chapter on Crop Rotations. Diversity lessens the risk of failure. It distributes both the income and the labor. No matter how profitable a crop may be, there is a limit to the amount that can be produced with a given amount of labor, because it requires seeding and harvesting in a limited season. This usually covers only a small portion of the year.

[1] Courtesy of The Macmillan Company, N. Y. From "How to Choose a Farm," by Hunt.

Diversified farming is most successful when the several enterprises chosen do not compete with each other for labor. Crops are said to be competing only when they demand the attention of the farmer at precisely the same portion of the growing season. Agricultural surveys in many localities have shown that diversity of enterprises increases the farmer's labor income.

Dairying, which is generally considered a specialty, is more successful when combined with the production of cash crops. The labor required to do the milking mornings and evenings is more than sufficient to care for the herd. If the time of the laborers can be utilized in the production of crops, profits are increased.

SIZE OF FARMS

The scale on which farm operations are conducted is determined by many factors. It is often fixed by the amount of available capital or the size of the farm on which one is already located. If large-scale farming were decidedly more profitable than small-scale farming there would be no difficulty in embarking in it. Investigations show that farming on a very large scale has certain disadvantages and more frequently results in failure than moderate or small-scale farming. Consequently, it is difficult to borrow capital on a large scale for this purpose. There are no hard and fast lines separating large, medium and small-scale farming. As already indicated, the acreage of the farm is not the true test. More capital and labor may be employed on ten acres intensively farmed than on a thousand acres most extensively farmed.

In North America the majority of farms fall into the class of medium-size farms. A study of statistics shows that there is a tendency for the size of farms to change with the change in efficiency of the unit of labor. The introduction of larger teams and large farm implements, thus increasing the area that one man can farm, has had a tendency to increase the size of the one-man farm over a considerable portion of the corn belt. Small farms and very large farms are decreasing in numbers, while medium-size farms are increasing in numbers.

Size Depends on Type of Farming.—For the production of cereals, 100 to 200 acres is considered a moderate size farm. For the raising of vegetables, 20 to 40 acres is a medium size. The size of the farm, therefore, is largely dependent upon the type of farming. If the crops grown are fed to dairy cows, the labor required and the gross income received are both increased. The combination of dairying and cash crops, therefore, need not be quite so large as when crops only are depended upon. The capital required will vary according to the region selected and the type of farming, and when comparing the labor income of different types of farms, capital invested is a better measure of size than land area. In the better portions of the corn belt $10,000 would be too small an investment to yield a fair income. Land values at $150 to $200 per acre would make the acreage too small to fully utilize the character of teams and equipment best adapted to the production of the crops of that region; 100 to 120 acres is about the minimum size that can be economically farmed there; 160 to 240 acres is a more economical size.

Bonanza Farms.—This term applies to the very large farms, most of which are located in the northwestern part of the United States and which frequently cover from ten to thirty or more square miles. Wheat is generally the leading product. Such farms usually consist of an aggregation of a number of farm units under one management. In some cases, however, several thousand acres may be farmed with one set of buildings.

A Small Farm Under Glass.[1]
Soil made to order; heat, light and moisture controlled.

Very large fields are used and the work is done either with farm tractors or with very large teams. The chief advantage in farming on such a large scale lies in the economy of skill, equipment, buying and selling. Large-scale farming, like manufacturing on a large scale, affords enough of one special kind of labor to fully occupy the time of one or more men. In this way, men are most highly skilled and confined to a performance of work for which they are specially trained. Unlike manufacturing, however, such large-scale farming covers so much area that close supervision by a superintendent is impossible. It also involves the loss of much time on the part of both manager and workmen in traveling to and from fields, or from one enterprise to another. It obviously enables the employment of a high-

[1] Courtesy of The Macmillan Company, N. Y. From "Farm Management," by Warren.

priced and competent manager. Such management is even more difficult than the management of large-scale manufacturing, because farming is subject to a great extent to weather conditions and includes a diversity of operations.

Manufacturing has increased enormously, and its development has seen a marked change from large numbers of small factories to smaller numbers of large factories. The increase in the size of manufacturing plants has been very great. There are such decided advantages in the large manufacturing plant that it is difficult for the small one to compete with it. The large area over which large farms must extend make factory methods inapplicable. The change in season, the variety of enterprises, the sudden approach of storms, the encroachment of insect enemies and plant diseases, all call for abrupt changes in the character of work and for reorganization of the farm forces.

Medium Size Farms Superior.—Farm surveys in many sections show that the labor income of farms increases with the size of business. With any particular type of farming, the size of business corresponds approximately with the area. By grouping farms according to capital invested or acres farmed, we find that the labor income increases, step by step, with each group as size is increased. This view is confined to comparison of small and moderate size farms. The exceedingly large farms are so small in number that but few have been investigated in these farm surveys. The few that have been investigated show conclusively that while farms of from 500 to 1000 acres may bring large labor incomes, they also offer opportunity for failure. The following table shows the labor income of farms in several farming districts as related to size of farms:

578 Farms, Livingston Co., N. Y.		586 Farms, Tompkins Co., N. Y.		410 Farms, Chester Co., Pa.	
Acres.	Income.	Acres.	Income.	Acres.	Income.
20	$24	30	$168	40	$297
21–40	257	31–60	254	41–80	611
41–60	400	61–100	373	81–120	963
61–80	481	101–150	436	121–160	1068
81–100	642	151–200	635	161–200	1630
101–140	937	Over 200	946	Over 200	1229
Over 140	1261				

The Family Size.—The typical American farm is, and probably always will be, the family size. It should be sufficiently large to employ advantageously the time of the farmer and his family. It will vary with the type of farming and the size of the family. The farmer with only one son should have sufficient land to employ the time of the son when not in school. If he has several sons his farm should be somewhat larger. Statistics show that less than one-half of the farms in the United States employ hired labor. There are about one and a half male workers engaged in agriculture for each farm. Some farms employ a number of workmen throughout the year. Others employ one workman and still others employ day labor in busy seasons.

The Economical Unit.—This provides for an area sufficiently large for economical production. The area depends on the type of farming. It should provide for the utilization of buildings and equipment essential to the type of farming. If the area is too small, labor is wasted, machinery is not fully utilized and the land is likely to be over-capitalized with respect to buildings. The 120-acre grain farm in the corn belt will require the same equipment in the way of machinery and cultural implements that a 200-acre farm would require. The 200-acre farm in this case would be the most economical unit, since it could be farmed equally as well without increased expense for equipment. The size most economical is, therefore, the largest unit that can be farmed satisfactorily with the implements that the crops call for. If one increases the size 40 or 80 acres beyond what a set of implements can manage, it calls for duplication of tools, and unless the extra tools are fully utilized the increase in size may be disadvantageous rather than otherwise. In order to farm most economically the logical step would be from one unit in size to two full units in size. Statistical studies show that a little increase in area generally increases efficiency greatly, but when the area of a 200 or 300-acre farm is increased, the increase in efficiency is scarcely noticeable. Usually 300 acres represent about the upper limit that can be most economically farmed as a unit. Larger areas result in much land being so remote from buildings that much time is wasted in going to and from fields, hauling products to the farmstead and manure to the fields.

Size Economizes on Buildings and Fences.—The farmer on an 80-acre farm usually desires as good a house as he would wish if he were on a 160-acre farm. The barn and outbuildings on the larger farm will generally be larger than on the smaller one, but the increase in size, cost and expense for upkeep will not be in proportion to the area farmed.

It costs twice as much per acre to fence a square 10-acre field as it does to fence a square 40-acre field.

Size Economizes on Equipment.—A survey of 586 farms operated by owners in Tompkins County, N. Y., showed that the investment in machinery per acre ranged from $3.50 on the larger farms to $6 on the smaller ones. The larger farms were generally best equipped. Very small areas of crops often prohibit the use of labor-saving machines, because the cost is too great to justify their employment. The annual cost of depreciation, interest, insurance, repairs, housing and oil for machinery is about 20 per cent of its value. This would make the annual cost per acre for these items 50 cents more on the small farms above mentioned than on the large ones.

In Livingston County, N. Y., the investment per acre in machinery ranged from $3.18 on large farms to $7.05 on small ones. In the United States as a whole, according to the census for 1900, the investment in machinery per acre on improved land ranged from $1.31 on farms ranging from 500 to 1000 acres each to $7.50 on farms ranging from three to nine acres each.

Size Economizes on Man and Horse Labor.—A comparison shows that the number of acres farmed per horse varies greatly with the size of farm. It will vary also with the type of farming, and a satisfactory comparison of efficiency can be made only when comparing farms of different size devoted to the same type of farming. Iowa is all devoted to general farming. The farms of Lyon County in the northwestern part of the state average 210 acres, and those in Henry County in the southeastern part of the state, 123 acres. In Lyon County there are 22 acres of crops per horse; in Henry County only 11 acres of crops per horse.

In Tompkins County, N. Y., on 586 farms operated by owners, the acres per horse ranged from 15 on farms of 30 acres or less to 49 on farms of over 200 acres. In most parts of the country it costs as much to keep a team of horses as it does to employ a hired man. Three or four horses is the smallest number that can be most economically employed in the use of machinery best adapted to the production of general farm crops. To employ a smaller number increases the cost of production, or to employ this number for too small a percentage of the time also increases the cost of production.

The same principles apply in the employment of man labor, especially in so far as this labor is associated with the use of horses and implements.

There are many farm operations that are more advantageously performed by two men working together than by one man. Among these may be mentioned the harvesting of grain and hay. In hauling these products from field to stack or barn, one man is required to pitch and the other to load, stack or mow the product. Very often, odd tasks such as changing of wagon racks and boxes are much more quickly and easily done by two men than by one.

Size Related to Crop Yields.—Small farms do not necessarily produce larger yields per acre than large farms. Farms in the two Iowa counties above referred to gave yields of the staple crops that were almost identical. In Tompkins County, N. Y., 586 farms show practically the same yield per acre of staple crops on large farms as on small ones. In one product only, namely, hay, did there seem to be any consistent decline in yield with increase in acreage. The average yield of hay was 1.38 tons per acre on farms of 30 acres or less, and 1.24 tons per acre on farms of 150 acres or more. Yield per acre will depend more upon the character of soil and value of land and the intensity of farming, than upon the size of farms.*

*The figures given for Livingston and Tompkins counties, New York, are from Warren's Farm Management.

Advantages of Buying and Selling.—Large farms, using larger quantities of supplies in the form of fertilizers and feed may often purchase at wholesale to good advantage. The manager can also afford to spend more time in investigating markets and market conditions, because of his larger business. There is also a gain when products can be shipped in carload lots. Not only may the price received be better, but there is a saving on transportation charges. One may be justified in some expenditure for advertising where the product is sufficiently large, for it generally costs no more to advertise large quantities of produce than small quantities.

Size of Fields.—Good size farms enable laying the farm out into fields of an economical size. Large fields are cultivated with less loss of time than small ones. They require less expense per acre in fencing. Less waste land is incurred about borders and for turning rows. Cost accounting has shown that it costs more to produce crops on very small fields than it does on good size fields.

Size Related to Capital.—All farms necessarily have some capital invested in unproductive ways. This is always relatively larger on small farms than on large farms. The chief item in this respect is generally the house and the ornamental features of the farm. In Livingston County, N. Y., the capital invested in the farmhouses ranged from nine per cent of the total investment on the large farms to 43 per cent on the small ones. The value of other buildings per animal unit ranged from $50 on large farms to $164 on small farms. Capital is more fully utilized on good size farms.

Size Related to Dairying.—The best size for a dairy farm will depend on location and the type of dairying. In Denmark, the most progressive dairy country in the world, dairy farms average 40 acres in size. The dairymen in Scotland favor a farm that will maintain a herd that can be milked by the dairyman and his family. In the United States dairy farms should generally range from 75 to 150 acres in extent. Grain and general farms should range from 120 to 300 acres in extent. A limited number in each of these cases, under favorable conditions, may be somewhat smaller.

Dairying will frequently prove advantageous for men located close to good markets and whose farms are too small for general farming. Dairying will enable them to increase the volume of business and permit them to remain in the same neighborhood without the purchase of more land.

Size of Farms in the United States.—The size of farms in the United States has changed somewhat during the past sixty years. In 1850 the average area per farm was 203 acres. It declined steadily until 1880, when it was 134 acres. Since then change has been slight. In 1910 it was 138 acres, 75 acres of which was improved land and 46 acres of which was in the principal crops. The smaller farms were found in the trucking regions and in the cotton belt. In Dallas County, Ala., farms average 44 acres. They are generally farmed by colored people. In Gloucester County, N. J., farms average 62 acres. Truck is the leading industry. Farms in an irrigated district in Utah average 82 acres per farm. McHenry County, Ill., in which dairying predominates, averages 129 acres per farm. In Shelby County, Iowa, farms devoted chiefly to grain, hogs and cattle, average 167 acres per farm. In Clay County, Neb., farms devoted largely to grain production average 182 acres. In Sherman County, Ore., farms devoted largely to the production of wheat average 799 acres. This gives some idea of the wide range in the size of farms in different localities.

Size Helps Prevent the Boys Leaving Farms.—Boys generally leave the farm because there is not sufficient work to make it pay to stay. Surveys in New York State show that 79 per cent of the sons of the smallest farmers had left home. On the largest farms only 16 per cent had left home.

Small Farms.—Small farms have the advantage in a better supervision of the farm work, less loss from waste of material and less difficulty with the labor problem from the standpoint of hired help. The chief disadvantage in the small farm lies in the lack of adequate equipment, which results in high cost of production and a small labor income. The many advantages that have been cited in favor of good size farms seldom apply to the very small farm.

A great deal has been written advocating the small farm, and so far as area is concerned, there is room in the vicinity of large cities for a considerable number of comparatively small farms. These must necessarily be devoted to intensive farming along the line of vegetables and fruits for human consumption. Any attempt to grow the staple crops on very small areas results in marked increase in cost of production or materially reduces the standard of farm life. In Belgium and other European countries comparatively little animal power is used in farming. Much of the work is done by hand. The areas are often so small that if a team of horses was employed, they would consume more than one-half of the products of the farm, thus leaving comparatively little for the farmer.

REFERENCES

New York Expt. Station Bulletin 178. "The Income of 178 New York Farms."

U. S. Dept. of Agriculture, Year-Book 1908, pages 311–320. "The Small Farm as a Remedy for Southern Rural Conditions."

U. S. Dept. of Agriculture, Bureau of Plant Industry, Circular 75.

U. S. Dept. of Agriculture, Bureau of Plant Industry, Bulletin 259.

Farmers' Bulletins, U. S. Dept. of Agriculture:

310. "A Successful Alabama Diversification Farm."
312. "A Successful Southern Hay Farm."
325. "Small Farms in the Corn Belt."
355. "A Successful Southern Dairy Farm."
364. "A Profitable Cotton Farm."
365. "Farm Management in Northern Potato Growing Sections."

Utilizing Stony and Bottom Land.[1]

[1] Courtesy of U. S. Dept. of Agriculture Bulletin 341.

CHAPTER 67

CROPPING AND FEEDING SYSTEMS

The character of cropping and feeding systems on a farm determines to a considerable extent the success of the farm. Cropping systems from the standpoint of crop production and maintenance of soil fertility are discussed in the chapter on "Crop Rotations." The discussion here will pertain more especially to the farm management phase of it and its relation to the other enterprises of the farm.

Feeding systems pertain to the rations and methods of feeding livestock and will be discussed especially from the farm management standpoint.

The Farm Scheme.—Success in farming does not rest on the results of a single year. It is not enough to be successful in the production of one crop, but one must continually grow a satisfactory crop at regular intervals. This calls for a cropping system in order that the chief crop or crops may be rotated with other crops to avoid the numerous difficulties mentioned in the chapter on rotations.

From the crop standpoint the two dominant factors are how to maintain the yield of cash crops year after year and at the same time prevent any decline in soil fertility. When animals enter into the farm scheme the cropping system must also meet the needs for animal feed, including necessary bedding. Profits necessitate considering the enterprises as a whole. The order in which crops may be grown and the feeding system to be adopted is a local question. It will be determined by a great many factors, such as character of soil, climatic conditions, price of land, markets, transportation and the personal preferences of the farmer. No definite system can be laid down that will be best under all conditions, but there is a philosophy underlying the question that will aid every farmer in working out the system best suited to his conditions.

Crops Related to Farm Management.—Crops are grown either to sell or to feed to livestock. One farmer may desire to grow corn principally, another cotton, a third one potatoes, and still another some other crop. Usually, several crops are grown. The problem from the standpoint of farm management will be that of determining how much of each of the different crops should be grown. From the standpoint of crop production the farmer is interested only in the method of growing the crop, and when the crop is harvested the task is completed. As a farm manager, it will be necessary for him to decide what to do with the crop. Will he profit more by feeding it or by selling it? If so, when should it be sold or to what class of livestock can it be most profitably fed?

Animals Related to Farm Management.—Most farmers raise some livestock. They should know the nature of animals and their requirements. The care, the character of feed and the breeding that will give best results are generally questions of animal husbandry. From the standpoint of farm management, farmers must decide what classes of stock they will raise. This will be determined by many factors. The class of livestock to be kept will depend largely upon the character of the crops to which the farm is best adapted. On some farms, horses, sheep and poultry may be most desirable. On others, dairy cows for market milk will prove most profitable. The problem resolves itself into making plans for a specific farm, arranging it into fields, selecting the kinds of crops and the classes of livestock that are best adapted to it, and deciding upon the proper proportion of each. The buildings, equipment and capital must all be considered in this connection.

Cropping and Feeding Systems are Related.—When crops are of prime importance on the farm, the livestock kept is generally selected chiefly for the utilization of by-products. Statistics show that four-fifths of the farms in the United States keep dairy cows. Two-thirds of the farms make butter. The small dairies are maintained largely on cornstalks, straw and hay of poor quality, and the cows are pastured on land that is not well adapted to the growing of crops.

On most farms a few swine and 50 to 100 head of poultry are kept largely as scavengers to utilize what would otherwise be wasted. In some instances a few sheep are kept in the same way, and more might be kept in small flocks at low cost, to the advantage of both the farmer and the consumer of meat.

On farms where livestock predominates, crops become subsidiary and the crops grown are those that meet the needs of the livestock.

Adaptation of Cropping and Feeding Systems.—Crop adaptation is discussed under the heading of "Soils" and also in the part of this book pertaining to crops, but there is a further adaptation involved in the cropping system as well as in the feeding system. These two are dependent upon each other. If crops are grown chiefly for livestock, consideration must be given not only to yields, but also to feeding values. Corn generally produces more digestible nutrients per acre than any of the small grains, clovers or grasses. For example, 12 tons of ensilage, a fair yield on an acre, contains approximately 3600 pounds of digestible nutrients. An acre of timothy yielding 2 tons contains only about 1700 pounds of digestible nutrients. An acre of clover yielding $2\frac{1}{2}$ tons of hay contains about 2300 pounds of digestible nutrients. The clover also contains much more protein than the timothy. Crops for feed must be compared in this way, and definite information from the standpoint of yield and feeding value may be ascertained from the chapters on "Crops" and from the tables given on "Feeding Values of Different Crops."

Adaptation must also be considered from the standpoint of cost of production and the relation to the labor problem of the farm as a whole. The relative cost per unit of digestible nutrients is the safest basis for comparison.

Usually the farms should provide sufficient pasture or the number of animals should be regulated in accordance with the available pasture. Generally the cropping system should provide all of the necessary pasture and roughage for livestock. To secure these from the outside usually entails much additional expense. In most cases the farm may also produce the major portion of the concentrates, and in many cases will produce all of the concentrates. This will generally be true in case of the production of swine, beef cattle and sheep. It is less frequently true in case of the

HOGGING DOWN CORN.[1]

production of dairy products. There are many factors that determine the proportion that should be produced and that should be purchased.

Cropping System Related to Future.—The plan of the cropping system should take into consideration the future productivity of the soil. Humus and nitrogen are most important in this connection. No cropping system will prove satisfactory for a long term of years that does not include at intervals of four to five years a leguminous crop such as clover, alfalfa or some of the annual legumes. Nitrogen in commercial form is much more expensive than that secured through the production of legumes. There is about $11,000,000 worth of nitrogen in the air resting on each acre of land. It is of prime importance to secure the soil nitrogen for crop production from this abundant supply. The humus will be maintained largely

[1] Courtesy of South Dakota Experiment Station, Brookins, S. D.

through the return of crop residues in the form of barnyard manure. This calls for the feeding of a considerable portion of the general farm crops.

Crop Rotations.—From the standpoint of the farm scheme, a rotation should ordinarily provide the roughage and pasture for the number of animals that are to be kept. It should include a sod and a legume for the supply of organic matter and nitrogen. It should also include as large an area of the profitable cash crops as can be produced advantageously. In nearly every region there is one cash crop or sometimes several that pay better than other crops. In the South it is cotton; in the corn belt it is corn; in the New England states it may be hay; in some other districts it is apples. These crops should dominate both from the standpoint of area and the care which they are given. Agricultural surveys show that the most successful farmers follow this practice.

Crops for Cash or for Feed.—The cash products may be either soil products or animal products. When animals dominate, the cropping system should be adjusted to meet their needs. Many small factors should be considered. The amount of bedding required should not be neglected. A crop that is of low value as a cash crop may be grown particularly for the straw it supplies for bedding.

Milk may be produced more cheaply by allowing the liquid excrements of the cows to go to waste rather than by going to the expense of securing sufficient bedding to absorb it. If, however, the saving of the liquid by the use of straw will increase the yield of corn for ensilage and result in 12 tons per acre instead of 8 and a corresponding increase in the other crops that are grown for feed, the enterprise as a whole will undoubtedly be much more profitable by providing the necessary straw for bedding purposes.

Straw contains a considerable part of the fertilizer constituents removed from the soil by a crop of grain. Prices for straw seldom justify selling it. If it cannot be used as bedding for livestock and returned to the fields in the manure, it should be returned in some other way. The practice of burning straw should be universally condemned. In cereal farming, the grain should be cut as high as possible, thus leaving the major portion of the straw on the land.

In some localities a cash crop may be grown, marketed and the proceeds enable the farmer to purchase twice as much of a given stock feed as he could produce on an equal area of land. Under such conditions the exchange is justifiable. In dairy districts farmers are often able to produce potatoes, and with the potatoes purchase more cow feed than they could possibly produce on the land devoted to potatoes.

Crops Related to Feed Requirements.—When grown chiefly for livestock, the proportion of the different crops should be determined to considerable extent by the requirements of the stock. One can ascertain how much of each crop should be produced by establishing what seems to be the best feeding system for the animals in question and calculating the year's requirements of silage, clover hay and grain as concentrates. Knowing the yield for these crops on the farm in question, the relative acreage of each can then be approximately determined.

Plenty of pasture may reduce the requirements for hay. Where corn does well, it is generally cheaper to feed than oats. A horse or mule without pasture generally requires about 3 tons of hay or its equivalent annually. He needs approximately 70 bushels of corn or 100 bushels of oats. Some of each is better than either alone.

Cows usually require about 1 to 1½ tons of grain, 1 ton of hay and 4 tons of silage per year. If silage is not available, about 2½ tons of hay per cow is needed.

Seven sheep require about as much feed as one cow. Hens eat about twice as much in proportion to their weight as other farm animals; 100 hens are equivalent to a 1000-pound cow and are considered an animal unit.

Buildings on a Dairy Farm.[1]

Changing Cropping System.—The development of the livestock enterprises on the farm will often necessitate changing the cropping system to meet the feed requirements. This will often require increasing the area of crops that produce the roughage. Such a change may be effected by substituting forage crops for cash crops or by increasing the length of the rotation. For example, the amount of hay produced could be very materially increased by increasing the ordinary four-crop rotation of equal areas of corn, oats, wheat and hay to a five-years' rotation of the same crops in which hay would remain for two years. This would increase the proportion of total cropped land in hay from one-fourth to two-fifths.

Two Rotations on the Same Farm.—The best development of the cropping system often calls for two rotations. This will be determined chiefly by the nature of the crops grown and the sequence that gives best

[1] Courtesy of Hoard's Dairyman.

results. Potatoes give good results when grown in a three-years' rotation consisting of potatoes, small grain and clover. At the same time a longer rotation of general farm crops may prove more advantageous for the major area of the farm.

On dairy farms where soiling crops are used, a subsidiary rotation on small fields close to the farmstead may prove advantageous for the production of a succession of suitable soiling crops. These will be supplementary to the general farm rotation that occupies the large fields for the remainder of the farm. The two systems can usually be worked together, so that a field is sometimes in one and sometimes in the other rotation.

Combining Fields.—Some farms are laid out in many small, irregular fields that cannot always be satisfactorily combined. The number and size of fields may not fit the most desirable rotation. This will involve a plan of cropping the fields in the most advantageous way to meet the desired cropping system.

Having decided upon the length of the rotation, ascertain the acreage of all fields and divide by the years in the rotation. This will give the area of each crop for each year. With this data, select the fields that are to be cropped the same in any given year in such a way that the acreage of two or possibly more, may equal as nearly as possible the desired acreage of the crop to be grown.

Fixed Rotations with Unequal Areas.—The livestock requirements often call for an unequal acreage of the several crops. This makes the cropping system more complicated, but never impossible of solution. Fields of equal area may be maintained without growing an equal area of each crop. Suppose the farmer wishes to grow 8 acres of potatoes, 20 acres of corn, 28 acres of oats, 16 acres of wheat and 40 acres of hay, making a total of 112 acres of crops. It will be necessary to decide on the number of fields that give best results in accommodating these crops. If the area is divided into four fields of equal size, there will be 28 acres in each. The following rotation and arrangement might be used:

Field.	1914.	1915.	1916.	1917.
1	8 acres potatoes 20 acres corn	28 acres oats	12 acres hay 16 acres wheat	28 acres hay
2	28 acres oats	12 acres hay 16 acres wheat	28 acres hay	8 acres potatoes 20 acres corn
3	12 acres hay 16 acres wheat	28 acres hay	8 acres potatoes 20 acres corn	28 acres oats
4	28 acres hay	8 acres potatoes 20 acres corn	28 acres oats	12 acres hay 16 acres wheat

Rotations for Dairy Farms.—The rotations on dairy farms will depend chiefly on the location and the relative cost of producing the dairy feeds as compared with purchasing. The cost of production may be relatively low, because the manure from the dairy is an important factor in the crop yields, and because labor will be available for a certain amount of field work and still fully meet the needs of the dairy. Corn as ensilage will prove an important crop wherever it can be successfully grown. Hay for supplementary roughage, and oats or wheat for the sake of the straw, will frequently be found advantageous.

Careful investigations of the success of dairymen show that a combina-

A Feed Lot Rack for Both Grain and Roughage.[1]

tion of dairy products and cash crops are generally more successful than dairying alone.

Corn, wheat, clover or clover and timothy mixed is a very common rotation. Where wheat does poorly, oats generally succeed and may supplant the wheat. In other districts both oats and wheat are advantageously grown. Alfalfa in limited acreage is generally advisable. It affords a most excellent crop, both for hay and soiling purposes.

Feeding Systems.—The feeding system for any particular farm should be based on the class of animals, their age and the chief purpose for which grown. The feeding system for the rearing of young stock is quite different than for dairy cows or stock that is being fattened. Many farmers find it advantageous to raise young stock and sell it for feeding purposes, while others are better equipped to purchase feeding stock and fatten it for

[1] Courtesy of The Pennsylvania Farmer.

market. The cheap feeds should be utilized to the fullest possible extent. Waste should be avoided.

Economy in feeding often calls for two or more classes of stock. Swine will follow steers and secure much feed from the droppings that otherwise would be wasted. They will also utilize the skim milk and buttermilk on farms that make butter. Under these conditions one may be justified in feeding steers whole grain in greater abundance than he would in the absence of swine.

The rearing of young stock generally necessitates depending chiefly on roughage and cheap feed. The roughage develops bone, and so long as the animal is kept thrifty and develops a good frame, the fat required for marketing can be secured by the use of concentrates during the feeding period. In this connection stockmen are cautioned to avoid the stunting of young stock by insufficient feed. The higher the grade and value of stock, the greater the necessity for quality in the feed consumed.

Feeding System Depends on Type of Farming.—Types of farming differ greatly in different sections of the country, depending on many factors previously mentioned in the chapter on this subject. Consequently, the feeding systems will vary greatly, depending on crops available. A type of farming that includes intensive crops, like tobacco, that respond abundantly to animal manures, may be justified in adopting a feeding system in which concentrates predominate. This results in more valuable manure which may increase the value of the cash crop to such an extent that stock can be fed on such a basis, even though there is no direct profit in the feeding enterprise.

The type of farming, however, will regulate the feeding system more largely from the standpoint of the products that are available for feed. On the grain farms in the corn belt, roughage in the form of stover and straw, supplemented with corn and oats, together with small amounts of hay, should constitute the chief products in the feeding system. In the cotton belt, corn and annual legumes which can be grown advantageously with cotton, should constitute the major portion of the livestock ration. This may be supplemented with cottonseed meal.

In the semi-arid belt, grazing combined with alfalfa and Kaffir corn would doubtless dominate the feeding system.

Feeding System Related to Cost of Production.—Agricultural surveys show that crops generally pay better than livestock for the time put upon them, but a combination of crops and livestock is generally more profitable than either alone. The feeding system for livestock produced in combination with cash crops will generally be more economical than that used when livestock alone is sold. When full time is spent upon crops, the waste products are not utilized and there is absence of manure to maintain yield; and when one devotes full time to livestock, time is spent in a line of production that is carried on at a very low margin of profit. It is a mistake either to overstock or understock on general farms. It is a good policy to keep enough livestock to consume all the by-products. When a farm is so heavily stocked that all the farm products are consumed, in years of low yields feed will have to be purchased, generally at such a high price as to make livestock an unprofitable enterprise.

Feed Units.—Feeds of different kinds are most readily compared by using a standard. Corn, being the leading crop in America, is the best standard to use. When corn is taken as 1, the equivalent value of a few other products is as follows: mixed hay .4, alfalfa .5, cottonseed meal

THE SCALE IS A NECESSARY ADJUNCT TO PROFITABLE FEEDING.[1]

1.25, wheat bran, oats, malt sprouts and similar feeds .91, corn silage .17, root crops .08. These equivalents vary somewhat, depending on the quality of the product in question. The value of other products is more definitely given in the feeding tables. Roughly, a cow or horse requires about 25 pounds of dry matter daily. This will generally contain from 18 to 20 feed units. The relation of protein, carbohydrates, etc., will depend on the work that the animal is doing or the product that is made.

Profits from Cheap Crop Products.—The cheapest stock feeds are products on which little labor has been expended and the cheapest way of

[1] Courtesy of The Pennsylvania Farmer.

feeding is to allow animals to harvest their own feed. The grazing of grass lands and the pasturing of cornstalk fields is typical of this process. It is further illustrated by chickens and pigs in small numbers that are allowed to forage for themselves about the farm premises. The farmer who produces pork on concentrates alone is at a disadvantage with the one who depends partly on hog pasture.

Livestock Gains in Relation to Feed.—Swine gain about 10 pounds in weight for each bushel of corn; steers require about 1000 pounds dry matter to make 100 pounds of gain; sheep require somewhat less food per pound of gain than steers; 100 pounds dry matter in dairy rations will produce about 74 pounds of milk containing 3½ per cent of fat.

Better gains are made with given amounts of feed during the early portions of the feeding period than toward its close. Young animals make a more profitable use of feed than older ones. It is seldom that the value of gain in fattening cattle is equal to the cost of the feed consumed. The profit is usually made on the increase in value of the total weight of the animal. Generally, a feeder weighing 1000 pounds can be purchased for from 1 to 3 cents per pound less than he will bring when in prime condition and weighing 1200 to 1400 pounds.

Corn Silage as Base for Ration.—In the corn belt, corn silage should form the base for feeding rations. It should be supplemented with dry roughage and nitrogenous concentrates in such a way as to meet the requirements of the stock raised. It has a wide adaptation and may be extensively used, either for the production of dairy products, the fattening of steers or the feeding of horses and young stock.

The following are a few rations which include corn ensilage as taken from an article by J. G. Grigsdale, published in the *Tribune Farmer:*

	Pounds.
For yearling heifers:	
Corn silage	25 to 35
Straw or chaff	4 to 6
Clover hay	4
Bran	2
For dry cows:	
Corn silage	50 to 60
Straw	8 to 10
Clover hay	4
Bran	1 to 2
For cows in milk:	
Corn silage	45
Straw	6
Clover hay	4 to 6
Meal mixture: Bran, oats, gluten or oilcake or cottonseed meal, equal parts. One pound of meal to three or four pounds of milk produced per diem.	
For steers running over winter (1000 pounds weight):	
Corn silage	60 to 75
Straw	8 to 12
Clover hay	2 to 4

For fattening steers (1000 pounds):

	Pounds.
Corn silage	50 to 60
Straw	6 to 10
Hay	3 to 6
Meal, starting at one pound, go up to 10 pounds per diem.	

Balanced Rations.—Animals require not only a sufficient amount of feed, but also enough of each of the different food elements as well. This pertains to the relationship of protein to carbohydrates and fat, and is spoken of as the nutritive ratio. The nutritive ratio is determined, as above indicated, by the character of animal and the work performed. It may vary somewhat within reasonable limits without seriously affecting the yield of animal products. The relative cost of protein and carbohydrates often justifies some modification in the ratio.

Standard rations for different classes of livestock will be found in the chapters pertaining to each class of animals. Methods of calculating rations are given in Chapter 45.

REFERENCES

Illinois Expt. Station Bulletin 125. "Thirty Years of Crop Rotation on a Prairie Soil."

Minnesota Expt. Station Bulletins:
104, 109. "The Rotation of Crops."

U. S. Dept. of Agriculture, Bureau of Plant Industry, Bulletin 102. "Planning a Cropping System." (In three parts.)

U. S. Dept. of Agriculture:
Year-Book 1902, pages 342–364. "Systems of Farm Management in the United States."
Year-Book 1907, pages 385–389. "Cropping Systems for Stock Farms."

Farmers' Bulletin 337, U. S. Dept. of Agriculture. "Cropping Systems for New England Dairy Farms."

CHAPTER 68

PLANNING THE FARM AND FARMSTEAD

Economy in the management of a farm and the pleasure in farm life depend to a considerable extent upon the plan of the farm, the arrangement of the farmstead and its position on the farm. A plan that will meet the needs of the farm for a long period of time calls for a plan of farm operations that is likely to be most successful under the conditions which prevail. These consist of crop adaptation, market demands and other factors discussed in preceding chapters.

Location of the Farmstead.—The farmstead includes that portion of the farm on which is located the farm buildings and feedlots, and generally includes the garden, orchard and ornamental features. The farmstead is the center of the business operations of the farm and from the business standpoint should be centrally located in order to economize in time in the performance of all farm operations. On the other hand, the farmstead is the home of the farm family, and many factors which contribute to the ideal home may be opposed to those which pertain strictly to the farm business. From the standpoint of the home, the farmstead should be near the public road. It should be in a sightly position and have a pleasing outlook. Proximity to a lake, stream or wooded area may add to the attractiveness of the home surroundings. These factors may not be available if the farmstead is centrally located.

The central location economizes time in getting to and from the fields. All four sides of the farmstead are directly connected with the fields. It makes possible minor rotations immediately adjacent for the pasturing of livestock so that they are near buildings, have access to water, shade and other protection. From the central location the manure is more easily returned to the fields and the crops are more easily brought to the barns.

The location will generally be a compromise between the business and the living requirements. If the farm is not too large, it will often be possible to locate the farmstead on the public highway in the center of one side of the farm. This brings three sides of it in contact with the fields and at the same time has the advantage of the public highway and nearness to the market, school and church.

The best position for the farmstead will be determined by the several factors that must be considered. If, for example, the trips to town are very numerous as a result of the character of farming, the children who go to school, etc., it may be equally as saving of time to have the farmstead located on the part of the farm that will bring it nearest to the place or places where members of the family must go.

Size of Farmstead.—The size of the farmstead should be in proportion to the size of the farm and will vary with the type of farming and the price of land. Livestock farming calls for more buildings and paddocks than most other types, and demands a relatively larger farmstead to provide these features. Where land is cheap more space may be given up to the farmstead than where it is high-priced. Land in feed-lots, paddocks and driveways is unproductive. A large farmstead generally results in scattering the buildings and increasing the distance between them, thus causing increased work in doing the chores.

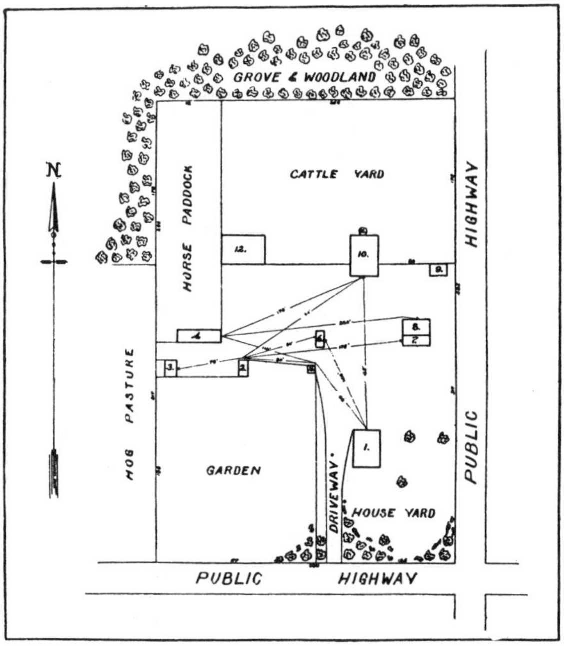

SKETCH OF FARMSTEAD THAT IS TOO LARGE.[1]

Buildings poorly located. 1—Farm-house. 2—Poultry-house. 3—Hog-house. 4—Horse-barn. 5—Smoke-house. 6—Milk and well-house. 7—Corn-crib. 8—Machine-shed. 9.—Ice-house. 10—Cow-barn and granary. 11—Silo. 12—Hog-shed. Distance from horse-barn to machine-shed 220 feet; from corn-crib to hog-pen 250 feet, and from well to hog-pen 155 feet. In one trip three times a day for a year between the corn-crib and the hog-house and between the well and the hog-house, 199.6 miles would be traveled. In one trip a day between the machine-shed and horse-barn 26.1 miles would be traveled, and going to the poultry-house three times a day for a year would require 78 miles of travel. A total of 264 miles would be traveled, which, at the rate of 15 miles a day, would require 18.1 days.

From four to six acres are usually sufficient for the farmstead on a diversified grain and stock farm of 160 acres. More can sometimes be used to advantage. The dimensions may be varied in any way desirable. The length and breadth will often be determined by the position on the farm as affected by topography, proximity to shelters of either hills or wooded areas and the direction which it faces. In any locality, the prevailing direction of the wind at different seasons of the year will determine the position of windbreaks and the direction of the barns from the house.

Extensive lawns and ornamental features, while very attractive,

[1]Courtesy of Lyons & Carnahan, Chicago. From "Farm Management," by Boss.

really have no place on the average farm. The farmer, as a rule, has less time to devote to keeping a lawn in good condition than does his village or city brother. The latter may not only have more time for such work, but finds the outdoor exercise a decided advantage to him, while the farmer already has enough of such exercise without mowing a lawn or trimming shrubbery. A lawn of moderate size planted with a limited number of ornamental trees and shrubs artistically grouped with plenty of open space for air circulation and good views, is desirable.

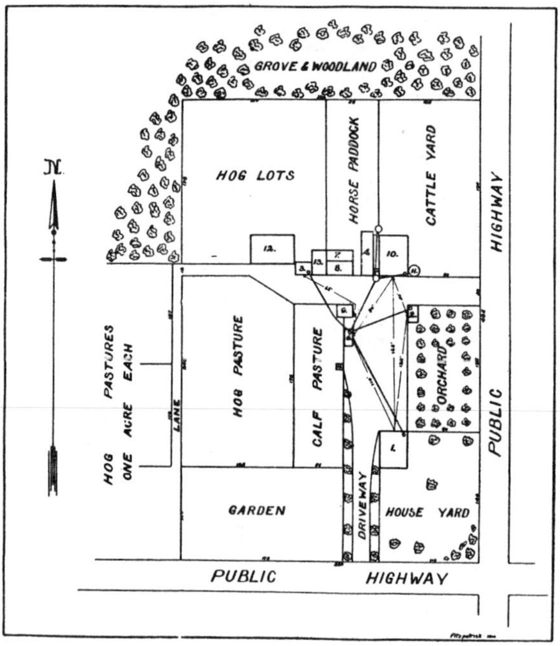

THE FARMSTEAD REARRANGED FOR ECONOMY.[1]

Buildings moved and more closely grouped. 1—Farmhouse. 2—Poultry-house. 3—Hog-house. 4—Horse-barn. 5—Smoke-house. 6—Milk and well-house. 7—Corn-crib. 8—Machine-shed. 9—Ice-house. 10—Cow-barn. 11—Silo. 12—Hog-shed. 13—Feeding-floor. In doing the same chores described under previous illustration on page 845, only a little over 30 miles need be traveled, and but 2.24 days would be required. The water is piped to the hog-house and barns, greatly reducing the labor. The feeding-floor joins the corn-crib and hog-house, and the machine-shed is only 30 feet distant from the barn.

Arrangement of Orchard, Garden and Lots.—The orchard may serve for fruit supply, ornamental features and wind protection. Its position need not necessarily be especially near the house. Often, however, it will be found advantageous to utilize it as runs for poultry, or as pasture for pigs, calves and lambs. These features should be considered in connection with its location and relation to the buildings.

It is more important to have the garden easy of access, and it may lie in any direction from the house so long as it is not brought conspicuously to the front. Vegetable gardens at certain seasons are, as a rule, weedy and more or less unsightly, and should, therefore, be kept in the background.

The position and size of the paddocks or feedlots will be determined

[1] Courtesy of Lyons & Carnahan, Chicago. From "Farm Management," by Boss.

by the class of stock that is to use them, the position of the buildings and the drainage of the land. Sheltered lots that drain away from the buildings are to be preferred. One-half acre to an acre in extent will generally meet the needs for most any class of stock on the average size farm. For the minor classes, smaller ones may be sufficient.

The garden calls for careful arrangement within itself, in order to make it as sightly as possible and at the same time economize in the garden work. The small fruits and all perennials, such as asparagus and rhubarb, should

General View of a Well-arranged Farmstead.[1]

be brought together in one portion of the garden, preferably with the taller growing ones in the further background. These should be arranged in rows as long as the dimensions of the garden will permit and with sufficient space between for horse cultivation.

The annual crops should occupy another portion, preferably in the foreground with the corn and tall-growing plants to one side or in the rear. It will be necessary, however, to plan the plantings of annual crops with a view of crop rotation. Such will lessen the difficulties with plant diseases and help to maintain satisfactory yields and high quality. The garden should be planned with this rotation in view, and the rotation should pro-

[1] Courtesy of The Macmillan Company, N. Y. From "How to Choose a Farm," by Hunt.

vide for legumes, such as beans and peas, every third or fourth year, so as to avoid the successive plantings of cabbage, cauliflower and other crops that are likely to be affected with diseases as a result of continuous cropping.

It is generally advisable to fence the garden against poultry. This is better than to confine the poultry within a limited enclosure. After the planting season is over and the garden crops are well established, there are periods of time when poultry may be more beneficial in the garden than harmful. Young chickens catch many insects and feed upon quantities of weeds and weed seeds, and except for injury to the fruits or berries, may do little harm in the garden.

Grouping the Buildings.—The farm buildings should be grouped with reference to economy in doing the chores and with the object of good appearance. Both of these features call for careful consideration. From the architectural standpoint, it may be wise to consult the landscape architect, but from the standpoint of economy in work, no one is better able to calculate the position of the several buildings and their distance from each other in relation to economy in the work than is the farmer himself. The horse barn, machine shed and shop should be located near each other and on the main lane that leads to the largest number of fields. This arrangement reduces to the minimum the time in caring for the horses and machinery and connecting them up with the field work. The buildings in which animals are sheltered, and those used for storing feed are generally visited several times daily throughout the year. A few yards added to the distance between those visited in this way means many miles in the course of a year. One needs only to make some calculations in order to ascertain how much time can be saved by a better arrangement and lessening the distance between buildings. It is always wise, however, to have the barns 100 feet or more distant from the house or any other buildings in which fire is maintained at any time during the year. This lessens the danger of loss by fires and furthermore reduces the insurance rates on barns and outbuildings.

All of the buildings should be in keeping with the farm. The size of the farm and the value of the land will determine in a large measure the size and quality of the buildings. It is seldom wise to build an expensive house or barn on a small farm or on cheap land. This principle holds even in city building where it is not considered good business to build an expensive house on a cheap lot, or vice versa.

In all parts of the country are found farms which are over-capitalized with buildings, and it is not unusual to see a farm advertised for sale, the buildings on which cost more than is asked for the farm. These are examples of farm buildings that were either illy adapted to the type of farming or too expensive to enable the farm to maintain them.

It is always wise to provide ample shelter and sufficient storage to meet the needs of the type of farming. Storage capacity often enables the farmer to hold his products for the best market prices. In starting the new farm, however, it is better to defer building until one is sure of what he needs than it is to build hastily. Energy and money spent in raising crops in the beginning will place the farm on a good financial basis sooner than to put all capital into buildings at the sacrifice of working capital.

Buildings should be located and constructed with reference to future needs. When a barn is being built a place should be provided for other buildings that will be needed in the future, in order that they may bear

A FARM OVER-CAPITALIZED WITH BUILDINGS.[1]
Buildings too extensive for average business farming.

proper relation to the barn. If enlargement of the business contemplates enlarging the barn, this should be provided for at the outset.

Numbers of farmhouses, the result of additions from time to time without reference to future needs, are striking illustrations of the lack of forethought in this respect.

Few buildings with ample capacity are generally more economical, both from the standpoint of construction and upkeep, than many small buildings having equal capacity. The less the number of buildings, the greater the economy in doing the chores.

Water Supply.—The well or water supply for the farm should be

[1] Courtesy of Doubleday, Page & Co., Garden City, N. Y. From "Farm Management," by Cord.

brought to a central location that will be easily accessible to the farm family, as well as to all livestock on the farm. Where gravity systems are feasible, or where wind or gasoline power can be provided, water may be forced through pipes to different points on the farmstead, and the time required in watering stock thus greatly reduced. As far as possible, water should be in all feedlots and paddocks, and running water in the house saves many steps on the part of the housewife or other members of the family.

Relation of Buildings to Farm.—The farmhouse should be set in the foreground of the farmstead, and, when possible, should occupy a prominence that will afford drainage and enable the farm family to have a good outlook in as many directions as possible. From the business standpoint,

Adequate but not Over-capitalized.

the more of the farm that can be seen directly from the house, the better. There is nearly always some member of the family there, and if stock escape from the pasture, or the neighbor's cattle break into the grain fields, the trouble may be detected and damage avoided.

In most localities an east or a south front for the farmhouse is preferred.

In type, a farmhouse differs from the city house. It should be built on broader lines and not so tall as the city house. Numerous gables or striking shapes are more noticeable in the country house and should be largely avoided. Since the back of the farmhouse is more extensively used than that of the city house, relatively more attention should be given to its construction, appearance and convenience. The back yard of the farmhouse frequently comes more into prominence than the front yard. Since all travel is done by team or automobile which find their housing in the rear, the farmer seldom leaves by the front door or through the front yard. Even visitors often drive directly to the barn, where their team is cared for, and enter the house at the rear.

The interior arrangement of farm buildings and plans for their construction will be found in Chapter 57.

Sightliness and Healthfulness.—The appearance of the farmstead and the buildings which it contains may be greatly enhanced by natural features. These should be taken advantage of in selecting the site. The arrangement of the buildings and the ornamental planting, together with the type of architecture and the use of paint of suitable shades should be considered also. The ornamental plantings are discussed in Chapter 41.

Healthfulness is often closely related to one's surroundings. Ill health may result from living on wet land or in close proximity to stagnant ponds of water. This should be avoided in locating the farmstead. Where surface water is depended upon, contamination of the water supply often results in sickness. A proper location of the well which is to provide a healthful water supply is quite as important as the location of the farmstead.

Size, Shape and Number of Fields.—The size and shape of the fields on a farm are determined to a considerable extent by the size and shape of the farm. For most economical cultivation, fields should be fairly large and rectangular in shape. Fields twice as long as they are broad are generally preferable to square fields. Long fields lessen the number of turns in the operation of farm machinery, thus avoiding wasted time. Triangular fields, or those of irregular shape, necessitate irregularity in the length of the rows, and are more expensive to plow, seed, till and harvest than rectangular fields.

The more horses one drives the greater the necessity for large fields. In most kinds of general farming, fields should be at least 40 rods long; 80 rods is much better and 160 rods is ample. Little is to be gained by having fields more than half a mile in length. Ordinarily teams should rest for a few minutes during each half mile of travel when doing field-work.

Large fields waste less land than small ones. Some space is required along the margins of the field for turning rows, especially when the fields are fenced. The smaller the field, the larger the percentage of land wasted in this way. Even though crops are planted close to the borders of the field, considerable will be wasted along the turning rows in both cultivation and harvesting.

Large fields economize in the cost of fencing, and the shape of the field is also a factor in the relative cost of fencing. A strip of land 1 rod wide and 160 rods long contains 1 acre and would require 322 rods of fence to enclose it. One acre in the form of a square requires about 50 rods of fence; 10 acres in the form of a square requires 16 rods of fence per acre; 40 acres in the form of a square requires only 8 rods of fence per acre; while a square mile requires only 2 rods of fence per acre. The most economical fence and the one occupying the minimum of space is the woven wire fence. This calls for very strong deeply-set and well-braced corner posts. Fields of irregular shape have more corners and increase the expense of brace posts.

It is much better to have a few large fields than many small ones. The number of fields should be determined by the crop rotations, there being one field for each crop or each year in the rotation. In some cases, streams, woodlots, roads or railways may interfere with the regularity in shape of

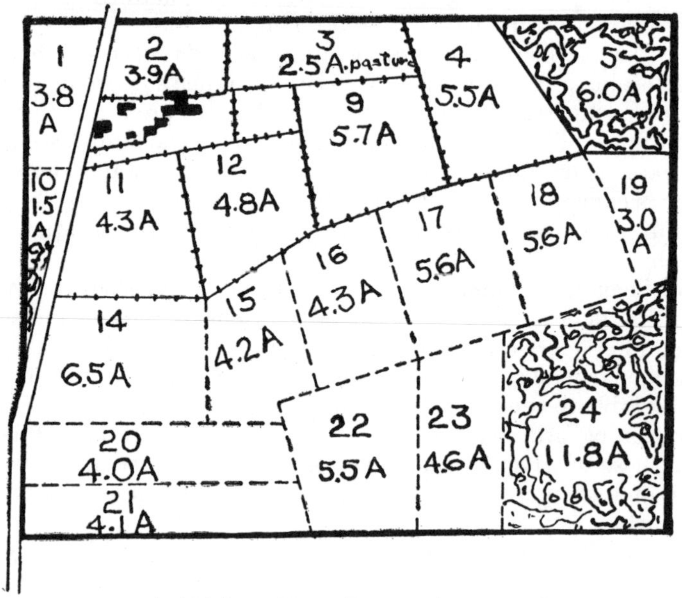

A 100-Acre Farm Poorly Arranged.[1]

fields and uniformity in size. The plan for the fields should be such as to obviate these difficulties as much as possible.

Distance to Fields.—The distance from the farmstead to fields should be as short as possible and the reduction in this distance will depend greatly on the size, shape and arrangement of the fields. Time spent in traveling to and from the fields is unproductive. The number of trips in the course of a year are many. If the average distance to fields is reduced by twenty rods as a result of proper planning, many miles of travel and many hours of time on the part of the men and teams will be saved in a year. Even though the average distance of the farmlands from the farmstead is not changed, bringing the nearest portion of the field close to the farmstead greatly facilitates the work. All tillage, seeding and harvesting operations should be so planned that the machine and team begin work at once upon

[1] Courtesy of The Macmillan Company, N. Y. From "Farm Management," by Warren.

reaching the nearest point of the field. It is unwise to finish the work of that portion of the field nearest the farmstead, and thus travel over it unoccupied to reach the more remote portions. The hauling of the products from the field and the manure to the field may, to a certain extent, be planned in the same way. Only when the field is so long that a load can be gathered or unloaded in passing less than twice across the field, is there any loss in this respect.

Variation in topography and character of soil will sometimes influence the shape and direction of the lines of the field. It is desirable as far as

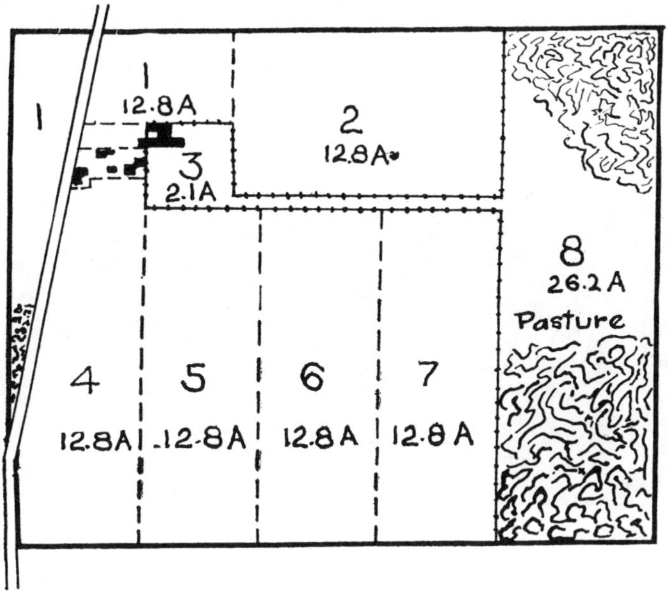

THE FARM REARRANGED FOR ECONOMY IN CULTIVATION AND SAVING IN WASTE LAND AND FENCES.[1]

possible to have fields uniform in character of soil, for this facilitates uniformity in treatment and rate of seeding.

Rotation Groups.—Under present conditions, the arrangement of fields on a farm most generally becomes a question of remodeling the farm. To lay out a farm from the beginning, the fields should be free from crops, but since this is usually not feasible, it is necessary to change the plan by degrees. It will usually require two or three years to establish the change and satisfactory rotation. Sometimes it will take four years to do it.

Where two rotations are advisable, two groups of fields are called for, one for the major rotation consisting of large fields, and the other for the minor, made up of small fields. The latter should lie adjacent to the

[1] Courtesy of The Macmillan Company, N. Y. From "Farm Management," by Warren.

farmstead. The small fields, if not immediately adjacent to the farmstead, should be connected with it by short lanes, and these should generally be fenced. Such fields will be most frequently used for summer forage and pasture purposes.

Farm Lanes, Roads and Fences.—Farm lanes should be sufficiently wide for the convenient passage of all types of farm machinery to and from the fields, and where gates are necessary, they should be sufficiently wide to permit the passage of wide machines, such as hay-rakes and binders. Seldom should a lane be less than two rods in width. Lanes that are fenced and through which livestock frequently pass may be considerably wider, providing they can be pastured. Such lanes are fully utilized and

A Good Farm Fence.[1]

provide for some latitude in the roadway in case the road becomes too bad and difficult for heavy hauling.

The ideal road arrangement is to have the public road pass through the center of the farm. While this takes more of the farm land than when the road passes along the border of the farm, the saving in time in reaching all parts of the farm over a good public road more than offsets the loss in land. Where such roads are available, farmers are advised to have their buildings all on one side. To have the house on one side and the barn on the other is both dangerous and inconvenient. Such an arrangement calls for fences and gates to keep the livestock within the farm fields and out of the public highway. If such provision is not made, stock may be injured by automobiles, or passersby may be injured as a result of the stock being in the highway.

Unnecessary fences are frequently provided at much expense. Many

[1] Courtesy of Wallace's Farmer.

farms and fields, expensively fenced, could be better utilized in the absence of fences. In the better and more prosperous grain-growing districts many farms have no fences other than those about the farmstead. On general and livestock farms it is advisable to have the farm as a whole enclosed by a good stock fence that will afford protection from the encroachment of stock of neighbors or from stock escaping on to neighbor's land. Aside from this, the only fences advisable are those enclosing the permanent pasture, the paddocks and the farmstead. It is true that having all fields fenced enables the farmer to pasture such fields temporarily when forage is available. Utilization of such material, however, can frequently be

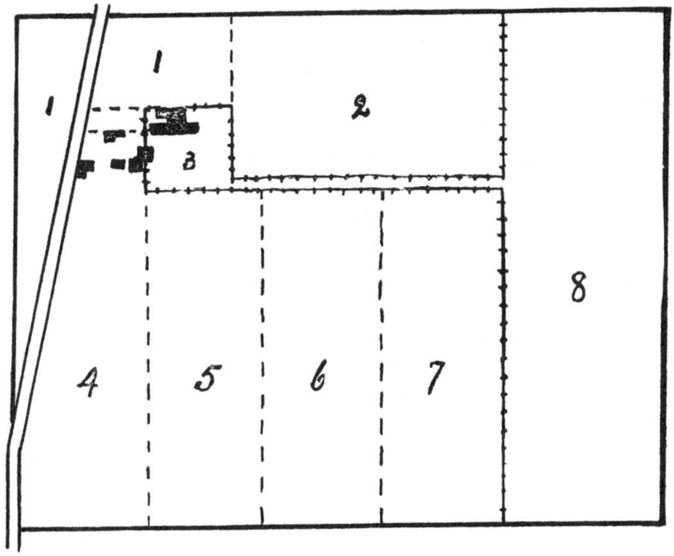

THE FARM SKETCH THAT IS USEFUL FOR RECORDING CROPS GROWN, YIELDS, PLACE OF APPLYING MANURE, LIME, ETC.[1]

arranged for without permanent fences. Stock may be herded a few hours each day during the short period when pasture is thus available at small cost, or temporary fences may sometimes be provided and shifted from year to year as the temporary grazing shifts from field to field in the rotation.

The type of fence is important from the standpoint of land occupied, effectiveness, durability and first cost. The old type of rail fence is no longer economical. Hedge fences once were quite desirable, but today have been universally discarded except in a small way and for ornamental purposes. The standard woven wire fence, well built, is both effective and economical. It occupies little land and calls for little work in keeping the fence row free from weeds.

[1] Courtesy of The Macmillan Company, New York. From "Farm Management," by Warren.

Rearranging Farms.—As a rule, farms in the older sections of the country contain many small, irregular fields. These should be combined in a systematic way and the number reduced to meet the crop rotation most desirable. Such revision calls for careful thought, considerable work and some expense. Often stone fences or rows of trees and shrubbery will have to be removed in order to unite small fields. Such revision brings into cultivation land that is now absolutely wasted and increases the economy in the farm operations. It enables the use of larger teams and bigger machines which cannot be economically used on very small, irregular fields. One needs to figure on the saving in time and land for the years to come as compared with the expense called for in the remodeling process. As a rule, the work of cleaning up old fence rows can be performed at odd times when the crops and animals of the farm do not call for all the time of the farmer. From this standpoint, the cost may be comparatively slight.

The accompanying sketches show a farm before and after revision in this way and bring out the decided advantages of the new plan

Crop Ledger Plan.—A plan of the farm drawn to scale, showing the size and shape of fields and location of buildings, woodlots, etc., is always desirable. Such a plan is necessary in connection with working up a new plan for the farm. After the new plan has been arranged, a number of copies may be provided so there may be one copy for each year on which may be entered the name of the crop grown on each field, the acres contained and the yield secured. This forms a permanent record, if kept, and gives a history of the cropping of the fields.

If desirable, two copies may be used, one to show the fields which have received manure from year to year, in order that manure may be equally distributed over the entire farm. The use of this system is illustrated in the accompanying sketch.

REFERENCES

"The Farmstead." Roberts.
Illinois Expt. Station Bulletin 105.
Wisconsin Expt. Station Bulletin 105. "Improvement of Home Grounds."
U. S. Dept. of Agriculture, Bureau of Plant Industry, Bulletin 236.
Farmers' Bulletin, U. S. Dept. of Agriculture:
317. Pages 5–10. "Planning a Home."
370. "Replanning a Farm for Profit."

CHAPTER 69

LAND RENTAL AND FARM TENANTRY

Methods of renting land are becoming of greater importance. According to census figures there were 2,354,676 tenants in the United States in 1910. This constitutes 37 per cent of all farms, as compared with 35.3 per cent in 1900 and 28.4 per cent in 1890.

TYPICAL FARM IMPROVEMENTS IN A FARM COMMUNITY WHERE TENANTRY PREVAILS.[1]

The report shows further that while 37 per cent of farms are rented, 22 per cent are on the share basis, 11.2 per cent on a cash basis and 2 per cent on a combination cash and share rent. The system of rental for the remaining 1.8 per cent is not given.

Is Tenantry Desirable?—Much has been written relative to the evils of farm tenantry, but it is not necessarily a curse to the country. The

[1] Courtesy of Doubleday, Page & Co., Garden City, N.Y. From "Farm Management," by Cord.

evils depend upon the land-owner and the tenant, rather than upon the system. Where both conspire to rob the soil, the landlord will ultimately be forced to sell and eventually the tenant will have to discontinue renting. The landlord should pursue a policy that will provide for the maintenance of permanent soil fertility, and should offer terms that will secure the co-operation of a tenant who can see the advantage of such a system.

There are marked advantages in the long-time lease for both landlord and tenant, but as yet relatively few land-holders are willing to enter into long contracts with tenants. Investigations show that permanency on the part of a tenant makes for a higher standard of farming than where tenants are changing annually from farm to farm.

Relation to Progress.—In these times of extensive co-operation, a shifting tenant population blocks progress along agricultural lines. Such movement, common in the South and in the prairie states, affects not only marketing, but the character of schools, churches and other helpful institutions.

Closer sympathy between owner and tenant is needed. Both should realize that a farm will not stand the old-time drain on fertility and continue to yield a profit. In England, where tenantry is usually common, the long-time lease prevails. There legislation has been enacted in the interest of both landlord and tenant, providing for the adjustment of capital invested in improvements or fertilizers, in case the tenant leaves.

Classes of Land-Owners.—There are three classes of land-owners that rent their land, viz., retired farmers, who think the rent of the farm will enable them to live in town; business men, who buy farms for an investment; and speculators, who buy because they expect a rise in land values. These all rent to men who cultivate the land. The contracts made between owner and tenant should be such as to provide for permanent improvements and for soil fertility better than would be necessary if the owner farmed the land. Two families, instead of one, must live from the farm returns.

Under the present system of tenantry, ownership on the part of the tiller of the soil has decided advantages. It stimulates interest in building up the farm and providing improvements.

Farming with Small Capital.—Land values have increased so rapidly that profitable farm ownership requires considerable capital. As a result, men with limited capital are obliged to farm as tenants. This enables them to put all their capital into livestock and equipment. Investigations show that they generally make more on rented farms than they possibly could on farms of such size as their limited capital would enable them to purchase. Recent farm surveys in Indiana, Illinois and Iowa show that tenants with $1758, $2867 and $2667, respectively, invested in machinery and livestock, made average labor incomes of $755, $1139 and $716, respectively. The average labor income of farm owners for these same districts in the same year was only $310, $622 and $291, respectively. The capitalization in case of the owners was $17,000, $51,000 and $23,000, respectively, 5 per cent interest on the investment being considered as part of the expense this percentage being deducted in order to find the labor income.

Starting as a Tenant.—It is seldom advantageous for a young man to start farming as a tenant with less than $1000 in capital. The good farm workman will generally make more as a farm laborer than he can make with a capital of less than $1000.

In most of the livestock, grain and hay regions, one should have from $1500 to $2000 worth of stock and equipment before one is ready to start as a tenant. Part of this may be borrowed. In such regions a tenant with $3000 can run a business with a total capital of about $15,000.

It is usually better to rent a good farm than a poor one. Half the crop on good land is frequently equal to the entire crop on poor land.

If land is likely to rise in value, a tenant is justified in purchasing a farm sooner than would otherwise be advantageous. Increase in land values are sometimes greater than rents or interest.

Basis of Rental.—The rent or share of crops demanded by the landlord should be in proportion to his risk and trouble. Cash rent involves little trouble on his part and no risk. It is, therefore, the cheapest form of rent.

THE HOME OF A NEGRO TENANT IN THE SOUTH.

The more the landlord furnishes, the larger is his risk. He takes a risk on weather conditions and upon the possible loss of livestock. The more the landlord furnishes, the greater the necessity on his part of looking after his interest. This demands considerable of his time, and he therefore should have a larger proportion of the products.

The extreme of this system occurs in the South with the negro tenants where the landlord supplies everything and trusts the tenant for his groceries.

Investigations in Tompkins County, N. Y., show that landlords who rent for cash averaged 5.2 per cent on their investment, while those owning part of the stock and equipment and receiving half the proceeds, made 9 per cent. In the same county, tenants renting for cash averaged $604 yearly for their labor, while those renting for a share averaged only $342 a year for their labor.

As a general principle, tenants who are good farmers had best rent for cash when possible. They assume all responsibility and all risk of loss, but by good management generally make more than by renting for shares.

One who is just starting as a farmer will often do better by renting for shares from a good landlord who will give close supervision to the management of the farm. A landlord's good judgment and experience in farming will be of much assistance to an inexperienced tenant.

Systems of Rental.—The method of renting land varies greatly, but there are three general methods, which are subject to variation: (1) the landlord may furnish only the land; (2) he may furnish land, half the livestock, other than work animals, and pay half of the feed, fertilizer and seed bills; and (3) he may furnish everything but the labor.

Renting for shares is more common than any other system. The division of the farm proceeds may be based on the gross receipts or upon net returns. The former is much more common. The basis of division is determined by the relative responsibility and interest of the landlord and tenant. The more working capital the landlord has invested, the larger will be his share of the farm products.

In most of the eastern states, the tenant and owner each pay one-half of the running expenses of the farm, such as feed, fertilizers, spraying material, binder twine, etc., and divide the proceeds equally. Under this system, the owner pays for all permanent repairs, improvements on buildings and land, taxes and sometimes half the school taxes. The tenant furnishes all the labor, all tools and implements, the work stock and pays the school taxes. Where dairy cows are kept, their ownership is frequently shared equally. Under this arrangement the equal division of all products between landlord and tenant is a fair division in most regions where general farming prevails. The more intensive the type of farming, the larger should be the tenant's share of the products. Tenants can seldom afford to spend most of their time milking cows and raising such crops as potatoes for one-half the products. In this kind of farming, labor constitutes a large part of the cost of production and the burden of this falls entirely on the tenant.

Cash Rental.—Good farms rented for cash often bring better returns than land ownership. This gives the tenant independence in the conduct of his methods of farming and he reaps the benefit of all special efforts that he puts forth. If better tillage increases crop yields, he receives the full benefit of his effort.

If a long lease can be secured, there is nearly as much inducement to build up the fertility of the soil as if he were the owner.

Advantages of Share Rental.—Renting land on shares lessens the investment on the part of the renter and makes it possible for some men to farm who would not be able to buy or rent for cash. The lower the value of the land rented, the larger should be the tenant's share of the farm products; the greater the intensity of the type of farming and the larger the cost on the part of the tenant, the larger should be his share of the products. The chief disadvantage in share renting lies in the fact that the tenant receives only a share of the increased yield of crops that may result from extra tillage. For example, if a tenant puts an extra dollar per acre into the preparation of land for wheat, and gets an increased yield of two bushels per acre, selling for 90 cents per bushel, he loses 10 cents per acre, although the landlord profits 90 cents per acre.

In the newer regions, where grain is the chief crop, tne tenant usually gets three-fifths of the crop, and usually agrees to deliver the landlord's two-fifths at the nearest market place. The tenant generally gets all the straw, corn stocks and other by-products, for use on the farm. Where land is very poor or crops uncertain, the tenant may get a larger share. On the high-priced corn lands in the corn belt, tenants are now getting one-half of the grain and hay or any other products that may be sold. They are allowed all straw, stalks and other by-products if they are fed on the farm.

In the eastern part of the United States, where a considerable part of the crops are fed to livestock, the landlord bears a certain portion of the expenses and furnishes part of the livestock as previously mentioned, and gets one-half of all receipts. The straight half-and-half division is fair when the tenant sells many cash crops, such as corn, wheat and hay. If nearly all receipts are from dairying and livestock products, he should receive three-fifths of all the receipts.

Personal Element.—It is to the interest of the landlord to have the good will of the tenant. This will generally be secured by giving him a fair deal. The proper understanding between the landlord and tenant can best be secured if each will put himself in the place of the other in their business dealings. In order to do the square thing with each other, each must understand that the right thing must be done by the soil. The maintenance of soil fertility is to the best interest of both of them, and is likewise a matter of interest to the general public. If this situation is thoroughly understood, there is little need of legislation.

Legisiation.—The increase in the percentage of farms in this country that are rented from decade to decade indicates an increased need for legislation that will protect landlords, tenants, farm lands and the public. Legislation relative to land tenantry should be preceded by education. Before we can intelligently adopt the measures of older nations, we must ascertain whether or not they will meet local requirements.

The merchant who rents a building in the city uses it, but he does not improve it. He may be liable for damages beyond reasonable wear and tear, but he never adds to its value. Good tenant farming necessitates fertility being at least maintained, and it should be improved. The burden of this falls upon the tenant, and the character of the lease should be such that he can afford to do it. The tenant uses up the fertility of the soil, hence there is some necessity that the landlord also be protected. If regulations necessitate the tenant applying fertilizers to the land, the landlord should be prevented from confiscating the benefits, as he may do if he finds some rival tenant that will pay a larger rent. Legislation, therefore, should be framed for the protection of both parties.

What the Lease Should Contain.—Bulletin 198 of the Wisconsin Agricultural Experiment Station gives the results of investigations relative to methods of renting farm land in that state. As a result of these investigations the author gives the following considerations:

"A. The contract should vary with the farm, the tenant and the landlord. The three chief requisites of the tenant are honesty, efficiency and capital. The capable and fair-minded landlord gets better tenants, higher rents and has many more friends than the grasping type of landlord. In choosing a farm, fairness of rent, suitability to type of farming most congenial to tenants, location with respect to markets, schools, etc., are of primary importance to tenant.

"B. What the lease should contain: A dozen or more phases of the contract are discussed in detail. Some of the phases will be merely enumerated, while a brief summary of others will be given:

"1. A description of the farm as found in the deed.

"2. Uses of property should be stated specifically.

"3. Disposition of products regulates crops that are to be produced and the forms in which they are sold. Restrictive clauses should be accompanied by supply of credit to comply with restrictions.

"4. Manure should not be removed from the farm.

"5. Purchase of feed and fertilizers should be encouraged by compensating the tenant for exhausted fertility added to the land.

"6. Noxious weeds should be destroyed before maturity.

"7. Duration of lease should be longer than one year. Extensions of one year or more make it possible for less loss in division each year, and afford the tenant time to realize more fully on labor and investments he has made. Compensation for unexhausted improvements is the remedy for many of the evils of the short term lease.

"8. The amount of rent to be paid depends upon the usefulness of the land and the scarcity of land of this grade.

"9. Time of paying rent should be adjusted to the time of the tenant's sales.

"10. Guarantee that the rent shall be paid should be obtained either by statutory law or by agreement.

"11. The agreement should provide for closing the tenancy at the end of the current year in case either party fails to perform his part.

"12. Notice to terminate a lease is used where a contract remains in force as long as it is mutually agreeable."

Time of Lease.—The tenant on the one-year lease has little interest in the fertility of the soil. His chief object is to realize on what fertility is there. It is not to his interest to turn from grain to livestock farming. The land, on the other hand, is entitled to stock and the pasture on which stock should graze. On the one-year lease, it seldom gets it.

The long-term lease increases the interest of the tenant both with reference to keeping up a good appearance of the farm buildings and maintaining the fertility of the soil. It makes it to his interest to destroy noxious weeds and haul out all stable manure. Long-time leases enable him to plan ahead and to establish crop rotations on which he may realize the most in the long run. They enable him to build up a herd of livestock of such a kind as may be best adapted to the conditions. Neither landlords nor tenants can profit in the long run through short-term leases, especially if this results in a change of tenant from year to year.

Profits Under Different Methods of Renting.—Investigations show conclusively that the average tenant makes larger profits when he rents for cash than he does when renting for shares of either the products or the net returns. The cash tenant assumes all risk and the largest possible supervision of the farm and labor, while the tenant who rents for share, shares the risks with the landlord and frequently is subject to the direction of the landlord with reference to the manner of farming. In Tompkins County, N. Y., it was found that cash tenants made an average labor income of $604, as compared with $342, the average made by share tenants.

Systems of rental should be flexible and leases drawn to fit the type of farming engaged in. The equitable division of the farm products should be determined by the relative cost of labor and expenses on the part of the tenant as compared with the interest value on the capital on the part of the landlord.

REFERENCES

"Agricultural Economics." Taylor.
Wisconsin Expt. Station Bulletin 198. "Methods of Renting Farm Lands in Wisconsin."
Farmers' Bulletin 437, U. S. Dept. of Agriculture. "A System of Tenant Farming and Its Results."

CHAPTER 70

FARM LABOR

Labor is an important factor in the production of all wealth. The labor of men and teams is the chief item in the cost of producing general farm products. Agriculture, therefore, has its labor problem the same as all other productive industries. Farming, however, is less industrialized than other industries, and the labor problem on farms should be less difficult.

Social Relation of Farm Labor.—It is often necessary for the farm laborer to eat at the same table with the farm owner and mingle freely with the farm family. In this respect farm labor differs from most other kinds. For this reason, there can be less difference in the social standing of employer and employee than in most other industries. Industrial corporations do not discriminate against employment of southern European immigrants, Hungarians or negroes, so long as they are fitted for the work to be done. The reason these people are not more extensively employed in farm labor is because the farm hand is so frequently a member of the farmer's family.

Why is Farm Labor Scarce?—Most of the farm labor has always been done by the farmer and his family. The increasing importance of education, raising the age limit of employment, the lessening tendency for women to work in fields, together with increased farm tenancy, has lessened the number of farm laborers. Foreign immigrants are not going to farms as much as formerly, because the development of farm machinery and improved methods in agriculture call for a higher class of workmen. The unskilled labor can be used only on farms where much hand labor is required.

Farming does not sort men as closely as most other occupations, consequently, the farm laborer has less opportunity for advancement. In many industries there are many grades of men and this gives opportunities for advancement from one grade to another, to positions of considerable responsibility. The farm laborer must either look upon his employment as temporary or be willing to serve without much advancement. Such service is essential to the young man who intends to become a farmer. It prepares him first as a farm tenant, and paves the way to ownership.

Extent of Employment.—On the average American farm there is not much opportunity to profit by the employment of labor. According to the last census, there was employed less than 5,000,000 people, exclusive of owners, on something over 6,000,000 farms in the United States. Over half of those employed were members of the farmers' families, so there was less than one hired employee to every two farms. A considerable number of farms employ more than one person, so the majority of farms employ no help other than members of the family.

The development and introduction of farm machinery have more than offset the increased need of farm workmen, resulting from more intensive agriculture. This has increased the difficulty in keeping farm labor continuously employed.

Solution of Farm Labor Problem.—It is not possible to settle once for all the labor problem on farms or in any other occupation. Changed economic conditions will raise new questions and result in new phases of the old problem. The problem is not to be solved by having more and cheaper labor, but by a better direction of the efforts of labor, and by providing homes and a continuous employment for it.

Many farmers have reported favorably on the employment of married men. This necessitates a tenant house in which the man and his family can live. This arrangement relieves the farmer's household from the housing and feeding of farm workmen, and often enables the housewife to secure assistance. Often children, members of the workman's family, can assist in the household or on the farm during rush seasons. Often, too, the wife of the farm laborer is glad of the opportunity to furnish meals, lodging and do the work of the laundry for extra day labor at reasonable compensation. Such an arrangement usually calls for the allotment of a small parcel of land to the laborer for a garden. Milk, butter and eggs may be supplied by the farm. These, together with the home, constitute the perquisites in addition to the monthly wage.

The distribution of labor throughout the year will aid materially in the solution of the problem. Continuous and effective employment enables the farmer to pay wages equal to those of other industries.

Since so large a part of the labor on the average farm is performed by the farmer and his family, his income is largest when labor is high-priced. Under these conditions, much of his profit is on the labor of himself and family.

Demand for Labor.—The farm labor supply is not equal to the demand. In many sections of the country, and at rather frequent intervals, good land lies idle or crops are not harvested when they should be because of insufficient help. Such conditions necessitate long hours of labor on the part of the farmer and his inadequate supply of help. This condition often creates dissatisfaction on the part of labor and tends to aggravate rather than relieve the situation.

Many farmers become discouraged and sell or rent their farms and move to town. In other cases, they adopt a type of farming that will enable them to remain on the farm without hiring help. Often their farms are better adapted to more intensive agriculture than the type into which they are forced by the scarcity of labor.

Less than a century ago, more than four-fifths of the population of this country in gainful occupations were engaged in agriculture. At present less than one-third are so engaged. High wages, short hours and other advantages enjoyed by workmen in other industries are more attractive than the farm.

Hours of Work.—A day's work on the farm is usually longer than a day's work in the city. There are good reasons why it should be longer. Farm work is more varied and less monotonous than most kinds of city work. It is subject to interruptions as the result of unfavorable weather. The care and feeding of livestock often requires attention in the early and

INSUFFICIENT LABOR AND EQUIPMENT.[1]
A farmer seventy-five years old harvesting thirty acres of hay alone.

late portions of the day. Best results in the dairy are obtained by milking twice daily with equal intervals between milking periods. For these and other reasons, the farm work-day is long. There are seasons, however, when work is slack and the day's work is materially shortened. During the short winter days darkness prevents long hours for outside work.

While the eight-hour day is becoming common in many industries, many men live so far from their work that an hour or more is required daily going to and from their duties. The farmer usually loses less time in

[1] Courtesy of Doubleday, Page & Co., Garden City, N. Y. From "Farm Management," by Cord.

going to and from his work than the city man. He is at home for the midday meal.

In the majority of cases, the hours of labor during the busy season could be reduced to ten without reducing the work accomplished. In most cases men and teams could work more rapidly and accomplish just as much as by longer hours and slower work. There always will be the emergency demand for long hours for a few days now and then during seeding and harvesting periods.

Statistics indicate that the hours on farms are not as long as claimed. While the farmer may be out of bed sixteen hours a day, he does not work this many hours. In southeastern Minnesota, where crop raising and dairying is followed, the average work-day throughout the year was found to be 8.6 hours. In the northwestern part of the state, in a grain-growing district, the average work-day was 7.4 hours. The average hours on Sunday in these two districts were 3.4 and 2.2 respectively. (These data were gathered by the United States Department of Agriculture and the Minnesota Experiment Station.)

When long hours of labor prevail, the condition may be somewhat ameliorated by allowing the help holidays during the time when the work is not pressing.

Wage of Farm Labor.—In comparing farm wages with those of other occupations, the numerous perquisites of the farm laborer are underestimated. Unmarried men generally receive room, board and laundry, in addition to wages. The clothing requirements for farm work are generally less than in most other occupations, and the incentive and opportunities for spending for pleasure are also minimized.

A comparison of wages in different occupations shows that farm wages in this country have advanced relatively more than in any other line. United States census reports show that the amount of wage paid farm laborers in 1909 was over $650,000,000, or about double the amount reported by the census ten years earlier. Farm wages are now about 55 per cent higher than during the 80's, and 67 per cent higher than in 1894, a year of financial depression. The average monthly rate of farm wages in the United States, including board, is $21.38. When board is not included it is $30.31. It varies widely in different sections of the country, and ranged from $17.90 in South Carolina and $19.60 in Mississippi to $54 in Montana and $56.50 in Nevada. It is thus seen that the wage in Nevada is more than three times that in South Carolina. Since 1890, wages of farm labor have increased relatively more than those of manufacturing industries. The increase in farm wages was 37 per cent from 1900 to 1910 and 55 per cent from 1890 to 1910. For the same periods the increase in manufacturing wages was 22 and 23 per cent only. The relative increase of farm wage acts automatically in the distribution of labor.

It is customary to employ a farm laborer by the day, week, month or year. Frequently, a man will hire out early in the season for the year at a

The Harrow Cart Lightens the Work of Harrowing.[1]

[1] Courtesy of The International Harvester Company, Chicago.

comparatively low wage. In the rush season, when day wages are high, he will sometimes leave his employer unless his wages are raised. To avoid this difficulty, a sliding wage is suggested, the gradation depending on the variation in character and amount of work at different seasons. In general farming districts a wage of $20 per month and board during December to March, inclusive; $25 for April, May, June and November; and $35 per month from July to October, inclusive, gives a satisfactory adjustment and a fairly liberal wage.

Where records and accounts are kept and it is possible to figure closely on profits, there are conditions under which a living wage may be given, supplemented by a percentage of profits. This offers an incentive to the

THE TROUBLES OF A TENANT.[1]

laborer, increases his interest in the work and frequently proves advantageous for both employer and employee.

The wide range in farm wages previously mentioned is due both to the skill of labor and the equipment which is provided. On an average, the wage of labor is in proportion to the work accomplished. The farm laborer in Idaho, receiving more than three times the wage of the colored laborer in the cotton belt, generally drives from three to six horses in a team, whereas the colored man of the South generally drives one mule. In this comparison, the highest paid laborer is probably the cheapest and the one on which the employer makes the greatest profit.

Housing Farm Labor.—Whether the farm laborer is a single man or is married, whether he lives in the farm owner's house or in the tenant house, it generally pays to provide him with comfortable quarters and plenty of good food. There are many advantages in having a tenant house on the

[1] Courtesy of The Macmillan Company, N. Y. From "Farm Management," by Warren.

farm so that a married man can be employed. Such a man is less apt to be dissatisfied, or, if he does become dissatisfied, he cannot pack his suitcase and leave on a day's notice, and before he can leave his dissatisfaction may prove to be more imaginary than real.

The better the housing facilities for labor, the better the class of labor that can be secured. Laborers are always looking for good places, and the farmer who can offer the right inducements generally has a number of applicants from which to choose.

Interesting the Farm Hand.—The average farmer is not satisfied with unskilled labor or with labor of such a social status that it will prove undesirable on the farm under the close relationships which must exist. This is especially true when the laborer becomes a member of the farmer's family. Men of fair intelligence will not take an interest in the work or put forth their best efforts unless their employer shows an interest in them. This lack of interest is quite as often the fault of the farmer as it is of the laborer. Labor must be given as much responsibility as is consistent with its intelligence and ability. It must be advised of the plans for the next day's work, at least during the evening of the day preceding. Men are seldom opposed to hard work, providing work is well arranged, runs smoothly and something is accomplished at the day's close. The farmer who gets the most out of his labor is generally the one that most fully satisfies the labor and pays the best wage.

Skilled and Unskilled Labor.—The degree of skill that the farmer can afford to pay for will depend chiefly upon the character of work to be done and the degree of supervision. The man of mediocre intelligence who is faithful and industrious will often give better satisfaction under good supervision than the more intelligent workman. The man who knows more than his employer should be working for himself.

The immigrants from southern Europe, unskilled in most farm work, may be used advantageously in types of farming where much hand labor is required. The immigrants from northern Europe, on the other hand, are more skilled, and are accustomed to driving teams and caring for livestock. These men, with a little training and experience in this country, generally make desirable farm hands.

The rapid development of agricultural machinery and the adoption of up-to-date methods, together with specialization in certain lines, call for more skill on the part of the farm laborers than formerly. In general, farm work requires more skill than that of most any other industry. In the industries, labor is specialized. In agriculture, it is usually general.

The most satisfactory farm hand is usually a farmer's son. His natural training and his knowledge of local conditions and practices give him a decided advantage over men from foreign countries or from cities. A young man from an adjacent county is frequently better than one from the immediate neighborhood. The latter is apt to have social attachments that may interfere with his duties. A second choice will generally lie with men who have been farm laborers for a number of years, and who, because of their inability to direct their own efforts satisfactorily, have been unable to farm for themselves. Such men require careful direction, but are generally faithful and reliable. The hobo class is the least desirable. They find employment for short periods of time in busy seasons, and frequently do satisfactory work for a short time. Such men are restless and never satisfied to remain long in one locality. Furthermore, they are not dependable.

Permanency of Employment.—The satisfactory workman attaches much importance to the certainty of employment. This is especially true of all workmen who have families dependent upon them. In this fact lies the weakness of the farm opportunities for labor. In many instances a farmer wishes to employ labor for only a portion of the year. The keen competition for labor in other industries makes it essential for the farmer to so plan his business that he can employ labor continuously and at a wage which, including perquisites and opportunities, will be equal to that in other industries.

Permanency of employment applies not only to the year, but also to the duties of each day. Some farmers manage so badly that men are idle during a portion of many days. The farm laborer, if paid regardless of whether working or not, ordinarily will not object, but good workmen generally prefer to have their time during working hours fully employed. A good farm laborer, who left the farmer for whom he was working and engaged with another having the reputation of being strict, made this explanation: "Jones is a mighty nice fellow to visit with, but he has no management about his work. He put me at easy jobs that did not pay for doing. I suppose I ought not to kick so long as I got my pay, but I could not bear to see work so mismanaged, and I was afraid I would get so easy going that I would fall into bad habits and do my own work slack when I got on the farm I hope to buy some day. Sands keeps me hopping, but every lick of work counts, and when bedtime comes I can see I have accomplished something."

Management of Men.—The management of farm laborers depends largely on the character of the employees, but to some extent upon the nature of the work and the number employed. It is always best to hire with a definite understanding. A written agreement is desirable when hiring by the year. In any event, it is always a good plan to have a witness to the agreement, whether it be written or oral. Such an agreement should state the wages to be paid, the dates of payment and the time to be covered. It is also well to specify the character of duties of the employee and the obligations of the employer. If privileges are granted, such as keeping a horse, storing a buggy or furnishing a driving horse occasionally, it is well to have this understood at the outset.

The successful employer will assign certain chores and more or less definite procedure for the direction of his employees. With the more skilled laborers, it is often advantageous to discuss the work with them. The average farmer works with his men. He thus becomes the leader in the work and sets the pace. More is generally accomplished in this way than by assigning men to work, each at a different task. This is not applicable to chores and the care of livestock to the extent that it is in field operations, where men work advantageously in groups.

Where a number of men are employed on the farm, the business is sufficiently large to require much of the operator's time in directing the work. It may even call for his being absent quite frequently. Under such conditions he should employ one of his men as a foreman and make him responsible for the direction of the work in his absence. Such a man should have a wage correspondingly larger than those with less responsibility. The wage of such a man, however, need be very little larger than that of other workmen.

Successful management calls for a definite plan of work with orders or instructions given either the evening before or promptly on the morning of the day's work. Directions must be concise and definite. Much should not be merely understood.

Productive and Unproductive Work.—No matter what the price of labor, it is essential to employ it in productive enterprises to the greatest possible extent. There will always be some unproductive work on farms, such as cutting weeds, cleaning ditches, repairing fences, caring for work stock, and household items. These are necessary in the long run for the upkeep and the good appearance of the farm and home. Labor employed in this way should not interfere with the productive enterprises any more than possible. Many farmers have a weakness for tinkering around with old machinery and patching up old fences and buildings, thus materially reducing productive operations and failing to make the farm pay. We may go still further, and by knowing what crops and animals give the best returns per hour of labor, select those that are most profitable.

Many farmers postpone necessary repairs and adjustments to important machinery, such for example as a self-binder, until the day that it is to be used. It often requires a full day of time to put the machine in working order. A day or two in the winter, or during a rainy period, may have had little value, but in the heat of harvest it is worth much. The delay of one day may mean serious loss in connection with the harvesting of the crop.

Doing Work on Time.—It is important in farming to do the right thing at the right time. This never increases the amount of work to be done, whereas a delay often means added labor. Seeds not planted on time result in a late crop, the yields of which may be reduced. Crops not harvested on time result in loss or depreciation in quality. Weeds not destroyed in time are destroyed when larger only by added labor, and in the meantime reduce the yield. Best results are secured and the minimum amount of labor is required by doing work on time. This is one item in the solution of the labor problem of farms.

Winter Work.—Winter, being the slack period in nearly all kinds of farming, calls for careful consideration relative to the work that may be done in order to relieve the work situation during the crop season. The more nearly the winter work equalizes that of the summer, the less difficulty will be encountered when employing laborers the year around. Such a plan greatly relieves the labor problem on farms. Work that may be done during the winter includes the repair of buildings, fences and farm implements, preparation of seeds and the mixing of fertilizers, or any improvements possible to make under winter weather conditions.

The type of farming and the adjustment of farm enterprises may also be made to furnish work during the winter. Winter lambs may be raised and dairy cows made to freshen at the beginning of the winter, and other livestock features so adjusted as to demand the major portion of the work

THIS FORM OF GRAIN RACK SAVES LABOR.[1]

pertaining to them at this season of the year. By proper planning, crops may be marketed and manure hauled in the winter season.

A work-shop, suitably warmed, makes a place where harness can be oiled and repaired during the coldest and stormiest weather.

Work for Stormy Days.—In order that labor may be employed during stormy days it is advisable to keep a memorandum of things to be done. Oiling machinery, sharpening sickles, oiling harness and making repairs are some of the items that can be done on rainy days, and are likely to be overlooked unless a memorandum of this and other things is kept. The barn, the hen-house and the pig-pens may be cleaned on rainy days. Feed may be ground, rations mixed for cows and pigs, etc.

Economizing Time.—Time is economized in many ways, but largely through the arrangement of the fields and the position and interior arrangement of buildings. It is saved also by the proper equipment, teams of adequate size, by a careful planning of the work and by reducing the motions required in the performance of work. All useless motions should

[1]Courtesy of The Macmillan Company, N. Y. From "Farm Management," by Warren.

be eliminated. This principle comes prominently into play in the husking of corn, pitching of grain and in all work where many motions of the hands are called for. With the same effort, a skilled workman will often accomplish more than an unskilled one. Time is saved on the part of both men and teams by hauling large loads. The wagon constitutes a considerable percentage of the load hauled. The larger the load, the less the relative importance of the weight of the wagon. The movement of the large load seldom requires more time than the small one unless it involves a slower speed or resting the team. Much time may be saved by making the loads large. This is especially true when the haul is long.

Workman's Attitude.—One working for another should always endeavor to earn more than he receives. The employment of both capital and labor is supposed to be at some profit. The employee who recognizes this is the one first likely to receive promotion.

Saving Horse Labor.—The value of work-horses and the high cost of their maintenance make necessary their fullest possible employment on the farm in order to make the farm pay. This calls for a distribution of the horse work so as to increase the average hours of labor per horse as much as possible. Investigations show that on many farms the farm work-horse averages only about three hours of work for the work days in the year. Good management has enabled farmers to increase this in some cases to six or seven hours daily.

Horse labor may also be economized by using cheap horses, by feeding them on low-grade roughage when not being worked, and by using brood mares and raising colts.

REFERENCES

"Farm Management." Boss.
"Farm Management." Warren.
South Carolina Expt. Station Bulletin 84. "The One Horse Farm."
U. S. Dept. of Agriculture, Bureau of Statistics, Bulletin 26. "Wages of Farm Labor in the United States."
U. S. Dept. of Agriculture, Year-Book 1911, pages 269–284. "Seasonal Distribution of Labor."
U. S. Dept. of Agriculture, Bureau of Plant Industry, Bulletin 269, pages 16–23. "What is Farm Management?"

CHAPTER 71

THE FARMER'S CAPITAL

More capital is now required for farming than formerly. Land costs more, livestock is more valuable, labor is dearer and the necessary equipment more extensive and costly. During the decade 1900 to 1910, the average value of farms in the United States nearly doubled.

How to Secure Capital.—Capital is brought into existence by work, and its accumulation is the result of consuming less than is produced. If one has a dollar and spends it for an article of consumption, the process will not increase capital. If the dollar is used to purchase an implement such as a hoe or a spade, the purchaser becomes a capitalist to that extent. He now possesses articles of production that are as truly capital as was the dollar.

In farming, the easiest way to secure capital is to become a tenant. That this method is being taken advantage of is set forth in a preceding chapter. For several decades the percentage of tenant farmers has been on the increase. If one has a very small capital, it is usually undesirable to attempt to farm on the very small area that the money will purchase. It is much better to put all of the capital into equipment, livestock, labor and seeds, and operate a rented farm. In 1910 land made up 70 per cent and buildings 15 per cent of the average value of all farm property.

Cash Transactions.—Because of inadequate facilities to secure credit, farm implements, fertilizers and other necessities are often purchased on time, the manufacturer acting as banker as well as a dealer. As a rule, the price paid for purchases made in this way differs from the cash price by two or three times the usual rate of interest. Under such conditions, it is economy for the farmer to borrow the money at a bank if possible and pay cash for his purchases. One's credit is frequently jeopardized by allowing innumerable little bills to go unpaid. This may be avoided by establishing credit at a bank and borrowing enough to pay bills promptly.

One should not promise to pay an obligation too soon. A payment made in advance of the promised time strengthens one's credit much more than a payment delayed.

Payment by check is preferable to payment by cash. A check serves as a receipt. For this reason, it is well to indicate on the check the item for which it is drawn.

Agricultural Credit.—The increased capitalization of farming in recent years has emphasized the need for an adequate system for agricultural credit in this country that will meet the needs of farmers, just as such systems have met the needs of the farmers in many of the European countries.

Good farming necessitates improving the soil, providing buildings suitable for animals and crops, as well as a modern equipment of machinery. The farmer must have money to put in his crop, harvest it, market it and maintain his family in the meantime. If he is obliged to sell to a creditor or sell at a sacrifice to meet a debt, he is at a decided disadvantage. This represents the condition of many farmers. It is a condition under which railroads and manufactures could not develop.

At a general conference on rural credits, held in Washington, D. C., 1914, it was the consensus of opinion that the primary demand is for a system of credit that will recognize farm property of all kinds as security against loans on the same basis that other forms of real estate and property are recognized. There is no reason why farmers should not have as low rates of interest and as long time on loans as other lines of business. They do not ask for special privilege or a laxity in security precautions. The present banking laws favor manufactures and commerce. This is but natural, since the banking system was developed first in the cities and primarily for the purpose of aiding and doing business with commercial and manufacturing interests.

The problem which presents itself is that of securing a better system of credit. With this in view, a large commission visited Europe during the summer of 1914 to study credit systems. The finding of this committee has been reported in full by Congress, but as yet the foremost farmers and bankers of this country have failed to adapt European systems to our conditions. (This report, "Agricultural Co-operation and Rural Credit in Europe," is document No. 214, Sixty-third Congress, First Session, Washington, Government Printing Office.)

The credit of the American farmer is as good as that of farmers in any other country. The strength of the European credit system lies in the co-operation of a group of farmers sufficiently small that each is acquainted with the others. The combined credit of a group of farmers in any part of the world is security on which the money market will lend almost as cheaply as on government bonds. The nearest approach to this co-operative credit system in this country is found in the financial aid extended to Jewish farmers by organizations composed of their own race. There are a number of these in the State of New Jersey, the membership ranging from 20 to 40 each. The first of these was established in 1890, through the agency of Baron de Hirsch. Since 1900 nearly $2,000,000 have been loaned through this agency to Jewish farmers in this country.

While the credit system is faulty, farmers may remedy conditions if they will fully realize that tangible property, such as land, is not necessarily the best security. The character and ability of the man is of much greater importance than the land he may own. In many localities, farmers have little difficulty in borrowing of local banks most any reasonable amount without security. Where farmers are known to be capable of paying their debts, and do so without legal proceedings, credit conditions are good and desirable money lenders are attracted. Honest bankers avoid localities where legal pressure is necessary to collect loans; such men merely want their principal back with the stipulated rate of interest. It is true that the individual farmer can do little to establish the reputation of a neighborhood, but the co-operative effort of many farmers can establish confidence and credit.

The Raiffeinsen Bank has done much for the farmers of Germany, and under a different name has been equally effective in other European countries. It is a neighborhood bank and is a simple institution. Each member pledges his entire resources for the bank's debts. Each must be a shareholder, the minimum holdings seldom being more than $5 each. As soon as the bank is established, confidence is established and the wealthier members deposit their money to be loaned to their neighbors and draw interest. If this does not provide funds for all the would-be borrowers, more money is easily secured in the open market, generally from the largest bank in the nearest town. The organization is entitled to deal in credit like bankers in cities, with the exception that it may lend only to its own members, and then for productive purposes only, the purpose being stated in the application for the loan. The officers, except the cashier, serve without pay and are elected from the membership.

Borrowing Money.—The amount of money that one can borrow on a farm depends to a considerable extent on the character of the borrower. In most cases, however, the limit is placed at about one-half the actual value of the farm. Bankers will lend freely to men who are industrious and honest, even though they have very little property. Applications for loans are frequently refused to extensive property holders because of a reputation they may have of not paying their debts unless forced to do so.

Large insurance companies loan much money at the rate of 5 per cent, but the farmer cannot get it at this rate. He must pay his banker or some agent for securing the loan. The commission may range from .5 to 2 per cent of the loan. There are innumerable other small expenses, such as abstract of title, exchange in remitting, recorder's fee, etc. The tendency is to shorten the time of the loan and increase the frequency of these renewal expenses. This condition is unfair to the farmer and calls for a remedy, legislative or otherwise.

Farmers' Bulletin No. 593, issued by the United States Department of Agriculture, states five rules that should be observed in borrowing money. These are believed to be applicable under all systems of credit. They are as follows: "(1) Make sure that the purpose for which the borrowed money is to be used will produce a return greater than needed to pay the debt. (2) The contract should provide for the repayment of the principal at the most convenient time; that is, when the farmer is most likely to have the means wherewith to repay it. (3) The length of time the debt is to run should have a close relation to the productive life of the improvement for which the money is borrowed. (4) Provision should be made in the long-time loan for the gradual reduction of the principal. (5) As low interest rates as possible should be secured."

Farm Mortgages.—A farm mortgage is a legal document which usually transfers the title of land or an equity in it for a money consideration. It is redeemable upon fulfillment of the stipulated requirements. It is usually safe to mortgage a farm for one-half its actual value. The more stable land prices are, the greater the relative proportion that may be covered by a mortgage. In this, as in case of other credits, much will depend on the character and ability of the one giving the mortgage. Some men are able to secure mortgages covering three-quarters or more of the actual value of their property.

According to the census of 1910, 62 per cent of the farms in the United States were operated by owners. These are the only ones for which statistics relative to mortgages were obtained. Of these, 33.6 per cent were mortgaged. The average mortgage per farm was $1715, while the average value of mortgaged farms was $6289. The average mortgage is, therefore, 27.3 per cent of the actual value of the mortgaged farms. A study of the census figures shows that mortgaged indebtedness does not necessarily indicate a lack of prosperity. This form of indebtedness is higher in Iowa and Wisconsin than in most other states. These are prosperous agricultural states.

Farm mortgages are of prime interest chiefly to four classes: (1) owners of farms, who live on them and operate them; (2) absentee owners, who generally operate through tenants or by managers; (3) prospective owners, including tenants and farm laborers who hope to become owners; and (4) those individuals, institutions and organizations having funds to loan on farm real estate as security. Some of the questions in connection with farm mortgages are: For how long a period should the mortgage run? For what purpose should money be borrowed? How is the loan to be repaid? What institutions should be allowed to negotiate loans? Where is the money to be secured? These and many other questions arise in connection with farm mortgages.

The period for which mortgages run has been settled by practice in some localities. Occasionally they are renewed every second year; more frequently every third year. In other localities they run five years, and in some exceptional cases ten years. Seldom, if ever, are farm mortgages drawn for a period longer than this. In many instances, the period of time is too short. It should not be determined by custom, but should depend upon the purpose for which made. If a mortgage is negotiated for the construction of farm buildings, the time should be sufficiently long to make reasonably certain the payment for these out of farm profits. The duration of the buildings covers many years. If money is borrowed to equip the farm with machinery, the time of the mortgage should be shorter than if borrowed for buildings. The life of the machinery will be shorter than that of the buildings. In no case should the mortgage be drawn for a period of time longer than the life of the equipment or improvement for which it is negotiated.

One should be prepared to pay a mortgage when it comes due, or to renew it. If renewal is necessary, the agreement for such should be made at least a year in advance of the date of maturity.

Extent of Debt Permissible.—The extent of indebtedness depends partly on the man, but to a considerable extent on the character of land for which he goes in debt. Careful observations indicate that good land may justify a larger percentage of indebtedness than poor land, yet we more often find the man of very limited means purchasing poor land and going deeply in debt for it because it is cheap. The young man runs fewer risks in paying a big price for first-class land than he does in paying a low price for run-down land, especially if he must go heavily in debt for the purchase. The one will bring good returns without expenditure for improvements, while the other must be improved before it can be farmed with profit. In this case the poor land requires relatively more working capital than the good land.

Relation of Banker to Farmer.—In recent years bankers have begun to realize that farming is the first industry of the country, and that the needs of farmers should be cultivated by the bankers. As a result, conferences between bankers and farmers with a view of establishing a better relationship between them are frequent. It is interesting to note that the bankers always take the initiative in this matter. The agitation for better systems of rural credit have doubtless had their effect in this direction.

If farmers are not accorded fair treatment by banking institutions, it is quite possible for them to establish their own credit institutions, just as has been done in many of the European countries.

Building and loan associations have been successful in aiding home builders in cities and towns. These, with slight modifications, could doubtless be applied to farm conditions. Such co-operative institutions reduce to the minimum the incidental costs that now often seem exorbitant in negotiating loans for farmers.

Working Capital.—One of the greatest needs of farmers in most parts of the country is working capital; that is, money that can be used for the purchase of livestock, equipment, fertilizers, seeds and an adequate labor supply in order to increase to the maximum the efficiency of the farm. For example, the efficiency of labor has been enormously increased by the invention and use of labor-saving farm implements. The time required to produce nearly all of the staple crops has been reduced from 60 to as much as 95 per cent through the introduction of machinery. This has made necessary an increase in the working capital of the farmer.

It is not uncommon to find railroads discarding their old locomotives and purchasing costly new ones, abandoning old tracks and constructing new ones alongside of them with lesser grades. Such investment is for the purpose of increased efficiency. Farmers often work with antiquated tools and inefficient teams when, with a little more working capital, their efficiency could be materially increased by replacing these with a better equipment.

One buying a farm should have enough money to pay a reasonable part of the purchase price, to buy an adequate equipment, the necessary stock, and have some cash left on which to do business.

Investigations at the University of Wisconsin relative to the relationship between the percentage of capital on farms as working capital and the net profits are as follows: On fifteen farms on which only 13.5 per cent of the total capital was in the form of working capital, the average net profit was $167; on twelve farms where it was increased to 17.7 per cent the profit was $433; on six farms having 28 per cent as working capital the profit was $1628; and on six farms where one-third was working capital the net profit was $3511. The number of farms investigated is too small to draw sweeping conclusions, but the results are certainly very suggestive.

In Scotland and England, where 90 per cent of the farms are rented, the tenants regard $50 an acre as the necessary working capital in order to farm profitably. The amount of working capital necessary depends on the kind of farming and also on the fertility of the soil. The extensive purchase of commercial fertilizers very materially increases the need of a working capital.

Distribution of Capital.—The proportion of capital to be invested in land, stock and equipment varies with the price of land and kind of stock and the type of farming. There will always be a distribution between these various parts that will give best results. This distribution may be settled for a given time, but a marked rise in land values will necessitate changing it. Any change in the type of farming is also likely to disturb it. We have already found that the size of the farm is also a factor, and that the larger the farm the smaller the relative investment in buildings and equipment. The average distribution of capital on farms in the United States is given in the following table:

Average Capital and Its Distribution on Farms in the United States.
(U. S. Census, 1910.)

	1900		1910	
Land, exclusive of buildings	$2285	64%	$4476	70%
Buildings	620	17%	995	15%
Implements and machinery	133	4%	199	3%
Livestock	536	15%	774	12%
Tota	$3574	100%	$6444	100%

Capital Related to Area.—No matter how small a farm may be, the owner demands a respectable house, and on the small farms of the United

States this represents from 40 to 50 per cent of the total capital. Larger farms generally have better houses, and have only 5 to 10 per cent of the capital in the house. This to a less marked extent applies to the other buildings on the farm.

In the same way the investment in machinery and in work horses is relatively less per unit of area as the farm increases in size. This holds up to about 300 acres in extent, beyond which it becomes relatively small.

Capital Related to Labor Income.—In previous chapters the importance of farming on a scale sufficiently large to provide an adequate labor income has been pointed out. The size of the farm is measured both by extent and by capital invested. The importance of having sufficient capital, as an owner or as a tenant, to farm on a scale that will bring a satisfactory income has been emphasized. Under-capitalization limits decidedly the possibilities of the farmer. Investigations in many parts of the country show that in a general way the average labor income is approximately in proportion to the capital invested. Doubling the capital generally doubles the farmer's income. For this reason, one may purchase a comparatively large farm, make a living from it and pay for it out of the farm proceeds more easily and in less time than he could pay for a very small farm under the same conditions and with the same initial capital.

Capital Related to Type of Farming.—Types of farming that require the minimum of land usually require more capital in other forms, so that the investment required to bring a suitable labor income does not differ materially in different types of farming. With restricted capital, one had best engage mainly in grain farming or the production of truck crops. With plenty of capital, one will generally succeed best in a long term of years by giving considerable attention to livestock. Livestock farming calls for equally as large an investment in land, while the stock and buildings to shelter them increase the investment. The risk is also increased. These disadvantages, however, are more than counterbalanced by better distribution of labor, the better maintenance of soil fertility and increased returns.

Investigations of hundreds of farms in New York State show that farmers with less than $7500 in capital secured two-thirds of their receipts from cash crops, while those with more than $20,000 in capital secured three-fifths of their receipts from livestock. By examining those having less than $15,000 in capital, it was found that those having the greatest receipts received four-fifths of their income from crops.

Farming with Small Capital.—One with small capital or heavily in debt should avoid types of farming that require much time for returns. Among such types are orcharding, timber culture, the rearing of horses and other classes of animals requiring several years for maturity. One should avoid purchasing pure-bred stock or making extensive improvements when money is scarce. Debts will more quickly be cancelled by devoting one's efforts chiefly to cash crops.

Purchasing a Farm.—There are many items to consider in the purchase of a farm. One should begin with a careful estimate of his own inclination. The size and character of farm, and the nature of the buildings upon it, will depend largely on the type of farming to be followed. The experience and ability of the purchaser, together with his available capital, will also determine the size of the business. With the vast majority the available capital will likely be the determining factor, but in some cases there may be plenty of capital and a lack of both experience and ability, in which case it will be wise to start on a conservative basis.

In purchasing a farm, it is wise to take plenty of time and look at farms in different localities. It is also well to visit the farm at two seasons, once in the spring time when some of the fields are in the course of preparation for crops, and again late in the season as harvest is approaching. The first visit enables one to inspect the soil carefully with reference to its fertility, drainage and uniformity. The second one enables him to see what the soil has really done. In this connection one must, of course, consider the character of the season and know whether it is better or worse than the average season in that locality. One should also be able to judge as to the nature of the farming that is being done and how rapidly a run-down farm may be improved. It is a good idea to make a sketch of the farm, outlining the fields and making careful estimates of the acreage of woodlots, stream courses and waste lands of whatever nature. It is much safer to put a valuation on the different classes of land than it is to value the whole farm at so much per acre. The probable returns from the farm can be more closely calculated when one figures on the different classes of land and the crops to which they are likely to be adapted.

Most men will not base their selection on the monetary value of the farm alone, but will be influenced by the character of the neighborhood. Social and educational advantages should not be ignored. These count much in the environment of the family.

Land as an Investment.—The rise in land values in recent years has caused many persons not desiring to be farmers to invest money in land. Such investments are considered among the safest and the appreciation in land values is often as large and sometimes larger than the actual returns in land rents. Such investments frequently tend toward inflated prices, and these occasionally finally result in a decline in land values. Land investment for the investment only tends to increase land tenancy, and makes necessary a better tenant system than now generally prevails. Land tenancy ordinarily does not encourage increase in land values, because of the marked tendency for poor farming and for neglect to maintain the fertility of the soil.

When purchasing a farm, the farmer should always consider carefully the investment phase. Other things equal, there are decided advantages in the purchase of farms in localities where land values are rising. Under those conditions one is justified in purchasing a larger farm and in going more deeply in debt for it. Seldom is it possible under present regulations to mortgage a farm for more than one-half its actual value. This makes farm mortgages comparatively safe, both for the borrower and lender.

REFERENCES

"Principles of Rural Credit." Morman.
"Agricultural Credit." Bullock.
"Principles of Rural Economics." Carver.
"Rural Credits." Myron Herrick.
New York (Cornell) Expt. Station Bulletin 295. "Farm Management Survey of Tompkins County, N. Y."

CHAPTER 72

FARM RECORDS AND ACCOUNTS

In every farm community may be found a certain percentage of the farmers who are successful, but who are often unable to tell upon what their success is based. This lack of knowledge on the part of the farmer is due chiefly to his not keeping farm records and accounts, and, therefore, not attempting to carefully analyze his business. Investigations along this line have done much not only to show the labor income of farmers in different districts, but also to give rather definite information on the duty of farm machines and the average amount of work that should be expected from the farm laborer.

Those farmers who keep good farm records and accounts are not willing to do without them. The study of such accounts adds much to the interest of farm management. Farm accounts should not only show the profit or loss of the farm as a whole, but should show on which crops or animals the best profits are made. Such accounts do not make necessary a course in bookkeeping. Cost accounting is quite different from bookkeeping. Attempts to apply bookkeeping to farm cost accounting usually fails. The first essential to cost accounting is good judgment and a thorough knowledge of the farm business.

Object of Keeping Accounts.—The chief object of farm accounts is to learn how to improve the business of farming and make it more profitable. Another object is to have a record of everything bought or sold on time. This enables the farmer to know what he owes and what is due him. Having a record of the payment of a bill may prevent its payment a second time. Most of the disputes relative to debts are the result of forgetfulness rather than of dishonesty. When accounts are not kept, bills creep up in an astonishing way, and one is apt to buy more than one's resources warrant.

Many of the industries employ efficiency experts to study their business plants with a view of increasing their efficiency. This may result in a saving in the cost of production or increasing the output without increasing the investment. Such experts emphasize the importance of keeping accurate records of costs. This enables them to ascertain just where losses occur and where to apply the remedy. It is just as essential that farmers keep records that will enable them to know where the gains and losses occur. Such records would surprise many farmers.

Essential Records.—There are many kinds of farm records that are desirable. Those most useful will be determined by the kind of farming. Animal records, egg records and records of milk production are important where any of these enterprises are conducted on the farm. Weather records are helpful in most any kind of farming. Records of the application of manure or fertilizers to the different fields of the farm are essential in deciding where manure or fertilizers had best be used. Records in the form of maps showing the location of drains should never be neglected.

Records of milk production and butter-fat content enable the farmer to dispose of unprofitable cows. Such records enable him to select calves most likely to develop into productive adults.

An inventory once a year is a necessary part of farm accounting.

Blank Forms and Books Necessary.—Farm bookkeeping does not require that accounts be kept in a particular form so long as it is a logical selection and arrangement of facts, bearing upon the essentials of the business. So far as blank forms fit the records to be made, time is saved in using them. The kind of blank forms suitable for one farmer may not fit the conditions of another. It is generally best for each farmer to ascertain the character of books and what kind of blank forms best meet his requirements. Once started in farm accounting, he will ultimately find it advantageous to devise blanks that will fit his needs. He may then have a supply printed at small cost that will last for several years.

To begin with, one will usually need a day book small enough to be carried in the pocket. This should have one column ruled on the left of each page for the date and a double column at the right for receipts and expenses. The space between the columns may be used for entering the items bought or sold, or may be used for memorandum for work done. In addition to this, a journal should be kept. It should have two double columns on the right of each page, one for charges and the other for credits. Some prefer to use two pages for each account, devoting the right-hand page to credits and the left-hand page to charges.

How to Keep Accounts.—Cost accounts contain many estimates. It is foolish to spend time and refinement in methods of bookkeeping in hunting for insignificant errors or to check accounts to the last cent. It is these things that frequently disgust farmers with accounting. One should know what facts to record in order to accomplish the desired object. A full set of accounts will involve an annual inventory, a record of time accounts, a record of all receipts and disbursements, and a record of the time of men and teams for all of the farm enterprises, including each crop grown and each class of animals raised. It should also show the amount and value of crops transferred from bin or mow to the feed bins and feed lots. These data may be classified and distributed to the different accounts. These accounts should include interest, notes and accounts payable, notes and accounts receivable, equipment, labor, improvements and a personal account. In addition to these there should be one for each farm enterprise.

It is essential to record in the diary transactions of the day. Little difficulty will be found in remembering the day's transactions and amount of time employed in work items. If no record is made for several days, considerable guessing is involved. The transfer of daily transactions to the journal may be done at odd times. It is wise, however, to post at frequent intervals in order to avoid work of tedious length.

Time Required to Keep Accounts.—Farmers who have used the system of farm accounts here outlined report the time consumed in making daily entries to be from two to ten minutes, with an average of about five minutes. To this must be added a number of hours at the close of the year to close the accounts. This will vary with the size of the farm and the complexity of the business. Much of this work can be done evenings and on winter days when there is little to interfere. It is time most profitably spent.

Best Time to Start Accounts.—Farm accounts may be started any time after the last crop is harvested in the fall and before the first crop preparations are started in the spring. The exact date will vary with location and type of farming. Usually January 1st, March 1st or April 1st will be suitable dates. For the tenant farmer the date should correspond with the termination of his lease. In all cases the date should be early enough to enable the farmer to close his year's accounts and make his plans for the new year before the rush of the season's work begins.

Annual Inventory.—The time of making inventory should correspond with the date of opening farm accounts. It is the most essential item in the accounts. The inventory items should be the first entries in the accounts with the various enterprises on the farm. For example, the cow account should be charged at the beginning of the year by inventory for the number and value of the cows; also the amount and value of cow feed on hand. At the close of the year the accounts should be credited with the number and value of cows and the amounts of feed on hand. The purchase or sale of cows during the year may have materially increased or diminished the value of the herd. The herd itself may have increased in value because of the development of a number of heifers or decreased because of a predominance of old cows. The inventory at the beginning and close of the year shows the actual value of the herd, and is necessary for an accurate account with it.

The inventory should include the farm, the buildings, all stock and equipment, cash crops and feed on hand, cash and money in bank, and should also show accounts that are due. This gives the total assets. It should likewise show notes that have been given and bills to be paid. These represent the liabilities and when subtracted from the assets give the true worth of the farmer.

Values to Use.—In estimating values, the market price at the farm or the price at the nearest selling points, less cost of hauling, should be used. Hay in mow or stack and grain in bins may be closely estimated by measurements. It is customary to allow a reasonable depreciation on all machinery and implements, on buildings and on stock that is past its prime.

No other account will give so much information for the time spent upon it as the inventory. If no money has been taken from or added to the business during the year, by gift or transfer, the inventory will show the net gain or loss of the year's business. Every farmer should make an annual inventory, even though he keeps no further accounts. The inventory, however, will not show on what enterprise he has gained or lost.

Account No. 16 shows the summary of the inventory for one year. The inventory proper may be kept on separate sheets, so ruled that the inventory may be repeated for four or five years without rewriting the items. It is usually best to list each animal by name or number, stating age and value. Important machines should be listed separately. Small implements, like hoes, rakes and forks, need not be itemized.

Receipts and Expenditures.—A record of receipts and expenditures is necessary, but the receipts minus the expenditures seldom give a correct idea of the net returns for the year. Some permanent improvements may have been added or machinery and livestock purchased. This would make expenses run high and make the net receipts appear low. The improvements, machinery and stock would increase the farmer's net worth at the end of the year by their purchase price, less a very small amount for depreciation. On the other hand, stock may have been sold, thus increasing receipts and making the net receipts appear large. At the same time, the net worth at the end of the year might be reduced by an amount nearly equal to the sale of stock. It is, therefore, necessary that records of expenses and receipts be studied in connection with the inventory at the beginning and the end of the year.

Accounts with Farm Enterprises.—The farmer should have as definite a method of measuring the profits of his business as does any other business man. This calls for an account with each crop and each farm enterprise. The fields should be numbered or lettered and an account kept of the hours of man and horse labor, the cost of seed and fertilizer and any other expenses, such as twine for binding grain, and coal and oil for threshing.

Whether the product is sold directly from the field or placed in storage, the field or crop should be credited with the farm value of it.

Each class of livestock should be charged with the feed consumed, the labor required, including incidentals, such as bedding, medicine, equipment and a charge for interest and housing. In like manner, it should be credited with all animal products sold.

The important points in connection with the several farm enterprises are: What does each product really represent in cash, and what is the profit at the selling price?

Work Records.—A record of the time of men and horses is most conveniently kept by having blank forms, in which the man hours and horse hours may be recorded. These blanks may be either in the form of cards or blank books. Cards are suitable for weekly or monthly reports. If one does not wish to transfer the time records to the several enterprises, a blank may be provided for each and the hours of labor entered as they occur. The following record serves to show how such records can be kept:

SAMPLE WORK RECORD WITH CORN. FIELD NO. 3 (24 ACRES).

1916.	Operation.	Man.		Horse.[1]	
		Hours.	Minutes.	Hours.	Minutes.
March 30	Hauling manure	9	15	18	30
April 20	Plowing	10		30	
April 21	Harrowing	5		15	
April 21	Planting	5		10	

[1] Horse hours are recorded as hours worked times the number of horses in the team.

For chores a special page or form should be provided as follows:

TIME SPENT IN DOING CHORES.

1916.	Horses.		Cows.		Sheep.		Poultry.	
	Hours.	Minutes.	Hours	Minutes.	Hours.	Minutes.	Hours.	Minutes.
April 1	1	30	2	50		30		45
April 2	1	30	2	50		30		45
April 3	1	30	2	50		30		45
April 4	2		2	50	1			45

If horses are used in doing chores, extra columns must be provided. Time required for doing chores usually is about the same each day, and it may not be necessary to make the entries daily

Abbreviated Accounts.—If one does not wish to keep a full set of accounts, one may keep accounts with the most important enterprises only. This necessitates more estimating than full accounts. The cost of labor per hour can be accurately determined only by knowing the total number of hours and the total cost. If the time for each enterprise is kept, it can be accurately distributed. Otherwise, labor charges must be based on estimates. Such accounts, however, are better than mere guesses.

Classification of Troublesome Items.—The items most likely to puzzle the beginner in farm accounts are those relating to real estate and machinery. To the farm should be charged all items of repairs on buildings and fences, the construction of new buildings and fences, expenditures for drainage, taxes and insurance. These pertain to the farm and its permanent improvement. This account should be credited with rent of land and buildings; also with the sale of buildings or building materials, such as stone, from the farm.

The machinery account should be charged with all costs of machinery repairs, oil for lubricating machines and the purchase of new implements. There should also be a charge for housing. The machinery account should be credited for machinery rented or sold. The balance of the machinery account should be distributed in the same manner that man and horse labor is distributed. Since the important machinery is used with horse power, the distribution is based on the time records kept for horses.

Plan of Farm and Cropping Systems.—A sketch of the farm, showing the size, number and location of fields and the location of drains, is advisable. A number of copies may be provided so that one may be posted for each year. On this may be entered certain data, such as the crop in each field, its acreage and yield. If manure is applied to a portion of the field, the portion so treated should be indicated on the map. These maps, if kept, form a permanent record and make an efficient aid to the farmer in subsequent years.

Closing the Accounts.—Considerable time will be required in this operation. A definite order should be adhered to, which is as follows: The first step is to take the inventory and enter the inventory items after all regular posting has been done.

Items, such as the use of pasture and farm products used by the family, should be entered in the proper accounts. These may involve estimates. The value of board, produce and other allowances furnished to labor should be charged against labor and the proper accounts be credited with equal amounts. Animals should be credited with the value of manure produced and the same charged to the crops to which applied.

The crops, animals, the farmer and laborers should be charged with the proper amounts for the use of buildings. The several classes of live-stock should also be charged with straw used for bedding and the crops from which the straw was obtained credited with the same amount.

After all these entries have been made, the labor account should be balanced, the rate per hour obtained and the cost of labor distributed to all enterprises in proportion to the time put upon them. Some of this will be charged to the horse account. When this charge has been made the horse account may be closed in the same way and the cost of horse labor distributed to the several enterprises in proportion to the horse hours that each required.

Next in order will be the equipment account and the distribution of the cost of the equipment to the several enterprises in proportion to the hours of horse labor.

If the system is in double entry, that is if each item is entered in two accounts, the books may be closed and accurately balanced. In this case, the total gain or loss for the year, as shown by the account with "loss and gain," should check with the increase or decrease in inventory. After the system is thoroughly understood, there is no necessity for making double entries throughout. This will reduce the work by about one-third and obviate the difficulties frequently encountered in attempting to balance books to the last cent.

The following set of accounts, the details and daily entries of which are given for the first two months of the year, will show the method. The transactions for the remainder of the year are summarized and the account closed and balanced.

Those wishing further aid in this system of farm accounts are referred to Farmers' Bulletin 572 of the United States Department of Agriculture, or to Chapters 16 and 17 in Warren's "Farm Management," published by The Macmillan Company, New York City.

INDEX.

2

CASH.

1911		Dr.		Cr.	
April					
1	To inventory.....16	$536	54		
1	By plow.....9			$10	00
1	By allowance.....20			30	00
1	By copper sulphate.....19			5	70
1	By house rent, tenant.....14			6	00
4	By purchase 1 horse.....26			75	00
7	By *Courier* subscription.....20			1	50
7	By P. O. box rent.....20				10
7	By stamps, candy and tobacco.....20			1	35
8	By bull service.....17			1	00
8	By 40 eggs for incubating.....23			10	00
8	By thresher repairs.....28			2	00
10	To sale 6 cows.....7	180	90		
10	By telephone, horse doctor.....27				25
11	To sale 2 bbls. potatoes.....22	2	00		
12	By tickets to schoolhouse.....20				70
12	By church, haircut, tobacco.....20				70
13	By purchase, 3 cows and freight.....7			111	64
13	By pump casting and express.....9			2	76
18	To $3\frac{1}{2}$ bbls. seed potatoes.....22	8	50		
18	To sale 30 bales hay.....12	39	29		
18	By purchased feed.....7			2	80
18	By coal.....8			15	00
23	By use drill and help.....18			10	50
23	By purchased feed.....7			130	69
23	By purchased pair mules.....27			100	00
24	By $\frac{1}{2}$ bu. seed corn.....6				50
	Carried forward.....	$767	23	$518	19

4

Note: Page number 3 omitted.

Cash (*Continued*).

1911		Dr.		Cr.	
May					
15	Brought forward........................	$2,334	29	$690	40
15	Interest on note, 6 months...............17			37	50
15	To balance horse-trade..................27	25	00		
15	By rent on tenant house.................14			12	00
25	To sale, veal calf........................7	9	13		
26	To sale, 93 bales hay...................12	134	78		
26	By 1 bbl. of lime.......................19			1	50
26	By fert. for corn.........................6			4	00
26	By mat. for wagon bottom................9			1	40
26	By 8 ton coal and stamps...............20			40	04
28	By 6 days labor by Brown...............14			9	00
31	To sale eggs.............................23	1	07		
31	By amount paid Cremor...................8			25	00
1912					
March					
31	To cash receipts for balance year............	4,189	54		
31	By cash disbursements for balance year......			5,432	63
31	By inventory............................16			140	34
	By balance (loss)........................15			300	00
	Total..................................	$6,693	81	$6,693	81

6

Note : Page number 5, blank.

Corn (30 Acres).

1911		Dr.		Cr.	
April					
1	To inventory............................16	$125	00		
24	To purchase ½ bu. seed corn..............2		50		
30	By 300 bu. corn for teams...............27			$125	00
May					
14	To 6 bu. Leaming seed....................3	5	10		
14	To 2 bottles strychnine, for corn..........3		25		
26	To fertilizer..............................4	4	00		
1912					
March					
31	To expense balance year..................	47	00		
31	By produce balance year..................			642	25
31	By inventory............................16			240	00
31	To use of land..........................11	120	00		
31	To 1987 hr. labor @ 15.576 cents........14	309	50		
31	To 2974 hr. horse labor @ 17.385 cents...27	517	03		
31	To 2974 hr. equipment 9	88	60		
	By balance (loss)........................15			209	73
	Total..................................	$1,216	98	$1,216	98

7

Cows.

19—		Dr.		Cr.	
April					
1	To inventory (10 cows)....................16	$355	00		
8	To bull service..............................2	1	00		
8	To 800 lbs. rye.............................24	8	00		
10	By sale 6 cows..............................2			$180	90
13	To purchase 3 cows and freight............2	111	64		
18	To purchase of feed........................2	2	80		
23	To purchase of feed........................2	130	69		
25	To purchase of feed........................2	63	65		
30	By 94 qts. milk for house.................20			2	82
May					
14	By sale of milk..............................3			52	51
25	By sale of veal calf........................4			9	13
31	By 87 qts. milk for house.................20			2	61
1912 March					
31	To expense balance year....................	2,447	36		
31	By products balance year..................			2,009	67
31	By inventory (23 cows)....................16			975	00
31	To interest..................................13	33	25		
31	To use of buildings........................11	100	00		
31	To use of pasture..........................11	100	00		
31	To 2480 hrs. of labor.....................14	386	28		
31	To 576 hrs. horse labor...................27	100	14		
31	To 576 hrs. equipment......................9	17	16		
	By balance (loss)..........................15			624	33
	Total..	$3,856	97	$3,856	97

8

Charles Cremor.

19—		Dr.		Cr.	
April					
18	To coal...2	$15	00		
30	By labor for month........................14			$25	00
30	To cash...3	10	00		
May					
30	By labor for month........................14			25	00
30	To cash for month...........................4	25	00		
1912 March					
31	To amount paid..............................	281	00		
31	By labor for balance year..................			281	00
		$331	00	$331	00

9

Equipment.

1911		Dr.		Cr.	
April					
1	To inventory............................16	$602	15		
1	To 1 plow..............................2	10	00		
13	To pump-casting and express charges......2	2	76		
25	To repairs on wagon....................3		15		
27	By sale of 4 milk cans...................3			$2	00
May					
14	To purchase whip lash...................3	1	00		
26	To mat. for wagon-bottom................4	1	40		
1912 March					
31	To expense balance of year................	232	61		
31	By inventory...........................16			723	90
31	To interest............................13	33	13		
31	To use of buildings.....................11	50	00		
31	To 79 hrs. of labor.....................14	12	30		
31	To 40 hrs. of horse labor................27	6	96		
	By balance............................15			226	56
	Total..................................	$952	46	$952	46
	Distribution of Expense on Equipment.				
	By 2974 hrs. on corn @ 2.979 cents.......6			88	60
	By 576 hrs. on cows @ 2.979 cents........7			17	16
	By 156 hrs. on farm @ 2.979 cents.......11			4	64
	By 1408 hrs. on hay @ 2.979 cents.......12			41	94
	By 366 hrs. on oats @ 2.979 cents.......18			10	90
	By 300 hrs. on orchard @ 2.979 cents....19			8	94
	By 228 hrs. on personal @ 2.979 cents....20			6	79
				$178	97

Continued

10

EQUIPMENT (*Continued*).

	Dr.		Cr.	
Brought forward........................			$178	97
By 280 hrs. on potatoes @ 2.979 cents....22			8	34
By 18 hrs. on poultry @ 2.979 cents......23				53
By 920 hrs. on rye @ 2.979 cents........24			27	40
By 250 hrs. on thresher @ 2.979 cents....28			7	45
By 130 hrs. on wheat @ 2.979 cents.....25			3	87
Total................................			$226	56

Hours = 7606) $226.56 (2.979 = Cost per hour, cents.
152.12

74.440
68.454

5.9860
5.3242

.66180
.60848

11

Farm (200 Acres).

1911			Dr.		Cr.	
April 1	To inventory	16	$14,000	00		
May 1	To plastering tenant house	3		75		
14	To thread on pipe, 55 cents	3		85		
	To freight on tile and cement, 30 cents					
1912 March 31	To expense balance of year		480	84		
31	By returns balance of year				$48	15
31	By inventory	16			14,000	00
31	To interest	13	700	00		
31	By rent buildings (personal)	20			200	00
31	By use tenant house (labor)	14			50	00
31	By use buildings (equipment)	9			50	00
31	By use buildings (teams)	27			50	00
31	By use buildings (thresher)	28			15	00
31	By use buildings, cows, $100	7			200	00
	By use land, $100	7				
31	By use building and yard poultry	23			5	00
31	By use land (corn)	6			120	00
31	By use land (potatoes)	22			10	00
31	By use land (wheat)	25			15	00
31	By use land, hay, $225	12			325	00
	By use buildings, $100	12				
31	By use land, oats, $36	18			51	00
	By use buildings, $15	18				
31	By use land, rye, $60	24			75	00
	By use buildings, $15	24				
31	By use land (orchard)	19			20	00
31	To 614 hrs. labor	14	95	64		
31	To 156 hrs. horse labor	27	27	12		
31	To 156 hrs. equipment	9	4	64		
	By balance (loss)	15			75	69
	Total		$15,309	84	$15,309	84

12

HAY (75 ACRES).

1911		Dr.		Cr.	
April					
1	To inventory.....16	$445	00		
13	By sale 30 bales hay.....2			$39	29
29	To phone.....3		30		
May					
26	By sale 93 bales hay.....4			134	78
1912					
March					
31	To expense balance of year.....	44	06		
31	By product for balance of year.....			1,533	95
31	By inventory.....16			176	00
31	To use of land.....11	225	00		
31	To use of buildings.....11	100	00		
31	To 1550 hrs. labor.....14	241	43		
31	To 1408 hrs. horse labor.....27	244	78		
31	To 1408 hrs. equipment.....9	41	94		
	To balance (gain).....	541	51		
	Total.....	$1,884	02	$1,884	02

13

INTEREST.

1911		Dr.		Cr.	
May					
15	To 6 mo. interest on note $1500.....4	$37	50		
1912					
March					
31	To balance year.....	512	50		
31	By interest on cows.....7			$33	25
31	By interest on equipment.....9			33	13
31	By interest on farm.....11			700	00
31	By interest on poultry.....23			2	82
31	By interest on teams.....27			31	38
31	By interest on thresher.....28			47	50
	To balance (gain).....15	298	08		
	Total.....	$848	08	$848	08

14

Labor.

1911			Dr.		Cr.	
April						
1	To house rent	2	$6	00		
27	By labor of a man	3			$2	75
30	To cash	3	40	00		
30	To 1 mo. of labor	8	25	00		
May						
15	To house rent	4	12	00		
28	To 6 days labor by Brown	4	9	00		
30	To 1 mo. of labor	8	25	00		
1912						
March						
31	To labor balance of year		990	68		
31	By amount received balance of year				4	50
31	To use tenant house	11	50	00		
31	To use personal labor	20	500	00		
	To balance	15			1,650	43
	Total		$1,657	68	$1,657	68

Distribution of Labor.

Distribution of Labor.		Hrs.	Amount	
Corn	6	1,987	$309	50
Cows	7	2,480	386	28
Equipment	9	79	12	30
Farm	11	614	95	64
Hay	12	1,550	241	43
Oats	18	243	37	85
Orchard	19	483	75	23
Personal	20	285	44	39
Potatoes	22	368	57	32
Poultry	23	304	47	35
Rye	24	793	123	52
Teams	27	983	153	11
Thresher	28	330	51	40
Wheat	25	97	15	11
Total		10,596	$1,650	43

$1650.43 ÷ 10596 = 15.576 cents per hour.

15

LOSS AND GAIN.

	Dr. Loss		Cr. Gain	
Cash.....4	$300	00		
Corn.....6	209	73		
Cows.....7	624	33		
Farm.....11	75	69		
Hay.....12			$541	51
Interest.....13			298	08
Oats.....18	2	75		
Orchard.....19	57	58		
Personal.....20	329	37		
Potatoes.....22			1	51
Poultry.....23	49	53		
Rye.....24			179	83
Wheat.....25	4	69		
Thresher.....28		31		
By balance (loss).....15			633	05
Total.....	$1,652	98	$1,652	98

16

INVENTORY.

April	*Resources*	1911		1912	
1	Farm.....11	$14,000	00	$14,000	00
	Cash.....2	536	54	140	34
	Equipment.....9	602	15	723	90
	Teams { 6 horses in 1911, 7 horses in 1912 }.....27	580	00	675	00
	Cows { 10 in 1911, 23 in 1912 }.....7	355	00	975	00
	Poultry.....23	12	00	100	95
	Corn.....6	125	00	240	00
	Hay.....12	445	00	176	00
	Oats.....18	54	95		
	Potatoes.....22	18	50	4	25
	Rye.....24	18	00		
	Wheat.....25	30	00		
	Threshing machine.....28	900	00	1,000	00
	Notes and accounts receivable.....17			4	00
	Total resources.....	$17,677	14	$18,039	44
	Liabilities				
	Notes and accounts payable.....17	$10,004	65	$11,000	00
	Net assets, 1911.....	7,672	49		
	Net assets, 1912.....			7,039	44
	Balance (loss).....15			633	05
	Total.....	$7,672	49	$7,672	49

17

Notes and Accounts Payable.

1911		Dr.		Cr.	
April 1	By inventory........................16			$10,004	65
May 14	To bill, James Holden, paid..............3	$4	65		
15	By note given........................3			1,500	00
1912 March	To notes paid........................	500	00		
March	To inventory........................	11,000	00		
		$11,504	65	$11,504	65

Notes and Accounts Receivable.

1911		Dr.		Cr.	
April 1	To inventory........................16				
March 31	To amount paid on notes..................	$4	00		
31	By inventory........................16			$4	00
		$4	00	$4	00

18

Oats (12 Acres).

1911		Dr.		Cr.	
April 1	To inventory........................16	$54	95		
23	To use drill and help....................2	10	50		
30	By 157 bu. for teams...................27			$54	95
1912 March 31	To expense balance of year................	8	87		
31	By product balance of year................			180	00
31	To use of land........................11	36	00		
31	To use of buildings....................11	15	00		
31	To 243 hrs. of labor...................14	37	85		
31	To 366 hrs. horse labor.................27	63	63		
31	To 366 hrs. equipment...................9	10	90		
31	By balance (loss).....................15			2	75
		$237	70	$237	70

19

ORCHARD.

1911		Dr.		Cr.	
April					
1	60 lbs. copper sulphate.......................2	$3	30		
1	60 lbs. sulphur.................................2	2	40		
29	To 4 lbs. lead arsenate.................. } 3 To 16 lbs. sal. soda....................... }		77		
May					
15	To use wheels and tank..................28	10	00		
26	To 1 bbl. lime.................................4	1	50		
1912					
March					
31	To expense balance of year..................	16	80		
31	By products......................................			$133	52
31	To use of land..................................11	20	00		
31	To 483 hrs. labor 14	75	23		
31	To 300 hrs. horse labor 27	52	16		
31	To 300 hrs. equipment.......................9	8	94		
31	By balance (loss)..............................15			57	58
	Total..	$191	10	$191	10

20

PERSONAL.

1911			Dr.		Cr.	
April						
1	To allowance, household expense	2	$30	00		
7	To *Courier* subscription $1.50 To P. O. box rent, 10 cents	2	1	60		
7	To stamps, 90 cents candy, 25 cents tobacco, 20 cents	2	1	35		
12	To church, 30 cents tobacco, 20 cents haircut, 20 cents	2		70		
12	To tickets to school house	2		70		
30	To 15 doz. eggs	23	3	00		
30	To 2 hens	23	1	00		
30	To 94 qts. milk	7	2	82		
May						
1	To allowance, household expense	3	30	00		
6	To shoe repairs, 75 cents tobacco, 40 cents Masonic dues, $3.00	3	4	15		
14	To postage stamps	3	1	00		
26	To coal (8 tons) and stamps	4	40	04		
30	To 87 qts. milk	7	2	61		
31	To $13\frac{3}{4}$ doz. eggs	23	2	34		
1912						
March						
31	To expenses balance of year		436	59		
31	By receipts balance of year				$19	35
31	To rent buildings	11	200	00		
31	By labor for year	14			500	00
31	To 285 hrs. labor	14	44	39		
31	To 228 hrs. horse labor	27	39	64		
31	To 228 hrs. equipment	9	6	79		
31	By balance (loss)	15			329	37
	Total		$848	72	$848	72

Note: Page 21 blank.

22

POTATOES.

1911		Dr.		Cr.	
April					
1	To inventory 16	$18	50		
11	By sale 2 bbls. potatoes 2			$2	00
18	By sale 3½ bbls. seed potatoes 2			8	50
1912					
March					
31	By sale product			129	60
31	By inventory 16			4	25
31	To use of land 11	10	00		
31	To 368 hrs. labor 14	57	32		
31	To 280 hrs. horse labor 27	48	68		
31	To 280 hrs. equipment 10	8	34		
	To balance (gain) 15	1	51		
	Total	$144	35	$144	35

23

POULTRY.

1911		Dr.		Cr.	
April					
1	To inventory 16	$12	00		
8	To 40 eggs for incubating 2	10	00		
30	By sale of eggs 3			$2	05
30	By 15 doz. eggs used 20			3	00
30	By 2 hens used 20			1	00
May					
14	To chick food, wire and brooder 3	9	10		
31	By sale eggs 4			1	07
31	By 13¾ doz. eggs for house 20			2	34
1912					
March					
31	To expense, balance of year	108	75		
31	By receipts			38	74
31	By inventory 16			100	95
31	To interest 13	2	82		
31	To use buildings and yard 11	5	00		
31	To 304 hrs. labor 14	47	35		
31	To 18 hrs. horse labor 27	3	13		
31	To 18 hrs. equipment 10		53		
	By balance (loss) 15			49	53
	Total	$198	68	$198	68

24

Rye (20 Acres).

1911		Dr.		Cr.	
April					
1	To inventory....16	$18	00		
8	By 500 lbs. of rye....27			$5	00
8	By 800 lbs. rye....7			8	00
1912					
March					
31	To expense, balance of year....	26	65		
31	By product, balance of year....			597	34
31	To use of land....11	60	00		
31	To use of buildings....11	15	00		
31	To 793 hrs. of labor....14	123	52		
31	To 920 hrs. horse labor....27	159	94		
31	To 920 hrs. equipment....10	27	40		
	To balance (gain)....15	179	83		
	Total....	$610	34	$610	34

25

Wheat (5 Acres).

1911		Dr.		Cr.	
April					
1	To inventory....16	$30	00		
1912					
March					
31	To expense balance of year....	4	30		
31	By sale product....			$86	19
31	To use of land....11	15	00		
31	To 97 hrs. labor....14	15	11		
31	To 130 hrs. horse labor....27	22	60		
31	To 130 hrs. equipment....10	3	87		
	By balance (loss)....15			4	69
	Total....	$90	88	$90	88

26

TEAMS.

1911		Dr.		Cr.	
April					
1	To inventory, 6 horses..................16	$580	00		
4	To 1 horse purchased....................2	75	00		
8	To 500 lbs. rye.........................24	5	00		
10	To phone horse doctor...................2		25		
23	To purchase pair mules..................2	100	00		
27	By labor of teams..........................			$2	75
30	To 157 bu. oats from field...............18	54	95		
30	To 500 bu. corn from field................6	125	00		
May					
14	To purchase salt.........................3		49		
15	By cash to boot on horse trade............4			25	00
1912					
March					
31	To expense..............................	1,131	29		
31	By returns..............................			274	45
31	By inventory, 7 horses..................16			675	00
31	To interest.............................13	31	38		
31	To use buildings........................11	50	00		
31	To 983 hrs. labor.......................14	153	11		
	By balance..............................15			1,329	27
	Total...................................	$2,306	47	$2,306	47

27

TEAMS (*Continued*).

DISTRIBUTION OF EXPENSE ON HORSE LABOR.

By 2974 hrs. on corn @ 17.385 cents	6	$517.03
By 576 hrs. on cows @ 17.385 cents	7	100.14
By 40 hrs. on equipment @ 17.385 cents	9	6.96
By 156 hrs. on farm @ 17.385 cents	11	27.12
By 1408 hrs. on hay @ 17.385 cents	12	244.78
By 366 hrs. on oats @ 17.385 cents	18	63.63
By 300 hrs. on orchard @ 17.385 cents	19	52.16
By 228 hrs. on personal @ 17.385 cents	20	39.64
By 280 hrs. on potatoes @ 17.385 cents	22	48.68
By 18 hrs. on poultry @ 17.385 cents	23	3.13
By 920 hrs. on rye @ 17.385 cents	24	159.94
By 250 hrs. thresher @ 17.385 cents	28	43.46
By 130 hrs. wheat @ 17.385 cents	25	22.60
Total		$1,329.27

```
7646 ) $1329.27 ( 17.385+
        764.6
        -----
        564.67
        535.22
        ------
         29.450
         22.938
         ------
          6.5120
          6.1168
          ------
           .39520
           .38230
```

28

THRESHER.

1911		Dr.		Cr.	
April					
1	To inventory.........................16	$900	00		
8	To repairs.............................2	2	00		
May					
14	By use of thresher.....................3			$5	00
15	By use of wheels and tank..............19			10	00
1912					
March					
31	To expenses............................	58	20		
31	By returns.............................			109	70
31	By inventory..........................16			1,000	00
31	To interest...........................13	47	50		
31	To use building.......................11	15	00		
31	To 330 hrs. labor.....................14	51	40		
31	To 250 hrs. horse labor...............27	43	46		
31	To 250 hrs. equipment.................10	7	45		
	By balance (loss).....................15				31
	Total..................................	$1,125	01	$1,125	01

Interpretation of Results.—This is the most important phase of the work. If an account comes out even it means that the enterprise has paid rent on land, taxes, interest on inventory value, all expenses, and has also paid wages to men and teams for the time put upon it. If a more profitable crop can be found that will not interfere with the other successful crops of the farm, it should be substituted. If not, it may be good business to continue the old crop. It employs labor at actual cost when it would otherwise be idle.

A study of the account may reveal a way to make the crop pay. It may be by increasing the yield or it may be by reducing the cost per acre. One should judge whether the year is an average one and if average prices prevail.

A study of the accounts here given shows a heavy expense for teams. There are too many horses for the acres of land under plow; consequently, they averaged only about three hours of work daily for 300 days in the year. This made the horse labor cost 17.38 cents an hour. On a well-managed farm the cost should be not more than 10 cents an hour.

Of the crops, hay and rye gave good profits, while potatoes gave a very small profit. The potatoes could be made more profitable by increasing the acreage. Two and one-half acres are too small to justify the maintenance of a good equipment. Corn gave a fair gross return, but the hours of man and horse labor devoted to it were so large and costly that the crop shows a loss of $209.73. Six units of horse labor per acre of corn are usually sufficient. In this crop nearly ten units per acre were used.

The cows show a heavy loss. Without the detailed account for the year it is not quite clear why they lost money. The gross receipts appear fairly good, but expenses are too high. It is evident that one may lose money by feeding profitable crops to unprofitable cows.

The poultry shows a loss due very likely to the fact that it was just being established. The account is not fully satisfactory because it does not show the number of fowls either at the beginning of the year or in the final inventory. The hours of labor on poultry are rather large in comparison with the inventory at the close of the year. The expenses, as will be noted, are much greater than the actual receipts. From 100 to 300 hens can usually be kept on a farm with comparatively small expense for feed and with cheap labor. Such usually should give a good profit.

REFERENCES

"Farm Accounts." Vye.
"Farmers' Business Hand Book." Roberts.
"The Business Side of Farming." Bexel.
Illinois Expt. Station Circulars 77, 84, 102, 114, 115. "Records of Dairy Cows."
Indiana Expt. Station Bulletin 127. "Records of Dairy Cows."
Massachusetts Expt. Station Bulletin 120.
Farmers' Bulletins, U. S. Dept. of Agriculture:
511. "Farm Bookkeeping."
572. "A System of Farm Cost Accounting."

CHAPTER 73

MARKETS, MARKETING AND CO-OPERATION

The rapid increase in the size and number of cities has increased the need of markets; not only has the number of people that gain a livelihood in the market business increased, but the proportion of them to the total population has also increased. This is due to the increased complexity of the marketing system. At present, the cost of getting perishable products from the producer to the consumer is greater than the actual cost of production.

A reduction in the cost of production can not be hoped for. Living costs can be lowered only through more economical methods of distribution. This important problem is now engaging the attention of town and city organizations in many centers.

Cost of Distribution.—Careful investigations have been conducted in recent years in a number of large cities to ascertain through what hands products pass in transit from producer to consumer and the increase in cost resulting from each transfer. The following table from a report by Dr. C. L. King, of the University of Pennsylvania, to Mayor Rudolph Blankenburg of Philadelphia, shows the prices of various products as they passed from farmers in counties near Philadelphia to the consumer in the city:

TABLE GIVING THE PRICE RECEIVED BY THE PRODUCER AND EACH MIDDLEMAN AND THE PER CENT INCREASE OF EACH PRICE OVER THE PRECEDING PRICE, TOGETHER WITH THE TOTAL INCREASE OF CONSUMERS' PRICES OVER PRODUCERS' PRICES.

	FARMER.	PLUS FREIGHT TO TERMINAL.		JOBBER.		WHOLESALER.		RETAILER.		Per Cent Increase in Consumers' Prices over Producers' Prices.
	Price Received by.	Amount.	Per Cent Increase over Preceding Price.	Price Received by.	Per Cent Increase over Preceding Price.	Price Received by.	Per Cent Increase over Preceding Price.	Price Received by.	Per Cent Increase over Preceding Price.	
Butter (low-grade), per pound	$0.18½	$0.19	2	$0.21½	13	$0.24	11	$0.32–0.38	33–58	73–105
Butter (high-grade), per pound	.23	.23½	2	.26	10	.29	11	.40– .45	38–55	74–96
Potatoes (low-grade), per bushel	.53	.62	17	.68	9	.75	10	1.10–1.30	46–73	108–145
Potatoes (high-grade), per bushel	.63	.72	14	.80	11	.90	12	1.30–1.60	44–78	106–154
Eggs (low-grade), per dozen	.11	.12	9	.13½	12	.15	11	.25– .30	67–100	121–173
Eggs (high-grade), per dozen	.21	.22	4	.24	10	.27	11	.35	30	67
Huckleberries (low-grade), per quart	.04½	.05½	22	.06	9	.07	16	.12	71	166
Huckleberries (high-grade), per quart	.07	.08	14	.10	25	.11	10	.15	36	114
Blackberries (low-grade), per quart	.04	.05	25	.05½	10	.06	13	.12	100	200
Blackberries (high-grade), per quart	.06	.07	16	.08	14	.09	12	.15	66	150
Live poultry (low-grade), per pound	.06	.06½	7	.09	38	.11	22	.22	100	266
Corn, per dozen	.15	...	..	...	..	...	..	.40	...	167
Tomatoes, per peck	.32	.32½	1	.36	10	.40	11	.80	100	150

From this table it will be noted that the price paid by the consumer ranges from 67 to 266 per cent above that received by the producer, the average increase in price being 136 per cent. This represents the average condition for farmers who sell comparatively small quantities and for consumers who buy for daily needs.

The large cost of distribution tends towards increasing prices for the consumer and lowering prices for the producer. It is estimated that the people of New York City are paying over $150,000,000 annually to have their foodstuffs carried from railroad terminals to their kitchens. Of the $146,000,000 paid annually by the people of New York City for milk, eggs, potatoes and onions, less than $50,000,000 are received by the farmers who produced them.*

It is clear that a cheaper method of food distribution is of much concern to both producer and consumer.

Frederick O. Sibley makes the following report in regard to the cost of growing apples as compared with the cost of selling them:

SOME COSTS OF GROWING APPLES.

	Per Barrel.
Labor	$0.50 to $0.75
Cash expenses	.15 to .25
Interest and overhead charges	.10 to .15
The barrel	.35 to .40
Storage	.25 to .40
	$1.35 to $1.95

SOME COSTS OF SELLING APPLES.

	Per Barrel.
Freight	$0.10 to $0.15
Commission	.06 to .25
Cartage	.15 to .25
Storage	.25 to .25
Jobbers	.25 to .40
Retailers	.50 to 3.00
	$1.31 to $4.30

Middlemen.—There is a long list in this class. They consist of the local buyer and shipper, the transportation companies, the transfer companies, the commission merchant, the jobber and the retailer. Their purpose is to serve the producer and consumer. In this capacity they find a market for the farmers' goods. So long as this service lessens the work of the farmer, he may give more time to production and produce more cheaply.

Some of the transportation companies have established market bureaus to assist the farmer in finding a market for his produce. They have also taught the farmer how best to prepare his produce for sale. Some have also aided the producers in securing suitable packages in which to ship produce. In like manner other classes of middlemen have estab-

* These figures are from report by Dr. King.

lished markets and standardized produce so that farmers can better understand market quotations and know the price their goods will command.

The chief difficulty lies in the large number of middlemen and the complexity of the business, thus entailing an expense in getting produce from the farm to the consumer's table. This unwarranted expense is a burden to both producer and consumer.

The Consumer is helpless and must pay the price asked for produce by those with whom he deals. Dissatisfaction on his part has more recently given rise to public meetings with a view of forming associations for the purpose of protecting consumers. The chief difficulty encountered has been a lack of reliable data on which to base practical plans. In some places the consumers have organized and established a market place where

A Farmers' Retail Curb Market.[1]

they may meet the producer directly, thus eliminating the middlemen. In other instances co-operative associations have been established, and produce bought at wholesale and retailed to the members of the association at actual cost. In order to succeed, such an organization requires loyalty of all its members. The organized trade does everything to discourage such competition. For a time regular dealers will reduce prices even below that of the co-operative store, for the purpose of putting it out of business. This is often looked upon as a failure on the part of the co-operative store, and it receives no credit for having reduced the prices. If the co-operative store is forced to close, prices again rise, frequently above their former level, to enable the regular merchants to make up for their sacrifice in gaining their point. Co-operative associations should recognize this difficulty and hold out against it until the regular trade resumes normal prices; this critical period once passed, such organizations are usually in a position to render good service to a community.

[1] From Year-Book, U. S. Dept. of Agriculture, 1914.

Such organizations must be willing to pay the price necessary to secure a good manager. It is good business to purchase directly from producers so far as it is possible.

The Producer's Share.—The producer of food supplies is complaining bitterly of the small share of the consumer's dollar which he receives. Farmers producing vegetables, fruit, potatoes and milk suffer more in this respect than do those who produce the staple crops, such as corn and wheat. The latter pass through a less number of hands and are sold in larger quantities.

The remedy for the farmer will not be found until the farmers themselves organize for the purpose of solving their own problem and protecting their own interests. In numerous localities, the farmers have organized and are selling their produce through the organization, adopting a system of grading and a style of package found to be best suited to the market on which their goods are placed. In this way prices have frequently been increased from 50 to 100 per cent above what they were before they were organized.

Such organizations may also profit by purchasing fertilizers, seeds, feeds, coal, farm machinery and shipping packages by wholesale. In this way the goods are secured at lower prices and much is saved in freight by carload rates.

Aside from financial benefits, farmers come to know each other better, trust one another more fully and profit by the leadership of those best qualified to lead.

Legislative Regulations of Commission Business.—The commission business offers opportunities for gross irregularities. There is plenty of evidence to show that unscrupulous commission men have made false reports to the farmer, relative to the condition of his goods upon arrival at the market and the price for which they were sold. In such instances, the farmer is at a decided disadvantage and is usually unable to secure redress. It is important that farmers ship only to commission men who have a reputation for honest and fair dealing.

Commission merchants should be compelled by law to keep records and accounts of their business, and these should be open to inspection with reference to their regularity. Should a question of unfairness arise on the part of the farmer, the records of the merchant should show the condition of the goods when received, the price for which sold and the parties to whom sold. Such a requirement would provide for establishing the truth relative to the condition of goods and price received.

The law should also protect the farmer against the commission man who voluntarily goes into bankruptcy for the purpose of swindling the farmer.

Advertising.—When the farmer goes to town, he sees on the front of every business house the name of the firm and usually finds a display of goods to attract attention. This is advertising. As one drives through the country, one sometimes sees the farmer's name on the mail-box, more frequently it is painted out. Many farmers would find it to their advantage to have a farm bulletin board so placed that passersby could easily see it. A good one is shown in the accompanying picture. This one is in front of the farm home of Mr. W. W. Scott, Clinton County, Iowa. His name and the name of his farm appear on the upper part of the board; on the lower part is a blackboard on which may be written articles for sale and also articles wanted.

There are many ways of advertising. The neat appearance of a farm is always more or less of an advertisement. One may use letter-heads

The Farm Bulletin Board Brings Business.

giving the name of the farmer, the name of the farm and the special line of production. Produce sold in packages should be marked with the producer's name and address. Produce and animals exhibited at fairs are also advertisements.

The four essentials to a good advertisement are: (1) to attract attention, (2) excite interest, (3) convince buyers, and (4) to consummate sales.

Marketing the Farm Products.—The first essential to the successful marketing of farm products is to have an article of good quality put up in an attractive form. If package goods, the package should be of suitable size. Whatever the package, make sure that it is full measure or weight. Be ready to stand back of every sale to the extent that if the product is not up to what is claimed, the purchase price will be refunded or the package duplicated.

Purchasers do not like to find upon opening a box or a barrel of fruit that the interior is inferior to the surface layer. Goods packed in this way cannot establish a reputation for the producer, and, of course, are not branded. It pays to be honest.

Trend of Prices.—The prices of farm products fluctuate from month to month and from year to year. The rise and fall in prices is due to several factors, such as over and under production, the financial condition of the country and the extent of exportations.

For products that have a world-wide market, such as wheat and most animal products, the supply in foreign countries will also affect price. A shortage in the supply of any farm product results in a rise in price. Frequently the advance in price is such as to render a short crop more valuable on the market than a large crop sold at a price somewhat below normal. Nearly all crops are subject to high and low prices at rather regular intervals. When prices of any commodity are unusually good, farmers generally plant more extensively of it the following year and thus cause the price to decline. A decline in price is then followed by a reduced acreage and a consequent rise in price.

With annual crops these periods are of short duration. With crops that require several years for fruitage, these periods are much longer. With apples, for example, the periods of high prices and low prices occur at intervals of about twenty years. With horses, that require four or five years for maturity, these periods occur at intervals of eight to ten years, while with swine the intervals are about three years. It is good business on the part of the farmer to anticipate these periods of high prices and increase his output to meet the demand.

There is also a monthly trend of prices for nearly all farm products. As a rule, prices are lowest just at the close of the harvesting period of each crop and during the months that livestock is most conveniently marketed. If crops are held for an advance in price, one should compare the cost of holding with the probable rise in price. The cost of holding consists of shrinkage, storage, interest on value of crop, insurance and possible depreciation in quality, together with loss from vermin. The cost of marketing should also be taken into account. It is advantageous to market the crops when farm work is not pressing. This helps to distribute the work of the farm and keep men and teams more fully employed. It will pay the farmer to study market prices and the forecast of probable yields as reported by the U. S. Department of Agriculture and the daily and agricultural press.

Selling Directly to Consumer.—Milk, butter, poultry, eggs and nearly all classes of fruit and vegetables are frequently sold by the farmer directly to the consumer. While this method eliminates all middlemen and secures for the farmer the best price, the production side of his business generally suffers as a consequence. Whether or not a farmer should follow this method will depend on many factors, such as personal qualifications, the character of local markets, the distance from market, the character and condition of roads, the amount and kind of produce for sale and the assistance available to conduct both the production and marketing ends of the business.

The recent introduction of parcel post has made it possible for the farmer to market direct to the consumer without wagon service. Such

THE MOTOR TRUCK IN MARKETING.[1]
Milk delivery truck, making 67-mile trip daily. Carries 96 cans of 46 quarts each from Elmer, N. J., to Philadelphia.

marketing necessitates special packages to protect the produce and guarantee safe delivery. It is not applicable to very low-priced and bulky materials at present delivery rates. Its chief merit lies in the better quality secured through direct and quick delivery.

The Motor Truck in Marketing.—The motor truck in regions of good roads is a saver of time and horses in the marketing of produce. Such a truck frequently dispenses with one team of horses and at the same time reduces the time consumed in delivery.

In New Jersey, peaches from one neighborhood, hauled to market in a market truck netted about twenty cents per basket more than those

[1] Courtesy of The Pennsylvania Farmer.

THRESHING SCENE

hauled in the old way. This difference was due to the reduced cost of hauling and the better condition of the peaches upon their arrival.

It is estimated that it costs $7,500,000 annually to haul the freight to and from terminals in Philadelphia by teams. A motor truck of half the length of a wagon carries double the load and travels twice as fast. In cities, the motor truck is relatively more important than in the country. This is manifest by the extent to which it has already displaced the city horse.

Co-operation.—The tendency of the twentieth century is for all produce growers to increase their facilities for direct marketing by organizing co-operative associations. This is a world-wide movement and such societ es are doing business successfully in nearly every country of the world. Co-operation is best developed in little Denmark. There it has become almost a national trait. Denmark has more than a thousand co-operative dairies, 500 egg societies and numerous other co-operative associations. Through these associations the farmers of Denmark are exporting nearly $100,000,000 worth of butter, eggs and meat every year. Through such associations the Danish farmer purchases nearly $20,000,000 worth of machinery, fertilizers, etc., each year

Successful co-operation among farmers in the United States is manifest in such associations as the New England Cranberry Sa es Company, the Monmouth County Farmers' Exchange of Freeho d, N. J., the Citrus Association of Florida, the Dassel Co-operative Association of M nnesota, the Rockyford Melon Association of Colorado, the Hood River Apple

SHOWING CO-OPERATION.

Growers' Union of Oregon, the California Fruit Growers' Exchange, and more than a hundred others that could be mentioned.

Successful co-operation requires a study of local conditions and needs. While the principles of co-operation are similar for all places, their application must be modified to meet the needs of each locality. It is, therefore, necessary for the farmers to get together and thoroughly discuss the ways in which they can co-operate. The launching of a co-operative association requires leadership, and whether for the purpose of selling or buying, a good business manager is needed. The Citrus Growers' Association of California pay their manager $10,000 a year.

As co-operation develops in any region, there should be a central co-operative organization to assist and advise the local branches. Once such an organization is started, it makes easy the organization of many local branches which soon enlarge and increase the effectiveness of the organization as a whole. The more territory covered by the organization the more completely can it control the marketing and prices of the farm products.

In What Can Farmers Co-operate?—Farmers can co-operate in exchanging work, in the joint ownership of expensive machines and pure-bred sires. They may exchange fresh meats, fruits and vegetables to advantage. By co-operation, they can secure and maintain good roads, improve the schools, churches and country clubs. They may establish cow-testing associations and co-operative creameries; drainage districts may be formed and farm bureaus established for the betterment of agriculture. They may also co-operate in selling the farm products and in purchasing farm supplies. There are other ways in which to co-operate.

Exchanging Help.—There are many kinds of farm work that can be most advantageously done by two or more men working together. The most striking instances of this are the harvesting of grain and hay, the shelling and marketing of corn and threshing. The exchanging of work during the harvest season often meets the temporary demand for extra labor. Such co-operation brings neighbors closer together and promotes friendly relations. It should be free from abuse. It frequently happens that some farmers, through generosity, give much more than they receive. A value should be placed on the labor of men and teams and a record kept of the time given to the neighbor as well as that received from him. A settlement for the difference in time at the close of the year promotes good feeling and avoids dissatisfaction.

Co-operation in threshing grain is now the rule in the large grain districts. The threshing ring consists of fifteen to thirty farmers, the number depending on the size of the farms and the length of the threshing period. The members of the ring either purchase an outfit for their own use or contract to hire one. They hold meetings at stated times and agree upon terms, the price to be paid for threshing the different kinds of grain and the order in which the members' threshing shall be done. The order should change from year to year in order to be fair to all members. Each member furnishes help in proportion to the size of his farm or acreage of grain he grows.

Cow Testing Associations.—These associations have been of great service to dairymen in many districts. Such an association requires that there shall be fifteen or more dairymen living within an area sufficiently restricted so that all herds can be visited by one man once or twice each month, and that each dairyman own ten or more cows. The association members each agree to pay so much per cow, usually not more than $1.50 per year. They employ an official tester, who spends one day each month with each farmer; or, in case of small herds, two herds may be tested in one day. The tester weighs the milk of each cow morning and evening, and takes samples of it which are tested for butter-fat. He also weighs or measures the quantity of each kind of feed that is given the cows. These records, after being repeated for several months, enable the farmer to recognize which cows are the most productive and which should be sold in order to make the herd more profitable.

Many farmers milk cows year after year that do not pay for the feed consumed. The average annual production of cows in the United States is about 4000 pounds of milk and 160 pounds of butter-fat. Since this is the average, many cows must produce less than this. The best dairymen say there is no profit in cows of even average production. One association in Pennsylvania reports a gain valued at $4500 in one year as a result of cow testing.

The associations lead to a better understanding of the feed requirements, and have resulted in auxiliary associations for the purchase of feed at wholesale with much saving to members.

Marketing Dairy Products.—The intelligent marketing of dairy products necessitates an understanding of the value of the different milk products and the cost of making them as compared with the price for which the milk may be sold. Knowing the price of milk, cream, butter, cheese and ice cream, and the butter-fat content of the milk and cream, one can easily calculate the relative values of the several products as compared with the milk. This, in connection with the cost of production, will enable the farmer to determine which is most profitable for him.

Where there is little demand for market milk, co-operative creameries are advantageous. They enable the farmer to have his milk manufactured

A Full Load Reduces Cost of Hauling.[1]

into a first-class article and sold for a good price. Such creameries need not be expensive structures nor contain elaborate equipment. A creamery to be successful must have the product of a sufficient number of cows within reasonable distance to fully employ the time of the creameryman and utilize the equipment. The rules of the creamery in reference to the standard and condition milk received, should be enforced; otherwise a few careless patrons may impair the product of the whole. The essentials to success are the making of a first-class product and economy in the cost of production. There will be no difficulty in finding a gilt-edged market for the good product.

Marketing Livestock.—Under the old system of marketing livestock there were two or more local buyers at every small shipping point who made a living from this business. Very few farmers have enough of any one class of stock to make a carload. For this reason it is not feasible

[1] Courtesy of The Macmillan Company, N. Y. From "Farm Management," by Warren.

for them to make their own shipments. Furthermore, many a farmer would like to feed half a carload of cattle, but since he can buy from the stockyards economically only in carload lots, he concludes that less than this number is not worth while. In this way he fails to utilize roughage and other unmarketable products on the farm.

The above state of affairs has given rise to dissatisfaction and has resulted in the organization of co-operative livestock shipping associations. Minnesota has taken an active lead in this respect. In 1911 there were three of these associations in that state. In 1913 the number had increased to forty. Since then others have been added and a State Central Association has been formed, the chief purpose of which is to assist in organizing local associations and developing a uniform system of accounting.

No capital is required in this kind of an organization. A manager is employed to look after the shipping and selling of the stock. His chief qualifications must be honesty, a thorough knowledge of livestock and business methods. He should be a good salesman. Since he handles considerable sums of money for the patrons, he should be required to give a bond.

When the members of the association have stock to market they advise the manager. As soon as several members have enough to make a carload, he advises each of the date when shipment will be made. The stock of each member is given a number or other identification mark and weighed. The stock of each is sold on its merit at the stockyards, and the bill of sale shows what it brought. The manager may conduct the business on a percentage basis or on a salary. He is subject to the direction of the board of managers.

Marketing Eggs.—The bulk of eggs being produced as a side line, can be marketed co-operatively in connection with the co-operative marketing of other products to good advantage. In districts where co-operative creameries exist, eggs are found to be an ideal side line for the creamery. The teams that bring milk or cream to the creamery can also bring eggs produced by the patrons with practically no extra expense. In this way the creamery can secure eggs at frequent intervals and by having definite rules will secure nothing but strictly fresh eggs and guarantee them to consumers. It has the facilities for keeping eggs cool and for putting them upon the market at the minimum of expense.

The plan that has proven successful with some creameries is to deliver egg cartons and to supply a stamp to each patron. The eggs are stamped as soon as gathered, placed in the carton, sealed and kept cool until delivered. Patrons should gather eggs twice daily, especially during warm weather, should grade them to uniform size and keep white and brown eggs separate. If these eggs are stamped and guaranteed not to be over a certain number of days old, they have been found to bring on the market from two to five cents per dozen over the ruling price. In some cases advance in price is even greater.

In Prince Edward Island, within the last few years, a number of egg circles or associations have been formed for no other purpose than the marketing of eggs. The membership of each circle ranges from 35 to 100, and each elects a board of directors and hires a manager. It is the business of the manager to collect eggs and find a market for them, finally turning over the proceeds to each member. The eggs are gathered and graded according to rules which must be lived up to. The manager's salary is paid out of a commission which, during the summer, equals about one cent

Shipping Vegetables by Water.[1]

a dozen on all eggs, and during the winter amounts to about two cents a dozen.

The objects of the egg circles are to secure better prices for poultry products, to improve the quality and to buy at wholesale supplies for its members. It also aims to introduce pure-bred poultry and disseminate poultry information.

Marketing Vegetables.—In the marketing of vegetables, quality and appearance count for nearly as much as in case of fruit. The attractive package of the family size is in greatest demand. Several of the western states are taking the lead in this respect, as they have done in case of the marketing of fruit. They are teaching the eastern truck grower the value of putting upon the market a strictly fancy article.

The most successful vegetable growers are those that produce as high

[1] Courtesy of The Pennsylvania Farmer, Philadelphia, Pa.

a percentage of first-class product as possible and grade their product carefully, putting their first-class stuff on the market under their farm name or brand. Any low-grade material should go to the market as such where there is likely to be the greatest demand for it, and should not bear a brand.

Co-operative marketing of vegetables has proven eminently successful in many localities. A notable example of such marketing is that of the Eastern Virginia Produce Exchange, with the headquarters at Onley, Va.

LOADING PEACHES FOR AUTO TRUCK TRANSPORTATION AT GLASSBORO, N. J.[1]

The sales of this exchange amount to about $4,000,000 annually, and it carries from $50,000 to $100,000 to the surplus fund each year.

The sales of this exchange are made in the northern cities in carload lots by traveling salesmen. If there is any complaint on the arrival of produce, a man is sent at once to investigate, and if claim is just, settlement is made. It is the aim to market goods that come up to the standard claimed for them and to do a strictly honest business.

Marketing Fruit.—Reference has been made to the advantages frequently taken of the producer by commission merchants. On the other hand, producers frequently put upon the market all kinds of fruit, the package contents of which are not what is represented by the top layer. This is especially true in case of barreled apples. An instance is cited of a producer who visited a commission merchant and saw first-class Winesap

[1] Courtesy of The Pennsylvania Farmer, Philadelphia, Pa.

apples sold at $3.20 per barrel. These were purchased on the assumption that the interior of the barrel would be decidedly inferior to the surface layer. This gentleman followed the apples to the retailer and found that they were sold at the rate of $9.60 a barrel. The retailer stated that if he had believed the label true he would have willingly paid $5.00 a barrel for them. Further investigation among both retailers and wholesalers in that city brought out the fact that honest packing was estimated at about 5 per cent of the fruit product put upon the market, and that 95 per cent of it was more or less dishonest.

Until fruit producers correct this defect in their method of marketing their product, they cannot hope for the best results. Neither can they blame the commission merchant and other middlemen for taking what seems to be advantage of them. Everyone should aim to have his goods fully up to what they are claimed to be in quality and then insist on fair treatment by the dealer.

Co-operation in the marketing of fruit is becoming the rule rather than the exception. At a recent meeting of the fruit growers of Kansas, Nebraska, Iowa and Missouri for the purpose of forming a fruit growers' association of these states, it was stated on good authority that 90 per cent of the Missouri fruit growers not marketing co-operatively were not making the fruit business pay, while over 90 per cent of those selling co-operatively were making good money at it. Many of those present gave testimony to this effect and cited specific instances to substantiate their claims. It was shown that in 1912 grape growers along the Missouri River on one side received an average price of $1\frac{1}{2}$ cents per pound for grapes, whereas just across the river grapes marketed co-operatively brought an average price of $2\frac{3}{8}$ cents.

Some Successful Co-operative Associations.—The Farmers' Incorporated Co-operative Society of Rockville, Iowa, has been in existence for twenty-four years. It markets general farm products, particularly corn, wheat, oats and barley, and purchases for its members feed, fuel, building material, machinery, clothing and some other items. It does an annual business of about $600,000. In the spring wheat belt, including the states of Michigan, Wisconsin, Minnesota and North and South Dakota, co-operative associations are numbered by the hundreds. These consist of co-operative elevators, creameries, cow-testing associations, poultry associations, and organizations for buying farm supplies. Minnesota claims first rank in co-operative business enterprises. In January, 1914, it reports 2013 co-operative establishments. These did a total business of $60,760,000 in 1913. Co-operative creameries led with a total business of $21,675,000. There were 270 farmers' elevators with a membership of 34,500. To these enterprises may be added co-operative telephone and insurance companies. The total amount of insurance outstanding in January, 1914, was $342,000,000. The cost of each $100 worth of insurance in force was 18 cents as against 46 cents the rate of stock companies soliciting business in competition on three-year contracts.

Even among the conservative people of Lancaster County, Pa., there is a co-operative association which has been in existence for over eight years, having a membership of over a thousand, and doing an annual business of nearly $200,000. Enough has been stated to indicate that co-operation is growing rapidly and that it is successful.

Importance of Able Management.—Success in co-operation depends much upon the manager. It is necessary to leave a great deal of the business to the manager, and for this reason a man should have wide experience, ability and be of unquestioned integrity. Such a man is sure to be worth a good salary. Those trying to co-operate who underestimate the value of good management are likely to fail.

Supervision of Co-operation.—Several of the states have already passed laws regulating certain features of co-operation. Such a law passed a few years ago by New York, provides that five or more persons may become a co-operative corporation, company, association, exchange, society or union for the purpose of conducting a general producing, manufacturing and merchandising business on a co-operative plan. It limits the aggregate value of shares held by one person to not more than $5,000. Each stockholder is entitled to only one vote, regardless of the amount of his stock. It provides for the apportionment of the net earnings by first paying dividends at a rate not exceeding 6 per cent per annum on stock. Not less than 10 per cent of the net earnings are reserved until the reserve fund equals 30 per cent of the paid-up capital stock. Five per cent of the earnings must be devoted to an educational fund designed especially for the teaching of co-operation. The remainder of the net earnings shall be distributed to members of the first class, that is, stockholders, and those of the second class, non-stockholders. Dividends are paid on purchases and sales and in proportion to the amount purchased or sold.

A Bureau of Supervision of Co-operation has been created. The superintendent is appointed by the Commissioner of Agriculture. It is his duty to aid co-operative associations throughout the state.

REFERENCES

"Co-operation in Agriculture." Powell.
"Farm Management." Card.
"Farmers' Business Hand-Book." Roberts.
"Markets for People." Sullivan.
"Marketing of Farm Products." Weld.
Michigan Expt. Station Bulletin 191. "Shrinkage of Farm Products."
Illinois Expt. Station Bulletin 124. "Shrinkage of Corn in Cribs."
Kansas Expt. Station Bulletin 147. "Shrinkage of Corn in Cribs."
New Jersey Extension Circular No. 5. "Marketing White Potatoes in New Jersey."
Wisconsin Expt. Station Bulletin 209. "Prices of Farm Products."
Canadian Dept. of Agriculture Bulletins:
192. "Agricultural Co-operation."
216. "Box Packing of Apples."

U. S. Dept. of Agriculture, Bureau of Statistics:
Bulletin 21 "Rates of Charge for Transporting Garden Truck with Notes on Growth of the Industry."
Bulletin 49. "Cost of Hauling Crops from Farms to Shipping Points."
Bulletin 9. "Production and Price of Cotton for 100 Years."
U. S. Dept. of Agriculture:
Year-Books, Appendix. Transportation Rates, Yields, Prices, etc.
Year-Book 1906, pages 371–386. "Freight Costs and Market Values."
Farmers' Bulletins, U. S. Dept. of Agriculture:
321. "Use of the Split-Log Drag on Earth Roads."
445. "Marketing Eggs Through the Creamery."
594. "Shipping Eggs by Parcels Post."

BOOK VIII

PLANT AND ANIMAL DISEASES, INSECT ENEMIES AND THEIR CONTROL

CHAPTER 74

DISEASES OF ANIMALS AND THEIR MANAGEMENT

By Dr. S. S. Buckley

Professor of Veterinary Science and Pathology, Maryland Agricultural College

"You say you doctored me when lately ill;
To prove you didn't, I'm living still."

Domestic animals contribute largely to the benefits of country life, and, aside from house pets, these pleasures are denied the residents of towns and cities. Farms devoted to trucking and fruit growing may prove financially profitable, as do mercantile pursuits, but they fail to make the farm *a home* as do those which possess a varied assortment of species of live stock.

Domestic animals share our labor, contribute to our food supply and furnish the means for improving our soil and maintaining its fertility. While the different species of domestic animals are materially unlike in some respects, yet the general scheme on which their conformation and action is planned makes it possible to apply similar broad rules for the care and management of them all.

Animals in health are by nature intended to serve man's purposes and, according to the degree of impairment of health, so is the degree of their usefulness to man affected.

Strictly considered, there are not different degrees of health, since health signifies a normal condition of the body. Abnormal conditions of the body occur, however, which are variable in degree, and these constitute disease.

Disease, therefore, may range from slight unrecognizable disturbances of the body functions to extremely complex modifications which terminate life in death.

An animal is most highly profitable to its owner when in a normal or healthy condition, and its value to him diminishes according to the degree of abnormality or disease. It is for the stockman, therefore, to interest himself in maintaining animals in health, rather than in the study of the nature and treatment of their diseases, if he is to derive the greatest benefits from them.

The Essentials for Health.—In order to be most successful in the management of animals, a study should be made of the efficacy of sound, wholesome food and pure water; the necessity for pure air and proper exercise; the effects of proper dieting, over-feeding and abstinence; the necessity for comfortable quarters, and lastly the benefits of humane and intelligent treatment. This means familiarity with the laws of hygiene and as far as possible with the structures (anatomical parts) and the functions (normal actions) of the animal body. There is a general similarity of the organization of animal bodies and of the human body, and what is bad for mankind is most likely bad for animals.

Knowledge of Disease Should Precede Treatment.—No one should undertake the treatment of a disease of animals whose nature he is not familiar with, nor to administer medicines whose effects are unknown to him, any more than he should attempt to treat similar disorders in the human. The mere fact that one is animal and the other human does not alter the chances for success, nor prove more creditable to his intelligence.

Stockmen should exercise *common sense* in the management of animals in health and disease, and remember that there is always to be regarded the powerful effort on the part of nature to combat bodily disturbances and disease. Intelligent assistance would frequently restore, where indiscreet meddling will destroy.

There is a strong propensity on the part of stockmen to resort to the use of powerful remedies for all diseases without first deliberating on the *nature* of the disorder, its *cause*, its *symptoms*, its *course*, its *normal duration* and, finally, its *rational treatment*. Such deliberation would frequently indicate that the disorder was due to some *lapse* in management; that some of the symptoms were mere evidences of nature's effort to overcome the disorder; that its normal cause and duration was dependent upon the duration of mismanagement and that rational treatment should be directed towards assisting rather than in opposing nature's efforts. For example, an animal has been over-fed and diarrhea results. More frequently than otherwise, such a case is treated with opium preparations or astringents, to check the diarrhea, possibly with serious consequences; while on the other hand, rational treatment would consist in restricting the diet, perhaps modifying it, and administering a mild laxative, mashes, flaxseed tea, or raw linseed oil, to assist nature in her efforts at the expulsion of the offending material as shown by the condition of diarrhea. After the desired result has been secured, the animal is brought back, by gradually increased amounts of food, to the usual ration which had been fed.

Intelligent and judicious management is essential, both in preserving health and in restoring it when impaired.

GENERAL RULES FOR MAINTAINING HEALTH

1. Feed only sound, wholesome grain and fodder. Supply abundantly pure water, at short intervals.

2. Supply salt regularly to all animals. Rock salt is preferable to purified salt, as it contains other needed elements than soda. Hogs and poultry need little salt compared to other farm animals, excessive amounts being poisonous to them.

3. Charcoal may be given occasionally with benefit to all animals, and may be fed with salt.

4. Feed with extreme regularity, and according to the requirements of animals, in quantity and nutritive value.

5. Developing or growing animals, females with young, pregnant females, males for breeding purposes, work animals and animals not at work require different feeds, in quantity and quality.

6. Animals at pasture require attention. Pasturage may be adequate or it may need to be supplemented with additional feed.

7. Make all changes in rations gradually. Add any new variety of feed to the ration in small and successively increasing amounts until the desired addition is secured.

8. Unwholesome food is frequently produced on farms, and, being unmarketable, is kept for feeding purposes. Such foods may be fed safely if proper methods are employed.

Damaged grain, soft, rotten, mouldy, worm-eaten and otherwise unwholesome, may be made safe for feeding if it is first shelled from the cob or threshed from the straw and then carefully fanned to remove the light, badly damaged and unwholesome grains. By the same process, the spores of mold and poisonous dust are largely eliminated.

Damaged fodder and hay may be made less objectionable and safer by shaking out as much as possible the dust and must as it is removed from the stack. It should then be run through a cutting box and cut into convenient lengths. This cut fodder should be mixed with a proper amount of grain and salted at the rate of one pound of salt to the hundred pounds of chop. Moisten the entire mass and after macerating for several hours, it can be fed. Where this is practiced, the chop box should be kept scrupulously clean.

Comfort.—Animals may be well bred and well fed and yet not develop nor thrive properly if kept in uncomfortable surroundings.

Stables which are comfortable should be well lighted, but the light must be admitted into the building in such a way as not to subject the animals to a constant glare of bright sunlight and they should not face dark, unlighted walls. Stables, however, should be so arranged that all parts of the enclosure are well lighted with diffuse light. They should be devoid of dark recesses which might serve for the accumulation of filth, as breeding places for vermin or for the decomposition of feed and fodder.

Mangers and racks for feed should be convenient alike for feeder and animals and easy to clean. Refuse must not be allowed to accumulate, as when moistened with saliva it sticks to the mangers and affords an ideal place for decomposition processes and the development of attendant poisons.

Floors must be kept with even surfaces, and be clean. If hard and impervious, they should be well bedded. If porous, they must not be permitted to become foul. Foot and hoof troubles, lameness and foul skins develop in dirty stalls.

The air of stables must be pure. Any ventilating system which admits an abundance of pure air and allows the escape of foul air is a proper one. There is no one system suited to all stable designs. Muslin stretched across window openings, instead of glazed sash, makes a desirable covering. It at the same time allows the passage of air through its meshes and subdues the light from without.

Animals must be kept well groomed. It is an old adage that "grooming is half the feed." The skin of animals becomes dirty with dust from without and from the dried sweat and skin emanations from within the body. Unless accumulations are removed through grooming, the natural function of the skin is impaired and debility results. Dirty coats of animals afford desirable breeding places for vermin.

Proper light, pure air, suitable mangers and floors, together with cleanliness of stables and bodies, all tend toward the comfort of animals, and the less perfect these are, the more likely is it that the animals will be affected with abnormal sight, unhealthy skin, disordered respiration and impaired digestion, with all their consequent ills. It is necessary, therefore, to study carefully the comfort of animals, to insure good condition or physical fitness of their bodies.

Exercise.—Regularity of exercise in the open air is necessary for the health of all classes of animals. Animals closely confined in stables, even though well fed and watered, properly groomed and otherwise well cared for, will become soft, their body tissues more or less watery. They become less resistant to disease and less vigorous in every way.

The proper assimilation of food and bodily comfort is dependent upon proper exercise. The appearance of animals is deceptive in this respect. If a lot of young animals with similar treatment is divided and one part is allowed a paddock for exercise and the other part confined to stalls, the latter will usually appear to better advantage. They will be well rounded, smooth and apparently in prime condition, compared with the other lot, which is rough, rugged and more or less angular. As they mature, however, the lot which has been allowed to exercise in the open will continue a steady development to maturity, while the stalled lot will undergo a period of arrested development and fail totally in becoming large, robust, resistant animals. It is in the young and developing animals particularly that opportunity for exercise in the open should be given.

Failure to provide this has resulted, among other things, in the unnecessary susceptibility of horses to heaves, or cattle to tuberculosis, and of hogs to thumps, etc.

General Management.—Intelligent management of animals, therefore, may be said to consist of the following essentials:

1. An abundant supply of pure air at all times.
2. Proper food and water, regularly and judiciously provided.
3. Good grooming for all animals when stabled.
4. Proper exercise in the outside air.

Nursing.—In spite of intelligent management and due regard for the laws of hygiene, disorders and disease of the animal system will occur. Sick animals require intelligent care and greater attention to details of management even than do animals in health. Good nursing is of prime importance in the treatment of disease.

Sick animals should be placed in detached, well-ventilated and clean box stalls, conveniently located. Such stalls should be roomy, clean, cool and dry. In certain cases body clothing—blankets and bandages—are necessary.

All utensils, buckets, brooms, etc., used in the care of sick animals should be kept clean and should not be used in other parts of the stable.

All food not eaten should be removed from the sick animal and under no circumstances offered to other animals.

Bedding must be clean, sufficient in amount and comfortable for the patient.

Sick animals should be seen frequently, but should not be disturbed more nor oftener than is absolutely necessary.

Sick animals are more comfortable and improve more rapidly when the bowels are in a lax state. Mashes and soft feed tend to keep them in this condition. In addition to having laxatives, mashes, flaxseed tea, apples, carrots or potatoes are serviceable in catering to their appetites. Exposure for a short while daily to sunlight acts as a tonic to convalescent animals and enables them to regain strength rapidly.

Disease.—With the appearance of disease in an animal, it is essential that its true nature be speedily recognized or diagnosed. To this end there are observed the modifications in the external visible or otherwise accessible parts of the body which indicate the nature of the internal changes occurring.

These modifications are perceived through one or more of the special senses: sight, revealing alteration in size, conformation, color, etc.; sound, differentiating cavities and solid parts; touch, the texture, sensibility to pain, temperature variation, etc.; smell, the natural or modified odor and even the sense of taste, in milk examination for instance, serving an important end.

In addition to the immediate employment of the senses, the clinical thermometer gives accurately the internal temperature, and various tests are at the command of veterinarians for special examinations. It is necessary for the stockman to recognize health and the earliest approach of disease and be capable of applying the treatment prescribed. To do so, he must acquaint himself with a system of examination which will enable him to fairly well approximate the condition of the animal, as well as to secure information which, compared with later examinations, will show the *progress* of disease.

The modifications in form and function of the body are known as *symptoms*. By observing these the disease is located, and by them also its character is shown or a *diagnosis* made. For correctly diagnosing disease it is necessary that *all changes* be noted.

The following procedure is recommended to the stockman who should make written rather than mental notes, in order to have positive and complete information about the patient prior to a veterinarian's examination, if such proves to be necessary.

Examination of Sick Animals.—1. *Description oj Animal.*—This refers to the kind of animal, the sex, color, age, size and breed. This serves not only as a mark of identification, but such information may limit the diagnosis to certain diseases or may eliminate certain diseases from consideration.

2. *Characteristic Pose.*—The attitude of the patient, whether standing or lying down, and the particular positions assumed are to be noted. The mere pose of an animal is more or less significant in some diseases, *e. g.*, by rigidity of muscles, dilated nostrils, slightly extended tail and extension of the haw over the corners of the eyes in the standing horse, picture tetanus or lockjaw; the recumbent cow with muzzle at the flank, dull eyes, slow respiration and grating teeth, with history of calving within a few hours or days, designates calving fever or paralysis, etc.

The physical *condition* of the animal suggests the possibility of certain diseases, *e. g.*, azoturia, while *conformation* and *temperament* may point equally well to other diseases; *e. g.*, long-coupled, thin-barreled and long-legged horses are liable to scours.

3. *The Skin.*—The *condition* of the skin indicates in an accurate way the *condition* of the body. In its examination we must take into account the disposition of the hair, the action of the sweat glands, presence of enlargements or growths upon the skin, any changes in the color of skin and whether these are *confined* to the skin or are evidences of general disease.

4. *The Eye.*—An examination of the eye will indicate the volume and character of the blood, as seen in the visible capillary vessels. The color of the conjunctiva shows the condition of the animal and the character of its blood. The discharge of tears and swellings about the eyes should be noted as important to diagnosis.

5. *Temperature.*—The internal body temperature in health varies within certain narrow limits, the average being for—

Horses	100.0°–101.5° F.
Cattle	100.5°–102.5° F.
Sheep	102.5°–105.0° F.
Hogs	100.5°–104.0° F.

In diseases, these temperatures may range for—

Horses	102.0° F. and over
Cattle	103.5° F. and over
Sheep	104.0° F. and over
Hogs	104.0° F. and over

In all animals the temperature may rarely reach as high as 110° F., but life will soon terminate at such. The temperature must be accurately gotten with a thermometer inserted into the rectum for at least three minutes. The clinical thermometer registers only from 95° to 110° F. and is self-registering. This allows ample time for accurate reading and does away with the errors of estimating fever by the sensation of touch.

Temperatures should be taken throughout the course of the disease and should be taken at about the same hour, once or twice daily.

Fevers are measured by temperature and, in addition, by noting the accompanying chill, the uneven surface temperature, the alteration of pulse and respirations, the alteration of appetite and the general depression produced.

6. *The Pulse.*—The pulse or blood force in the arteries indicates the frequency or rapidity of circulation, its rhythm or regularity and its quality and character.

The normal pulse rate for animals is for—

Horses	30–40
Cattle	40–60
Swine	60–80
Sheep	70–80

The rapidity of circulation or pulse frequency varies and is easily influenced by age, sex, external temperatures, exercise, the digestive processes, and by disease.

The regularity of the pulse beat is greatly modified according to the state of health.

The quality or character of the pulse is determined by the resistance to pressure by the finger tips when placed over the accessible arteries.

7. *The Respirations.*—The examination of the respiratory system should be complete and thorough. The respirations are to be noted as to frequency, the manner in which they are produced and by the various chest sounds. The normal respiration of animals is as follows:

Horses	8–16
Cattle	10–30
Swine	10–20
Sheep	12–20

In health, the respirations are carried on noiselessly. There are certain physiological or normal noises, as the snort and the blowing sound made by horses when galloping.

On the other hand, with abnormal conditions, there is the *snoring* sound produced with the mouth partially open in semi-comatose animals from any cause; a *wheezing* sound from the nostril when polyps, tumors or thickening of the bones occur; *gargling* or *gurgling* sounds are produced when mucus is present; and, finally, grunting sounds occur when the abdomen is greatly distended.

The *breath* of animals in health is inoffensive. In disease it may become intensely disagreeable. It may indicate bad teeth, pus in the sinuses or chronic catarrhal conditions. Septic and gangrenous pneumonia is accompanied by foulness of breath.

The *nasal discharges* signify various conditions by their quantity, color, consistency, odor, and by the presence of particles of food, blood, etc.

They afford an excellent opportunity for examination of the quantity and the character of capillary blood circulation and characteristic evidences of particular diseases.

The cough is indicative of various conditions such as *heaves, bronchitis* or *pneumonia.*

8. *The Mouth.*—An examination of the mouth is of particular importance, inasmuch as it exposes to view mucous surfaces which are altered in some diseases. It allows an opportunity for judging age, by the characters upon the teeth; and further, the amounts of secretion present indicate the degree to which the secretory glands are disturbed.

9. *The Kidneys and Bowels.*—Direct examination of the kidneys and bowels is only safely conducted by experienced and trained men, but the stockman has an opportunity to examine the urine and the excrement. He should note the amount, color, consistency and any unusual odor of either. He should observe the frequency of the evacuations and whether they were made without causing distress.

Rational Measures for Treatment.—Not until after having made a critical examination of the sick patient is the stockman or attendant justified in the attempt to supply remedial measures.

If the condition of the patient justifies it, the services of a veterinarian should be secured promptly. If, on the other hand, there is no necessity for professional services, it is advisable that a comparison be made of the symptoms presented by the animal and the symptoms described in books on diseases of animals. When these are found to closely correspond, then, and only then, should the administration of medicines be begun. Many animals are destroyed or permanently ruined by unwise treatment. The eagerness "to do something" for these animals prevents proper deliberation and proper judgment, and the result is that the "cure is worse than the ill."

REFERENCES

"Common Diseases of Farm Animals." Craig.
"Care of Animals." Mayo.
"Diseases of Animals." Mayo.
Kentucky Expt. Station Circulars:
5. "A Remedy for Clover Bloat."
7. "Blackhead of Turkey."
Montana Expt. Station Bulletin 105. "Intradermal Test for Bovine Tuberculosis."
Ohio Expt. Station Bulletin 280. "Important Animal Parasites."
Canadian Dept. of Agriculture Bulletin. "A Plain Statement of Facts Concerning Tuberculosis."
Farmers' Bulletins, U. S. Dept. of Agriculture:
351. "Tuberculin Test."

Farmers' Bulletins, U. S. Dept. of Agriculture:
366. "Hookworm Disease of Cattle."
379. "Hog Cholera."
380. "Loco-Weed Disease."
439. "Anthrax."
449. "Rabies or Hydrophobia."
473. "Tuberculosis."
569. "Texas or Tick Fever."
666. "Foot and Mouth Disease."

CHAPTER 75

Diseases of Farm, Garden and Orchard Crops; and Their Remedies

By Dr. Mel. T. Cook

Plant Pathologist, New Jersey Agricultural Experiment Station

When any of the various parts of a plant are not doing their work properly the plant is said to be diseased. The disease frequently causes poor growth or poor fruit, or both; and in case of our cultivated plants, an unsatisfactory crop.

The most important causes of plant diseases are fungi, bacteria, slime moulds, parasitic flowering plants, insects, mites, nematodes, unsatisfactory soil, too much or too small amount of moisture, unfavorable temperature, gas fumes and smoke. Some plant diseases occur for which there are no satisfactory explanations.

Plant diseases may be detected by characteristic symptoms which readily distinguish the disease upon the healthy plants. The most common of these symptoms are: (*a*) a discoloration of the foliage and sometimes of the new growths; (*b*) wilting, frequently followed by yellowing and browning; (*c*) dropping of the foliage; (*d*) the formation of spots on foliage, stems or roots; (*e*) perforation of the foliage commonly called "shot hole;" (*f*) variegation of the foliage commonly called mosaic; (*g*) the "damping off" or dying which is especially common on seedling plants; (*h*) the blight or dying of leaves, twigs or stems; (*i*) the dwarfing of parts; (*j*) the increase in size of parts; (*k*) formation of galls, pustules or corky growths; (*l*) cankers on fruit, stems or roots; (*m*) abnormal fruits; (*n*) the formation of masses of small shoots called "witches' brooms;" (*o*) the curling of leaves; (*p*) the formation of leaf rosettes; (*q*) abnormal root growths commonly known as hairy root; (*r*) exudations of gums, resins, etc.; (*s*) the rotting of fruit, stems or other parts; and (*t*) sunburn of fruits and foliages.

Some diseases of the soil, such as "damping off," are very severe in seed-beds and in greenhouses, and can be controlled by sterilizing the soil. Diseases that occur in the soil in fields are frequently overcome by a rotation of crops, by improved drainage and sometimes by stimulating the plants with suitable fertilizer.

Many diseases are controlled by spraying, but in most cases spraying is used for the protection of plants against disease and not for curing them; therefore, it is a kind of insurance and must always be supplied in advance of the appearance of the disease. Spraying cannot be conducted in a satisfactory manner unless the grower is sufficiently familiar with the disease to understand when, why and how to give the necessary treatments. In recent years it has been found possible to overcome some diseases by growing plants that are disease-resistant and, therefore, do not need treatments.

In this chapter only the most common and important plant diseases in the United States and Canada are considered. Brief descriptions and condensed directions for treatment are given.

Farmers should always report the presence of disease on crops to the agricultural experiment station of the state in which they reside, and ask advice as to treatment. The treatment of some diseases will vary somewhat, dependent upon the part of the country in which it occurs.

APPLE

Bitter Rot or Ripe Rot (*Glomerella rufomaculans* [Berk.], Spaul and von Schrenk).—This rot is not confined to ripe apples and is not always bitter. It attacks both fruit and twig and occurs in orchard and in storage. On the fruit it appears as a brown, sometimes black, circular spot which gradually enlarges. It may be soft and wet or dry and corky, depending on variety of the fruit and age of the infection. The spore pustules start from the center of the spot and gradually spread over the surface, usually forming rather definite circles. They are pinkish in color and watery and spread the disease from fruit to fruit. Large spots become depressed and wrinkled and the entire fruit eventually becomes rotten, then dry and shrunken, and is finally known as a "mummy."

The disease may be carried from year to year on these mummies and also on the stems. On the twigs and branches it causes rough spots known as cankers. These cankers are rough and vary in size with age. The fungous spores from these infect the growing crop.

Treatment.—Remove and burn the mummied fruit and twig cankers. Spray with lime-sulphur before the buds open. After the petals fall, spray with self-boiled lime-sulphur or Bordeaux mixture. (See spray table on page 943.)

Black Rot (*Sphæropsis malorum*, Peck).—The rotten spot on the fruit is usually blacker and drier than the bitter rot spot and can be readily distinguished by the numerous black dots or papillæ from which masses of black spores emerge.

It also causes a stem canker in which the twigs become swollen, rough and black. On the trunk and larger branches it causes peculiar cankers. On young trees it causes a blight which is somewhat similar to the fire blight of the pear, but which can be readily distinguished by the presence of numerous small black dots. It also attacks the leaves, causing peculiar spots frequently spoken of as "frog eye."

Treatment.—Same as for bitter rot.

Brown Rot.—Usually not severe on the apple. (See Peach.)

Storage Rots.—The rots which occur in storage may be due to the preceding fungi or to a number of others. Thorough spraying of the orchards, careful handling of the fruit, regulation of temperature and humidity will reduce these rots to a minimum.

Scab (*Venturia inœqualsis* [Cke.], Wint.).—This is one of the most injurious diseases of the apple. It causes a dry, black spotting of the fruit which is well characterized by the name "scab." As the season advances the seriously infected fruits become distorted and cracked. Affected fruits are especially susceptible to storage rots.

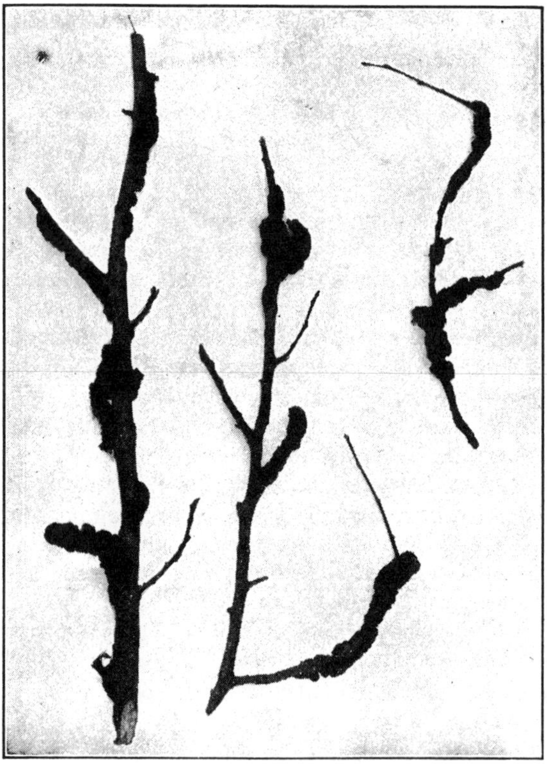

Apple Scab.[1]
Photograph by Prof. M. A. Blake.

The disease also attacks the leaves and twigs, causing a more or less thick, velvet-like covering, varying in color from olive-green to black.

Treatment.—Spray with concentrated lime-sulphur (5 quarts in 50 gallons of water) or Bordeaux mixture when the pink shows, but just before the blossom opens.

Blotch (*Phyllosticta solitaria,* Ell. and Ev.).—This disease causes dark, irregular blotches on the fruit and, when severe, causes a cracking. In the older spots a number of small, black, fruiting dots are formed. It also attacks the twigs, causing small tan-colored cankers. In the old cankers the bark becomes cracked and roughened.

Treatment.—Spray with lime-sulphur or Bordeaux mixture. (See table for apples, pears and quinces.)

Rust (*Gymnosporangium macropus,* Link.).—This disease attacks foliage, fruit and twig, causing a yellowish orange-colored spot which is not readily confused with other diseases. On the upper surface these spots show numerous small yellow pustules becoming black. On the

[1]Courtesy of New Jersey Agricultural Station.

under surface, in the late stages, are produced small, fringed, cup-like structures containing great masses of spores. These spores will not re-infect the apple, but are carried by the wind to neighboring red cedar trees, where they cause the formation of the familiar cedar apples.

These large brown cedar apples of the cedar, occurring in the spring, produce gelatinous, horn-like projections, bearing masses of spores. These spores are borne by the wind to the apple tree, which is re-infected with the disease.

Treatment.—Remove the cedar apples, or still better, remove the cedar trees. Spraying the apple trees as for scab will reduce the disease to some extent.

Fire Blight.—See Pear.

Other Foliage Spots and Twig Cankers.—There are leaf spots and twig cankers due to other causes which cannot be enumerated in this brief discussion. These diseases are all more or less injurious, but can be controlled by the regular spraying methods and sanitation.

Apple Tree with Typical Collar Blight.[1]
Showing proper method of cutting back into healthy bark before treating with paint.

Mildew (*Sphærotheca mali* [Duby], Burr.).—This fungus grows on the surface of the leaf, causing a grayish or whitish covering. Usually it is not severe and can be controlled by the regular spraying or by spraying with potassium sulphide.

Crown Gall and Hairy Root (*Bacterium tumefaciens,* Smith and Townsend).—These two diseases are due to the same organism. The crown galls or root galls occur at the crown or on the roots and sometimes on the stems. They are more or less spherical, with irregular, roughened surfaces. Some are hard and others soft, but they are all probably due to the same cause. They are most severe on red raspberries, are very injurious to peach

[1] Courtesy of Pennsylvania Agricultural Experiment Station, State College, Pa.

trees and more or less injurious to apple trees, dependent somewhat on the varieties. They also occur on pears, quinces, cherries, plums, grapes, roses and many other plants. The diseased tissues extend throughout a considerable part of the plant which makes cutting off of these malformations a very uncertain treatment.

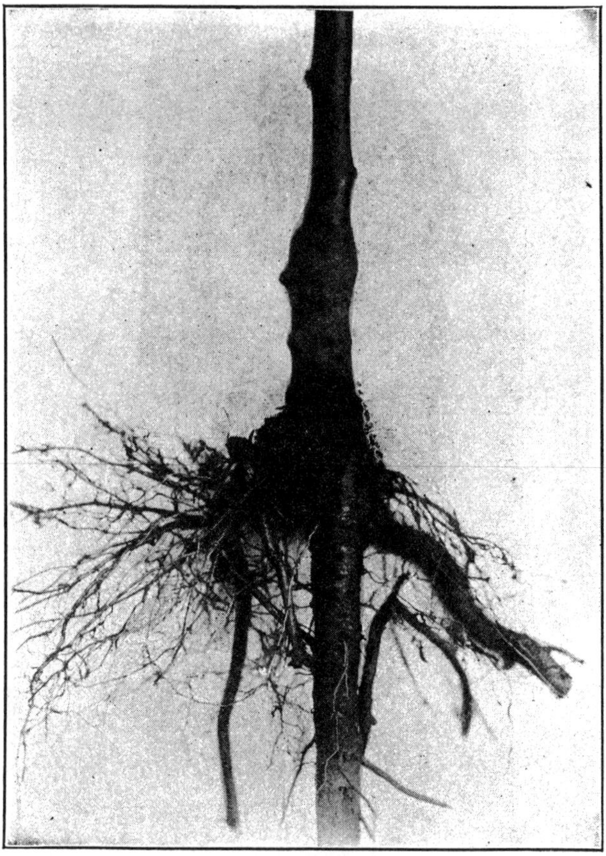

YOUNG APPLE TREE FROM NURSERY.[1]
Showing the disease known as Root Gall.

The hairy root appears underground as a mass of fibrous roots and above ground as warty knots on trunk and branch, and is sometimes mistaken for cankers, due to other causes.

Treatment.—The organism which causes this disease lives in the soil for several years, and cannot be eradicated except by a long rotation of crops. It is unwise to set orchards, especially peach orchards, in old berry fields or other fields known to be infected or to use berries as inter-row crops in orchards. Nursery stock known to be infected should be destroyed.

PEAR

Blight (*Bacillus amylovorus* [Burr], De Toni).—This very familiar disease causes the leaves and young twigs to die and blacken very much as though injured by fire. These dead leaves hang on the trees during the winter instead of falling in the autumn, as is the case with healthy leaves. The disease also attacks the branches, causing black, sunken cankers from

[1] Courtesy of The Field, New York.

which a sticky, milky fluid oozes in the early spring, and from which the disease is spread, by means of insects, to the opening blossoms. If the weather conditions are favorable the blossoms and fruit spurs die and blacken and very frequently considerable quantities of the young fruit are destroyed. The disease also occurs on the apple, crab, hawthorn and other related trees.

Treatment.—Prune and burn the diseased twigs on young trees; clean out the cankers on old trees, dipping the knife from time to time in formaldehyde (1 part in 20 parts water). Paint these wounds with formaldehyde and then with white lead paint or coal tar. Do not over-fertilize or over-cultivate the orchard.

SPRAY TABLE FOR APPLES, PEARS AND QUINCES.

TIME.	MATERIAL.	PURPOSE.
1. Before the buds swell.	Concentrated commercial lime-sulphur, 1 part in 9 parts water or home-made concentrated lime-sulphur diluted to a specific gravity of 1.03.	For fungous diseases and for San José scale.
2. As soon as the flower buds show the pink color.	Concentrated commercial lime-sulphur, diluted to 5 quarts in 50 gallons of water or 1–40, or home-made concentrated lime-sulphur diluted to specific gravity of 1.007. (Bordeaux mixture can be used for this treatment.)	For scale and fungi.
3. Immediately after petals fall.	Same as 2.	Same as 2.
4. Ten days after blossoms fall.	Same as 2.	Same as 2.

Two additional sprayings are frequently necessary for fall and winter varieties.
Arsenical poisons for chewing insects and tobacco extracts for sucking insects may be added to treatments 2 and 3.

Rust (*Gymnosporangium juniperi-virginianae* and *G. blasdaleanum* [D. and H.], Kern).—Similar to apple rust.

Scab (*V. pyrina*, Aderh.).—Similar to apple scab.

Leaf Spot (*Septoria pyricola*, Desm.) appears as numerous small, well-defined, angular, ashy-colored spots with minute black dots. It is not often severe.

Leaf Spot (*Entomosporium maculatum*, Lev.) occurs on the leaf, causing small, circular spots with dull, red centers and dark borders. When severe it causes the leaves to become yellow or brown and fall. It also attacks the fruit, causing spots which are at first red, becoming dark and in severe cases causing the fruit to crack. It is carried over the winter on the fallen leaves.

Treatment.—This disease can be controlled by spraying with Bordeaux mixture, beginning when the leaves are about half or two-thirds full grown and repeating at intervals of three weeks until four treatments have been given.

Rots.—The black rot and brown rot also occur on the pear. (See Apple.)

Crown Gall.—See Apple.

QUINCE

Rust (*G. clavipes*, C. and P.).—This disease is very similar to the rusts on apples and pears, but is more severe on the fruit and twigs than on either of the preceding. It also has the cedar for its alternate host.

Blight.—See Pear.

Leaf Spot.—See Pear.

Rots.—See Apple.

Crown Gall.—See Apple.

PEACH

Brown Rot (*Sclerotinia fructigena* [Pers.], Schroet.).—This is one of the most destructive diseases of the peach. It attacks the fruit as

PEACHES ENTIRELY DESTROYED BY BROWN ROT.[1]
Showing gray masses of spores of fungus.

it is approaching maturity, causing it to rot, become brown, soft and useless. The fungus produces an abundance of spores which form a dense brown, powdery mass over the fruit. It also attacks the blossoms, causing them to die, turn brown and fall soon after opening. It then spreads to the twigs, causing death of the young shoots and causing cankers on the older branches.

Treatment.—See spray table for peach.

Scab or Freckles (*Cladosporium carpophilum*, Thuem.).—This extremely common disease attacks the fruit, causing sooty, black specks or blotches

[1] From Farmers' Bulletin 440, U. S. Dept. of Agriculture.

which, when severe, may prevent normal ripening and cause the fruit to be irregular in shape and to crack.

Treatment.—See spray table for peach.

Leaf Curl (*Exoascus deformans* [Berk.] Fckl.).—This very familiar and very injurious disease causes the leaves to curl, reduces their value to the tree and finally causes them to fall. With the appearance of the second crop of leaves, the growers frequently suppose the tree to have recovered. However, it has lost in vitality and vigor, which results in a reduction or complete loss of the crop.

Treatment.—Spray with lime-sulphur before the buds open.

Shot Holes (*Cercospora circumscissa,* Sacc., *Phyllosticta circumscissa,* Cke.).—These "shot hole" diseases are quite common, but readily controlled by the regular spraying treatments.

Another shot hole (*Bacterium pruni,* Smith) is very common in the Southern states and especially on Elbertas. It frequently causes the foliage to fall in midsummer. It also attacks the fruit, causing a spotting somewhat similar to the scab. It cannot be controlled by spraying.

Crown Gall.—See Apple.

Mildew (*Sphærotheca pannosa* [Wallr.], Lev.).—Similar to the mildew of the apple. It is of little importance and can be controlled by the regular spray treatment. (See table for peach.)

Yellows.—The cause of this very destructive disease remains a mystery. In its earlier stages it causes a premature ripening of the fruit, accompanied by a red blotching over the surface and through the flesh which is usually insipid and frequently bitter. (Prematuring may also be caused by borers or winter injury.) In its later stages it causes the so-called "willowing" or formation of slender yellowish-green shoots on the trunk and larger branches. The leaves on these shoots are small, narrow and greenish-yellow. The foliage is frequently greenish-yellow, but when supplied with nitrogenous fertilizers will not show this character. In its earlier stages, one part of the tree may show the disease and the other parts appear perfectly healthy, but in fact the entire tree is diseased. It can be transmitted from tree to tree by contact and to young trees by budding. Buds from the apparently healthy parts of very slightly diseased trees will transmit the disease. Healthy nursery stock is of the greatest importance.

Treatment.—Dig and burn the trees as soon as the disease appears, using care to prevent the tree coming in contact with others. Young trees can be set in the places from which the old ones were removed; the disease does not persist in the soil. The greatest care should be used in the selection of bud wood, to insure its freedom from disease.

Little Peach.—The cause of this disease is also unknown, but it is of the same nature as yellows. The fruit of diseased trees is small, ripens late, is inferior in quality, frequently insipid and watery. The leaves are frequently lighter than normal leaves or yellowish-green and often rolled and drooping.

Treatment.—Same as for yellows.

Peach Rosette.—The cause of this disease of the peach in the Southern states is also unknown. It is very similar to yellows, but the leaves tend to cluster, giving the general appearance of green roses.

Treatment.—Same as for yellows.

SPRAY TABLE FOR PEACH.

TIME.	MATERIAL.	PURPOSE.
1. Same as first treatment for apple. (See page 943.)		
2. Just as the husks fall from the small fruit.	Self-boiled lime-sulphur.	For brown rot, scab and other diseases.
3. Three weeks after 2.	Same as 2.	Same as 2.
4. Three weeks after 3, for late varieties.	Same as 2.	Same as 2.
5. Same as 2 for very late varieties.		

NOTE.—Arsenical poisons may be added to No. 2 for curculio. Tobacco extrac' s and soap can also be added for sucking insects.

PLUM

Black Knot (*Plowrightia morbosa* [Schw.], Sacc.).—This very common and well-known disease causes swollen growths on the branches which are at first olivaceous in color, but finally become deep black and very hard and brittle. It will spread over the greater part of a tree, interfere with its growth and finally cause its death.

BLACK KNOT ON THE CHERRY.[1]
Photograph by Prof. J. P. Helyar.

Treatment.—The diseased parts should be cut out and burned and the trees should be sprayed with lime-sulphur in the spring before the opening of the buds. (See table for plum.)

Leaf Spot.—See Cherry. **Mildew.**—See Cherry. **Yellows.**—See Peach. **Brown Rot.**—See Peach. **Crown Gall.**—See Apple.

[1]Courtesy of New Jersey Agricultural Experiment Station.

SPRAY TABLE FOR PLUM.

TIME.	MATERIAL.	PURPOSE.
1. Same as for apple and peach.		
2. Immediately after the petals fall.	Self-boiled lime-sulphur.	For brown rot.
3. When fruit s about the size of green peas.	Same as 2.	Same as 2.
4. Three weeks after 3.	Same as 2.	Same as 2.

NOTE.—Arsenical poisons for control of curculio may be added to No. 2. Tobacco extracts and soap may be added for control of plant lice.

CHERRY

Leaf Spot (*Cylindrosporium padi*, Karst.).—This disease, which is also called "shot hole," causes discolored circular spots usually with reddish or purple border, eventually becoming brown and breaking into a hole and often resulting in a defoliation of the tree. When severe it is very injurious to the growth and health of the tree.

Treatment.—Spray with lime-sulphur or with Bordeaux mixture.

Black Knot.—See Plum.

Crown Gall.—See Apple.

Brown Rot.—See Peach.

Powdery Mildew (*Podosphæra oxyacanthæ* [D. C.], DeBy.).—Similar to *S. mali* of the apple.

SPRAY TABLE FOR SWEET CHERRIES.

1. Treatment same as first treatment for apple.
2. Treatment same as second treatment for peach, but given immediately after the petals fall.
3 Repeat treatment 2 when fruit is about the size of small pea.
4. Repeat treatment 2 after the fruit is picked.

CITRUS FRUITS

Brown Rot (*Pythiacystis citrophthora*, Sm. and Sm.).—This disease of the California fruit is sometimes the cause of heavy losses. It is caused by a fungus which is white in mass. It occurs in the orchard and packing houses causing a rot with a peculiar rancid odor. Very slightly infected fruits rot during transportation. It is most abundant in wet weather or on low ground after irrigation.

Treatment.—It is spread by contact and can be controlled by using a heavy straw mulch or cover crop under the trees and by disinfecting the fruit in the packing house.

Black Rot (*Alternaria citri*, Pierce).—This disease of the navel orange causes a premature ripening. It usually enters the fruit through the navel, causing an internal rot accompanied by a reddish color. The diseased fruit should be burned or buried.

Stem End Rot and Melanose (*Phomosis citri*, Fawcett).—This disease is most common on mature packed fruit, causing a circular patch of soft rot at the stem end which can be detected by a pressure of the finger even though there may be no discoloration. The presence of scale insects and warm, damp weather tend to increase the disease.

This organism also causes the disease known as melanose of the fruit, twig and leaf. This form of the disease appears as a raised brownish area forming dots, lines and crosses, varying from yellow to brown and black. The cutting out of the dead wood is an important factor in the control of this disease.

Other Rots (*Penicillium italicum,* Wehm., and *P. digitatum* [Fr.], Sacc.).—These rots are covered by the fungus and appear as blue moulds. They are the causes of heavy losses in transportation. The fungus enters the fruits through slight wounds and therefore the fruit should always be handled carefully.

Sooty Mould (*Meliola camelliæ* [Catt.], Sacc.).—In this case the fungus covers the fruit with a black velvety coating which can usually be removed. It is not nearly so serious as some other diseases. It really grows on the exudations (honey dew) of plant lice and its control depends on their destruction.

Black Pit of the Lemon (*Bacterium citriputeale,* Sm.).—This disease appears as circular or oval, well-defined, reddish-brown, brown or black spots or pits on the fruit. They are caused by bacteria which gain entrance through wounds.

Anthracnose or Wither Tip (*Colletotrichum glœosporioides,* Penz.).—This disease attacks the young leaves, twig tips and fruits. It causes a yellowish spotting of the leaves, a withering and dying of the new shoots and canker-like spots on the fruit. It is one of the most common diseases of the citrus fruits.

Scab (*Cladosporium citri,* Mass.).—This very common disease attacks leaves, twigs and fruits, causing prominent warty or corky outgrowths. The leaves are frequently twisted and twigs are frequently cracked.

Canker.—This is a comparatively new disease in America and there is some difference of opinion as to the cause. It is very destructive and a very vigorous campaign is being made against its spread. It occurs on leaf, twig and fruit, causing dead, circular spots which are usually raised. They are light-colored when young, but become brown and corky and frequently marked with small cracks.

Other important diseases of the citrus fruits are the scaly bark or nail head rust (*Cladosporium herbarum var. citricolum*) of Florida, the citrus knot (*Sphæropsis tumefaciens,* H. and T.) of the West Indies, and the gummosis, which is very widely distributed.

Treatment of Diseases of Citrus Fruits.—So much progress is being made in the study of these diseases at this time that it is inadvisable to attempt a discussion as to treatment. Those interested in these diseases should consult with the agricultural experiment station in the state in which the disease occurs.

FIG

Rust (*Kuehneoal fici* [Cast.], Butler).—This very common disease causes numerous rusty red spots on the lower surface of the leaves. When severe the trees are almost defoliated. It can be controlled by frequent spraying with Bordeaux mixture.

Cankers (*Libertella ulcerata,* Massee).—This disease is sometimes severe where figs are grown under glass. It starts as small radiating cracks which develop into cankers, sometimes completely girdling the branch and causing the death of the parts beyond the point of attack.

Another canker (*Tubercularia fici*) causes a shrinking and drying out of the tissues surrounding the fruit scars, followed by a drooping of the dead parts.

Fruit Rots (*Glomerella rufomaculans* [Clint], Sacc.).—This disease is due to the same organism as the bitter rot of the apple. It causes sunken, rotten spots, usually covered with a whitish growth and later by numerous pustules of salmon-pink colored spores. If the fruit is attacked when young, it becomes dry and hangs on the tree.

Ripe fruit rots may be due to various fungous organisms.

Other important diseases of the fig are the yellow rot (*Fusarium roseum,* Lint), the leaf spot (*Cercospori fici,* H. and W.) and the limb blight (*Corticuem lœtum,* Karsten).

PINEAPPLE

The pineapple is subject to several diseases, all of which should receive more attention. Growers who have reason to complain of these troubles should consult with the state agricultural experiment station.

MANGO

Bloom Blight (*Colletotrichum glœosporioides,* Penz).—This most severe disease of the mango is due to the same organism that causes the wither top of the orange. It attacks the blossoms, causing them to turn black and fall. Unfortunately, the blooming is during the rainy season, which makes spraying impractical.

AVOCADO

Leaf Spot (*Colletotrichum glœosporioides*).—This is due to the same organism as the wither top of the orange. It is frequently so severe as to cause a heavy loss of foliage. It also attacks the fruit, frequently causing a pronounced cracking.

OLIVE

Olive Knot (*Bacterium savastanoi,* Smith).—This disease originates as irregular, more or less hemispherical swellings on trunk, branches and leaves. They are firm and fleshy, but finally become woody and crack. Badly infected trees frequently die as a result of this disease.

BLACKBERRY, DEWBERRY AND RASPBERRY

Crown Gall.—See Apple.

Leaf Spot (*Septoria rubi*, West) occurs on the leaves of these bush fruits, causing small white or ash-colored spots with brown or reddish margins. Close examination shows very small black dots in each spot. It is frequently the cause of considerable damage.

Treatment.—Spray in the spring with Bordeaux mixture.

Anthracnose (*Glœosporium venetum*, Speg.).—This disease attacks the young canes of these fruits, causing small purplish spots which enlarge and become grayish or dirty white in the centers. When severe, it causes the canes to crack and die, the leaves to be dwarfed and the fruit to ripen prematurely. The disease also occurs on the leaves, causing them to develop unequally.

Treatment.—Cut and burn the diseased and dead canes soon after picking the fruit. Spray with Bordeaux mixture in the spring and also as soon as possible after the berry season.

Orange Rust (*Gymnoconia interstitialis* [Schlecht], Lagh.).—This disease is very abundant in the spring of the year, causing a dense coating of red rust (spores) on the under surface of the leaves. The fungus grows within and may spread throughout the entire plant.

Treatment.—Dig and burn the entire plant. Spraying with Bordeaux mixture will prevent the infection of healthy plants, but is not a practical treatment.

Double Blossom (*Fusarium rubi*, Wint.).—This disease is especially abundant on the Lucretia dewberry, the black diamond or Brazil blackberry and also occurs upon other varieties of blackberries and dewberries. The fungus lives within the buds, causing them to form witches' brooms of slender shoots with deformed or double flowers producing little or no fruit. The infection of the new buds which are forming for the next year occurs when the diseased flower buds are opening.

Treatment.—The disease can be greatly reduced by picking these deformed leaf buds soon after they open and before the opening of the flower buds. Select plants so far as possible from fields free from the disease.

Cane Blight (*Coniothyrium fuckelii*, Sacc.).—This disease of the raspberries attacks the canes, causing them to be lighter in color, with smoke-colored patches. The foliage of diseased canes wilts and dies very much as from drought. The disease penetrates wounds, frequently those made by pruning. It is readily distributed in nursery stock and will persist in the soil for several years.

Treatment.—Rotate the crops and use only healthy plants for setting.

Yellows.—This disease is confined to the raspberries. It resembles peach yellows and should be treated in the same manner.

STRAWBERRY

Leaf Spot (*Sphærella fragariæ* [Tul.], Sacc.).—This is the most prominent of the diseases of the strawberry. It causes small leaf spots with white or ashy centers and purple or red borders. These spots frequently unite, forming irregular blotches. It reduces the vigor of the plant and, therefore, the quality and quantity of the fruit.

Treatment.—The soil should be well drained and rotation of crops practiced. Cutting over the beds and burning of the tops is advantageous. Spraying with Bordeaux mixture will reduce the disease to the minimum, but there is doubt as to whether this treatment will prove profitable.

CRANBERRY

Scald or Blast (*Guignardia vaccinii*, Shear)—This fungus attacks the blossoms or the fruit soon after the falling of the blossoms, causing the berry to shrivel and turn black. This form of the disease is known as the "blast." The form known as "scald" causes small, soft, light-colored, watery spots which spread over the entire fruit, sometimes causing a zone effect. The disease also attacks the leaves, causing irregular, reddish-brown spots. The disease is likely to become worse from year to year and to prove very destructive.

Treatment.—Careful irrigation, raking and burning of the dead material and sanding the bogs are advantageous. Selection of resistant strains for planting, and spraying with Bordeaux mixture about five times during the season will control the disease.

Rot (*Acanthoshynchus vaccinii*, Shear).—This fungus causes a rot very similar to the scald and can be controlled by the same treatment.

Anthracnose (*Glomerella rufomaculans* [Berk.] Spaul. and von Schrenk; *vacinii*, Shear).—This fungus is the same as the one causing the bitter rot of the apple. It causes a rot very similar to and almost indistinguishable from the scald and can be controlled by the same treatment.

GOOSEBERRY

Powdery Mildew (*Sphærotheca mors-uvæ* [Schw.] Berk. and Curt.)—Very similar to the mildew on the apple and other fruits and should receive the same treatment.

CURRANT

Anthracnose (*Pseudopeziza ribes*, Kleb.).—This very common disease of the currant attacks the leaves, causing many small brown or black spots, followed by a yellowing of the entire leaf which falls prematurely. The disease also attacks the canes and the fruit, causing small black, sunken areas. It is carried from year to year on the canes.

Treatment.—Remove the old canes and spray with Bordeaux mixture in the spring just before the buds open, again after the leaf buds open and then at intervals of three weeks until the fruit is two-thirds grown.

NOTE.—There are several other leaf diseases of the gooseberry and currant that may occasionally prove injurious or destructive and should be treated as the circumstances may demand.

GRAPE

Black Rot (*Guignardia bidwellii* [Ell.], V. and R.).—This very common disease of the grape attacks the fruit, causing a black rot followed by a shriveling and drying into a hard, wrinkled mummy. It occurs on the leaves and young shoots earlier than on the fruit and causes tan-colored spots with minute black dots in the centers.

Treatment.—Spray with Bordeaux mixture before the opening of the flower-buds, and again after the setting of the fruit. Additional spraying should depend on the weather; in dry seasons it may be necessary to spray every three weeks until the fruit is two-thirds grown.

Bird's Eye or Anthracnose (*Sphaceloma ampelinum,* DeBy.).—This disease is not nearly so severe as the black rot. It attacks the fruit, causing brown or black spots with sunken centers and red borders. On the canes it causes similar spots, but as they approach maturity the centers become ashy in color and the edge dark.

Treatment.—Badly diseased canes should be cut out and burned. Spray treatment same as for black rot.

Bitter Rot or Ripe Rot.—This is the same as on the apple but its attack is confined to the ripe fruit.

Treatment.—Spray with ammoniacal copper carbonate solution.

Downy Mildew (*Plasmopara viticola* [Berk. and Curt.], Berl. and De Toni.).—This fungus causes whitish and finally brownish areas in the leaf, followed by a very perceptible downy growth on the lower surface. It sometimes causes the death of the entire leaf, shoot or vine. It is especially severe on the European varieties. It sometimes attacks the fruit, causing the gray or brown rot.

Treatment.—Same as for black rot.

Powdery Mildew (*Uncinula necator* [Schw.], Burr.).—This fungus is very similar to the powdery mildew of the apples and other fruits. It attacks all parts of the plant above ground, and occurs on both upper and lower surfaces of the leaves, causing circular, whitish, powdery spots which frequently unite and cover the entire leaf. It also attacks the fruit, causing it to develop irregularly, fail to develop or to fall. It is especially common in vineyards where the vines are too closely set and on vines grown under glass.

Treatment.—Spray with potassium sulphide or when the temperature is above 75° F., sprinkle the vines with flowers of sulphur.

Necrosis (*Fusicoccum viticolum,* Reddick).—This disease causes a dwarfing of the new shoots and leaves and when severe a shriveling up and dying.

Treatment.—Dig and burn all diseased vines, and spray with Bordeaux mixture.

Crown Gall.—See Apple.

NOTE.—A number of other minor diseases will be controlled by the treatment prescribed for the rot.

ASPARAGUS

Rust (*Puccinia asparagi*, D. C.).—This fungus causes the tops of the plants to redden soon after blossoming. The leaves turn yellow and fall, and the stems show numerous small blisters containing masses of rust-colored powder (spores). Later in the season these pustules break and become black in color. The disease spreads rapidly and causes heavy losses.

Treatment.—Cut and burn diseased plants as soon as observed. Spray with Bordeaux mixture.

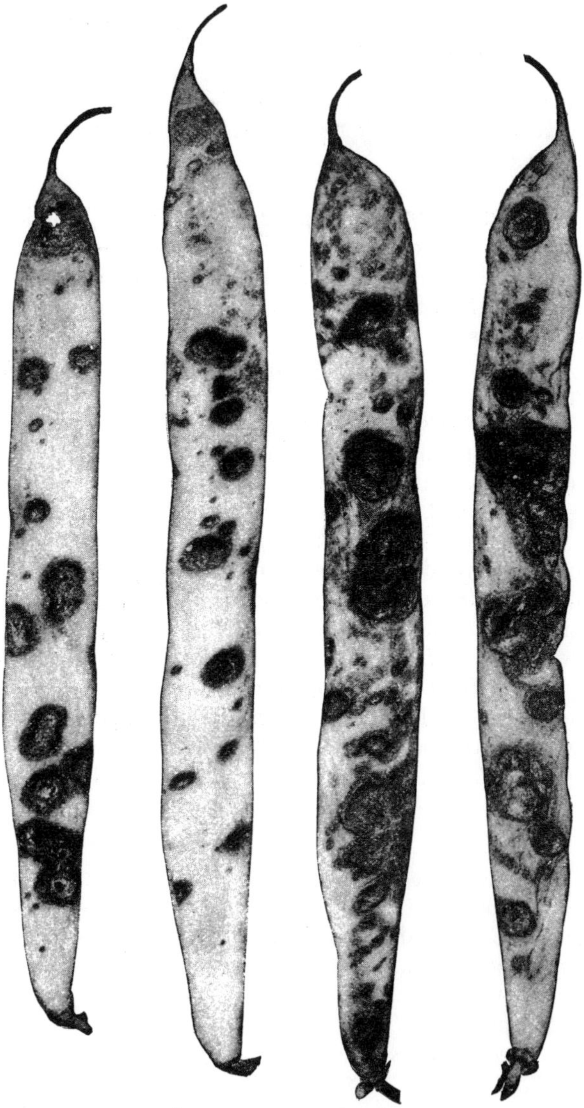

ANTHRACNOSE OF BEAN.[1]

The brown spots occur on both the pods and plants. They are caused by spores coming in contact with the tender plant tissues, where they germinate and give rise to serious damage.

[1] Cornell Agricultural Experiment Station Bulletin 255.

BEAN

Anthracnose (*Colletotrichum lindemuthianum* [Sacc. and Magn.], B. and C.).—This disease is most severe on the wax beans. It occurs on the pods, causing unsightly, dark-colored, sunken, canker-like spots. It also attacks the leaves and stems, producing similar spots and frequently causing the death of the plants. The fungus is carried in the seed and one diseased seed in a thousand is enough to infect a large number of growing plants.

Treatment.—Select clean seed.

Rust (*Uromyces appendiculatus* [Pers.], Link).—This fungus causes minute rusty spots or blisters on the under surface of the leaves and occasionally on the pods. These blisters break and set free great quantities of the reddish or rust-colored spores. It is not so severe as the anthracnose.

Treatment.—Practice clean cultivation and burn all old vines in the fall.

Blight (*Pseudomonas phaseoli*, Smith).—This disease attacks leaves, stems and pods, causing large watery areas, which later become dry, brown and papery. It is carried from year to year in the seed.

Treatment.—Use seed from healthy plants.

Downy Mildew (*Phytophthora phaseoli*, Thaxt.).—This disease is unlike the mildew on the fruits. It attacks the pods of lima beans, causing irregular areas of dense, woolly-white growth. It also occurs on other parts of the plant, causing dwarfing and irregular growths.

Treatment.—Spray with Bordeaux mixture.

Leaf Spot (*Phyllosticta sp.*).—This disease is most severe on the pole lima beans. It causes an irregular spotting of the leaves and to some extent of the pods. It is carried from season to season in the seed.

Treatment.—Spray with Bordeaux mixture.

PEA

Spot (*Ascochyta pisi*, Lib.).—This disease causes spots on stems, leaves and pods which are most conspicuous on the latter. On the pods they are circular, sunken with dark borders and pale centers, becoming pink when mature. The spots on the leaves are oval and usually show concentric circles. When severe on the stems it causes wilting and death of the plant.

Treatment.—Select clean seed and rotate crops.

BEET

Leaf Spot (*Cercospora beticola*, Sacc.).—This fungus causes the very common circular, brown, purple-bordered spots with ash-colored centers.

Treatment.—Spraying with Bordeaux mixture will control this disease.

Root Rot (*Rhizoctonia betæ,* Kuhn).—This disease causes the outer leaves to turn black and fall. As the disease advances the roots crack and then rot from the crown downward.

Treatment.—Use lime and rotate crops.

Scab.—See Potato.

CABBAGE, CAULIFLOWER, TURNIP, ETC.

Black Rot (*Pseudomonas campestris,* Pammel).—This is a bacterial disease which attacks all of the above and many related plants. It starts at the edges of the leaves, causing a blackening of the veins, gradually working downward to the main stalk and then upward and outward until the entire plant is affected. The affected leaves become yellow, wilt and then dry. In advanced stages the disease is accompanied by other rot organisms which cause a pronounced odor.

Treatment.—When once in the soil it is extremely difficult to eradicate. Prevent infection by using clean seed, which as a precautionary measure should be soaked for fifteen minutes in formaldehyde (1 part formalin to 30 parts water).

ENLARGED ROOTS OF CABBAGE CAUSED BY NEMATODES.[1]

Club Root or Finger and Toe Disease (*Plasmodiophora brassicæ,* Wor.).—This very destructive and well-known disease attacks cabbage and related plants, causing unsightly knotted roots. The diseased plants are dwarfed and fail to develop heads.

[1] From Farmers' Bulletin 488, U. S. Dept. of Agriculture.

Treatment.—Use nothing but absolutely clean soil in the seed-beds; use lime in the fields; rotate crops.

CANTALOUPES AND MELONS

Leaf Blight (*Alternaria brassicæ* [Berk.], Sacc. *var. nigrescens*, Pegl.).—This disease starts as small brown spots with concentric rings, which enlarge, unite and frequently cause the destruction of the entire leaf. The melons ripen prematurely and are soft, wilted and insipid.

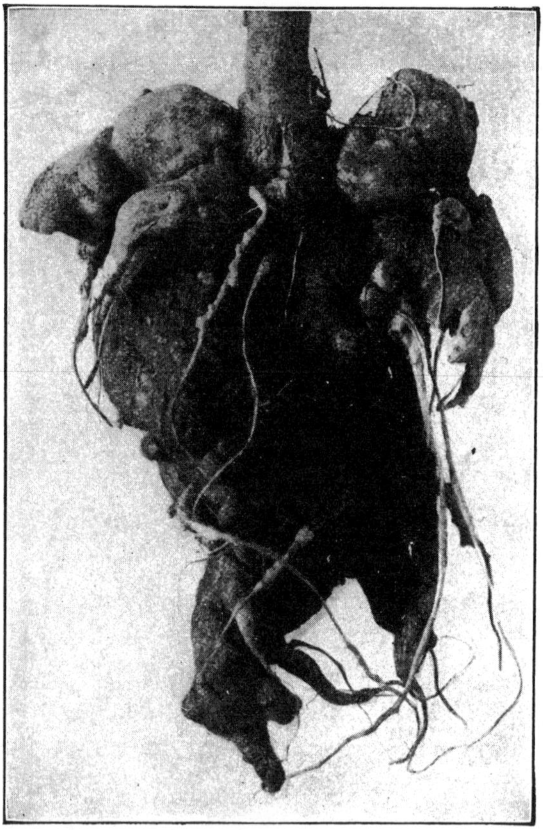

Enlarged Roots of Cauliflower Caused by the Club-root Organism.[1]

Treatment.—Rotate crops and spray with Bordeaux mixture.

Downy Mildew.—See Cucumber.

Anthracnose (*Colletotrichum lagenarium* [Pass.], Ell. and Halst.).—This disease attacks all parts of the vines of cucumber and squash, but is most injurious to watermelons, causing the buds and tendrils to die and turn black and the leaves to turn yellow. It is very noticeable on the fruit, causing sunken canker-like spots with pinkish centers. When the young fruit is attacked it is likely to have a bitter flavor.

Treatment.—Spray with Bordeaux mixture.

Wilt (*Bacillus tracheiphilus*, E. F. Smith).—This disease of melons and cucumbers, and sometimes of pumpkins and squash, may start with the central stem, causing the entire vine to wilt and die quickly, or it may start with a branch and work slowly back to the central stem.

Treatment.—Rotation of crops; avoid those that are susceptible.

[1] From Farmers' Bulletin 488, U. S. Dept. of Agriculture.

CUCUMBER

Downy Mildew (*Pseudoperonospora cubensis*, B. and C. Rost.).—This disease causes yellow, angular spots on the older leaves and eventually causes the entire leaf to turn pale and die. Considerable quantity of white growth appears on the under surface.

Treatment.—Spray with Bordeaux mixture.

Anthracnose.—See Melon.

Leaf Blight and Fruit Spot (*Cladosporium cucumerinum*, Ell. and Arth.).—This disease appears on the leaves as water spots and finally causes the entire leaf to wilt and rot. On the fruit it appears as minute gray, sunken, velvety spots which frequently unite and finally become black.

Treatment.—Prompt spraying with Bordeaux mixture.

Wilt.—See Melon.

CELERY

Leaf Spots (*Cerospora apii*, Fr., and *Septoria petroselini*, Desm., *var. apii*).—There are two leaf spot diseases which can be controlled by spraying with Bordeaux mixture. The first treatment should be while the plants are in the seed-bed and should be given whether the plants do or do not show the disease. Other sprayings should be at intervals of two weeks and with a high pressure sprayer. If necessary to spray late in the season, the last treatment should be with ammoniacal copper carbonate solution.

ONION

Smut (*Urocystise cepulæ*, Frost.).—This very destructive disease attacks the young plants, causing dark opaque spots on the leaves. The leaves finally die and dry up and the spots burst and permit the escape of masses of spores. The mature bulbs show black masses of spores in the outer and sometimes in the inner leaves, and when badly infected dry and rot. Sets and young onions when well started are practically immune from the disease. The spores persist in the soil for many years.

Treatment.—Prevent the introduction by using clean sets. Use lime and long crop rotations for infected soil. In small plantings disinfect the soil with formaldehyde.

Downy Mildew or Blight (*Peronospora schleideni*, Ung.).—Diseased plants have a tendency to develop a violet tint by which they can be recognized at a distance. As the disease advances, they become covered with a mouldy coating and finally collapse. Slightly affected plants may recover under suitable weather conditions. The disease spreads rapidly in damp, warm weather and on wet land.

Treatment.—See that lands are well drained. Rotate crops. Spray with Bordeaux mixture.

CARROT

Soft Rot (*Bacillus carotovorus*, Jones).—This is a bacterial disease which causes a soft rotting of the roots. It also attacks turnips, radishes, parsnips, onions, celery, beets and many other plants. The only satisfactory treatment lies in the rotation of crops.

PARSNIP

Blight (*Cercospora apii*, Fr.).—See Celery.

POTATO

Late Blight or Downy Mildew (*Phytopthora infestans* [Mont.], De By.).—This disease usually starts near the tip or margin of the leaf, but causes the infected area to die and blacken. In cool, wet, cloudy weather it spreads very rapidly and causes an offensive odor. The diseased tubers may show slightly depressed, dark-colored areas and a dirty brown color within. The disease is frequently the cause of heavy losses by rotting.

Treatment.—Spray with Bordeaux mixture, beginning when the plants are about six inches in height and repeat about every two or three weeks throughout the growing season.

Early Blight (*Alternaria solani* [E. and M.], J. and G.).—This disease appears earlier in the season than the late blight. It causes brown, brittle, irregular, more or less circular leaf spots with rather definite concentric circles. These spots frequently unite and the plant dies very much as though from natural causes.

Treatment.—Same as for late blight.

Wilt, Stem Rot and Dry Rot (*Fusarium oxysporum*, Schlecht).—The plant assumes an unhealthy appearance, the leaves roll and curl and the plant falls and dies prematurely. The stems are partly or entirely black and dead near the base and frequently show a white or pink mould. When stems are cut across below the ground they show discolorations just below the surface. This field form of the disease is known as "wilt" or "stem rot."

In storage the tubers undergo a "dry rot" beginning at the stem end, which causes them to shrivel and become light in weight. When cut across, these tubers show black discolorations just below the surface. The disease can be carried on the seed and will also persist in the soil.

Treatment.—Select seed potatoes which are free from surface cankers and are perfectly white when cut. When the soil becomes infected use rotation of crops for from three to five years.

Black Leg (*Bacillus phytophthorus*, Appel).—This disease causes the plants to be dwarfed, erect, pale in color and to die early. The stems become brown or black near the ground and the disease works downward. It is carried in the seed.

Treatment.—Soak the seed in formaldehyde or corrosive sublimate as recommended for potato scab.

Scab (*Oospora scabies*, Thaxt.).—This well-known disease is readily recognized by the rough, pitted character of the tubers and is the cause of heavy losses. It can be carried on the seed and will persist in the soil for several years.

Treatment.—Soak the seed potatoes for two hours in formaldehyde (1 pound in 30 gallons of water) or in corrosive sublimate (4 ounces in 30 gallons of water) for one and one-half hours. When the land becomes infected, avoid the use of stable manure and lime, and rotate crops for three to five years.

Little Potato, Rosette, Stem Rot, Scurf (*Rhizoctonia* or *Corticium vagum*, B. and C., *var. solani*, Burt.).—This disease assumes different forms, varying with the climatic conditions, soils and varieties. In very severe cases many of the young plants fail to get through the ground. Many that do get through are dwarfed and show a peculiar crinkling of the foliage. The part of the stems below ground shows peculiar brownish or black cankers. In some cases the leaves tend to roll upward; many small tubers are formed just below the surface of the ground and just above a very pronounced canker, and aerial potatoes along the stem above ground. The fungus can be readily detected on the tubers; it appears as small black spots, which do not wash off, but can be readily removed by rubbing. However, the presence of these spots on the tubers does not necessarily mean a severe outbreak of the disease.

A Potato Affected with Russet Scab.[1]
Showing the russeting and cracking, associated with the fungus Rhizoctonia.

Treatment.—Soak seed potatoes in corrosive sublimate as recommended for scab.

Bacterial Wilt (*Bacillus solanacearum*, Smith).—The plants wilt prematurely, become yellow, then black and dry. This disease attacks tomatoes, tobacco, peppers and eggplants.

Treatment.—Rotate crops, avoiding those that are susceptible.

Tipburn.—This disease is due entirely to hot, dry weather. It causes the leaves to dry at the tips and margins, roll up and break off.

Note.—There are a number of other diseases of the potato which cannot be included in this brief discussion.

TOMATO

Early Blight.—See Potato.

Leaf Blight (*Septoria lycopersici*, Speg.).—This disease appears as

[1] From Farmers' Bulletin 544, U. S. Dept. of Agriculture.

numerous small spots over the surface of the leaves, beginning with the lower and older leaves, causing them to turn yellow and fall. It is one of the most severe diseases and the cause of heavy losses.

Treatment.—Spray with Bordeaux mixture.

Fusarium Wilt (*Fusarium lycopersici*, Sacc.).—This is a disease in which the fungus works on the inside of the plant, causing it to wilt and die. It cannot be controlled by spraying. Rotation of crops is advantageous.

Bacterial Wilt.—See Potato.

Blossom-End Rot or Point Rot.—The cause of this disease is disputed, but it is now generally believed to be due to drought, although it may also be due to other causes. It is a dry black rot starting at the blossom end of the fruit and is often very destructive. It is more serious in dry weather and in dry soils.

Treatment.—Practice thorough cultivation of the soil and remove diseased fruit.

Anthracnose (*Colletotrichum phomoides* [Sacc.], Chester).—This disease causes discolored, sunken spots which become centers of decay. It is likely to be very severe in wet weather.

Treatment.—Spray with Bordeaux mixture.

Fruit Rot (*Phoma destructiva*, Plowr.).—This disease causes a spotting of the leaves and a fruit rot. Can probably be controlled by spraying with Bordeaux mixture.

EGGPLANT

Attacked by several fruit rots and leaf spots which sometimes prove destructive. They can be controlled by the use of Bordeaux mixture.

PEPPER

Susceptible to several fruit and stem rots, leaf spots and wilts which can be controlled by rotation of crops and treatment with Bordeaux mixture.

LETTUCE

Mildew (*Bremia lactucæ*, Regel).—This disease is frequently very destructive. It causes rather large, pale spots, which become yellowish above and fuzzy below.

Treatment.—Good cultural methods for outdoors. Ventilation for crops grown under glass.

Drop or Wilt (*Sclerotinia libertiana*, Fckl.).—This causes a very pronounced wilting and drooping, beginning with the lower leaves and gradually spreading throughout the entire plant.

Treatment.—Removal and destruction of the diseased plants and disinfection of soil at that point with Bordeaux mixture.

Note.—There are several other diseases of the lettuce more or less important.

SWEET POTATO

Soft Rot (*Rhizopus nigricans*, Ehrbg.).—This storage rot is caused by the bread mould fungus and can be readily recognized. It is accompanied by a sweetish odor and dense growth of white mould which becomes black. It spreads rapidly, but can be controlled by proper ventilation and regulation of temperature.

Black Rot (*Sphæronema fimbriatum* [Ell. and Halst.], Sacc.).—This disease occurs in both field and storage house. It appears as dark-brown or black, irregular, dry patches on the potatoes, sometimes causing breaking or cracking near the center of the diseased area. On the young sprouts and stems it causes black patches and frequently kills the entire plant.

Treatment.—Do not use diseased plants for setting. Do not use stable manure. Grow seeds from slips. These slips should be cut from the old plants and set as early in July as possible.

Stem Rot (*Nectria ipomoeæ*, Halst.).—This disease attacks the stem near the surface of the ground and spreads in both directions, frequently causing the death of the plant. The interior of the stem shows a yellow discoloration.

Treatment.—Rotate the crops and use slip seed.

NOTE.—There are a number of other rots and diseases which will not be taken up in this discussion.

PEANUT

Peanuts are subject to several foliage and root diseases of more or less importance. Growers of this crop should consult with their state agricultural experiment station.

TOBACCO

Granville Tobacco Wilt (*B. solenacerarum*, Smith).—This is due to the same organism as the wilt of the potato, tomato, peppers and eggplants. (See Potato.)

Mosaic, Calico or Mottle Top.—The cause of this disease is still somewhat uncertain. The leaves of the diseased plants show dark and light areas and frequently irregular thickenings or twistings.

Treatment.—Remove the diseased plants. Be careful not to touch healthy plants while working with the diseased plants. The disease can be communicated by contact.

Leaf Spots.—There are a number of leaf spot diseases and also mildews which cause more or less trouble.

Root Rots (*Thielavia basicola*, Zopf.).—This disease is a rotting of the roots, accompanied by the production of numerous new roots. The affected plants are dwarfed and frequently killed.

Treatment.—Sterilize seed-bed. Rotate crops. Avoid liming and acid fertilizers.

CORN

Smut (*Ustilago Zeae* [Beckm.], Ung.).—Corn smut on ear, tassel and leaves is so common that it is not necessary to give a description. It is frequently very destructive, especially on sweet corn.

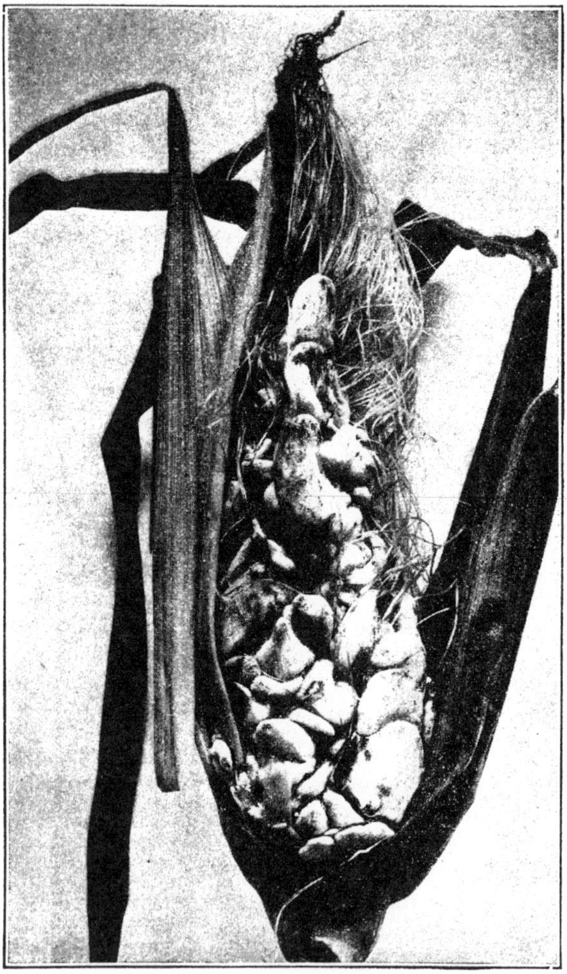

SMUT OF CORN.[1]
Showing a young smutted ear.

Treatment.—The treatment will depend largely on the severity of the disease. Do not use manure from animals which have been fed on smutted cornfodder. Rotate crops. If growing sweet corn on the same land year after year, it is advisable to remove the smut balls as soon as they appear.

WHEAT

Rust (*Puccinia graminis*, Pers., *P. rubigovera*, Wint.).—This crop is affected with the familiar rust diseases, all of which appear to be very much the same to the inexperienced student of plant diseases. They are very difficult to control and in fact comparatively little effort is made to protect the crop. Resistant varieties should be used and if possible spring wheat should be sown early.

Loose Smut (*Ustilago tritici* [Pers.], Jens.).—This very common and familiar disease is the cause of much greater loss than the growers realize.

Treatment.—It can be controlled by treating with hot water. Clean

[1] From Farmers' Bulletin 507, U. S. Dept. of Agriculture.

LOOSE SMUT OF WHEAT.[1]

Showing four smutted heads of various stages of development, and for comparison a sound wheat head.

[1] From Farmers' Bulletin 507, U. S. Dept. of Agriculture.

the seed and sack for five to seven hours in water at a temperature of 63° to 72° F. Then put into loose bags or wire baskets holding about one peck each and plunge into water ranging between 126° and 129° F. for ten minutes. Care should be taken to keep the water at the proper temperature and to keep the grain well stirred. The seed can be dried on a barn floor or canvas.

Smut of Oats.[1]
Showing a smutted head, and for comparison a sound oat head.

Stinking Smut or Bunt (*Tilletia fœtans* [B. and C.], Trel.).—This disease is very different from the loose smut. The diseased grains are shriveled, greenish tinted, filled with a mass of black spores and have a disagreeable odor. Badly infested crops are worthless for milling or for stock feed.

Treatment.—Put one pound of formaldehyde in fifty gallons of water and sprinkle on the grains at the rate of one gallon to each bushel of grain. Shovel the wet grain into a pile and cover with canvas or burlap for six to twelve hours. Spread and dry. (See Chapter on Wheat.)

OATS

Rust (*Puccinia coronata*, Cda.).—Also the two species found on wheat.

Treatment.—Same as for wheat.

Smut (*Ustilago avena* [Per.], Jens.).—Very similar in appearance to the loose smut of wheat.

Treatment.—Formaldehyde treatment same as for stinking smut of wheat.

---

[1]From Farmers' Bulletin 507, U. S. Dept. of Agriculture.

SUGAR CANE

Red Rot (*Colletotrichum falcatum*, Went.).—This is one of the most destructive diseases of the sugar cane. It causes the plants to wilt and finally a yellowing of the upper leaves. This is followed by a blackening and dying of the eyes and a gradual discoloration on the outside extending from the nodes. Upon splitting the canes, the fibro-vascular bundles are found to show reddish discolorations.

Rind Disease (*Trichosphaeria sacchari*, Massee).—Although this disease is not so severe as the preceding one, it causes a premature yellowing and dying of the plants. The joints become discolored and shrunken and the entire plant loses weight. Finally, small black eruptions which are thread-like in appearance appear over the canes.

The Pineapple Disease (*Thielaviopsis ethacetica*, Went.).—This disease is of comparatively little importance, but it sometimes attacks the cuttings which have been prepared for planting and prevents their growth or causes weak, unhealthy plants.

Treatment.—The most satisfactory treatment for these three diseases is care in selection of good healthy plants for cutting and the treatment of these cuttings with Bordeaux mixture before planting.

A Young Cotton Plant Affected by Cotton Wilt.[1]

When the grower has any reason to suspect the appearance of these or other diseases, he should consult with the plant pathologist of the experiment station in the state in which he is located.

Other Diseases.—There are a number of other diseases of minor importance.

COTTON

Anthracnose (*Golmerella gossypii*, Southworth Edg.).—This disease attacks stem, boll and leaves, causing dull, reddish-brown spots which are

[1]Courtesy of Bureau of Plant Industry, U. S. Dept. of Agriculture

slightly depressed. In advanced stages these spots are covered with a dirty gray or pinkish powder which is the spores of the fungus. This disease is carried in the seed and is the cause of heavy losses.

Damping Off, Sore Shin, Seeding Rot.—These diseases may be due to any one of several organisms. They attack the young plants at or just below the surface of the ground, causing them to rot off and die. They are sometimes the cause of heavy losses.

There are a number of other diseases of the cotton. The most satisfactory remedy for most diseases is the selection of seed from healthy plants. Where growers experience much difficulty, they should consult with the authorities at the state agricultural experiment station.

FLAX

Wilt (*Fusarium lini*, Bolley).—This is one of the most severe diseases of the flax. Sometimes the organism causing this disease is so abundant in the soil that it leads to the term "flax sick soil." The new plants affected with this disease wilt and die and fields are very frequently seen in which there are large bare spots due to the ravages of this disease. When the older plants are attacked they wilt and gradually turn yellow and die.

The grower who has any difficulty with this or other diseases should consult with the state agricultural experiment station.

REFERENCES

"Diseases of Tropical Plants." Cooke.
"Fungous Diseases of Plants." Duggar.
"Spraying of Plants." Lodeman.
"Minnesota Plant Diseases." Freeman.
"Diseases of Economic Plants." Stevens and Hall.
"Diseases of Cultivated Plants and Trees." Massee.
California Expt. Station Bulletin 262. "Citrus Diseases of Florida and Cuba Compared with California."
Michigan Expt. Station (Technical Bulletin 20). "Control of Root Knot Nematode."
New Jersey Expt. Station Circulars:
44. "Diseases of Apples, Pears and Quinces."
45. "Diseases of Peach, Plum and Cherry."
Ohio Expt. Station Bulletin 265. "Cob Rot of Corn."
Pennsylvania Expt. Station Bulletin 136. "Collar-Blight and Related Forms of Fire Blight."
Washington Expt. Station Bulletin 126. "Bunt or Smut of Wheat."
Canadian Dept. of Agriculture Bulletin 229. "Smuts and Rusts."
U. S. Dept. of Agriculture Bulletins:
64. "Potato Wilt and Other Diseases."
203. "Field Studies of the Crown Gall of Sugar Beets."
216. "Rust of Grain in United States" (Bureau of Plant Industry).
Farmers' Bulletins, U. S. Dept. of Agriculture:
333. "Cotton Wilt."
507. "The Smuts of Wheat, Oats, Barley and Corn."
544. "Potato Tuber Diseases."
555. "Cotton Anthracnose and Its Control."
618. "Leaf Spot: A Disease of the Sugar Beet."
625. "Cotton Wilt and Root Knot."
648. "The Control of Root Knot."

CHAPTER 76

INSECT PESTS AND THEIR CONTROL

By W. B. Wood

Scientific Assistant, Bureau of Entomology, U. S. Department of Agriculture

Insects are, without doubt, the greatest enemies of the farmer, for they destroy the crops of field and garden and render the fruit on the trees unfit for use; they injure the domestic animals by constant irritation, causing them to lose weight and even to die. Stored grains, tobacco and other farm products also suffer from their attacks.

After studying their life histories and habits, methods of control have been devised by which they can be combated with a reasonable amount of success. Many species can be held in absolute control by thorough and timely applications of the proper remedies, while others are only partially held in check.

In order to intelligently apply a treatment for the control of an insect, something of its habits must be known, especially in regard to its manner of feeding. Most of the important pests fall within two great groups, namely, biting or chewing insects and sucking insects, depending on whether the mouth parts are chisel or pincher-like in the first class, or beak-like and made for piercing and sucking in the second class. A number of these pests will fall in certain special groups which require a definite treatment, indicated by their manner of living or by the injury they do. Some of these special classes are internal feeders, as boring insects, subterranean insects and insects affecting stored products.

The external feeders, which have biting mouth parts, usually feed upon plants by gnawing out small pieces of the plant tissue which are swallowed. This group includes the larvæ or caterpillars of moths and butterflies, the larvæ of beetles and the adults, grasshoppers and crickets, and the larvæ of some species of *Hymenoptera* or the wasp group. Such insects may usually be controlled by applying a poison to the plant, either as a fine spray or as a powder dusted or blown over its surface. The arsenicals have been found to be the best remedy for this group.

The sucking insects feed by piercing the skin or epidermis of plants with their sharp beaks and sucking the sap. This group of insects is represented by the tree bugs or *Hemiptera,* to which order belong the squash bug, scale insects, plant lice and leaf hoppers. It is evident that a stomach poison on the surface of the plant would not affect insects of this class, so it is necessary to use what is known as a contact insecticide, which should be applied as a spray or wash directly to the insect's body. Such remedies kill by their suffocating or corrosive action. The most common of these insecticides are nicotine solutions, kerosene or oil emulsions, lime-sulphur wash and fish-oil soap.

In the following pages will be found listed the principal insect pests of farm crops under the class of crops to which they are most injurious. Only a very brief description of each insect can be given, and in most cases nothing of their life histories, in the limited space devoted to the subject. The treatments which have given the best results in each individual case are indicated briefly and reference is made to publications which give a more extended account of the insects. The abbreviations which are used in the references are as follows:

Bur. Ent. Bull.—U. S. Department of Agriculture, Bureau of Entomology Bulletin.
Bur. Ent. Cir.—U. S. Department of Agriculture, Bureau of Entomology Circular.
Farm. Bull.—U. S. Department of Agriculture, Farmers' Bulletin.
Dept. Bull.—U. S. Department of Agriculture Bulletin.

GENERAL CROP INSECTS

Caterpillars (leaf-eating).—Many plants are attacked by caterpillars which feed upon the leaves. These worms are the larvæ of *Lepidopterous* insects, or moths and butterflies.

Treatment.—Spray with an arsenical, preferably arsenate of lead, or dust with powdered arsenate of lead or Paris green. If the spray gathers in drops and does not adhere well to the surface of the leaves, use a resin fish-oil soap sticker.

Cutworms.—Various species of the family *Noctuidæ*, usually feeding at night upon the roots, crowns or foliage of plants. The worms may be found in daytime lying curled up in ground about an inch below surface.

Treatment.—Broadcast poison bran mash about the garden in the spring just before the plants come up. Make other applications later if the cutworms are still found. Cultivate the ground thoroughly in late summer and early in the spring to prevent the growth of grasses and weeds, thus starving out worms if present.

Grasshoppers or Locusts.—A number of species feed on corn, wheat, sorghum and other field crops, also on many garden crops and at times on fruit trees.

Treatment.—Cultivate the fields and stony fence rows in the fall to break up the egg masses deposited one to two inches below the surface of the ground. Broadcast Criddle mixture or poison bran mash flavored with juice of orange or lemon in fields where grasshoppers are plentiful.

Leaf Beetles (*Chrysomelidæ*).—Crops of many kinds are injured by beetles which feed upon the leaves as adults and sometimes as larvæ.

Treatment.—Spray or dust the affected plants with arsenicals.

Plant Lice (*Aphididæ*).—Many species of plant lice are found attacking field, garden and orchard crops. They feed by sucking the juices of the host plant and cannot be controlled by a poison spray.

Treatment.—Use sprays of nicotine or tobacco extract, kerosene emulsion (5 to 10 per cent strength) or fish-oil soap just after the aphids appear and at such other times as may be necessary. Spray thoroughly, being sure to wet all plant lice. If spray does not adhere to the bodies of insects, add 2 or 3 pounds of laundry soap to 50 gallons of spray solution, or preferably an equal amount of resin fish-oil soap as a sticker. For underground forms practice rotation of crops or use soil fumigants.

White Grubs (*Lachnosterna spp.*).—White grubs or grubworms are the larvæ of the common brown beetles known as May beetles or June bugs, commonly seen around lights and on the screens in the spring and summer. Their natural breeding place is grass lands, but they are found in fields and gardens feeding upon the roots of many plants.

Treatment.—No successful treatment is known. Practice crop rotation when necessary. Fall plowing will be of some benefit. Do not plant crops liable to be injured, as strawberries, on recently broken sod land.

Wire Worms (*Elateridæ*).—Slender, brown, hard, shining larvæ, ½ inch to 1½ inches long, body divided into several segments which show plainly three pairs of small legs near front end of body. Their natural breeding place is grass lands, but they feed on or in the roots of many garden and field crops. Two years or more are required for development.

Treatment.—No satisfactory treatment has been found. Rotation of crops, preventing ground from remaining long in grass, and late fall plowing followed by repeated harrowing for a month or two are the best means of preventing their increase. Seeds might be protected by the use of some substance as a repellent which would not injure germination.

GENERAL CROP INSECTS

The Army Worm (*Leucania unipuncta*, Haworth).—In general appearance it resembles cutworms. About 1½ inches long, dark in color, with three yellowish stripes down the back. The adult insect is a dull brown moth, often seen about lights in the spring. The worm feeds naturally on wild grasses, but when it is abundant marches across fields, destroying many crops, including corn, wheat, oats and related crops, as well as many truck crops.

Treatment.—The march of the worms to uninfested fields may be checked by a deep dust furrow through which a log is dragged occasionally to crush the worms and to maintain a thick coat of dust on the sides. Scattering poison bran mash through infested fields will often prove very effective. Late fall plowing and cultivating will help in destroying overwintering worms.

The Alfalfa Leaf Weevil (*Phytonomus posticus*, Fab.).—This insect, which has been accidentally introduced into the United States from Europe, now threatens the alfalfa industry of the country. From a small field near Salt Lake, where it was first found, it has spread through the surrounding country until it has gone as far as Wyoming and Idaho. In the spring the adult insect punctures the stems of the plants as they are coming up, and deposits its eggs in the wounds. The grubs hatch and feed upon the tender leaves until they are fully developed. Transformation then takes place and the adult beetle begins to destroy the foliage.

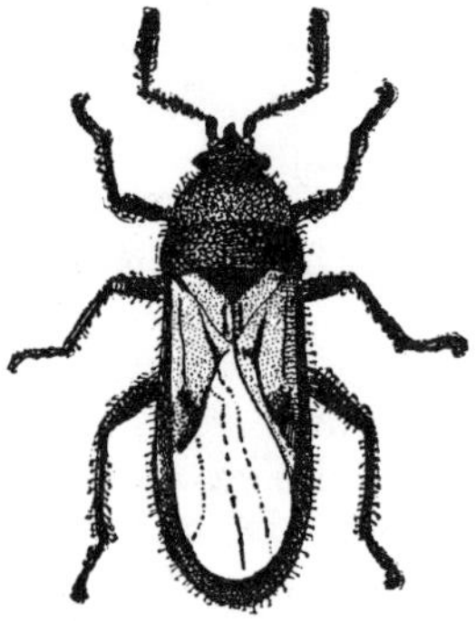

CHINCH BUG (*Blissus leucopterus*).[1]
Adult of long-winged form, much enlarged.

Treatment.—Breaking up the ground in the spring with a disk harrow to stimulate a rapid growth has been found to be beneficial. Clean up all trash and rubbish which might form hiding places for the insect. Immediately after first cutting use a spike-tooth harrow, followed closely by a brush drag to knock off and kill the grubs.

Bur. Ent. Bull. 112; Utah Exp. Sta. Bull. 110.

The Chinch Bug (*Blissus leucopterus*, Say.).—Throughout the Middle states this is the worst enemy of all kinds of grains. It hibernates for the most part in clumps of grass, but may be found in weeds and rubbish along fence rows. The bug injures the plant by sucking the sap from the stalks.

Treatment.—Concerted action by the farmers in a large area, in burning the bunch grass late in the fall or in early winter, is the best means of control. The grass should be burned close to the ground when it is perfectly dry, thus destroying many of the insects and leaving others unprotected against the storms of winter. When bugs are migrating from small grains to corn or other crops in summer, they may be stopped by dusty ditches with post holes in bottom, by dust ridges or coal tar barriers.

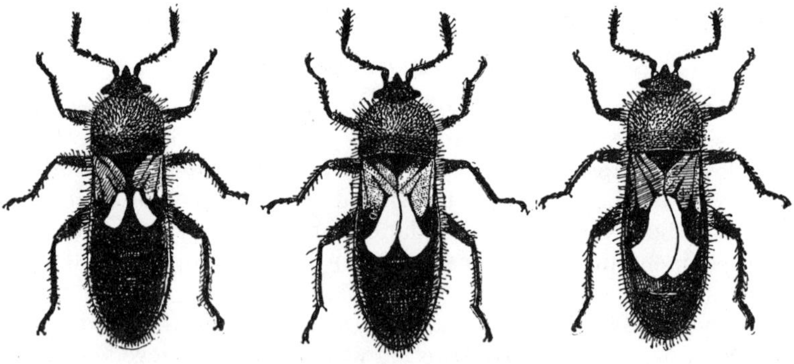

CHINCH BUG (*Blissus leucopterus*).[1]
Adults of short-winged form, much enlarged.

Farm. Bull. 657.

[1] Bur. Ent. Cir. 113.

Clover Mite (*Bryobia prætensis*, Garm.).—A common red mite on many plants, including clover, alfalfa and a number of varieties of fruit.

Treatment.—Dust the plants with sulphur and lime at rate of 1–4, or spray with either 10 per cent kerosene emulsion or sulphur in water, 1 pound to 4 gallons. Destroy eggs on fruit trees in winter with 20 per cent kerosene emulsion or with lime-sulphur.

Bur. Ent. Cir. 158.

Clover Root Borer (*Hylastinus obscurus*, Marsham).—The beetle winters over in clover roots; emerges in the spring and lays eggs in the larger roots. The grubs, on hatching, bore through central part, destroying plants.

Treatment.—Plow the fields after haying, allowing the roots to dry. Pasturing checks the injury. Infested field should not be allowed to stand over the second season.

Bur. Ent. Cir. 119.

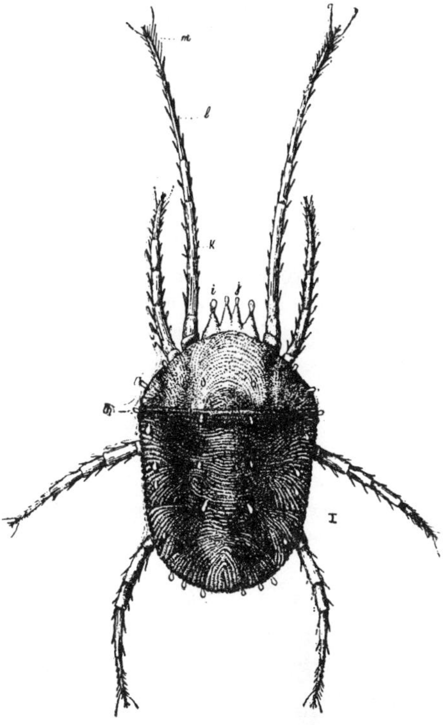

Clover Mite (*Bryobia prætensis*).[1] Enlarged; natural size shown by line at right.

Corn Ear Worm (*Heliothis obsoleta*, Fab.).—This insect is also known as the cotton boll worm, the tobacco bud worm or the tomato fruit worm. It has a long list of other food plants, but on many causes no serious injury. On corn the eggs are laid by the moths upon the silk. The larvæ upon hatching enter the ear and begin to feed on the immature grains. Cotton is not seriously attacked until the corn silks are drying up, as corn is much preferred by the worms. The adults lay their eggs upon the cotton leaves and the larvæ, after feeding for a short time upon the foliage, enter the bolls. They attack tobacco by eating into the buds, and tomatoes are injured by attacks upon the fruit.

Treatment.—For all crops the injury is materially lessened by late fall plowing and cultivation which crushes many pupæ in the soil and exposes others to the winter. On cotton the insect may be well controlled by two applications of an arsenical dust or spray at the time the eggs are hatching. Tobacco may be protected by dropping into the buds a little

[1] Bur. Ent. Cir. 158.

corn meal, poisoned with powdered arsenate of lead, using 2 or 3 spoonfuls to a quart of meal. Early maturing varieties of corn or cotton will not be so seriously injured as the later kinds.

Farm. Bull. 290; Bur. Ent. Bull. 50.

The Corn Root Aphis (*Aphis maidi-radicis*, Forbes).—A bluish-green plant louse found on the roots of corn, broom corn, sorghum and on several weeds. It weakens the plant, causing it to be stunted and poorly nourished.

Treatment.—One year rotation to other crops than corn, clean cultivation and liberal use of fertilizers, winter plowing to break up nests of ants where aphis eggs are stored.

Bur. Ent. Cir. 86; Bur. Ent. Bull. 85, Pt. 6.

Cotton Boll Worm (*Heliothis obsoleta*, Fab.).—See Corn Ear Worm.

Cotton Worm (*Alabama argillacea*, Hbn.).—A dark-greenish caterpillar, striped with black, the larva of a grayish-brown moth marked on the fore wings with irregular darker bands. They feed on the under side of leaves when young, later feeding on the entire leaf and when abundant on buds and tender stalks. Adults make strong flights, going as far north as Canada. They feed at times on ripe fruit, which they are able to puncture with strong mouth parts.

Treatment.—Dust the plants with powdered arsenate of lead when the worms appear.

Bur. Ent. Cir. 153.

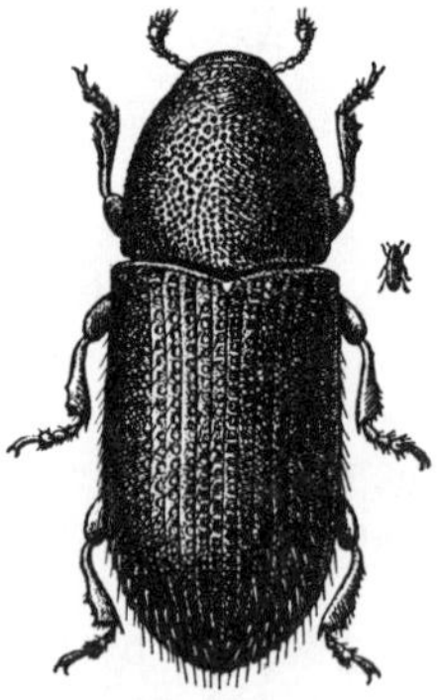
1

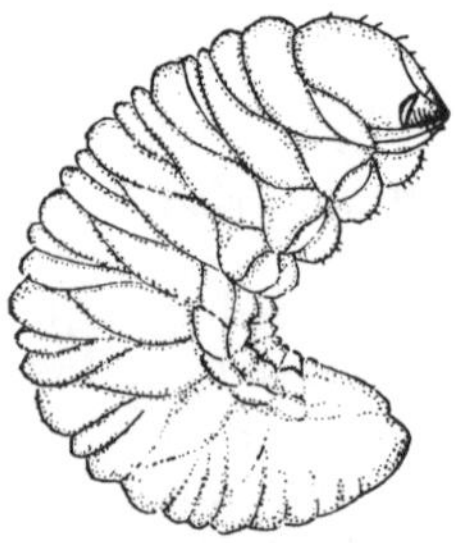
2

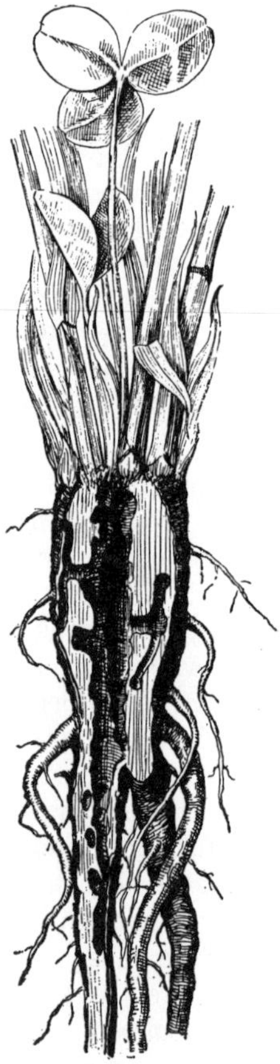
3

CLOVER ROOT BORER (*Hylastinus obscurus*).[1]

1—Adult beetle, natural size at right. 2—Larva or grub, much enlarged. 3—Showing work of the borer.

[1]Bur. Ent. Cir. 119.

Cotton Worm (*Alabama argillacea*).[1]
Stages and work.

[1]Bur. Ent. Cir. 153.

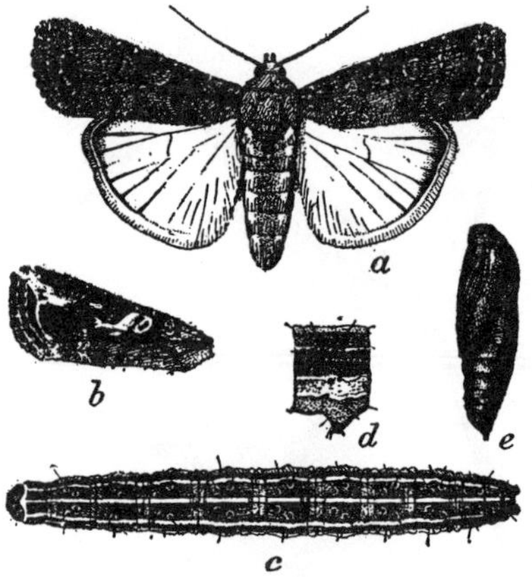

FALL ARMY WORM
(*Laphygma frugiperda*).[1]

A—Moth, plain gray form. B—Fore wing of prodenia-like form. C—Larva extended. D—Abdominal segment of larva, lateral view; twice natural size. E—Pupa, lateral view.

The Cotton Red Spider (*Tetranychus bimaculatus,* Harvey).—This small red mite is common on cotton and on several other plants, especially pokeweed and violet. It causes the leaves of cotton to turn red and fall off. It kills plants if abundant.

Treatment.—Prevent the mites from starting on the cotton by clean culture, being sure to eradicate all pokeweed and violets near the fields. If found in cotton fields, spray the affected plants with potassium sulphide 3 pounds and water 100 gallons; make two applications one week apart.

Bur. Ent. Cir. 172.

The Fall Army Worm (*Laphygma frugiperda,* S. and A.).—In general appearance is similar to the common army worm, but with different markings. It has wide range of food plants, including many forage and truck crops.

Treatment.—Practice fall plowing to break up the pupæ cells in the ground. Scatter poison bran mash when the caterpillars appear, or spray or dust with arsenicals.

Bur. Ent. Bull. 29.

The Green Bug, or Spring Grain Aphis (*Toxoptera graminum,* Rond.).—A small green plant louse which attacks oats, wheat, barley and other grains. It appears very early in the spring.

Treatment.—No satisfactory method of control is known. Attacks may be partially pre-

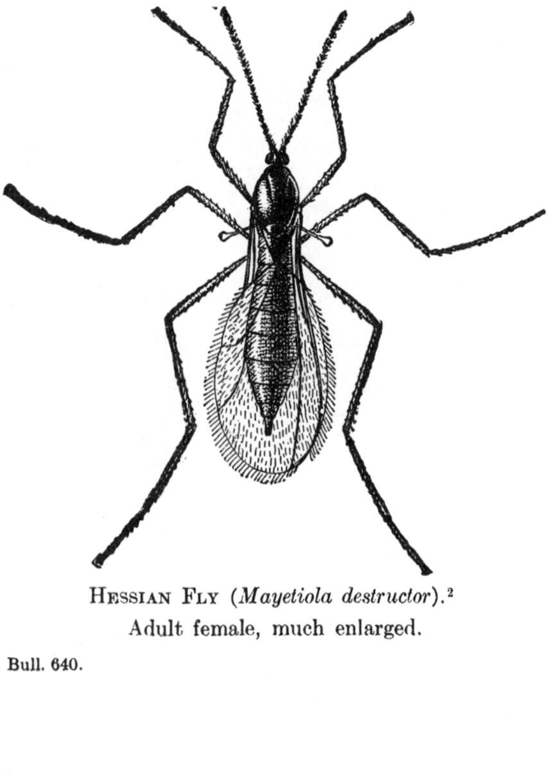

HESSIAN FLY (*Mayetiola destructor*).[2]
Adult female, much enlarged.

[1]Bur. Ent. Bull. 29. [2] Farm. Bull. 640.

Green Bug, or Spring Grain Aphis (*Toxoptera graminum*).[1]

Wheat plant showing winged and wingless viviparous females with their young clustered on leaves, and a few parasitized individuals on lower leaves. About natural size.

[1] Bur. Ent. Bull. 110.

vented by late planting and by the destruction of volunteer wheat and oats in the fall.

Bur. Ent. Bull. 110.

The Hessian Fly (*Mayetiola destructor,* Say.).—This small two-winged fly is one of the most destructive insects of growing wheat, causing the plants to be stunted and to break down near harvest time.

Treatment.—Burn the stubble or plow it under as soon after harvest as possible. Destroy all volunteer wheat just before sowing. Delay the sowing until ten days or two weeks after usual time. The two latter precautions should prevent most of usual injury.

U. S. Dept. Agri. Cir. 51, Office of Secretary; Farm. Bull. 640.

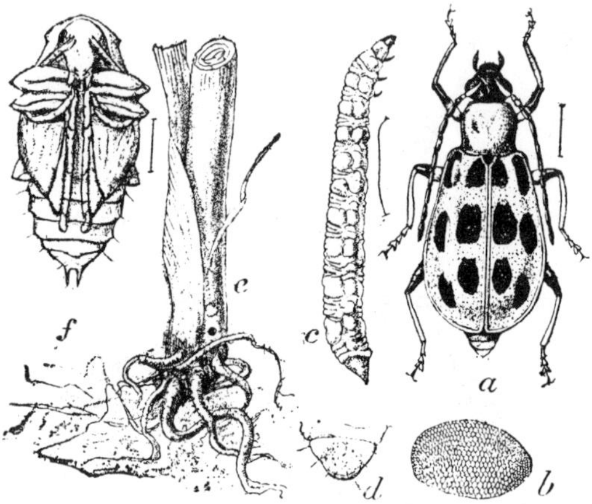

SOUTHERN CORN ROOT WORM
(*Diabrotica duodecimpunctata*).[1]

A—Beetle. B—Egg. C—Larva. D—Anal segment of larva. E—Work of larva at base of corn stalk. F—Pupa. All much enlarged except E, which is reduced.

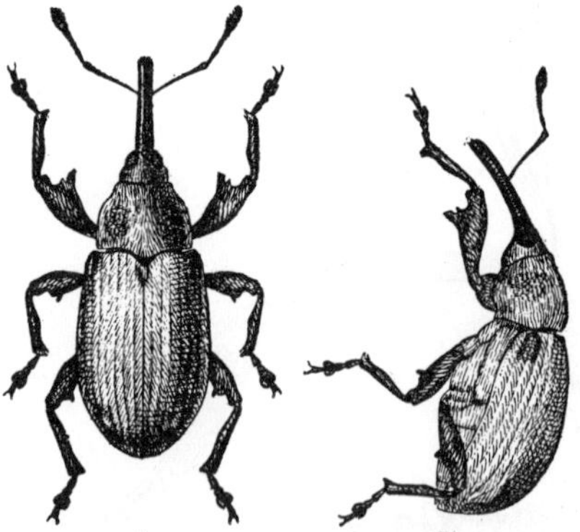

COTTON BOLL WEEVIL
(*Anthonomus grandis*).[2]

A—Beetle, from above. B—Same from side. About five times natural size.

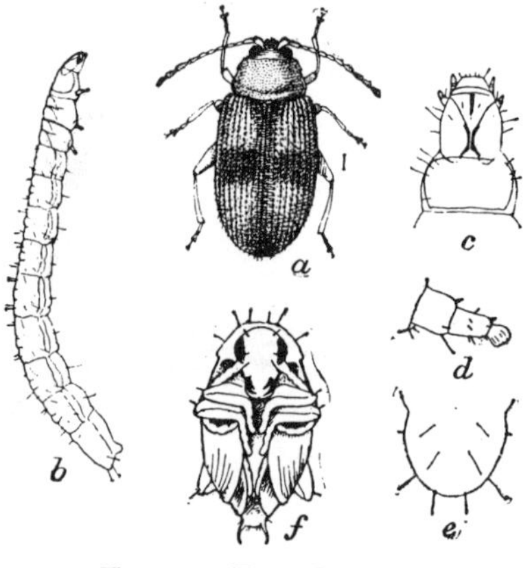

TOBACCO FLEA BEETLE
(*Epitrix parvula*).[3]

A—Adult beetle. B—Larva, side view. C—Head of larva. D—Hind leg of same. E—Anal segment of same. F—Pupa. A, B, F—Enlarged about fifteen times. C, D, E—More enlarged.

[1] Bur. Ent. Bull. 43. [2] Farm. Bull. 344. [3] Bur. Ent. Cir. 123.

Mexican Cotton Boll Weevil (*Anthonomus grandis*, Boh.).—No pest of cotton has caused so much injury as this small brown beetle. Both the adult insects and the larvæ feed upon the squares and the bolls, injuring the fiber.

Treatment.—Clean up and destroy all stalks, dead bolls and crop remnants as soon as cotton is picked, either by burning or burying. Plow under or burn in the fall and winter all trash in neighboring fields and

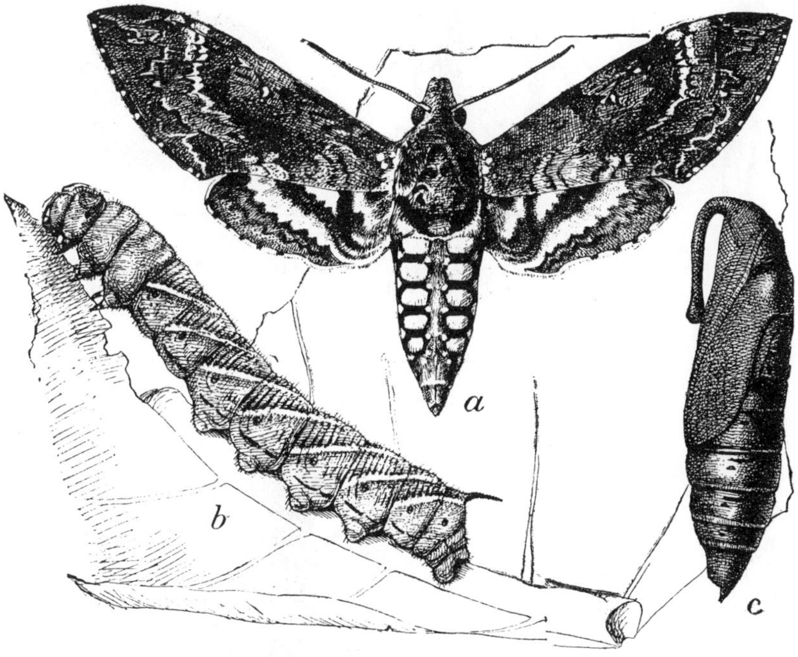

SOUTHERN TOBACCO HORN WORM (*Phlegethontius sexta*).[1]
A—Adult. B—Larva. C—Pupa.

hedgerows where the insect might hibernate. Prepare the land early, plant early and fertilize heavily to secure an early crop.

Farm. Bull. 344, Senate Document No. 305, 62d Congress.

Spring Grain Aphis (*Toxoptera graminum*, Rond.).—See Green Bug.

Southern Corn Root Worm, or Bud Worm (*Diabrotica duodecimpunctata*, Oliv.).—Greenish-yellow beetle marked on the back with twelve black spots. Feeds on a variety of plants. Larva or grub feeds on roots of corn after boring into roots and stem.

Treatment.—No satisfactory insecticidal treatment is known. The worst of the injury may be prevented in Southern states by planting about

[1] Bur. Ent. Cir. 123.

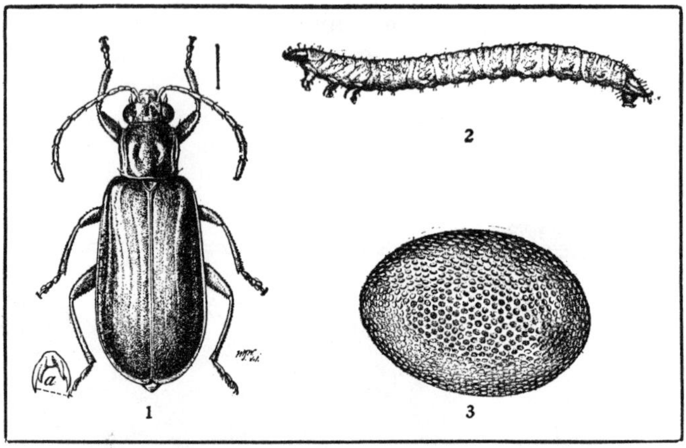

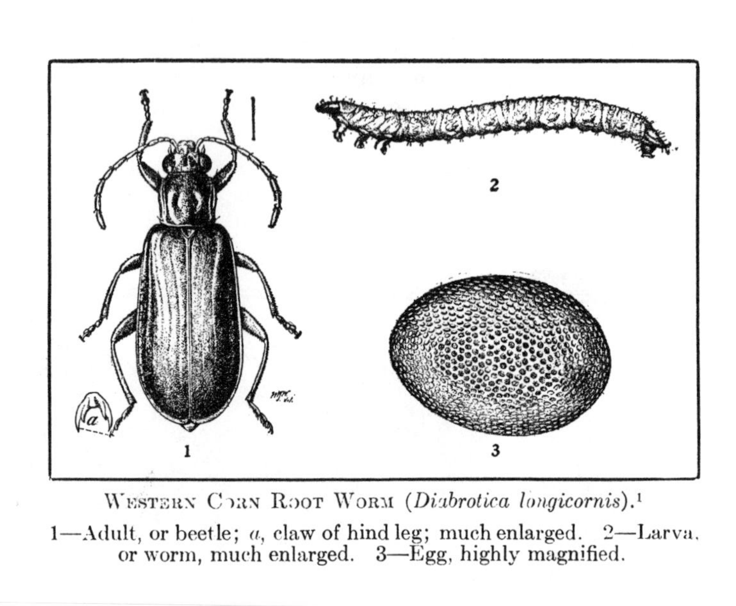

Western Corn Root Worm (*Diabrotica longicornis*).[1]

1—Adult, or beetle; *a*, claw of hind leg; much enlarged. 2—Larva, or worm, much enlarged. 3—Egg, highly magnified.

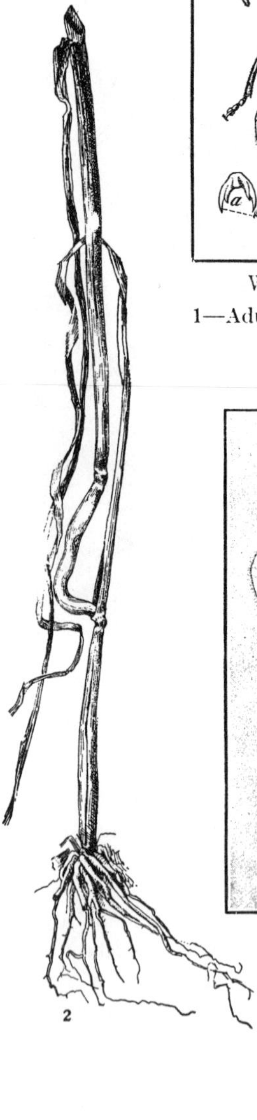

2

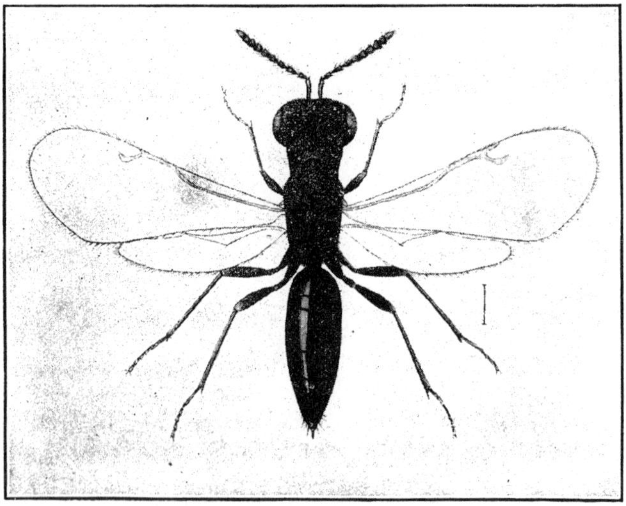

1

Wheat Joint Worm (*Isosoma tritici*).[2]

1—Adult of the joint worm much enlarged. 2—One effect of the joint worm in wheat straw.

[1] Dept. Bull. 8. [2] Bur. Ent. Cir. 66.

three weeks later than usual or after most of the first brood eggs have been deposited.

Dept. Bull. No. 5.

Tobacco Flea Beetle (*Epitrix parvula*, Fab.).—A small dark-colored beetle, eating holes in the leaves of tobacco. The beetle is a very active jumper and cannot be readily captured.

Treatment.—Apply arsenicals by spraying or as dust when the injury is first noticed and again a few days later, if the beetles are still present.

Bur. Ent. Cir. 123; Year-Book 1910, pp. 281–296.

Tobacco Worms, or Horn Worms (*Phlegethontius quinquemaculata*, Haw., and *P. sexta*, Johan.).—These two pests are the most destructive of the tobacco insects. They feed on the leaves and buds.

Treatment.—Hand picking or the use of arsenicals will prevent serious injury.

Bur. Ent. Cir. 123; Bur. Ent. Cir. 173

Western Corn Root Worm (*Diabrotica longicornis*, Say.).—A yellowish green beetle, the larva of which feeds on the roots of corn. There is only one generation of the insect each year.

Treatment.—The only successful way of combating the pest is to rotate crops from corn to one of the small grains.

Dept. Bull. No. 8.

Wheat Joint Worm (*Isosoma tritici*, Fitch).—Most of the injury from this insect has been found in the wheat-growing regions east of the Mississippi River. The adult is a small black insect somewhat resembling a small winged ant. Eggs are laid in the straw of growing wheat after several joints have been formed. The larvæ develop in the joints and emerge in the following spring.

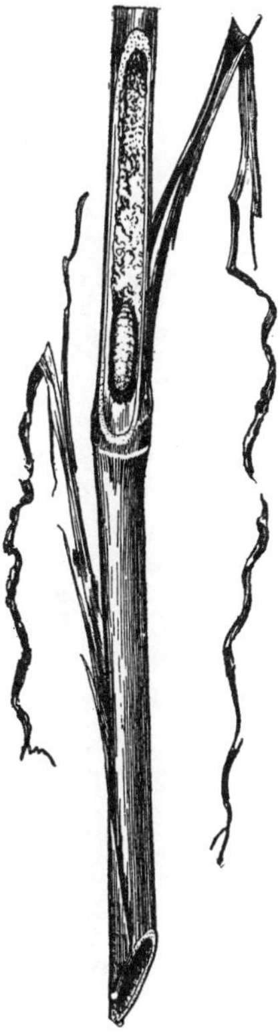

LARVA OF *Isosoma grande* IN WHEAT STRAW.[1]

Treatment.—Burn or plow under all stubble in the fall. Burn all outstanding straw in spring. Do not scatter green manure in fields to be planted in wheat in spring if infested straw was used for bedding. Fertilize liberally. Practice rotation of crops.

Bur. Ent. Cir. 66.

Wheat Straw Worm (*Isosoma grande*, Riley).—West of the Mississippi River this insect is often a very serious enemy of wheat. The larva works inside the young shoots early in the spring and the later generation in straw.

Treatment.—Injury can be largely prevented by a rotation of crops

[1] Bur. Ent. Cir. 106.

according to Prof. F. M. Webster. Wheat should not be planted on the same ground two years in succession.

Bur. Ent. Cir. 106.

TRUCK CROP INSECTS

The Asparagus Beetle (*Crioceris asparagi*, Linn.).—This beetle is about one-fourth of an inch long, with dark-colored body, red thorax and yellow wing covers marked and bordered with blue. The adults and grubs feed on the stems and tender shoots of asparagus.

Spray of Asparagus, with Common Asparagus Beetle in its Different Stages.[1]

Asparagus tip at right, showing eggs and injury. Natural size.

Treatment.—Apply arsenical sprays. Air-slaked lime will kill the grubs.

Bur. Ent. Cir. 102.

Bean Aphis (*Aphis rumicis*, Linn.).—A small black plant louse with pale shanks. It attacks beans, dock, shepherd's purse, pigweed, "burning bush" and snowball bush.

Treatment.—Spray the plants thoroughly with nicotine solution.

Bur. Ent. Bull. 33, p. 109.

Bean Weevil, The Common (*Bruchus obtectus*, Say.).—A small gray or brown beetle with mottled wing covers, about one-eighth of an inch long. It lay its eggs on or in beans in the field, also breeds in stored beans. The grub eats its way into the bean and develops there, sometimes several to one bean.

Treatment.—Heat the infested seed or fumigate with carbon bisulphide.

Bur. Ent. Bull. 96; Year-Book, U. S. D. A., 1898, p. 239.

Other Bean Weevils.—Several other weevils affect the bean in March, in the same way as the common bean weevil.

Treatment.—See Bean Weevil, the Common.

The Beet Army Worm (*Laphygma exigua*, Hbn.).—Beets are sometimes attacked in the Western states by this insect at about the same time the fall army worm is making its attacks in other sections. Several

[1] Bur. Ent. Cir. 102.

other food plants are known, including a few garden crops and a number of weeds.

Treatment.—Spray or dust arsenicals upon the leaves. Poison bran mash may also be of value.

Bur. Ent. Bull. 43.

Beet Leaf Beetle, The Larger (*Monoxia puncticollis*, Say.).—This leaf beetle, known also locally as the alkali bug and the French bug, resembles somewhat the elm-leaf beetle. It causes considerable injury to the sugar-beet in Colorado and nearby states.

BROAD-BEAN WEEVIL (*Laria rufimana*).[1] Adult, or beetle, enlarged.

Treatment.—Dust or spray foliage with arsenicals.

The Beet Leaf Hopper (*Eutettix tenella*, Baker).—The beet in the Western states is often troubled with a condition known as "curly leaf," caused by the above-named leaf hopper, a light yellowish green species about one-eighth of an inch long.

Treatment.—Spray the beets thoroughly with a 40 per cent nicotine sulphate solution in water, diluted 1 part to 600; or spray with 5 per cent kerosene emulsion. Many hoppers may be captured on a shield smeared with tanglefoot or covered with sticky fly paper if it is pushed up and down between the rows. A wire or rod should be fastened in front of the shield at the proper distance to stir out the hoppers.

Bur. Ent. Bull. 66, Pt. 4.

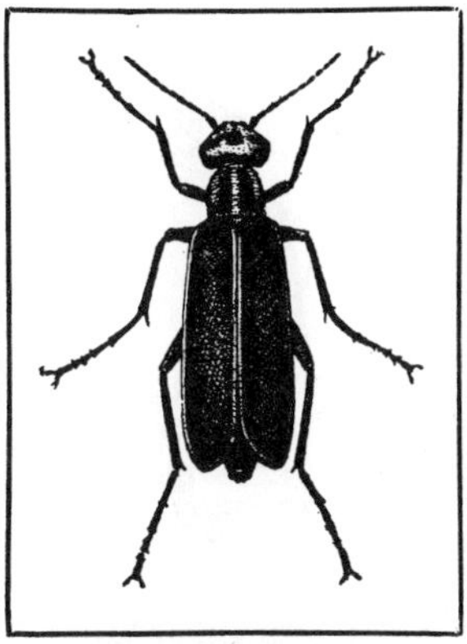

BLISTER BEETLE (*Epicauta marginata*).[2] Enlarged.

Blister Beetles (*Meloidæ*).—At times a number of crops are badly damaged by the insects known as blister beetles or "old-fashioned potato bugs." These beetles are rather large, long-legged and are variously colored, the usual colors being black, gray or striped with yellow and black.

Treatment.—Apply arsenate of lead or other arsenicals to the affected plants as a spray or dust. Several treatments may be necessary if the beetles swarm on crops from other localities.

Bur. Ent. Bull. 43, pp. 21–27.

The Cabbage Looper (*Autographa brassicæ*, Riley).—The looper is a light-green worm often referred to as a measuring worm because of its looping movement when crawling. It feeds on the leaves of cabbage.

[1] Bur. Ent. Bull. 96, Pt. 5. [2] Bur. Ent. Bull. 43.

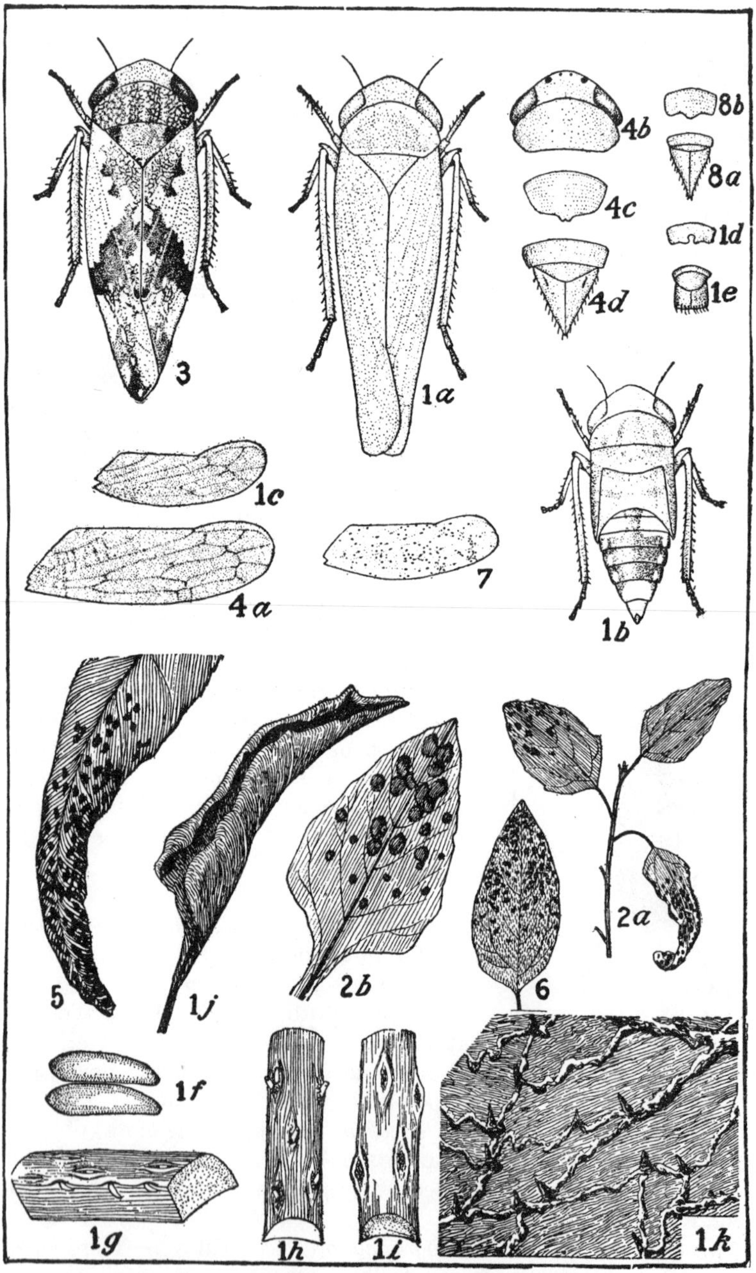

LEAF HOPPERS (*Eutettix, Spp.*) AND THEIR WORK.[1]
Explanation of illustration on page 983.

[1] Bur. Ent. Bull. 66, Pt. 4.

1—*Eutettix tenella:* *a*, adult; *b*, nymph; *c*, wing; *d*, *e*, genitalia; *f*, eggs, greatly enlarged; *g*, section of beet stem, showing fresh eggs in place; *h*, same, showing eggs ready to hatch; *i*, old egg-scars on beet stems; *j*, small leaf of sugar-beet, showing characteristic "curly-leaf" condition; *k*, enlarged section of back of an extreme case of "curly-leaf," showing "warty" condition of veins. 2—*Eutettix strobi:* *a*, work of nymphs on sugar-beets; *b*, leaf enlarged. 3—*Eutettix scitula:* adult. 4—*Eutettix clarivida:* *a*, wing; *b*, head and pronotum; *c*, *d*, genitalia. 5—*Eutettix nigridorsum:* work of nymphs on leaf of Helianthus. 6—*Eutettix straminea:* work of nymphs on leaf of another Helianthus. 7—*Eutettix insana:* wing. 8—*Eutettix stricta:* *a*, *b*, genitalia.

Treatment.—Apply arsenicals until the cabbage head is half grown. If spray is used, add resin fish-oil soap as a sticker.

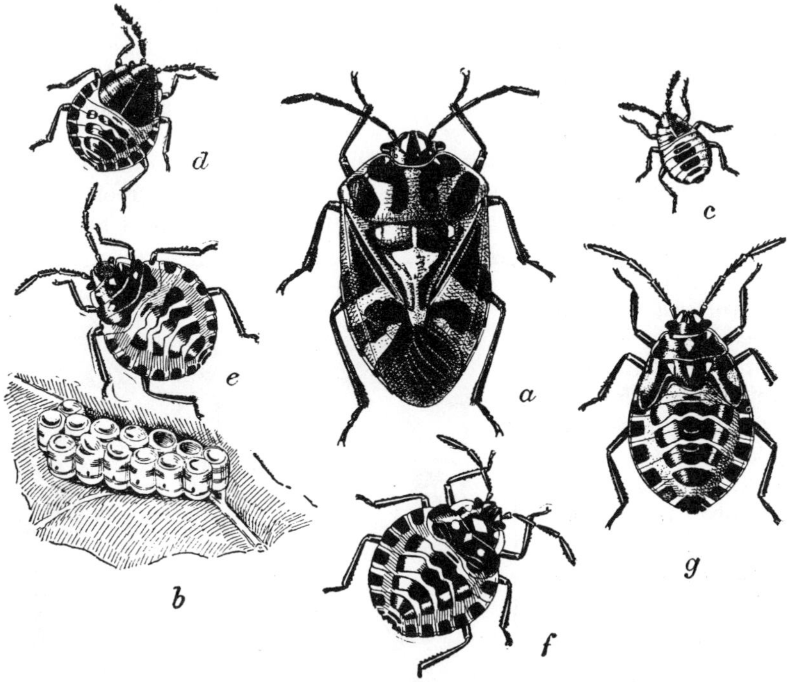

HARLEQUIN CABBAGE BUG (*Murgantia histrionica*).[1]

A—Adult. B—Egg mass. C—First stage of nymph. D—Second stage. E—Third stage. F—Fourth stage. G—Fifth stage. All enlarged.

The Cabbage Maggot (*Pegomya brassicæ*, Bouché).—Soft white maggots work in the roots of cabbage, turnip and cauliflower, eating away the root hairs and scarring the surface of the larger roots. This maggot is the larva of a two-winged fly which lays its eggs in the ground near the plants.

Treatment.—Fit a disk of tarred paper about four inches in diameter around the stem of each plant, letting it lie flat on the ground to keep the

[1] Bur. Ent. Cir. 103.

maggots from reaching the roots. Clean up all cabbage stumps in the fall and plow deeply. Rotate crops.

The Colorado Potato Beetle (*Leptinotarsa decemlineata*, Say.).—Without doubt the worst enemy of the potato is the robust yellow-striped beetle which, together with its larvæ or slugs, feeds upon the leaves. The insect is too well known to need description.

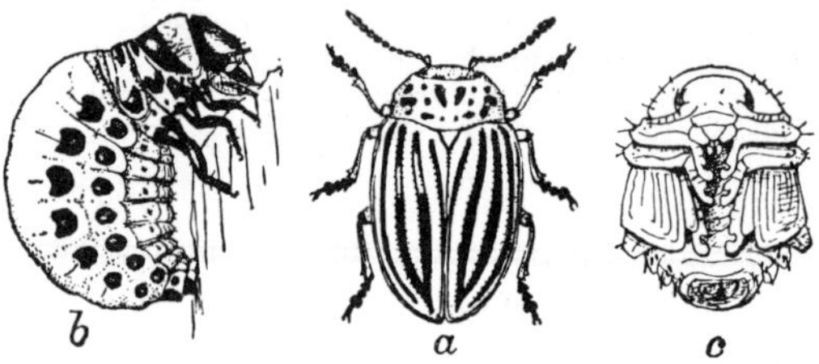

COLORADO POTATO BEETLE
(*Leptinotarsa decemlineata*).[1]
A—Beetle. B—Larva. C—Pupa. Enlarged.

Treatment.—Apply arsenicals either as a dust or as a spray. Hand picking or "bugging "may be resorted to in a small garden patch.

Bur. Ent. Cir. 87, Bull. 82, Pt. 1.

Flea Beetles.—Small dark-colored insects which as adults feed upon the foliage of many truck crops and weeds. The larvæ feed upon the roots. The name is derived from the active way in which the insect hops about. In this respect it resembles a flea.

Treatment.—Clean up weeds about the garden that may form a breeding place for the pests. Apply arsenicals to plants as a spray, using Bordeaux mixture preferably, which acts as a repellent.

Harlequin Cabbage Bug (*Murgantia histrionica*, Hahn.).—This gaudily marked bug is easily recognized by its bright colors of red, yellow and blue. It feeds upon cabbage, cauliflower, mustard and other related plants.

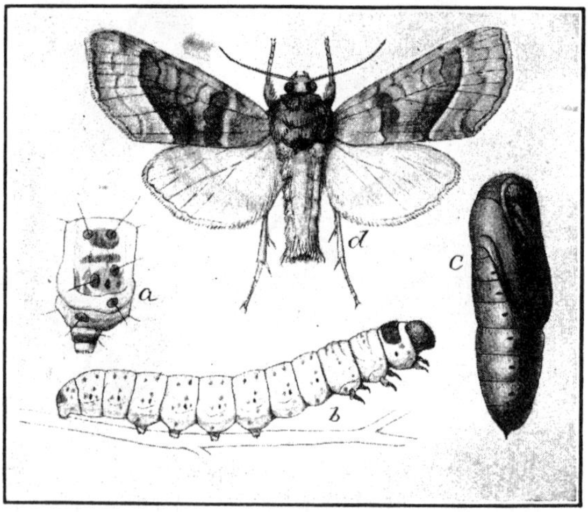

HOP PLANT BORER (*Hydrœcia immanis*).[2]
A—Enlarged segment of larva. B—Larva. C—Pupa. D—Adult. Natural size.

Treatment.—Plant a trap crop of mustard or turnips in the spring and fall and when the bugs have become numerous spray with pure kerosene. Hand picking may be profitable in the spring. Methods of clean culture should be practiced, especially in the fall, tall cabbage stalks and weeds

[1]Bur. Ent. Cir. 87 [2] Bur. Ent. Bull. 7

being destroyed in order to starve out the bug. Destroy trash where it might hibernate.

Bur. Ent. Cir. 103.

The Hop Aphis (*Phorodon humuli*, Schr.).—This plant louse is found on the plum in spring, but flies to the hop plant in early summer, where it turns the leaves yellow, causing them to fall.

WINGLESS PROGENY OF WINGED HOP APHIDS FROM ALTERNATE HOST.[1]

Treatment.—When aphids appear spray thoroughly with 40 per cent nicotine solution diluted 1 part to 800 parts water.

Bur. Ent. Bull. 111.

The Hop Plant Borer (*Hydrœcia immanis*, Get.).—The hop plant is attacked in three places by this borer during the period of development of the insect. Early in the season it bores into the tender tips, causing them to droop; after a short time it falls to the ground and bores into the stem at the crown. Later it bores out of the stem and goes below the ground, feeding just above the old roots, where it nearly severs the plant.

Treatment.—In the spring search for the affected tips and crush the insects in the stem.

Bur. Ent. Bull. 7, p. 40.

[1] Bur. Ent. Bull. 111.

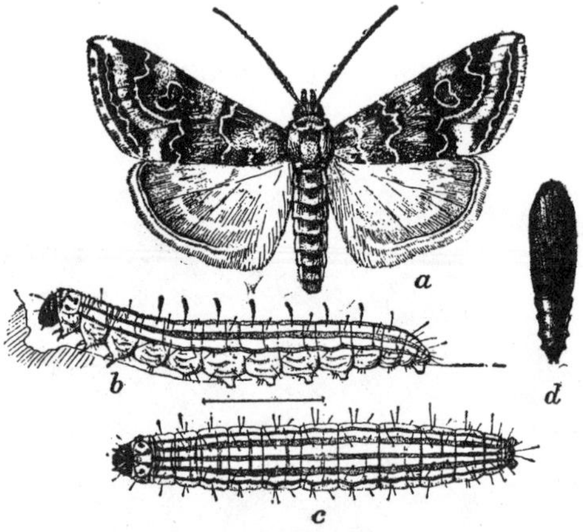

Imported Cabbage Web Worm
(*Hellula undalis*).[1]

A—Mature moth. B—Larva, lateral view. C—Larva, dorsal view. D—Pupa. All three times natural size.

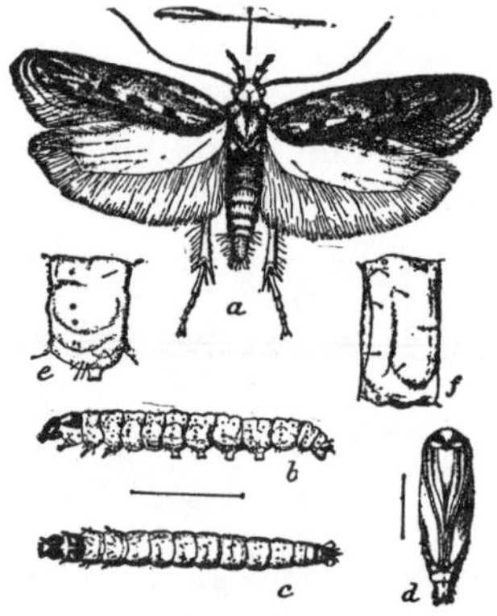

Potato Tuber Moth
(*Phthorimœa operculella*).[2]

A—Moth. B—Larva, lateral view. C—Larva, dorsal view. D—Pupa. E, F—Segments of larva, enlarged. (Redrawn from Riley and Howard.)

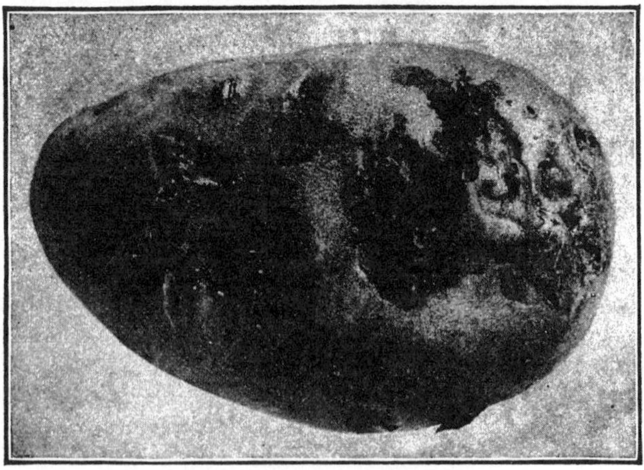

Work of the Potato-Tuber Moth.[2]
Exterior view of potato.

[1] Bur. Ent. Bull. 19. [2] Farm. Bull. 557.

The Imported Cabbage Web Worm (*Hellula undalis*, Fab.).—This worm feeds upon cabbage, turnips and other similar crops, spinning a web under which it retires when not feeding.

Treatment.—Same as for cabbage looper.

Ent. Bull. 23, p. 54.

The Imported Cabbage Worm (*Pontia rapæ*, Linn.).—Of all the insects on cabbage, this is the worst pest. It is the larvæ of the white butterfly seen fluttering about over fields of cabbage during spring and summer.

Treatment.—Same as for cabbage looper.

Bur. Ent. Cir. 60.

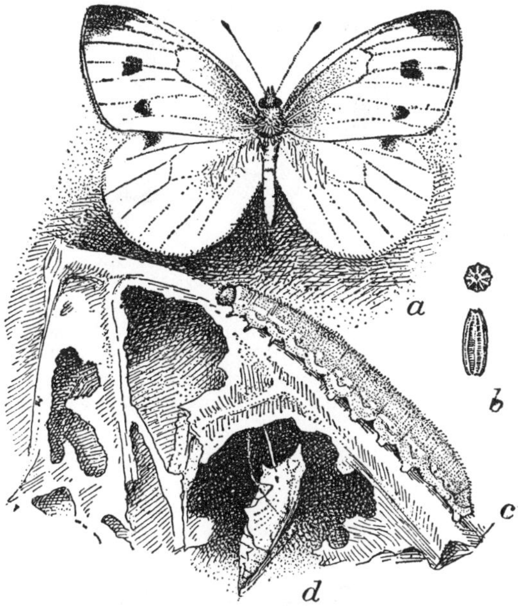

IMPORTED CABBAGE WORM (*Pontia rapæ*).[1]

A—Female butterfly. B—Above, egg as seen from above; below, egg as seen from side. C—Larva in natural position on cabbage leaf. D—Suspended chrysalis. A, C, D—Are slightly enlarged. B—More enlarged.

The Melon Aphis (*Aphis gossypii*, Glov.).—This plant louse feeds not only on melons but on cotton, strawberries and a number of other plants.

Treatment.—Before the leaves are badly curled spray them with nicotine solution, turning the vines over if necessary, so as to hit the under sides of the leaves. In small gardens fumigate under tub with carbon bisulphide, using about a teaspoonful to each cubic foot of space. Tobacco fumes may also be used.

Bur. Ent. Cir. 80.

The Potato Tuber Moth (*Phthorimæa operculella*, Zell.).—Potato growing is now menaced in California, Washington and southern Texas by this insect, which bores into the vines and tubers of potatoes. It also feeds upon tomato, eggplant and tobacco, and on the latter plant is known as the split worm.

Treatment.—No satisfactory method of treatment is known, but the injury may be partly prevented by clean methods of cultivation, crop rotation and fumigation of infested tubers. The latter is by far the best remedy. For a full discussion of methods of control see Farm. Bull. 557.

[1] Bur. Ent. Cir. 60.

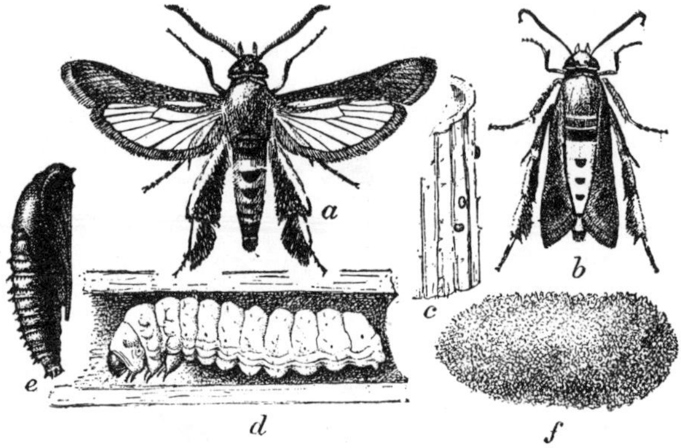

Squash Vine Borer
(*Melittia satyriniformis*).[1]

A—Male moth. B—Female moth with wings folded in natural position when at rest. C—Eggs shown on bit of squash stem. D—Full grown larva—*in situ* in vine. E—Pupa. F—Pupal cell. All ⅓ larger than natural size.

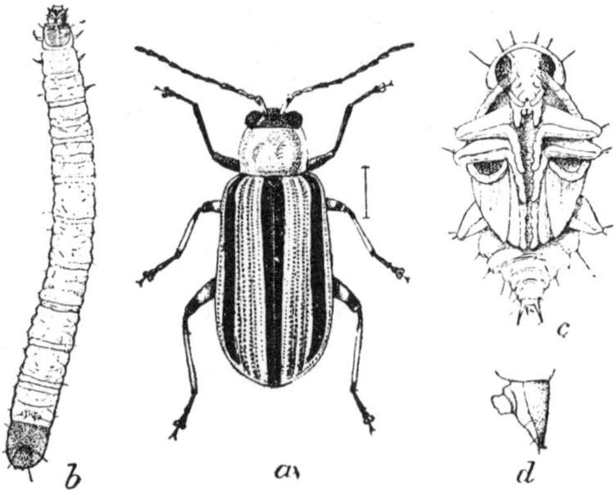

Striped Cucumber Beetle
(*Diabrotica vittata*).[2]

A—Beetle. B—Larva. C—Pupa. D—Anal proleg. A, B, C—Much enlarged. D—More enlarged.

Sugar-Beet Web Worm
(*Loxostege sticticalis*).[3]

Moth twice natural size.

[1] Bur. Ent. Cir. 38. [2] Bur. Ent. Cir 31. [3] Bur. Ent. Bull. 109, Part 2.

The Squash Bug (*Anasa tristis,* De G.).—This well-known insect is often a serious pest of squashes and pumpkins and can nearly always be found upon the vines during the summer.

Treatment.—Pick off and destroy the eggs in the spring. Trap the

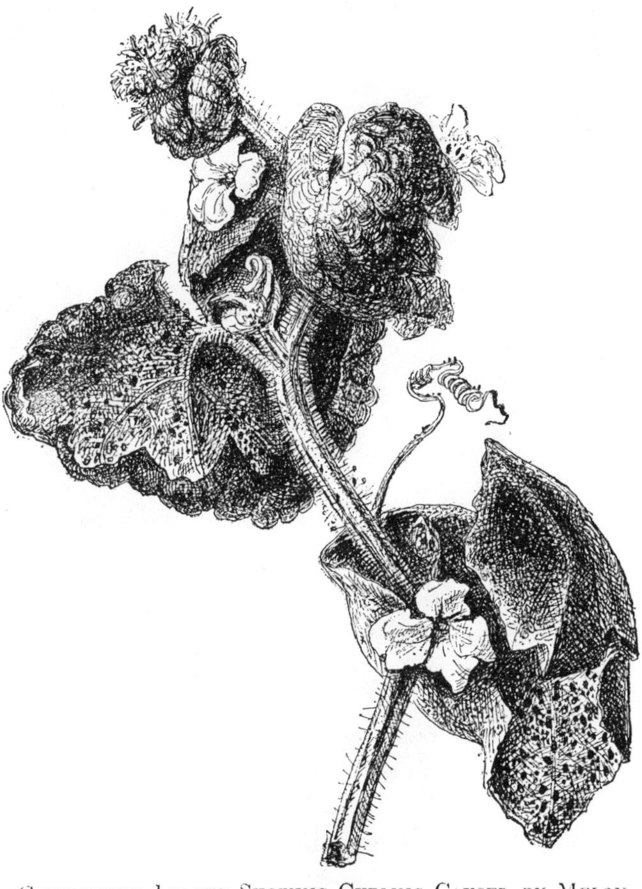

Cantaloupe Leaves, Showing Curling Caused by Melon Aphis; Aphides on Lower Surface.[1]

Slightly reduced.

bugs under boards placed near the vines and gather them up in the morning. Protect cucumbers and melons by planting early squashes among them, from which adults should be picked. Spray with kerosene emulsion.

Bur. Ent. Cir. 39.

Squash Vine Borer (*Melittia satyriniformis,* Hbn.).—In many localities

[1] Bur. Ent. Cir. 80.

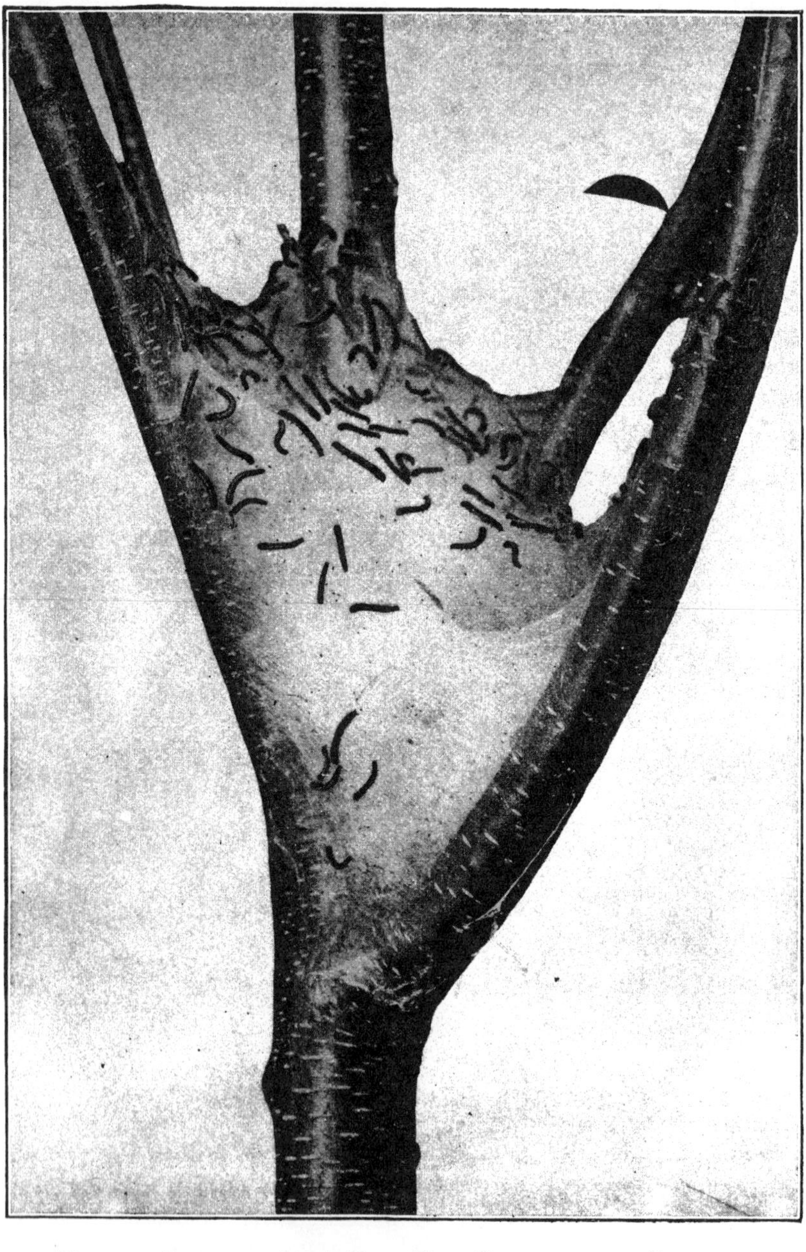

NEST AND LARVÆ OF APPLE TREE TENT CATERPILLAR IN CROTCH OF WILD CHERRY TREE.[1]

[1] Farm. Bull. 662.

this is the most serious pest of squash vines. The larvæ bore into the vines, causing them to rot and break off easily.

Treatment.—Rake up and destroy vines as soon as possible in the fall. Plow deeply in the spring. Rotate crops; plant early squashes among other vines as a trap crop.

Bur. Ent. Cir. 38.

The Striped Cucumber Beetle (*Diabrotica vittata,* Fab.).—A black-and-yellow striped beetle two-fifths of an inch long, injuring cucumbers, squashes and melons by feeding on the young plants as they come up.

Treatment.—Cover the hills of young plants with nets to protect them from beetles. Dust heavily with air-slaked lime and tobacco dust while the dew is on. Spray the plants with arsenate of lead 3 to 5 pounds to 50 gallons.

Bur. Ent. Cir. 31.

Sugar Beet Web Worm (*Loxostege sticticalis,* Linn.).—This insect defoliates beets and webs them together at times, causing notable injury. It also feeds on onions, cabbage, alfalfa, pigweed and careless weed.

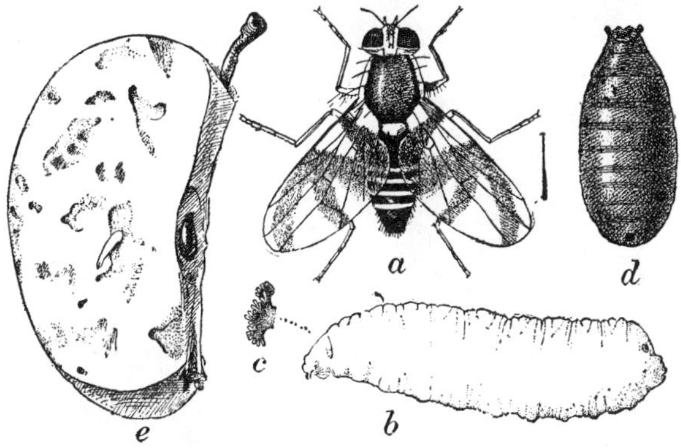

APPLE MAGGOT, OR RAILROAD WORM (*Rhagoletis pomonella*).[1]

A—Adult. B—Larva or maggot. C—Funnel of cephalic spiracle. D—Puparium. E—Portion of apple showing injury by maggots. A, B, C—Enlarged. D—Still more enlarged. E—Reduced.

Treatment.—Plow the infested land in late fall or winter. Spray or dust the plants with arsenicals.

Bur. Ent. Bull. 109, Pt. 6.

FRUIT INSECTS

Apple Maggot, or Railroad Worm (*Rhagoletis pomonella,* Walsh.).—The larva of a two-winged fly. It infests summer and early fall apples and occasionally winter apples, tunneling through the flesh of fruit and causing it to fall.

Treatment.—Spray the trees during the first week in July with arsenate of lead, 4 pounds to 100 gallons. Pick up infested fruit every two or three days and feed it to hogs or bury it deeply.

Bur. Ent. Cir. 101.

Apple Tree Tent Caterpillar (*Malacosoma Americana,* Fab.).—The

[1] Bur. Ent. Cir. 101.

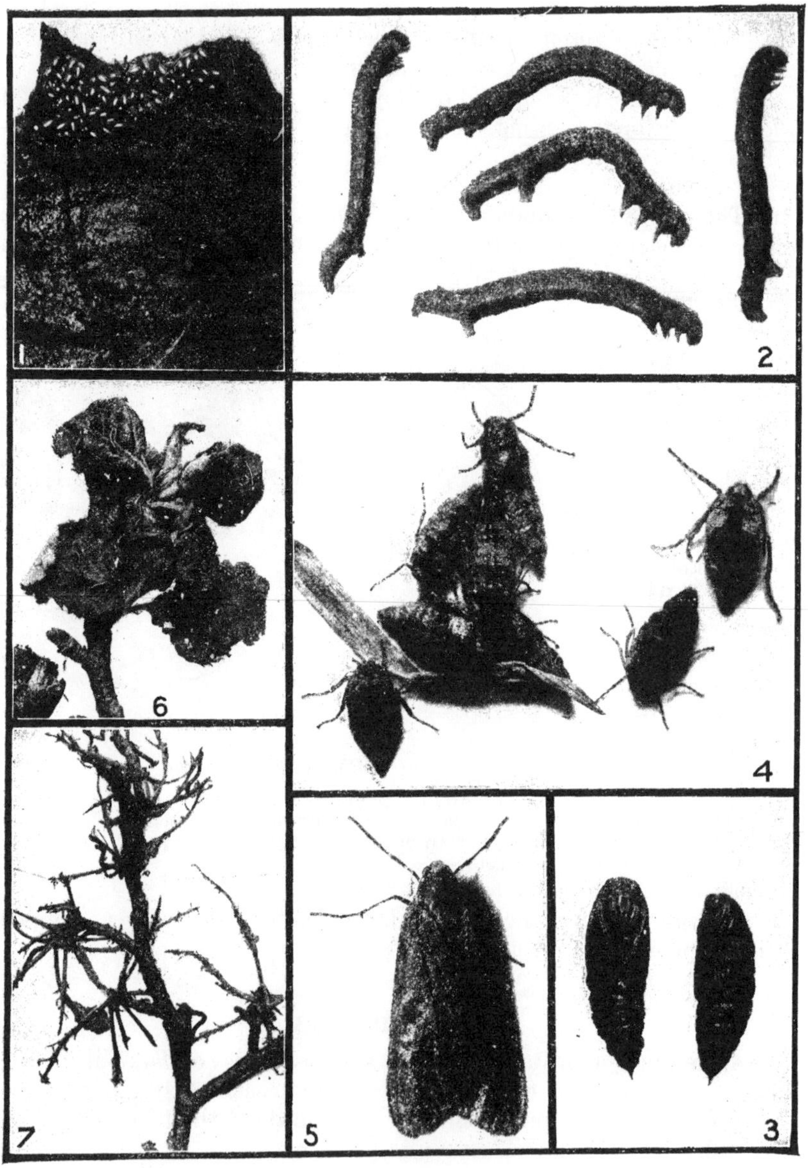

STAGES AND WORK OF SPRING CANKER-WORM (*Paleacrita vernata*).[1]

1—Egg mass on bark scale. 2—The larvæ or canker-worms. 3—Pupæ. 4—Female moths. 5—Male moth. 6—Work of canker-worms on apple leaves when small. 7—Later work of the larvæ, only the midribs of leaves being left. 1-5—Considerably enlarged. 6, 7—Reduced.

[1] Bur. Ent. Bull. 68 Pt. 2.

larva of moth. It feeds on the foliage of apple and a number of other trees. It makes large nests or web tents in which caterpillars stay when not feeding.

Treatment.—Spray the trees with arsenate of lead when the nests first appear. Make later application if necessary.

Farm. Bull. 662.

The Brown-Tail Moth (*Euproctis chrysorrhœa*, Linn.).—This well-known caterpillar was accidentally introduced into Massachusetts from Europe. It has now spread over a large part of New England and is still extending its territory. The moths appear early in July and the female deposits masses of eggs on the under side of leaves. The young caterpillars web terminal leaves together and spend the winter in those nests in partially grown condition. They resume feeding in the spring and soon reach their full development.

Treatment.—Cut out and burn all the winter nests before the buds start. Spray the trees with arsenate of lead, 4 pounds to 100 gallons. Band the trees with tanglefoot to prevent the ascent of caterpillars from other trees.

Farm. Bull. 264.

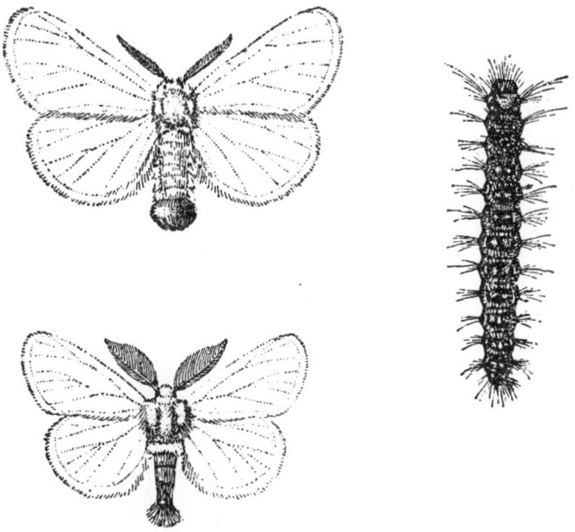

BROWN-TAIL MOTH (*Euproctis chrysorrhœa.*)[1]

Female moth above, male moth below, larva or caterpillar at right, slightly enlarged.

Canker-Worm, The Spring (*Paleacrita vernata*, Peck), **and The Fall** (*Alsophila pometaria*, Harris).—The larvæ of canker-worm moths are measuring worms about an inch long, dark-colored and variously striped. The adult males are winged, females wingless. They defoliate apple trees.

Treatment.—Cultivate orchards well in summer to destroy pupæ. Apply sprays of arsenate of lead 4 or 5 pounds to 100 gallons water, first before the blossoms open; second, just after petals fall. Apply barriers of tanglefoot or cotton batting to the trunks of trees to prevent the ascent of the moths to lay eggs.

Bur. Ent. Cir. 9; Bur. Ent. Bull. 68, Pt. 2.

The Cherry Fruit Flies (*Rhagoletis cingulata*, Loew, and *R. fausta*, O. S.).—Two-winged flies deposit eggs in cherries. Maggots develop in the fruit on the tree, causing it to rot on one side. They enter ground to pupate.

[1] Farm. Bull. 264.

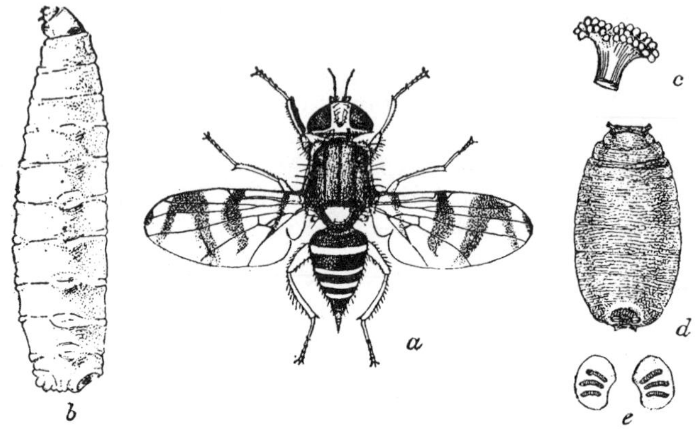

CHERRY FRUIT FLY
(*Rhagoletis cingulata*).[1]

A—Fly. B—Maggot from side. C—Anterior spiracles of same. D—Puparium. E—Posterior spiracular plates of pupa. All enlarged.

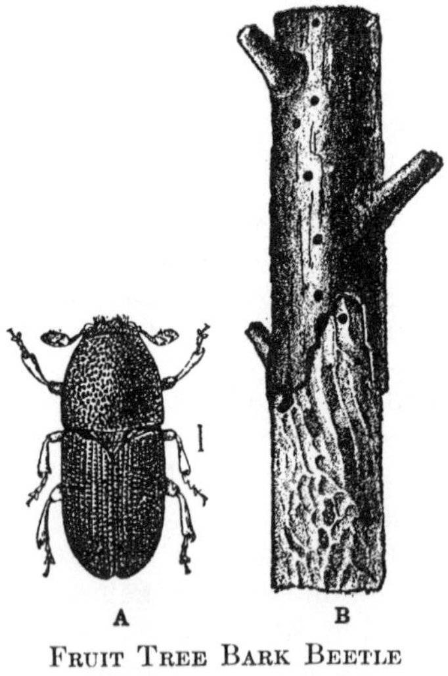

FRUIT TREE BARK BEETLE
(*Scolytus rugulosus*).[2]

A—Adult beetle. B—Work in twig of apple. Natural size.

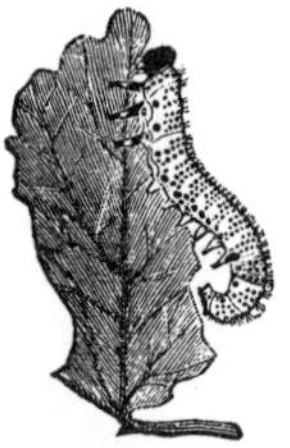

AN IMPORTED CURRANT WORM.[3]

[1] Bur Ent. Bull. 44. [2] Bur. Ent. Cir. 29.
[3] Courtesy of Connecticut Agricultural Experiment Station.

Treatment.—Apply arsenate of lead to the trees, either with or without sweetening, 4 or 5 pounds to 100 gallons, at the time the flies are emerging. Two applications usually necessary.

Bur. Ent. Bull. 44, pp. 70–75; Cornell Agricultural Exp. Sta. Bull. 325.

The Codling Moth, or Apple Worm (*Carpocapsa pomonella*, Linn.). —This is the insect the larva of which is responsible for most of the wormy apples, pears and quinces. The female moths lay their eggs upon the leaves and fruit. The larvæ upon hatching begin at once to hunt for the fruit, which they enter mostly through the calyx cup. In spraying it is very necessary to fill this cup with the poison, as it is here that the larvæ mostly take their first meal. The time when this first and most important spray should be made is just after the petals have fallen and before the calyx cup closes. In most parts of the country there are two broods of insects, but in the South there may be three and in places even four.

Treatment.—Spray with arsenate of lead, 4 pounds to 100 gallons, first just as the petals have fallen; second, three to four weeks after the petals have fallen; third, eight to nine weeks after petals have fallen.

Bur. Ent. Bull. 115, Pts. 1 and 2; Farm. Bull. 492.

Currant Worm, The Imported (*Pteronus ribesii*, Scop.).—This currant worm is the most destructive insect enemy of the currant, but is easily controlled.

Treatment—At the time the worms begin to appear spray or dust with an arsenical.

Report of the Conn. State Entomologist, 1902, pp. 170–172.

The Flat-Headed Apple Tree Borer (*Chrysobothris femorata*, Fab.).—A larva about one inch long, slightly flattened. The front end much enlarged. It usually attacks trees partly dead or in poor condition, rarely sound trees. As a preventive measure, keep trees healthy by use of fertilizers and thorough cultivation.

Treatment.—Dig the borers from burrows with sharp instrument.

Bur. Ent. Cir. 32.

The Fruit Tree Bark Beetle (*Scolytus rugulosus*, Ratz.).—The small dark-brown beetle which bores shot holes in fruit trees of nearly all kinds, like the flat-headed borer, works only in dead or dying wood. As a preventive, keep the trees healthy; clean up all dead wood about orchards; cut out and burn all infested wood.

Treatment.—No satisfactory treatment is known.

Bur. Ent. Cir. 29, Revised.

The Gipsy Moth (*Porthetria dispar*, Linn.).—The gipsy moth, like the brown-tail, is a serious enemy of forest and fruit trees. Egg masses are deposited in the fall on trunks of trees, on fences or wherever a roughened surface can be found. They hatch in the spring and the larvæ feed on the foliage of various trees.

Treatment.—Hunt out the egg masses in winter time and soak with coal tar creosote. Spray the trees in the spring with arsenate of lead as soon as the eggs hatch, using 10 pounds to 100 gallons of water.

Bur. Ent. Bull. 87; Farm. Bull. 564.

The Grape Berry Moth (*Polychrosis viteana*, Clem.).—A larva about

GRAPE BERRY MOTH
(*Polychrosis viteana*).[1]
1 and 2—Adult, or moth. 3—Full grown larvæ. 4—Pupæ.
All greatly enlarged.

one-fourth of an inch long, works in the berry of grape, webbing several together. It is the cause of most of the wormy grapes in the eastern sections of the country.

Treatment.—Spray with arsenate of lead, 6 pounds to 100 gallons.

[1] Bur. Ent. Bull. 115.

First application shortly after fruit sets; second, about ten days later, and third, when the fruit is about half grown or when the second brood eggs are hatching.

Bur. Ent. Bull. 116, Pt. 2, Farm. Bull. 284.

Grape Leaf Hopper (*Typhlocyba comes*, Say.).—This active little

INJURY TO GRAPES BY LARVÆ OF SECOND BROOD OF GRAPE-BERRY MOTH.[1]

Just previous to harvesting of fruit.

hopper is known in all parts of the country where grapes are grown. It is yellowish in color, marked with green stripes. The leaves of the grapes are injured by the puncture made by the hopper in feeding on the under side of the leaf, causing them to turn spotted and yellow and finally fall off.

Treatment.—Spray the vines thoroughly about the first week in July, when the maximum number of young hoppers are on the leaf, with a

[1] Bur. Ent. Bull. 116, Pt. 2.

Grape Leaf Hopper (*Typhlocyba comes*).[1]

Adult, winter form. Greatly enlarged.

solution of 40 per cent nicotine, diluted 1 part to 1000 parts water. Clean up trash and weeds in fence corners and practice clean culture generally.

Dept. Bull. 19.

The Grape Vine Flea Beetle (*Haltica chalybea*, Ill.).—A blue metallic beetle about one-fourth of an inch long. It feeds on buds and tender shoots in early spring. The grubs feed later upon the leaves.

Treatment.—Spray with arsenate of lead to kill the adults and grubs on the leaves during May and June. The beetles may be captured in sheets or pans by jarring the vines.

New York (Geneva) Exp. Sta. Bull. 331, pp. 494–514.

The Lesser Apple Worm (*Enarmonia prunivora*, Walsh).—This insect is closely related to the codling moth and has very much the same life history.

Treatment.—Spray as for the codling moth, but take especial pains to make the second spray very thorough, three to four weeks after petals have fallen.

Bur. Ent. Bull. 68, Pt. 5; Bur. Ent. Bull. 80, Pt. 3.

The Peach Tree Borer (*Sanninoidea exitiosa*, Say.).—The larvæ of this insect are found at the crown of peach, plum and cherry trees, boring

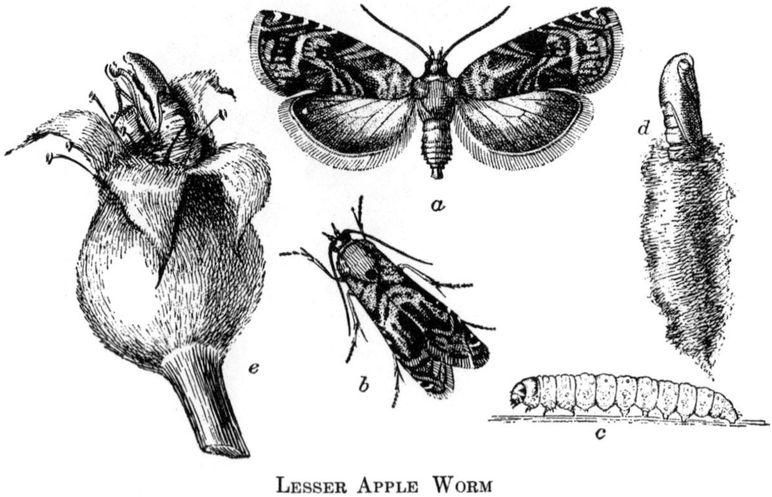

Lesser Apple Worm (*Enarmonia prunivora*).[2]

A—Adult, or moth. B—Same with wings folded. C—Larva. D—Pupa in cocoon ready for transformation to adult. E—Young apple, showing at calyx end empty pupa skin from which moth has emerged. Enlarged about three times.

[1] Dept. Bull. 19. [2] Bur. Ent. Bull. 68.

beneath the bark. The external indications of their presence are the mass of sap which is commonly seen at the base of the tree and the frass or wormwood that has been worked out through holes in the bark. Preventive means that have given some measure of success are various styles of protectors placed around the base of the trees, and coating washes applied to the trunk. The latter are not satisfactory.

Treatment.—Remove the ground from the crown of the tree in the spring and fall and dig out the borers with a sharp knife.

Georgia Agri. Exp. Sta. Bull. 73; N. J. Agri. Exp. Sta. Bull. 235.

Pear Leaf Blister Mite (*Eriophyes pyri*, Pagenstecher).—This small mite, only $\frac{1}{125}$-inch in length, is the cause of the rough, blistered surface of pear and apples leaves. When the attack is severe the trees become so brown that they have the appearance at a distance of having been swept by fire.

Treatment.—Spray in the spring or fall with concentrated commercial lime-sulphur testing 33° Baumé, diluted at the rate of 1–10 or 11.

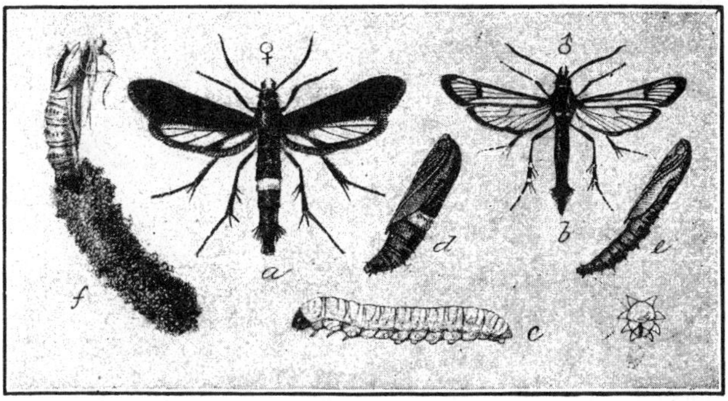

Peach Tree Borer
(*Sanninoidea exitiosa*).[1]

A—Adult female. B—Adult male. C—Full grown larva. D—Female pupa. E—Male pupa. F—Pupa skin extruded partially from cocoon. All natural size.

Plant Lice (*Aphididæ*).—Many species of plant lice are found upon the various fruit trees grown in this country. They feed by sucking the sap from the leaves and stems and thus do considerable injury at times. Some species curl the leaves about them so that they are very difficult to reach with a spray unless the treatment is made before the attack becomes severe. The treatment for all aerial forms is practically the same.

Treatment.—Spray carefully with a 40 per cent nicotine sulphate solution diluted at the rate of 1 part to 800 parts of water, being sure to touch all insects with the spray. A kerosene emulsion spray is also good if used at the 8 or 10 per cent strength.

Plum Curculio (*Conotrachelus nenuphar*, Hubst.).—On apples this insect injures the fruit by deforming or scarring it by its feeding and egg-laying punctures.

Treatment.—Spray as for codling moth, except that one additional

[1] Bur. Ent. Cir. 17

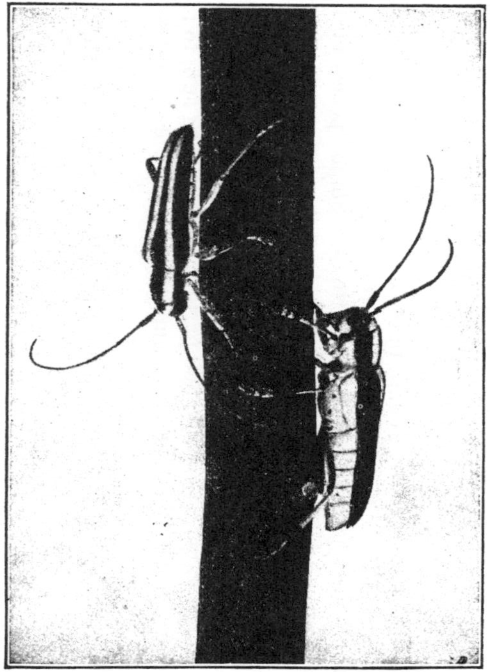

Adult Male and Female Roundheaded Apple Tree Borer.[1]

Male on left, female on right, slightly enlarged.

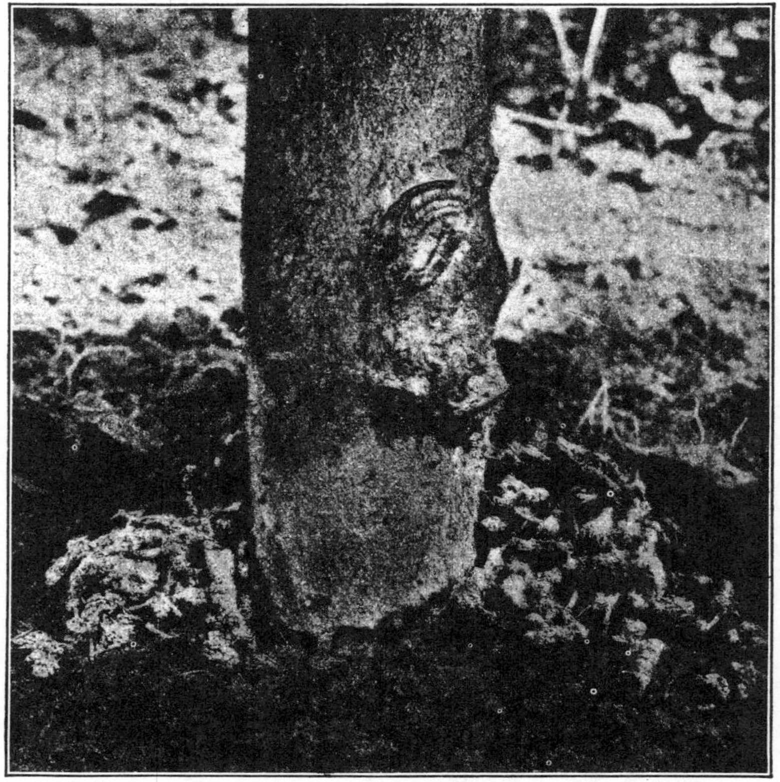

Castings of Roundheaded Apple Tree Borer at Base of Young Apple Tree.[2]

[1] Farm. Bull. 675. [2] Farm. Bull. 657.

spray should be given before the blossoms open or at the time the cluster buds have opened out.

On plum, peach and cherry trees most of the injury is caused by the grubs inside the fruit.

Treatment.—For plums, spray with arsenate of lead, two pounds to 50 gallons; first, soon after petals fall; second, a week or ten days later.

For cherries, same as for plum.

For peaches, first, spray just as calyxes or shucks are shedding;

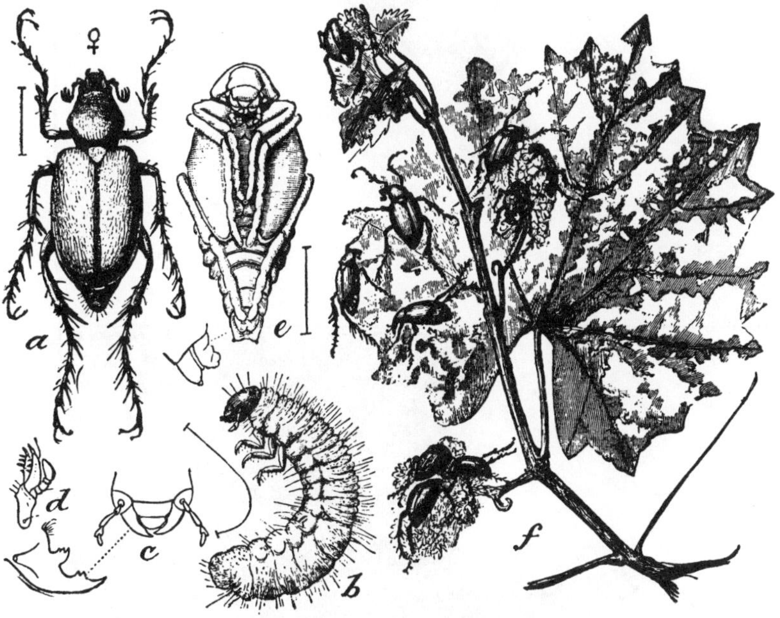

ROSE CHAFER
(*Macrodactylus subspinosus*).[1]

A—Adult, or beetle. B—Larva. C, D—Mouth parts of larva. E—Pupa. F—Injury to leaves and blossoms of grape, with beetles at work. A, B, E—Much enlarged. C, D—More enlarged. F—Slightly reduced.

second, spray three weeks later. When spraying peaches, self-boiled lime-sulphur is usually added to prevent fungous troubles.

Farm. Bull. 440, Farm. Bull. 492.

The Rose Chafer (*Macrodactylus subspinosus*, Fab.).—This beetle is recognized by his long legs and yellowish-gray color. Often in sandy regions the beetles swarm upon the grapes in great numbers, causing serious injury.

[1] Bur. Ent. Bull. 97.

1

2

SAN JOSÉ SCALE[1]

1—Peach tree with top killed by scale. 2—Peach twig moderately infected, showing male and female scale. Enlarged four times.

[1] Bur. Ent. Bull. 62.

Treatment.—Spray with arsenate of lead at the rate of 8 or 10 pounds to 100 gallons of water, to which is added 1 or 2 gallons of cheap molasses.

Bur. Ent. Bull. 97, Pt. 3.

Round-Headed Apple Tree Borer (*Saperda candida,* Fab.).—The adult of the round-headed borer is a handsome striped beetle. It lays its eggs on the bark at the base of apple trees and the young grubs eat through and enter the wood. Their presence can be detected by dark areas or discolored places on the bark and usually by the presence of small chips or frass worked out through the holes.

Treatment.—When the borer is located it should be removed with a sharp knife. Many washes and protectors have been tried to prevent injury from this insect. White lead has been strongly recommended for this purpose.

Bur. Ent. Cir. 32, Farm. Bull. 675.

The San José Scale (*Aspidiotus perniciosus,* Comst.).—This insect has a wide distribution throughout the country and is a serious enemy of fruit trees in many states. The scale is nearly circular in outline and about the size of the head of a pin. When it is plentiful upon trees it becomes encrusted on the trunk and branches, giving the tree a scurfy appearance. The insect under the protecting scale feeds by sucking the sap of the tree, so a contact insecticide is necessary for its control.

Treatment.—Spray the trees during the dormant season with concentrated lime-sulphur giving a Baumé test of 33°, diluted at the rate of 1 gallon to 8 or 9 of water. The so-called miscible oils (mineral oils which have been so treated that they may be readily mixed with water) are also used successfully.

Bur. Ent. Bull. 62.

REFERENCES

"Insects Injurious to the Household and Annoying to Man." Herrick.
"Injurious Insects: How to Recognize and Control." O'Kane.
"Manual of Fruit Insects." Slingerland and Crosby.
"Insects Injurious to Vegetables." Chittenden.
"Manual for Study of Insects." Comstock.
"American Insects." Kellogg.
"Insect Pests of Farm, Garden and Orchard." Sanderson.
California Expt. Station Bulletin 255. "The Citricola Scale."
California Expt. Station Bulletin 258. "Mealy Bugs of Citrus Tree."
Connecticut Expt. Station Bulletin 186. "Gipsy Moth."
Iowa Expt. Station Bulletin 162. "Strawberry Slugs."
Illinois Expt. Station Bulletin 174. "Method for Controlling Melon Lice."
Maine Expt. Station Bulletin 242. "Pink and Green Aphid of Potato."
Missouri Expt. Station Bulletin 134. "Insect Pests of Field Crops."
New Jersey Expt. Station Circular 46. "Hessian Fly."
New York Expt. Station Bulletin 402. "Controlling Plant Lice in Apple Orchard."
Ohio Expt. Station Bulletins:
154. "Important Greenhouse Pests."
264. "Orchard Bark Beetles and Borers."
Utah Expt. Station Bulletin 138. "Control of Grasshoppers."
Canadian Dept. of Agriculture Bulletins:
150. "Common Fungous and Insect Pests."

Canadian Dept. of Agriculture Bulletins:
187. "The Codling Moth."
219. "San José Scale and Oyster Shell Scales."
227. "Cherry Fruit Flies."
Farmers' Bulletins, U. S. Dept. of Agriculture:
344. "Boll-Weevil Problem."
492. "Insect Enemies of Fruit and Foliage of the Apple."
498. "Methods of Exterminating the Texas Fever Tick."
512. "Boll-Weevil Problem."
543. "Common White Grubs."
557. "Potato Tuber Moth."
564. "The Gipsy and Brown Tail Moth: Their Control."
634. "The Larger Corn Stalk Borer."
636. "The Chalcis-Fly in Alfalfa Seed."
637. "The Grass Hopper Problem and the Alfalfa Culture."
639. "Eradication of the Cattle Tick."
640. "The Hessian Fly."
650. "The San José Scale and Its Control."
658. "Cockroaches."
659. "The True Clothes Moth."
662. "The Apple Tree Tent Caterpillar."
668. "The Squash Vine Borer."
671. "Harvest Mites or Chiggers."
674. "Control of Citrus Thrips in California and Arizona."
675. "The Round-Headed Apple Tree Borer."

CHAPTER 77

Insecticides and Fungicides

By H. Garman
Professor of Entomology, University of Kentucky

The word insecticide has come to mean any chemical or other substance used to destroy insects that are hurtful or objectionable in any way to man. This definition excludes substances such as sticky fly-paper that may be employed to entrap pests and would, according to some entomologists, exclude also simple deterrents, such as oil of citronella, used to keep insects away by their offensive odors. In a general way, however, every substance employed to prevent the injuries of insects is an insecticide and in this view it does not matter whether or not they kill, deter or entrap.

The insecticides most used and valued by practical men either kill as poisons when eaten with food, or else destroy when brought in contact with the bodies of insects, in which case they are sometimes called contact insecticides.

A group of insecticides of which the effective ingredient is arsenic has proved especially popular and useful in suppressing insects which feed by gnawing away and devouring the leaves of plants.

Paris Green.—Of these the one best known and most used is Paris green, Schweinfurth green, or Imperial green, French green and Emerald green. It was first used in the arts, and because of its cheapness and poisonous properties was early tried on the Colorado potato beetle (about 1868) proving a very satisfactory means of suppressing the pest when used either as a dry powder or when stirred into water. It contains a little soluble arsenic however, and in water this is liable to burn leaves to which it is applied, hence care must be exercised not to use too much. Four to five ounces of the powder in a barrel of water is commonly regarded as enough; if more is used a pound or two of freshly-slaked lime may be added to neutralize its caustic effect.

Arsenate of Lead.—Paris green has two defects: Its burning action is often hard to guard against, and its weight causes it to settle quickly when used in water, rendering the spray produced uneven in strength. Stirrers connected with spray pumps obviate the latter trouble, but sometimes increase the labor of operating pumps. The addition of lime, as already suggested, lessens the burning action, though the lime may, if care is not exercised, increase the labor of applying.

Arsenate of lead has neither of these defects. It is practically insoluble in water, does not burn foliage, and it is so finely subdivided that it remains suspended much better than Paris green. It has the additional advantage of adhering to leaves longer than Paris green, and thus fewer applications are required. A single spraying with this substance, if applied at the right time, is for some plants sufficient for a whole season. The arsenate of lead paste is commonly used with water in the proportion of 2½ to 3 pounds in 50 gallons. As found in the market it contains about 50 per cent of water.

It requires more by weight to destroy insects than Paris green, but the cost per pound is less and hence the actual cost for materials amounts to about the same, whichever poison is used. Its advantages are so decided in other directions that it is now supplanting Paris green in popular favor. For the injuries of most gnawing insects working on foliage this insecticide may be safely recommended.

To meet the objections sometimes made to arsenate of lead paste, a powdered arsenate of lead has recently been offered to the public by manufacturers of insecticides. The paste when dried out is lumpy and is not in this condition easy to mix with water. In the powdered form it is not open to this objection and may, besides, be dusted over plants without the addition of water.

There are serious objections to the use of poisonous dusts, however, though in practice they have advantages that always commend them to workmen. The weight of the water to be carried when using liquid sprays increases the labor, of course, and this ought to be lessened if it can be done without diminishing the effectiveness of the applications, and also without increasing the danger to those making the applications. The inhaling of either dry Paris green or arsenate of lead is a serious matter, and if continued long is certain to lead to ill health. Liquid sprays go more directly and evenly to the plants and stay there. They may be made just as promptly effective as the dusts if used when the injury is beginning. They are not so likely to be inhaled.

Arsenite of Zinc.—This poison has somewhat recently been recommended as a substitute for Paris green and arsenate of lead, and appears to be about equally good and somewhat cheaper than either. It is a finely divided white powder as put on the market and remains in suspension about as well as arsenate of lead, having thus some advantage over Paris green. It contains a little water-soluble arsenic and has been claimed to be less injurious to foliage even than arsenate of lead, possessing at the same time about the same killing power. For use it is stirred first into a little water and allowed to soak for a time, then is stirred into the water in which it is to be used, about one pound of the powder being added to 50 gallons of water. It contains nearly the same quantity of arsenic as Paris green. Like arsenate of lead, it remains in suspension better if a little soap is dissolved in the water into which it has been stirred. It has of late been quoted by dealers at from 20 to 25 cents per pound.

London Purple.—This arsenite came into use for injurious insects somewhat later than Paris green (about 1878), but is less used now than formerly because of its lack of uniformity in composition and its excessive burning of foliage. Its affective ingredient as an insecticide is arsenic in the form of lime arsenite and lime arsenate, of which it contains about 40 per cent, nearly half of which is soluble. It is the soluble arsenious and arsenic oxides that make this insecticide so injurious to the foliage and render necessary the addition of lime. The amount of pure arsenic present has been found to be about 29 per cent. For use it is customary to recommend about one-quarter pound each of London purple and fresh lime in from 50 to 75 gallons of water.

White Arsenic.—The use of this poison has been recommended from time to time for gnawing insects, but the time and labor required in boiling it with milk of lime (thus producing an arsenite of lime) in order to avoid its burning effect on foliage has prevented its general employment as an insecticide. It can be made to accomplish the same purpose as Paris green and arsenate of lead, without injury to foliage, by boiling for a half hour 1 pound of commercial arsenic and 2 pounds of fresh lime in 4 gallons of water, diluting with water finally to make 100 gallons.

Sulphur.—Flowers of sulphur has been used for many years as an insecticide, especially for mites infesting hothouse plants. When dusted on plants it does no harm to the leaves, but is not as effective as could be desired. When burned in hothouses it may do severe injury to plants. These defects have led to its neglect by entomologists. When sulphur is boiled with lime, however, it produces a lime sulphide, in which condition it becomes one of the best of insecticides for use in winter against scale insects.

Lime-sulphur Wash.—In this condition thousands of barrels of the boiled sulphur and lime are sold to fruit growers every year, who use it largely as a remedy against San José scale. A concentrated solution is prepared by boiling in large iron kettles, tanks or other vessels, 50 pounds of fresh lime, 100 pounds of sulphur and 50 gallons of water. Part of the water is heated, then the lime is added and is followed by the sulphur, the whole being stirred continually while boiling, the time employed being from fifty minutes to an hour. Finally, after adding enough hot water to make 50 gallons, the solution is strained and set aside until ready to use. Home-made solutions may not test higher than 27 to 30° Baumé, but when carefully made go higher and may even reach 34 or 35° Baumé, the differences being apparently due to differences in the quality of limes used.

The manufacturers now follow about the same formula in producing their concentrated products, but because of having better facilities will perhaps average higher in concentration than the fruit grower, although analysis of samples bought in the market have sometimes shown that they did not test as high as good home-made lime-sulphur.

These concentrated solutions are of a deep reddish-yellow color and for use must be greatly diluted with water. It is customary in spraying for San José scale to use one part of the solution to eight or ten of water and to apply during open weather in February or March, while the trees are still dormant. For summer use they must be diluted with from 30 to 50 parts of water to avoid injury to the foliage, but lose much of their value as insecticides when thus weakened. The concentrated solution is regarded as the most effective remedy for scale injury now in use.

It should be added that there has somewhat recently appeared a so-called "soluble sulphur" which is recommended for the same uses as

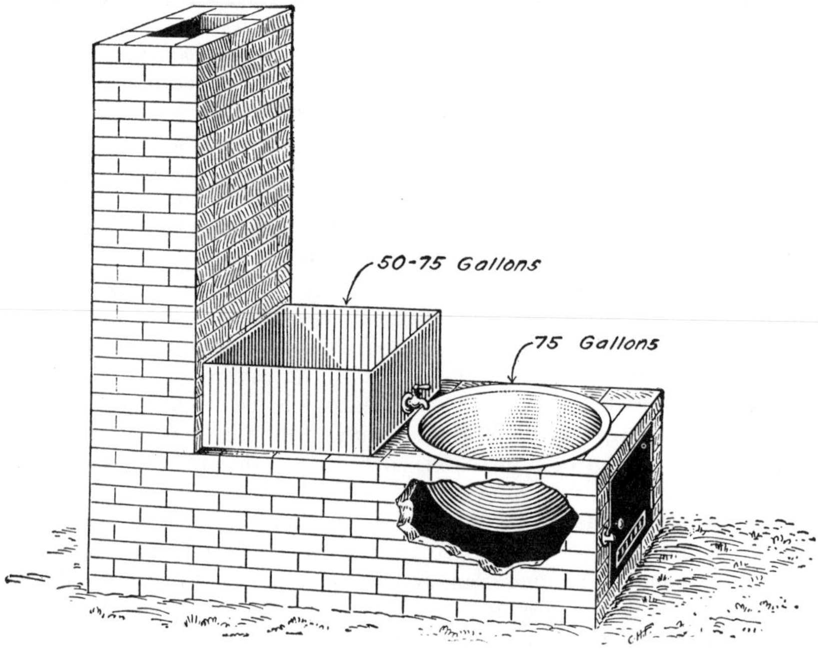

A LIME-SULPHUR COOKING OUTFIT.[1]

lime-sulphur. It promises well, but has not been tested long enough and carefully enough to justify very positive statements as to its merits.

Tobacco Extracts.—For use against soft-bodied insects such as plant lice there is no more useful insecticide than extracts made from the midribs of tobacco leaves. These extracts contain as their effective ingredient nicotine and differ widely in the percentage of nicotine they contain. Home-made extracts or decoctions are made by placing a couple of pounds of the midribs in a wooden bucket full of boiling hot water, allowing it to stand over night. The percentage of nicotine under such treatment will probably not be more than 0.07 per cent, but it is a very useful wash for plants infested with aphides, does no harm at all to leaves, and where

[1] From Farmers' Bulletin 650, U. S. Dept. of Agriculture.

tobacco is grown and the midribs can be easily secured is one of the very best insecticides for uses of this sort. The whole leaf makes a somewhat stronger extract (0.12 per cent) as determined by tests recently made at the Virginia Station. Soaking seems to extract as much of the nicotine as boiling. When plants are to be treated on a larger scale it becomes important to know just how much nicotine is present in a wash, and manufactured extracts, some of them containing 40 per cent of nicotine, are demanded. For the apple leaf louse, the lettuce louse, the rose aphis and other similar pests, these extracts are safe and effective. For thick-skinned insects they are not so satisfactory.

Tobacco is often used in other ways as a remedy for insect injuries, but is open to some objections when so employed. Florists have long used the midribs (often called "stems") for making a smudge for the destruction of plant lice. The tobacco is simply burned in a perforated iron vessel. The smoke leaves a strong smell of tobacco on flowers, which is sometimes objected to by buyers. The odor can be avoided by using the extract diluted with water and driven off as a vapor by dropping a hot iron into a pan containing it.

Pyrethrum.—Under the name Persian insect powder or simply insect powder this insecticide is to be obtained from most dealers in drugs. It is a brown powder made from the flowers of a rather handsome plant of the sunflower family (*Compositæ*). Its beauty leads florists to propagate it, though few who grow the plant know that it has any relation to the powder sold in drug stores. It comes to us from the East, and the powder commonly sold here is imported, though an effort has been made in the west coast states to manufacture the powder in this country.

The powder is thought to give off a volatile oil which penetrates the breathing tubes of insects and thus by some irritating or suffocating effect overpowers them. It is effective either dry, in water or when burned to produce a smudge, but must be fresh. It loses much of its effectiveness if kept in open packages. Though rather costly for use on field crops, it has a place in the household at times, and may sometimes be profitably resorted to for limited outbreaks of garden pests. Unlike most other insecticides, this one is not hurtful to man; at any rate, not more so than snuff.

White Hellebore.—This is another vegetable product, being the pulverized rootstocks of a plant (*Veratrum album*) of the lily family, occurring in Europe and northern Africa. It is used in this country for the rose slug, either dry or in water, in the latter case about two heaping tablespoonfuls being stirred into a wooden bucketful (2½ gallons) of water. It is a stomach poison and also a contact insecticide.

Old samples when not kept in airtight receptacles lose their virtue and tend to discredit this vegetable poison as a remedy for pests.

Coal Oil.—This oil has become well known as an insecticide in the form of an emulsion. It is a good contact insecticide, serving the same purpose as lime-sulphur wash in the destruction of scale insects, and having the advantage of remaining effective when diluted. It can, therefore, be used on foliage in summer for both scale insects and plant lice, and being quicker in its action than tobacco extract, has advantages under some circumstances over the extract for the prompt destruction of soft-bodied insects. It is, however, more likely to do injury to plants, especially if the emulsion is badly prepared, and this, together with the work required in making it, leads practical men to neglect it whenever they can use something else.

The standard emulsion is made of one-half pound of whale oil or laundry soap dissolved in a gallon of boiling hot water, this to be added to two gallons of coal oil, and the whole churned for ten minutes by passing rapidly through a force pump. As thus made it can be diluted for use, one part to ten of water.

Crude Oils.—These are sometimes used for the same purposes as the refined oil, and to render them easily mixed with water are sometimes mixed with caustic potash, fish oil and crude carbolic acid, producing a so-called miscible or soluble oil, which can be diluted with water for use like the coal oil emulsion.

Soaps.—Many of the soaps sold in our market can be used at times as a means of lessening the injuries of insect pests. A good soapsuds frequently and freely used on plants infested with aphides or scale insects has a good effect, though not a very prompt one. Stronger solutions must be used with some caution to avoid injury to foliage. When trees are dormant very strong solutions (one or two pounds to a gallon of water) are sometimes used on the trunks for scale and other insects.

Whale oil soap or fish oil soap, as it is sometimes called, is to be preferred to most others because of its more even composition. It is particularly good for use in making coal oil emulsion.

Coal Tar.—In the early days of fruit growing in America this substance was much used on the trunks of trees to prevent the ascent in the spring of the wingless female canker worm moth. It proved to have an injurious effect on the trees after a time, and hardened on exposure, so that the insects could pass over the barrier. It was then used on bands of tin, and by frequent renewal proved a useful check on the insect. But with the introduction of arsenites and spraying machinery, it was given up for the more convenient treatment. It is still used as a barrier, poured along the ground, for chinch-bugs which are migrating from small grains to corn. Seed corn may be treated with it before planting to deter wire worms and the seed corn maggot from attacking the germinating seeds. The corn is first immersed in warm water for a minute or two, then a couple of teaspoonfuls of the tar are stirred quickly among the grain so as to bring a little in contact with each seed. It dries over night so as to be ready for planting the following day. The application does no harm to the germ, as has been determined by germination tests of treated seeds.

Borax.—This substance has often been recommended for roaches in dwellings, and is sometimes found with an arsenite as an ingredient of proprietary roach pastes. Recent work done with a view to destroying the larvæ of house flies in manure indicates that this is one of the best of insecticides for the purpose, excelling for this use, coal oil, pyroligneus acid, formalin and Paris green. Sodium borate and crude calcium borate were both found effective in killing the larvæ, either when used dry or in solution. It was recommended as a result of the work done that about 0.62 pound of borax be used in 8 bushels of manure. Larger amounts of borax are believed to be injurious to plants when the manure is spread on land. The cost was estimated at one cent per horse per day.

Other Insecticides.—Numerous other insecticides have been recommended, and have had a limited use, but, excepting the fumigants considered later, they have not been generally adopted by practical men. Among them may be mentioned benzene, which is sometimes applied to fabrics to destroy clothes moth; carbolized plaster, sometimes recommended as a remedy for fleas about stables; fir-tree oil, lemon oil and oil of citronella, the latter often employed as a deterrent against the attacks of mosquitoes and also as a preventive of injury to seed corn in the soil. Quassia, the effective ingredient of which is quassiin, is obtained from the wood of the Jamaican *Picrasma excelsa.* It is an old insecticide that has been perhaps most used in solutions for the hop aphis in the West. The extract is made from the "chips" by either soaking or boiling.

Bisulphide of Carbon.—As sold by druggists and manufacturers, this is a brownish fluid which quickly disappears in the air when exposed in an open vessel. Its disagreeable odor is due to impurities, since the odor of pure bisulphide of carbon is not unpleasant. The fumes are not only poisonous, but are inflammable, so that some care must be exercised in handling the fluid. It has proved of special service as a remedy for grain weevil, bean weevil and other insects attacking stored seeds, and for the phylloxera of grapevines in Europe, for the woolly aphis, for ants, and even for the clothes moth. Its great value for such purposes comes not only from its effectiveness in destroying all insects, but also because it is not corrosive and is otherwise not injurious to seeds, fabrics and other objects fumigated. The offensive odor is soon gone if objects that have been exposed to the fumes are thoroughly aired. It cannot be used for fumigating plants infested with insects because of its destructive effect on the plants themselves.

About one fluid ounce should be used on each bushel of seed, and may be poured over the seeds or simply placed in a saucer or other open vessel set on their surface. It is absolutely necessary that the seeds be enclosed in a tight box or bin to get satisfactory results, and the time of exposure should not be less than two hours.

Carbon Tetrachlorid.—The disagreeable odor of commercial bisulphide of carbon renders it objectionable to some people for use on fabrics infested with moth, and has led to the suggestion that carbon tetrachlorid, which has a rather pleasant odor, be used in its stead. This also is a fluid, and is used in the same way as carbon bisulphide, namely, by pouring it into open dishes or crocks and allowing it to evaporate in a box, bin or room.

It is not nearly as effective in small quantities as either cyanide of potassium or carbon bisulphide, and the large quantities that must be used increase the cost of treatment.

Para-dichlorobenzene.—This is a recently proposed fumigant and is not yet in general use, because of its cost. It is not evil-smelling like

MAKING PREPARATIONS TO FUMIGATE WITH HYDROCYANIC GAS.[1]
Front edge of sheet tent and top of derrick ready to be pulled over tree.

carbon bisulphide, and appears to be quite effective in destroying weevils in grain and clothes moth. Since it is not inflammable, it can be more safely used about dwellings, though its fumes have wonderful penetrating power and escape in some quantity even from tightly stoppered bottles. From a limited experience with it the writer is disposed to regard it very favorably for fumigating seeds and fabrics, though more extended tests may show it to have defects that are not now apparent.

Hydrocyanic Acid Gas.—This gas is made from cyanide of potassium (98 per cent), commercial sulphuric acid of good grade and water. The

[1]Courtesy of U. S. Dept. of Agriculture.

gas produced is very poisonous, as are also the cyanide of potassium and sulphuric acid. When fumigating it is well to place a notice on the room or house warning people not to enter. After the fumigating is accomplished it is advisable to open up doors and windows and air out for ten minutes or more before entering.

The dose to be used depends upon the space to be fumigated and upon the character of the plants to be treated. Dormant trees can be exposed for

Fumigating with Hydrocyanic Gas.[1]
Sheet tent ready for introduction of chemicals.

a time to very strong fumes. Growing plants must be treated cautiously with very mild doses. Some of them are very sensitive to the gas and will be slightly burned with any dose calculated to be of value in destroying insects. The condition of the air as to moisture may influence the results, since dampness favors injury from the gas.

For nursery stock it is customary to employ for each 100 cubic feet enclosed, the following:

Cyanide of potassium	1 ounce
Sulphuric acid	1 25 fluid ounces
Water	3 fluid ounces

[1] Courtesy of U. S. Dept. of Agriculture.

The box or house should be as nearly gas-tight as possible, with a very tight-fitting door. The water and sulphuric acid are placed in a deep open crock, then the cyanide of potassium, broken up into pieces about as large as a hickory nut, is poured into the crock and the door shut as quickly as possible. The fumes must be left about the trees not less than forty minutes, and fifty minutes or an hour is better. Short exposures in badly constructed houses have sometimes resulted in the sending out of living San José scale on trees.

In the hothouse the gas must be used with very great care to avoid injury to plants. Plants of the grass family (*Gramineæ*) endure more gas than most others tested by the writer. Corn, timothy, blue grass and the like are not very sensitive. The leguminous plants, such as clover, sweet pea and cowpea, are very likely to suffer some injury with any but very light doses, and on this account it is best to use the less hurtful tobacco extract when practicable. The extract will not, however, destroy the adults of all hothouse pests, and has no effect at all on the scale insects nor on the immature white fly.

FUNGICIDES

When all has been said the number of fungicides approved by the experience of practical and scientific men is very small. Many have been recommended, but comparatively few have stood all the tests as to effectiveness, convenience of application and cheapness. Some are cheap and only slightly effective; some are difficult to prepare; others are too costly for extensive use.

Copper Sulpnate.—At the head of the list stands copper sulphate, a cheap, effective fungicide, commonly known as bluestone. This is the active and most essential ingredient of Bordeaux mixture. Concentrated solutions of it cannot be used alone on foliage because of their caustic action. In winter on dormant trees it is sometimes used for fungous troubles, about two pounds being dissolved in a barrel of water. A weaker solution—1 pound in 200 gallons of water—may be used on foliage in summer when fruit is well matured and it is not desirable to use sprays like Bordeaux mixture, which leave a residue. The bluestone may be quickly dissolved by pouring boiling hot water over it. When one is not hurried it may be dissolved by suspending in a loose sack in the water. It dissolves slowly if simply thrown in the water and allowed to settle.

To avoid to some extent the delays involved in dissolving bluestone it is well to buy a finely powdered grade now manufactured for the making of fungicidal preparations.

Bordeaux Mixture.—A standard formula for the preparation of this valuable mixture is the following:

Bluestone	4 pounds
Fresh lime	4 pounds
Water	50 gallons

Dissolve the bluestone in 25 gallons of water, slake the lime separately, and add water to make 25 gallons; then pour the two, bucket by bucket, into a third barrel so as to mix thoroughly. For peach and plum, which are more tender than apple and grape, the above formula may be changed to the following:

Bluestone	2.5	pounds
Fresh lime	2.5	pounds
Water	50	gallons

These are the best preparations known for mildews, rots, scabs, smuts and the like, and where one is dealing with a fungous trouble and is uncertain as to how to proceed, the chances are that he will accomplish as much

Efficiency of Bordeaux Mixture on Potatoes. One Row Not Sprayed.[1]

by using this preparation as with anything that could be recommended. It is the best general-purpose fungicide we have at present.

Copperas, or Iron Sulphate.—While this is less often used than bluestone, yet it has decided fungicide and antiseptic value, and because of its cheapness may sometimes be found serviceable. As now used it generally comes to the market as a waste product in the manufacture of steel wire, and may be bought for a cent or less per pound.

Formalin, or Formaldehyde.—This very valuable preservative and antiseptic has been much used of late as a remedy for potato scab and to some extent for wheat smut. It is sold as a fluid containing forty per cent of formalin. In this condition it is very acrid, and gives off fumes that affect the eyes and nostrils unpleasantly. Used on the hands, it quickly destroys the outer skin. It cannot, therefore, be employed except when

[1] Courtesy of New York Agricultural Experiment Station, Geneva, N. Y.

greatly diluted. But since it retains its active fungicide and bactericidal properties even when very greatly diluted, and is not so dangerous a poison in this condition as are corrosive sublimate and other antiseptic agents, it becomes very useful in the hands of those who wish to disinfect quarters in which have been lodged people, or animals, affected with communicable diseases. The wash or spray of the dilute formalin has always seemed to the writer much better for such uses than the fumes of formalin as generally produced.

On plants the action of even dilute sprays is very quickly destructive, and I doubt if it has a value for their treatment. But for seed wheat,

TREATING GRAIN WITH FORMALIN FOR SMUT.[1]

likely to produce smutted heads and for potato scab it has proved very convenient and useful. A pint of the 40 per cent formalin may be poured into a barrel containing 30 gallons of water, stirred thoroughly, and the potatoes in a sack can be set in the barrel for disinfection. They should be left in the fluid for two hours and may then be removed and spread out on grass or on a clean plank floor to dry, when another sack may be placed in the barrel. The treated potatoes must not be put in barrels or sacks that have not been treated with the formalin. By having a number of barrels at hand, the work proceeds rapidly.

Oats and wheat liable to smut may be treated by sprinkling the seed with dilute formalin (1 pint in a barrel of water) until every seed is moist, not wet, then leaving for several hours in a heap, finally spreading out to dry.

[1] Courtesy of H. L. Bolley and M. L. Wilson, North Dakota Agricultural Experiment Station.

Fumes of formalin produced either by heat or by the use of permanganate of potash have been recommended as a remedy for potato scab, but the writer's experience with the fumes has not been such as to warrant him in recommending them for this or for other purposes.

Bichloride of Mercury.—A very poisonous chemical, valuable in dilute solutions (1 part in 1000) as a disinfectant, and particularly good as a remedy for potato scab. The whitish, crystalline, very heavy material is very dangerous to have about, since it may attract the attention of children or animals. It should of course always be kept labeled as a poison. It dissolves slowly in cold water, and it is best, therefore, to make use of heat, afterward turning the dissolved poison into the larger quantity of water required, best kept in a barrel. Good results have been obtained in checking potato scab with this disinfectant, using 4 ounces in 30 gallons of water and soaking the seed potatoes one hour. They were placed in the fluid in gunny sacks and afterward spread out on a barn floor to dry.

It is very essential that poisoned potatoes be not left where stock will eat them, and the poisonous fluid must be disposed of after treating the seed, so that it will do no harm.

Lime-Sulphur Wash.—This preparation of sulphur and lime has already been mentioned under insecticides. It has undoubted fungicide value both in concentrated and dilute preparations. For foliage the latter must always be used. Even the sulphur alone thickly strewn over leaves is a fairly good remedy for mildew. A very small quantity of the sulphur dissolved in the presence of lime renders it more effective both as an insecticide and as a fungicide.

COMBINED INSECTICIDES AND FUNGICIDES

The cost of treatment for pests is greatly increased by the necessity for frequent spraying when insecticides and fungicides are used separately. They have been combined in some cases with no loss in the effectiveness of either, and one of the important problems of both entomologists and plant pathologists at the present time is the finding of ways and means of reducing the number of sprayings still further.

Some work in determining the compatibility of different mixtures has already been done, and it may be said that the following mix without less and in some cases with a gain in effectiveness:

Arsenate of lead (acid) and Bordeaux mixture.
Arsenate of lead and tobacco.
Arsenate of lead and acids.
Arsenate of lead (neutral) and Bordeaux mixture.
Arsenate of lead (neutral) and lime-sulphur.
Arsenate of lead (neutral) and tobacco.
Paris green and Bordeaux mixture.
Arsenite of lime and Bordeaux mixture.
Arsenite of lime and tobacco.
Lime-sulphur and tobacco.
Soaps and Bordeaux mixture.

Soaps and tobacco.
Soaps and emulsions.
Tobacco and lime-sulphur.
Tobacco and soaps.
Tobacco and emulsions.
Tobacco and alkalies.

Some dangerous combinations are the following:

Arsenate of lead (acid) and soaps.
Arsenate of lead (acid) and emulsions.
Arsenate of lead (acid) and alkalies.
Arsenate of lead (neutral) and acids.
Arsenite of zinc and lime-sulphur.
Arsenite of zinc and soaps.
Arsenite of zinc and emulsions.
Arsenite of zinc and alkalies.
Arsenite of zinc and acids.
Hydrocyanic acid gas and Bordeaux mixture.

REFERENCES

The literature dealing with the subject is very extensive and cannot be cited at all adequately in a brief résumé such as this. The few recent papers given will help the reader to an understanding of the range and character of work being done to throw light on this important subject:

"The Spraying of Plants." Lodeman.

"The Chemical Composition of Insecticides and Fungicides." Bul. 68, Bur. of Chem., U. S. Dept. Agr., 1902.

"An Investigation of Lime-sulphur Injury—Its Causes and Prevention." Bul. No. 2, Ore. Agr. College Exp. Sta., 1913.

"Fumigation and Spraying," Bul. No. 172, Kentucky Agr. Exper. Sta., 1913.

"Chemical Studies of Lime-sulphur—Lead-arsenate Spray Mixture." Research Bul. No. 12, Iowa Exper. Sta., 1913.

"Analyses of Materials Sold as Insecticides." Bul. No. 262. New Jersey Exper. Sta., 1913.

"Preparation of Nicotin Extracts on the Farm." Bul. No. 218, Virginia Agr. Exper. Sta., 1914.

"The Compatibility of Insecticides and Fungicides." By George P. Gray, Vol. III, No. 7, Monthly Bulletin, Cal. State Comm. of Horticulture, 1914.

"A Report of Chemical Investigations on the Lime-sulphur Spray." Research Bul. No. 3, Oregon Exper. Sta., 1914.

"Experiments in the Dusting and Spraying of Apples." Bul. No. 340, N. Y. Agr. Exper. Station (Ithaca), 1914.

"Analyses of Materials Sold as Insecticides and Fungicides." Bul. No. 384, N. Y. Agr. Exper. Station (Geneva), 1914.

"Experiments in the Destruction of Fly Larvæ in Horse Manure." Bul. No. 118, U. S. Dept. Agr., 1914.

"Quassiin as a Contact Insecticide." Bul. No. 165, U. S. Dept. Agr., 1914.

"Notes on the Preparation of Bordeaux Mixture." Circular No. 15, New Hampshire Agr. Exper. Station, 1914.

"Bordeaux Mixture." Technical Bul. No. 8, New Hampshire Agr. Exper. Station, 1914.

"The Nicotin-Sulfate-Bordeaux Combination." By V. I. Safro, *Journal of Economic Entomology*, Vol. 8, No. 2, 1915.

"Homemade Lime-sulphur Concentrate." Bul. No. 197, U. S. Dept. Agr., 1915.

"Para-Dichlorobenzene as an Insect Fumigant." Bul. No. 167, U. S. Dept. Agr., 1915.

"Cactus Solution as an Adhesive in Arsenical Sprays for Insects." Bul. No. 160, U. S. Dept. Agr., 1915.

"Further Experiments in the Dusting and Spraying of Apples." Bul. No. 354, New York Agr. Exper. Station (Ithaca), 1915.

California Expt. Station Bulletin 257. "Dosage Tables;" "Fumigation Studies No. 1."

Kansas Expt. Station Bulletin 203. "Orchard Spraying."

Michigan Expt. Station Bulletin (Technical) 21. "How Contact Insecticides Kill."

New Jersey Expt. Station Bulletin 48. "Bordeaux Mixture."

Canadian Dept. of Agriculture Bulletin 198. "Lime-Sulphur Wash."

U. S. Dept. of Agriculture, Plant Industry Bureau, Bulletin 265. "Some Factors Influencing the Efficiency of Bordeaux Mixture."

Farmers' Bulletins, U. S. Dept. of Agriculture:

440. "Self Boiled Lime Sulphur Mixture as a Promising Fungicide."
492. "Fungous Enemies of the Fruit and Foliage of Apple Tree."
595. "Arsenate of Lead, An Insecticide Against Tobacco Worms."
603. "Arsenical Cattle Dips."

BOOK IX

HOME ECONOMICS AND AGRICULTURAL EDUCATION

CHAPTER 78

Food Materials and Their Functions

By Miss Nellie E. Goldthwaite, Ph.D.
Dean of Women, New Hampshire College

Woman, from the savage state on, has been more fundamentally interested in foods and food-products than man. This has been true because of the dependence of the young upon her. She who bore the little ones must find food for them when her breast no longer furnishes sustenance. For the children who walked by her side she was still responsible. It was she who gathered fruits and berries for their subsistence. Perhaps it was she who first noted the sprouting of seeds and thus learned how she could increase the food-supply of her family. Possibly she was the first agriculturist. At least, we know that among all primitive peoples it was woman who tilled the soil. So, from the most primitive times on, it has been woman's work and pleasure to be interested in foods. Sad will it be for the race if she ever lose that interest.

In these days when science is rapidly unlocking the mysteries concerning the ways in which food sustains life, subjects of vital interest to humanity, let not the woman be ignorant. It is still her province to mother the young and to look into the ways of her household. Today she may, if she will, profit by the researches of a multitude of workers. But there is no royal road to learning. It takes time and patient work to grasp the underlying principles in any field of knowledge. Let the woman once grasp the fact that food not only preserves life, but that the general condition of the body depends largely upon the food taken into it, and she will be quick to make herself intelligent concerning the principles of human nutrition. Planning the family dietary is a work worthy of the best thought.

Elements of the Body.—Scientists tell us that our earth is composed of about eighty different elements, *i. e.*, simplest forms of matter. Our bodies consist of about a dozen of these elements that have been taken from the earth and so combined with each other that we should not recognize any one of them as elements. Combinations of elements in which each one has lost its own peculiar identity, we call compounds.

Chemists generally agree that our bodies are composed of the following elements, and that each is present in about the percentage given:*

* Sherman, "Food and Nutrition," page 260.

Oxygen	about	65	Potassium	about	.35
Carbon	"	18	Sulphur	"	.25
Hydrogen	"	10	Sodium	"	.15
Nitrogen	"	3	Chlorine	"	.15
Calcium	"	2	Magnesium	"	.05
Phosphorus	"	1	Iron	"	.004

Iodine, Fluorine, Silicon } Very minute quantities.

These percentages mean that the body of an adult weighing 154 pounds contains about—

100 pounds of oxygen		$\frac{1}{2}$ pound of potassium.	
27 " " carbon		$\frac{1}{3}$ " " sulphur.	
15 " " hydrogen		$\frac{1}{4}$ " " sodium.	
$4\frac{1}{2}$ " " nitrogen		$\frac{1}{4}$ " " chlorine.	
3 " " calcium		1 ounce " magnesium.	
$1\frac{1}{2}$ " " phosphorus		$\frac{1}{10}$ " " iron.	

Description of Body Elements.—Let us familiarize ourselves a little with these body elements.

As every one knows, the air we breathe is made up mainly of a *mixture* (not compound) of *oxygen* and *nitrogen,* each of which is a colorless, odorless gas. About one-fifth the air by volume is oxygen.*

Water is a chemical combination of *oxygen* and *hydrogen,* another colorless, odorless gas. In water both oxygen and hydrogen have lost their identity and have formed a compound.

Carbon is familiar in its elemental form as soot, lampblack, coal, charcoal. We realize its presence in our foods by the blackening that occurs when we scorch or burn them.

Calcium, Potassium and *Sodium* never occur in nature in their elemental forms. When freed from their compounds they are soft, silvery-white metals that are dangerous to handle.

Phosphorus, also, is never in nature in its elemental form. When pure it may be a yellow solid that burns spontaneously in the air.

Sulphur, a yellow solid, is well known. When set on fire it burns with the familiar blue flame of the sulphur match.

Chlorine is a heavy, yellow, suffocating gas. Our common table salt is composed of chlorine and sodium chemically combined. Chlorine so combined with hydrogen forms hydrochloric acid, the essential acid of the gastric juice of the stomach.

Magnesium is a silvery-white metal that burns with a brilliant white flame.

Iron we are all familiar with.

Body Compounds.—From the foregoing it is needless to add that no one of the elements that make up the human body occurs in the body in its elemental form. These elements are chemically combined with each other in

*The carbon dioxide found in the air is very important in plant life. It occurs to the extent of about 3 parts per 10,000.

divers ways to form the various types of compounds which make up the different parts of the body, *e. g.*, the bones, teeth, muscles, vital organs, fat, etc.

Body Oxidation Products, or Final Metabolic Products.—All of the body compounds seem to be in a state of more or less constant chemical change, *i. e.*, new combinations of the elements are being produced. So long as life lasts the body is constantly breathing in air and, in a sense, is constantly burning itself up—oxidizing itself. The mere act of living involves constant chemical changes in the body. Chemical changes keep the body warm. Each motion of the body involves chemical change; the more strenuous the motions, the more rapid the chemical changes.

In the process of these changes, some of the body compounds are constantly uniting with the oxygen of the air and are producing so-called highly oxidized waste products, or final metabolic products, which the body must finally rid itself of. Thus, the body finally burns its carbon mainly to carbon dioxide, a colorless gas which it excretes by way of the lungs; its hydrogen mainly to water; its nitrogen mainly to urea, a white crystalline solid, which is held in solution and excreted in the urine and perspiration; its sulphur and phosphorus to sulphates and phosphates of calcium, sodium, potassium, magnesium, which also are white solids excreted in the same manner as urea. Some other solid metabolic products are similarly excreted.

All these waste products the body proceeds to get rid of as rapidly as possible, so all such products are ultimately returned to the air or to the soil. Other things being equal, the quantity of these products which the body excretes per 24 hours is in proportion to the amount of its physical exertion per 24 hours.

It should be mentioned that small amounts of certain metabolic products are excreted in the feces, but in the main, the feces consist of any indigestible materials taken in with the food, the remains of the food that has escaped digestion, residues of digestive juices and bodies of bacteria.

Need of Food.—Because of these constant oxidizing processes which are going on in the body, and because of the quantity of metabolic products thus produced and excreted, the body soon wastes away if food be not supplied. Hence, we eat to make up for the constant oxidation which the body compounds are undergoing. More explicitly, the adult eats to repair the body tissues constantly being torn down through the life processes, and also to supply energy for physical motions; the child eats not only to repair its body tissues and to supply energy for its intense activities, but also for *growth.* This last fact must not be forgotten in feeding the child.

Elements Needed in Food.—Obviously, the food furnished must contain the same elements as those which constitute the body. In the case of the healthy adult, the amount of these elements furnished daily should correspond to the daily losses of these elements; *e. g.*, if the body loss of carbon per 24 hours is 12 ounces, then the food intake of carbon per 24 hours should be 12 ounces; and the same must be true of the twelve or more body elements. In the case of the growing child, the amount of each element furnished in the daily food must exceed that daily excreted from his body. In any case, food must be presented to the body in an organic form, *i. e.*, a form which has been produced by *life*, either vegetable or animal. Table salt is a noted exception to this rule.

Nature's Preparation of Food Materials.—All the essential elements of food are found in soil or air. The plant apparently builds its parts from this dead world of matter and serves as food for the animal. Both plant and animal may serve as food for man. In nature's scheme the plant kingdom is intermediate between the inorganic kingdom and the animal kingdom. Always the animal economy returns to the inorganic kingdom its metabolic products to serve again and again in ceaseless cycles of life forms.

Man's Selection of Food Materials.—In the ages past man has selected food materials from both plant and animal kingdoms. Instinctively, he has selected the most appetizing and nutritious parts of either plant or animal. He has gradually added to his diet, till now it is possible for civilized man to choose from a wide variety of food materials.

Foodstuffs, their Composition and Functions.—Scientific investigation of these food materials shows that man has unerringly chosen substances that are composed of water and of various types of so-called foodstuffs: proteins, fats, carbohydrates and mineral matter. The *proteins* are composed of carbon, hydrogen, oxygen, nitrogen and sulphur (phosphorus and iron sometimes). The *fats* and the *carbohydrates* are composed of carbon, hydrogen and oxygen, the fats being much the richer in carbon. The *mineral matter* of the food includes sulphur, phosphorus, calcium, sodium, potassium, magnesium, iron, in various combinations.

The term *protein* means "I take the first place;" the name is given because the proteins appear to be the most important of the foodstuffs. Their nitrogen and sulphur seem necessary to every living cell of the body. These living cells apparently are constantly breaking down; hence, proteins (nitrogenous foodstuffs) are always needed to repair the worn-out tissues or to build new tissues, as in the growth of the child. The white of egg and the curd of milk are typical protein foodstuffs; the muscle fiber of meat is largely protein; the gluten of wheat (the sticky stuff that permits the making of a dough) consists of proteins. Dried beans and peas are rich in protein.

Typical animal *fats* are butter-fat of milk, lard, beef tallow, mutton tallow, fat of chickens; while typical vegetable fats are olive oil, cottonseed oil and oils of various nuts.

Typical *carbohydrates* are the various sugars and starches; *e. g.*, cane sugar, milk sugar, glucose, cornstarch, starch of wheat flour, starch of potatoes.

Both fats and carbohydrates are essentially energy foods; *i. e.*, they do not appear to be repairers of tissue, but they are burned in the body in proportion to the physical exertion that the body puts forth; in the main, they furnish the energy that enables the body to do work, although protein may serve as an energy food. It is very interesting that fats and carbohydrates, which are eaten when not immediately needed as energy foods, seem to be stored as body-fat till sometimes the body becomes much over-weight from the amount of stored fat it is forced to carry about. The carbon of an over-supply of food-protein may also be converted into fat and stored in the body. In a certain sense the fats are more concentrated foods than the carbohydrates. Fats are about 76 per cent carbon, whereas carbohydrates are 40 to 44 per cent carbon. The fats furnish more than twice as much energy, pound for pound, as the carbohydrates; but apparently the fats are digested, assimilated and used by the body less easily.

The *mineral matters* contained in the food are needed, like the proteins, in the repair and growth of tissues—they are needed in every living cell; hence the great necessity of their presence in the diet of both adult and child. They are often called body regulators. Fruits and vegetables are very important sources of food mineral matter.

Proportions of Foodstuffs in Food Materials.—Most food materials contain varying proportions of the different foodstuffs along with water. Consideration of the constituents of some typical food materials will make this clear.

Average cow's milk consists of 87 per cent water, 3.3 per cent protein, 4 per cent fat, 5 per cent carbohydrate and .7 per cent mineral matter; or cow's milk contains 13 per cent of foodstuffs. A quart of standard milk weighs 34.4 ounces. Hence, a quart of whole (unskimmed) milk contains 4.46 ounces of foodstuffs; a pint 2.23 ounces and a cup 1.12 ounces. In other words, a quart of milk contains some more than a quarter of a pound of total foodstuffs; while a pint and a cup each contain proportional amounts.

First patent flour consists of 10.55 per cent water, 11.08 per cent protein, 1.15 per cent fat, 76.85 per cent carbohydrates and .37 per cent mineral matter. Hence a pound (about one quart sifted) of this flour consists of 14$\frac{1}{3}$ ounces of foodstuffs, and a pint or a cup contains proportional amounts. Note that fine wheat flour is a very concentrated food, since it contains nearly 90 per cent of foodstuffs.

Whole eggs consist of 11.2 per cent refuse (shell), 65.5 per cent water, 11.9 per cent protein, 9.3 per cent fat, no carbohydrate and .9 per cent mineral matter; or whole eggs consist of about 22 per cent foodstuffs. An egg of average size weighs about 2 ounces. Such an egg contains nearly $\frac{1}{4}$ ounce protein, a little less than $\frac{1}{4}$ ounce fat and about $\frac{1}{50}$ ounce mineral matter. It will be discovered that in eggs we have to consider a

Table I.

Food Materials (Purchased).	Composition of Food Materials, per cent. Refuse.	Water.	Protein	Fat.	Carbo-hydrate.	Mineral Matter.	Number of Ounces Foodstuffs per Pound of Food Material. Protein.	Fat.	Carbo-hydrate.	Mineral Matter.	Total.
I. Animal Foods.											
(a) Meats:											
Bacon (smoked)	8.7	18.4	9.5	59.4	....	4.5	1.52	9.50	.....	.72	11.74
Beef:											
Porterhouse	12.7	52.4	19.1	17.9	....	.8	3.05	2.86	....	.12	6.03
Rib rolls	....	64.8	19.4	15.5	....	.9	3.10	2.48	....	.14	5.72
Round	8.5	62.5	19.2	9.2	....	1.0	3.07	1.47	....	.16	4.70
Chicken, broilers	41.6	43.7	12.8	1.4	....	.7	2.04	.22	....	.11	2.37
Mutton, leg	17.7	51.9	15.4	14.5	....	.8	2.46	2.32	....	.12	4.90
Pork, chops	19.3	40.8	13.2	26.0	....	.8	2.11	4.16	....	.12	6.39
(b) Fish:											
Mackerel	44.7	40.4	10.2	4.2	....	.7	1.63	.67	....	.11	2.41
Lake trout	48.5	36.6	9.1	5.1	....	.6	1.45	.81	....	.09	2.35
(c) Milk Products:											
Butter	....	11.1	1.0	85.0	....	3.0	.16	13.60	....	.48	14.24
Cream cheese	....	34.2	25.9	33.7	2.4	3.8	4.14	5.39	.38	.60	10.51
Whole milk	....	87.0	3.3	4.0	5.0	.7	.52	.64	.80	.11	2.07
Skim milk	....	90.5	3.4	.3	5.1	.7	.54	.04	.81	.11	1.50
Cream	....	74.0	2.5	18.5	4.5	.5	.40	2.96	.72	.08	4.16
(d) Other Animal Products:											
Eggs	11.2	65.5	11.9	9.3	....	.9	1.90	1.48	....	.14	3.52
Lard	....	....	....	100.0	....	....	....	16.00	....	...	16.00
II. Cereal Products.											
Bread, white	....	35.3	9.2	1.3	53.1	1.1	1.47	.20	8.49	.17	10.33
Bread, whole wheat	....	38.4	9.7	.9	49.7	1.3	1.55	.14	7.95	.20	9.84
Corn meal	....	12.5	9.2	1.9	75.4	1.0	1.47	.30	12.06	.16	14.00
Flour, first (patent)	....	10.5	11.0	1.5	76.8	.3	1.77	.24	12.28	.04	14.33
Flour, whole wheat	....	11.4	13.8	1.9	71.9	1.0	2.20	.30	11.50	.16	14.16
Oatmeal	....	7.3	16.1	7.2	67.5	1.9	2.57	1.15	10.80	.30	14.82
Rice	....	12.3	8.0	.3	79.0	.4	1.28	.05	12.64	.06	13.82
Shredded wheat	....	9.6	12.1	1.8	75.2	1.3	1.93	.30	12.03	.20	14.46
III. Fruits.											
(a) Fresh:											
Apples	25.0	63.3	.3	.3	10.8	.3	.04	.04	1.72	.04	1.84
Cherries	5.0	76.8	.9	.8	15.9	.6	.14	.12	2.54	.09	2.89
Grapes	25.0	58.0	1.0	1.2	14.4	.4	.16	.19	2.30	.06	2.71
Oranges	27.0	63.4	.6	.1	8.5	.4	.09	.01	1.36	06	1.52
Peaches	18.0	73.3	.5	.1	7.7	.3	.08	.01	1.23	.04	1.36
Strawberries	5.0	85.9	.9	.6	7.0	.6	.14	.09	1.12	.09	1.44
Watermelon	59.4	37.5	.2	.1	2.7	.1	.03	.01	.43	.01	.48
(b) Dried:											
Apples	....	28.1	1.6	2.2	66.1	2.0	.25	.35	10.57	.32	11.49
Apricots	....	29.4	4.7	1.0	62.5	2.4	.75	.16	10.00	.38	11.29
Prunes	15.0	19.0	1.8	....	62.2	2.0	.28	....	9.95	.32	10.55
IV. Vegetables.											
(a) Fresh:											
String beans	7.0	83.0	2.1	.3	6.9	.7	.33	.04	1.10	.11	1.58
Beets	20.0	70.0	1.3	.1	7.7	.9	.20	.01	1.23	.14	1.58
Cabbage	15.0	77.7	1.4	.2	4.8	.9	.22	.03	.76	.14	1.15
Green corn	61.0	29.4	1.2	.4	7.7	.3	.19	.06	1.23	.04	1.52
Lettuce	15.0	80.5	1.0	.2	2.5	.8	.16	.03	.40	.12	.71
Onions	10.0	78.9	1.4	.3	8.9	.5	.22	.04	1.42	.08	1.76
Green peas	45.0	40.8	3.6	.2	9.8	.6	.57	.03	1.56	.09	2.25
Potatoes	20.0	62.6	1.8	.1	14.7	.8	.28	.01	2.35	.12	2.76
Spinach	....	92.3	2.1	.3	3.2	2.1	.33	.04	.51	.33	1.21
(b) Dried:											
Beans	....	12.6	22.5	1.8	59.6	3.5	3.60	.28	9.53	.56	13.97
Peas	....	9.5	24.6	1.0	62.0	2.9	3.93	.17	9.92	.46	14.47
(c) Vegetable Products:											
Molasses	....	2.51	2.4	....	69.3	3.2	.38	....	11.08	.51	11.97
Olive oil	....	....	....	100.0	....	....	....	16.00	....	...	16.00
Sugar	....	....	....	....	100.0	....	....	....	16.00	...	16.00

certain amount of "refuse." This is true in many foods, as apple and potato parings, corn cobs, bones of meat, etc.

Table I gives the percentage composition and the number of ounces of foodstuffs per pound of some of our more common food materials.

Discussion of Table I.—The data concerning these food materials refer to each "as purchased." The first column under "per cent composition" gives the average amount of refuse to be expected from the food indicated. It should be noted that the percentage of refuse is generally high in meats, fish and fresh vegetables. The prepared food products, such as butter, cheese, flours, sugar, contain little or no refuse; these are the most concentrated foods that the ingenuity of man has produced.

The percentages or proportions of water, protein, fat, carbohydrate and mineral matter that make up the "edible portion" of foods are very instructive.

Every housewife knows, even if the chemist did not tell her, that fresh fruits and vegetables contain the most water of any of our food materials, but she may not realize that meats also contain the astonishing amount of water indicated in the second column.

Examination of the third and fourth columns shows that the animal foods are essentially our protein and fat foods, with the cereal products a close second in proteins. Note, however, that olive oil, as well as refined fat, is 100 per cent fat.

The fifth column shows that animal foods, with the exception of milk and cheese, contain no carbohydrate; that the vegetable kingdom furnishes our carbohydrate foodstuffs, with the cereal products containing from 50 to 75 per cent carbohydrate.

The sixth column shows the percentages of mineral matter contained in these foods. It should be noted that the high percentages of mineral matter in bacon, butter, cheese, are due to the added salt in their manufacture. Animal foods generally are more or less rich in mineral matter; but the mineral matter contained in fruits and vegetables is considerable, and seems best adapted to keep the body in health. Witness the dread disease, scurvy, which ensues when vegetables are excluded from the diet.

The number of ounces of foodstuffs per pound yielded by each food material is given in columns 7 to 11. These data have been worked out in terms of protein, fat, carbohydrate, mineral matter and the total. The results are given for the convenience of the housewife who wishes to calculate just how many ounces of the various foodstuffs, per pound of food material, she is purchasing and is feeding to her family in any given length of time.

Fuel Value of Foodstuffs.—As has already been said, protein is essentially a tissue builder, but it may be burned in the body as an energy food; fats and carbohydrates are essentially energy foods.

Just as the fuel value of coal is measured by the quantity of heat that is given off when a given weight of it is burned, so the fuel value of a foodstuff is measured by the quantity of heat given off when a given weight of the foodstuff is burned. Such measurements are made by burning a small weighed quantity of the foodstuff in such a way that all the heat produced is used to heat a given weight of water. From the results obtained, the quantity of heat given off, when a pound of the foodstuff is burned, is readily calculated.

Very careful experiments have been made further to determine the quantity of heat that is *yielded to the body* when it burns a given weight of a foodstuff. Such experiments have shown that when a pound of protein or of carbohydrate is *burned in the body*, the body derives 1814 calories* of heat, while it derives from the burning of a pound of fat 4082 calories of heat. In terms of ounces these figures mean that—

1 ounce protein	yields 113.5 calories of heat to the body.
1 ounce fat	yields 255.1 calories of heat to the body.
1 ounce carbohydrate	yields 113.5 calories of heat to the body.

Thus, a given weight of fat yields to the body more than double as much heat or energy as the same weight of carbohydrates. Given weights of protein and of carbohydrate yield the body equal amounts of heat or energy.

Fuel Value per Pound of Food Material.—In addition to a knowledge of the amount of each foodstuff contained in a given weight of a food material, the woman who would feed her family intelligently should know how much heat these foodstuffs will yield the body. Knowing the number of ounces of each foodstuff contained in a pound of the food material, and also the fuel value per ounce of each foodstuff, she can easily calculate the fuel value of each foodstuff contained in the pound, and readily obtain the total. Thus, a pound of milk (Table I) contains .52 ounce protein, .64 ounce fat and .80 ounce carbohydrate; then the fuel-value of each foodstuff is as follows:

Protein	.52 × 113.5 =	59	calories
Fat	.64 × 255.1 =	163	"
Carbohydrate	.80 × 113.5 =	92	"
Total		314	"

Hence, the fuel value per pound of milk equals 314 calories. By this method Table II has been prepared. (See following pages.)

Discussion of Table II.—The first three columns of figures in Table II show the number of calories of heat which the protein, fat and carbohydrate contained in a pound of the food material yield respectively when burned in the body. The last column shows the total number of calories of heat the pound yields when burned in the body.

It should be noted that the pure fats, *i.e.*, lard and olive oil, yield the greatest number of calories (4082) per pound, while butter is a close second,

*A calorie is the amount of heat necessary to raise the temperature of 1000 grams (or 2.2 pounds) of water 1° Centigrade or 1°.8 Fahrenheit.

TABLE II.

Food Materials (Purchased).	Calories per Pound.			
	Protein.	Fat.	Carbo-hydrate.	Total.
I. ANIMAL FOODS.				
(*a*) Meats:				
Bacon, smoked	173	2423	..	2596
Beef:				
Porterhouse	346	729	..	1075
Rib rolls	352	633	..	985
Round	349	375	..	724
Chicken, broilers	232	56	..	288
Mutton, leg	279	592	..	871
Pork, chops	239	1061	..	1300
(*b*) Fish:				
Mackerel, whole	185	171	..	356
(*c*) Milk Products:				
Butter	18	3469	..	3487
Cheese, cream	470	1375	44	1889
Milk, whole	59	163	92	314
Milk, skim	61	10	92	163
Cream	45	756	82	883
(*d*) Other Animal Foods:				
Eggs	216	378	..	594
Lard	..	4082	..	4082
II. CEREAL PRODUCTS.				
Bread, white, average	167	51	964	1182
Bread, whole wheat	176	36	902	1114
Corn meal	167	77	1369	1613
Flour, first patent	201	61	1394	1656
Flour, whole wheat	249	76	1305	1630
Oatmeal	292	293	1226	1811
Rice	145	10	1435	1590
Shredded wheat	219	76	1365	1660
III. FRUITS.				
(*a*) Fresh:				
Apples	5	10	195	210
Cherries	16	31	288	335
Grapes	18	49	261	328
Oranges	10	3	154	167
Peaches	9	3	140	152
Strawberries	16	23	127	166
Watermelon	4	3	49	56
(*b*) Dried:				
Apples	28	89	1200	1317
Apricots	85	41	1135	1261
Prunes	32	..	1129	1161

TABLE II (*Continued*).

Food Materials (Purchased).	Calories per Pound.			
	Protein.	Fat.	Carbo-hydrate.	Total.
IV. VEGETABLES.				
(*a*) Fresh:				
Beans, string	38	10	125	173
Beets	23	3	140	166
Cabbage	25	8	86	119
Corn, green	23	15	140	178
Lettuce	18	8	46	72
Onions	25	10	161	196
Peas	65	8	177	250
Potatoes	32	3	267	302
Spinach	38	10	58	106
(*b*) Dried:				
Beans	409	72	1082	1563
Peas	446	41	1126	1613
V. VEGETABLE PRODUCTS.				
Molasses	43	....	1258	1301
Olive oil	..	4082	....	4082
Sugar	..	....	1816	1816

with its 3487 calories per pound. In general, the foods containing notable quantities of fat are the richest in fuel value. This has long been known in a practical way.

It is interesting to note that although cheese, the richest protein food material of the list, yields 470 calories from its protein per pound, that dried peas and beans are close seconds, with 408 and 446 calories per pound respectively. Fully one-quarter the total fuel value of cheese, peas and beans is due to protein.

Note that, exclusive of breads, the cereal products yield 1600 to 1800 calories per pound; breads are much less because of the proportion of water used in making them.

While the fuel value per pound of fresh fruits and fresh vegetables is not high, their fuel value when dried is much increased, 1100 to 1600 calories per pound.

The first and last columns of Table II are of particular interest to the housekeeper. This is because she wishes primarily to know how many calories a pound of a given food material will yield the body; and secondarily, because she wishes to know how many of these calories are due to protein.

Amount of Food Needed for Twenty-four Hours.—It has already been pointed out that the quantity of food needed by the body per twenty-four hours is in direct proportion to the amount of its physical exertion per twenty-four hours. This amount of food needed is measured in calories per kilogram (2.2 pounds) of body weight. Young to middle-aged men and women are said to require:

	Per kg.
At complete rest (*i. e.*, for life processes only)	30–35 calories
With light exercise	35–40 "
With moderate exercise	40–45 "
With hard muscular labor	45–60 "

Translated into terms of calories per pound of body weight, these figures become:

	Per pound body weight.
At complete rest	13.6–16 calories
With light exercise	16.0–18 "
With moderate exercise	18.0–20 "
With hard muscular labor	20.0–27 "

These figures mean that if your body weighs 154 pounds, and you are doing light exercise, you need $16 \times 154 = 2464$ calories, to $18 \times 154 = 2772$ calories of food per day.

Authorities disagree as to the proportion of these calories which should be yielded by protein. The high protein advocate considers the ration

TABLE III.

Food Materials.	Weight in Ounces.	Fuel Value in Calories.	
		Total.	Protein.
Orange	12	124	7
Shredded wheat	1	104	13
Bread, white	6	342	62
Egg	2	75	27
Potato	8	151	16
Round steak	4	181	87
Beets	4	41	5
Corn, green	8	89	11
Whole milk	16	314	60
Butter	2	436	2
Prunes	4	282	8
Cheese	1	117	29
Sugar	2	227	0
Cream	2	110	5
Total	....	2593	332

"balanced" when 1 calorie out of 6 calories is obtained from protein; the medium protein advocate, 1 out of 9 or 10; and the low protein advocate, 1 out of 12 or more.

Suppose you eat during twenty-four hours the kinds and quantities of foods indicated in Table III. The fuel value of these food materials (computed from Table II) is given in column 2, and the fuel value of the protein in column 3. Evidently the total fuel value of this food is 2593 calories, of which 332 calories are yielded by protein. The amounts mean that one calorie in 7.8 (*i.e.*, 2436 ÷ 302) is yielded by protein. This means that the dietary is intermediate between *high* and *medium* protein. If a

Window Box for Storage of Food.[1]

A north window is to be preferred. It should exactly fit the width of the window and come half way up the lower sash. When sash is raised the contents are easily accessible. In freezing weather the sash may be left up so that box becomes a part of the room.

greater or a less proportion of protein is desired, it can most readily be secured by increasing or diminishing the quantity of meat or cheese eaten. If you weigh 154 pounds, this day's dietary, 16.8 calories per pound (*i. e.*, 2593 ÷ 154) corresponds very well to that calculated for light exercise.

Since the child must eat not only for repair and for energy, but also for growth, it follows that he needs much more food in proportion to his body

[1]Courtesy of U. S. Dept. of Agriculture. From Farmers' Bulletin 375.

weight than the adult. Authorities say that the needs of children are as follows:

Under 1 year......	100 calories per kg.	(or 45 calories per pound body-weight).
1–2 years........	100–90 calories per kg.	(or 45–40 calories per pound body-weight).
2–5 years........	90–80 calories per kg.	(or 40–36 calories per pound body-weight).
6–9 years........	80–70 calories per kg.	(or 36–32 calories per pound body-weight).
10–13 years.......	70–60 calories per kg.	(or 32–27 calories per pound body-weight).
14–17 years.......	60–45 calories per kg.	(or 27–20 calories per pound body-weight).

Comparison of these food requirements of the child with those of the adult shows strikingly that the child needs a liberal diet)—much more in proportion to his body weight than the adult.

Reasons for Cooking Food Materials.—Food materials are cooked for the purpose of developing flavor, of making foods more easily digestible and of killing bacteria. Doubtless, the use of fire in the preparation of food was first discovered by accident; the flavor developed by the action of heat on meat appealed to the palate, and so the action of heat on various other food materials was tried gradually. Also, it was learned that some foods keep better when cooked. But it was reserved for the last century to explain scientifically the effects of heat on the palatability, the digestibility and the keeping qualities of food.

Effects of Heat on Foodstuffs.—One of the first lessons to be learned in the cooking of foods is that all food materials conduct heat slowly—that is, it takes time for heat to penetrate them.

The *proteins* are coagulated, hardened by heat, and the higher the temperature the more they are hardened. Hence the difficulty of cooking meats "to a turn" and of cooking eggs so that they shall be tender. Proteins should be cooked at a comparatively low temperature—below the boiling point of water; otherwise they will be tough and indigestible, *i. e.*, not easily dissolved by the digestive juices. High heat should be used at first with meats to sear the outside in order to keep the juices in; then the cooking should proceed at a lower temperature. In re-heating meats, the protein is much hardened. Soft-cooked eggs are best when put into cold water, covered, and heated just to the boiling point. Eggs may be cooked hard, and yet be tender, if put into boiling water, covered, then set where water will keep hot, but not boil, for 30 to 45 minutes.

The *fats* melt when heated, but otherwise seem to be unchanged except at high temperatures. Then they are decomposed, and at least one of the products formed is very irritating to the mucous membrane. This is one reason why people find difficulty in digesting fried foods. Butter scorches easily because of the small amount of protein in it.

Of the *carbohydrates*, *sugar* is melted, but not otherwise changed by heat unless heated to caramelization, or to the burning point. *Starch*, when cooked by dry heat, changes into dextrine, a more soluble and hence more digestible form of carbohydrate. When starch is cooked in the presence of water, the granules swell and finally burst open. This renders the starch more readily accessible to the digestive juices. Cereals to be well cooked should be brought to the boiling temperature and kept there for some time.

Cooking of Combinations of Foodstuffs.—The baking of breads consists in hardening the protein (gluten) and thus enclosing the gas bubbles in cooking the starch and in killing the yeast plants. The baking of cake is essentially the same thing, except that cake contains the protein of egg as well as gluten, and it also contains more fat, but does not contain yeast ordinarily. As the inside of the cake (or bread) never goes above the boiling point of water, there is little danger of decomposing the fat. All foodstuffs carbonize if subjected to dry heat when cooked; this explains the browning of baked foods. In baking doughs, the smaller the mass, the hotter should be the oven; the larger the mass, the lower the oven temperature in order to give time for the heat to penetrate.

In these days much is said of preventive medicine. The study of the human body is demanding the attention of some of the finest minds of the age. Physicians are seeking to learn and to teach fundamental laws of health. Better than in any past generation may the woman of today understand her own body and that of her child. Better than ever before may she know how best to care for and to feed each member of the family. When she fully appreciates that only food well chosen, well cared for and properly prepared should be presented to the wonderful human laboratory to be made over into body substance, then will the physician find his most efficient co-worker in woman, the mother of the race.

(The woman who wishes to know the subject of foods should write to the Department of Agriculture, Washington, D. C., for a list of bulletins published on food and nutrition. She should get some of these, especially the Farmers' Bulletins. After these have been studied, she should send for others, and so gradually build up her own food library.)

REFERENCES

"Principles of Human Nutrition." Jordan.
"Things Mother Used to Make." Gurney.
"Household Bacteriology." Buchanan.
"Foods and Household Management." Cooley.
"Food for the Invalid and Convalescent." Gibbs.
New Jersey Home Economics Bulletin 2. "Milk and Eggs."
South Carolina Circular 27 (Expt. Station). "Home Canning of Fruits and Vegetables."
Canadian Dept. of Agrculture Bulletin 221. "Food Value of Milk and Its Products."
Farmers' Bulletins, U. S. Dept. of Agriculture:
332. "Nuts and Their Uses as Food."
359. "Canning Vegetables in the Home."
363. "Use of Milk as Food."
375. "Care of Food in the Home."
389. "Bread and Bread Making."
391. "Economical Use of Meat in the Home."
526. "Mutton and Its Value in the Diet."
535. "Sugar and Its Value as Food."
559. "Uses of Corn, Kaffir and Cowpeas in the Home."
565. "Corn Meal as a Food: Ways of Using It."

CHAPTER 79

Housing and Clothing

By Mrs. Cecil Baker
Textile and Clothing Expert, Chicago, Ill.

Among the many improvements that have come to the farm in the last generation, those to the farm-house have been the slowest. Many a farm woman still carries on her work in the same sort of kitchen, with the same tools, in the same way that her grandmother did. Her husband has new barns designed according to the latest ideas of farm building, filled with the most improved machinery. His high-bred stock drink from a trough filled by a windmill or a gasoline engine, while his wife must often carry the water necessary for her operations, by the pailful, sometimes a considerable distance. If the farmer's wife and children are to have the full advantage of the new life made possible to them by better roads, automobiles, telephones, better schools and community associations of all sorts, the woman must have more time and energy for the new things. She must have her share of the improvements that come to the farm.

Leisure to the farm woman can come only through better facilities for work. The amount of work to be done has not markedly decreased, help from the outside is difficult to obtain, but the conveniences which make for efficiency have multiplied. Some expense is involved in obtaining labor-saving appliances, but this must be considered in the house as it is in the barn or the field, as an investment, the profit to come to be in the form of a freer, fuller life for the woman, which will mean in the end a fuller, saner life for the family.

House Plan Essentials.—Not only labor-saving devices are needed, but more thought in planning and arranging the house and its furnishings. Health is a vital factor; better sanitation and better facilities for personal hygiene aid in conserving it.

Finally, the more intangible psychological effect of the home on the family must be dealt with. This relates to the provision for the individual needs of the members of the family and the esthetic requirements of the home, and is more important than is sometimes understood.

The first requisite of a good house plan is that it shall fulfil the needs of the family that is to live in it. A farm-house must be planned for the farm, and not be a copy of the town house; the needs are not the same. A four-room cottage may be planned to meet the requirements of a certain family, but it is not likely to do so if taken from a builder's catalog. Four walls with a roof will give shelter, but only with intelligent thought will it give convenience and comfort. Generous porches for work, rest and sleeping, and plenty of windows to let in air and sunshine are blessings that should not be forgotten. The water supply which brings drink to the cattle should bring water to the house, to the sinks in the kitchen and milk room, to the laundry, to the family bath and to the wash, toilet and bathroom for the help; all of which must have the proper sewage disposal.

The Basement.—In constructing the house the basement will be the

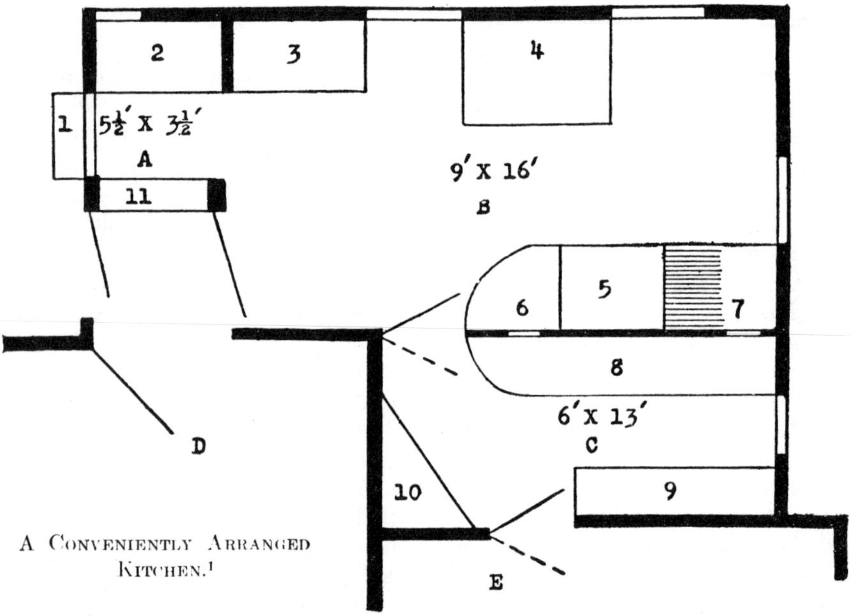

A Conveniently Arranged Kitchen.[1]

A—Cold pantry. B—Kitchen. C—Butler's pantry. D—Back hall. E—Dining room. 1—Window box to be used to keep food material in cold weather. 2—Refrigerator, with outside door for icing. 3—Kitchen cabinet. 4—Range. 5—Sink with shelves and drainage board on either side. 6—Door for passage of soiled dishes to be passed back. 7—Shelves in pantry. 8, 9 and 10—Shelves with glass doors. 11—Shelves in cold pantry.

starting point and should be planned as carefully as any other part. The walls should be well constructed of stone, brick or concrete. The floor should be of cement and there should be enough windows to insure plenty of light and air. The heating plant should go in the basement, a separate room being planned for it and the fuel. A hot-air furnace is the cheapest to install and in a small house is fairly satisfactory. The initial expense of hot water is greater, but it is most satisfactory, especially in larger houses; it is less affected by wind, is more easily regulated and causes less dirt in the house. As far from the furnace as possible, and near the stairs, should

[1] Courtesy of U. S. Dept. of Agriculture.

be a storage room for vegetables, canned goods and extra provisions; this should be light and have shelves or cupboards. A dumb-waiter to carry wood, coal and food supplies to the kitchen will add little expense and save many steps and much back-breaking labor. The cellar stairs, not too steep or winding, may lead directly to the kitchen, with a landing at the grade level.

The Kitchen is a place for preparing food and not a laundry, men's wash room or nursery. If well arranged, 8 by 10 feet or 10 by 12 feet will give ample room for stove, sink, cupboards with working counter and work table. Utensils used every day may hang on the wall over stove and sink; open shelves above these give place for salt, spices, cereals, coffee, etc. The things in constant use will not be any trouble to care for when kept in this way and the steps saved are well worth counting; the writer knows this from experience with both types of kitchen. The sink and tables should be high enough so the worker does not have to stoop over them, and if a high stool is nearby, it will many times be found a comfort to tired feet and back. A table on large casters, which will move easily to any part of the room and even to the dining room, is very useful. With good cross ventilation, air from two sides at least, good light, a floor and walls which may be easily cleaned, and the whole cheerful in color, a beginning has at least been made towards a comfortable work room.

The Pantry.—If the kitchen is fitted with cupboards for the dishes and staple food supplies, and plenty of working space, the pantry then becomes a place for keeping food cold, a place for the storing of large utensils and those used occasionally, and possibly for food which must be bought in quantities on the farm. The ice box may be here and should be so arranged that it will be filled from outdoors through a window at the back. Whether there will be one or more pantries is a question for the individual to decide. The main point to be emphasized is, that the food supplies, the dishes, the working area embraced by the stove, sink and table, and the ice box, should be as close together as possible.

Dining Room.—The entrance to the dining room should be near the working area, though possibly through a pantry or passage, to cut off the odors of cooking. The dining room itself should be large enough to hold the table, extended to accommodate the seasonal help, with room when the persons are seated to pass easily around when serving. A serving table or buffet will add to the convenience and attractiveness of the room. If one room must serve the double purpose of kitchen and dining room, the two parts should be kept as much separated as possible; the kitchen may be an alcove.

Wash Room.—A small room with a sink for the care of the milk may be near the kitchen, in the house or in a separate building, which latter may also contain the laundry and summer kitchen, and the wash room for the men help. In more pretentious houses there should be a sitting room and porch for the men, opening into the dining room and provided with washing facilities.

The Living Room.—The living room is very important in the life of the family. If it is attractive and amusement is here provided for all, the boys and girls will be more content to stay at home. The piano, book case, sewing table, writing table, magazines and the toys all belong here and each should have its place. A comfortable couch, a good light for reading and a small chair for the child are more essential than a gay rug, lace curtains or vases on the mantelpiece. An open fire adds cheer to the room, spares the furnace in the first chill mornings of fall and in spring, and aids in ventilation.

The Office.—A most desirable room, even though it be a small one, has a place for the farmer's desk, account books, catalogs and other business papers. If access to it from the front or side door is easy, business callers

A Cheerful Living Room.[1]

may be received here apart from the family life. This room is much more important on the farm, where there are almost no formal calls, than a reception hall or parlor.

The Hall.—The chief use of the hall is to serve as a passage from outdoors to the living rooms and to the upstairs. Here there should be a good closet for wraps, possibly under the stairs. It is not desirable to have the front entrance direct into the living room, as the privacy of the room is thus destroyed and cold and dirt from outdoors come in. Neither should the kitchen entrance be the most used one, as the passing interferes greatly with the work done there.

Sleeping Rooms.—Bedrooms are the strictly individual rooms of the house and here, above all, the individual should have privacy. A painted or hardwood floor, simple washable curtains and rugs, a comfortable bed, a good clothes closet and plenty of fresh air are the essentials of a bedroom.

[1] Courtesy of The Macmillan Company, N. Y. From "Shelter and Clothing," by Cooley.

In case there are children or old people, and one woman does all of the work, a bedroom on the first floor is very desirable.

Bathroom facilities on a farm, where hard physical exercise makes frequent baths necessary, are too often inadequate. Tub, toilet and wash bowl of enameled iron may be purchased for from $60 to $100. The toilet should be a syphon closet and the tub one which sets close to the floor, leaving no room for dirt underneath. The fixtures should be of good quality and without ornament.

Interior Finish.—Bathroom and kitchen walls painted with oil paint or papered with a washable paper are easily cared for. Inlaid linoleum on the floor wears well, printed oilcloth does not. Woodwork may be painted or varnished, or white enamel is an attractive finish and is not as much work to care for as is usually thought. Light, cheerful colors which show when they need washing are desirable. Walls in other rooms may be calcimined, a cheap but temporary finish; painted with oil paint, which is more permanent and may be cleaned, or papered. Plain colors or simple figures with not too contrasting colors are good backgrounds for pictures. The soft greens and browns of the woods, buff and warm grays, yellows in rooms inclined to be dark, are all good colors to live with. Red is exciting, blue apt to look cold and gold far from restful. There are many papers on the market bad in color and design; seek to avoid them.

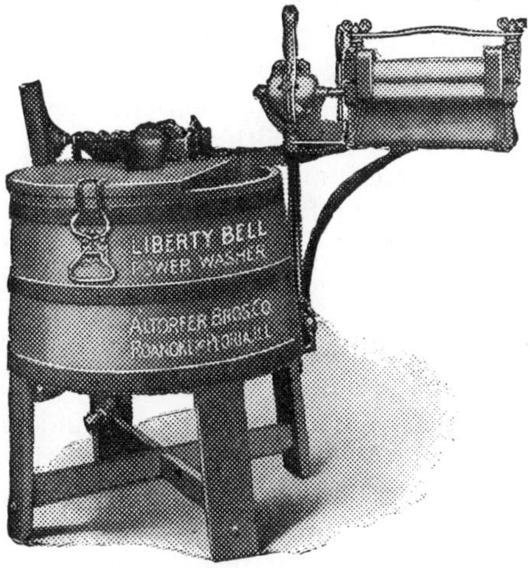

A POWER WASHING MACHINE.[1]

The Furnishings.—Simple, substantial furniture should be chosen, with comfort and usefulness its first requirements. Cheap machine carving only mars the beauty of wood. A dull wax finish on both furniture and woodwork is more restful than a shiny varnish and mars less easily. Rugs should keep their place on the floor and not, because of their giddy color and design, seem to rise and hit one upon entering the rooms. Hardwood or painted floors are easier to care for and more sanitary than carpets. Washable rugs are good for bedrooms and bathrooms. Wilton, Brussels, Smyrna, Scotch and grass rugs each have their good features for general wear; they are named in the order of their cost.

Draperies and Decoration.—Simple white or ecru curtains of scrim, net, swiss or madras soften the windows, yet let in sunlight and air, and

[1] Courtesy of Altorfer Bros., Roanoke, Ill.

are easily washed. They should hang straight to the window sill or be draped back simply. Pictures and bric-a-brac give the finishing touches to the room, but should be used sparingly. A few good pictures, one or two pieces of pottery, and a good clock add character to the room. Embroidered satin sofa pillows set against the table leg have no excuse for being, but two or three plain washable ones will make the couch more comfortable. Whether the home be a cottage or a mansion, the same principles of good taste hold. Simplicity is a good guide, and inexpensive things are often more beautiful than the costly.

Household Appliances.—To aid in caring for the house there are now many mechanical devices; the vacuum cleaner, the power washing machine, the mangle, the bread mixer, the fireless cooker and the meat grinder are all good investments. Before buying labor-saving devices be sure they save more work than they make.

Ventilation in most houses is merely a question of opening and shutting windows. The health derived from sleeping outdoors has proved that plenty of fresh air at all times is more to be sought than avoided. The house sealed up from fall until spring is more likely to breed consumption than is cold, fresh air. Let in the air and sunshine; if dust and faded carpets result, they are less to be feared than air which has been breathed over and over and contains many impurities. In winter boards 6 to 8 inches high fitted under the lower sash make an opening between the two sashes and let in air without draft. In bedrooms a window screen covered with a coarse muslin allows the window to be kept open, winter and summer, regardless of the weather. Artificially heated air is usually too dry; flat pans or buckets hanging in the registers, kept filled with water, and plenty of fresh air from the outside will help. It is a mistaken idea of economy to re-circulate air through the house and furnace again and again; open the windows wide and let in the fresh air; it will heat much more quickly than foul air.

A house planned and furnished with thought for all of these details, provided love dwells there, will be truly a home, a haven for old and young alike, and a joy to the guest who enters its doors.

CLOTHING

To the farm woman, because of the simpler social demands, the problem of clothing her family is less trying than to her city sister. It is her duty, nevertheless, to see that those of her household are dressed as well and as economically as can be; and in accordance with the needs of their social environment. She must do the planning, the buying, perhaps the making, as well as care for the clothes. She must set the standard of dress for her own family, and thus help set that of the community.

Bodily protection, and through it the preservation of health, should be the first consideration in the choice of dress. Clothing to be most healthful must allow perfect bodily freedom, must protect the body from extreme cold and from sudden changes in temperature, and must take care of the perspiration. Our ideas of clothing that will best accomplish this, change as we gain knowledge of the physiology of the human body and of the properties of the different materials. Also, the changes which have come about in living conditions (warmer houses, overheated trains and street cars, and poorly ventilated buildings of all sorts) demand a different type of clothing from that worn by our forefathers.

The entrance of women into outdoor sports, into business and professional life, has done much towards the development of saner and more varied dress for women, such as the tailored suit, the shirtwaist, the house dress and the sporting costume.

Undergarments.—In considering clothing for health the undergarment is of first importance. It must perform three functions: (1) regulate the body heat, (2) absorb perspiration, (3) allow ventilation, so that the perspiration may be evaporated. It is important that the materials chosen be capable of fulfilling these requirements. Loosely woven materials absorb moisture into the meshes readily without seeming wet, these same meshes allow ventilation, and the still air held in them, being a poor conductor of heat, makes them warm, provided the outer garment is woven sufficiently close to prevent too sudden changing of the air. Several layers of light-weight material, holding air between them, are warmer than a single layer of heavier material. For summer, the loosely woven undergarment takes care of the perspiration and a thin outer garment allows the constant changing of air which cools the skin.

Character of Material.—Cotton and linen, closely woven, absorb perspiration, but get very wet, because of the lack of meshes to hold air to hasten evaporation. Woolens, because of the nature of the fiber, contain many air spaces; however, with washing the fibers mat or felt together and this characteristic is lost. Wool also is irritating to some skins and is too warm for people in active life. As the skin must be clean, so must the clothing be clean if it is to perform its functions. Cotton and linen wash more easily than wool, another point in their favor.

Amount of Clothing.—The actual amount of clothing required for warmth will vary with the individual. Old age and infancy require more than active youth, while an invalid or a person leading a sedentary life requires more than one exercising or possessing great vitality. The old custom of putting the woolens on in October and not taking them off until May was not a good one. Clothing should be regulated by the weather; with summer temperature in the house, summer underwear may be worn all the year and the outer garments changed when greater warmth is desired.

When the body is over-heated and the ventilation cut off, one becomes over-sensitive to cold. The throat swathed in furs or a muffler is more likely to suffer when exposed than one left bare; therefore when such neckwear is worn it should be left slightly open at the throat. This does not mean, however, that the whole chest need be exposed, as fashion sometimes decrees it should be. A great criticism of women's clothing is due to its weight and uneven distribution over the body. Union suits and one-piece garments of all kinds are desirable, because they do away with extra layers about the waist and throw some of the weight on the shoulders. Tight clothing is of course undesirable, but garments fitting close to the body give more protection from cold; tights in preference to petticoats is a striking example of this.

The Outer Garments.—In considering the outer garments, the desire for beauty plays a large part. We owe it to ourselves, our family and our neighbors, to look as well as we can. Becoming colors are no more expensive than unbecoming ones, and careful observation in the mirror will tell better than fashion which are the ones suited to us. The color of the skin, the hair and the eyes should be considered; if the skin is sallow and the hair colorless, do not emphasize the fact by matching them in the clothes. It is not possible to give rules, as each individual is a law unto herself and must be studied as such.

In choosing the pattern for a dress, the figure of the individual should be considered. A tall, slender woman should not wear stripes or tucks up and down, nor the fat one tucks or flounces around. Ruffles, fullness, the kind of material, all should conform to the figure.

Extremes of Fashion.—Our ideas of modesty are purely conventional, but change from time to time as fashions change, though not so rapidly. The ball gown is immodest if worn on the street, and the bathing suit shocking in the parlor. The extremes of fashion are often immodest from the standpoint of the more conservative. Fashion in its greatest follies makes it difficult at times to live up to our highest ideals in clothes, but we must not deviate too far from convention. There seem to be some indications that thinking woman is dictating slightly to fashion; at least, she insists that for work and recreation she have suitable clothes. To be sure, the idle rich are willing to give most of their time to the pursuit of this capricious queen, and, unfortunately, many of those less able to do so ape them, but the better class of working, thinking women, be they rich or poor, do not follow the greatest extremes.

If one is to be well dressed, one must not be conspicuous, and to be entirely out of fashion makes one so. Therefore, consider the prevailing style, modify it to suit the individual figure, the individual needs and the occasion for which it is to be worn. Consider the material to be used and do not copy a design suited for filmy chiffon in heavy woolen. The trimming should harmonize with the material and excessive trimming is in all cases to be avoided. Do not overdress; this is as great a sin as to be not well enough dressed; in fact, it usually makes one more conspicuous. One is always at greater ease when one's clothes are suitable. An old woolen skirt worn in the kitchen does not hurt one socially, but a clean, simple wash dress worn there increases one's self-respect.

Footwear.—Shoes are the cause of more discomfort than is usually attributed to them, not being restricted to that of the feet. Shoes should be designed on lines more nearly corresponding to those of the foot; straight on the inside, with broad toes and low heels broad and long enough to support the arch of the foot. Very high heels throw the whole body out of balance, cause backache and legache, and affect the nerves. Most women wear shoes too small for them; they should fit, with the large joint opposite the broad part of the sole, with ample room for the toes and at the same time snug at the heel.

Children's Dress.—In dressing children, it should be remembered that they are at the playtime of life, that bodily freedom is of more importance to them than fine clothes. Children like cheerful colors and look well in bright ones; they should not be made somber before their years. Cleanliness should be instilled in their minds early; not the cleanliness which forbids making mud pies, but the kind which means frequent baths and clean underwear.

The infant needs the softest wool, or silk and wool, as protection about the stomach, and flannel shirts and skirts in cool weather; they should all be light in weight, very soft and clean. Woolen dresses are very undesirable, being too warm and not easily washed. The habit of bundling babies up until they cannot move, and then covering their faces if there is a breath of air about, is greatly to be deplored.

Economy in Clothing.—The question of economy in clothes comes close home to most people. Changes in fashion are so frequent, and the manufacture of materials has become so cheap that fabrics are often not made for wear, but merely for fleeting appearance. To buy wisely a woman needs some knowledge of textiles. She should, for example, recognize cotton when it is skilfully mixed with wool or finished to look like linen; she should understand that silk can be loaded with metallic salts, and that a heavy silk is not necessarily a good one. With fresh accessories, such as girdles and collars, garments made from good materials in not too marked a style may be worn for several seasons. Poor materials, though perhaps attractive to begin with, are disappointing in wear, and the time spent in making them up is often lost.

Ready-made garments are much more attractive than formerly, and if carefully chosen save time and may prove economical. One must consider the cost of making as well as the cost of materials, when comparing them with homemade garments. If much alteration is necessary it will probably not pay to buy them. The greatest fault in much ready-made clothing is the amount of cheap trimming used, a good garment often being spoiled thus.

In conclusion, much thought is necessary if we are to get the best results for the least money. More and more intelligent thought is needed, and when women as a whole are willing to give this, perhaps they will be masters over fashion, instead of allowing fashion to master them.

REFERENCES

"From Kitchen to Garret." Van De Water."
"Shelter and Clothing." Cooley.
Farmers' Bulletin 317, U. S. Dept. of Agriculture. "The Farm Home."

CHAPTER 80

Education and Information for the Farmer

Having read the preceding chapters of this book, one must be impressed with the breadth of the field of agriculture and its close relationship with or dependence upon the laws of nature. No other occupation calls for so great a knowledge in the sciences. Science applied to farming does not make the occupation less difficult. On the other hand, it gives the man with a trained mind a decided advantage over the man who depends chiefly upon physical exertion and set rules.

North America needs a type of agriculture that will provide for the future generations. During the nineteenth century, the population of the United States doubled every twenty-five years. In the normal course of events, this rate of increase should continue throughout the twentieth century. If population continues to increase at this rate, what will be the population at the close of the twentieth century, and what kind of an agriculture must exist in order to provide a cheap and abundant food supply for all the people? Furthermore, what kind of men will be needed on the farm in order to meet the needs of the high civilization?

The extent to which the national and state governments are fostering agricultural research and education is an index of the importance of agriculture and the necessity of making it more productive and profitable.

The character of agriculture can be no better than the man who follows the occupation. It is evident that the agriculture of the future will need the brightest and best trained men. In order to train these men in adequate numbers, agricultural subjects are being introduced into the secondary schools in every state in the Union. The agricultural departments of our colleges are being taxed to the limit of their capacity.

While vocational training dominates our entire educational system, it is the hope that it will not be developed at the expense of non-vocational education. To maintain a strong and virile nation demands that all be educated for citizenship. The training for this function is the same for the man on the farm as it is for the man in the city. All should be educated for one citizenship.

Agriculture in Secondary Schools.—It was about thirty years after the first agricultural college was opened in the United States, before the first successful agricultural high school was opened. This was established in Minnesota in 1888. A Farmers' High School had been established in Pennsylvania in 1855, but was afterwards merged with The Pennsylvania State College. For twenty years after the establishment of the Minnesota school the development of agriculture in secondary schools, as well as that in institutions of higher learning, was slow. During the past ten years however, it has gained a wonderful momentum, and has exceeded in development and support any other branch of vocational training in an equal length of time. According to the annual report of Director Claxton, of the United States Bureau of Education, for the year 1914, there were in the United States 1792 secondary schools teaching agriculture to 53,367 pupils.

At the present time the greatest need in these secondary schools is for teachers with suitable agricultural training. To meet this demand, many of the normal schools have established courses in general agriculture, and

School Wagons Returning Pupils to their Homes.[1]

the agricultural colleges have introduced courses in agricultural education, running parallel with the courses in agronomy, animal husbandry, horticulture, etc.

Teachers of agriculture in secondary schools should have a liberal education in the general cultural subjects and in the physical and natural sciences, particularly as related to agriculture. They should also have sufficient training in psychology and pedagogy to make them efficient teachers. It is also essential that they have the technical and practical knowledge needed on the farm. Familiarity with farm life, acquired by having lived on a farm, strengthens the sympathy between teacher and pupils, and makes possible the most effective teaching.

There is a large field of service for young men and young women trained

[1]Courtesy of U. S. Dept. of Agriculture.

for the vocation of teacher of agriculture in secondary schools. The full fruits of such service will be realized in the years to come.

There are several types of secondary schools, chief among which are the congressional district agricultural high schools, county agricultural high schools, and the high school proper having a course in agriculture. No attempt will here be made to outline a course of study in agriculture for the secondary school. The growth of these schools has been so marked that the courses first outlined have been greatly modified and further improvement will doubtless be made from year to year. The guiding principle should be that the technical agricultural studies fit local conditions and meet the needs of the rural community to the fullest possible extent.

In many of these schools, the teacher is expected to serve the farm community in an advisory capacity. Vacation periods are spent in visiting farmers, studying their methods and conducting tests that bear directly on their soil, crop and livestock problems.

Those seeking detailed information on the introduction of agriculture into secondary schools and the most effective way of conducting it are referred to bulletins issued by the States Relation Service of the United States Department of Agriculture, the United States Department of Education, and State Department of Education, and the State Agricultural College and Experiment Station.

Agricultural Colleges.—In 1821, R. H. Gardner of Maine secured from the state legislature an appropriation of $1000 to aid in maintaining an institution that would give farmers and mechanics a scientific training for their occupation. Students were first received in this institution in 1823. Three years later there was established at Derby, Conn., an agricultural school that proved successful and was obliged to increase its accommodations for students. Between 1845 and 1850 a number of agricultural colleges were established as private enterprises in New York State and Connecticut. In 1846 Yale College established a chair of Agricultural Chemistry and Vegetable and Animal Physiology, and the demand for teachers in these subjects became sufficient in 1848 to justify establishing a course for their preparation at this institution.

Agricultural colleges which have proven permanent were opened in Michigan in 1857 and in Maryland and Pennsylvania in 1859. State agricultural colleges were opened in Iowa and Minnesota in 1858.

In 1862, Senator Morrill of Vermont, after several unsuccessful attempts, secured the passage of a bill in the United States Congress, establishing land grant colleges. This bill bestowed 30,000 acres of land for each member of Congress upon the several states, the proceeds of the land by sale or rental to be used in maintaining courses of learning related to agriculture and the mechanic arts.

A second bill, also introduced by Senator Morrill, passed Congress in 1890, provided for an annual appropriation of $15,000 to each state and territory, to be used for instruction in these colleges. This amount was increased $1000 annually for ten years, since which time it has remained at $25,000.

At present there is a strong agricultural and mechanical college in every state in the Union and in sixteen states an additional institution is maintained for colored students. In one state only, viz., Massachusetts, is there a college teaching agriculture only. In a number of the states, such as Ohio, Indiana, Illinois, Wisconsin, Missouri, California and a few others, the college of agriculture is affiliated with the state university. In most of the states, the Federal appropriation is multiplied several times by state appropriations. In a few states, support of the institutions is provided for by a mill tax. Most of them, however, are dependent on direct appropriations for buildings and maintenance. These are made annually or biennially at the pleasure of the legislature.

Agricultural Experiment Stations.—The first experiment station in North America receiving state aid was established in 1875, at Meriden, Conn. The success attending the first station attracted sufficient attention throughout the country to warrant the establishment of similar stations in other states. The second station was established at Chapel Hill, N. C., in 1877 in connection with the state university. Two years later one was established at Cornell, N. Y. New Jersey followed in 1880 by establishing a station in connection with Rutgers College.

The Hatch Act passed the United States Congress in 1887. This provided $15,000 annually for each state to establish an agricultural experiment station in connection with the land grant colleges, except in those states where experiment stations had already been established as separate institutions.

In 1907, the Hatch appropriation was supplemented by the passage of the Adams bill, which provided $15,000 annually to each state or territory, this sum to be used in agricultural research.

The forces of the agricultural colleges and experiment stations should be utilized as fully as possible by farmers, in order that their sons and daughters may be educated to a better citizenship and higher efficiency.

Farmers' Institutes.—These grew out of meetings held at a comparatively early date under the auspices of local agricultural societies. This form of education did not become regularly established until the organization of the agricultural colleges in 1862. At this time they began to receive state aid. These institutes are conducted under the direction of the state board of agriculture or the state agricultural college. They are usually held during the winter months and, as a rule, each institute continues from two to four days.

In 1885, Wisconsin appointed a Superintendent of Farmers' Institutes. Other states soon followed the example of Wisconsin and now nearly every state has a Director or Superintendent of Farmers' Institutes. In 1910, there were 5651 regular institutes having an aggregate of over 16,000 sessions and a total attendance of nearly 3,000,000 persons. For the same year the appropriations for institutes aggregated $432,000.

The weakest point in the institute lies in the failure to put the platform teaching into practice. Another defect has been the absence of an attempt to do for the woman in the farm home what it has so long tried to do for the man in the field. There is as great need for work of this kind for the woman as for the man in the country, and the institute that serves both sexes more than doubles its efficiency. Happily, these two defects are now being corrected in a number of the states. If the institute is to continue to be a living force among farmers, it must meet the needs of all rural communities as fully as possible. It is well to forget these defects and improve the institute work in every possible way.

The teachings of the institute force should be interesting and apply to the problems of the farm, farm home and rural community. It may serve a good purpose by carrying to the farmer the findings of the state experiment station. For this reason there are advantages in placing this work under the direction of the agricultural college and experiment station. It is a form of extension work that fits well with the research and teaching of the college and experiment station.

Agricultural Fairs.—This form of agricultural education and entertainment had its inception in this country in the fairs held on market days in Washington, D. C., as early as 1804. A similar fair was started at Pittsfield, Mass., in 1807.

An investigation in 1909 showed over 1200 county agricultural fair associations in the United States at that time. The membership was approximately 250,000. The annual gross receipts was $6,500,000, and expenditures for premiums $2,500,000.

In addition to the county fairs, most of the states maintain a state fair. These are usually on a larger scale, having extensive grounds and numerous and commodious buildings for the housing of livestock and exhibits.

Fairs serve as effective educational, stimulating and advertising agencies for both the farmer and merchant. They also afford social and amusement features. Often the last-mentioned feature is allowed to become too prominent and sometimes lowers the tone and reflects discredit on the fair.

Agricultural Societies.—The first agricultural societies were formed between 1785 and 1794, as follows: 1785, Society for Promotion of Agriculture, Charleston, S. C.; same year, Philadelphia Society for Promotion of Agriculture; 1791, New York Society for Promotion of Agriculture, Arts and Manufactures; 1792, Massachusetts Society for Agricultural Promotion; 1794, Society for the Promotion of Agriculture in Connecticut. These early agricultural societies were soon followed by the formation of many others in different sections of the country. In the majority of cases, city men took the initiative in the formation of societies for the promotion of agriculture. Then, as now, far-sighted men realized that the welfare of the country was based largely on the prosperity of the farmers.

The first attempt to form a national agricultural organization was made in Washington, in 1841. This attempt was unsuccessful, and it was not until after the Civil War that the first national agricultural society was successfully launched. This was established in 1866 and known as the National Grange or Patrons of Husbandry. The organization has for its chief objects the promotion of the welfare of the agricultural classes through better legislation, better systems of agricultural education and co-operation along all lines, especially with reference to buying and selling products. Under its auspices many co-operative stores have been established. For various reasons many of these have failed, although some have been successful. Co-operative creameries and co-operative wholesale buying and selling through local agents have been more successful.

At this time the country has need of the conservative advice and constructive criticism of the conservative farmer on many of the live issues of the day. We find the grange in both its state and national meetings discussing such questions as public roads, taxation, rural education, co-operation, woman suffrage and many other of the issues that conern the nation at large.

In recent years farmers' clubs in large numbers have been organized in all sections of the country and these have proven successful chiefly because the membership covers a small area. This enables the club to consider matters that are of practical local interest to all its members, and affords an opportunity to discuss topics at close range. Such clubs have been instrumental in advancing the welfare of the neighborhood through securing postal routes, telephone lines, the introduction of pure-bred stock and pure-bred seeds, and various other improvements.

These various agricultural societies frequently carry with them social advantages, and the bringing of farmers together helps to overcome the conservativism for which they are frequently criticised.

Extension Work.—This consists in conveying information to the people at large, whether it be from the university, the college, the experiment station or other sources of information. In no field has extension developed more rapidly than in agriculture. This information may be conveyed through circulars and bulletins, by means of correspondence lessons, through lectures delivered by college and experiment station members or by means of the farmers' institutes. More recently these methods have been very effectively supplemented by means of the county farm advisor. While the old method carried information, it too frequently failed to get it put into operation. The last-mentioned method of extension overcomes this more fully than any of those previously used.

The passage of the Smith-Lever Bill by the United States Congress has given the farm advisory work financial support that far exceeds that ever before given by the Federal Government to any form of education. This, within a few years, will amount to over $4,000,000 annually, which must, according to the act, be met with an equal amount from the several states. This will ultimately make possible the placing of trained agriculturists in every one of the three thousand counties in the United States.

Extension Representatives.—In farming, knowledge is quite as important as experience. There are many things that the farmer should know which he has not learned by experience, no matter how many years he may have been farming. There are many valuable things that he could learn from those who are not farmers. Dr. Babcock was not a farmer, but he gave to the farmer the milk test which bears his name, and which has done more than any one thing to improve the yield of milk and butter-fat per cow all over the United States. This simple test that may be used by any farmer, enables him to accurately measure the yield of each cow with but little time and expense. With the knowledge thus secured, the farmer is able to decide which cows are profitable and which are unprofitable.

The advantages of the extension representative, located permanently in a county, lie in his being able to become acquainted with the people and their problems. If he is the right man for the position, he soon secures the confidence of the people with whom he works, and this is the first essential in his success. He cannot be expected to solve all of the problems that may be presented, but he should know where to secure expert advice on those which he is unable to solve by himself.

It has so far been a principle in the development of this type of extension that men would be placed only in the counties where the people were ready for this kind of work and willing to aid in financing it. To attempt to force this type of advice on unwilling people only invites failure. So far, many counties have requested the work, and the great difficulty has been to find the type of men who could successfully fill the positions. There are many lines of work that such a representative can undertake. Among these may be mentioned tests of varieties of the different crops best adapted to local conditions, methods of testing seed corn and other seeds for germination, the treatment of seeds for smut and potatoes for scab. Much good can be done by showing the farmers how to compute feeding rations and advising them relative to the home mixing of fertilizers and the amount and form of lime that may be most economically used. Cow testing associations may be formed and boys' and girls' corn clubs, pig clubs, etc., organized.

Agricultural Publications.—The earliest publications in the interest of agriculture in this country consisted of reports of the early agricultural societies. As early as 1792 the New York Society published a volume on its transactions and five years later the trustees of the Massachusetts Society began a series of pamphlets on agricultural topics which afterwards developed into a journal.

The first distinctively agricultural periodical in America was the *American Farmer*, started in Baltimore in 1819. The century has seen a wonderful growth in the development of publications bearing upon the occupation of farming. In fact, there has been such a wealth of literature that it is often a problem to know what to select for one's library.

Every farmer should take one or more agricultural journals that deal most directly with the enterprises in which he is engaged. He should also secure and make use of many of the free publications that are issued by his state experiment station and by the National Department of Agriculture.

EXHIBIT OF CORN AND VEGETABLES GROWN BY MEMBERS OF A BOYS' CLUB.

He should see that his name is on the mailing list of the experiment station of the state in which he resides. The bulletins and circulars issued by his home station are more likely to deal directly with his problems than those from other states. Frequently, he will learn of particular bulletins from stations in other states that bear directly on his enterprises and these he may secure by addressing a letter to the director of the station concerned.

The publications of the national department are classified into *Journal of Research*, department bulletins and farmers' bulletins. The farmers' bulletins are issued in large editions and are free to all farmers. The department bulletins are printed in small editions and are sent free to farmers especially interested as long as the department supply lasts. After this, they may be purchased from the Superintendent of Public

Documents. The *Journal of Research* is technical and designed only for libraries and college and experiment station workers. In addition to these mentioned, the department issues a year-book annually which is also free. It also issues the crop reporter monthly which gives statistics on the condition, acreage and farm price of farm products, and also gives each year the numbers of different classes of livestock and their farm value.

All farmers are privileged to have their names placed on the department monthly mailing list of publications. This brings them the number and title of all department bulletins and circulars issued during the month. Having this list, they may send for those that will be of interest to them.

Libraries.—Farmers, as well as men in other occupations, should have access to a library. It is more difficult to establish and maintain libraries of much consequence in the country than it is in villages or cities. In the country the school or grange should be the agency through which small libraries could be established. With the large number of available free publications a library of much value can be assembled with little cost.

To overcome the difficulties in establishing libraries in the country, the moving library has recently been inaugurated. This is a case of books varying in number and consisting of those pertaining to various phases of farming, as well as some of fiction. A set of books is supplied to a neighborhood, where it is placed in charge of an individual. The books may be secured by neighbors, returned and others secured until the set has been read by most of those interested. It is then returned to the central library and another set secured.

Boys' and Girls' Clubs.—The object of these clubs is to interest the boys and girls in the business affairs of the farm, and in order to create interest the members are induced to compete for prizes. This is the stimulus which induces the boy or girl to study the problem in which he is engaged. Prizes are offered for the largest yield of corn or potatoes on an acre or fraction of an acre of ground. This induces the boy to use the best seed that he can secure, to prepare the ground in the best possible way, and use the manure or kind of fertilizer that will give best results. The problem of economic production, as well as large yield, is generally taken into consideration in this connection. A careful presentation of the method of procedure is also taken into account. In a similar way, girls compete in the growing of tomatoes or other vegetables.

These clubs to be effective should be under the direction of a competent teacher who can guide and train the participants. The boys and girls in the country are excellent material on which those interested in agricultural betterment can work. These clubs have not only a direct stimulating and educational effect, but they often pave the way for a higher education. "Knowledge is power" applies to the man on the farm with as much force as to the man in any other occupation. Farm surveys have shown that farmers with college training are making larger incomes than those whose schooling ended in the high school. Those who completed the high school are making more than those whose school days ended with the grades. In one county in New York, the average labor income of 165 farmers who had attended high school was $304 more than the average for 398 farmers none of whom had gone beyond the district school. Three hundred dollars is 5 per cent interest on $6000. This investigation indicates that a high school training is worth $6000 to one engaged in

MEMBERS OF A BOYS' CORN CLUB AT TYLER, TEX.[1]
A real school of agriculture.

farming. Which would you prefer for your boy, a high school training or $6000?

Education is much more essential now in farming than it was twenty-five years ago. One preparing for farming is preparing not only for the present but for forty years of active service, each succeeding year of which will demonstrate the greater need of education.

REFERENCES

"Country Life Movement;" "Outlook to Nature;" "State and the Farmer." Bailey.
"Farm Boys and Girls;" "Training the Boy;" "Training the Girl;" "The Industrial Education of the Boy;" "The Industrial Education of the Girl." McKeever.

[1] Courtesy of the U. S. Dept. of Agriculture. From Year-Book 1909.

"Farmer of Tomorrow." Anderson.
"Soils and Crops." (A Manual for Secondary Schools.) Hunt and Burkett.
"Training of Farmers." Bailey.
"Law for the American Farmer." Green.
"Farmers' Manual of Law." Willis.
"New Methods in Education." Todd.
University of California Pamphlet. "How Can a Young Man Become a Farmer?"
Nebraska Extension Bulletin 32. "Lunches for the Rural School."
Missouri Expt. Station Circular 77. "Value of Education to the Farmer."
Wisconsin Research Bulletin 34. "Social Anatomy of Agricultural Community."
U. S. Dept. of Agriculture Bulletin 213. "Use of Land in Teaching Agriculture in Secondary Schools."
U. S. Dept. of Agriculture, Year Book 1914. "Organization of Rural Communities."
Farmers' Bulletins, U. S. Dept. of Agriculture:
385. "Boys' and Girls' Agricultural Clubs."
408. "School Exercises in Plant Production."
409. "School Lessons on Corn."
422. "Demonstration Work on Southern Farms."
428. "Testing Farm Seeds in the Home and in the Rural School."
468. "Forestry in Nature Study."
521. "Canning Tomatoes in Clubs."
562. "Organization of Boys' and Girls' Poultry Clubs."
566. "Boys' Pig Clubs."
586. "Plant Material for Use in Study of Agriculture."
617. "School Lessons in Corn."
638. "Laboratory Exercises in Farm Mechanics."

BOOK X

TABLES OF WEIGHTS, MEASURES AND AGRICULTURAL STATISTICS

TABLE I.—PERCENTAGE OF TOTAL DRY MATTER AND DIGESTIBLE NUTRIENTS IN FEEDING-STUFFS.

GRAINS AND SEEDS.

Feeding-stuff.	Total Dry Matter, per cent.	Digestible Nutrients.			Nutritive Ratio, 1:
		Protein, per cent.	Carbohydrates, per cent.	Fat, per cent.	
Cereals:					
Dent corn	89.4	7.8	66.8	4.3	9.8
Corn meal	85.0	6.1	64.3	3.5	10.8
Corn and cob meal	84.9	4.4	60.0	2.9	15.1
Wheat	89.5	8.8	67.5	1.5	8.1
Oats	89.6	8.8	49.2	4.3	6.7
Barley	89.2	8.4	65.3	1.6	8.2
Rye	91.3	9.5	69.4	1.2	7.6
Rice	87.6	6.4	79.2	0.4	12.5
Emmer (Spelt)	92.0	10.0	70.3	2.0	7.5
Legumes:					
Field pea	85.0	19.7	49.3	0.4	2.5
Cowpea	85.4	16.8	54.9	1.1	3.4
Soy bean	88.3	29.1	23.3	14.6	1.9
Peanut	92.5	25.1	13.7	35.6	3.7
Oil-bearing seeds:					
Flax seed	90.8	20.6	17.1	29.0	4.0
Cotton seed	89.7	12.5	30.0	17.3	5.5
Sunflower seed	91.4	14.8	29.7	18.2	4.8

CEREAL BY-PRODUCTS.

Gluten meal	90.5	29.7	42.5	6.1	1.9
Gluten feed	90.8	21.3	52.8	2.9	2.8
Germ oil meal	91.4	15.8	38.8	10.8	4.0
Corn bran	90.6	6.0	52.5	4.8	10.5
Hominy feed	90.4	6.8	60.5	7.4	11.3
Corncobs	89.3	0.5	44.8	...	89.6
Wheat bran	88.1	11.9	42.0	2.5	4.0
Wheat middlings (standard)	88.8	13.0	45.7	4.5	4.3
Wheat middlings (flour)	90.0	16.9	53.6	4.1	3.7
Red Dog flour	90.1	16.2	57.0	3.4	4.0
Oat hulls	92.6	1.3	38.5	0.6	30.6
Oat dust	93.5	5.1	32.8	2.3	7.5
Dried brewers' grains	91.3	20.0	32.2	6.0	2.3
Wet brewers' grains	23.0	4.9	9.4	1.7	2.7
Malt sprouts	90.5	20.3	46.0	1.4	2.4
Dried distillers' grains	92.4	22.8	39.7	11.6	2.9

OIL BY-PRODUCTS.

Linseed oil meal (O. P.)	90.2	30.2	32.0	6.9	1.6
Linseed oil meal (N. P.)	91.0	31.5	35.7	2.4	1.3
Cottonseed meal	93.0	37.6	21.4	9.6	1.1
Peanut cake	89.3	42.8	20.4	7.2	0.9
Corn germ cake	91.4	15.8	38.8	10.8	4.0

TABLE I.—PERCENTAGE OF TOTAL DRY MATTER AND DIGESTIBLE NUTRIENTS IN FEEDING-STUFFS (*Continued*).

PACKING HOUSE BY-PRODUCTS.

Feeding-stuff.	Total Dry Matter, per cent.	Digestible Nutrients. Protein, per cent.	Carbohydrates, per cent.	Fat, per cent.	Nutritive Ratio, 1:
Dried blood	91.5	70.9	....	2.5	0.09
Tankage	93.0	50.1	....	11.6	0.52
Meat scrap	89.3	66.2	....	13.4	0.45

MISCELLANEOUS CONCENTRATES.

Beet molasses	79.2	4.7	54.1	...	11.5
Cane molasses	74.1	1.4	59.2	...	42.3
Molasses beet pulp	92.0	6.1	68.7	...	11.3
Dried beet pulp	91.6	4.1	64.9	...	15.8
Molasses alfalfa feed	90.9	9.8	40.8	0.9	4.4
Cows' milk	12.8	3.4	4.8	3.7	3.9
Skim milk	9.4	2.9	5.3	0.3	2.1
Buttermilk	9.9	3.8	3.9	1.0	1.6

HAYS.

Legumes:					
Red clover	84.7	7.1	37.8	1.8	5.9
Mammoth clover	78.8	6.2	34.7	2.1	6.4
Alsike clover	90.3	8.4	39.7	1.1	5.0
Alfalfa	91.9	10.5	40.5	0.9	4.1
Soy bean	88.2	10.6	40.9	1.2	4.1
Cowpea	89.5	9.2	39.3	1.3	7.3
Grasses:					
Timothy	86.8	2.8	42.4	1.3	16.2
Redtop	91.1	4.8	46.9	1.0	10.2
Blue grass	86.0	4.4	40.2	0.7	9.5
Bermuda grass	92.9	6.4	44.9	1.6	7.6
Prairie grass	90.8	3.0	42.9	1.6	15.5
Cereals:					
Oat	86.0	4.7	36.7	1.7	8.6
Barley	85.0	5.7	43.6	1.0	8.0
Millet	86.0	5.2	38.6	0.8	7.8

FODDER AND STOVER.

Corn fodder	83.3	2.4	50.4	1.2	22.1
Corn stover	81.8	1.9	43.9	0.5	23.7

STRAWS.

Oat	90.8	1.3	39.5	0.8	31.8
Wheat	90.4	0.8	35.2	0.4	45.1
Barley	85.8	0.9	40.1	0.6	46.0
Rye	92.9	0.7	39.6	0.4	57.9

TABLE I.—PERCENTAGE OF TOTAL DRY MATTER AND DIGESTIBLE NUTRIENTS IN FEEDING-STUFFS (*Continued*).

Feeding-stuff.	Total Dry Matter, per cent.	Digestible Nutrients.			Nutritive Ratio, 1:
		Protein, per cent.	Carbohydrates. per cent.	Fat, per cent.	
PASTURE OR FORAGE, AND SOILING CROPS.					
Grasses:					
Blue grass	34.9	2.8	19.7	0.8	7.7
Timothy	38.4	1.5	19.9	0.6	14.1
Orchard grass	27.0	1.2	13.4	0.5	12.1
Bermuda grass	28.3	1.3	13.4	0.4	11.0
Green corn	21.0	0.9	12.2	0.4	14.6
Sorghum	20.6	0.6	11.6	0.3	20.5
Rye	23.4	2.1	14.1	0.4	7.1
Rape	14.3	2.0	8.2	0.2	4.3
Legumes:					
Red clover	29.2	2.9	13.6	0.7	5.7
Alsike clover	25.2	2.6	11.4	0.5	4.8
Alfalfa	28.2	3.6	12.1	0.4	3.6
Cowpea	16.4	1.8	8.7	0.2	5.1
Soy bean	24.9	3.1	11.0	0.5	3.9
SILAGE.					
Corn	26.4	1.4	14.2	0.7	11.3
Corn and soy bean	24.0	1.6	13.2	0.7	9.3
ROOTS.					
Mangel	9.1	1.0	5.5	0.2	5.9
Rutabaga	11.4	1.0	8.1	0.2	8.5
Sugar beet	13.5	1.3	9.8	0.1	7.7
Carrot	11.4	0.8	7.7	0.3	10.5
Potato	20.9	1.1	15.7	0.1	14.5

TABLE II.—DRY MATTER, DIGESTIBLE PROTEIN, AND NET ENERGY PER 100 POUNDS OF FEED. (ARMSBY.)

Feeding-stuff.	Total Dry Matter, pounds.	Digestible Protein, pounds.	Net Energy, therms.
Green fodder and silage:			
Alfalfa	28.2	2.50	12.45
Clover, red	29.2	2.21	16.17
Corn fodder, green	20.7	0.41	12.44
Corn silage	25.6	1.21	16.56
Hungarian grass	28.9	1.33	14.76
Rape	14.3	2.16	11.43
Rye	23.4	1.44	11.63
Timothy	38.4	1.04	19.08
Hay and dry coarse fodders:			
Alfalfa hay	91.6	6.93	34.41
Clover hay, red	84.7	5.41	34.74
Corn forage, field cured	57.8	2.13	30.53
Corn stover, field cured	59.5	1.80	26.53
Cowpea hay	89.3	8.57	40.76
Hungarian hay	92.3	3.00	44.03
Oat hay	84.0	2.59	26.97
Soy bean hay	88.7	7.68	38.65
Timothy hay	86.8	2.05	33.56
Straws:			
Oat straw	90.8	1.09	21.21
Rye straw	92.9	0.63	20.87
Wheat straw	90.4	0.37	16.56
Roots and tubers:			
Carrots	11.4	0.37	7.82
Mangels	9.1	0.14	4.62
Potatoes	21.1	0.45	18.05
Rutabagas	11.4	0.88	8.00
Turnips	9.4	0.22	5.74
Grains:			
Barley	89.1	8.37	80.75
Corn	89.1	6.79	88.84
Corn and cob meal	84.9	4.53	72.05
Oats	89.0	8.36	66.27
Pea meal	89.5	16.77	71.75
Rye	88.4	8.12	81.72
Wheat	89.5	8.90	82.63
By-products:			
Brewers' grains, dried	92.0	19.04	60.01
Brewers' grains, wet	24.3	3.81	14.82
Buckwheat middlings	88.2	22.34	75.92
Cottonseed meal	91.8	35.15	84.20
Distillers' grains, dried:			
Principally corn	93.0	21.93	79.23
Principally rye	93.2	10.38	60.93
Gluten feed, dry	91.9	19.95	79.32
Gluten meal, Buffalo	91.8	21.56	88.80
Gluten meal, Chicago	90.5	33.09	78.49
Linseed meal, O. P	90.8	27.54	78.92
Linseed meal, N. P	90.1	29.26	74.67
Malt sprouts	89.8	12.36	46.33
Rye bran	88.2	11.35	56.65
Sugar beet pulp, fresh	10.1	0.63	7.77
Sugar beet pulp, dried	93.6	6.80	60.10
Wheat bran	88.1	10.21	48.23
Wheat middlings	84.0	12.79	77.65

TABLE III.—WOLFF-LEHMANN FEEDING STANDARDS.
(Showing amounts of nutrients per day per 1000 pounds live weight.)

Animal.	Total Dry Matter, pounds.	Digestible Protein, pounds.	Digestible Carbohydrates, pounds.	Digestible Fat, pounds.	Nutritive Ratio, 1:
Oxen, at rest in stall	18	0.7	8.0	0.1	11.8
Fattening cattle:					
First period	30	2.5	15.0	0.5	6.5
Second period	30	3.0	14.5	0.7	5.4
Third period	26	2.7	15.0	0.7	6.2
Milch cows, when yielding daily:					
11.0 pounds of milk	25	1.6	10.0	0.3	6.7
16.6 pounds of milk	27	2.0	11.0	0.4	6.0
22.0 pounds of milk	29	2.5	13.0	0.5	5.7
27.5 pounds of milk	32	3.3	13.0	0.8	4.5
Sheep:					
Coarse wool	20	1.2	10.5	0.2	9.1
Fine wool	23	1.5	12.0	0.3	8.5
Breeding ewes, with lambs	25	2.9	15.0	0.5	5.6
Fattening sheep:					
First period	30	3.0	15.0	0.5	5.4
Second period	28	3.5	14.5	0.6	4.5
Horses:					
Light work	20	1.5	9.5	0.4	7.0
Medium work	24	2.0	11.0	0.6	6.2
Heavy work	26	2.5	13.3	0.8	6.0
Brood sows	22	2.5	15.5	0.4	6.6
Fattening swine:					
First period	36	4.5	25.0	0.7	5.0
Second period	32	4.0	24.0	0.5	6.3
Third period	25	2.7	18.0	0.4	7.0
Growing cattle (dairy breeds):					
2–3 months, 150 pounds	23	4.0	13.0	2.0	4.5
3–6 months, 300 pounds	24	3.0	12.8	1.0	5.1
6–12 months, 500 pounds	27	2.0	12.5	0.5	6.8
12–18 months, 700 pounds	26	1.8	12.5	0.4	7.5
18–24 months, 900 pounds	26	1.5	12.0	0.3	8.5
Growing cattle (beef breeds):					
2–3 months, 160 pounds	23	4.2	13.0	2.0	4.2
3–6 months, 330 pounds	24	3.5	12.8	1.5	4.7
6–12 months, 550 pounds	25	2.5	13.2	0.7	6.0
12–18 months, 750 pounds	24	2.0	12.5	0.5	6.8
18–24 months, 950 pounds	24	1.8	12.0	0.4	7.2
Growing sheep (mutton breeds):					
4–6 months, 60 pounds	26	4.4	15.5	0.9	4.0
6–8 months, 80 pounds	26	3.5	15.0	0.7	4.8
8–11 months, 100 pounds	24	3.0	14.3	0.5	5.2
11–15 months, 120 pounds	23	2.2	12.6	0.5	6.3
15–20 months, 150 pounds	22	2.0	12.0	0.4	6.5

TABLE III.—WOLFF-LEHMANN FEEDING STANDARDS (*Continued*).

Animal.	Total Dry Matter, pounds.	Digestible Protein, pounds.	Digestible Carbohydrates, pounds.	Digestible Fat, pounds.	Nutritive Ratio, 1:
Growing sheep (wool breeds):					
4–6 months, 60 pounds.....	25	3.4	15.4	0.7	5.0
6–8 months, 75 pounds.....	25	2.8	13.8	0.6	5.4
8–11 months, 80 pounds....	23	2.1	11.5	0.5	6.0
11–15 months, 90 pounds....	22	1.8	11.2	0.4	7.0
15–20 months, 100 pounds...	22	1.5	10.8	0.3	7.7
Growing swine (breeding stock):					
2–3 months, 50 pounds......	44	7.6	28.0	1.0	4.0
3–5 months, 100 pounds.....	35	4.8	22.5	0.7	5.0
5–6 months, 120 pounds.....	32	3.7	21.3	0.4	6.0
6–8 months, 200 pounds.....	28	2.8	18.7	0.3	7.0
8–12 months, 250 pounds....	25	2.1	15.3	0.2	7.5
Growing fattening swine:					
2–3 months, 50 pounds......	44	7.6	28.0	1.0	4.0
3–5 months, 100 pounds.....	35	5.0	23.1	0.8	5.0
5–6 months, 150 pounds.....	33	4.3	22.3	0.6	5.5
6–8 months, 200 pounds.....	30	3.6	20.5	0.4	6.0
9–12 months, 300 pounds....	26	3.0	18.3	0.3	6.4

TABLE IV.—ARMSBY FEEDING STANDARDS.*

FOR MAINTENANCE.

Cattle.			Horses.		Sheep.		
Live Weight, pounds.	Digestible Protein, pounds.	Net Energy, therms.	Digestible Protein, pounds.	Net Energy, therms.	Live Weight, pounds.	Digestible Protein, pounds.	Net Energy, therms.
150	0.15	1.7	0.3	2.0	20	0.03	0.30
250	0.20	2.4	0.4	2.8	40	0.05	0.54
500	0.30	3.8	0.6	4.4	60	0.07	0.71
750	0.40	4.95	0.8	5.8	80	0.09	0.87
1000	0.50	6.0	1.0	7.0	100	0.10	1.00
1250	0.60	7.0	1.2	8.15	120	0.11	1.13
1500	0.65	7.9	1.3	9.2	140	0.13	1.25

FOR GROWTH.

Cattle.				Sheep.			
Age, months.	Live Weight, pounds.	Digestible Protein, pounds.	Energy Value, therms.	Age, months.	Live Weight, pounds.	Digestible Protein, pounds.	Energy Value, therms.
3	275	1.10	5.0	6	70	0.30	1.30
6	425	1.30	6.0	9	90	0.25	1.40
12	650	1.65	7.0	12	110	0.23	1.40
18	850	1.70	7.5	15	130	0.23	1.50
24	1000	1.75	8.0	18	145	0.22	1.60
30	1100	1.65	8.0				

* Modified from Armsby's original table for the sake of simplicity.

TABLE IV.—ARMSBY FEEDING STANDARDS* (*Continued*).

FOR FATTENING.

Cattle.			Sheep.		
Live Weight, pounds.	Digestible Protein, pounds.	Net Energy, therms.	Live Weight, pounds.	Digestible Protein, pounds.	Net Energy, therms.
250	1.1	2.4+(3.5×daily gain)	40	....	0.54+(3.5×daily gain)
425	1.3	3.4+ "	60	....	0.70+ "
500	1.5	3.8+ "	70	0.30	0.79+ "
650	1.7	4.5+ "	80	0.28	0.87+ "
750	1.7	5.0+ "	90	0.25	0.94+ "
850	1.7	5.4+ "	100	0.24	1.00+ "
1000	1.8	6.0+ "	110	0.23	1.06+ "
1100	1.7	6.4+ "	120	0.23	1.13+ "
1250	1.6	7.0+ "	130	0.23	1.19+ "
1500	1.5	7.9+ "	140	0.22	1.25+ "
			145	0.22	1.28+ "

*Modified from Armsby's original table for the sake of simplicity.

TABLE V.—HAECKER'S STANDARD FOR MILK PRODUCTION.
Digestible nutrients for the production of one pound of milk.

Fat in Milk, per cent.	Protein, pounds.	Carbohydrates, pounds.	Fat, pounds.	Fat in Milk, per cent.	Protein, pounds.	Carbohydrates, pounds.	Fat, pounds.
2.5	0.0446	0.176	0.0151	4.8	0.0591	0.276	0.0236
2.6	0.0451	0.180	0.0155	4.9	0.0597	0.280	0.0240
2.7	0.0455	0.185	0.0159	5.0	0.0604	0.284	0.0243
2.8	0.0460	0.190	0.0163	5.1	0.0611	0.288	0.0247
2.9	0.0464	0.194	0.0166	5.2	0.0618	0.291	0.0250
3.0	0.0469	0.199	0.0170	5.3	0.0625	0.295	0.0253
3.1	0.0474	0.203	0.0174	5.4	0.0632	0.299	0.0256
3.2	0.0478	0.207	0.0178	5.5	0.0639	0.302	0.0259
3.3	0.0483	0.212	0.0181	5.6	0.0644	0.307	0.0263
3.4	0.0486	0.216	0.0185	5.7	0.0651	0.310	0.0266
3.5	0.0492	0.221	0.0189	5.8	0.0656	0.314	0.0269
3.6	0.0501	0.225	0.0193	5.9	0.0663	0.318	0.0273
3.7	0.0511	0.229	0.0196	6.0	0.0668	0.322	0.0276
3.8	0.0520	0.234	0.0200	6.1	0.0679	0.326	0.0279
3.9	0.0530	0.238	0.0204	6.2	0.0689	0.330	0.0283
4.0	0.0539	0.242	0.0208	6.3	0.0700	0.334	0.0286
4.1	0.0546	0.247	0.0211	6.4	0.0710	0.338	0.0289
4.2	0.0553	0.251	0.0215	6.5	0.0721	0.342	0.0293
4.3	0.0558	0.255	0.0218	6.6	0.0724	0.345	0.0296
4.4	0.0565	0.260	0.0222	6.7	0.0728	0.349	0.0299
4.5	0.0572	0.264	0.0226	6.8	0.0731	0.353	0.0302
4.6	0.0579	0.268	0.0230	6.9	0.0735	0.357	0.0305
4.7	0.0584	0.272	0.0233	7.0	0.0738	0.359	0.0308

TABLE VI.—PERCENTAGE COMPOSITION OF AGRICULTURAL PRODUCTS.

Crop.	Water.	Ash.	Protein.	Crude Fiber.	Nitrogen-Free Extract.	Ether Extract.
Corn, dent	10.6	1.5	10.3	2.2	70.4	5.0
Corn, flint	11.3	1.4	10.5	1.7	70.1	5.0
Corn, sweet	8.8	1.9	11.6	2.8	66.8	8.1
Corn meal	15.0	1.4	9.2	1.9	68.7	3.8
Corn cob	10.7	1.4	2.4	30.1	54.9	0.5
Corn and cob meal	15.1	1.5	8.5	6.6	64.8	3.5
Corn bran	9.1	1.3	9.0	12.7	62.2	5.8
Corn germ	10.7	4.0	9.8	4.1	64.0	7.4
Hominy chops	11.1	2.5	9.8	3.8	64.5	8.3
Germ meal	8.1	1.3	11.1	9.9	62.5	7.1
Dried starch and sugar feed	10.9	0.9	19.7	4.7	54.8	9.0
Starch feed, wet	65.4	0.3	6.1	3.1	22.0	3.1
Maize feed, Chicago	9.1	0.9	22.8	7.6	52.7	6.9
Grano-gluten	5.8	2.8	31.1	12.0	33.4	14.9
Cream gluten	8.1	0.7	36.1	1.3	39.0	14.8
Gluten meal	8.2	0.9	29.3	3.3	46.5	11.8
Gluten feed	7.8	1.1	24.0	5.3	51.2	10.6
Wheat, all analyses	10.5	1.8	11.9	1.8	71.9	2.1
Wheat, spring	10.4	1.9	12.5	1.8	71.2	2.2
Wheat, winter	10.5	1.8	11.8	1.8	72.0	2.1
Flour, high grade	12.2	0.6	14.9	0.3	70.0	2.0
Flour, low grade	12.0	2.0	18.0	0.9	63.3	3.9
Flour, dark feeding	9.7	4.3	19.9	3.8	56.2	6.2
Bran, all analyses	11.9	5.8	15.4	9.0	53.9	4.0
Bran, spring wheat	11.5	5.4	16.1	8.0	54.5	4.5
Bran, winter wheat	12.3	5.9	16.0	8.1	53.7	4.0
Middlings	12.1	3.3	15.6	4.6	60.4	4.0
Shorts	11.8	4.6	14.9	7.4	56.8	4.5
Wheat screenings	11.6	2.9	12.5	4.9	65.1	3.0
Rye	11.6	1.9	10.6	1.7	72.5	1.7
Rye flour	13.1	0.7	6.7	0.4	78.3	0.8
Rye bran	11.6	3.6	14.7	3.5	63.8	2.8
Rye shorts	9.3	5.9	18.0	5.1	59.9	2.8
Barley	10.9	2.4	12.4	2.7	69.8	1.8
Barley meal	11.9	2.6	10.5	6.5	66.3	2.2
Barley screenings	12.2	3.6	12.3	7.3	61.8	2.8
Brewers' grains, wet	75.7	1.0	5.4	3.8	12.5	1.6
Brewers' grains, dried	8.2	3.6	19.9	11.0	51.7	5.6
Malt sprouts	10.2	5.7	23.2	10.7	48.5	1.7
Oats	11.0	3.0	11.8	9.5	59.7	5.0
Oat meal	7.9	2.0	14.7	0.9	67.4	7.1
Oat feed	7.7	3.7	16.0	6.1	59.4	7.1
Oat dust	6.5	6.9	13.5	18.2	50.2	4.8
Oat hulls	7.3	6.7	3.3	29.7	52.1	1.0
Rice	12.4	0.4	7.4	0.2	79.2	0.4
Rice meal	10.2	8.1	12.0	5.4	51.2	13.1
Rice hulls	8.2	13.2	3.6	35.7	38.6	0.7
Rice bran	9.7	10.0	12.1	9.5	49.9	8.8
Rice polish	10.0	6.7	11.7	6.3	58.0	7.3
Buckwheat	12.6	2.0	10.0	8.7	64.5	2.2
Buckwheat flour	14.6	1.0	6.9	0.3	75.8	1.4
Buckwheat hulls	13.2	2.2	4.6	43.5	35.3	1.1
Buckwheat bran	10.5	3.0	12.4	31.9	38.8	3.3
Buckwheat shorts	11.1	5.1	27.1	8.3	40.8	7.6
Buckwheat middlings	13.2	4.8	28.9	4.1	41.9	7.1
Sorghum seed	12.8	2.1	9.1	2.6	69.8	3.6

TABLE VI.—PERCENTAGE COMPOSITION OF AGRICULTURAL PRODUCTS (*Continued*).

Crop.	Water.	Ash.	Protein.	Crude Fiber.	Nitrogen-Free Extract.	Ether Extract.
Broom-corn seed	11.5	3.4	10.2	7.1	63.6	3.0
Kaffir seed	9.3	1.5	9.9	1.4	74.9	3.0
Millet seed	14.0	3.3	11.8	9.5	57.4	4.0
Hungarian grass seed	9.5	5.0	9.9	7.7	63.2	4.7
Flaxseed	9.2	4.3	22.6	7.1	23.2	33.7
Flaxseed, ground	8.1	4.7	21.6	7.3	27.9	30.4
Linseed meal, old process	9.2	5.7	32.9	8.9	35.4	7.9
Linseed meal, new process	10.1	5.8	33.2	9.5	38.4	3.0
Cotton seed	10.3	3.5	18.4	23.2	24.7	19.9
Cotton seed, roasted	6.1	5.5	16.8	20.4	23.5	27.7
Cottonseed meal	8.2	7.2	42.3	5.6	23.6	13.1
Cottonseed hulls	11.1	2.8	4.2	46.3	33.4	2.2
Cottonseed kernels (no hulls)	6.2	4.7	31.2	3.7	17.6	36.6
Cocoanut cake	10.3	5.9	19.7	14.4	38.7	11.0
Palm nut meal	10.4	4.3	16.8	24.0	35.0	9.5
Sunflower seed	8.6	2.6	16.3	29.9	21.4	21.2
Sunflower seed cake	10.8	6.7	32.8	13.5	27.1	9.1
Peanut kernels (no hulls)	7.5	2.4	27.9	7.0	15.6	39.6
Peanut meal	10.7	4.9	47.6	5.1	23.7	8.0
Rape seed cake	10.0	7.9	31.2	11.3	30.0	9.6
Pea meal	10.5	2.6	20.2	14.4	51.1	1.2
Soy bean	10.8	4.7	34.0	4.8	28.8	16.9
Cowpea	14.8	3.2	20.8	4.1	55.7	1.4
Horse bean	11.3	3.8	26.6	7.2	50.1	1.0
Corn fodder, field cured	42.2	2.7	4.5	14.3	34.7	1.6
Corn stover, field cured	40.5	3.4	3.8	19.7	31.5	1.1
Corn husks, field cured	50.9	1.8	2.5	15.8	28.3	0.7
Corn leaves, field cured	30.0	5.5	6.0	21.4	35.7	1.4
Corn fodder, green	79.3	1.2	1.8	5.0	12.2	0.5
Dent varieties, green	79.0	1.2	1.7	5.6	12.0	0.5
Dent, kernels glazed green	73.4	1.5	2.0	6.7	15.5	0.9
Flint varieties, green	79.8	1.1	2.0	4.3	12.1	0.7
Flint, kernels glazed green	77.1	1.1	2.7	4.3	14.6	0.8
Sweet varieties, green	79.1	1.3	1.9	4.4	12.8	0.5
Leaves and husks, green	66.2	2.9	2.1	8.7	19.0	1.1
Stripped stalks, green	76.1	0.7	0.5	7.3	14.9	0.5
HAY FROM GRASSES:						
Mixed grasses	15.3	5.5	7.4	27.2	42.1	2.5
Timothy, all analyses	13.2	4.4	5.9	29.0	45.0	2.5
Timothy, cut in full bloom	15.0	4.5	6.0	29.6	41.9	3.0
Timothy, cut soon after bloom	14.2	4.4	5.7	28.1	44.6	3.0
Timothy, cut when near ripe	14.1	3.9	5.0	31.1	43.7	2.2
Orchard grass	9.9	6.0	8.1	32.4	41.0	2.6
Redtop, cut at different stages	8.9	5.2	7.9	28.6	47.5	1.9
Redtop, cut in full bloom	8.7	4.9	8.0	29.9	46.4	2.1
Kentucky blue grass	21.2	6.3	7.8	23.0	37.8	3.9
Kentucky blue grass, cut when seed is in milk	24.4	7.0	6.3	24.5	34.2	3.6
Kentucky blue grass, cut when seed is ripe	27.8	6.4	5.8	23.8	33.2	3.0
Hungarian grass	7.7	6.0	7.5	27.7	49.0	2.1
Meadow fescue	20.0	6.8	7.0	25.9	38.4	2.7
Indian rye grass	8.5	6.9	7.5	30.5	45.0	1.7
Perennial rye grass	14.0	7.9	10.1	25.4	40.5	2.1
Rowen (mixed)	16.6	6.8	11.6	22.5	39.4	3.1

TABLE VI.—PERCENTAGE COMPOSITION OF AGRICULTURAL PRODUCTS (*Continued*).

Crop.	Water.	Ash.	Protein.	Crude Fiber.	Nitrogen-Free Extract.	Ether Extract.
HAY FROM GRASSES (*Continued*):						
Mixed grasses and clovers	12.9	5.5	10.1	27.6	41.3	2.6
Barley hay, cut in milk	15.0	4.2	8.8	24.7	44.9	2.4
Oat hay, cut in milk	15.0	5.2	9.3	29.2	39.0	2.3
Swamp hay	11.6	6.7	7.2	26.6	45.9	2.0
Salt marsh hay	10.4	7.7	5.5	30.0	44.1	2.4
Wild oat grass	14.3	3.8	5.0	25.0	48.8	3.3
Buttercups	9.3	5.6	9.9	30.6	41.1	3.5
White daisy	10.3	6.6	7.7	30.0	42.0	3.4
Johnson grass	10.2	6.1	7.2	28.5	45.9	2.1
FRESH GRASS:						
Pasture grass	80.0	2.0	3.5	4.0	9.7	0.09
Kentucky blue grass	65.1	2.8	4.1	9.1	17.6	1.3
Timothy, different stages	61.6	2.1	3.1	11.8	20.2	1.2
Orchard grass, in bloom	73.0	2.0	2.6	8.2	13.3	0.9
Redtop, in bloom	65.3	2.3	2.8	11.0	17.7	0.9
Oat fodder	62.2	2.5	3.4	11.2	19.3	1.4
Rye fodder	76.6	1.8	2.6	11.6	6.8	0.6
Sorghum fodder	79.4	1.1	1.3	6.1	11.6	0.5
Barley fodder	79.0	1.8	2.7	7.9	8.0	0.6
Hungarian grass	71.1	1.7	3.1	9.2	14.2	0.7
Meadow fescue, in bloom	69.9	1.8	2.4	10.8	14.3	0.8
Italian rye grass, coming in bloom	73.2	2.5	3.1	6.8	13.3	1.3
Tall oat grass, in bloom	69.5	2.0	2.4	9.4	15.8	0.9
Japanese millet	75.0	1.5	2.1	7.8	13.1	0.5
Barnyard millet	75.0	1.9	2.4	7.0	13.1	0.6
HAY FROM LEGUMES:						
Red clover	15.3	6.2	12.3	24.8	38.1	3.3
Red clover in bloom	20.8	6.6	12.4	21.9	33.8	4.5
Red clover, mammoth	21.2	6.1	10.7	24.5	33.6	3.9
Alsike clover	9.7	8.3	12.8	25.6	40.7	2.9
White clover	9.7	8.3	15.7	24.1	39.3	2.9
Crimson clover	9.6	8.6	15.2	27.2	36.6	2.8
Japan clover	11.0	8.5	13.8	24.0	39.0	3.7
Alfalfa	8.4	7.4	14.3	25.0	42.7	2.2
Cowpea	10.7	7.5	16.6	20.1	42.2	2.2
Soy bean	11.3	7.2	15.4	22.3	38.6	5.2
Pea vine	15.0	6.7	13.7	24.7	37.6	2.3
Vetch	11.3	7.9	17.0	25.4	36.1	2.3
Serradella	9.2	7.2	15.2	21.6	44.2	2.6
Flat pea	8.4	7.9	22.9	26.2	31.4	3.2
Peanut vines (no nuts)	7.6	10.8	10.7	23.6	42.7	4.6
Sainfoin	15.0	7.3	14.8	20.4	39.5	3.0
FRESH LEGUMES:						
Red clover, different stages	70.8	2.1	4.4	8.1	13.5	1.1
Alsike clover	74.8	2.0	3.9	7.4	11.0	0.9
Crimson clover	80.9	1.7	3.1	5.2	8.4	0.7
Alfalfa	71.8	2.7	4.8	7.4	12.3	1.0
Cowpea	83.6	1.7	2.4	4.8	7.1	0.4
Soy bean	75.1	2.6	4.0	6.7	10.6	1.0
Serradella	79.5	3.2	2.7	5.4	8.6	0.7
Horse bean	84.2	1.2	2.8	4.9	6.5	0.4
Flat pea	66.7	2.9	8.7	7.9	12.2	1.6

TABLE VI.—PERCENTAGE COMPOSITION OF AGRICULTURAL PRODUCTS (*Continued*).

Crop.	Water.	Ash.	Protein.	Crude Fiber.	Nitrogen-Free Extract.	Ether Extrac
STRAW:						
Wheat	9.6	4.2	3.4	38.1	40.4	1.3
Rye	7.1	3.2	3.0	38.9	46.6	1.2
Oat	9.2	5.1	4.0	37.0	42.4	2.3
Barley	14.2	5.7	3.5	36.0	39.0	1.5
Wheat chaff	14.3	9.2	4.5	36.0	34.6	1.4
Oat chaff	14.3	10.0	4.0	34.0	36.2	1.5
Buckwheat straw	9.9	5.5	5.2	43.0	35.1	1.3
Soy bean	10.1	5.8	4.6	40.4	37.4	1.7
Horse bean	9.2	8.7	8.8	37.6	34.3	1.4
SILAGE:						
Corn	79.1	1.4	1.7	6.0	11.0	0.8
Sorghum	76.1	1.1	0.8	6.4	15.3	0.3
Red clover	72.0	2.6	4.2	8.4	11.6	1.2
Soy bean	74.2	2.8	4.1	9.7	6.9	2.2
Apple pomace	85.0	0.6	1.2	3.3	8.8	1.1
Cowpea vine	79.3	2.9	2.7	6.0	7.6	1.5
Cow and soy bean vines mixed	69.8	4.5	3.8	9.5	11.1	1.3
Field pea vine	50.1	3.5	5.9	13.0	26.0	1.6
Barnyard millet and soy bean	79.0	2.8	2.8	7.2	7.2	1.0
Corn and soy bean	76.0	2.4	2.5	7.2	11.1	0.8
Rye	80.8	1.6	2.4	5.8	9.2	0.3
ROOTS AND TUBERS:						
Potato	78.9	1.0	2.1	0.6	17.3	0.1
Common beets	88.5	1.0	1.5	0.9	8.0	0.1
Sugar beets	86.5	0.9	1.8	0.9	9.8	0.1
Mangels	90.9	1.1	1.4	0.9	5.5	0.2
Turnip	90.5	0.8	1.1	1.2	6.2	0.2
Rutabaga	88.6	1.2	1.2	1.3	7.5	0.2
Carrot	88.6	1.0	1.1	1.3	7.6	0.4
Parsnip	88.3	0.7	1.6	1.0	10.2	0.2
Artichoke	79.5	1.0	2.6	0.8	15.9	0.2
Sweet potato	71.1	1.0	1.5	1.3	24.7	0.4
MISCELLANEOUS:						
Cabbage	90.5	1.4	2.4	1.5	3.9	0.4
Spurry	75.7	4.0	2.0	4.9	12.7	0.8
Sugar beet leaves	88.0	2.4	2.6	2.2	4.4	0.4
Pumpkin, field	90.9	0.5	1.3	1.7	5.2	0.4
Pumpkin, garden	80.8	0.9	1.8	1.8	7.9	0.8
Prickly comfrey	88.4	2.2	2.4	1.6	5.1	0.3
Rape	84.5	2.0	2.3	2.6	5.4	0.5
Acorns, fresh	55.3	1.0	2.5	4.4	34.8	1.9
Apples	80.8	0.4	0.7	1.2	16.6	0.4
Cow's milk	87.2	0.7	3.6	...	4.9	3.7
Cow's colustrum	74.6	1.6	17.6	...	2.7	3.6
Mare's milk	91.0	0.4	2.1	...	5.3	1.2
Ewe's milk	81.3	0.8	6.3	...	4.7	6.8
Goat's milk	86.9	0.9	3.7	...	4.4	4.1
Sow's milk	80.8	1.1	6.2	...	4.4	7.1
Skim milk, gravity	90.4	0.7	3.3	...	4.7	0.9
Skim milk, centrifugal	90.6	0.7	3.1	...	5.3	0.3
Buttermilk	90.1	0.7	4.0	...	4.0	1.1

TABLE VI.—PERCENTAGE COMPOSITION OF AGRICULTURAL PRODUCTS (*Continued*).

Crop.	Water.	Ash.	Protein.	Crude Fiber.	Nitrogen-Free Extract.	Ether Extract.
MISCELLANEOUS (*Continued*):						
Whey	93.8	0.4	0.6	...	5.1	0.1
Dried blood	8.5	4.7	84.4	...	...	2.5
Meat scrap	10.7	4.1	71.2	...	0.3	13.7
Dried fish	10.8	29.2	48.4	...	...	11.6
Beet pulp	89.8	0.6	0.9	2.4	6.3	...
Beet molasses	20.8	10.6	9.1	...	59.5	...
Apple pomace	76.7	0.5	1.4	3.9	16.2	1.3
Distillery slops	93.7	0.2	1.7	0.6	2.8	0.9
Dried sediment from distillery slops	5.0	11.3	27.4	8.0	36.1	12.3

TABLE VII.—FERTILITY IN FARM PRODUCE.

Produce.	Amount.	Nitrogen. pounds.	Phosphorus, pounds.	Potassium, pounds.
Corn, grain	100 bushels	100	17	19
Corn, stover	3 tons	48	6	52
Corn crop	....	148	23	71
Oats, grain	100 bushels	66	11	16
Oats, straw	2½ tons	31	5	52
Oat crop	....	97	16	68
Wheat, grain	50 bushels	71	12	13
Wheat, straw	2½ tons	25	4	45
Wheat crop	....	96	16	58
Soy beans	25 bushels	80	13	24
Soy bean straw	2¼ tons	79	8	49
Soy bean crop	....	159	21	73
Timothy hay	3 tons	72	9	71
Clover seed	4 bushels	7	2	3
Clover hay	4 tons	160	20	120
Cowpea hay	3 tons	130	14	98
Alfalfa hay	8 tons	400	36	192
Cotton, lint	1000 pounds	3	0.4	4
Cotton, seed	2000 pounds	63	11	19
Cotton, stalks	4000 pounds	102	18	59
Cotton crop	....	168	29.4	82
Potatoes	300 bushels	63	13	90
Sugar beets	20 tons	100	18	157
Apples	600 bushels	47	5	57
Leaves	4 tons	59	7	47
Wood growth	1 tree	6	2	5
Total crop	....	112	14	109
Fat cattle	1000 pounds	25	7	1
Fat hogs	1000 pounds	18	3	1
Milk	10,000 pounds	57	7	12
Butter	400 pounds	0.8	0.2	0.1
Rye, grain	1470 pounds	28	12	9
Rye, straw	3500 pounds	12	4	27
Rye crop	4970 pounds	40	16	36
Beets, roots	36,800 pounds	88	22	158
Beets, tops	9200 pounds	26	11	69
Beets, crop	46,000 pounds	114	33	227
Grass	4000 pounds	53	13	58
Cotton cake, decoraticated	1000 pounds	66	31.2	15
Rape cake	1000 pounds	48	24.6	13.2
Linseed cake	1000 pounds	45	19.6	14.7
Cotton cake, undecorticated	1000 pounds	39	22.9	20.1
Linseed	1000 pounds	36	15.4	12.3
Palm kernel meal, English	1000 pounds	25	12.2	5.5
Malt dust	1000 pounds	38	17.2	19.5
Bran	1000 pounds	22	32.3	14.8
Mangels	1000 pounds	1.9	0.7	3.9
Swedes	1000 pounds	2.4	0.6	2.0
Carrots	1000 pounds	1.6	1.0	3.2
Turnips	1000 pounds	1.8	0.6	2.9

TABLE VIII.—COMPOSITION OF VARIOUS EXTENSIVE TYPE SOILS OF UNITED STATES.

Locality.	Soil Formation.	Soil Type.	Pounds in 2,000,000.*			
			P	K	Mg	Ca
California, Fresno	Volcanic ash (arid)	Fresno fine sandy loam	1,830	68,200	17,400	50,100
California, Indio	Coastal plain (arid)	Indio fine sandy loam	2,090	50,300	40,560	195,000
Maryland, St. Marys Co.	Coastal plain	Leonardtown loam	160	18,500	3,480	1,000
Maryland, Prince George Co.	Coastal plain	Norfolk sand	520	10,000	2,280	1,860
Maryland, Worcester Co.	Coastal plain	Norfolk loam	610	13,200	1,200	3,430
Alabama, Montgomery Co.	Coastal plain	Orangeburg sandy loam	520	4,700	600	1,400
Louisiana, Acadia Parish	Coastal plain	Crowley silt loam (prairie)	1,220	15,400	3,480	2,720
Texas, Nacogdoches	Coastal plain	Orangeburg fine sandy loam	960	7,000	1,200	1,860
Missouri, Shelby Co.	Glacial (loessial)	Shelby silt loam (prairie)	1,920	32,000	5,160	8,280
Minnesota, Marshall	Glacial (loessial)	Marshall loam (prairie)	1,830	27,700	12,000	26,700
Wisconsin, Janesville	Glacial (loessial)	Marshall silt loam (prairie)	2,430	41,000	10,440	12,300
Wisconsin, Janesville	Glacial (loessial)	Miami silt loam (timber)	2,100	32,700	8,160	13,000
Illinois, McLean Co.	Glacial (loessial)	Marshall black clay loam (prairie)	2,970	38,500	18,440	46,200
Illinois, Clay Co.	Glacial (loessial)	Marion silt loam (prairie)	1,050	24,900	3,480	8,000
Ohio, Toledo	Glacial sand	Miami sand, upland	2,360	33,300	4,920	13,860
Ohio, Toledo	Glacial	Wabash loam, valley land	1,570	45,500	9,960	26,600
Ohio, Wooster	Glacial	Volusia silt loam, upland	1,480	38,300	6,840	4,850
Connecticut, Connecticut valley	Glacial valley land	Podunk fine sandy loam	1,920	22,500	15,600	33,430
Kansas, Parsons	Shale and limestone, residual	Oswego silt loam (prairie)	1,050	20,700	4,320	5,710
Kansas, Riley Co.	Shale and limestone, residual	Wabash silt loam, valley	1,140	49,500	8,520	16,860
Kentucky, Scott Co.	Limestone, residual	Hagerstown clay, upland	3,490	71,700	18,480	44,900
Tennessee, Pikeville	Limestone, residual	Hagerstown loam, upland	1,050	12,800	6,720	5,710
Alabama, Dallas Co.	Limestone, residual	Houston clay, coastal plain	5,150	16,800	14,520	400,300
North Carolina, Statesville	Piedmont residual	Cecil clay	960	22,700	3,480	2,000
Virginia, Albemarle	Piedmont residual	Porter's black loam	4,630	48,300	12,360	23,700
Maryland, Harford Co	Granite and gneiss, residual.	Chester mica loam, upland	1,130	34,400	10,200	3,290
New Jersey, Salem	Greensand, residual	Collington sandy loam, coastal plain	260	13,200	1,560	1,720
New Jersey, Salem	Greensand, marl	Collington lower subsoil	27,600	18,200	38,760	18,300

* 2,000,000 pounds is the average weight of an acre of soil to the depth of 6½ inches.

TABLE IX.—WEIGHT PER BUSHEL, SEEDING RATE PER ACRE, NUMBER OF SEEDS PER POUND AND DEPTH TO COVER FARM SEEDS.

Crop.	Weight per Bushel, pounds.	Rate of Seeding.	Number of Seeds per Pound.	Depth to Cover, inches.
GRASSES.				
Bermuda	36	5 pecks	180,000	¼–½
Canada blue	14–20	15 pounds	2,583,000	¼–½
Creeping bent	15–20	....	8,000,000	¼–½
Crested dog's tail	26–30	....	897,000	¼–¾
Erect brome	14–15	4–6 pecks	162,000	½–1
Fowl meadow	12–15	30 pounds	....	¼–½
Hard fescue	10	....	578,000	¼
Italian rye	17–24	30 pounds	275,000	¼–¾
Johnson	28	4–6 pecks	....	½–1½
Kentucky blue	6–28	25 pounds	2,637,000	¼–½
Meadow fescue	12–28	12–15 pounds	264,000	½
Meadow foxtail	6–14	40 pounds	769,000	½
Orchard	12–21	20 pounds	457,000	¼
Perennial rye	18–30	30 pounds	280,000	¼–½
Redtop	12–40	12–15 pounds	4,135,000	.....
Reed canary	14–48	20–25 pounds	632,000	.....
Rough stalked meadow	12–28	26 pounds	2,706,000	¼
Sheep's fescue	12–28	30 pounds	802,000	¼
Smooth brome	12–14	15–20 pounds	120,000	½–1
Sweet vernal	6–15	30 pounds	837,000	½
Tall meadow fescue	14–25	12–20 pounds	246,000	¼
Tall meadow oat	7–14	30–40 pounds	151,000	½–¾
Timothy	44–50	15 pounds	1,146,000	¼
Velvet	6–7	20 pounds	1,268,000	¼–½
Yellow oat	12–14	30 pounds	1,540,000	¼
LEGUMES.				
Alfalfa	60–63	15–25 pounds	210,000	½–1½
Alsike clover	60–66	4–8 pounds	692,000	½
Bird's foot trefoil	60	11 pounds	367,000	¼–½
Bur clover	60	15 pounds	......	½–1
Common vetch	60	60 pounds	......	.....
Cowpeas	60	4–6 pecks	......	1–2
Crimson clover	60	12–15 pounds	129,000	½–1½
Field peas	52–68	2½–3½ bushels	2,400–4,000	1½–3
Garden peas	60	3 bushels	800–2,400	1–3
Hairy vetch	60	40–60 pounds	75,000	1½–2
Horse bean	56	4 bushels	......	.....
Japan clover	25	15–25 pounds	370,000	¼–¾
Kidney beans	60	....	3,200–4,000	1–2
Kidney vetch	60–64	18–22 pounds	169,000	½–1½
Red clover	60–64	8–14 pounds	304,000	½
Soy beans	60	2–3 pecks	2,000–7,000	1–2
Sweet clover	.....	2–4 pecks	......	½–1
Velvet beans	60	2–6 pecks	......	.....
White clover	60–63	3–6 pounds	739,000	¼
White lupine	50–60	1½–2 bushels	......	1–2
Yellow trefoil	64–66	4–6 pounds	305,000	¼–½
ANNUAL FORAGE CROPS.				
Barnyard millet, Japanese	35	1–2 pecks	212,000	.....
Broom corn millet	60	2–4 pecks	212,000	½–1½

TABLE IX.—WEIGHT PER BUSHEL, SEEDING RATE PER ACRE, NUMBER OF SEEDS PER POUND AND DEPTH TO COVER FARM SEEDS (*Continued*).

Crop.	Weight per Bushel, pounds.	Rate of Seeding.	Number of Seeds, per pound.	Depth to Cover, inches.
ANNUAL FORAGE CROPS (*Continued*).				
Millet, common / Millet, Hungarian / Millet, German / Millet, Golden Wonder	50	2–3 pecks	212,000	¼–½
Rape	50–60	3–8 pounds	......	½–1
Sainfoin	26	40 pounds	22,500	¾–1
Serradella	28–36	40–50 pounds	......	1–2
Sorghum	56	1½–2 bushels	23,000–35,000	½–1½
Sunflower	24–50	10–15 pounds	......	1½–2½
CEREALS.				
Barley	48	7–9 pecks	......	1–2½
Buckwheat	42–50	3–5 pecks	......	1–2
Flax	56	2–8 pecks	......	½–1
Kaffir corn	50–60	3–12 quarts	......	1–2
Milo	50–60	5 quarts	......	1–2
Maize, shelled / Maize, on cob	56 / 70	5–16 quarts	......	1½–4
Rice	43–45	1–3 bushels	......	1½–3
Rye	56	5–10 pecks	......	¾–2
Spelt	40–60	....	......	1–3
Wheat	60	5–8 pecks	......	1–3
Oats	32	8–10 pecks	......	1–2½
VEGETABLES AND ROOTS.				
Artichokes	.....	6–8 bushels	......	.....
Beets	50–60*	4–6 pounds	25,000	.....
Carrots	50*	3–4 pounds	384,000	¼–¾
Mangels	50–60*	5–8 pounds	......	½–1
Parsnip	45–50*	4–8 pounds	112,000	½–1
Potato	60	8–15 bushels	......	2–4
Turnip	55–60*	2–4 pounds	208,000	½–1
Rutabaga	50–60*	3–5 pounds	......	½–1
Sugar beets	50–60*	15–20 pounds	......	½–1
Sweet potato	50–55	1½–4 bushels	......	2–4
FIBER.				
Broom corn	30–48	3 pecks	......	1–2
Cotton, Sea Island / Cotton, upland	44 / 30	1–3 bushels	......	1½–3
Hemp	44	3½–4 pecks	......	1–2

* Roots.

TABLE X.—WATER REQUIREMENTS OF VARIOUS STANDARD CROPS.

Crop.	Location.	Experimenter.	Pounds Water per Pound Dry Matter.		
			Maximum.	Minimum.	Mean.*
Wheat	Germany	Sorauer	708	...	708
	Germany	Hellriegel	390	328	339
	Germany	Von Seelhorst	333	...	333
	India	Leather	544	...	544
	Akron, Col	Briggs and Shantz	534	468	507
	England	Lawes	235	...	235
	Logan, Utah	Widstoe	489	427	458
	Davis, Cal	Fortier and Beckett	359	286	326
	Bozeman, Mont	Fortier and Gieseker	334	226	271
	Reno, Nev	Fortier and Peterson	395	309	360
Oats	Germany	Wollny	...	...	665
	Germany	Sorauer	...	...	600
	Germany	Hellriegel	464	339	401
	India	Leather	...	...	469
	Wisconsin	King	526	502	514
	Akron, Col	Briggs and Shantz	639	598	614
Barley	England	Lawes	262	258	260
	Germany	Wollny	...	...	774
	Germany	Sorauer	...	...	490
	Germany	Hellriegel	366	263	297
	Germany	Von Seelhorst	454	295	365
	India	Leather	...	...	468
	Wisconsin	King	401	375	388
	Akron, Col	Briggs and Shantz	544	527	539
Corn	Germany	Wollny	...	...	233
	India	Leather	...	...	337
	Wisconsin	King	390	305	348
	Akron, Col	Briggs and Shantz	420	319	369
Rye	Germany	Hellriegel	438	315	377
	Germany	Von Seelhorst	700	343	469
	Akron, Col	Briggs and Shantz	...	...	724
Peas	England	Lawes	...	...	235
	Germany	Wollny	...	...	416
	Germany	Hellriegel	353	231	292
	India	Leather	...	...	563
	Wisconsin	King	...	...	477
	Akron, Col	Briggs and Shantz	...	...	800
Potatoes	Germany	Von Seelhorst	294	268	281
	Wisconsin	King	...	...	423
	Akron, Col	Briggs and Shantz	...	...	448
Alfalfa, 1 year	Davis, Cal	Fortier and Beckett	1265	1005	1102
Alfalfa, 2 years	....	....	971	522	761
	State College, N. M.	....	889	757	823
	Akron, Col	Briggs and Shantz	...	...	1068
Clover, red	England	Lawes	...	...	251
	Germany	Hellriegel	363	297	330
	Wisconsin	King	564	398	481
Sugar beets	Logan, Utah	Widstoe	...	...	497
	Akron, Col	Briggs and Shantz	...	...	377
Rice	India	Leather	...	...	811

* This column represents the average of all reliable and comparable tests.

TABLE XI.—COST PER ACRE, PRODUCING CROPS.*

Crop.	Average Cost.
Barley, fall plowed	$8.21
Clover, cut for seed	6.50
Corn, ears husked from standing stalks	10.44
Corn, cut, shocked and shredded	15.30
Corn, cut, shocked and hauled in from field	10.26
Corn, grown thickly and siloed	19.89
Flaxseed, threshed from windrow	7.50
Flaxseed, stacked from windrow	7.85
Flaxseed, bound, shocked, stacked, threshed	7.28
Fodder corn, cut and shocked in field	9.65
Fodder corn, cut, shocked and stacked	12.36
Hay, timothy and clover, first crop	5.59
Hay, timothy and clover, two cuttings	7.18
Hay, millet	7.10
Hay, wild grasses	4.04
Hay, timothy	3.39
Hemp	6.74
Mangels	32.68
Oats, fall plowed	8.86
Oats, on disked corn stubble	8.88
Potatoes, machine production	26.37
Potatoes, machine production, use of fertilizer	37.72
Timothy, cut for seed	4.43
Wheat, fall plowed	7.25

*Minnesota Experiment Station, Bulletin No. 117, page 29.

TABLE XII.—COST OF FARM HORSE POWER.*

Agricultural Region.	Total Annual Cost of Keeping One Horse. Average 5 Years, 1908–12.	Actual Cost per Hour of Work for One Horse. Average 9 Years, 1904–12.
Southeastern Minnesota	$103.27	9.72 cents
Southwestern Minnesota	100.64	8.64 cents†
Northwestern Minnesota	84.67	8.05 cents

NOTE.—The cost figures shown in this table have been selected from the statistical data of the Division of Farm Management of the Minnesota Agricultural Experiment Station. These figures are not estimates, but actual records from a large number of Minnesota farms. The averages are based on records of about 450 horses in each region. The annual cost includes interest on investment, depreciation, harness depreciation, shoeing, feed, labor and miscellaneous expense. Feed is the largest item in the cost of farm horse power, representing on the average ⅔ to ¾ of the total cost. The cost of horse power per hour is computed by dividing the total annual cost by the actual number of hours worked.

* Taken from "Field Management and Crop Rotation," by Parker.
† Seven-year average.

TABLE XIII.—WORK CAPACITY OF FARM MACHINES.*

Kind of Machine.	Size of Machine.	Horse Power Required.	Speed per hr. in Miles, or Revolutions per Minute.	Acre Capacity per Hour.	Ton Capacity per Hour.
Binder, small grain	6-foot cut	3	2½	1.5–1.8	....
Binder, small grain	7-foot cut	4	2½	1.7–2.1	....
Binder, small grain	8-foot cut	4	2½	2.0–2.4	....
Binder, corn	....	3–4	2½	0.8–1.0	....
Cultivator, single row (42-inch rows)	....	1	2	0.5–0.8	....
Cultivator, riding (42-inch rows)	....	2	2	0.5–0.8	....
Cultivator, 2 row riding (42-inch rows)	....	3–4	2	1.0–1.6	....
Drill, small grain	12 tube	2	2½	1.5–1.8	....
Drill, small grain	16 tube	3	2½	2.0–2.4	....
Drill, small grain	20 tube	4	2½	2.5–3.0	....
Ensilage cutter, with flywheel diameter of	42 inch	15–20	...	....	9–15
Ensilage cutter, with flywheel diameter of	36 inch	12–15	...	....	8–12
Ensilage cutter, with flywheel diameter of	30 inch	8–12	...	....	5–8
Harrow, disk (½ lapped)	4 foot	2	2	0.4–0.5	....
Harrow, disk (½ lapped)	6 foot	3	2	0.6–0.7	....
Harrow, disk (½ lapped)	8 foot	4	2	0.8–1.0	....
Harrow, spring tooth	6 foot	3	2	1.0–1.4	....
Harrow, spring tooth	8 foot	4	2	1.5–2.0	....
Harrow, spike tooth	3 section	2–3	2	3.0–3.6	....
Harrow, spike tooth	5 section	4	2	5.0–6.0	....
Header, small grain	12 foot	6	2½	3.0–3.6	....
Mower	5 foot	2	2½	1.2–1.5	....
Mower	6 foot	2	2½	1.5–1.8	....
Packer	10 foot	4	2	2.0–2.4	....
Planter, beet (18-inch rows)	4 row	2	2½	1.5–1.8	....
Planter, corn, 1 row (42-inch rows)	....	1	2½	0.5–1.0	....
Planter, corn, 2 rows (42-inch rows)	....	1	2½	1.0–2.0	....
Planter, potato, 1 row (40-inch rows)	....	2	2½	0.6–1.0	....
Planter, potato, 2 rows (40-inch rows)	....	4	2½	1.2–2.0	....
Plow, walking	14-inch cut	2	2½	0.25–0.35	....
Plow, walking	16-inch cut	3	2½	0.3–0.4	....
Plow, sulky	16-inch cut	3	2½	0.3–0.4	....
Plow, sulky gang	28-inch cut	4–5	2½	0.5–0.7	....
Plow, engine gang, 4 plows	56-inch cut	14–18	2	0.9–1.1	....
Plow, engine gang, 6 plows	84-inch cut	20–25	2	1.4–1.6	....
Plow, engine gang, 8 plows	112-inch cut	25–30	2	1.9–2.2	....
Plow, deep tillage, 2 disk	20-inch cut	6	2	0.34–0.4	....
Potato digger, 40-inch rows	....	4	2½	0.7–1.0	....
Rake, self-dump	10 foot	2	2½	2.5–3.0	....
Rake, side-delivery	8 foot	2	2½	2.0–2.4	
					Bushel
Shredder and husker, corn	4 roll	10–12	...	....	25–50
Shredder and husker, corn	6 roll	15–20	...	....	50–75
Shredder and husker, corn	8 roll	25		....	80–100

* From "Field Management and Crop Rotation," by Parker.

TABLE XIII.—WORK CAPACITY OF FARM MACHINES* (*Continued*).

Kind of Machine.	Size of Machine.	Horse Power Required.	Speed per hr. in Miles, or Revolutions per Minute.	Acre Capacity per Hour.	Bushel Capacity per Hour.
Threshing separator, pea and bean special................	12 inch.....	2–4	300–350	....	8–10
Threshing separator, pea and bean special................	20 x 32 in...	6–8	300–350	....	35–50
Threshing separator, pea and bean special................	26 x 44 in...	10–14	300–350	....	50–80
Threshing separator, pea and bean special................	36 x 54 in...	14–18	300–350	....	80–100
Threshing separator, small grain (wheat and flax)............	18 x 36 in...	15–18	1050–1150	....	60
Threshing separator, oats and barley.....................	18 x 36 in...	15–18	1050–1150	....	220
Threshing separator, wheat and flax......................	28 x 50 in...	30–40	750–800	....	75
Threshing separator, oats and barley.....................	28 x 50 in...	30–40	750–800	....	275
Threshing separator, wheat and flax......................	32 x 54 in...	40–50	750–800	....	125
Threshing separator, oats and barley.....................	32 x 54 in...	40–50	750–800	....	300
Threshing separator, wheat and flax......................	36 x 58 in...	50–60	750–800	....	160
Threshing separator, oats and barley.....................	36 x 58 in...	50–60	750–800	....	350
Threshing separator, wheat and flax......................	40 x 62 in...	60–80	750–800	....	200
Threshing separator, oats and barley.....................	40 x 62 in...	60–80	750–800	....	375

NOTE.—Data on ensilage cutters and shredders by courtesy of The International Harvester Co.; pea threshers, J. L. Owens Manufacturing Co.; grain separators, J. I. Case Threshing Machine Co.; all other data by the author.

Horse power for engine plows is horse power at the draw-bar; for threshing machines, shredders and ensilage cutters, horse power on the belt.

The work capacity of farm machines varies through very wide limits, due to soil and crop conditions, speed and stamina of horses, size and shape of fields, condition of the machine to stand steady work, and the experience and character of the operator. In this table there is shown the maximum capacity per hour for the common tillage, planting and harvesting machines at the standard speeds for best work, also the average capacity per hour based on observations of the actual average daily capacity of farm machines.

The actual average work capacity of any farm machine may be determined very closely by subtracting 15 to 20 per cent from the maximum capacity at a given speed—this deduction being made for time lost in turning, resting horses, oiling, adjusting, filling seed hoppers, etc.; or, in case of power machinery, for oiling, adjusting and taking fuel. The capacity of certain machines such as the corn binder and the potato digger are especially subject to variation. For best results these machines must be driven at comparatively high speed ($2\frac{1}{2}$ to 3 miles per hour) and this speed quickly tires the horses. In order to maintain maximum capacity it is necessary to change horses once or twice a day. If the horses are not changed the capacity varies greatly according to the amount of rest allowed.

* From "Field Management and Crop Rotation," by Parker.

TABLE XIV.—COMPOSITION AND AMOUNTS OF MANURE PRODUCED BY DIFFERENT KINDS OF FARM ANIMALS.*

Kind of Animal and Kinds of Food Fed.	Analysis.				Amount per 1000 Pounds Live Weight.			
	Water.	Nitrogen.	Phosphorus.	Potassium.	Pounds per Day.	Pounds per Year.	Pounds Absorbents per Year.	Total Tons Farm Manure per Year.
Sheep.—Fed hay, corn, oats; or hay, wheat bran, cottonseed meal and linseed meal	59.52	0.77	4.10	0.59	34.1	12,446	5,000	8.7
Swine.—Fed skim milk, corn meal, meat scraps; or corn meal, wheat bran and linseed meal...............	74.13	0.84	0.17	0.32	83.6	30,514	5,000	17.7
Cattle.—Fed hay, silage, beets, wheat bran, corn meal and cottonseed meal	75.25	0.43	0.127	0.44	74.1	27,046	3,000	15.0
Horses.—Fed hay, oats, corn meal and wheat bran.....	48.69	0.49	0.114	0.48	48.8	17,812	3,000	10.5

NOTE.—The analyses and amounts of manure produced by farm animals, as shown in this table, are from the Cornell Experiment Station, and the estimates of pounds absorbents per year from "Farm Management," by Andrew Boss. It is estimated that under average farm conditions 50 per cent of the elements of fertility in farm manures is lost by leaching and fermentation. Direct hauling of manure to the field or composting in concerete pits will prevent much of this loss.

* From "Field Management and Crop Rotation," by Parker.

TABLE XV.—PRICES OF FARM PRODUCTS. AVERAGE FARM VALUE PER HEAD, FIVE-YEAR PERIODS, UNITED STATES.*

Period.	Horses.	Mules.	Milch Cows.	Other Cattle.	Swine.	Sheep.
1866–1870.............	$60.83	$73.16	$29.29	$17.11	$4.44	$1.98
1871–1875.............	66.24	83.50	28.29	18.28	4.41	2.49
1876–1880.............	55.37	61.96	24.36	16.24	4.79	2.20
1881–1885.............	67.18	77.45	28.22	21.16	5.60	2.36
1886–1890.............	71.19	79.21	24.84	18.20	4.85	2.07
1891–1895.............	55.47	66.77	21.70	14.78	5.22	2.26
1896–1900.............	36.17	45.87	26.88	20.24	4.45	2.33
1901–1905.............	62.40	74.04	29.22	17.72	6.63	2.73
1906–1910.............	94.29	109.18	31.85	17.35	7.11	3.75
1911–1915.............	108.16	121.39	46.73	26.52	9.50	3.97

* From United States Year-Book, 1914.

Table XVI.—Average Farm Prices for the United States. Five-Year Periods, 1866–1915.*

Period.	Corn per Bushel, cents.	Barley per Bushel, cents.	Cotton per Pound, cents.	Buckwheat per Bushel, cents.	Oats per Bushel, cents.	Rye per Bushel, cents.	Wheat per Bushel, cents.	Hay per Ton, dollars.	Potatoes per Bushel, cents.	Tobacco per Pound, cents.
1866–1870....	52.1	79.8	....	73.3	39.7	85.5	115.5	10.62	56.1	9.4
1871–1875....	43.6	78.2	14.5	71.6	36.0	70.7	101.7	12.50	53.7	8.9
1876–1880....	35.5	61.9	9.7	61.1	30.9	62.5	97.2	9.10	51.2	6.6
1881–1885....	44.6	61.8	9.3	71.3	34.6	64.5	88.1	9.32	54.6	8.6
1886–1890....	38.8	53.8	8.4	56.4	30.7	54.5	76.6	8.42	53.3	8.1
1891–1895....	37.5	43.8	7.0	53.6	29.0	55.4	60.0	8.38	48.3	8.0
1896–1900....	28.5	38.5	6.5	47.6	23.2	46.8	66.4	7.07	41.4	6.4
1901–1905....	45.7	43.9	10.1	59.5	33.0	58.2	72.4	9.08	58.4	7.5
1906–1910....	51.9	55.3	11.5	68.2	39.6	70.2	86.8	10.78	58.8	10.1
1911–1915....	60.3	59.4	10.2	73.9	39.2	76.7	86.8	12.07	61.9	10.4

* Taken from United States Year-Books, 1914-15.

Table XVII.—Capacity of Round Silos in Tons.

Inside Height of Silo, feet.	Inside Diameter of Silo, feet.												
	8	10	11	12	13	14	15	16	17	18	19	20	22
20	17	26	30	38	44	51	59	67	..	...	...	...	...
21	18	28	33	41	47	55	63	72	..	...	...	...	...
22	19	30	36	43	50	59	67	77	86	...	...	...	...
23	20	32	39	46	54	63	72	81	92	103	...	...	...
24	22	34	41	49	57	67	76	86	98	110	122	...	...
25	23	36	43	52	60	71	80	91	104	116	129	143	...
26	24	38	46	55	64	75	85	97	110	123	137	152	...
27	25	40	49	58	68	79	90	102	116	130	145	160	...
28	27	42	51	61	71	83	95	109	122	137	152	169	205
29	28	44	54	64	75	87	100	114	128	144	160	178	216
30	30	47	56	67	79	91	105	119	135	151	168	187	226
31	31	49	59	70	83	96	110	125	141	158	176	196	237
32	33	51	62	74	86	100	115	131	148	166	184	205	248
33	35	53	65	77	90	105	121	137	155	174	192	215	260
34	36	56	68	80	94	109	126	143	162	181	200	224	271
35	37	58	70	84	98	114	132	149	169	189	209	234	282
36	39	61	73	87	102	118	136	155	176	196	218	243	293
37	40	63	76	90	106	123	142	161	183	204	227	252	305
38	41	66	79	94	110	128	148	167	190	212	236	262	316
39	43	68	82	97	115	133	154	173	197	220	245	272	328
40	45	70	85	101	119	138	160	180	204	228	255	282	340
41	..	72	88	105	124	143	166	187	211	236	262	291	352
42	..	74	91	109	128	148	172	193	218	244	270	300	363
43	..	..	..	113	133	154	179	201	225	252	280	310	375
44	..	..	..	117	137	159	184	207	233	261	289	320	387
45	..	..	..	...	...	165	191	215	240	269	298	330	399
46	..	..	..	...	...	170	197	222	247	277	307	340	412
47	..	..	..	...	...	...	...	229	254	285	316	350	424
48	..	..	..	...	...	...	...	236	261	293	325	361	436
49	..	..	..	...	...	...	...	...	...	301	334	371	449
50	..	..	..	...	...	...	...	...	...	310	344	382	462

TABLE XVIII.—SPOUTING VELOCITY OF WATER, IN FEET PER SECOND, IN HEADS OF FROM 5 TO 1000 FEET.*

Head.	Velocity.	Head.	Velocity.	Head.	Velocity.	Head.	Velocity.
5	17.9	17.5	33.6	29	43.2	48	55.6
6	19.7	18	34.0	30	43.9	49	56.2
7	21.2	18.5	34.5	31	44.7	50	56.7
8	22.7	19	35.0	32	45.4	55	59.5
9	24.1	19.5	35.4	33	46.1	60	62.1
10	25.4	20	35.9	34	46.7	65	64.7
11	26.6	20.5	36.3	35	47.4	70	67.1
11.5	27.2	21	36.8	36	48.1	75	69.5
12	27.8	21.5	37.2	37	48.8	80	71.8
12.5	28.4	22	37.6	38	49.5	85	74.0
13	28.9	22.5	38.1	39	50.1	90	76.1
13.5	29.5	23	38.5	40	50.7	95	78.2
14	30.0	23.5	38.9	41	51.3	100	80.3
14.5	30.5	24	39.3	42	52.0	200	114.0
15	31.3	24.5	39.7	43	52.6	300	139.0
15.5	31.6	25	40.1	44	53.2	400	160.0
16	32.1	26	40.9	45	53.8	500	179.0
16.5	32.6	27	41.7	46	54.4	1000	254.0
17	33.1	28	42.5	47	55.0	....	.....

* Taken from "Electricity for the Farm," by F. I. Anderson.

TABLE XIX.—WEIGHTS AND MEASURES.

AVOIRDUPOIS WEIGHT.

16 ounces (oz.) . . . . =1 pound (lb.).
100 pounds . . . . =1 hundredweight (cwt.).
20 cwt . . . . =1 ton (T.).
1 ton . . . . =20 cwt. or 2000 lbs. or 32,000 oz.

DRY MEASURE

2 pints (pt.) . . . . =1 quart (qt.).
8 qts . . . . =1 peck (pk.).
4 pks . . . . =1 bushel (bu.).
1 bu . . . . =2150.42 cu. in.

LIQUID MEASURE

4 gills (gi.) . . . . =1 pint (pt.).
2 pints . . . . =1 quart (qt.).
4 quarts . . . . =1 gallon (gal.).
31½ gallons . . . . =1 barrel (bbl.).
U. S. gallon . . . . =231 cu. in.
7½ gallons water . . . . =1 cu. ft. approximately.

LINEAR MEASURE.

12 inches (in.) . . . . =1 foot (ft.).
3 feet (ft.) . . . . =1 yard (yd.).
5½ yds. or 16½ ft . . . . =1 rod (rd.).
320 rds . . . . =1 mile (mi.).
1 mile or 320 rds. or 1760 yds. or 5280 ft. or 63,360 ins.

TABLE XIX.—WEIGHTS AND MEASURES (*Continued*).

SQUARE MEASURE.

144 square inches (sq. in.)	=1 square foot (sq. ft.).
9 square feet (sq. ft.)	=1 square yard (sq. yd.).
30¼ sq. yds	=1 square rod (sq. rd.).
160 sq. rds	=1 acre (a.).
640 acres	=1 square mile (sq. mi.).
1 sq. mi	=1 section.
36 sections	=1 township (twp.).
43,560 sq. ft.	=1 acre.

SOLID OR CUBIC MEASURE.

1728 cubic inches (cu. in.)	=1 cubic foot (cu. ft.).
27 cu. ft	=1 cubic yard (cu. yd.).
1 cu. yd	=27 cu. ft. or 46,656 cu. in.
1 cu. yd	=1 load.
24¾ cu. ft	=1 perch.
128 cu. ft. or 8 ft. × 4 ft. × 4 ft.	=1 cord.
1 ft. x 12 in. x 1 in	=1 board foot.

SURVEYOR'S LINEAR MEASURE.

7.92 inches	=1 link.
100 links	=1 chain.
80 chains	=1 mile.

Gunter's chain is the unit and is 66 feet long.

SURVEYOR'S SQUARE MEASURE.

10,000 sq. links	=1 square chain.
10 sq. chains	=1 acre.
10 chains square	=10 acres.

WEIGHT OF DAIRY PRODUCTS.

Article.	Specific Gravity.	Weight of Gallon, pounds.	Weight of Quart, pounds.
Pure water	1.000	8.342	2.085
Skim milk	1.036	8.642	2.160
Whole milk	1.032	8.609	2.152
20 per cent cream	1.022	8.525	2.131
24 per cent cream	1.009	8.417	2.104
30 per cent cream	1.001	8.350	2.087
Pure butter-fat	0.910	7.591	1.898

TABLE XX.—LIST OF AGRICULTURAL COLLEGES AND EXPERIMENT STATIONS IN THE UNITED STATES.

State.	Name of Institution.	Location of College.	Location of Experiment Station.
Alabama	Alabama Polytechnic Institute	Auburn	Auburn
	Agricultural School of the Tuskegee Normal and Industrial Institute	Tuskegee Institute	Tuskegee Institute
	Agricultural and Mechanical College for Negroes	Normal	Uniontown
Arizona	College of Agriculture of University of Arizona	Tucson	Tucson
Arkansas	College of Agriculture of University of Arkansas	Fayetteville	Fayetteville
California	College of Agriculture of University of California	Berkeley	Berkeley
Colorado	The State Agricultural College of Colorado	Fort Collins	Fort Collins
Connecticut	Connecticut Agricultural College	Storrs	Storrs, New Haven
Delaware	Delaware College	Newark	Newark
Florida	College of Agriculture of University of Florida	Gainesville	Gainesville
	Florida Agricultural and Mechanical College for Negroes	Tallahassee	
Georgia	Georgia State College of Agriculture	Athens	Experiment
	Georgia State Industrial College	Savannah	
Hawaii	College of Hawaii	Honolulu	Honolulu
Idaho	College of Agriculture of University of Idaho	Moscow	Moscow
Illinois	College of Agriculture of University of Illinois	Urbana	Urbana
Indiana	School of Agriculture of Purdue University	La Fayette	La Fayette
Iowa	Iowa State College of Agriculture and Mechanic Arts	Ames	Ames
Kansas	Kansas State Agricultural College	Manhattan	Manhattan
Kentucky	The College of Agriculture of State University	Lexington	Lexington
	The Kentucky Normal and Industrial Institute for Colored Persons	Frankfort	
Louisiana	Louisiana State University and Agricultural and Mechanical College	Baton Rouge	Baton Rouge New Orleans (sugar) Crowley (rice) Calhoun, North
	Southern University and Agricultural and Mechanical College of the State of Louisiana	Scotland Heights, Baton Rouge	
Maine	College of Agriculture of University of Maine	Orono	Orono
Maryland	Maryland Agricultural College	College Park	College Park
	Princess Anne Academy, Eastern Branch of the Maryland Agricultural College	Princess Anne	
Massachusetts	Massachusetts Agricultural College	Amherst	Amherst
Michigan	Michigan Agricultural College	East Lansing	East Lansing
Minnesota	College of Agriculture of University of Minnesota	University Farm, St. Paul	University Farm, St. Paul
Mississippi	Mississippi Agricultural and Mechanical College	Agricultural College	Agricultural College
	Alcorn Agricultural and Mechanical College	Alcorn	
Missouri	College of Agriculture of University of Missouri	Columbia	Columbia, College Mountain Grove, (fruit)
Montana	Montana State College of Agriculture and Mechanic Arts	Bozeman	Bozeman
Nebraska	College of Agriculture of University of Nebraska	Lincoln	Lincoln
Nevada	College of Agriculture of University of Nevada	Reno	Reno
New Hampshire	New Hampshire College of Agriculture and Mechanic Arts	Durham	Durham
New Jersey	Rutgers College	New Brunswick	New Brunswick
New Mexico	New Mexico College of Agriculture and Mechanic Arts	State College	State College
New York	New York State College of Agriculture	Ithaca	Ithaca (Cornell) Geneva (State)
North Carolina	The North Carolina College of Agriculture and Mechanic Arts	West Raleigh	Raleigh and West Raleigh
	The Agricultural College for the Colored Race	Greensboro	
North Dakota	North Dakota Agricultural College	Agricultural College	Agricultural College
Ohio	College of Agriculture of Ohio State University	Columbus	Wooster
Oklahoma	Oklahoma Agricultural and Mechanical College	Stillwater	Stillwater
	Agricultural and Normal University	Langston	
Oregon	Oregon State Agricultural College	Corvallis	Corvallis
Pennsylvania	School of Agriculture of Pennsylvania State College	State College	State College
Porto Rico	College of Agriculture and Mechanic Arts of University of Porto Rico	Mayaquez	Mayaquez (Federal) Rio Piedras (Insular)
Rhode Island	Rhode Island State College	Kingston	Kingston

Table XX.—List of Agricultural Colleges and Experiment Stations in the United States (*Continued*).

State.	Name of Institution.	Location of College.	Location of Experiment Station.
South Carolina.	The Clemson Agricultural College of South Carolina	Clemson College	Clemson College
	The Colored Normal Industrial Agricultural and Mechanical College of South Carolina	Orangeburg	
South Dakota	South Dakota State College of Agricultural and Mechanic Arts	Brookings	Brookings
Tennessee	College of Agriculture, University of Tennessee	Knoxville	Knoxville
Texas	Agricultural and Mechanical College of Texas	College Station	College Station
	Prairie View State Normal and Industrial College	Prairie View	
Utah	The Agricultural College of Utah	Logan	Logan
Vermont	College of Agriculture of University of Vermont	Burlington	Burlington
Virginia	The Virginia Agricultural and Mechanical College and Polytechnic Institute	Blacksburg	Blacksburg(College) Norfolk (truck)
	The Hampton Normal and Agricultural Institute	Hampton	
Washington	State College of Washington	Pullman	Pullman
West Virginia	College of Agriculture of West Virginia University	Morgantown	Morgantown
	The West Virginia Colored Institute	Institute	
Wisconsin	College of Agriculture of University of Wisconsin	Madison	Madison
Wyoming	College of Agriculture of University of Wyoming	Laramie	Laramie

Table XXI.—How to Estimate Amount of Grain in Bins and Hay in Mow or Stack.

SMALL GRAIN AND SHELLED CORN.

Length multiplied by width multiplied by average depth in feet gives the cubic feet of grain. This multiplied by 8 divided by 10 equals the bushels.

Example:—A bin of wheat is 8 feet wide by 16 feet long and the average depth of wheat is 6 feet.

$$\frac{8\times16\times6\times8}{10}=614.4 \text{ bushels.}$$

FOR BUSHELS OF EAR CORN.

Multiply the cubic feet occupied by ear corn by 4 and divide by 10.

FOR TONS OF HAY.

If hay has stood for 60 days or more and mow or stack is deep, divide cubic contents in feet by 400. For shallow mows or stacks that have stood only 30 days or less, divide by 600. For intermediate conditions, divide by 500 more or less, depending on conditions. The cubic feet in a stack may be obtained as follows: Subtract the width from the over (the "over" is the distance from the ground on one side over the stack to the ground on the other side), divide by the height, then multiply successively by the over, the width the length, and by .225.

SOURCES OF INFORMATION

Circulars and Bulletins.—Every state in the Union maintains an agricultural college and experiment station. Canada has one at Ottawa, Ontario. Those for the several states are located as stated in Table XX. Each experiment station maintains a mailing list of names to which the station publications are mailed without charge. Any citizen of the state is entitled to the publications.

Information by Letter.—Information other then given in publications may be secured by letter of inquiry. Unless the writer knows the individual or department of the college or station to whom or which he should apply, letters should be addressed to the director of the experiment station. The stations are qualified to answer inquiries relative to the character of soil and crop adaptation for different sections of the state, the fertilizer and lime requirement of the soils, varieties of crops that will succeed best, best sources of seed, how to compound feeding rations and mix fertilizers. Insect enemies and plant and animal diseases will be identified and remedies advised.

Inspection by Experts.—Some of the experiment stations are able to furnish experts along several lines who can visit farmers and advise them relative to improving the farm business. This may be done free or there may be a charge for transportation and subsistence, depending on the federal and state aid the several institutions receive.

County Extension Representatives.—Through the support of the Lever Act, every state now has a number of county farm advisers, and these are being rapidly multiplied. Within a very few years every one of the 3000 counties in the United States should have such a representative on whom to call for any kind of a farm problem. These representatives are selected with much care and are available for counties that will organize a County Farm Bureau and assist in financing the enterprise.

United States Department of Agriculture Publications.—Several classes of bulletins and reports are issued. Non-technical bulletins are issued in large editions and are free to all farmers. The Department Year-Book is also free to all. Technical bulletins are issued in smaller editions and are free to libraries, experiment station workers and others as long as the supply lasts. The *Crop Reporter* is issued monthly and reports acreage, condition and price of farm crops. It is free to anyone.

Anyone may have his name placed on the mailing list for monthly lists of publications. This will give the number, title and character of all publications issued during the month. Knowing what has been issued, the farmer can write for those in which he is interested.

The Weather Bureau reports may be secured by writing directly to the Weather Bureau. For other department publications one should address the Divisions of Publications, U. S. Department of Agriculture, Washington, D. C.

For census figures, one should write to the Bureau of Census, Department of Commerce and Labor.

Each state also maintains a State Department of Agriculture. The duties generally pertain to the administration and police control of fertilizers, feeds, livestock regulations, etc. In many states other work is done, such as the control of Farmers' Institutes. Bulletins are frequently issued free.

GLOSSARY

Protoplasm.—The slimy, granular, semi-fluid content of vegetable cells. This substance is the living portion of the plant, the active, vital thing which gives to it its sensibility to heat, cold and other agents, and the power of moving, of appropriating food, and of increasing its size.

Chlorophyll.—The green coloring matter of plants.

Inoculation (Soil).—Introduction of a definite species of bacteria into the soil or on to the roots of a leguminous plant. This may be accomplished by transferring soil from fields already inoculated or by the use of artificial cultures applied either to the seed or the soil.

Manure Salt.—Double sulphate of potash and magnesia, sometimes referred to as low-grade sulphate of potash and kainit.

Denitrification.—A bacterial change which results in the breaking down of nitrogenous compounds, such as nitrates and nitrites, giving rise to free nitrogen or nitrogen in the form in which it is found in the air.

Concentrates.—These include all the grain and mill products used for feeding stock.

Humates.—These are compounds of humus and salts, such as lime.

Tri-Calcium Phosphate.—A definite compound of calcium, phosphorus and oxygen, containing 19 per cent phosphorus.

Rachis.—The continuation of the stem along which the spikelets, as in wheat for instance, are arranged.

Mycelium.—The thread-like growth of fungi which penetrates the host plant in case of such diseases as smut on grain or moulds on fruit. In the case of mushrooms, it takes the form of a thread-like growth in the soil.

Calcium Cyanimid.—A compound of calcium, carbon and nitrogen manufactured by an electrical process and used as a nitrogenous fertilizer.

"In Kase."—A technical phrase. When the tobacco leaf becomes dry and brittle in the curing barn, it will, during damp or rainy weather, become soft and pliable. When in this condition it is called "in kase."